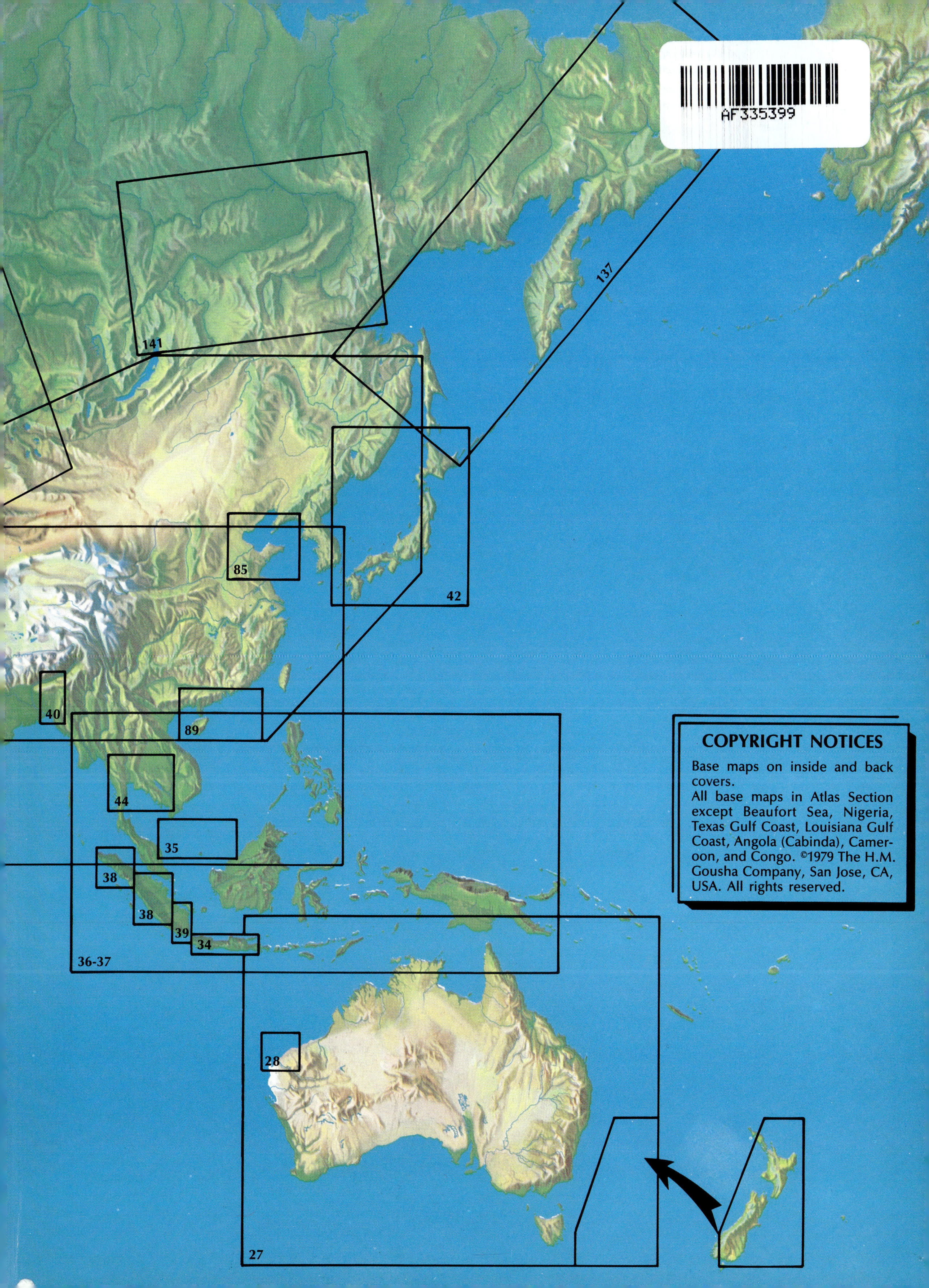

141
137
85
42
40
89
44
35
38
38
39
34
36-37
28
27
AF335399

International
Petroleum
Encyclopedia

INTERNATIONAL PETROLEUM ENCYCLOPEDIA

TABLE OF CONTENTS

Controlling OPEC oil prices........5

Oil hunters probe
Offshore Alaska 19

ATLAS

SPECIAL FEATURES

International Petroleum Encyclopedia

Publisher
John Kennedy

Editor
John C. McCaslin

Editorial coordinator
Robert G. Lair

Presentation editor
Max L. Batchelder

Artist
Kay Wayne

Atlas editors
Patrick Crow
Jim Stilwell
Jim West

Staff editors
Bob Beck
Debra Gwyn
Richard Corbett
Rick Hagar
Sandra Nobbe
G. Alan Petzet
James Redden
Bob Tippee
Warren True
Roger Vielvoye
Bob Williams

International Petroleum Encyclopedia is published annually by the Energy Group of PennWell Publishing Co., Box 1260, Tulsa, Okla. 74101. Phone: (918) 835-3161. Telex: 20-3604.

NEW, ADVANCED digital equipment controls Saudi Arabia's Zuluf/Marjan onshore gas-oil separator plant at Safaniya. Photo courtesy of Arabian American Oil Co.

Saudi Arabia's clout helps control OPEC oil prices and production

THE 13 MEMBERS OF THE ORGANIZATION OF PETRO-leum Exporting Countries entered 1989 with a new oil production quota system in place.

OPEC, with Saudi Arabia playing its accustomed powerful role, aimed to rescue crude oil prices from what threatened to become a collapse like the one that rocked the upstream segment of the international industry in 1986.

World oil markets were watching for signs that OPEC members had started the 3 million b/d production cut promised by the new quota agreement signed in Vienna late in November 1988.

After a week of classic OPEC haggling, oil ministers conjured up another compromise designed to reduce combined production to 18.5 million b/d (Table 1).

The accord brought Iraq back into the quota system at parity with its old enemy, Iran.

New quotas went into effect Jan. 1, 1989, with OPEC optimists hoping the production agreement would boost oil prices back to $18/bbl.

Initial relief that OPEC had escaped from the Iran-Iraq parity impasse pushed North Sea Brent blend prices to $14.60/bbl from $12.50/bbl. But there was uncertainty among oil traders because of the United Arab Emirates' apparent rejection of its quota. Oil Minister Mana Said al-Otaiba told the official emirates news agency after the meeting the 988,000 b/d allotted to the U.A.E. did not represent "an official quota." Otaiba did, however, promise to cooperate with OPEC.

West Texas intermediate futures for January 1989 delivery closed Nov. 30 at $15.32/bbl. U.S. posted prices jumped 75¢-$1.25/bbl, bringing the posting to $14/bbl for West Texas intermediate.

Those and other prices had wavered during a big part of 1988, largely because of large volumes of oil on world markets.

The 1989 production ceiling, which allowed modest increases to all members, was about the best compromise

OPEC could expect. And the maneuvering to persuade Iran to accept production parity with Iraq showed ministerial negotiating ingenuity at its best.

Analysts said the 18.5 million b/d ceiling was about 1 million b/d more than the volume needed to balance supply and demand in first quarter 1989. That took into account a massive inventory buildup that took place during the previous 6 months, when OPEC production consistently topped 20 million b/d.

Another make or break gathering

The November 1988 meeting in Vienna was one of the make-or-break gatherings that litter OPEC history.

Only market expectations that a full ministerial meeting would take firm steps required to prevent further declines put a temporary floor under prices throughout November.

The message to ministers at the start of the meeting was clear. Unless OPEC emerged from Vienna with a credible production quota agreement that included Iran, Iraq, and the U.A.E., the industry could expect single digit oil prices before late December.

OPEC ministers and officials rarely worked harder to prepare the ground for a successful outcome.

The powerful Gulf Cooperation Council opened a round of debate in October 1988 with a suggested 17.5 million b/d ceiling on production with Iraq brought back into the quota

HISTORIC MEETING of OPEC and non-OPEC oil officials Apr. 26-27 in Vienna failed to reach agreement on production limits prior to OPEC oil

fold at parity with Iran.

A series of ministerial committee meetings before the November conference failed to make real progress.

Like the full ministerial meeting that followed, ministers always returned to the same stumbling block. Iraq would accept only a production quota equal to Iran, which was unwilling to make any sweeping concessions to its hated neighbor.

A compromise based on export parity was suggested but rejected, leaving OPEC's mainstream members on the verge of despair and staring at the prospect of a new price collapse.

A breakthrough
in negotiations

The breakthrough in negotiations occurred with a proposal to raise the production ceiling to 18.5 million b/d and allow Iran to maintain its 14.27% share of production. Using this formula, Iran's quota rose to 2.64 million b/d from 2.369 million b/d.

Giving Iraq parity required a small cut in its November production—to 2.64 million b/d. No official quota was fixed for Iraq. It had a notional limit, based on previous levels, of 1.54 million b/d, only 9.28% of OPEC production.

That proved to be the kind of face-saver Iran could accept. After Oil Minister Gholamreza Aghazadeh returned to Tehran for high level consultations, the principle of parity with Iraq was accepted.

The deal required sacrifices on the part of the other 11 members. Their share of OPEC production fell, but the higher total quota was designed to allow small increases in official export levels.

In reality, most of the 11 members had been overproducing to the point that they were to be required to make production cuts ranging from just a few hundred thousand barrels to more than 1 million b/d in the case of Saudi Arabia and the U.A.E. Only Indonesia, Libya, and Qatar were scheduled to see an appreciable increase in production as a result of the 1989 quotas.

A glut of oil on world markets sent oil prices into a steep slide shortly before OPEC oil ministers began their November meeting. They gathered in the wake of Saudi Arabia's reprisals against other OPEC members that were cheating on oil production quotas.

In a determined effort to punish quota breakers, the Saudis stepped up production to 5.7 million b/d early in October from 5 million b/d at the end of September. That pushed OPEC production to 20.6 million b/d, with a larger volume still to come for the rest of October.

As a result, Brent blend fell more than $1 with crude for October delivery, trading at $11.275/bbl and November delivery at $11.375/bbl at closing Oct. 6.

West Texas intermediate futures price for November delivery dropped $1.55 on the week to an Oct. 5 closing of $12.60/bbl, a 2 year low.

Dubai crude, widely used in market related pricing formulas operated by many OPEC members, slumped briefly to $9.17/bbl, lowest since the 1986 collapse.

ministers' midyear meeting. It was the first time OPEC and non-OPEC officials had negotiated on the subject. Photo courtesy of OPEC.

There was a limited recovery in oil prices in 1987. Even so, an average of about $17/bbl for the year was only about half of that reached in 1980 in current dollars (Fig. 1).

**Saudi action
amounts to "cramdown"**

Petroleum analysts watched world markets closely late in 1988 as Saudi Arabia used its production muscle to whip other OPEC members into line and force an agreement on quotas.

The Saudi oil production hike amounted to a "cramdown," said Philip K. Verleger Jr., an analyst with Charles River Associates Inc., Boston and Washington.

"A cramdown," Verleger explained, "is one of those wonderful terms invented by investment bankers. Investors engage in a cramdown when they force some type of reorganization on a beleaguered company.

"In the oil market, the term cramdown will one day be associated with actions taken by Saudi Arabia in the fall of 1988. Specifically, it will be used to describe the Saudi attempt to cram new production quotas down the throats of OPEC and large non-OPEC oil exporting nations.

"Saudi Arabia has once again decided to pursue an increased share of the oil market."

"However, the motive behind this attempt is different from that in 1985-86, and as a consequence the effort is likely to succeed."

Three key facotrs made the 1988 effort different from the earlier one:

• There was a need to accommodate increased production from Iran and Iraq when the combatants returned to normal operations after agreeing to an Aug. 20, 1988, cease-fire in their 8 year old Persian Gulf war.

• The Saudi financial situation was more precarious.

• The effort was aimed at a more vulnerable target: non-OPEC production.

Verleger termed the need to increase OPEC's market to absorb increased production from Iran and Iraq the main reason for a rapidly developing price war.

"Some producers will be forced to cut production, and Saudi Arabia is attempting to shift the burden to non-OPEC producers," he said.

In addition, Saudi Arabia wanted to stop a decline in its financial resources. Verleger estimated the Saudis appeared to have about $30 billion in liquid assets in October 1988, but assets were falling at a rate of $1 billion/month before the oil price collapse.

**Non-OPEC producers
seek stability**

An unprecedented move in the spring of 1988 by half a dozen non-OPEC oil producing nations afforded a measure of world concern during most of the year over the stability of prices and markets.

In talks that began late in April in Vienna, six non-OPEC members offered cuts in production in return for reciprocal action by OPEC nations. But the non-OPEC initiative had exactly opposite the desired effect.

How OPEC oil quotas, production compare — Table 1

| | Old quotas | | October 1988 production* | | 1989 quotas | |
	Million b/d	% share	Million b/d	% share	Million b/d	% share
Algeria	0.667	4.02	0.7	3.302	0.695	3.76
Ecuador	0.221	1.33	0.3	1.415	0.230	1.24
Gabon	0.159	0.96	0.2	0.944	0.166	0.90
Indonesia	1.190	7.17	1.2	5.660	1.240	6.70
Iran	2.369	14.27	2.3	10.849	2.640	14.27
Iraq	1.540	9.28	2.7	12.736	2.640	14.27
Kuwait†	0.996	6.00	1.6	7.547	1.037	5.61
Libya	0.996	6.00	1.0	4.717	1.037	5.61
Nigeria	1.301	7.84	1.5	7.075	1.355	7.32
Qatar	0.299	1.80	0.3	1.415	0.312	1.69
Saudi Arabia†	4.343	26.16	5.7	26.887	4.524	24.45
U.A.E.	0.948	5.71	2.0	9.434	0.988	5.34
Venezuela	1.571	9.46	1.7	8.019	1.636	8.84
Total	**16.600**	**100.00**	**21.2**	**100.00**	**18.500**	**100.00**

Estimated. †Includes one half of production from Neutral Zone, which averaged 300,000 b/d in October 1988.

OPEC's drive for market share

Weak oil prices beginning in 1986 reflected a resolve among OPEC members to maintain market share at the expense of revenues.

For example, figures compiled by the Royal Dutch/Shell Group showed that world oil trade remained virtually unchanged at about 24.5 million b/d in 1987, the latest full year for which data are available. That consolidated the sharp increase from 21.9 million b/d in 1985.

Within the 1987 total, OPEC's contribution fell by 500,000 b/d, or 3%, to 15.7 million b/d. Meanwhile, net exports from non-OPEC, non-Communist countries rose to 6.3 million b/d from 5.9 million b/d.

Crude prices started to slide again as OPEC members, after four long and often bitter meetings, failed to respond to the non-OPEC offer with new quotas based on a lower production ceiling.

The price of North Sea Brent blend, about $17.50/bbl when the non-OPEC offer of a 183,000 b/d production cut was put to the OPEC consultative meeting, began to drop as it became clear that major exporters were having difficulty matching the offer.

The North Sea marker crude rapidly fell to less than $17/bbl and was heading for the $16/bbl mark. U.S. West Texas intermediate futures for June delivery, after trading at $18.30-18.60/bbl just before the meeting, fell to $17.15-17.30 shortly after talks broke off.

The six non-OPEC members—Angola, Egypt, China, Mexico, Oman, and Malaysia—offered to cut exports 5% in return for a 5% cut by OPEC members, excluding Iraq. The non-OPEC cuts were to have continued through May and June.

The cut would have totaled 183,000 b/d with Mexico providing 68,000 b/d, China 30,000 b/d, Oman 27,500 b/d, Egypt 22,500 b/d, Angola 20,000 b/d, and Malaysia 15,000 b/d.

Reducing OPEC exports by a reciprocal 5% found little favor, but there was support for making more than a barrel-for-barrel response. Iran pushed for a 300,000 b/d reduction with an offer of incentives to persuade the non-OPEC group to draw other producers like Colombia and the Soviet Union into export reductions.

Saudi Arabia, on behalf of the Gulf Cooperation Council (GCC) made up of itself, Kuwait, United Arab Emirates, and Qatar, suggested that the four gulf members cut production by a combined 56,000 b/d. The remaining 127,000 b/d would be allocated equally among eight other members, excluding Iraq.

The majority group had been looking for a bigger contribution from the gulf producers, and the meeting broke up without agreement.

The June 1988 ministerial meeting failed to produce an agreement that could shore up OPEC's ailing prices and production situation.

So the midyear meeting in Vienna was forced to roll over the first half 1988 production and price regime for the rest of the year.

Only about 20% of non-OPEC, non-Communist oil production moves in international trade, compared with more than 80% of OPEC production. As a result, although OPEC's share of international oil exports fell slightly in 1987, it still contributed nearly two thirds of total trade. However, that is well below the OPEC share of more than 90% in 1973, when its exports were about 30 million b/d.

Countries around the Persian Gulf account for the biggest volume of OPEC oil exports, and Saudi Arabia usually supplies the largest share of that oil. In 1987, total OPEC exports from the gulf were virtually unchanged at 10.3 million b/d (Fig. 2).

Outside the gulf, OPEC exports declined to 5.4 million b/d in 1987 from 5.8 million b/d in 1986, with the biggest declines in Nigeria, off 11%, and Indonesia, off 13%.

The main increases in non-OPEC, non-Communist oil exports in 1987 occurred in Norway, with a gain to 850,000

Distribution of oil export revenue loss — Table 2

| | Income | | |
| | 1985 | 1987 | % decline |
	Billion $		
Algeria	9,170	8,000	12.7
Ecuador	1,927	1,350	29.9
Gabon	1,668	950	43.0
Indonesia	9,083	6,144	32.3
Iran	13,115	9,200	29.8
Iraq	11,380	11,300	0.7
Kuwait	9,729	7,800	19.8
Libya	10,520	5,600	46.7
Nigeria	12,338	6,900	44.0
Qatar	3,355	2,100	37.4
Saudi Arabia	25,936	21,500	17.1
U.A.E.	13,395	8,800	34.3
Venezuela	10,325	7,200	30.2
Total	**131,941**	**96,844**	**26.6**

Source: OPEC Review

b/d from 700,000 b/d, and Egypt, with a similar volume increase (Fig. 3). Exports from the U.K. declined 5% to 1.1 million b/d in 1987, reversing an upward trend.

Since 1973, exports from non-OPEC, non-Communist countries have risen from 1.5 million b/d from 12 countries to 6.3 million b/d from 20 countries in 1987. So in spite of OPEC's recent vow to maintain market share, the long term trend has been toward increased reliance on non-OPEC oil supplies spawned by hefty prices of the early to mid-1980s.

1987 oil export revenues rebound

Despite a slight drop in volume, OPEC's income from oil exports increased to $97 billion in 1987 from $75 billion in 1986. The latest figure was far short of the $282 billion peak of 1980 (Fig. 4). The rebound resulted from a rise in the average price of a barrel of OPEC crude oil from $12.71 in 1986 to $17 in 1987 following a return to the official pricing system in January 1987.

Oil export revenues increased in almost all OPEC member countries, with Iran and Iraq the main beneficiaries of increased volumes and higher prices. In both countries, revenues increased by about $5 billion, restoring for Iraq most of its losses of the previous year.

Even with a decline in export volume, Saudi Arabia's oil export revenues at $23 billion were about $3 billion higher than they were in 1986.

Higher income from oil also improved OPEC's net current account balance, reducing the deficit from about $37 billion in 1986 to less than $20 billion in 1987 (Fig. 5).

The financial situation of OPEC members varied considerably, but at least 10 of the members were running a current account deficit in 1987. Some had seen their income from oil exports plunge by more than 40% from 1985 levels, one calculation showed (Table 2).

When is condensate crude oil?

An issue that OPEC officials tried—and failed—to resolve in 1988 was the definition of condensate. The issue revolved around whether condensate should be included in production quotas, thus reducing the volume of crude oil that could be produced if quotas were observed.

Before condensate became an issue with OPEC, the industry definition was generally accepted as hydrocarbons gaseous under reservoir conditions but liquid at atmospheric pressure and temperature.

The Illustrated Petroleum Reference Dictionary by Robert D. Langenkamp, published by PennWell Publishing Co., says condensate generally is 50-120° gravity and water-white, straw, or bluish in color.

Venezuela's definition of condensate appeared in a notice published in its Official Gazette of Jan. 20, 1982.

This said naturally occurring condensates are hydrocarbons that exist in gaseous phase in natural underground reservoirs at original reservoir conditions but are liquids at atmospheric pressure and temperature and have a gravity that is more than 40.2° at 15.56° C.

The Venezuelan definition excluded liquids obtained from absorption, adsorption, compression, and cryogenics.

In OPEC ministerial sessions Venezuela suggested another loose definition of condensate similar to its 1982 official definition but without mention of gravity.

Saudi Arabia and Kuwait pointed out that OPEC's auditors would be unable to distinguish between light crude and condensate using this imprecise definition.

They put forward their own proposal, which was similar to the generally accepted industry standard, using a minimum of 50° gravity at 15.56° C., a GOR higher than 5,000:1, light

KEY PROCESS UNITS at Ras Tanura, Saudi Arabia, refinery are reflected in this view from the crude oil tank farm. Photo courtesy of Arabian American Oil Co.

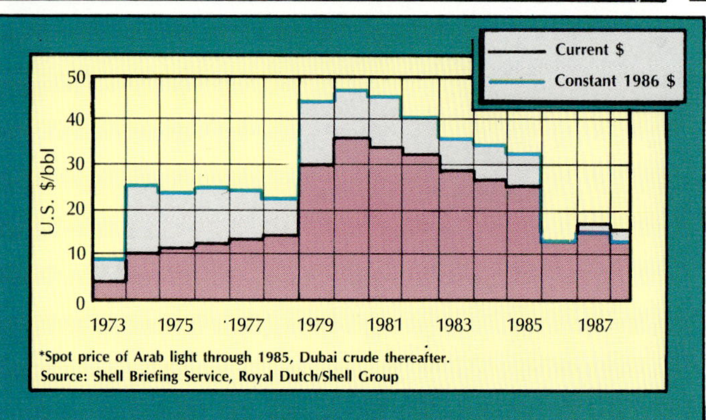

AMMONIA PLANT in Saudi Arabia's Al-Jubail Industrial City is rated at 500,000 metric tons/year. It's a joint venture of two affiliates of Saudi Basic Industries Corp. Photo courtesy of Saudi Basic Industries.

straw colored or colorless, and with a C_7+ content of not more than 3.8 mole % in the total gas stream.

A series of compromises was then put forward that progressively took condensate down from the Saudi-Kuwaiti-general industry definition toward the Venezuelan position.

One attempt at compromise would have established a minimum gravity of 45°, minimum GOR 5,000:1, C_7+ less than 5 mole %, and color of light straw or light tea.

A final compromise, also rejected, would have set gravity at a minimum of 43°, GOR above 4,000:1, C_1 70-85 mole %, C_7 less than 12.5 mole %, with a color of light straw to light tea.

With the issue of what is crude oil and what is condensate unresolved, OPEC members continued to produce condensate without regard to oil production quotas.

Saudi Arabia plays a key OPEC role

A founding member in 1960, Saudi Arabia with its big oil reserves has long played a key role in OPEC policy and operations. It is among the pacesetters in a drive by oil exporters to move into downstream operations at home and abroad.

In the early 1980s Saudi Arabia acted as a swing producer, balancing OPEC production with decreasing demand and increasing non-OPEC production. But the sharp drop in oil revenues in 1986 increased financial pressures and made Saudi Arabia reluctant to resume the role of swing producer.

Operations of Arabian American Oil Co. (Aramco) reflect the changes in Saudi oil fortunes (Fig. 6).

Aramco accounts for 91% of oil production in Saudi Arabia, its sole operating area.

In 1987 Aramco adjusted further to higher nonassociated gas production and lower crude oil and associated gas production.

Working on behalf of the Saudi government, Aramco increased its capacity to produce nonassociated gas from deep Permian Khuff limestone nearly 10% to 1.7 bcfd during 1987.

Khuff gas had become an important element in meeting summer domestic gas demand peaks.

Saudi Arabia's OPEC quota on crude oil production limited the volume of associated gas supply.

Aramco can produce an additional 450 MMcfd of gas from Abqaiq field to assist in periods of extremely high gas demand. The company previously injected the gas into Abqaiq to form a gas cap.

Aramco also nearly completed by yearend 1987 a major project to "preserve onshore and offshore production plants until a time when increased market demand requires their use."

The company mothballed four offshore gas-oil separator plants (GOSPs) in Zuluf and Marjan fields, one onshore GOSP, wet crude handling facilities at Tanajib, and four accommodation platforms during the year.

It used a method of fully covering whole facilities or plant components in a polyvinylchloride coating to shield them against the highly corrosive marine/coastal environment.

Aramco's 1987 field operations

Aramco achieved the 150 MMcfd increase in Khuff nonassociated gas productive capacity by drilling new wells and expanding capacity of existing wells.

Three flow lines to new Khuff wells and three flow lines to existing Khuff wells were to be completed in 1988, raising Khuff gas productive capacity to 1.95 billion cfd.

Production from the Shedgum and Uthmaniyah gas plants averaged 2.7 bcfd during 1987. Nonassociated gas accounted for about 34% of the supply.

Exploratory drilling for gas in 1987 found new Khuff reserves in Khursaniyah field and upgraded Khuff reserves at Shedgum, Uthmaniyah, and Ain Dar.

Aramco in January 1989 disclosed a big hike in its oil and gas reserves.

Based on 1988 production and evaluation of reservoir performance, Aramco jumped its reserves figures to 252.384 billion bbl of oil and 177.294 tcf of gas as of yearend 1988. By contrast, the government earlier pegged yearend 1988 reserves at 169.97 billion bbl of oil and 145.848 tcf of gas for the entire country, only slight increases from the 166.98 billion bbl and 139.9 tcf figures the government reported for yearend 1987.

What's more, Aramco said, with added development of today's oil fields and recent expansion of exploration to all areas of the kingdom, Saudi Arabia's remaining resources could be as much as 315 billion bbl of oil and 253 tcf of gas.

Aramco, at government request, undertook exploration outside its retained areas in 1987. It conducted seismic surveys north and northwest of Riyadh and south and west of Khurais field.

Three seismic crews were active at yearend.

Aramco's crude oil production fell 14.9% during 1987 to 3.991 million b/d.

Crude production peaked at 9,631,366 b/d in 1980 and fell as low as 3,041,104 b/d in 1985, the lowest since 1969.

NGL production rose 13.4% to 344,921 b/d in 1987. NGL exports rose to 333,578 b/d from 298,354 b/d.

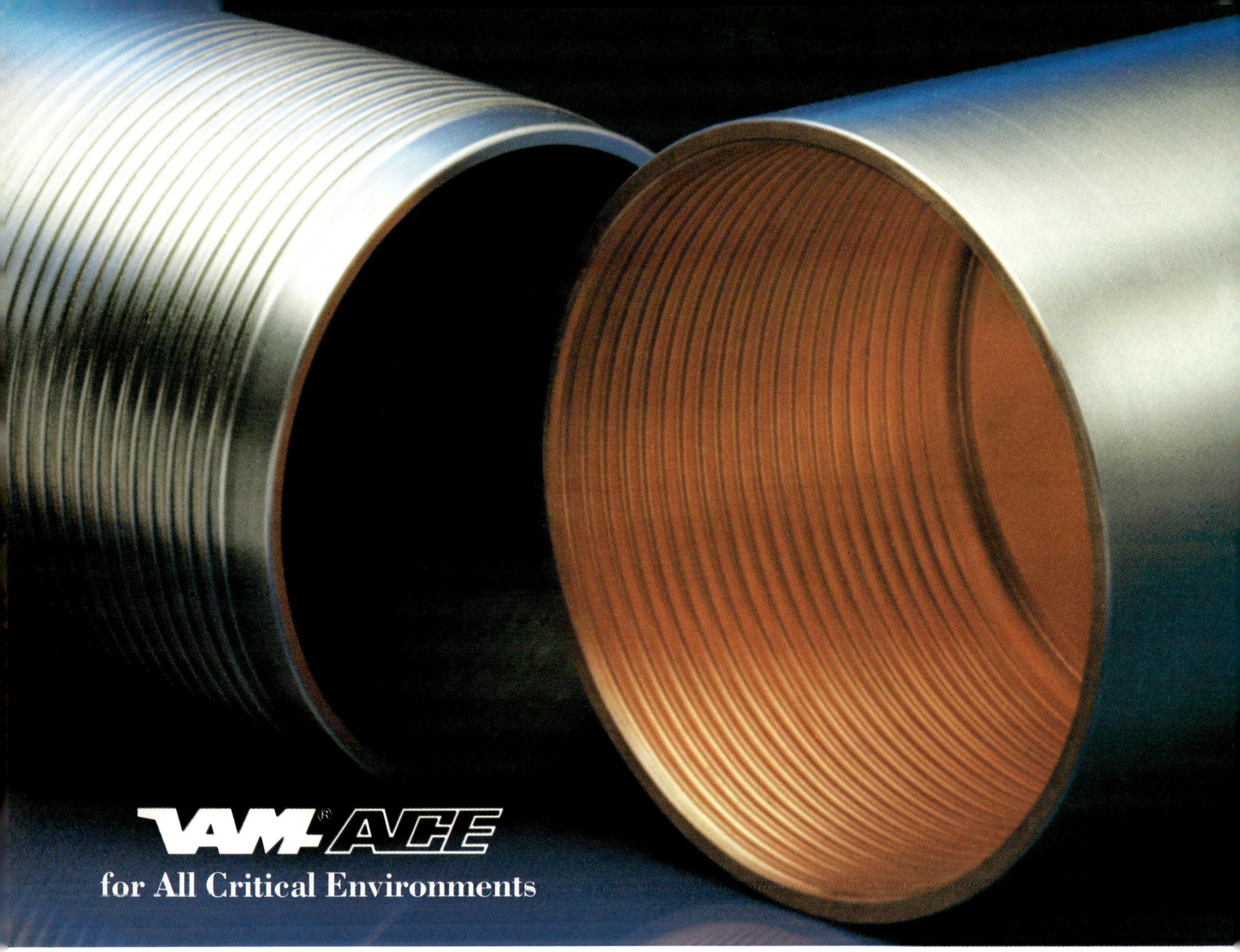

The Ideal "Super Premium" Connection

Here is a new breed of connections designed to give super premium performance on CRA tubing and casing in All Critical Environments.

VAM ACE has been selected for deep, high-pressure, corrosive wells the world over—including the North Sea, the Middle East, and locations in the USA such as Mobile Bay.

It is backed by a strong and continuing support—from the design and testing, to the technical assistance for accessory manufacturing and the running of the string.

Look at these benefits:

1. Less joint rejection—the geometry avoids high local-contact pressure on the seal during stabbing and early make-up, and the couplings have a proprietary copper plating to prevent mating surfaces from galling.

The result is that running CRA is almost as easy as carbon steel!

2. Higher pressure integrity—a new metal-to-metal sealing system, further energized by the well-known VAM reverse-angle torque shoulder, provides pressure integrity that exceeds the burst and collapse of the pipe!

3. 100% coupling efficiency—a coupling design with new hook threads provides greater strength with rigorous resistance to jump-out.

4. Smoother flow—a streamlined bore with only a 6° chamfer and tight tolerances on the coupling ID permits high velocity, corrosive gas flows.

VAM ACE is backed by "The VAM System." Your assurance of continuing support.

Contact us for complete details and our VAM Catalog 870.

vallourec industries

Oil and Gas Market Division

130 Rue de Silly
92100 Boulogne-Billancourt
France
Tel: (33-1) 49.09.35.00
Telecopy: (33-1) 49.09.37.13

11

EMPLOYEE INSPECTS the 56 in. scraper-launcher outlet valve on Saudi Arabia's east-west crude oil pipeline. Photo courtesy of Arabian American Oil Co.

Ras Tanura
refinery operations

Crude processing at Aramco's 415,000 b/d Ras Tanura refinery averaged 381,902 b/d in 1987, down from 390,251 b/d the previous year.

NGL runs at Ras Tanura averaged 76,880 b/d, the highest since 1984's 93,648 b/d.

NGL runs reached a high of 303,226 b/d in 1979.

Combined crude and NGL runs reached a high of 701,518 b/d in 1980. The lowest volume of combined runs in recent years was the 1986 average of 442,002 b/d.

Aramco installed a computer controlled gasoline blending system at Ras Tanura that will boost gasoline production capacity by as much as 460,000 bbl/year.

The company also reactivated a mothballed gas processing plant in the refinery and reallocated pipelines to transfer NGL from Ras Tanura to Ju'aymah for storage and export.

Turnaround time for the 1,975 ships that called at Ras Tanura, Ju'aymah, and Yanbu terminals to take on crude, NGL, and products in 1987 averaged 27 hr.

Sabic expands
downstream operations

Saudi Arabia in 1988 continued its program of expanding downstream facilities in a campaign aimed at exporting high value commodities instead of raw material.

The vehicle assigned to accomplish that goal is Saudi Basic Industries Corp. (Sabic), a joint stock company established in 1976. By 1988 Sabic had become involved in 15 manufacturing operations supported by subsidiary marketing companies. Sabic also was a partner in three industrial and marketing ventures based in Bahrain (Table 3).

Many of Sabic's operations are joint ventures with international companies. Sabic regards such partnerships as bridges for the transfer of advanced technology to Saudi Arabia and a means to help prepare Saudi nationals to manage, operate, and maintain the country's expanding industrial base.

Sabic companies in 1987 produced 9.8 million metric tons of more than a score of commodities sold in more than 65 countries. Sabic is convinced it is positioned to ride the crest of increasing demand for all sorts of petrochemicals—methyl tertiary butyl ether (MTBE), for example.

Demand for MTBE could account for as much as 10% of the world gasoline pool by 1995, Sabic predicted in August 1988. Feedstock supply could become a problem for MTBE producers as capacity for methanol becomes squeezed in the mid-1990s.

"Methanol already consumes 1 out of every 7 gal of methanol made, and that percentage will continue to rise," said Sabic Marketing Ltd. Pres. Abdullah S. Nojaidi. "We believe that during the 1990s, MTBE will be one of the fastest growing chemicals in the world,"

Demand for MTBE, fueled by a need to provide octane to replace lead in gasoline, has sparked a building boom. Forty plants are running, and 30 are under construction or planned around the world.

Nojaidi said he expected MTBE operating rates to stay less than 70% of nameplate capacity until they begin to rise gradually in 1990. By 1995 MTBE plants will be running at 87% of nameplate, and feedstock could become a problem.

Sabic predicted methanol operating rates will hit 90% of capacity in 1990 and stay there through 1995.

"Companies with multiple methanol sources would then be in the best position to weather the storm," it said. "Producers also may have to find new isobutylene sources. Refiners with hydrocarbon streams they can upgrade to isobutylene are already building relatively inexpensive MTBE plants."

Sabic said MTBE plants of the future will have to be near sources of inexpensive field butanes, the approach it has taken.

Where Sabic''s operations are — Table 3

Companies	Location	Feedstock	Products	Capacity (1,000 metric tons/year)	Start-up
PETROCHEMICALS					
Saudi Methanol Co.	Al-Jubail	Methane	Methanol	640	1983
National Methanol Co.	Al-Jubail	Methane	Methanol	770	1984
Saudi Petrochemical Co.	Al-Jubail	Ethylene	Ethane	760	1985
		salt	Ethylene dichloride	560	
		Benzene	Styrene	360	
			Ethanol	300	
			Caustic soda	450	
Al-Jubail Petrochemical Co.	Al-Jubail	Ethylene	Lldpe	332	1984
Saudi Yanbu Petrochemical Co.	Yanbu	Ethane	Ethylene	560	1985
			Lldpe, HDPE	430	
			Ethylene glycol	250	
Arabian Petrochemical Co.	Al-Jubail	Ethane	Ethylene	650	1985
		Styrene	Butene 1	50	1988
			Polystyrene	100	1988
Eastern Petrochemical Co.	Al-Jubail	Ethylene	Lldpe	140	1985
			Ethylene glycol	330	
INTERMEDIATES					
National Plastic Co.	Al-Jubail	Ethylene	Vinyl chloride monomer	300	1986
			Polyvinyl chloride	200	
Saudi European Petrochemical Co.	Al-Jubail	Ethylene dichloride Methanol Butane	MTBE	500	1988
SUPPORT					
National Industrial Gases Co.	Al-Jubail	Air	Nitrogen	146	1984
			Oxygen	438	1985
FERTILIZERS					
Saudi Arabian Fertilizer Co.	Dammam	Methane	Urea	330	1969
			Sulfuric acid	100	
			Melamine	20	1985
Al-Jubail Fertilizer Co.	Al-Jubail	Methane	Urea	600	1983
National Chemical Fertilizer Co.	Al-Jubail	Methane	Ammonia	500	1987
			Compound and phosphate fertilizers	1,300	1988
MINERALS					
Saudi Iron & Steel Co.	Al-Jubail	Iron ore, limestone, natural gas, scrap iron	Rods and bars	1,200	1983
Steel Rolling Co.	Jeddah	Steel billets	Rods and bars	140	1962
GULF INDUSTRIES					
Aluminum Smelter of Bahrain	Bahrain	Alumina	Ingots, rolling slabs, billets, liquid metal	180	1971
Aluminum Rolling Mill	Bahrain	Aluminum	Sheets, corrugated sheets, can stocks	40	1985
Gulf Petrochemical Industries Co.	Bahrain	Natural gas	Ammonia	330	1985
			Methanol	330	

Source: Saudi Basic Industries Corp. 1987 annual report

The success of its venture could spawn similar projects in other nations that have no distribution system or market for their natural gas, Nojaidi said.

"Unless MTBE makers have secure feedstocks, they could find themselves in a bind as methanol operating rates rise and isobutylenes grow scarce."

Sabic unit starts shipments of MTBE

The international flavor of Sabic operations was apparent in start-up of the company's Saudi European Petroleum Co. MTBE plant at Al Jubail, the newest of the parent company's plants.

The 500,000 metric ton/year plant serves markets in the U.S. and western Europe, where MTBE is seeing increasing use as an octane enhancer in unleaded gasoline.

The plant shipped its first cargo Oct. 5, 1988, about 2½ months after start-up. The 20,000 metric ton shipment went to Rotterdam.

Sabic owns a 70% interest in the Saudi European plant, while Ecofuel of Italy, Neste Oy of Finland, and Arab Petroleum Investment Corp., an Arab regional financing firm, each own 10%.

The plant manufactures MTBE via an unusual configuration of technology licensed by Snamporgetti of Italy and the U.S.

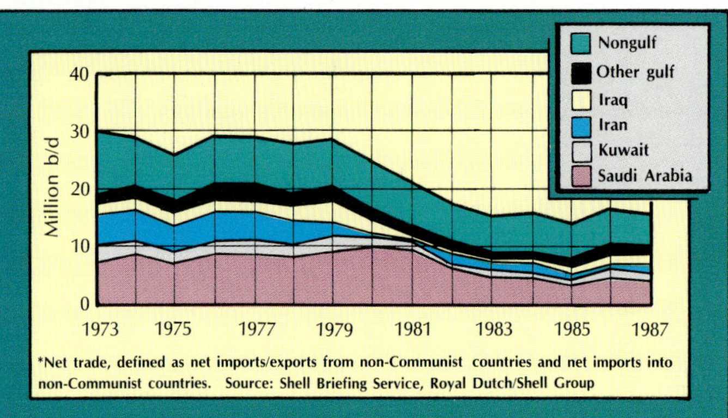

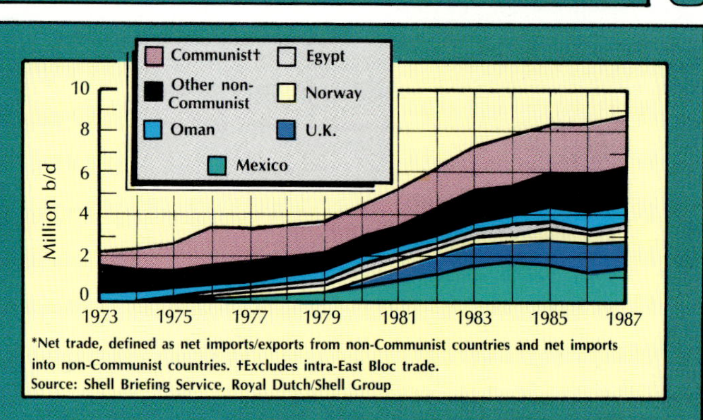

firms Air Products & Chemicals Inc. and UOP Processes International Inc.

It makes its own isobutylene feedstock by processing butane into isobutane, then dehydrogenating the isobutane to make isobutylene. Most other MTBE producers obtain isobutylene by cracking naphtha.

Downstream venture with Texaco

Saudi Arabia in November 1988 completed arrangements on a deal calling for a big downstream move abroad.

The Saudis and Texaco Inc. signed a definitive agreement selling Saudi Arabia a 50% interest in Texaco's refining/marketing business in 23 U.S. East and Gulf coast states and the District of Columbia. The nearly $2 billion joint venture agreement, plans for which were announced the preceding June 16, was signed in London by Saudi Petroleum Minister Hisham Nazer and Texaco Pres. James Kinnear. Nazer signed on behalf of Saudi Arabian Oil Co., the new state oil company created by a major reorganization of the Saudi oil industry that started after former Petroleum Minister Ahmed Zaki Yamani left almost 2 years ago.

Purchase price was $812 million. By sharing inventory and working capital, Texaco hoped to realize a total of about $1.8 billion in cash and savings. The exact amount was to depend on crude and product prices when the deal was scheduled to become final Dec. 31, 1988. The joint venture has the right to purchase as much as 600,000 b/d of Saudi crude at prevailing market prices and the right to use the Texaco trademark in the region covered by the deal.

Saudi Refining Inc., a wholly owned subsidiary of Aramco Services Co. (ASC), agreed to provide 75% of the joint venture's initial oil inventory, about 30 million bbl.

Texaco's interest in the venture is held by Texaco Refining & Marketing (East) Inc., a wholly owned subsidiary of Texaco Refining & Marketing Inc. (TRMI).

In addition to oil inventories, each partner agreed to contribute 50% of working capital. The venture, Star Enterprise, is based in Houston with marketing activities based in Dallas, Atlanta, Philadelphia, and Orlando, Fla.

Assets included in the deal were Texaco refineries in Port Arthur, Tex., with 250,000 b/cd capacity; Convent, La., 225,000 b/cd; and Delaware City, Del., 140,000 b/cd; 50 product distribution terminals with combined storage capacity of about 300,000 b/cd; and 11,400 Texaco branded stations, 1,400 of which were owned and leased and the rest franchised. Texaco's total wholesale and retail sales in the region averaged 390,000 b/d. Star Enterprise has the third largest gasoline sales volume in the region behind Exxon Corp. and Mobil Corp. and the 10th largest in the U.S.

The new company said it does not expect significant changes in operations, the 5,000 employee staffing level, or products and services. Nazer described the joint venture as a major achievement in Saudi Arabia's program of more active involvement in the downstream business.

Kinnear said formation of the joint venture was one of several major steps to restructure Texaco and boost shareholder value. "Not only will this joint venture produce significant cash benefits and savings for Texaco, it also will strengthen our nation's energy security by providing access to more than 4 billion bbl of crude oil over 20 years at market prices," he said. The two parent companies each have three representatives on a management committee.

Saudi budget reflects nonoil growth

Saudi Arabia's King Fahd Bin Abdul Aziz unveiled his country's budget for 1988 early in the year, calling it "a new phase of development in the national economy," reflecting the growth of the nonoil economy.

He also said the budget depended on internal resources "to achieve welfare and prosperity for the kingdom's citizens."

The budget projected revenues of 105.9 billion riyals ($28.24 billion U.S.) and expenditures of 141.2 billion riyals ($37.65 billion). To make up the shortfall of 35.9 billion riyals ($9.57 billion), the government planned to draw 8 billion riyals ($2.13 billion) from national reserves and issue treasury bonds with varying dates of maturity in amounts not to exceed 30 billion riyals ($8 billion).

Earmarked for domestic projects was 60 billion riyals ($16 billion), including 9 billion riyals ($2.4 billion) for 51 new projects. In announcing the budget, the king said, "In years past, our policies have directed oil revenues to meet requirements of national development and assist our citizens financially in improving living standards and establishing industrial and agricultrual projects, public utilities, and buildings for housing or investment.

"This assistance has achieved its mission."

Noting that government oil revenues had dropped to one fifth of earlier levels because of the decline in oil prices, the king cited budget revisions designed to maintain a level of expenditures "that will ensure the kindgdom's continued economic growth..."

Meantime, the Ministry of Finance and the National Economy released a summary that highlighted the growth of the Saudi economy. In 1987 Saudi Arabia's nonoil economy grew for the first time in 3 years. The real nonoil gross product increased by 0.8%, compared with declines of 5.7% in 1985 and 2.3% in 1986.

The country's trade surplus rose to 14 billion riyals ($3.73 billion) in 1987 from 4 billion riyals ($1.06 billion) in 1986. The value of total exports grew by 16% in 1987, paced by a 25% rise in nonoil products. Nonoil revenues for 1988 were projected to surpass those of 1987 by 7.8 billion riyals ($2.08 billion), reinforcing the signs that the country is changing

the past

1 - Conventional derrick lay barge
2 - Manual welding
3 - Jacket launching
4 - Conventional platform in shallow water
5 - Flare

the future

1 - The "Reed" deep water compliant platform (2000 ft)
2 - "Saturne" automatic welding system
3 - Arctic platform new concept
4 - Subsea pipeline trenching plough
5 - Derrick lay barge ETPM DLB 1601

ETPM is present

WORLD HEADQUARTERS: 57, AVENUE JULES-QUENTIN
B.P. 207 - 92002 NANTERRE CEDEX - FRANCE - Tel:(1) 40.97.63.00 - FAX:(1) 40.97.63.33 - Télex: 612021 F

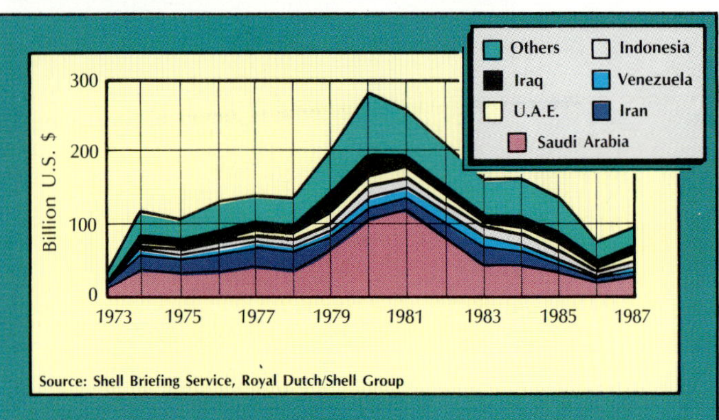

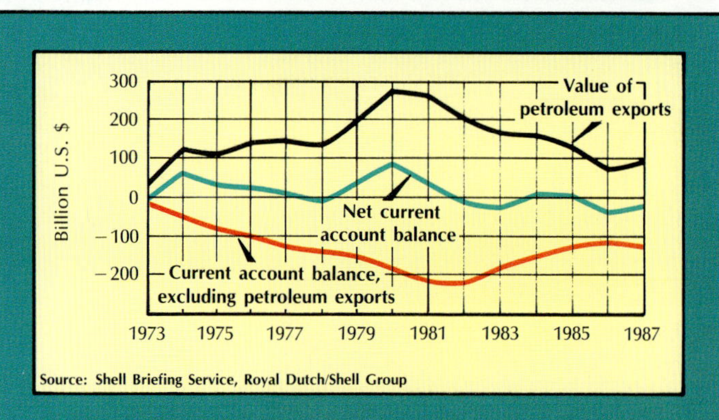

from an oil based economy to a more diversified one.

Analysts ponder direction of prices

As 1988 drew to a close, the oil market's mixed signals—price declines, slight recoveries, production increases, and promises of decreases—spawned a broad spectrum of analyses that predicted anything from another oil price collapse to oil price increases in 1989 and beyond.

Oil prices are likely to collapse in the near term despite current efforts within OPEC to prop them up by playing to crude oil buyer psychology, said Charles River Associates' Verleger.

"Pious appeals to producers to abide by quotas...should have only transitory impacts on prices," he said just before the November 1988 meeting of OPEC oil ministers in Vienna. Length and size of the price decline will be determined by actions of Kuwait, Saudi Arabia, and the U.A.E., which hiked production well above their OPEC quotas during July-August, he contended.

Verleger called Saudi Arabia the main culprit because its production in August 1988 was 698,000 b/d above its average level of production for the first 6 months of the year. "This sudden surge in output represents a preemptive bid for control of the market," he said.

In turn, the late 1988 downward pressure on prices stemmed from fears that overproduction would lead to a sharp increase in stocks by yearend. If continued, the excess production will cut into demand for OPEC oil in first half 1989, leaving OPEC with a market for crude oil that is less than 18 million b/d.

Verleger figured Saudi Arabia's rationale like this:

• The slide in non-OPEC production was offset fully by cheating on the part of other exporters, particularly the U.A.E.

• Expected future demands for higher quotas by Iran and Iraq would threaten to absorb any increase in demand.

• Rebuilding programs in Iran and Iraq could lead to a loss of Saudi and Kuwaiti oil exports to Japan and Korea if the programs were paid for through barter deals.

• The Saudis may have concluded that an increase in income was required.

Paul Mlotok of Salomon Bros. Inc., New York, cited $14/bbl for West Texas intermediate as a floor price.

"Below that level," he explained, "producers are unhappy because of the loss of revenues, consumers recognize that their dependence will rise too rapidly, and the world financial system recognizes that several heavily indebted oil producing nations are endangered.

"Looking back at better times, when the price was in the high teens, Saudi Arabia and Kuwait have tended to increase production to keep the price from rising much over $18/bbl. Hence, we conclude that the price range currently desired by these key OPEC nations would be within the limits of $14-18/bbl, with perhaps $16-17/bbl as the target."

Dean Witter Reynolds Inc.'s Eugene Nowak pegged the vagaries of diplomacy among Iran, Iraq, and Saudi Arabia as a key to the direction of oil prices in the short term. In the meantime, OPEC's inability to prevent cheating on production quotas will cause West Texas intermediate to fluctuate at $14-19/bbl in 1989-90. Nowak saw a note of optimism in Saudi diplomatic overtures to Iran, notably Saudi Prince Bandar's role in helping to arrange the cease-fire with Iraq, that could entail Saudi cooperation on OPEC policy in return for political concessions from Iran.

That cooperation might involve Saudi guarantees for a floor of $15-16/bbl for West Texas intermediate in exchange for Iranian promises to abandon efforts to subvert the Saudi monarchy and other gulf states.

Merrill Lynch cited $18/bbl as an ideal price for OPEC because it is low enough to spur demand but not high enough to head off declines in non-OPEC production.

Moreover, $18/bbl means a level of demand for OPEC oil that yields optimum revenues for OPEC members, satisfying OPEC needs while taking into account the price elasticity of supply/demand. At $18/bbl, demand for OPEC oil would reach about 19.5 million b/d (Fig. 8).

Of that volume, 16.5 million b/d would be exported to earn almost $110 billion/year. Merrill Lynch estimated nonoil revenues of another $20 billion.

Those revenues would just about meet OPEC's needs without incurring a deficit. Other combinations of price and production could produce the same revenues but would not be as workable, said Merrill Lynch.

"In fact, as oil prices decline, OPEC's revenue shortfall increases geometrically with the gap expanding at its fastest rate once prices drop under $15/bbl. If oil prices were to remain in the $17-22/bbl range, OPEC could basically generate the same amount of revenues if it adjusted its production accordingly. For prices of more than $22/bbl, OPEC would again be worse off." Thus Merrill Lynch saw $15/bbl as the "absolute minimum price tolerable" to OPEC and $18/bbl as highly desirable for the organization.

Low prices seen through 2000

Resilience of oil production growth outside OPEC likely will force the industry to live with low prices through 2000, said a study released in October 1988 by Arthur Andersen & Co., Houston, and Cambridge Energy Research Associates (CERA), Cambridge, Mass.

The industry is changing profoundly to adapt to a low oil price environment. And the Andersen-CERA study concludes the outlook for continued low prices is forcing industry adjustments in these major directions:

• Repositioning via consolidation, technological innova-

Ten years of Aramco producing, refining operations 

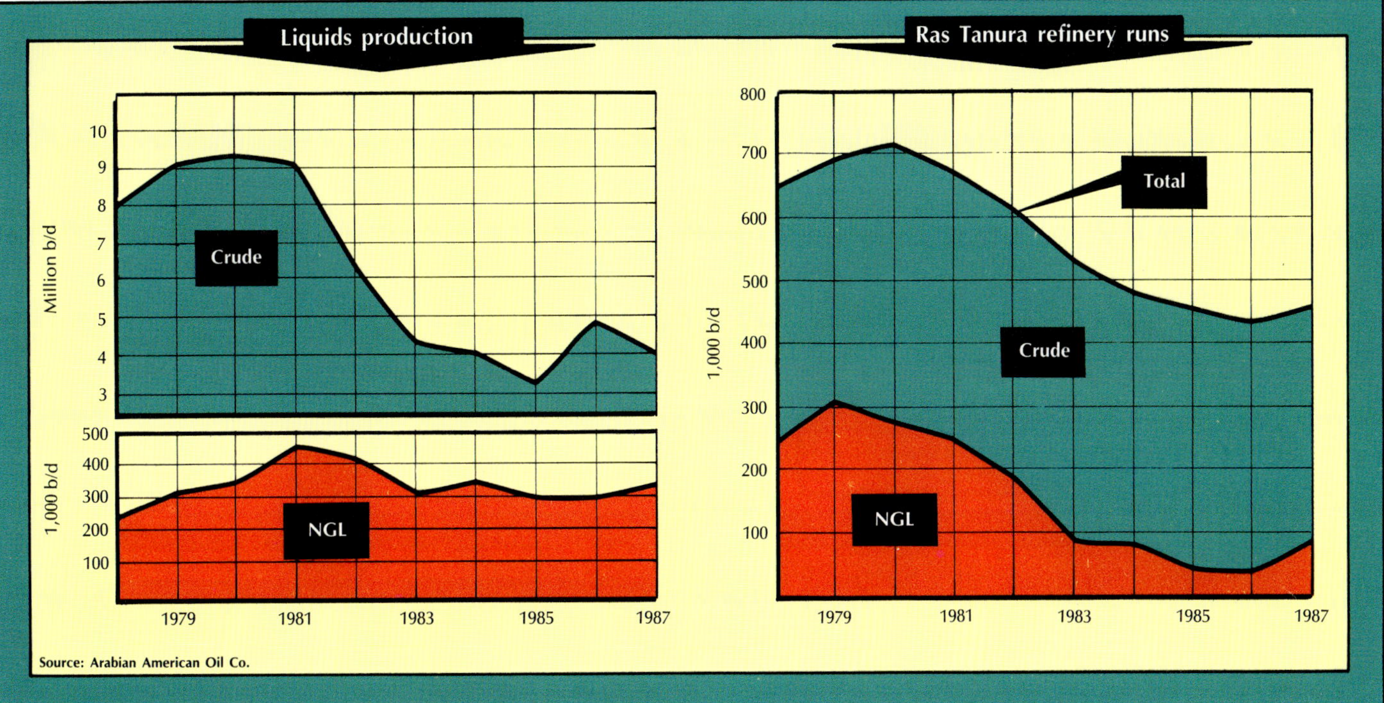

tion, and operating effiency, exploration/development activities, buying and selling reserves, international capital flows, and downstream capacity.

• Reintegration, as powerful economic reasons provide impetus to link upstream and downstream segments.

• Realignment, as resilience and further growth in non-OPEC production results in intensified competition in oil markets and in efforts to create new relationships among non-OPEC and OPEC producers.

The view of more robust prices in the early 1990s persists in some circles of the industry, the study acknowledged. But the intensity of activity apparent in the three trends "reflects a lowering of expectations, or at least a lessening of conviction, about how the price lines will point in the 1990s."

Since early 1986, non-OPEC production rose by 1 million b/d while U.S. production fell by about 1 million b/d, Andersen-CERA pointed out.

"This means that non-OPEC production outside the U.S. has increased by 2 million b/d since the [1986] price collapse," the study said.

Repositioning, reintegration, and realignment represent industry's attempts to prosper under low prices, Andersen-CERA said. Technological innovation is a major force in repositioned companies. For example, with better technology North Sea fields of 50-100 million bbl judged uneconomic at $28-30/bbl 4 years ago were being developed in 1988 on the basis of $15/bbl oil. Surviving companies underspent 1988 exploration and development budgets and boosted acquisitions of producing properties. Reintegration is driven by the quest for secure, adequate markets, but its pace will be limited by political factors in producing and consuming countries, the study said.

Surprising growth of non-OPEC production was spearheading creation of relationships between non-OPEC exporters and OPEC nations.

Non-OPEC production growth sharply eroded OPEC's market share, and the end of the Iran-Iraq conflict will intensify the surplus, Andersen-CERA predicted.

**OPEC in control
for the long term**

Never mind the short term gyrations of oil prices and shifting market shares, said Salomon Bros.' Mlotok in a 1988 analysis. Asked to cite the greatest misperception about international oil equities, he replied, "That OPEC is incapable of ever holding together long enough to exert control over world oil markets.

"In fact, we are convinced OPEC will remain an effective force for managing world oil markets, and its influence will grow through the rest of the century. While it may periodically lose control because of internal problems, these lapses will be of limited duration and magnitude."

OPEC is frequently criticized for its squabbling and inability to keep enough oil back from world markets to manage oil prices tightly. But Mlotok pointed out that OPEC has consistently kept 12-14 million b/d of productive capacity in check during a period of more than 6 years, with nearly all its members producing substantially below capacity.

What's more, since 1973 the organization seriously lost control of oil markets only once—and then for only 8 months—during January-August 1986.

"We do not mean to picture the organization as a well oiled, perfectly functioning machine," Mlotok said. "On average, however, it works, and it works better than most observers expect it to."

The mechanism for this is a mix of meetings, prices, and quotas confused by national and regional politics and saved in the end by pragmatic recognition of economic realities and ad hoc midstream ajustments.

Mlotok called OPEC "a problematic melding of 13 sovereign states" with often disparate interests and conflicting politics.

"Fundamentally, however, they are all nations that are extremely dependent on oil revenues," he said.

"Hence there is a great degree of common interest in stable oil prices. OPEC periodically will behave in a manner that leads to downspikes in oil prices, but self-interest will dictate that such lapses are limited and short lived."

For Mlotok, the picture that emerges from OPEC's history of jolting price increases, price discounting, netback pricing, production quota cheating, and the Saudis' willingness—then refusal—to play the role of swing producer is this:

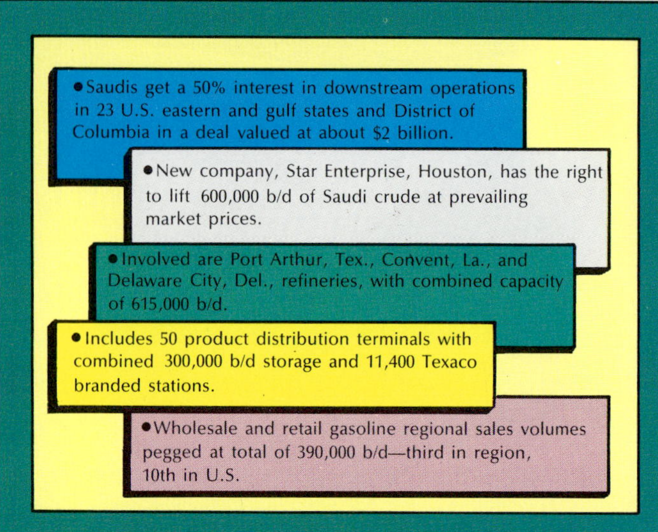

Highlights of the Texaco-Saudi deal | 7

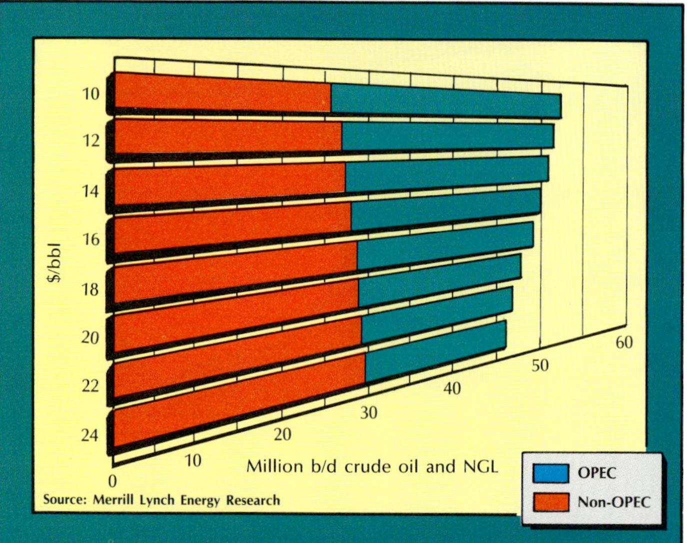

Oil demand at various prices | 8

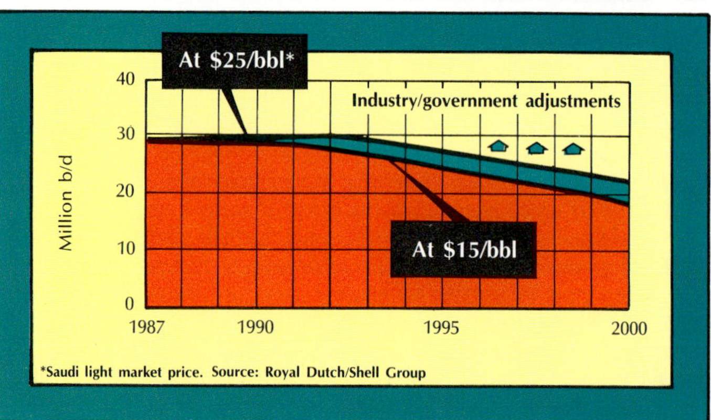

Outlook for non-OPEC oil flow | 9

A group of nations that usually has a fair idea of what's happening in world oil markets will make agreements designed to regulate markets, then ignore those agreements individually and collectivley whenever they think they can get away with it, and respond fairly rapidly when they realize they have gotten things wrong.

Mlotok said, "This will lead to ongoing volatility but not to collapse. And in a market where supplies and demand are converging, a modestly increasing real price of oil can be sustained."

Non-OPEC production seen sustainable

Non-OPEC oil production need not decline significantly for years, depending on what average prices the industry believes OPEC will be able to defend in the short and medium term, a Royal Dutch/Shell official predicted in August 1988. But the end of low, volatile oil prices could be further away than many in the petroleum industry expect, said John Jennings, a group managing director and exploration/production coordinator.

Successful defense of an $18/bbl marker price by OPEC in the medium term will help sustain non-OPEC production well into the 1990s, Jennings predicted.

Especially if demand grows slowly and research and development cut the cost of recovering large non-OPEC unconventional oil reserves and synthesizing transportation fuels from natural gas, "the eventual strengthening of oil prices may not be as dramatic as many seem to assume," Jennings told a British Institute of Energy Economics conference in London. But he warned that even modestly assertive predictions are almost always wrong. Non-OPEC production may decline quite slowly if Saudi light prices are maintained at $15-25/bbl, Jennings said (Fig. 9).

"It is likely that both these volume forecasts will prove pessimistic because they include only a small contribution from new discoveries, very little improvement in recovery efficiency in existing fields, and assume no further innovative cost cutting by the industry," Jennings said.

"They also neglect any further relaxation—adjustments—of fiscal regimes by governments, who I think will neither wish to see their indigenous industry wither nor become unduly dependent on imported oil.

"Thus, if OPEC defends a price of $18/bbl, it could be at least the mid-1990s before surplus production capacity begins to reduce significantly—and it may be later."

Only if OPEC fails to defend $18/bbl and retreats to a marker below $15/bbl is non-OPEC production likely to decline rapidly. Retreat to less than $15/bbl would cause most OPEC members financial difficulty and raise the possibility of destabilization.

"Unless some unexpected event disturbs the balance and either imposes restrictions on supply or somehow stimulates demand, this industry and governments may well have to come to terms with low, unstable prices for rather longer than we first thought," he said.

High oil prices of the late 1970s and early 1980s stimulated exploration in almost every sedimentary basin, Jennings pointed out.

But results for the most part were disappointing, and no new North Seas or Alaskan North Slopes turned up.

As a result, the industry is becoming more selective about major new commitments to frontier exploration and is increasingly being forced back to established, non-OPEC oil provinces with new emphases on maximizing recovery, "reexploration," and cost reduction.

The combination of improved industry operating effectiveness and reduced government takes has, at a 15% discount rate, halved the breakeven cost of many projects to about $15/bbl, Jennings said.

Shell companies' research and development targets these areas for conventional oil in the short term: cost reduction particularly offshore, seismic technology, enhanced waterflooding, integrated reservoir simulation and management, and enhanced oil recovery.

BEAUFORT SEA WILDCAT was under way in 1988 by Tenneco Oil Exploration & Production at a site 6 miles off the Arctic National Wildlife Refuge in Alaska. The Canmar/Reading & Bates Joint Venture Single Steel Drilling Caisson/MAT submersible rig was drilling the well to a projected 12,500 ft. Photo by Carlos Carpenter, courtesy of Tenneco.

Declining oil production, rising imports focus attention on frontier regions off Alaska

PERSISTENT FORECASTS OF DECLINING OIL PRO-duction and rising imports have helped focus petroleum industry and government attention on frontier regions off northern Alaska, as well as on the state's oil riches on its North Slope.

Industry executives in 1988 continued their efforts to find an answer to big questions:

How much oil is left to be discovered in areas like the Beaufort Sea?

And will the Chukchi Sea, when drilled, live up to high expectations?

What's more, will industry ever overcome environmental-ists' opposition and be allowed to explore federal acreage on the Coastal Plain of the onshore Alaskan National Wildlife Refuge (ANWR)? If so, will those prospects live up to industry's expectations?

Answers to those questions could help determine whether the U.S. has a chance to avoid undue dependence on imported oil.

Industry has the technology to find, develop, and produce arctic offshore oil. But whether oil prices will remain firm long enough to allow further development in the Beaufort Sea and a thorough exploration program in the Chuckchi Sea is a troubling question.

Alaska's role in U.S. oil supply

Conant & Associates Ltd., Washington, D.C., in October 1988 published an analysis of the oil supply importance of the region on and off Alaska's North Slope.

A report written by William E. Westermeyer for the October 1988 issue of Conant's Geopolitics of Energy said if a relatively rapid decline in North Slope production is to be forestalled, significant discoveries will have to be made very soon and brought on stream quickly.

Westermeyer, an analyst with the Ocean and Environment Program in Congress's Office of Technology Assessment, pointed out that by any measure Alaska figures prominently in calculations of U.S. oil supply.

In 1978, the first full year of operation of the Trans Alaska Pipeline System (TAPS) from supergiant Prudhoe Bay field to the Valdez tanker terminal, Alaska supplied an average 1.23 million b/d of oil—14% of U.S. production.

Alaska's share of U.S. production has increased about 1%/year since then, resulting in a 1988 flow of 2.05 million b/d, or 24% of U.S. crude.

"But what of the future of Alaskan production?" Westermeyer asked.

As recently as 4 years ago, virtually everyone was optimistic about the potential for substantial additional oil production from arctic Alaska, he said.

However, 4 years of largely unsuccessful exploration changed that perception. New assessments by Alaska, the Interior Department, Energy Information Administration, and operators suggest that Alaska's potential is far less promising that was once believed.

"To prevent North Slope production from falling, additional production must come either from more intensive development of existing fields, from discovered but undeveloped fields, or from undiscovered resources," Westermeyer said.

In late 1988, TAPS was operating near its capacity of 2.2 million b/d carrying about 1.6 million b/d from Prudhoe Bay field, 300,000 b/d from Kuparuk River, 100,000 b/d from Endicott, and 50,000 b/d from Lisburne.

Those four fields accounted for almost all of Alaska's oil. And all were expected to go into decline within the next 12 years despite added production from things like horizontal drilling to tap thinner pay zones and use of enhanced oil recovery methods.

North Slope flow was expected to begin dropping in 1990 and fall to 450,000-930,000 b/d by 2000, or about 22-46% of 1988 production.

Alaska's Department of Natural Resources (DNR) predicted Prudhoe Bay production in 2000 will be about 350,000 b/d, Kuparuk River 75,000 b/d, Endicott 40,000 b/d, and Lisburne 25,000 b/d.

"Production from developed fields cannot be expected to keep TAPS full much longer," Westermeyer said.

However, he warned against placing rigid faith in forecasts. Estimates of the onset of a decline in Prudhoe Bay field have been extended several times, Westermeyer pointed out, and the effect of new technologies is hard to gauge.

More available at a price

Uncertainty about the outlook for North Slope and Beaufort Sea oil stems from the direction prices will take, Westermeyer said.

DNR estimated that if low oil prices of $15/bbl in 1987 dollars prevail, only about 5.25 billion bbl is likely to be recovered from existing North Slope fields. But prices of more than $24/bbl, also in 1987 dollars, coupled with optimistic assumptions about technology could yield about twice as much oil as the low price scenario.

Two thirds of the difference between those low and high cases would come from infill drilling and use of EOR methods in Prudhoe Bay field and from development of West Sak field. The rest would come from development of new fields.

Standard's figures showed that beginning recoverable reserves amounted to 10.5 billion bbl of oil in Prudhoe Bay field alone. In March 1987 the field was half depleted.

Westermeyer said still more North Slope oil production could come from other known fields which, because of their smaller size, greater distance from TAPS, or more challenging reservoir characteristics than currently producing fields, were too costly to develop as 1989 approached.

Most of the undeveloped fields—Niakuk, Colville Delta, Gwydyr Bay, Seal Island, Point Thomson, and Hammerhead, for example—are small by North Slope standards, Westermeyer said. Even if developed they would make only a marginal contribution to production—not enough to make much of a difference in production forecasts.

Development of those fields is sensitive to the price of oil. Most are not likely to be developed until the price rises and stabilizes at $20-25/bbl.

One small developed field, Conoco Inc.'s Milne Point, was temporarily shut down because low oil prices did not allow economic operations.

West Sak, unlike the smaller undeveloped fields, could potentially make an important difference in North Slope production.

With an estimated 15-25 billion bbl of original oil in place, West Sak is by far the largest undeveloped North Slope field. But the outlook for development is clouded by things such as high viscosity of the oil and a complex structure. Unless currently unforeseen developments in EOR technology are made, recovery will be limited to about 5% of oil in place, Westermeyer said.

He called ANWR's Coastal Plain in the northeast corner of Alaska "the most promising prospect for major new discoveries."

Interior estimated there is a 19% chance that economically recovereable oil will be found in ANWR. And if ANWR holds any recoverable oil, a mean of 3.2 billion bbl is likely to exist.

Operators consider those odds encouraging and favor leasing of the Coastal Plain for exploration and development. However, there is spirited opposition by environmentalists to that move, which requires an act of Congress.

Even if oil operations are allowed on the Coastal Plain, production probably would not begin before 2000. So that region would go on stream too late to halt a decline in

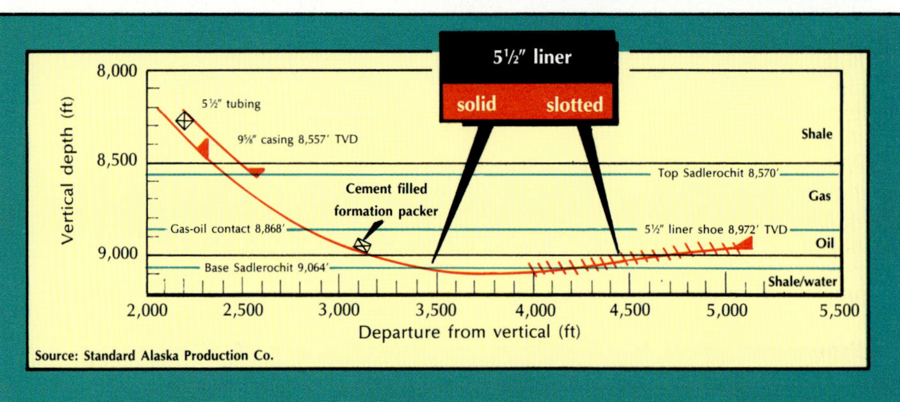

Alaskan production.

Except for ANWR, Westermeyer said, the outlook for major North Slope discoveries has dimmed considerably.

Preliminary data from a new assessment by the U.S. Geological Survey and Minerals Management Service (MMS) of undiscovered oil and gas resources suggest that there is less oil likely to be found in Alaska's onshore and offshore areas than was previously forecast.

For the federal Outer Continental Shelf, for example, MMS reduced its estimate of undiscovered, economically recoverable oil to 900 million bbl from 3.3 billion bbl.

Exploration since an MMS assessment in 1985 was disappointing in the Beaufort and Bering seas. Only the Chukchi Sea looked more promising in 1988 than it did in 1985, Westermeyer said.

Prudhoe Bay boost planned

Prudhoe Bay field operators in summer 1988 disclosed plans to begin work on a $350 million expansion of gas handling facilities in the big oil field.

Project goal is to boost production by about 90,000 b/d in 1990, partly offsetting a field decline. The project will hike total field recovery by at least 400 million bbl.

Plans called for the addition of two compressors at Prudhoe Bay's central compression plant (CCP), increasing gas handling capability to 5.1 bcfd from 3.6 bcfd.

The main Prudhoe Bay Unit owners—ARCO Alaska Inc., Standard Alaska Production Co., and Exxon Co. U.S.A.— were studying the feasibility of a further expansion of gas handling capability with a similar project.

A decision on that project could come in 1989, depending on technical factors, world oil prices, and Alaskan taxes.

In addition to boosting oil recovery through improved gas handling and greater use of dry gas for reservoir sweep in a reinjection program, the project will hike yield of miscible gas and NGL for Prudhoe Bay's miscible enriched gas enhanced oil recovery project.

The field's central gas facility (CGF) was handling 3 bcfd of gas and providing miscible injectant for the EOR project and about 40,000 b/d of NGL commingled with Prudhoe Bay crude and shipped through TAPS.

Without the expansion, field production would be limited by capability to handle associated gas. About 10% of the raw gas stream is miscible injectant and NGL, and the rest is sent to the CCP for compression and reinjection into the Prudhoe Bay gas cap pending marketability.

The project's resulting increase in oil production will allow wells to remain on stream longer and thus partly offset the expected natural decline in production.

Compressors will be shipped on the 1990 sealift of equipment to the North Slope. Other work related to the project will involve construction of pipelines to the CCP and CGF and some modifications to increase gas handling capability at Prudhoe Bay flow stations and gathering centers.

Field work was expected to get under way in the 1988-89 winter.

ARCO eyes hike in North Slope reserves

ARCO revealed in April 1988 that it expects to boost its proved North Slope oil reserves by almost 900 million bbl by 1997. In addition, the company said it expects to maintain its North Slope liquids production at more than 425,000 b/d through 1992, 2 years longer than it previously thought.

To achieve those goals, ARCO Alaska was prepared to spend more than $7 billion in the next 10 years, assuming reasonable oil prices and stable taxes.

In the Lower 48, ARCO's liquids production will dip slightly in the mid-1990s, then rebound in the late 1990s. Its natural gas production in the Lower 48 probably will increase slightly

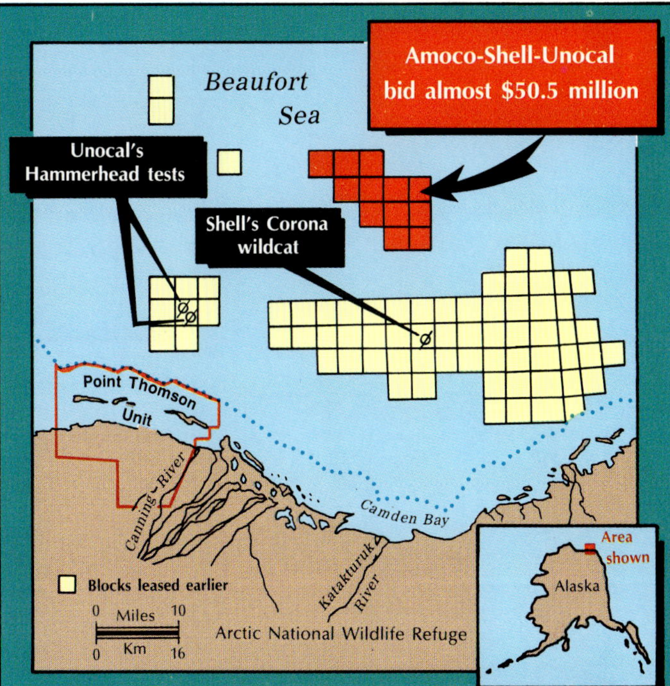

in the early 1990s.

In foreign operations, ARCO planned selective expansions, notably in the U.K. North Sea.

However, much of ARCO's upstream emphasis in the 1990s will be on Alaska, which holds 70% of its worldwide liquids reserves.

ARCO's North Slope reserves additions target is a 100 million bbl hike from its 1987 estimate.

The latest estimate did not include reserve additions that may come from exploration or the full potential of West Sak field, where ARCO plans extensive development in the late 1990s.

ARCO booked 260 million bbl of proved oil reserves on the North Slope in 1987.

That broke out to 146 million bbl in Prudhoe Bay field, 144 million bbl in Kuparuk River field, and a loss of 30 million bbl in Lisburne field.

The Lisburne decline occurred because of a reservoir that was not performing as well as expected.

During 1988-97, ARCO will spend about $4 billion for capital items associated with continuing development involving about 1,000 wells—plus continuing enhanced oil recovery efforts—in Prudhoe Bay and Kuparuk River fields and start-up of West Sak.

First phase development of West Sak is to start in the mid-1990s, although expected production levels were not disclosed.

Also on tap was the 1988 start-up of production from the Eileen West End portion of Prudhoe Bay field. Total Eileen development costs will be about $300 million.

ARCO's reserves share will be about 33 million gross bbl of crude and condensate—already mostly booked as proved reserves—out of Eileen's total contribution of about 170 million gross bbl.

It was prepared to spend another $3 billion on further exploration and development on the North Slope.

ARCO planned to jump its North Slope exploration spending in 1988 to $40 million from $18 million in 1987 and $10 million in 1986.

The 1988 figure was shy of the $78 million ARCO spent for North Slope exploration in 1985.

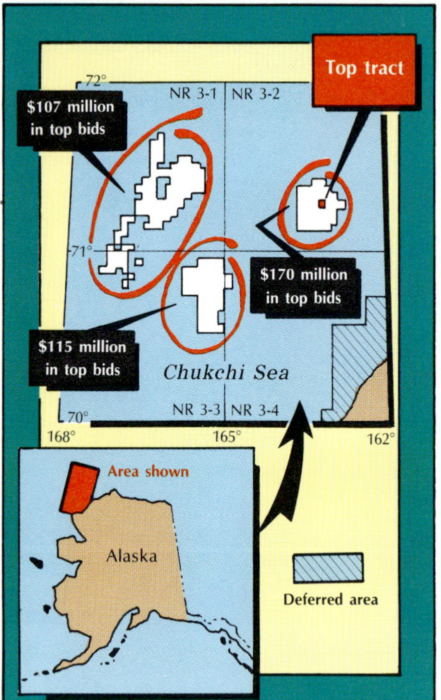

How Angts cost estimates compare

Segment	1982 $ 1982 Baseline	— Jan. 1, 1988 $ — 1982 Baseline	1988 Reestimate	Current $* 1988 Reestimate
	Billion U.S. $ (Billion Canadian $)			
Alaska	14.2 (18.0)	15.6 (20.8)	7.2 (9.6)	10.2 (13.6)
Canada	7.9 (10.0)	8.2 (11.0)	5.6 (7.5)	9.5 (12.7)
Lower 48	2.1 (2.7)	2.3 (3.1)	1.8 (2.4)	2.8 (3.7)
Total	**24.2 (30.7)**	**26.1 (34.9)**	**14.6 (19.5)**	**22.5 (30.0)**

*In service Nov. 1, 1995.

Source: Foothills Pipe Lines (Yukon) Ltd.

Standard's "inverted" Prudhoe Bay wells

Standard's operations in Prudhoe Bay field include a new twist in horizontal drilling: turning the wellbore past 90°. In a report in the Apr. 11, 1988, issue of Oil & Gas Journal, the company called two of its Prudhoe Bay wells, E-28 and E-25, high angle, "inverted" wells.

Standard had drilled 15 horizontal and high angle wells on the North Slope that were producing oil at rates three to four times that of conventionally drilled wells.

In Prudhoe Bay field, the company accessed 800-1,200 ft of Permian Sadlerochit sandstone through slotted liner in long horizontal holes, compared with 50 ft accessed in a vertical well. Sadlerochit is about 9,000 ft deep and 450 ft thick.

Standard tested its E-28 Prudhoe Bay Unit well at a rate of 15,300 b/d of oil, about four times that of a conventional well drilled in similar reservoir conditions.

The inverted wells were drilled vertically, then built angle while penetrating the Sadlerochit. The well path was designed to reach 90° near or just below the base of the reservoir. The wellbore was then turned upward, and the interval to be produced was drilled at an angle of 92-97° (Fig. 1).

The technique leaves a greater distance, or "standoff," between the reservoir gas-oil contact and the well's perforations. That extends the well's long term productivity and practically eliminates gas channeling.

In some cases, the inverted well can be converted to a conventional completion after the inverted interval is abandoned. Standard can maintain the standoff distance from the gas-oil contact as the gas cap expands.

The depth of the contact drops an average 10 ft/year.

Gas coning near the deepest measured depth can be restricted by setting a downhole choke to produce that part of the hole at lower rates.

The company set a cement filled formation packer along the unperforated part of the liner. The packer is insurance against gas from the gas cap channeling along the solid liner toward the perforations in the inverted hole section.

Liner cement jobs have been excellent, practically eliminating squeeze jobs, due to simultaneous reciprocating and rotating during cementing in wellbores inclined as much as 97°.

Standard said rotation mechanically assures that cement reaches under the low side of the pipe and that reciprocation provides fluid velocity and pressure surges to help break up gelled mud in washout zones.

The company had cemented 4 1/2-in. slotted liner in 8 1/2-in. hole, 5 1/2-in. and 7 in. slotted liner in 8 1/2-in. hole, and in one well 4 1/2-in. slotted liner in 6 in. hole.

Standard said costs of its inverted high angle wells in Alaska were no higher than conventional high angle wells at 82°-87°. Slotted liner completions can be used as long as the formation does not contain vertical permeability barriers.

Standard plans second offshore development

Standard in first quarter 1988 started permitting efforts for its second commercial project to produce oil from arctic waters, Niakuk field in the Beaufort Sea.

Niakuk lies northeast of Prudhoe Bay and west of Endicott field, where Standard on Oct. 3, 1987, started the world's first commercial production of arctic offshore oil.

The Niakuk project will cost about $130 million in 1987 dollars. Production is to start in fourth quarter 1991 and reach about 20,000 b/d of oil and 12 MMcfd of gas. Except for field use, the gas will be reinjected.

Plans call for Standard to develop the 58 million bbl, 30 bcf reservoir with 14 producers and injectors, including three existing wells on the Niakuk 4 exploration island. Drilling will begin in early 1991 and take about 13 months.

The project called for expansion of the Niakuk 4 gravel island in 4½ ft of water and linking it to Heald Point with a 1¼ mile gravel causeway. The causeway would have a 350 ft breach to accommodate marine life passage.

Also involved was a new road and pipelines from the tip of Heald Point to Lisburne drillsite L-5 and new pipelines to existing Prudhoe Bay Unit roads from L-5 to either the Lisburne field production center or Prudhoe Flow Station 2.

Standard will have minimodules fabricated in winter 1989-90 and trucked to the island. Niakuk construction, if permits are timely, would begin in winter 1989-90 and take about 22 months.

Standard had drilled nine wells in the Niakuk area since 1985. A 1986 step-out and a 1985 3D seismic survey spurred the development decision.

Niakuk is productive from a faulted extension of Kuparuk River sands. Standard called its sediments highly variable in quality but very permeable in the core development area—in some places capable of producing 5,000 b/d.

Unocal discovery disclosed off ANWR

Unocal Corp. in May 1988 disclosed a sizable oil discovery off ANWR's Coastal Plain.

The strike buttressed industry's view of the plain's high petroleum potential, which is crucial to efforts to lease and drill there. The well was drilled in the deepest water among disclosed discoveries in the U.S. share of the Beaufort Sea.

Unocal's 1 OCS P-0849 wildcat, a test of the Hammerhead prospect, flowed at a combined rate of 1,580 b/d from two Tertiary zones below 5,000 ft. Total depth was 8,034 ft. Site was on Block 624 in 103 ft of water 12 miles northeast of giant Point Thomson Unit gas/condensate field.

Unocal drilled a successful confirmation, 2 OCS P-0849, which it didn't test, on the same block in 107 ft of water 1 mile north.

The Canmar Explorer II rig drilled the wells, the first use of a drillship on the U.S. side of the Beaufort Sea. It drilled the discovery well in fall 1985, the step-out in fall 1986.

Unocal's partners in the discovery were Amoco Production Co. and Shell Western Exploration & Production Inc.

Unocal-Amoco-Shell, each with one third interest, acquired the Hammerhead prospect tracts in Beaufort Sea Sale 87. Block 624 netted more than $55 million. The same combine later acquired 12 parcels in Sale 97 of March 1988 for $50.5 million, accounting for almost half of the sale's total high bids.

Shell in 1986-87 drilled a wildcat, 1 OCS P-0871 on its Corona prospect, to 10,000 ft in 119 ft of water on Block 678 in a cluster of tracts that netted more than $100 million in Sale 87, conducted in August 1984. It has not disclosed results.

The award of Sale 97 leases more than doubled Unocal's holdings in the Beaufort Sea. In 1988 Unocal held interests of 25-100% in 70 U.S. Beaufort Sea blocks for a total of 145,365 net acres.

Little interest shown in latest Beaufort tracts

Sale 97 yielded the poorest showing in high bids of any U.S. Beaufort Sea federal sale to its time.

Various combinations of 13 oil companies submitted high bids totaling $118,574,936 for 218 blocks scattered from the edge of the Chukchi Sea to the Canadian border.

The sale's sole hotspot of interest in terms of very high bids was off ANWR. A combine of Shell, Amoco, and Unocal offered high bids totaling $50.5 million for 12 parcels about 40 miles north of ANWR (Fig. 2).

In offering Sale 97's highest bids, the combine returned to an area in which it had offered the highest bids in Beaufort Sea Sale 87.

In all, companies submitted 276 bids to expose only $139 million in Sale 97. That showed minimal competition for most tracts and generally tight concurrence of how companies assessed lease values in the bidding.

Most of the bids in Sale 97 were about half the minimum bid for leases in Sale 87, which itself drew less than half the total high bids of Sale 71's $2 billion in October 1982.

The lack of opposition many bidders encountered as well as the scattering of bids across the sale area showed the broad diversity of geological plays in the offering. Although much of the bidding was for buffer acreage, heavy bidding occurred for acreage in the deepest and remotest U.S. Beaufort waters offered to date.

Groups led by ARCO acquired the most tracts in the sale, 66, but Amoco and Shell separately, together, and in combination with other companies participated in 99 winning bids.

Other companies solo or leading groups with winning bids were Exxon, Texaco Producing Inc., Mobil Corp., Amerada Hess Corp., Conoco Inc., and Phillips Petroleum Co. Other bidders were Chevron Corp., Murphy Oil Corp., Union Texas Petroleum Corp., Petrofina Delaware Inc., Odeco Oil Co., and Elf Aquitaine Inc.

Drawing the sale's top bid was Tract 412 at the heart of a big structure off the Kaktovik Inupiat Corp. native lands portion of ANWR where Chevron drilled the only onshore ANWR test to date.

Shell-Amoco-Unocal offered $23.388 million for 412, $11.787 million for 457, $7.65 million for 413, $4.73 million for 456, and $2.94 million for the remaining eight tracts in a southeast-northwest trending feature off Camden Bay.

Chukchi Sea's geologic provinces

Chukchi Sea sale sparks leasing interest

In May 1988, a group led by Shell overpowered the competition in a record federal oil and gas lease sale, the first in the Chukchi Sea off Northwest Alaska.

Chukchi Sea Sale 109, offering the most remote and severe environment acreage off the U.S., netted more bids—653 for 351 blocks—than any other federal sale outside the Gulf of Mexico. The total offering was 4,700 blocks covering about 26 million acres in arctic waters.

It was the second largest offering of acreage off Alaska, covering parcels 3-240 miles offshore in 60-250 ft of water.

Sale 109 garnered $478,177,948 in high bids out of $666,232,384 exposed. That made it the fifth biggest sale off Alaska in the sum of high bids, following only the Navarin basin sale and the first three Beaufort Sea sales.

Sale 109 and Beaufort Sea Sale 97 were the first federal sales outside the Gulf of Mexico in more than 3 years. Sale 109 originally had been scheduled for February 1985 but was postponed at Alaska's request.

Sale results showed a surprisingly strong belief, especially by the Shell group, in the Chukchi Sea's hydrocarbon potential. Not even a continental offshore stratigraphic test had been drilled in the sea. And only a few wildcats were drilled and plugged along the Chukchi coast.

Further, the Chukchi Sea could prove to be the most difficult petroleum province in terms of operating conditions that the U.S. industry has experienced: severe storms, massive ice forces, and extreme remoteness.

What was especially striking about the results, other than Shell's dominance, was the widespread nature of geological plays. About 40 structures underlie Chukchi tracts that received bids.

There were frequent instances of one or two tract features

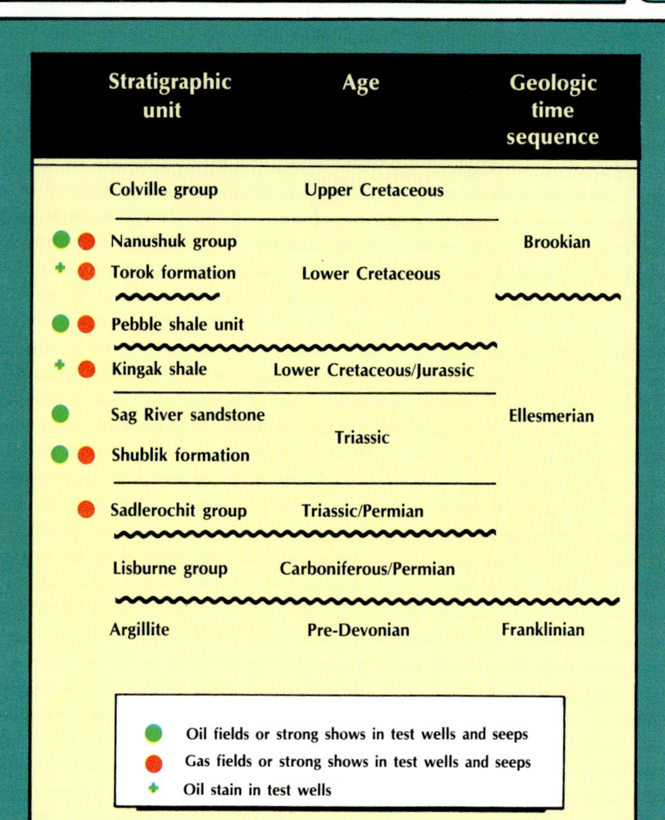

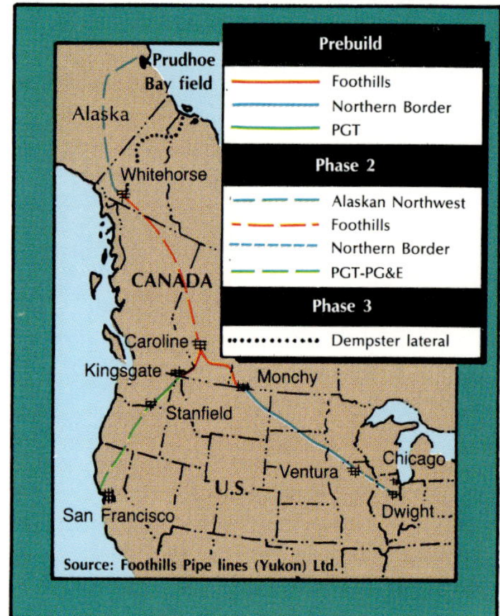

receiving single bids. And there were several big structures underlying parcels receiving multiple bids. Three such trends accounted for about three fourths of the sale's bidding, in dollars and number of bids (Fig. 3).

Bids ranged from about $30 to $3,000/acre.

Shell, bidding alone or in various combines with Conoco Inc. and Elf Aquitaine Inc., was high bidder on 164 tracts with bids totaling about $368.3 million.

That far outstripped the effort by a runnerup Texaco Inc. group. Texaco, bidding alone or in combines with Murphy Oil Corp., Ocean Drilling & Exploration Co., and Petrofina Delaware Inc. was high bidder on 87 parcels for about $47.5 million in the sale.

The Texaco group was as active as the Shell group, netting parcels in perhaps 13 plays.

Murphy later said the Chukchi Sea holds structures of a size seldom seen outside the Middle East. It revealed no further information.

The Shell group also accounted for all but two of the sale's tracts netting at least a $5 million high bid, leaving many millions of dollars on the table in the process. Seldom did another bidder offer more than $5 million, even on the most sought after tracts.

Often a Shell group bid was tenfold or more that of the next highest bid or for several million dollars when there were no other bids.

A combine of Amoco and Exxon offered about $22.2 million in high bids for 19 parcels. Its high bid of more than $11 million for a key tract in the sale's biggest prize, a massive structure in about 130 ft of water 70 miles northwest of Wainwright, topped a Shell group offer of $7.7 million.

Amoco-Exxon carved out a seven tract portion of the structure the Shell group otherwise dominated with apparent high bids totaling $151.7 million for 30 tracts. Many of the bids topped $2,000/acre.

The sale's top tract, NR 3-2 718, in the same area, went to a Shell group for more than $16 million, or about $2,800/acre.

MMS said the best chances for discoveries were on structures and in stratigraphic traps in the northeastern and western parts of the sale area: the Barrow arch, Arctic platform, Chukchi platform, and North Chukchi basin (Fig. 4).

Structural traps predominate in the Hanna trough, North Chukchi basin, and Hope basin, while stratigraphic traps predominate in the Barrow arch and Chukchi platform provinces. MMS said the most prospective reservoirs are expected to be found in the Lisburne and Sadlerochit groups and Shublik and Sag River formations (Fig. 5). All are productive in Prudhoe Bay field. Best potential source rocks are the Shublik formation, Kingak and Pebble shales, and Torok formation.

Shell's shift to Chukchi Sea

Shell Western's dominance of Sale 109 signaled a shift of some of parent Shell Oil Co.'s frontier exploration thrust to the Chukchi Sea from the deepwater Gulf of Mexico. The company said it was not leaving the deepwater gulf, however. In the gulf, Shell Oil Co. affiliate Shell Offshore Inc. in August 1988 gave notice to Reading & Bates Drilling Co. that it planned to end its multiyear contract for the Zane Barnes semisubmersible.

The rig was drilling in about 2,700 ft of water on Garden Banks Block 426 under a contract that required a 6 month notice of cancellation. So the unit was scheduled to continue under contract to Shell until late February 1989.

Shell Western signed a contract with the Canmar/Reading & Bates joint venture for use of the Explorer III arctic drillship in the Chukchi Sea. The rig was being retrofitted for Chukchi service with a top drive drilling system in Victoria, B.C.

Shell Western and partners Elf Aquitaine and Conoco planned to drill the Chukchi Sea's first wildcats beginning in July 1989 at a cost of about $70 million. The first prospect to be tested, Ivan, was in the northeast corner of Block NR 3-3 in about 100 ft of water. The combine expected a 90-100 day drilling window, during which it planned to drill one and one-third wells in 1989. If the first well is a discovery, the second probably won't be spudded.

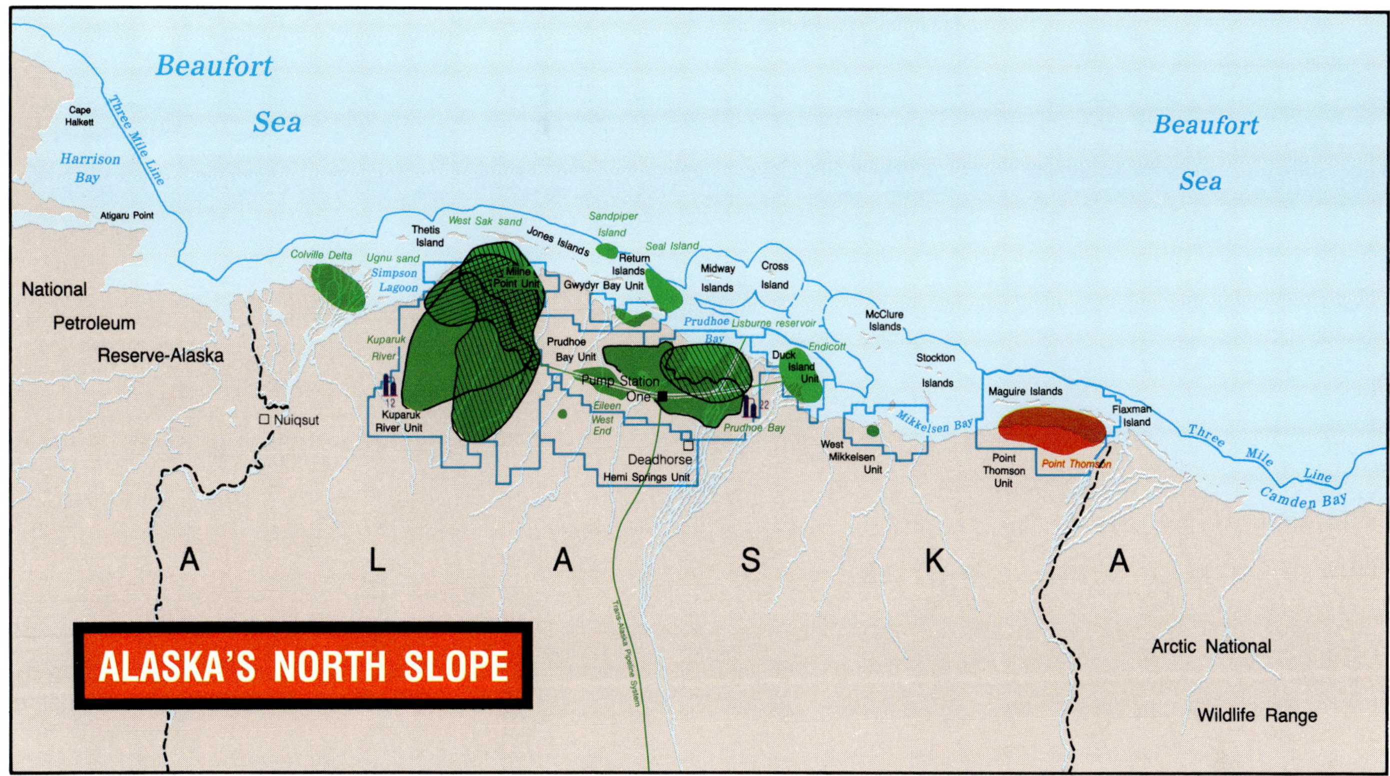

The partners hoped to prove that Chukchi Sea geology is an extension of the Alaskan North Slope's.

A spokesman said Shell Oil Co. decided to shift some of its exploratory drilling budget to the Chukchi Sea from the deepwater gulf because its 1988 exploration/production budget, although big at $2.4 billion, was not big enough to carry both programs.

A spokesman pointed out that Chukchi exploration will require a very long time because of the limited number of months exploratory drilling can take place. Because of ice, even with arctic drilling equipment, access to drillsites will average only 2-3 months/year. And in some years, some sites will not be accessible at all. Development lead times will be lengthy, too. Shell Oil Co. estimated it will require 10-15 years from the time of discovery until production starts.

The company said Shell Offshore's notice of cancellation of the Zane Barnes contract did not mean a long term shift away from the deepwater gulf. It planned to continue its multiyear contract to use Sonat Inc.'s Discoverer Seven Seas drillship, which was drilling for Shell Offshore in about 5,700 ft of water on Mississippi Canyon Block 730.

Shell Offshore used the previous 3 years of area-wide lease sales in the gulf to become the region's largest deepwater lessee.

Lower cost seen for gas pipeline

In 1988, 20 years after discovery of Prudhoe Bay field and 11 years after the field went on stream, the North Slope still had no pipeline outlet for remaining gas reserves of 20.5 tcf.

The key reasons: lack of a pressing need for the gas in the Lower 48 and a pipeline project's tremendous construction costs. However, the Alaska Natural Gas Transportation System (Angts) said its proposed northern section could be built for $14.6 billion in 1988 U.S. dollars, only 56% of the capital cost in an earlier estimate in 1982 (see table).

Based on the lower capital costs, the cost of shipping Alaska North Slope gas to the Lower 48 would average $4.80/MMBTU the first 5 years and $5.35/MMBTU the first 10 years of operation, said Northwest Alaskan Pipeline Co., Salt Lake City, and Foothills Pipe Lines (Yukon) Ltd., Calgary, chief sponsors of Angts.

Key factors in the reduced estimate of capital costs were faster construction, incorporation of results from project test sites, pipeline design advancements, improved welding productivity, more traditional owner/contractor functions, use of existing communication systems, and reduced contingency.

The latest estimate, unveiled in June 1988, also took into account more modest inflation and interest rates than the 1982 estimate. Under latest assumptions, project cost would reach $22.5 billion by the time the line went into service Nov. 1, 1995. In 1982 the companies estimated an in service cost of $24.2 billion in 1982 dollars.

The 1988 reestimate called for flow of 2.3 bcfd through 42-48 in. pipe, up from 2.1 bcfd through 48-56 in. pipe in the 1982 estimate. Pressures would be 1,680-2,160 psig, up from 1,260-1,080 psig. Angts initial throughput would be the equivalent of 400,000 b/d and increase to the equivalent of 600,000 b/d, the sponsors said.

The updated figures resulted from a detailed project the sponsors began early in 1987. Foothills said it was convinced more U.S. gas supplies will be required by the mid to late 1990s, when the pipeline could be operating, and the supplies can be delivered at market prices.

The system's southern legs had been delivering Canadian gas to the U.S. since 1981-82. They made up the so-called prebuild segment of the system (Fig. 6).

The partners' disclosure of the reestimate said nothing about a decision to proceed. Construction time was understood to be about 3 years, excluding time to obtain financing and sign gas purchase contracts.

ARCO withdrew from the Angts combine in December 1987, declaring the project uneconomic.

ARCO based its decision mainly on the belief that no Lower 48 market will exist for some time for gas at Angts delivered prices. That belief stemmed from updated figures similar to the Northwest-Foothills reestimate, an ARCO spokesman said.

Foothills in January 1988 blasted a determination by the Reagan administration that owners of Alaskan gas be allowed to seek sales in Asia.

Yukon Pacific Corp., which proposed to export Alaskan North Slope gas to Asia as LNG, said there is enough Alaskan gas to serve Asian and U.S. markets. IPE

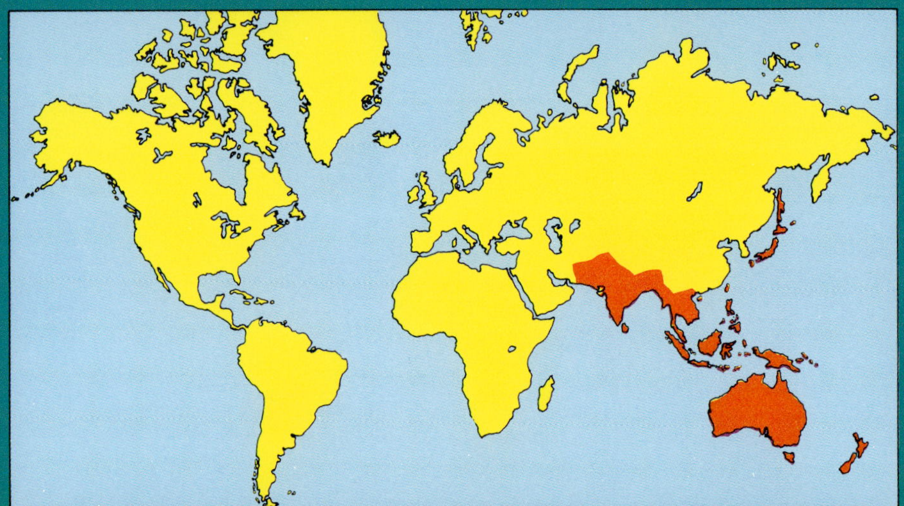

ASIA—PACIFIC

AUSTRALIA

CAPITAL: Canberra
MONETARY UNIT: Dollars
REFINING CAPACITY: 644,100 b/cd
PRODUCTION: 552.3 Mb/d
RESERVES: 1,673,089 Mbbl

AT MIDYEAR 1988 AUSTRALIAN ENERGY OFFICIALS expected a sharp rebound in offshore exploratory drilling for the year.

Sen. Peter Cook, minister for resources, said as many as eight rigs would be active offshore during 1988, twice the 1987 maximum. Six rigs at work then would remain under contract at least through early 1989, and two operators had plans to move one offshore rig each to Australia soon.

As many as five rigs would drill in the Timor Sea/Bonaparte Gulf region, where BHP Petroleum Pty. Ltd. and partners were boosting capacity of Jabiru oil field to 60,000 b/d, planned to develop Challis oil field, and scored other discoveries.

Cook cited a June 1988 survey by the Australian Petroleum Exploration Association (APEA) anticipating drilling of 32-47 exploratory wells offshore in 1988, compared with an estimated 15 in 1987. Onshore exploratory drilling was expected to be steady with 1987's level at 201-224 wells. According to the survey, operators would drill 38 onshore development wells and 22 offshore, both figures up two from 1987.

The offshore outlook brightened during the course of the year. In March 1988, APEA's survey anticipated drilling of 6-21 fewer exploratory offshore wells than it did in June. Of the new expected total, 27-36 would be be wildcats and the rest appraisal wells. Operators were committed to drill 16 offshore exploratory wells in 1988 and 24 in 1989 on work program permits. Of these, two in 1988 and 11 in 1989 are guaranteed wells.

The federal Department of Primary Industries and Energy says companies planned significant activity beyond commitment wells, especially in the Timor Sea. Minimum work program commitments called for seven wells in the area in 1988, and as many as 16 more wells might be drilled.

On the Northwest Shelf, another active area, work commitments called for five exploratory wells. Three additional wells probably would be drilled. Three commitment wells would be drilled in the Gippsland basin.

According to APEA's survey, operators would conduct 26,000-34,000 line km of seismic surveys offshore in 1988, compared with 24,889 line km in 1987. Recent seismic peak was 42,448 line km in 1985.

APEA's onshore exploratory drilling projection in June was slightly more optimistic than the 190 well projection in March. Onshore exploratory drilling has been buoyed in recent years by efforts to add gas reserves in the Cooper basin. APEA says 67 gas exploratory wells were drilled in 1987, and 65 were expected in 1988. The onshore active rig count had dropped to 15 from a maximum of 30 in 1987 and 22 in early 1986. Australian onshore drilling is concentrated in the Cooper/Eromanga and Bowen/Surat basins.

APEA expected onshore seismic activity to total 15,000-17,000 line km, about even with 1987's level.

Timor Sea activity

Australia and Indonesia took the first step toward resolving an offshore territorial dispute over the so-called Timor

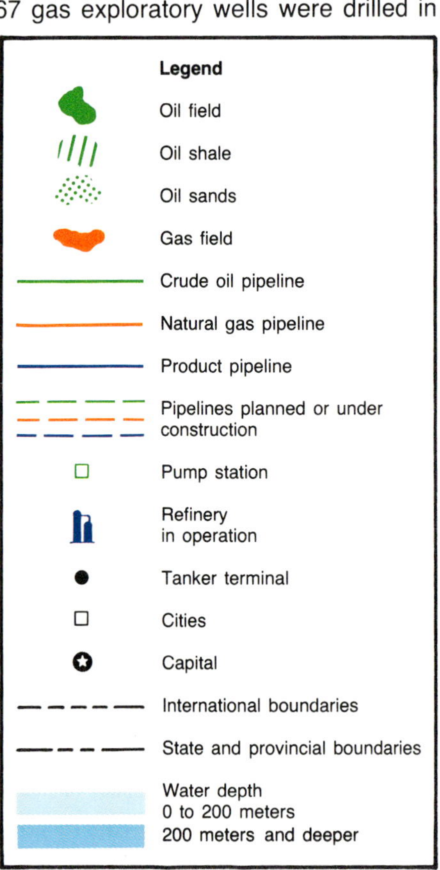

Gap.

Proposals for a joint zone of cooperation between the Indonesian island of Timor and Australia were disclosed after a meeting between officials from both countries. The proposals require approval by both governments before they can be implemented.

The disputed area lies northeast of the Jabiru and Challis oil discoveries by BHP Petroleum Pty. Ltd. and its partners in the Timor Sea. Australian and Indonesian officials proposed that the region under dispute be divided into three areas.

In Area C, closest to Timor, Indonesia's tax and legal regime would be in force, but Australia would have the right to 10% of revenues. Area B, closest to Australia, would be subject to the Australian regime. Indonesia would have the right to 10% of revenues. Area A, between Areas B and C, would be administered by the two countries through a joint authority and a ministerial council. This zone contains a large structure known as Kelp.

The areas would be delineated on the northwest side by a simplified bathymetric line and on the southeast by a 200 nautical mile line measured from Indonesian archipelagic base lines. Eastern and western sides would be set by equidistant lines.

Meanwhile, BHP Petroleum Pty. Ltd., Melbourne, was stepping up its drilling program in the Timor Sea frontier off the northwest coast of Australia. Peter Willcox, BHP chief executive officer, said his company would drill as many wells in the Timor Sea as it has in the past 6 years.

Four rigs—three semisubmersibles and a drillship—were to drill 25-26 wells for BHP and its partners in various permits. BHP is producing Jabiru oil field, the first developed in the area, through a floating production system. It is building a custom designed floating system for nearby Challis oil field, expected on stream in 1989 as the second development project in the Timor Sea.

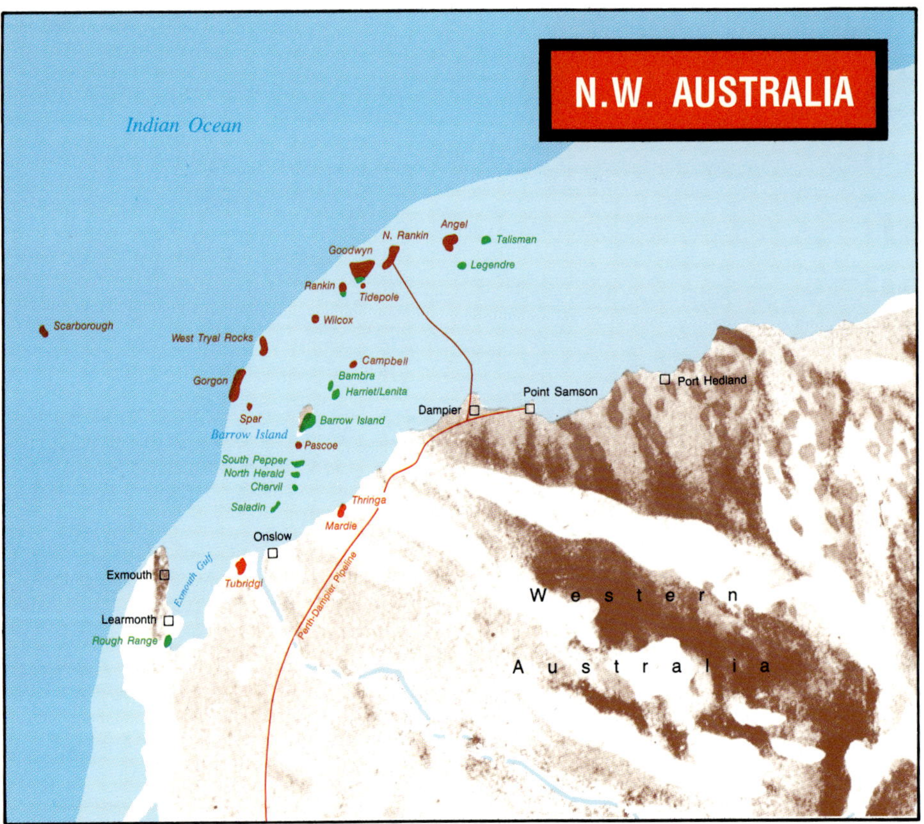

Jabiru and Challis, along with small Northeast Challis and Cassini that will be tied into Challis, hold reserves of 75 million bbl. Jabiru alone holds 50 million bbl, a big increase from an original estimate of 11 million bbl. Jabiru and Challis are 1984 discoveries. The 115,000 dwt production storage barge for Challis was under construction in Japan. It will be permanently moored by a single anchor leg rigid arm mooring (Salram). The $136 million (Australian) system will be able to handle as many as 12 wells in four fields. Production is to start in third quarter 1989 from 1, 2A, and 3 Challis at an initial rate of 24,000 b/d. Reserves there are estimated at 22 million bbl.

Bass Strait
activity

The Esso Exploration & Production Australia Inc.-BHP Petroleum (Australia) Pty. Ltd. combine plans to develop two small oil fields in the Bass Strait off Australia with steel monotowers at a cost of $85 million (Australian).

Esso-BHP will place a double deck monotower each over the discovery wells of Dolphin field, a 1967 find, and Perch field, a 1968 strike. The combine estimates reserves for Dolphin at 6 million bbl and Perch at 13 million bbl.

Top deck of the monotower will be a helipad, below which will be the equipment deck and wellhead. Iron ore gravel ballast, pumped into bases at installation, will hold the monotowers to the seafloor.

Esso-BHP will connect the fields to a processing plant at Longford, Victoria, with a 29 mile pipeline. A separate pipeline will return associated gas to the fields for reinjection. Field development became economic when Australia's government exempted from excise taxes the first 30 million bbl of production. The project remains sensitive to oil price changes, combine officials said. First production is due in September 1989. Meanwhile, Santa Fe International Corp.'s Southern Cross semisubmersible was to begin a Bass Strait exploration program for Esso-BHP. The rig, stacked the last 2 years, was to begin work in license areas near existing fields.

And the Esso-BHP partnership planned to develop two more small oil fields in Bass Strait off Victoria, Australia.

It will spend $88 million (Australian) to develop Seahorse and Tarwhine fields, near the planned Whiting development announced earlier. Seahorse and Tarwhine production will start in 1990 at an initial combined rate of 8,500 b/d. Esso-BHP estimates reserves at 7 million bbl for Tarwhine and 4.4 million bbl for Seahorse.

Western Australian
activity

Marathon Petroleum Australia Ltd.'s development of Talisman oil field off Western Australia will be similar to that of larger Jabiru oil field in the Timor Sea, 740 miles northeast.

Marathon and partners were to start production early in 1989 at about 10,000 b/d from a single seabed completion in 262 ft of water. A riser will connect the wellhead to a leased production/storage tanker. The operator will produce the field for a year before deciding whether to drill more wells.

Marathon had not disclosed reserves estimates for Talisman. Outside sources estimate 10 million bbl. The field is 75 miles off Dampier and about 43 miles east of North Rankin gas field.

Northwest Shelf
development

Horizontal wells have aided development of two marginal fields on Australia's Northwest Shelf.

Production is averaging 9,500 b/d of oil from North Herald and South Pepper fields off Western Australia in the Indian Ocean, 19 miles south of Barrow Island.

The fields, operated by Western Mining Corp. Ltd., Perth, were discovered in the early 1980s but considered uneconomic until early 1987. Combined reserves are estimated at 4-6 million bbl. Western believes the wells are the first horizontal penetrations in Australia. It chose horizontal drilling in an attempt to reduce gas and water coning and increase oil production rates.

North Herald, placed on stream Dec. 27, 1987, has one horizontal well, 3 North Herald, on production. Full flow from South Pepper started last Jan. 30, 1988.

North Herald has a 39 ft oil column and a water drive, and

South Pepper has a 33 ft oil column with a gas cap and water drive.

The South Pepper wells were planned to be horizontal in the oil zone so 9⅝ in. casing could be set to isolate the gas cap. Western believes the wells may be the first in which 9⅝ in. casing is set at 90° inclination.

The company drilled 3 North Herald in fall 1987 in 59 ft of water, setting 20 in. casing at 230 ft and 13⅜ in. casing at 1,650 ft. It drilled a 12¼ in. hole vertically from the 13⅜ in. shoe to 2,394 ft, the kickoff point. Western built angle to 89° at 4,625 ft measured depth, 3,916 ft true vertical depth, where deviation from vertical was 1,322 ft. Western set 9⅝ in. casing at 4,608 ft. It drilled an 8½ in. hole holding the 89° angle from the 9⅝ in. shoe to 5,691 ft measured depth, 3,939 ft true vertical depth, for a deviation of 2,388 ft. The 8½ in. hole was drilled with this inclination to penetrate various sections in the oil zone. Western then ran a 7 in. slotted liner to total depth.

The 6 and 7 South Pepper, spudded in December 1987, were similarly drilled to 90° inclinations. Horizontal sections are at 5,576-6,613 ft measured depth in 6 South Pepper and 5,543-6,629 ft in 7 South Pepper.

Queensland pipeline

Queensland, Australia, called for bids for construction of a 330 mile, 12.2 in. gas pipeline to be owned by the state.

The proposed Timor Gap solution

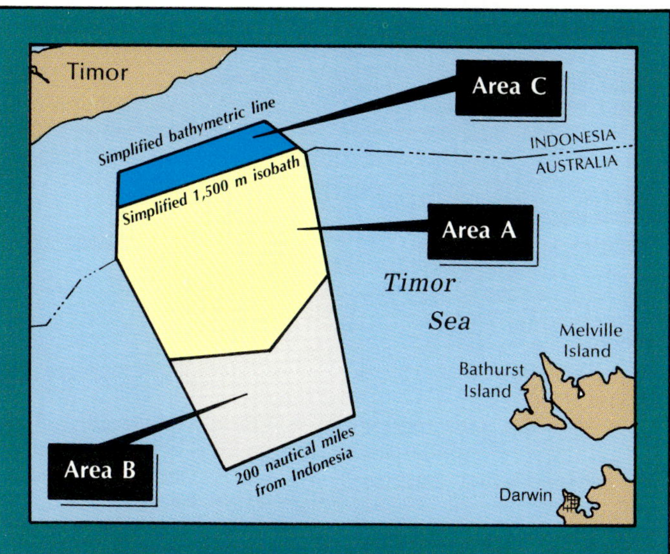

Bass Strait monotower

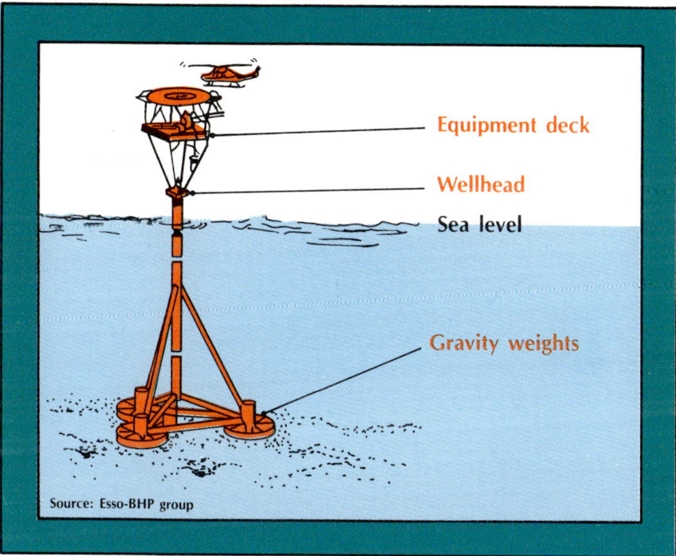

The line with connect Surat basin and new Denison Trough gas fields in Southeast Queensland with the central east coast industrial center at Gladstone. It will cost about $100 million (Australian).

A gas sales contract essential to the project had not been signed. It was expected to involve producers CSR Ltd. and Oil Co. of Australia NL and the buyer, Queensland Alumina Ltd., which runs an aluminum smelter at Gladstone.

AFGHANISTAN

CAPITAL: Kabul
MONETARY UNIT: Afghanis
REFINING CAPACITY: NA
PRODUCTION: NA
RESERVES: NA

THE U.S.S.R. EXPECTED TO WIELD PERVASIVE, long term influence in northern Afghanistan's petroliferous region after the withdrawal of Soviet occupation forces and the political chaos that may follow.

As the only purchaser of Afghanistan's major export, natural gas, the U.S.S.R. supplies most of its southern neighbor's foreign exchange. Because Afghan revenues from natural gas sales are in rubles, any new Kabul government must continue to accept the Soviet Union as its most important trading partner or obtain large loans from non-Communist sources to prevent economic collapse.

The Kabul correspondent of the Moscow newspaper Sovetskaya Rossiya reports that joint Soviet-Afghan exploration for oil and gas is slated to continue after the Soviet troop withdrawal.

Further electrical power development based on gas and increased nitrogen fertilizer production in northern Afghanistan depend largely on Soviet credits, materials, equipment, and technical personnel.

Gas production, exports

Most of the gas produced by northern Afghanistan's two large fields—Hodja-Gugerdag (Khoja-Gugerdak) and Djar-Kuduk (Jarkuduk)—moves by pipeline to the U.S.S.R. Development of both fields, construction of twin 63 mile pipelines from Shibarghan area fields to the Soviet border, and laying of a 55 mile domestic pipeline east to the Afghan industrial center of Mazar-i-Sharif were made possible by extensive Soviet aid.

It is not known how severely Afghan gas production was reduced by rebel activity or how much gas transmission was curtailed by pipeline sabotage. What is known is that Afghan gas exports to the U.S.S.R., which normally would account for at least two thirds of production, have slipped below the level achieved before the Soviet invasion in 1979.

The U.S.S.R. bought 2.65 billion cu m of Afghan gas in 1975 and had expected to take more than 3 billion cu m in 1981. Instead, Afghanistan sold only 2.3 billion cu m of gas to the Soviets in 1981 and 1982. Official Moscow figures place Soviet imports of Afghan gas at 2.422 billion cu m in 1985 and 2.22 billion cu m in 1986, the last year for which data are available.

Afghanistan's gas production has not been officially reported since the 1979 Soviet military takeover. But gas flow probably is close to 3 billion cu m/year, far short of the 4 billion cu m Moscow originally targeted for 1981.

The two 32 in. pipelines from Shibarghan area gas fields to the Soviet border have operated far below their capacity of 4 billion cu m/year ever since their construction. Gas flow through the 13 in. pipeline to Mazar-i-Sharif's 36,000 kw electrical power station and 105,000 ton/year fertilizer plant has repeatedly been disrupted by rebel action.

Soviet reports in the 1970s estimated Afghanistan's "explored" (proved plus probable) gas reserves at about 5 tcf. Hodja-Gugerdag's initial reserves were placed at slightly more than 2 tcf.

Discovered in 1962 and on production since 1967, Hodja-Gugerdag has had falling production during the 1980s.

Anticipating the decline in flow from Afghanistan's first commercial gas field, Russian and Afghan personnel placed nearby Djar-Kuduk on stream in May 1980. Djar-Kuduk's flow was expected to boost Afghanistan's total gas production "immediately" from about 2.3 billion cu m/year to nearly 4 billion cu m/year.

Soviet imports

Domestic oil and especially gas development will be highly important to Afghanistan's economy after the Russian troop withdrawal. The U.S.S.R., on the other hand, can easily do without imports of Afghan gas, which equal less than 0.3% of soaring Soviet production.

In contrast to the Soviet Union, Afghanistan would be dealt a severe blow if its gas sales to the U.S.S.R. stopped.

Afghanistan's 1986 gas exports to the Soviet Union represented about 82% of total exports to the U.S.S.R.

Value of Afghanistan's gas sales to the Soviet Union peaked at almost 262 million rubles in 1985. With volume and prices falling, Afghanistan received less than 201 million rubles from gas exports to the U.S.S.R. in 1986.

Afghan oil development

Development of Afghanistan's small oil reserves and construction of the nation's first refinery, with a rated capacity of 10,000 b/d, have lagged badly since a 1979 Soviet-Afghan treaty called for work on those projects.

First oil strike was made at Angot (Angoth) in 1977 at 3,182 ft. Soon afterward, another oil discovery was reported at Ak-Darya, and 2 years later a third oil field was found at Kashkar.

In 1979, Moscow announced that "the time is not far off when commercial production and refining of oil in Afghanistan will begin." The Soviets have been silent since 1979 regarding specific progress toward development of an Afghan oil industry, saying only that work is under way.

BANGLADESH

CAPITAL: Dhaka
MONETARY UNIT: Taka
REFINING CAPACITY: 31,200 b/cd
PRODUCTION: 0.9 Mb/d
RESERVES: 500 Mbbl

ROYAL DUTCH/SHELL GROUP, AFTER 8 YEARS OF geophysical work interrupted by tribal hostilities, was to begin its first drilling program in Bangladesh.

It planned wells on both of its large production contract areas. It was to begin drilling at Shalbanhat on its 5,000 sq mile area in the north. It has shot 1,214 line km of seismic survey in the area since signing the production sharing contract in April 1986. Shalbanhat is 6 miles from the Indian border. Shell was moving 10,000 tons of material and equipment into the area, including a rig from North Yemen.

On its 5,170 sq mile area in the Chittagong hills of southeastern Bangladesh, Shell was reactivating a program idle since January 1984, when three employees were kidnapped and held for ransom during a tribal insurrection.

Before then, the company had shot about 350 line km of seismic survey. It planned to begin drilling at Sitapahar near Kaptai Dam, the country's only hydroelectric project. The program includes horizontal drilling in the Tripura fold belt. Insurrections in the area had diminished, and the government beefed up security. Shell acquired the Chittagong contract in May 1980. Until Scimitar Exploration Co. signed a production sharing contract late in 1987, Shell was the only foreign operator at work in Bangladesh. Since then, the country has offered contracts covering all open acreage.

INDIA

CAPITAL: New Delhi
MONETARY UNIT: Rupees
REFINING CAPACITY: 1,051,441 b/cd
PRODUCTION: 631.8 Mb/d
RESERVES: 6,354,200 Mbbl

WITH OIL CONSUMPTION CLIMBING STEADILY AND production growth stalling, India was trying to find and develop oil and gas reserves to rein its dependence on imports. Results were spotty at the beginning of 1988.

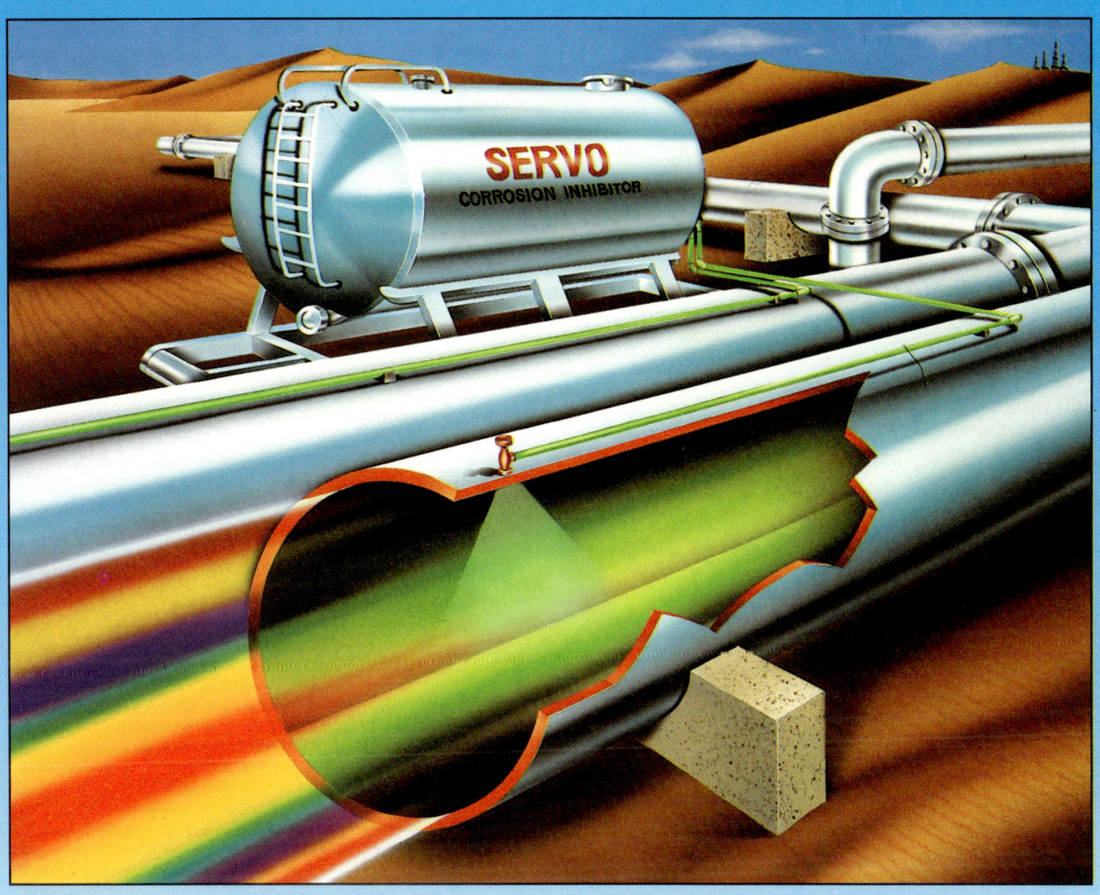

GAS CORROSION?
SERVO HAS THE INHIBITORS AND THE EXPERIENCE

We have been involved for many years in a world-wide campaign against corrosion. Our experience is at your disposal. Please contact:

Head office:
P.O.Box 1,
7490 AA Delden,
The Netherlands

Phone: 05407-63535*
Telex: 44347
Telefax: 05407-64125

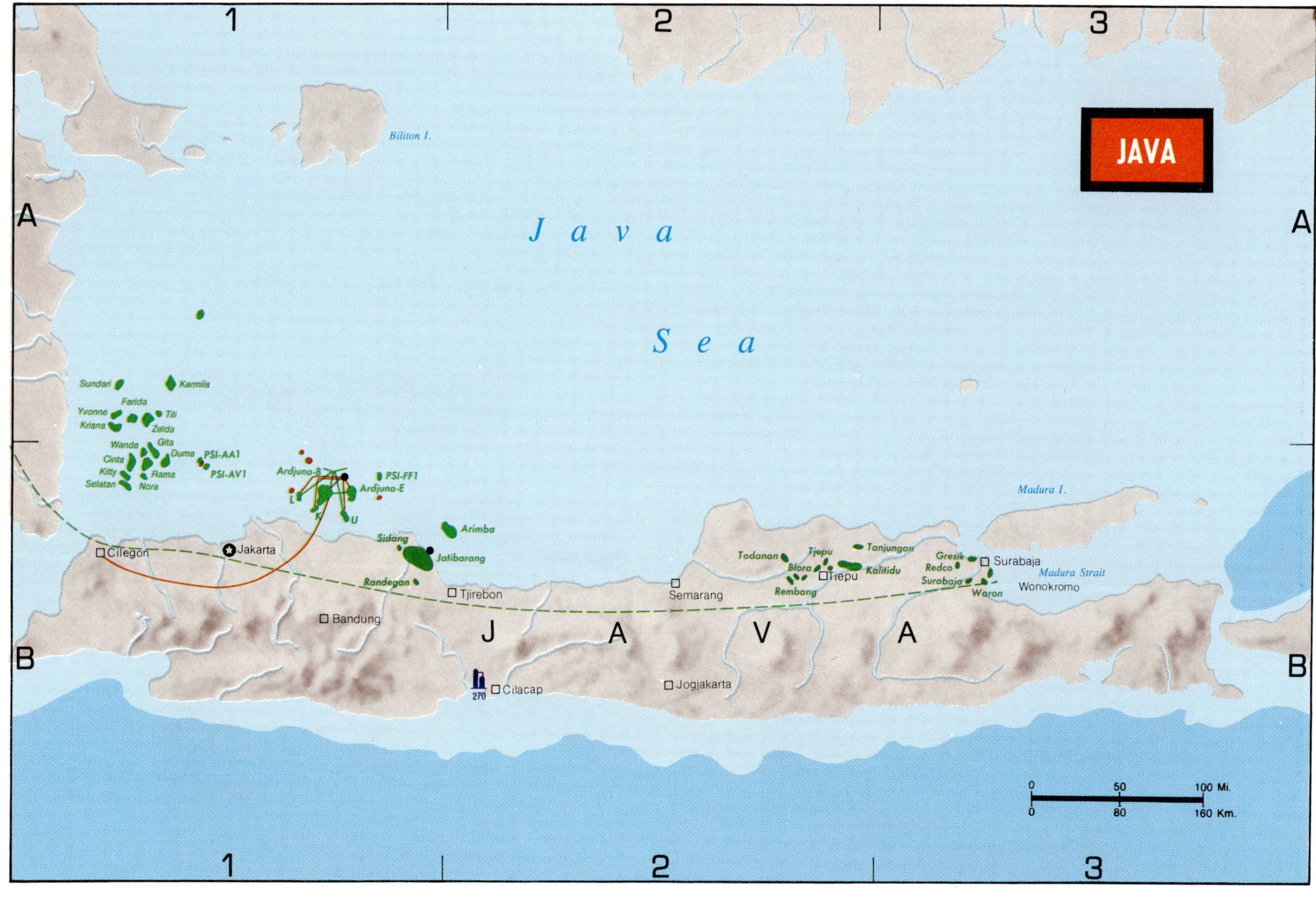

Bombay High oil field, discovered late in the 1970s off the western coast, remains India's last major discovery. Estimates of country's oil reserves are 4.25 billion bbl and gas reserves, 17.55 tcf.

Production in the fiscal year ended April 1987 totaled about 30 million metric tons, about its level of 1985-86 and only slightly more than the 29.99 million tons of 1984-85. Consumption for 1986-87 totaled 43.5 million tons, compared with 40.9 million tons the previous year, according to Indian Oil Corp. Ltd. It's expected to reach 50 million tons/year soon.

The government's two main oil and gas agencies—Oil & Natural Gas Commission (ONGC) and the smaller Oil India Ltd. (OIL)—have been trying to find and develop reserves quickly.

ONGC, which produced 27.86 million tons of crude in 1986-87, wants to add 2.3 million tons to its production total by 1990. Recent exploration successes indicate it may achieve the goal. OIL isn't faring as well. Its 1986-87 production total of 2.35 million tons lagged the prior year's total of 2.65 million tons and its target of 2.84 million tons. It was doubtful OIL will meet the unchanged target for 1987-88.

Even with shortfalls by OIL, however, marginally higher flow from ONGC fields could enable India to achieve national production targets of 30.46 million tons for 1987-88.

Total natural gas production is expected to increase to 231.7 bcf from 167.7 bcf in 1986-87, but much is still flared.

Exploration expands

ONGC and OIL are exploring throughout India, but production remains concentrated offshore around Bombay High and onshore in Assam and Gujarat.

Production from the western offshore area—including Heera, Ratna, and Panna fields in addition to Bombay High—averaged about 430,000 b/d.

At Bombay High, ONGC late in 1987 commissioned four major processing platform complexes and two water injection platforms, but the facilities are expected only to keep production at plateau levels.

Since 1980, ONGC has drilled 13 oil, six gas, and five oil and gas strikes in the area, where it operates 21 rigs. Western offshore fields under development include Tapti gas field and South Bassein, mainly a gas field, where ONGC is using horizontal drilling to try to recover oil from thin pay.

Among other western offshore fields under development are structures designated D-18, D-1, D-11, B-57, B-131, B-132, B-134, B-178, and B-179.

Elsewhere, ONGC has made promising discoveries in the southern onshore region, especially in the Cauvery basin. It has put its Narimanam oil find on production and expects the latest strike on the Bhuvanagiri-II structure near Chidambaram, Tamil Nadu, to add significantly to the country's oil flow. OIL's fields are in Assam.

The agency has explored in other states with limited success. Its only significant discovery in recent years was in Arunachal Pradesh, where production is not expected for some time.

Other activity

The Krishna-Godavari basin yielded a gas discovery in a

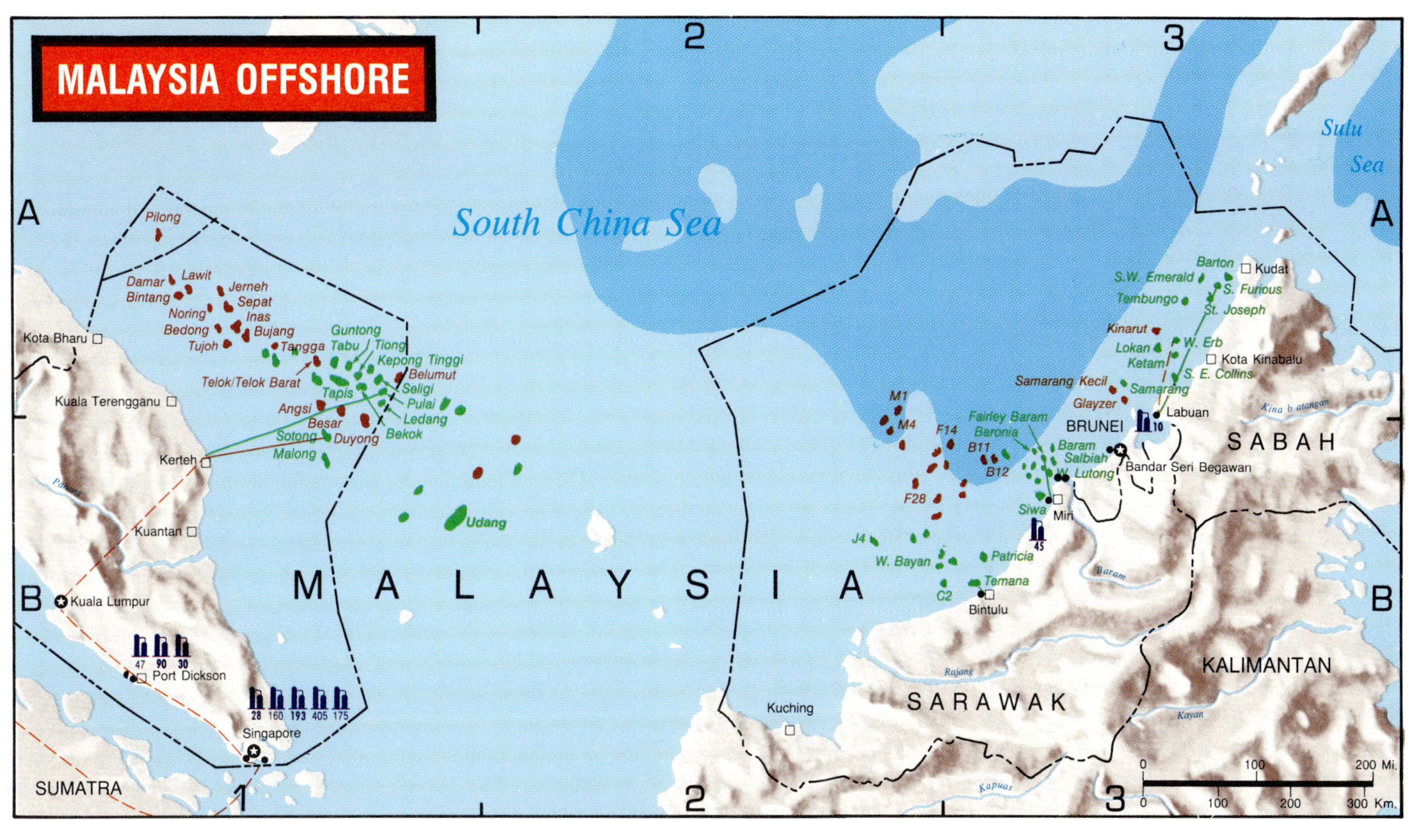

Prospects in disputed Gulf of Thailand

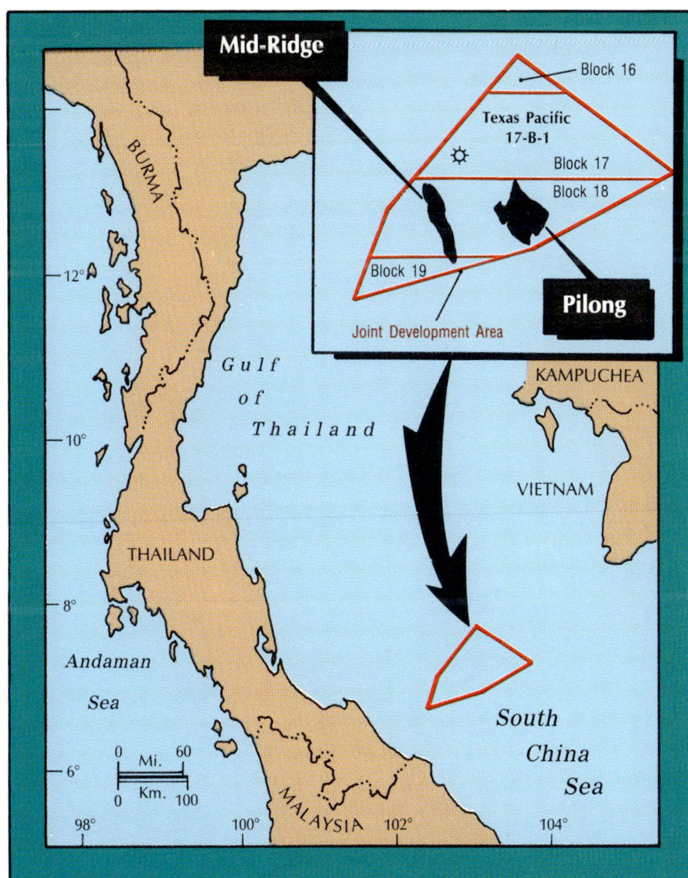

previously unproductive area in basin's north.

ONGC 1 Mandapetta, 15 miles southeast of Rajamundry in Andhra Pradesh, encountered gas at a depth of 14,114 ft.

Test results weren't available.

The onshore part of the basin is gas-prone. Petroleum Ministry officials plan to develop onshore discoveries in the next 2 years to supply 106 MMcfd of gas to commercial users in the region.

Elsewhere in eastern India, ONGC tapped oil in the first well on the Sonari structure in Assam's Sibsagar District.

Drilled to 14,928 ft, the well tested 700 b/d of oil and 706 Mcfd of gas. It's in the Assam-Arakan basin, which covers the far northeastern Indian states of Assam, Arunachal Pradesh, Nagaland, Mizoram, Manipur, Tripura, and Meghalaya.

Amoco Corp. took its first exploratory step into India.

And the Soviet Union will drill its first Indian well outside the Cambay and Cauvery basins.

The Indian government and ONGC signed a production sharing contract with Amoco India Petroleum Co. for Krishna-Godavari basin Block KG-OV-5, covering 580 sq miles in the Bay of Bengal off Andhra Pradesh.

Amoco will bear exploration costs and, at ONGC's option, share with the government 40% of any production. India has the right to buy all remaining production until it's self-sufficient in oil.

Under a contract signed in 1986 with ONGC, the Soviet Union was to drill its first well in the Bengal basin early in 1989. Site will be at Sonarpur. ONGC will bear costs of Bengal basin drilling.

India will develop an offshore oil field believed capable of producing 80,000 b/d. Petroleum Minister Brahm Dutt told a consultative committee of Parliament the field, Neelam, is the country's most promising discovery since Bombay High.

Like that 1974 discovery, Neelam is off western India. It has thick oil pay, high permeability, and adequate reservoir pressure, Dutt said. The field covers 33 sq miles earlier thought to involve two structures designated B-131 and B-132.

Dutt estimated recoverable oil reserves at 314 million bbl. He estimated "geological reserves" of oil and gas at 1.13 billion bbl of oil equivalent. ONGC will begin flow with a floating early production system while a platform is being

MALAYSIA-INDONESIA
BURMA
LAO PEOPLE'S DEMOCRATIC REPUBLIC
SOCIALIST REPUBLIC OF VIET NAM
Hanoi
Haik'ou
HAINAN
Fang
Vientiane
THAILAND
Hue
Da Nang
Payagon
Mae Sot
Sirikit
Pru Krathiam
BP1
Bangkok
Sriracha
64 63
65
Andaman Sea
Gulf of Thailand
DEMOCRATIC CAMPUCHEA
Tonle Sap
Mekong
Phnom Penh
Sihanoukville
Ho Chi Minh City
Long Xuyen
Mouths of the Mekong
South China Sea
Luzon I.
Subic Bay
Limay
Manila
Quezon City
Batangas
63
155
Mindoro I.
Platong
Trat
Pakarang
Satun
Erawan
Baanpot
Khanom
Jakrawan
Funan
Si Thammarat
12
12-8
15-B-4X
16-8-5X
16-B-1
17-E-1
Pilong
Dua-1X
Goloc
Matinloc
Pandan
Cadlao
Nido
Libro
Panay I.
Bacolod
Ce
Palawan I.
Negros I.
Sulu Sea
Kudat
S.W. Emerald
Tembungo
Barton
S. Furious
Kinarut
St. Joseph
Kota Kinabalu
Labuan
SABAH
Sandakan
MALAYSIA
Kamar
Lawit
Jerneh
Bintang
Sepat
Noring
Inas
Bedong
Bujang
Guntong
Tabu
Tujoh
Tangga
Kerteh
Tekok/Telok
Belumut
Barat
Besar
Tapis
Sotong
Anggi
Malong
Puyong
Terebuk
Belanak
Udang
Kutaradja
Pinang
Arun
Perlak fields
North fields
Rantau fields
Pangkalanberandon
Belawan
MALAYA
Kuala Lumpur
Port Dickson
Natuna I.
Anambas I.
M1
M4
Samarang Kecil
Samarang
Glayzer
Bandar Seri Begawan
Lutong
Miri
BRUNEI
F28
F14
J4
Patricia
W. Bayan
Temtina
Bintulu
C2
SARAWAK
Mengatal
Bunju
Pamusian
Bunju I.
Tawau
Simeulue I.
Sibolga
30
99
Dumai
Sungei Pakning
28
760
113
435
175
90
SINGAPORE
Batam
Nias I.
Puncak
Jingga
Gajom
Rintis
Minas fields
Beruk
Lirik fields
Kampar
Central fields
Bukittinggi
Padang
SUMATRA
Siberut I.
Djambi
Djambi fields
B. Hari
Mari
Bangka I.
Plaju
Sungei Gerong
61
Palembang
Limau fields
Talang Akar fields
Radja fields
Muraraönim fields
South fields
Mentawai I.
Billiton I.
Pontianak
Kuching
BORNEO (KALIMANTAN)
Kapuas
Makaham
Batuputih
Sangkulirang
Badak
Kerindingan
Melahin
Santan
Attako
Dian
Tanjilatan
Nilan
Pamaguan
Handil
Badak
Senga-Sanga
Bekapi
Kenindan
Louise
Sepinggan
Sambodja
Tengah
230
Balikpapan
Bangkerak
Sanggur
Tandjung
Palangkaraya
Bandjarmasin
SULAWESI (CELEBES)
Makasar
Indian Ocean
Telukbetung
Sundari
Karmila
Krisna
Cinta
Nora
Rama
PSI AA1
PSI-AV1
PSI FF1
Ardjuna-B
Ardjuna-E
Jakarta
Sindang
Arimbi
Balongan
Tjibarang
Tjedugan
Semarang
Rembang
Bandung
Cilacap
Jogjakarta
Tjepu
Rembang fields
Surabaja
WonoKromo
270
JAVA
Java Sea
Singaradja
Bali I.
Sumbawa I.
Raba
Flores
INDO
Strait of Malacca
South
China
Sea
200
100
300
400 Mi.
200
400
600 Km.
1
2
3
A
B
C
D

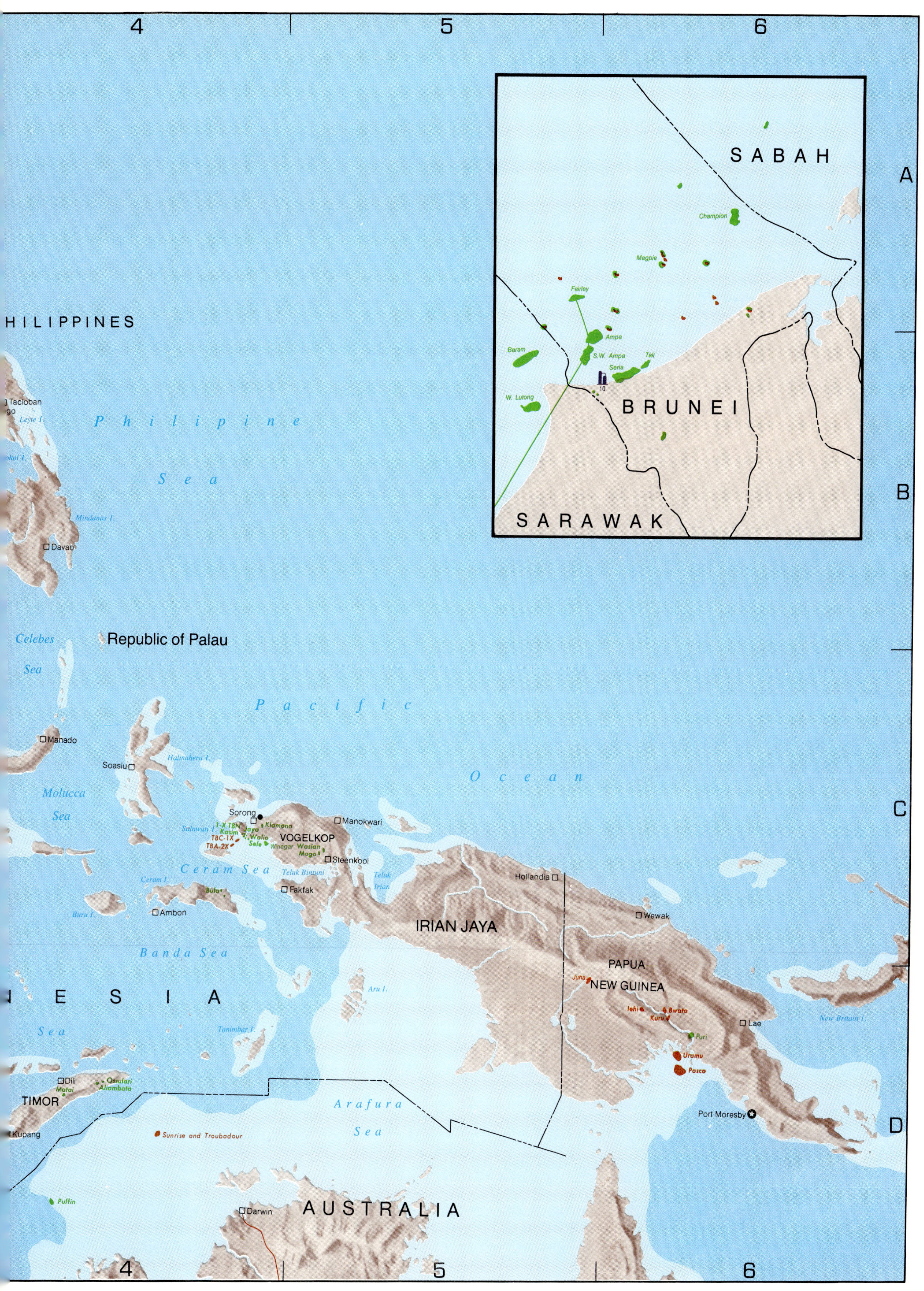

4
5
6
SABAH
Champion
Magpie
Fairley
Ampa
Baram
S.W. Ampa
Tali
Seria
W. Lutong
10
BRUNEI
A
B
SARAWAK
PHILIPPINES
Tacloban
go
Leyte I.
ohol I.
Philipine
Sea
Mindanas I.
Davao
Celebes
Sea
Republic of Palau
Manado
Pacific
Soasiu
Halmahera I.
Ocean
C
Molucca
Sea
Sorong
Klamono
1-X T8N
Jaya
Manokwari
Salawati I.
Kasim
Wolio
T8C-1X
Sele
VOGELKOP
T8A-2X
Winagar
Wasian
Mogo
Ceram Sea
Steenkool
Ceram I.
Teluk Bintuni
Buru I.
Bula
Teluk
Irian
Fakfak
Hollandia
Ambon
Wewak
IRIAN JAYA
Banda Sea
PAPUA
Juha
NEW GUINEA
NESIA
Aru I.
Sea
Iehi
Bwata
Kuru
Lae
Tanimbar I.
New Britain I.
Puri
Dili
Oasulari
Uramu
Matai
Aliambata
Pasca
TIMOR
Kupang
Arafura
Port Moresby
D
Sunrise and Troubadour
Sea
Puffin
AUSTRALIA
Darwin
4
5
6

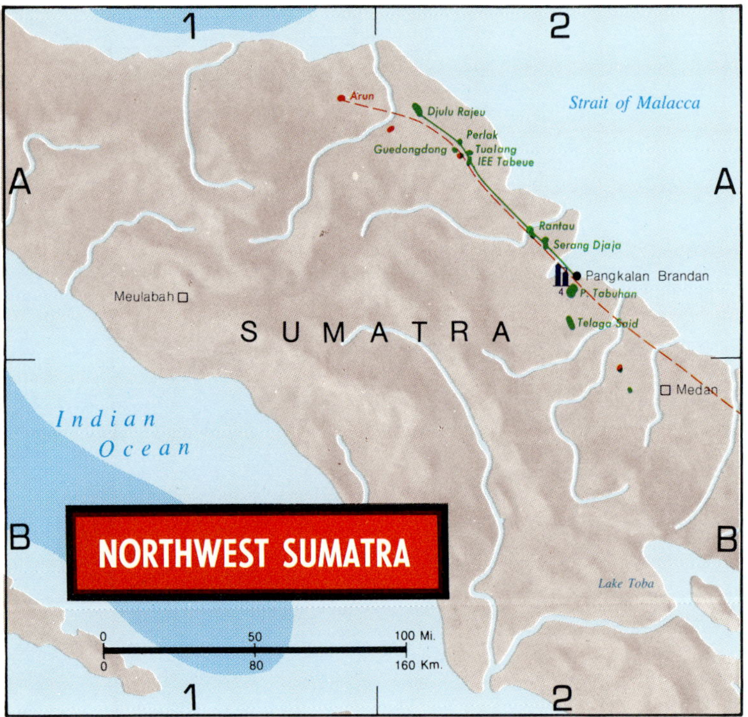

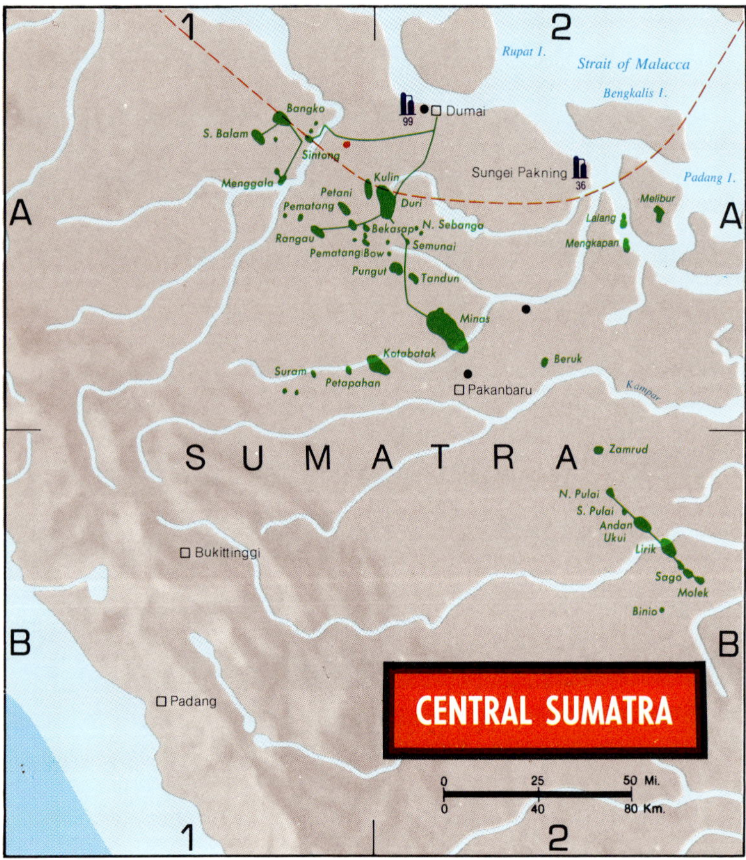

installed.

Gas production

India's efforts to replace imported petroleum and other products with domestically produced natural gas are paying off.

A study by ONGC says fuel switching saved the country $1.125 billion in foreign exchange in each of fiscal years 1986-87 and 1987-88. The savings will climb to $1.38 billion in fiscal 1988-89.

Industries increasing their use of gas include fertilizer, ceramics, glass, textiles, paper, petrochemicals, and power generation.

In far eastern India, gas is finding markets in power generation, fertilizer plants, and tea estates in Assam and in power generation in Tripura, although local productive potential exceeds demand. Gas from India's major producing area, the Bombay High offshore area, is used mainly to produce fertilizer and generate power.

Recent gas discoveries in the Krishna-Godavari basin soon will supply eight industries in Andhra Pradesh. And ONGC has started supplying gas to a steel rolling mill from a recent discovery at Narimanam in the Cauvery basin.

Cryogenic natural gas complex

India commissioned its first big cryogenic natural gas processing and treatment complex.

The plant at Hazira, operated by ONGC, can treat and extract liquids from 706.3 MMcfd of sour gas and 103 metric tons/hr of condensate from South Bassein field off western India. In addition, it can process 176.6 MMcfd of sweet associated gas and 20 tons/hr of condensate from Bombay High oil field north of South Bassein.

South Bassein production undergoes initial treatment offshore. A 135 mile, 36 in. pipeline carries it ashore to Umbhrat. From there it travels 8.7 miles via a 36 in. line to Hazira. The Hazira complex involves the sweet gas processing plant and gas treatment plant, built in two phases with identical capacities. Sweet gas from the complex will supply the Kribhco fertilizer plant at Hazira as well as six other fertilizer plants and three power plants along Gas Authority of India Ltd.'s 1,094 mile Hazira-Bijaipur-Jagdishpur gas pipeline. Plans for Hazira include a crude oil stabilization unit and a 1 million ton/year NGL dearomatization and fractionation unit.

New crude pipeline

India has approved construction of an 850 mile pipeline to carry crude oil from the port of Kandla in Gujarat to the 120,000 b/d Karnal refinery at Baholi, Haryana, and to transport products northwest from there.

Most of the crude will be imported, although some production from the Bombay High offshore complex may be shipped to Kandla to be fed into the pipeline.

Indian Oil Corp. Ltd. will lay the line. It hopes to commission it before completion of the Karnal refinery it is building with India's Tata Group. From the refinery, the pipeline will carry products to Bhatinda, Punjab. The terminus is in India's northwestern region, which accounts for nearly one third of the country's total petroleum products consumption.

INDONESIA

CAPITAL: Jakarta
MONETARY UNIT: Rupiahs
REFINING CAPACITY: 714,200 b/cd
PRODUCTION: 1,137.5 Mb/d
RESERVES: 8,250,000 Mbbl

INDONESIA IS PUSHING A PROGRAM DESIGNED TO wean its economy from heavy dependence on oil and gas exports.

For the first time, Indonesia's earnings from nonoil and gas exports have outstripped those from oil and gas exports. But income from oil and gas exports still will be a mainstay of the country's economy in the years to come.

Indonesia's earnings from nonoil and gas exports in first

quarter 1988 rose more than 50% to $3.33 billion from the same period a year earlier, its Ministry of Trade reports. The ministry noted that the range and value of exported products are rising. Trade Minister Arifin Siregar expects Indonesia to achieve its goal of $10.8 billion in earnings from nonoil exports.

Until 1986, oil and gas were Indonesia's chief sources of revenue. They now account for slightly less than half of total exports, down from 82.4% in 1982.

Japan with 43.1% and the U.S. with 19.5% are Indonesia's main export markets, focusing on oil and LNG to Japan and oil to the U.S.

The 1986 collapse in oil prices spurred Jakarta to begin a radical restructuring of the economy, emphasizing deregulation and cutting bureaucracy, along with a devaluation of the currency. Those efforts have spawned more private investment and boosted trade. In first quarter 1988 Indonesia approved $1.8 billion worth of foreign investment vs. $1.46 billion in all of 1987 and $826 million in 1986.

Indonesia energy options

How well Indonesia manages its petroleum policies will have a major effect on its economic future, writes Charles K. Ebinger in Geopolitics of Energy, a publication of Conant & Associates Ltd., Washington, D.C.

"Over the last 5 years, Indonesia's internal pricing policy, as well as the commercial terms of access offered to the international oil companies, has led to rising demand and a drop in energy prices," Ebinger writes.

Indonesia needs to find about 400 million bbl/year just to replace production, but reserve additions have been only about 100 million bbl/year since 1983.

OPEC's low quota for Indonesia masks the rate at which the situation is deteriorating. Future production levels will depend on the size of reserve additions in existing reservoirs, new field reserve additions, and OPEC production restraints. Oil prices and incentives to foreign oil and gas companies are keys to the first two.

Four of the eight petroleum products sold in Indonesia—kerosine, automobile and industrial diesel fuel, and resid—are subsidized. Demand for those four products has remained strong without production costs being fully reflected in the government price, Ebinger contends.

What's critical for Indonesia is whether energy pricing and contract terms may depress foreign investment in the petroleum sector.

There is a common assumption that a 15-20% pretax rate of return is enough incentive to justify development of Indonesia's marginal oil and gas reserves. It is important for Indonesia to understand that better terms, 25-30%, are offered elsewhere in the world, and consequently a competitive market will draw capital away from Indonesia, Ebinger points out.

Another key is Indonesia's effort to back out oil in domestic consumption with natural gas and coal. The country's gas reserves have jumped to 80 tcf in 1988 from 30 tcf in 1980. At the same time, falling oil prices have led Jakarta to realize that LNG exports may have plateaued and that further use of gas will be in the domestic market. An important question is whether Indonesia's gas potential should be handled by the public or private sector.

Improved E/D terms

Indonesia, trying to revive exploration/development, has improved the terms it offers foreign operators.

It has relaxed terms in production sharing contracts for frontier areas. And it will take steps to encourage development by foreign companies of marginal oil and gas fields.

For frontier areas, the country will implement a sliding scale for production sharing. The government's share will increase incrementally from 80% for production of as much as 50,000 b/d to 90% for production of more than 150,000 b/d.

Among other measures, the government will:
• Base oil company taxes on market prices for oil rather than official OPEC prices, which have been higher.
• Allow drilling costs to be written off in the year they are incurred.
• Permit use of depreciation and depletion accounting methods that let oil companies recover capital faster.
• Hike investment credits.

Production activity

Maxus Energy Corp. estimates production from its recent discoveries off Indonesia will reach 220,000 b/d of oil by 1991.

Cost to develop the recent Intan and Widuri discoveries in the Southeast Sumatra production sharing contract area is tagged at a combined $240 million. Maxus pegs the fields' reserves at a combined 275 million bbl. It expects first production from Intan in mid-1989 from an accelerated development program. Meantime, Maxus has identified other prospects nearby and planned an extensive seismic program. Plans call for two platforms and 15 wells in Intan

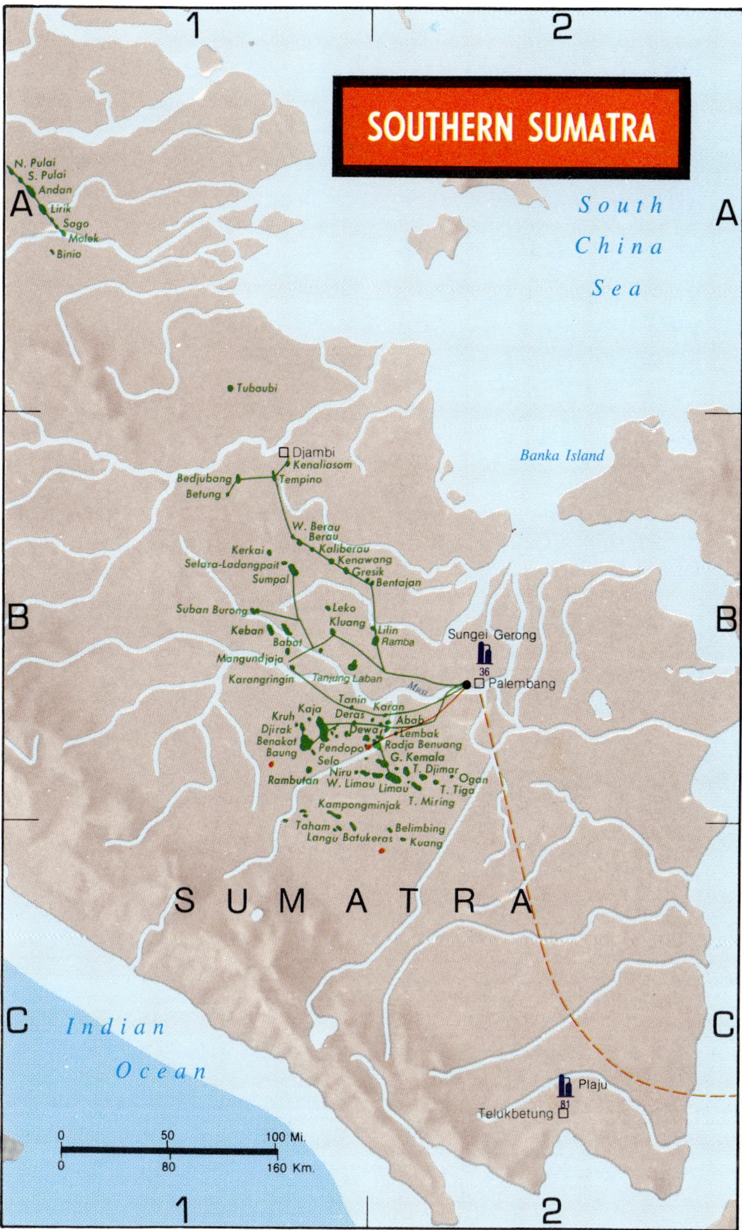

field and five platforms and 45 wells in Widuri. Maxus followed up the Intan strike with six step-outs, each of which flowed at rates of 3,000-5,800 b/d of oil. It drilled four step-outs to the 1 Widuri.

Indonesia's Pertamina was considering proposals from seven companies to boost oil recovery from fields in its Limau area of South Sumatra.

The state owned company wants to form a 50-50 venture with a contractor that would conduct secondary and tertiary recovery programs in the fields.

Wood Mackenzie says only less-exotic recovery techniques are likely to be used at Limauo.
Waterflood could recover 10-20 million bbl, and an added 10 million bbl could be produced by gas lift.

Initial development probably will consist of a pilot project in Belimbing, Limau Niru, and Karangan fields, which have the lowest expected primary recovery.

Production rose to 81,300 b/d of oil in Indonesia's Duri field, site of the world's biggest steamflood. About 63,000 b/d of the 22.7° gravity oil is due to steam drive, said PT Caltex Pacific Indonesia, a 50-50 partnership between Texaco Inc. and Chevron Corp.

The project is successful, profitable, and being expanded, Caltex said.

Caltex has 1,252 producing wells and 198 injecton wells in the project.

They tap Pertama-Kedua sand at 625 ft, which has average porosity of 32-36% and permeability of 1,550 md. Oil viscosity is 157 cp.

The company expects the project to cover 15,100 acres when completed, compared with the present 1,400 acres. Caltex said more than 4,000 wells will be drilled and more than 300 steam generators used during project life.

Asamera (Overseas) Ltd. and Bow Valley Industries Ltd. began production from their Corridor Block production sharing contract (PSC) area off South Sumatra, Indonesia.

And Asamera, operator, said Corridor Block exploration has been successful.

The companies also produce oil on the Corridor Block technical assistance contract (TAC) area. They expected combined gross production from the areas to average 35,000 b/d in 1988. PSC flow began at the rate of 1,200 b/d and would climb to 8,000 b/d gross as production facilities are completed at Rawa, South Rawa/Keri, and Supat fields.

A delineation well, North Rawa 2, on a separate structure discovered oil associated with the 1985 North Rawa 1 Batu Raja gas discovery, Asamera reports. The delineation well, ½ mile southeast of North Rawa 1, encountered 19 ft of net oil pay in the Batu Raja limestone.

A west step-out, North Rawa 3, ¾ mile from the discovery well, encountered 36 ft of net gas pay in Batu Raja. The North Rawa 3 encountered oil shows in nonreservoir rocks at the same depth that North Rawa 2 produced oil, "indicating a productive oil rim could exist on the west flank of the structure," Asamera says.

MALAYSIA

CAPITAL: Kuala Lumpur
MONETARY UNIT: Ringgit
REFINING CAPACITY: 209,300 b/cd
PRODUCTION: 540 Mb/d
RESERVES: 2,922,000 Mbbl

MALAYSIA EXPECTED CRUDE OIL and natural gas production to increase in the fiscal year beginning Apr. 1, 1988. State owned Petroliam Nasional Bhd. (Petronas) set targets of 540,000 b/d of crude oil and condensate and 1.13 bcfd of natural gas.

Targets for the fiscal year ending Mar. 31, 1988, were 508,300 b/d of oil and 1.11 bcfd of gas.

Malay-Thai disputed area

Malaysia's Cabinet endorsed principles that may revive operations in the South China Sea's disputed Malay-Thai joint development area (JDA). Acceptance of the principles indicates Malaysia honors Thailand's right over the 1,088 sq mile Block 18 within the JDA held by Triton Oil Co. of Thailand. The principles state:

• Triton's acreage is a single area for exploration and development.

• Production from the entire disputed area is to be split equally between

Where IPC tested its gas discovery

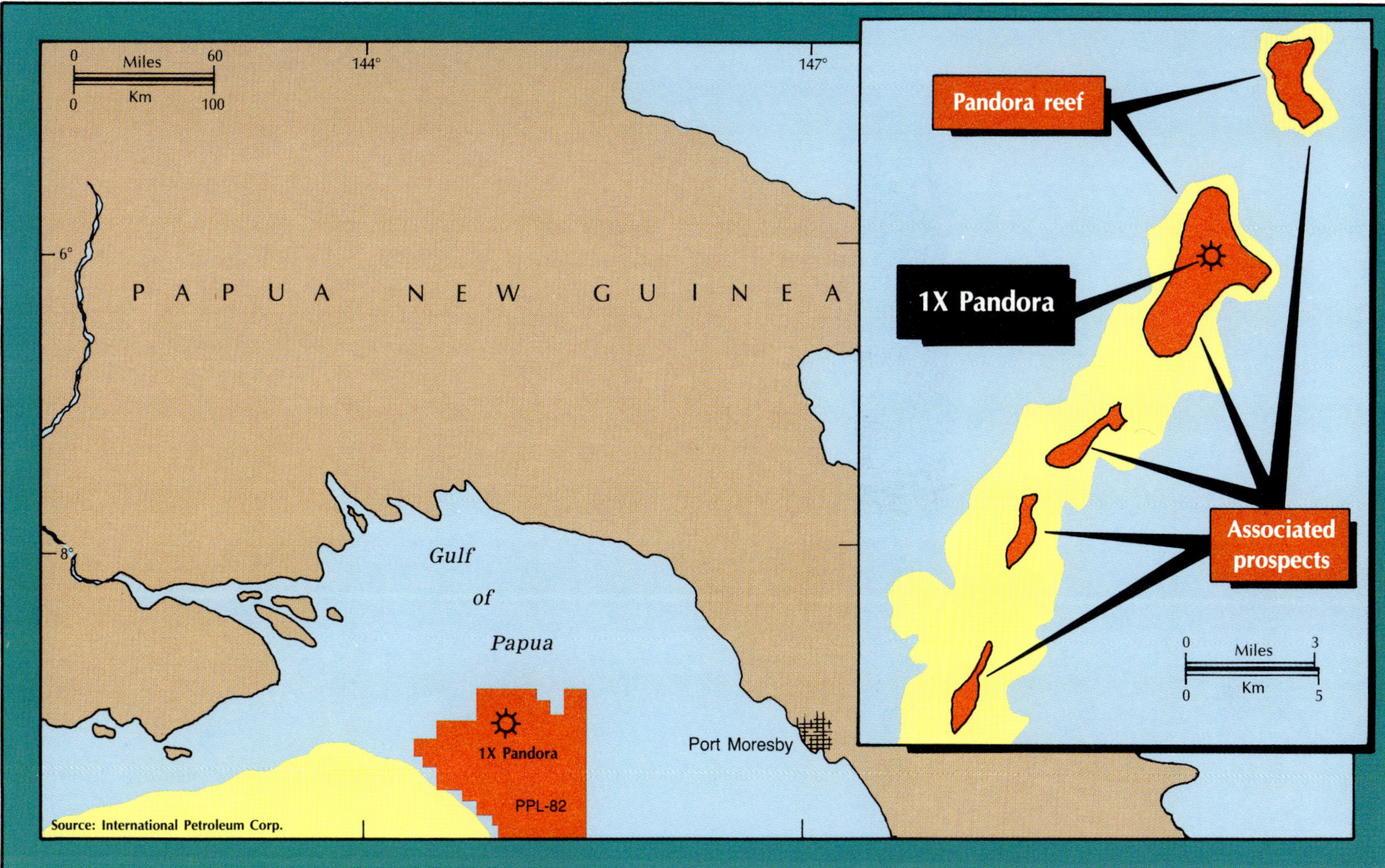

Thailand and Malaysia, with the portions subject to the countries' respective terms.

• Triton and Petronas will select an operator for the area.

• A joint operating committee involving representatives of Thailand and Malaysia will monitor and supervise operations.

Triton describes Block 18's Pilong and Mid-Ridge prospects as "major structures." It says a 1971 well, 1 Pilong, flowed at a combined rate of more than 29 MMcfd of gas from four pay intervals between 4,000 and 6,600 ft.

Based on geophysical data, it believes the Mid-Ridge prospect may hold oil. High temperatures, indicating natural gas, characterize the Pilong prospect and a structural trend northwest from there. The trend includes the Texas Pacific Thailand Inc. 17-B-1 natural gas discovery on Block 17 and undeveloped gas discoveries to the north.

Besides having a structural genesis different from the Pilong structure, the Mid-Ridge area is interpreted to be situated away from the extreme temperature gradient areas. It is for this reason the Mid-Ridge structure should produce oil. Triton says the Pilong structure could hold 4.5-6 tcf of gas in place. The Mid-Ridge structure could hold 2 billion bbl of recoverable oil or 2.5-3.5 tcf of gas in place.

Production sharing contracts

Petronas signed more production sharing contracts with units of U.S. companies. It signed a contract with Sun Malaysia Petroleum Co. and partners for Block PM15 in the Strait of Malacca. With Block PM1 to the north, Sun now holds most of the Malaysian part of the strait.

Block SK6 off Sarawak went to Occidental Petroleum (Malaysia) Ltd. Oxy agreed to spend $23 million during 5 years to shoot 7,500 line km of seismic surveys and drill five wildcats.

Sarawak Shell Bhd. signed a new production sharing contract for its Block SK 5 off Sarawak, Malaysia. Shell held 4,440 sq mile Block SK 5 under a concession agreement before 1976 and under a production sharing contract after 1976.

Shell will operate the block, where water depths are 0-200 ft. It will spend at least $100 million (Malaysian), about $39 million in U.S. dollars, drill six wildcats, and shoot 7,000 line km of seismic survey.

Petronas signed a contract for Block SK9 off Sarawak with Italy's Agip (Overseas) Ltd. and wholly owned subsidiary Petronas Carigali Sdn. Bhd.

Agip proposes to spend at least $16.5 million to drill three wildcats and shoot 5,000 line km of seismic surveys. Block SK9, covering about 6,700 sq km, has a maximum water depth of 197 ft.

NEW ZEALAND

CAPITAL: Wellington
MONETARY UNIT: Dollars
REFINING CAPACITY: 88,400 b/cd
PRODUCTION: 28 Mb/d
RESERVES: 182,650 Mbbl

IN THE ONSHORE TARANAKI BASIN AT YEAREND 1988, Petrocorp Exploration Ltd. gauged oil on four of five drillstem tests of undisclosed intervals in 2 Waihapa, a Waihapa oil field development well on PML 38140.

In the first test, the well flowed at rates of as much as 939 b/d of oil and 1.09 MMscfd of gas through a ½ in. choke with 490 psi flowing tubing pressure.

Mechanical problems hampered the second test. A third

Japan

Location	No. Refineries		Location	No. Refineries
Sakaide	1		Kawasaki	5
Yokohoma	2		Mizushima	2
Chiba	3		Okinawa	3
Matsuyama	1		Toyama	1
Sakai	2		Chita	1
Yokkaichi	2		Funakawa	1
Sodegaura	1		Niigata	2
Hyogo	1		Muroran	1
Tokuyama	1		Negishi	1
Kainan	1		Yamaguchi	1
Kashima	1		Ehime	1
Marifu	1		Wakayama	1
Osaka	1		Owase	1
Oita	1		Sendai	1
Tomakomai	1			

test of the second and third intervals gauged a combined 160 b/d of oil and 690 Mscfd through a ½ in. choke with 275 psi flowing tubing pressure. On a separate test, the same intervals flowed at rates of as much as 1,634 b/d of oil and 1.85 MMscfd of gas through the same size choke with 1,100 psi flowing tubing pressure. And a fourth interval flowed as much as 1,754 b/d of oil and 2 MMscfd of gas through a ½ in. choke. Pressure wasn't reported.

On extended production tests, Petrocorp 1 Waihapa was flowing 2,940 b/d of oil and 4.3 MMscfd of gas from Oligocene Tikorangi limestone.

Petrocorp had spudded 3 Waihapa about 4,600 ft south of 1 Waihapa.

On PLM 38091, Petrocorp 2 Kaimiro flowed at rates averaging 75 b/d of oil through a ½ in. choke during initial tests of the higher of two pay zones. Petrocorp injected nitrogen into the well, drilled to 7,021 ft.

Drilling planned

In other activity, New Zealand Oil & Gas Ltd. (NZOG) was to begin a five well drilling program on separate onshore licenses. It was likely to start with 1 Rotokare on PPL 38084 about 2½ miles southeast of Waihapa. A well then would be drilled on PPL 38083.

Also, NZOG planned to spud 1 Pigeon Creek on South Island's PPL 38075 east of Greymouth. It planned to drill one well each on PPL 38702 and PPL 38706, Taranaki. NZOG was to have begun shooting seismic surveys on PPL 38702, PPL 38706, and PPL 38707.

Interest surges

Encouraging test results outside established fields on the Taranaki Peninsula and offshore have boosted Taranaki basin interest.

Focus of the attention has been TCPL Resources Ltd. 3B Kupe South, an offshore appraisal well that flowed a total of more than 10,000 b/d during three drill stem tests. The well tested the eastern flank of Kupe South field on PPL 38116. TCPL reported these results of three drill stem tests of selected intervals between 10,564 ft and 11,060 ft:

• 4,400 b/d of oil and 9.7 MMscfd of gas through a ½ in. choke with 2,400 psi wellhead pressure.

• 2,370 b/d of oil and 20.6 MMscfd of gas through a ¾ in. choke with 1,820 psi wellhead pressure.

• 3,525 b/d of oil and 23.69 MMscfd of gas through a 1 in. choke with 1,318 psi wellhead pressure.

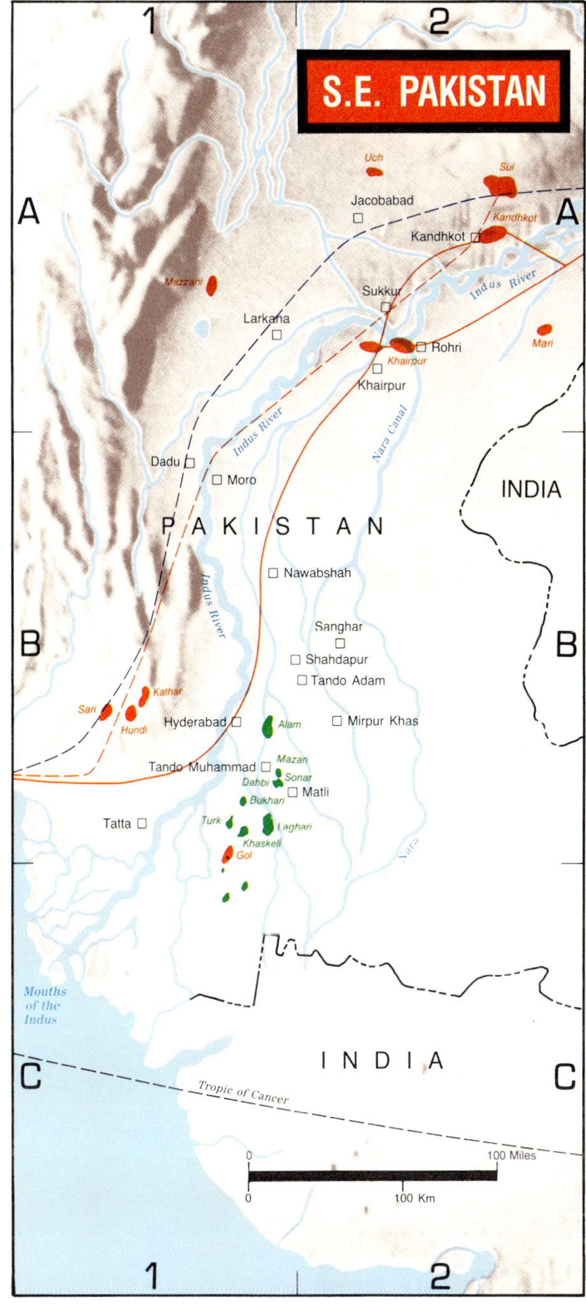

PAKISTAN

CAPITAL: Islamabad
MONETARY UNIT: Rupees
REFINING CAPACITY: 130,050 b/cd
PRODUCTION: 47 Mb/d
RESERVES: 170,438 Mbbl

WORLD BANK APPROVED FUNDS FOR A $150 MILLION loan to finance private sector energy projects in Pakistan.

National Development Finance Corp. will administer the funds, which will be augmented by loans from the U.S., U.K., Italy, Canada, and Japan totaling $450 million. Including private equity and commercial loans, the energy projects are expected to total nearly $2 billion.

The project's goal is to reduce Pakistan's dependence on imported oil and gas by producing an additional 2,300 mw of power generating capacity, 2 million tons/year of coal, and 132 MMcfd of gas by 1993. Most of the funds are expected to go for power plant construction, with three such projects proposed for funding. The project marks the first time Pakistan has allowed private companies to participate in energy production.

The government is allowing investors to build and operate power generation plants under the project, although it may buy them later.

Union Texas discovery

A group led by Union Texas Petroleum tested its 18th discovery since 1981 on the Badin concession in Southeast Pakistan.

The company also started production from South Mazari field on the Badin block, boosting its gross oil production in Pakistan to about 18,500 b/d. A 1985 discovery in the northern part of the block, South Mazari can produce about 2,500 b/d of oil. Production from other Badin fields—Khaskeli, Laghari, Dabhi, and Mazari—reached about 17,500 b/d in 1987.

Union Texas' strike, 1 Ghunghro, flowed 3,212 b/d of oil through a 48/64 in. choke with 470 psi flowing tubing pressure from the lower Goru at 4,615-25 ft. Site is about 110 miles southeast of Karachi.

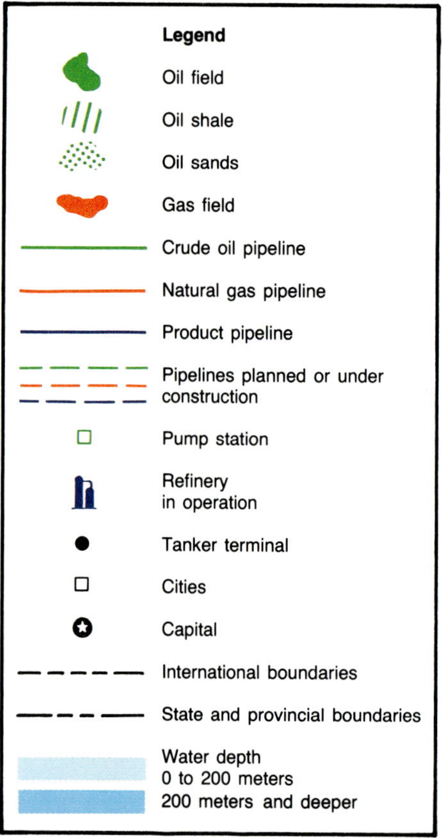

The Union Texas group has tallied 11 successes out of 13 wildcats in the past 3 years on the 2.2 million acre concession in Sind Province.

PAPUA NEW GUINEA

CAPITAL: Port Moresby
MONETARY UNIT: Kina
REFINING CAPACITY: 0
PRODUCTION: 0
RESERVES: 200,000 Mbbl

A GAS DISCOVERY OFF PAPUA NEW GUINEA confirms the presence of a "substantial hydrocarbon resource," says International Petroleum Corp., Geneva.

IPC tested 1X Pandora in the Gulf of Papua on four chokes from perforations at 4,800-25 ft and 4,925-50 ft in Miocene Darai reef. Best rate was 57.1 MMcfd of gas through an $^{80}/_{64}$ in. choke with 1,389 psi wellhead flowing pressure. It flowed 12.6 MMcfd of gas through a $^{32}/_{64}$ in. choke with 2,139 psi flowing pressure. Flows were 47.3 MMcfd through a 1 in. choke with 1,640 psi flowing pressure and 27 MMcfd through a $^{48}/_{64}$ in. choke with 1,962 psi flowing pressure. Equipment capacity restricted production rates. There was no significant liquid production. Logs showed the Darai gas column to be 904 ft thick.

Geophysical data indicate the size of the structure to be about 5.4 sq miles in Block PPL-82. The company has identified four other structures in the reef trend that holds the Pandora structure. The well is 160 miles west of Port Moresby in 385 ft of water. IPC planned to suspend the well, drilled by the Chancellorville drillship, and analyze the gas samples.

Chevron test

A group led by Chevron Niugini Pty. Ltd. tested good flows of oil and gas from three zones in an outpost well in the central highlands frontier.

Combined gauge was 8,787 b/d and 7.35 MMcfd from 1X Hedinia at a bottomhole location 2½ km south of the group's undeveloped Iagifu oil and gas field. Pay is the Upper Jurassic-Lower Cretaceous Toro sandstone, also productive in Iagifu field.

Merlin Petroleum Co., San Francisco, a member of the Chevron group, said it was not clear whether 1X Hedinia is an extension to Iagifu field or on a separate feature. Merlin describes the Iagifu structure as a thrust cored anticline trending northwest-southeast and holding about 100 million bbl of reserves. The area lies northeast of the Papuan foldbelt thrust front.

Drilled to 8,500 ft true vertical depth at a surface location 3 km south of Iagifu field, 1X Hedinia found a gas cap in the target Toro sandstone. Chevron then deviated the hole to the north to intersect the oil and gas zone of the reservoir, drilling to 9,421 ft measured depth, 8,700 ft true vertical depth.

**BP Petroleum
test**

BP Petroleum Development Ltd. tested a significant gas discovery in the southern highlands of Papua New Guinea.

After testing the Hides-1 well, BP issued a preliminary reserve estimate of 1.5 tcf of gas. There also is a poorly defined volume of condensate.

Hides-1 tested four zones in the Cretaceous Toro sandstone. Maximum flow rate was 15.9 MMcfd of gas and 39 b/d of condensate through a $^3/_8$ in. choke from pay at 9,836-10,013 ft. A deeper zone flowed 12.07 MMcfd and 408 b/d of condensate through the same choke size.

Talks are under way between BP and Papua New Guinea's Department of Minerals and Energy on the possibility of using the gas to generate electricity for mineral mines planned in the southern highlands. Hides-1 was the second well drilled on the permit. The first, Lavani-1, was dry total depth 9,794 ft in 1982.

PHILIPPINES

CAPITAL: Manila
MONETARY UNIT: Pesos
REFINING CAPACITY: 254,000 b/cd
PRODUCTION: 9.1 Mb/d
RESERVES: 15,800 Mbbl

ALCORN INTERNATIONAL INC., HOUSTON, WAS increasing oil production off the Philippines.

Alcorn hoped to start production from its 1 North Matinloc discovery. It planned to lay a 6 in. line from a piled wellhead to the Matinloc platform.

Alcorn's 1 North Matinloc wildcat flowed 43° gravity oil from three intervals in a reef reservoir. Stabilized flows through ¾ in. chokes were 3,060 b/d of oil from 6,832-52 ft, 2,475 b/d of oil from 6,908-28 ft, and 1,740 b/d of oil from 7,001-21 ft. The shallowest zone flowed at a maximum rate of 6,121 b/d of oil through a 1¾ in. choke with 200:1 gas:oil ratio.

The discovery, in Service Contract 14 off Palawan Island, is 2½ miles north of Matinloc field.

The Maersk Giant jack up, which drilled the discovery, was moving 3 miles west of Matinloc field to drill 1 Verde, a wildcat projected to 11,500 ft. The rig then will return to drill 2 North Matinloc.

Alcorn was evaluating further development of the Galoc sandstone reservoir. The 2 Galoc well was flowing 3,300-3,400 b/d of oil.

THAILAND

CAPITAL: Bangkok
MONETARY UNIT: Baht
REFINING CAPACITY: 191,045 b/cd
PRODUCTION: 38.4 Mb/d
RESERVES: 8,200 Mbbl

THAILAND HAS TAKEN STEPS TO BOOST NATURAL gas development and consumption.

After 4 years of negotiation, state owned Petroleum Authority of Thailand (PTT) and Esso Exploration & Production Khorat Inc. reached a final gas sales agreement clearing the way for development of Esso's gas discoveries in the northeast.

And PTT began a $10 million program to lay two spur lines to deliver Gulf of Thailand gas to power plants planned on the eastern seaboard by state owned Electricity Generating Authority of Thailand (EGAT).

**Esso-PTT
deal**

Esso and PTT reached final agreement on gas sales after resolving a dispute over the price of gas during an interim production period, during which Esso will further assess its gas reserves.

During that period, Esso will sell PTT 40-60 MMcfd for as many as 4 years. Price will be discounted 50-80% from the long term price, which is based on fuel oil prices in Singapore. Esso assured PTT a supply of 1 tcf, mostly from Nam Phong field in Khon Kaen Province, about 310 miles northeast of Bangkok. It has estimated the field's reserves at 1.5 tcf. The company has another gas discovery at Phu Horm, north of Nam Phong, and an indicated gas discovery to the south. First production is expected within 2 years.

Esso will spend $75 million to drill five more wildcats on Block E5, where it made the Nam Phong and Phu Horm discoveries. Depending on results, Esso will sell PTT as much as 250 MMcfd of gas for 20 years under the long term part of its new agreement. PTT during the interim period will sell the gas it buys from Esso to EGAT, which will build two sets of 105,000 kw combined cycle power plants near Nam Phong. The plants will begin operation in 1990-91.

**Spur lines
planned**

The spur lines PTT plans on the eastern seaboard will deliver gas to seven planned EGAT power plants with capacities totaling 2.7 million kw. The plants are due on stream in 1990-93.

EGAT is adding its third and fourth combined cycle generating units, with 300,000 kw capacity each, along with its third and fourth thermal plants, with 600,000 kw capacity each, to its Bang Pakong complex in Chachoengsao Province. Present capacity is 1.82 million kw. It also will build three combined cycle plants, with 300,000 kw each, in Rayong Province.

One 28 in. spur line will run 2.6 miles from PTT's Gulf of Thailand trunkline to Bang Pakong, about 25 miles southeast of Thailand. The line, estimated to cost $7.2 million, will increase gas deliveries to the power complex to more than 700 MMcfd from 400 MMcfd.

The other spur—16 in. in diameter, 2.5 miles long, and costing an estimated $2.8 million—will run from PTT's 350 MMcfd natural gas processing plant at Mab Ta Phud, Rayong, to the Rayong substation where EGAT plans the three combined cycle plants. It will deliver 120-180 MMcfd of gas.

**Exploration,
production activity**

Exploration is brighting in Thailand.

Gopher Oil Ltd., wholly owned Calgary subsidiary of Gulf Exploration Consultants of Houston, spudded the first of two rank wildcats on Block GO2 in the southern peninsula.

Elsewhere:

• Premier Oil Pacific Ltd., Hong Kong, planned its first wildcat on Gulf of Thailand Block B11/27.

• Petrocorp (Exploration) Ltd. of New Zealand planned more exploration on Central Plains Block SW 1, where it discovered oil and gas earlier in 1988.

• Thai Shell Exploration & Production Co. planned additional Chumphon basin seismic surveys following suspension of production from troubled Nang Nuan oil field on Gulf of Thailand Block B6/27.

Gopher's two tests will be the first major activity on 2,930 sq mile Block GO2. Rio Grande (Thailand) Ltd. was drilling Gopher 1 Khian Sa to a target depth of 8,000 ft. It was to follow with 1 Phun Phin about 12 miles northeast, targeted to

6,000 ft. Gulf Exploration took over Gopher in 1987. Commitments include 2,400 line km of induced polarization survey, one well, and expenditures of at least $6 million. Gopher had spent about $2.5 million on the block. Jack Copeland, chairman of Gopher and Gulf Exploration, said drilling outlays will satisfy the spending requirements. He pointed out that Thailand's southern peninsula is relatively unexplored and the least understood geological province in the country. Gopher 1 Khian Sa is on a structure designated "B." Target rocks are Permian limestone and Tertiary sandstone.

Petro-Canada (Thailand) Inc. will help fund drilling of the tests and earn an option to acquire an interest in the block. Rio Grande will acquire a 5% interest in the block when drilling is complete. Gopher plans to renew the GO2 concession agreement into the 5 year second obligation period. Commitments are more geophysical surveys, drilling of at least one well, and total spending of $7 million.

Premier's 1 Songkhla will be the first well on Block II/27.

The Maersk Voyager jack up was drilling the well in 77 ft of water. Target depth was 10,827 ft. The well will satisfy drilling commitments of an initial 3 year obligation period under an agreement signed in February 1986. Premier also was to shoot 1,100 line km of seismic survey. In a 5 year second obligation period, it would shoot more seismic survey and drill two wells. Petrocorp and Thai Shell planned more seismic work before yearend 1988.

The New Zealand company was to run 1,000 line km more seismic survey on its Block SW 1. It planned to drill another well by the middle of 1989, satisfying its remaining commitments.

Petrocorp 1 Wichianburi, drilled to 7,560 ft, gauged 500 b/d of oil and 3.6 MMcfd of natural gas.

Petrocorp received supplementary concessions from the Department of Mineral Resources to gain ownership of the block from previous concessionaires Promet Bhd. of Malaysia and Ultramar Oil & Gas (U.K.) plc.

On offshore Block B6/27, Thai Shell planned to shoot 1,200 line km of seismic survey, an increase from 850 line km planned earlier in 1988. The survey will identify locations for two or three wells, drilling of which was to begin in 1989.

Thai Shell has shot 6,500 line km of 2D seismic survey and 235 sq km of 3D survey and drilled four wildcats on the block. One of the tests, 1 Nang Nuan, found Nang Nuan oil field, which began production early in 1988. It flowed as much as 6,000 b/d and encountered water problems. Shell shut in the one well field. It hopes to revive production early in 1990 and drill another production well.

Kaphong field
platform

The first offshore production platform fabricated in Thailand has been set in Unocal Thailand Ltd.'s Kaphong gas/condensate field.

Thai Nippon Steel Engineering & Construction Corp. built the four leg KWA platform at its Samut Prakan fabrication yard near Bangkok. The company was due to complete the KWB platform, destined for the same Gulf of Thailand field.

Thai Nippon is building six platforms for Unocal under three contracts. Major interests in Thai Nippon are Thailand's Italthai Group 50% and Japan's Nippon Steel 38%. Fabrication of the KWA platform took about 6 months, less time than scheduled. Most of the raw material came from Japan.

Unocal plans to drill as many as 12 wells from each new platform to develop Kaphong field. It will be the company's fifth gas field on production in the area.

Satun field
platforms

Unocal Thailand Ltd. plans to install four more production platforms in the Gulf of Thailand.

The Satun field units will bring to 45 the number of platforms in Unocal's gas producing area.

Thai Nippon Steel Engineering & Construction Corp. (TNS), a joint venture of Thailand's Italthai Group and Nippon Steel Corp., will fabricate the Satun platforms in Thailand. Installation was to be finished in third quarter 1989. The platforms will be 330 ft high and weigh 2,100 metric tons each. Fields produce 530 MMcfd and 17,700 b/d of condensate.

PTT gas
development hopes

PTT wants to develop a natural gas discovery on offshore acreage it bought from Texas Pacific Thailand Inc. (TP) and partners.

A subsidiary, PTT Exploration & Production Co. Ltd., paid TP, Canadian Superior (Thailand) Ltd., and Highland Thailand Inc. $83.75 million for Blocks 14-17 covering 973 sq miles. It was seeking partners for development of the "B" structure, with reserves estimated by Texas Pacific at 1.8 tcf. PTT E&P says reserves may reach 6 tcf. The state owned company hopes to begin producing 150 MMcfd of gas in 5-6 years, rising to 250 MMcfd. It anticipates costs of $500 million for development and $150 million for a 105 mile, 28-32 in. pipeline to the eastern seaboard.

Erawan field
platforms

The 14th and 15th production platforms were installed in Unocal Thailand Ltd.'s Erawan gas field.

Nippon Steel Corp.'s Kuroshio II derrick and pipelay barge completed installation of the Erawan EWN and EWO platforms in 260 ft of water.

EWN is 0.44 nautical mile west of Erawan's central processing platform, EWO 1¼ miles south. Drilling was to begin at midyear 1988 from EWN and EWO.

Wildcat
near Bangkok

Takeover of its parent hasn't stalled exploration in Thailand by Britoil (Alpha) Ltd. The company planned to spud its first Thai wildcat near Bangkok.

On Block BT, Britoil was to drill the B1-1 wildcat, about 16 miles south of Bangkok in Samut Prakan Province, to 7,300 ft at a cost of $2-3 million. The well is part of Britoil's second 5 year obligation period, which began early in 1988. Commitments include drilling of one well/year and spending of $5 million/year.

Processing
activity

Thailand has taken a step toward construction of its third gas processing plant.

Thai Shell Exploration & Production Co. picked Randall Process Systems, U.S., as contractor for the Sirikit oil field facility in the northern Central Plains.

Construction was expected to start early in 1989. It's expected to take 13 months. The plant, at Sirikit's main flow station in northern Kamphaeng Phet Province, will be able to process 40 MMcfd of associated gas. That's an adjustment from the originally planned 30 MMcfd. Output capacity will be 350 metric tons/day of LPG and 700 b/d of condensate. Propane recovery is to be 95%. PTT will distribute the LPG locally. Condensate will be mixed with Phet crude oil in the pipeline from Sirikit, which produces about 21,000 b/d.

The gas processing project received a boost when the new Thai government reaffirmed an earlier ruling that excise tax should not be levied atop the 12.5% natural gas royalty and 50% income tax on LPG.

PTT operates a 350 MMcfd gas processing plant at Rayong on the eastern seaboard. It has let contract to Japan's NKK to build a 200 MMcfd plant there. In addition,

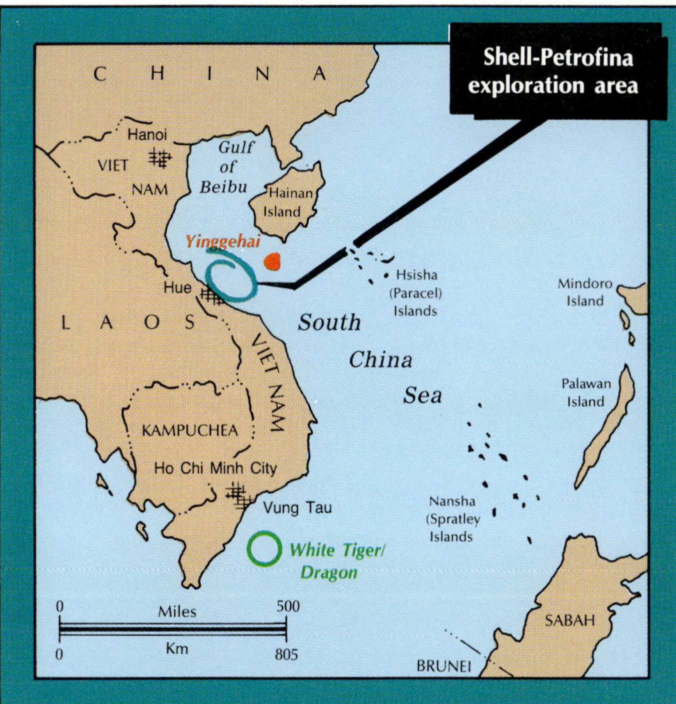

France has offered to study feasibility of a third gas processing plant near Rayong, where the petrochemical industry is developing rapidly.

Refinery expansions

Thailand approved expansion projects at two government controlled refineries.

The decisions by the National Energy Policy Committee, chaired by Prime Minister Prem Tinsulanonda, may result in suspension of private sector plans for refinery expansions and grassroots construction.

The committee approved a $150 million project by partly state owned Thai Oil Co. that would boost capacity of its 65,000 b/d refinery at Si Racha, now being expanded to 83,500 b/d, by another 100,000 b/d. It also approved a debottlenecking plan by state owned Bangchak Petroleum Co. that will boost capacity of its 65,000 b/d Bangkok refinery by 20,000 b/d.

Bangchak wants to merge the plan with a modernization program in progress. Merger, it says, would reduce costs of the debottlenecking to $30 million from $40 million.

Also under consideration were expansion of Esso Standard Thailand Ltd.'s 63,000 b/d refinery at Si Racha to 110,000 b/d and construction by Shell Co. of Thailand Ltd. of an 85,000 b/d refinery on the eastern seaboard.

Other companies had expressed interest in refining projects in Thailand, where consumption, estimated at 283,000 b/d, is growing at about 10%/year.

VIET NAM

CAPITAL: Hanoi
MONETARY UNIT: Dong
REFINING CAPACITY: NA
PRODUCTION: NA
RESERVES: NA

VIET NAM WAS RISKING INCREASED TENSION WITH China by opening its central coastal shelf to offshore exploration and development joint ventures with western oil companies.

The area near the port of Hue where Shell Exploration BV of Holland and Petrofina SA of Belgium was to conduct seismic surveys and wildcat drilling is less than 100 miles southwest of China's big Yinggehai gas field south of Hainan Island. The Shell-Petrofina combine will operate under a contract with Petrovietnam, Viet Nam's state oil company.

By contrast, the South China Sea sector where the joint Vietnamese/Soviet enterprise Vietsovpetro is developing two oil fields is more than 500 miles south of Hue and far from the nearest Chinese exploration areas.

For many years China and Viet Nam have squabbled and issued threats regarding sovereignty over the South China Sea's Nansha (Spratley) and Hsisha (Paracel) islands. Most of those tiny islands are hundreds of miles from China and Viet Nam.

Early in 1988, Chinese warships fired on two Vietnamese transports near the Nansha Archipelago.

Hanoi also has claimed sovereignty over Gulf of Beibu (Tonkin) areas northwest of Hainan Island, where China has had modest exploration success. Meanwhile, official Chinese maps continue to show China's territorial waters extending very close to Viet Nam's central coastline.

In announcing Petrovietnam's agreement with Shell-Petrofina for exploration and development, Hanoi said the contract term is 25 years, with the western companies providing financing and equipment.

Hanoi also said Shell-Petrofina is obligated to spend $70 million for exploration during the first 5 years of the contract. Shell-Petrofina pegged the obligation at $16.5 million in the first 3 years of exploration.

In any event, Hanoi said if exploration off Hue is successful, further capital investments will be made by Shell-Petrofina, "with all earnings to be distributed among participants on a mutually profitable basis."

Vietsovpetro operations

Moscow reports that Vietsovpetro, a joint venture between Viet Nam and the Soviet Union, continues to expand exploration on the southern Vietnamese coastal shelf.

The Soviets' Shelf-6 semisubmersible at midyear 1988 was drilling at a new South China Sea location, Danhung, off the Vietnamese coast.

Of the two Vietsovpetro discoveries on Viet Nam's southern coastal shelf, only White Tiger field, about 75 miles offshore between the port of Vung Tau and Kon Dao Island to the south, is on production. Moscow still describes White Tiger development as in the "pilot-commercial stage." No date has been disclosed for start of production from nearby Dragon field.

White Tiger oil flow during the first 4 months of 1988 was about 13,500 b/d. Vietsovpetro increased White Tiger production from less than 1,000 b/d in 1986 to 5,400 b/d in 1987. Production target for 1990 is 40,000 b/d.

First Viet Nam refinery

Viet Nam reports the nation's first refinery has been built "on the basis of free enterprise."

Financed by Vietnamese emigres, the 800 b/d Saigonpetro plant near Ho Chi Minh City (Saigon) was completed in 7 months with equipment bought in France. Private funds have been raised to double capacity. Crude for the refinery is purchased from Vietsovpetro.

Meantime, Viet Nam was building a 120,000 b/d, government owned refinery halfway between Ho Chi Minh City and the South China Sea port of Vung Tay, but no completion date has been disclosed.

It will process production from Viet Nam's offshore White Tiger and Dragon fields.

LATIN AMERICA

ARGENTINA

CAPITAL: Buenos Aires
MONETARY UNIT: Australes
REFINING CAPACITY: 690,400 b/cd
PRODUCTION: 449.6 Mb/d
RESERVES: 2,268,000 Mbbl

ARGENTINA HOPES FOREIGN COMPANIES CAN HELP it slow a production decline it estimates at 16%/year.

Rodolfo Terragno, Argentina's minister of public works, presented initial details of the plan to oil companies.

Argentina would allow contractors that hike production from existing contract areas to receive 80% of the f.o.b. international oil price for all incremental production.

Under another part of the plan, state owned Yacimientos Petroliferos Fiscales would enter into joint ventures with contractors for the first time in some of its existing fields.

YPF has 289 oil fields, and 12% of the company's production comes from 247 of them, Terragno said.

YPF would invite contractors to submit bids to enter joint ventures with it for as long as 20 years in an unspecified number of the 247 marginal, or underexploited, fields.

The bidder would be responsible for all investment, technology, exploitation, and taxes.

YPF would use revenue from the underexploited areas to increase operations in the more profitable fields, or central areas.

New Texaco block

Texaco Inc. agreed to acquire a 40% interest in Bridas Sapic's 1.5 million acre exploration/production contract area in Northwest Argentina.

Under terms of the deal, Texaco was obliged to drill at least one wildcat on the Hickmann block.

Bridas, Buenos Aires, has held the block with state owned YPF since 1987.

A subsidiary of Texaco was to begin seismic surveys on the block by late summer 1988.

The block is located near existing production and oil and gas pipelines in the far northern corner of Salta Province near the Bolivian border. YPF must approve the assignment of interest.

First Apache well outside U.S.

Apache Corp., Denver, planned to spud its first non-U.S. well in the summer of 1988 in Argentina.

The Argentine government formally approved Apache's application for a 2.2 million acre block. Apache, with a 75% interest, will be operator. Argerado Inc./Riumasa SA holds the remaining interest.

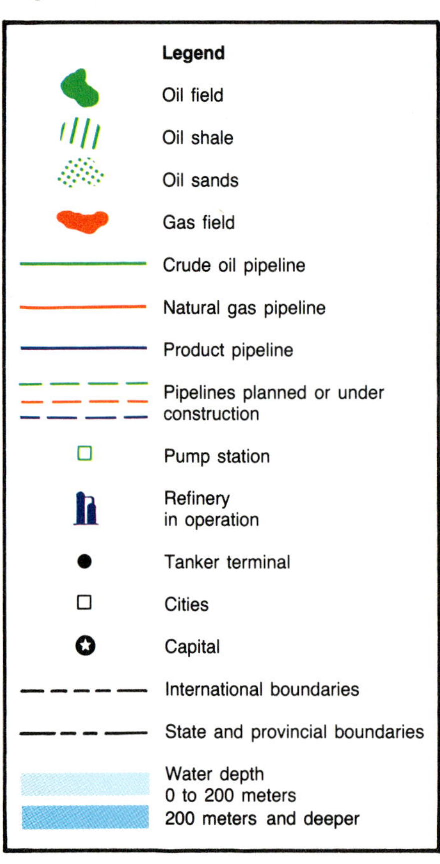

Arenal Block CNO-13 is in the Cretaceous Northwest basin of northern Argentina, one of the country's five oil producing basins. Apache expected to spud an 11,155 ft wildcat to Cretaceous Yacoraite on the concession.

Results of earlier drilling on the block and seismic studies conducted by Apache delineated several large structural closures, the company said.

BRAZIL

CAPITAL: Brasilia
MONETARY UNIT: Cruzados
REFINING CAPACITY: 1,407,300 b/cd
PRODUCTION: 555.6 Mb/d
RESERVES: 2,550,000 Mbbl

OIL WORKERS IN BRAZIL RETURNED TO THEIR JOBS after a strike that began in November 1988.

They reached an accord with the government calling for wage increases averaging 15%/month plus 4%/month compensation for productivity gains.

The strike, by employees of state owned oil company Petroleo Brasileiro SA (Petrobras), shut nine of Brazil's 11 refineries and reduced oil production from levels averaging 560,000 b/d. Petrobras denied reports that the strike shut operations on some offshore production platforms.

An official said the company reduced production because so much refining capacity had been idled. The Association of Petrobras Engineers had estimated the strike cut output by 80,000 b/d.

Petrobras expected its plants to be handling their usual 1.24 million b/d of crude shortly after the strike ended.

Workers won their demand that Petrobras within 60 days recognize the 6 hr work day prescribed in Brazil's new constitution.

Petrobras said Brazil's product supplies were secure except for LPG. It expected to be supplying the normal 9,090 metric tons/day of the product when its refineries returned to full operation. But stock drawdowns during the strike might interfere with deliveries in some regions, especially those supplied by refineries in Sao Paulo.

Those regions, mainly Sao Paulo and Matto Grosso do Sul states, account for 5,000 metric tons/day of the country's total 13,000 tons/day of LPG consumption. Imports account for about one-fourth of Brazil's LPG supply. Petrobras usually maintains LPG inventories sufficient for 10 days of consumption. It expected to rebuild stocks to that level in about a month.

The state owned company spent $65 million on unplanned product imports during the strike. Volumes totaled 110,000 metric tons of LPG, 153,000 tons of diesel, 60,000 tons each of kerosine jet fuel and petrochemical naphtha, and 185,000 tons of fuel oil.

Petrobras lost exports totaling 1.8 million bbl of products worth an estimated $28.8 million. It will be able to make up about one-half that amount. Brazil has export contracts for gasoline and fuel oil with the U.S. and for diesel fuel with Zaire.

New Brazilian
gas pipeline

Petrobras commissioned a 201 mile, 22 in. gas pipeline from Volta Redonda, Rio de Janeiro, to Capuava, Sao Paulo, Brazil.

The pipeline started operations with a capacity of 38.85 MMcfd of associated gas from offshore Campos basin fields. It extends a 60 mile, 18 in. pipeline from the state oil company's 226,600 b/d Duque de Caixis refinery in Rio de Janeiro to Volta Redonda.

The new pipeline crosses 18 major waterways, 24 high-

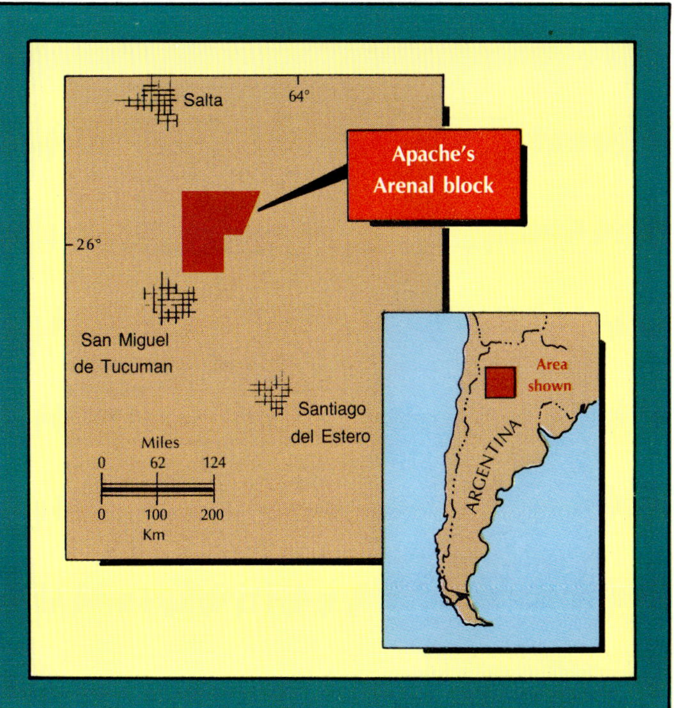

ways, and three railways. Compression facilities at the refinery will boost capacity to 105.9 MMcfd. Capacity will later increase to 176.57 MMcfd in a second project phase involving construction of a compressor station at Volta Redonda, expansion of the refinery compressor station, and looping of 22 miles of line. First phase costs total about $110 million. Petrobras operates a 32,500 b/d refinery near the pipeline terminus at Capuava.

New oil in
Santos basin

Petrobras discovered oil in the Santos basin off the southern coast and tested oil in a third well south of giant Marlim oil field in the offshore Campos basin.

The company also tested oil at a delineation well in a gas prone area of Potiguar basin off Rio Grande do Norte state.

And it added to its list of successes in the onshore Solimoes basin of the Upper Amazon.

The results brighten Brazil's chances for achieving its goal of petroleum self-sufficiency by 1993. The economically troubled country produced an average 578,573 b/d of crude oil and NGL at midyear 1988. First half 1988 products consumption averaged 1.089 million b/d, up 0.3% from the same period in 1987. The June 1988 production average was down 1.1% from June 1987 but up 34,700 b/d from the preceding month, when production was hurt by a blowout and fire on the Enchova platform in the Campos basin. Petrobras tested oil and gas from two pays in the Santos basin discovery well, 1-PRS-4.

The productive intervals are between 14,682 ft and 15,092 ft. One interval flowed at rates of 3,535 b/d of oil and 10.45 MMcfd of gas. Test rates from the other interval were much smaller. Gas contained some hydrogen sulfide.

The Petrobras XVI semisubmersible, which drilled 1-PRS-4, moved 3,600 ft northeast to drill a confirmation well.

Petrobras planned a 3D seismic survey of the structure and may move another rig to the area, which it named Tubarao field.

Exploration has been under way in the Santos basin since the early 1970s. Petrobras has drilled almost 30 wells in the basin and identified a number of limestone structures, each

7-19 sq miles in area. The structures are similar in size and lithology to those of Pampo and Garoupa oil fields in the Campos basin.

Pecten do Brasil, drilling under a risk service contract, discovered gas and condensate in the basin in 1984. It plans to place the field, Merluza, on production by yearend 1989. Petrobras estimates the field will produce at least 53 MMcfd for 13 years.

In the Campos basin, Petrobras tested 28° gravity crude in 3-MRL-4-RJS, a second delineation of the 1-RJS-382 discovery south of Marlim field. The oil in all three wells is lighter than that of Marlim, which averages 20° gravity.

In the Potiguar basin, Petrobras tested about 400 b/d of 41° gravity crude at 3-RNS-117A, a south delineation of a gas/condensate field discovered by Petrobras 1-RNS-89A.

Upper Amazon
sizzles

Brazil's onshore hotspot, along the Urucu River in the Upper Amazon region, has yielded another discovery.

It's Petrobras 1-SUC-1, drilled on a 2½ sq mile structure 7½ miles southwest of the Urucu area's discovery well, 1-RUC-1. Drilled to 8,760 ft, the 1-SUC-1 flowed 7 MMcfd of gas and 250 b/d of condensate from pay at 8,353-56 ft. And it flowed 925 b/d of 47° gravity oil from 8,271-81 ft. Pays are Pennsylvanian Itaituba and Monte Alegre.

In the same area, Petrobras completed 3-RUC-8, a delineation well 1½ miles northeast of the 4-RUC-2 discovery. The delineation, drilled to 8,580 ft, flowed 1,350 b/d of 41° gravity crude from an interval below 8,176 ft.

Petrobras started production from its Urucu fields in July 1988.

Enchova field
production restoration

Petrobras planned to use an early production system to restore oil and gas flow from offshore Enchova field, where a platform well blew out in April 1988.

Fire from the Campos basin blowout damaged the field's central platform, designated PCE-1. The jacket remained in place. An investigative commission told Petrobras the blowout resulted from a combination of failures during recement-ing of oil well No. 19, which was being recompleted as a gas producer. Petrobras expected to temporarily produce Enchova wells through a semisubmersible platform installed near PCE-1. Oil and gas would be separated on the semi and carried ashore via existing pipelines.

First production was expected in November 1988 from three nonassociated gas wells with total capacity of 26.5 MMcfd. The wells, north of PCE-1, have wet trees that will be connected with the semi.

Oil production was to start about May 1989 when debris is cleared from PCE-1 and the new equipment is installed.

The permanent system will involve a new fixed central process platform, designated PPE, linked to production facilities on PCE-1. PPE will have oil separation, gas compression, and water treatment facilities, handling production and injection for Enchova and Enchova West fields.

A fixed satellite platform, PEO-1, will be installed in Enchova West field. The semi will become unnecessary as the permanent system comes on stream.

Petrobras officials estimated Campos basin production fell by 62,000-75,000 b/d of oil and 60 MMcfd of gas as a result of the incident. The platform handled 27,500 b/d of production from Enchova field, 11,000 b/d from Bicudo, 10,000 b/d from Bonito, 14,000 b/d from Pirauna, 20,000 b/d from Marimba, and 50,000 b/d from Pampo.

The company expected to be able to make up oil requirements from crude stocks and about 38,000 b/d of additional imports. Brazil uses about 1.2 million b/d and produces 625,000 b/d, of which 365,000 b/d comes from the offshore Campos basin.

Petrobras says the Campos basin area may hold reserves of 5 billion bbl under water as deep as 6,560 ft.

The company had near term plans to drill five wildcats and five development wells in the area. One well, Petrobras 3-RJS-395, tested as much as 3,000 b/d of crude, confirming the discovery south of the giant Marlim field.

Marlim was to go on production in second half 1989 at a rate of 50,000 b/d from six wells. Petrobras estimates Marlim's probable reserves at 2.75 billion bbl of oil.

The Campos basin's first giant discovery, Albacora field, with reserves of 1.1 billion bbl, went on stream late in 1987 and was producing 18,000 b/d at yearend 1987.

Upper Amazon
Urucu production

Petrobras began producing 3,000 b/d of crude oil from its remote Upper Amazon Urucu area discoveries.

At first, it will carry production by barge and tanker to its 10,000 b/d refinery at Manaus, replacing some of the 8,000 b/d now reaching the plant from northeastern Brazil.

Petrobras has drilled eight Urucu wells. It estimates reserves at 150 million bbl of 40° gravity oil.

The state owned company plans to spend $60 million to develop the field with directional drilling, reducing the number of locations and cutting costs.

It called "completely successful" an experiment in 3-RUC-9D, the first directional well drilled in the Amazon region. The well initially flowed 1,700 b/d of 44° gravity oil.

Petrobras reported three other wells under way in the area, including a wildcat, 1-SUC-1, south of the productive area.

Urucu production will travel by barge down the Urucu River to the Solimoes River, where it will be transferred to small tankers for shipment to Manaus. Because navigation of the Urucu can be troublesome during the June-December dry season, Petrobras is considering construction of a 28 mile, 4½ in. pipeline from Urucu to Port Moura on the Tefe River. The line could carry 6,500 b/d of oil. A loop might be built later. At Port Moura, 200 ton barges would load the oil for shipment to Manaus.

Petrobras plans later to lay a 112 mile pipeline from Urucu to the Solimoes River. Numerous bayous and abrupt soil

SOUTH AMERICA
MEXICO
GUATEMALA
BELIZE
HONDURAS
EL SALVADOR
NICARAGUA
COSTA RICA
PANAMA
Pto. Barrios
Pt Cortes
Managua
San Jose
Pto. Limon
Havana
Cabaiguán
CUBA
Santiago de Cuba
JAMAICA
Kingston
HAITI
Port-au-Prince
DOMINICA REPUBLIC
Bonao
Nigua
Santo Domingo
PUERTO RICO
Ponce
Penuelas
Yabucoa
Bayamón
Virgin I.
St. Croix
Antigua I.
St. John's
Fort de France
Martinique I.
Barbados I.
Bridgetown
Fisher Pond
Caribbean Sea
Yucatán Channel
Coiba I.
Colón
Panamá
Cartagena
Barranquilla
Santa Marta
Caracas
Maracaibo
Pointe-a-Pierre
Forest Reserve
Port of Spain
Port Fortin
TRINIDAD AND TOBAGO
Oficina
Barrancabermeja
VENEZUELA
Cartago
Manizales
Medellín
Bogotá
Buenaventura
El Guamo
COLOMBIA
Georgetown
Paramaribo
GUYANA
SURINAME
FRENCH GUIANA
Pirapema
Tumaco
Esmeraldas
Quito
La Libertad
Guayaquil
Ancón
Talara
Bayovar
Lobitos
ECUADOR
Iquitos
Concordia
San Jacinto
Belém
Marajó I.
Manáus
Nova Olinda
Fortaleza
Recife
Ceara basin
Xareu
Ubarana
Potiguar
Agulha
Contamana
Pucallpa
La Oroya
La Pampilla
Conchán
Lima
PERU
BRAZIL
Maceió
Aracaju
Matáripe
Candeias
Salvador
La Paz
Cochabamba
Santa Cruz
Oruro
Sucre
Camiri
Arica
BOLIVIA
Lago Titicaca
Salar de Uyuni
Corumbá
Brasília
Belo Horizonte
Betím
Paulínia
Vitória
Campos
Rio de Janeiro
São Paulo
São Sebastião
São Mateus do Sul
Campo Duran
Villa Elisa
Asunción
PARAGUAY
Salta
Metán
Apucarana
Canoas
Tramandaí
Porto Alegre
Rio Grande do Sul
URUGUAY
Montevideo
Córdoba
Mendoza
San Lorenzo
Buenos Aires
La Plata
Olavarría
Concón
Valparaiso
Santiago
CHILE
ARGENTINA
Plaza Huincul
Concepción
Bahía Blanca
San Antonio
Golfo de San Matías
Chiloé I.
Comodoro Rivadavia
Golfo de San Jorge
Manantiales
San Sebastián
TIERRA DEL FUEGO
Cape Horn
Falkland I.
Pacific Ocean
Atlantic Ocean
BRAZIL
ALAGOAS
SERGIPE
BAHIA
Maceió
Aracaju
Salvador
Matáripe
Atlantic Ocean
BRAZIL
ESPIRITO SANTO
RIO DE JANEIRO
Vitória
Campos
Duque de Caxias
Niterói
Rio de Janeiro
Atlantic Ocean
0 50 100 Mi.
0 50 100 Km.
0 100 200 300 400 Mi.
0 200 400 600 Km.

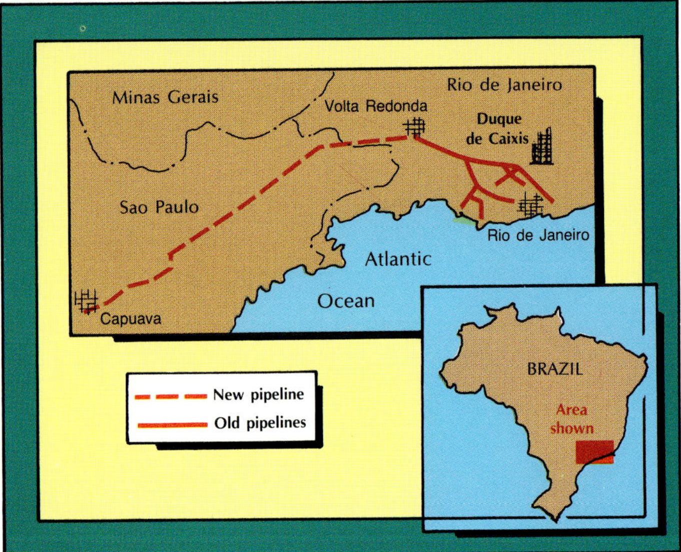

changes will make pipeline construction difficult.

Water depth production record

Petrobras again shattered the world water depth production record.

The company early in 1988 installed a subsea tree on its 3-RJS-376-D well in 1,613 ft of water in Marimba field off Rio de Janeiro. The previous record was 1,348 ft, set in 1987 at 3-RJS-294 in the same field.

After 3 days of well stimulation, testing, wireline work, and installation of the tree mandrel protector, 3-RJS-376-D started production at a rate of 5,095 b/d through the Petrobras XV semisubmersible production unit in Pirauna field 4.3 miles away. A second diverless layaway completion was installed on 4-RJS-330A in 1,067 ft of water in Albacora field, one of the six wells producing to a manifold in first phase development of the field.

A 4 in. by 2 in. flow line connection base was introduced on the project. The flow line terminations have swiveling elbows that can be manipulated to make precise alignment during makeup and allow flexible flow lines and umbilical attachments to be made up to the flow line base while the equipment is in the moonpool. Flow lines were passed from the dynamically positioned lay vessel to the moonpool of the completion rig and connected to the flow line base terminations.

Connections included the production line, annulus access line, hydraulic controls, and electrical systems. All subsea equipment was run on guidelines without the use of divers but monitored by a remotely operated vehicle.

CHILE

CAPITAL: Santiago
MONETARY UNIT: Pesos
REFINING CAPACITY: 146,800 b/cd
PRODUCTION: 24.8 Mb/d
RESERVES: 287,000 Mbbl

CHILE'S NEW METHANOL PLANT WAS RUNNING AT yearend at rates of as much as 110% of its 750,000 metric ton/year design capacity.

Cape Horn Methanol Inc. (CHM), operating Chile's biggest hydrocarbon processing plant, was producing chemical grade methanol that exceeds specifications, reports M.W. Kellogg Co., Houston.

Dedication at the site near Punta Arenas on the Strait of Magellan took place in the summer of 1988. It marked Chile's entry into the methanol market.

Kellogg designed, engineered, procured, built, and started up the $300 million plant for CHM.

Kellogg also will operate and maintain the plant under a separate contract.

Kellogg used computer aided design and engineering systems to build the plant. Three dimensional modeling techniques were used for the entire process area and steam systems and for more than 80% of the overall complex.

Procurement methods, which linked computer generated drawings with Kellogg's material control system, virtually eliminated surplus purchasing, "a major accomplishment in such a remote location," Kellogg said.

At full capacity, CHM's production will amount to 6% of world methanol supply and 13% of Western Hemisphere production, Kellogg said. CHM production will go to major markets, including the U.S., Europe, and Japan, with potential exports valued at $130 million/year. First shipment was due in November 1988.

Surging world demand for methanol as a chemical intermediate feedstock and for production of the octane booster methyl tertiary butyl ether is pressing industry capacity to the limit.

Chile, increasingly dependent on crude oil imports, is pressing efforts to tap its abundant gas reserves as feedstock in a petrochemical export industry. Two 41,000 dwt dedicated tankers will transport more than 90% of the plant's production to markets in North America, Europe, Japan, and Latin America. The plant is closer to the U.S. Gulf Coast than are plants in Canada, East Indies, New Zealand, Europe, and Middle East. Product is marketed in the U.S. through General Chemical Corp., Parsippany, N.J.

CHM, owner and operator of the plant, is a joint venture of Chilean and foreign investors. Henley, of La Jolla, Calif., and Hampton, N.H., is majority owner.

The plant uses natural gas delivered by pipeline from nearby reserves held by state owned Empresa Nacional del Petroleo. ENAP supplies the gas under a 20 year, fixed price plus escalation contract. The plant produces chemical grade methanol by a low pressure synthesis process, using spherical reactors instead of cylindrical ones to improve catalyst efficiency. It has all utilities on site, including a desalinization plant and a nitrogen plant.

Utilities and offsites include water treatment and power generation, four storage tanks, each with capacity totaling about 15 days of plant production, docks, and shiploading systems rated at 2,500 metric tons/hr of methanol.

COLOMBIA

CAPITAL: Bogota
MONETARY UNIT: Pesos
REFINING CAPACITY: 227,400 b/cd
PRODUCTION: 346.7 Mb/d
RESERVES: 2,028,000 Mbbl

THE CONSTRUCTION OF A MAJOR GRASSROOTS power generating and distribution system was an important element in producing and moving crude oil from Colombia's Caño Limón field to world markets, said Combustion Engineering Resource's J.P. Reynolds.

The system is the first source of continuous, reliable power in the region and also will provide significant socioeconomic benefits to the public sector.

The Caño Limón oil field in the Eastern Llanos region of Colombia was initially brought on stream in December 1985. The field is operated by Occidental on behalf of the Cravo

SOUTH AMERICA, SOUTHERN PORTION

Pacific Ocean
Atlantic Ocean

BOLIVIA
La Paz
Pto. Villarroel
Sicasica
Oruro
Cochabamba
Sucre
Santa Cruz
La Peña
Rio Grande
Camiri
Villa Montes
Sanandita
Campo Durán
Embarcacion
JUJUY
Jujuy
Salta
SALTA
Metán
Tucumán
CATAMARCA
Catamarca
LA RIOJA
La Rioja
Antofagasta
Vallenar

PARAGUAY
Asunción
Concepción
FORMOSA
Formosa
Villa Oliva
Villarrica
CHACO
SANTIAGO DEL ESTERO
Santiago del Estero
Resistencia
Corrientes
Posadas
MISIONES
CORRIENTES
Monte Caseros
Uruguaiana
Artigas
Santa Maria
São Gabriel
RIO GRANDE DO SUL
Santo Angelo
São Luís Gonzaga

BRAZIL
MATO GROSSO
Corumbá
Campo Grande
Aquidauana
Uberaba
GOIAS
Goiânia
Montes Claros
MINAS GERAIS
Belém
Belo Horizonte
Campos
Duque de Caxias
Rio de Janeiro
São Sebastião
SAO PAULO
Paulínia
Mauá
São Paulo
Cubatão
Santos
PARANA
Apucarana
Curitiba
Araucária
São Francisco do Sul
SANTA CATARINA
Florianópolis
Laguna
Canoas
Porto Alegre
Tramandaí
Rio Grande

ARGENTINA
Córdoba
CORDOBA
Rio Cuarto
San Luis
SAN LUIS
San Juan
SAN JUAN
Mendoza
Luján de Cuyo
MENDOZA
Quintero
Concón
Valparaiso
Santiago
S. Fernando
Talca
Chillán
Concepción
Los Angeles
Valdivia
Puerto Varas
Castro
Chiloé I.
SANTA FE
San Justo
Santa Fe
Paraná
ENTRE RIOS
La Paz
URUGUAY
Mercedes
Colonia
Montevideo
La Teja
José Ignacio
Minas
Melo
Rio Negro
San Lorenzo
Rosario
Pergamino
Campana
Buenos Aires
Quilmes
Lomas
La Plata
Ensenada
Dock Sud
Las Flores
Azul
Olavarría
Tandil
BUENOS AIRES
Santa Rosa
Cnl. Pringles
Bahia Blanca
Mar del Plata
Puerto Rosales
Punta Alta
Puerto Galván
NEUQUEN
Plaza Huincul
Neuquen
General Roca
Chañacal
Cerro Bandera
RIO NEGRO
San Antonio
Viedma
Puerto Madryn
Pen. Valdés
Rawson
CHUBUT
Rio Chubut
Sarmiento
Caleta Cordoba
Comodoro Rivadavia
Caleta Olivia
Pico Truncado
SANTA CRUZ
Santa Cruz
Rio Gallegos
Cóndor
Hidra
Punta Arenas
Magallanes
San Sebastián
Ushuaia
Estrecho Le Maire

Malvinas Islands
FALKLAND I.

100 200 300 400 Mi
100 200 400 600 Km

TIERRA DEL FUEGO
Monte Aymond
Dicky
Pampa Larga
Punta Delgada
Kimiri Aike
Cóndor
Daniel Este
Posesión
Cullen
Pta. Dúngeness
Dúngeness
Anegada
Estrecho de Magallanes
San Gregorio
Gregorio
Angostura
Manantiales
Spiniful
PTA. SAN ISIDRO
Puerto Percy
Cabo Nombre
Cabo Negro
Punta Arenas
Chañarcillo
Catalán
Chilán
Cullen
Tres Lagos
Bahía San Sebastián
Hidra
San Sebastián
Toma
Estancia Sara
Los Lagunas
Arroyo Gama
La Sara
Laguna Escondida
Rio Avilés
Rio Chico
Rio Grande
Arroyo Candelaria
CHILE
ARGENTINA
Ushuaia
Beagle Channel
Estrecho de Magallanes
Dawson I.
B. Inútil

25 50 Mi
50 Km

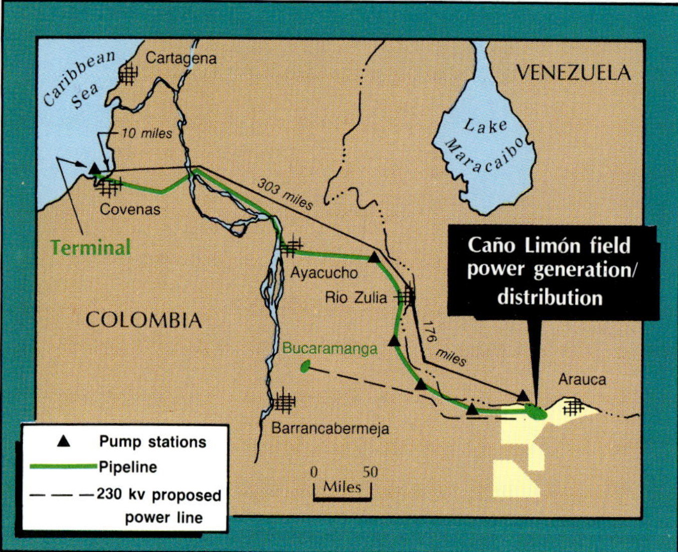

Norte Association which comprises Occidental de Colombia, Shell de Colombia, and the state oil company Ecopetrol.

The power generation system utilizes crude-burning reciprocating engines, which are believed to be the largest in use by the oil industry in South America. Overhead power lines were constructed in the most difficult of swamp conditions to distribute power to the wellsites and production facilities. Colombian contractors, labor, equipment, and materials were used as much as possible throughout the project.

Detailed design was carried out by contractors based on conceptual designs done in-house by Oxy.

Construction of the power system was carried out in stages to meet the power demands of the phased production buildup. For the early stages of production, the artificial lift electrical submersible pump (ESP) installations were powered individually by small portable generator sets.

The Caño Limón centralized power generating system, connected by the overhead distribution system, is designed to accept the planned 230 kv national grid connection. This arrangement was deemed to be the most reliable and long-term, cost-effective option.

The system is now essentially complete and able to meet all power demands which will reach steady state in 1989.

The power plants in each of the two production facilities are equipped with generator sets using crude-burning reciprocating engines, each rated at site conditions 2,400 kw output electrical.

The drivers are four stroke, turbocharged, eight cylinder, in-line 720 rpm 3,520 hp crude-burning reciprocating engines manufactured by W.H. Allen of U.K. The generators were made by GEC - U.K.

The central power plant was started up in the second half 1987 under a turnkey contract awarded in April 1986 to a consortium of Hawker Siddeley Power Engineering (HSPE) U.K. and Distral S.A., Colombia. The engines were manufactured by Mirrlees Blackstone and the generators by Brush Electrical Machines, both subsidiaries of the Hawker Siddeley group.

The central facility machines are each rated at site conditions for 8,404 kw output electrical. The drivers are four-stroke, turbocharged, 16-cylinder VEE form 600-rpm, 11,680-bhp crude-burning reciprocating engines.

Logistics

Some difficult construction logistics were involved which were further compounded by a compressed schedule. All roads and plant sites in Caño Limón had to be raised above swamp water levels that vary rapidly with the rainfall in the mountains and swamp regions. However, the central power plant benefited from having an infrastructure, including roads and camps, that had already been established during earlier construction of the production facilities.

Basic airport facilities were available at the regional capital town of Arauca which is located 48 km from the Caño Limón field on the Colombia-Venezuela border.

Each central plant generator/engine set weighs 160 tons assembled, with the heaviest single components of the engine being the generator rotor at 19.5 tons and the engine column assembly at 21 tons shipping weights.

The original plan was to ship the engines direct to Caño Limón via the Orinoco River through Venezuela and the Arauca River in Colombia, but seasonal low water levels exposed rapids and prevented access. Air freight was considered unnecessary and the costs prohibitive for the central plant. The engines were broken down and shipped by sea to Puerto Cabello, Venezuela, and then by road to Caño Limón. Transportation took 21 days from the U.K. port to the site. Reassembly was carried out in situ by the manufacturer's technicians utilizing a 40-ton overhead crane installed in the powerhouse.

To meet the earliest stages of the construction and production schedule, the generator sets for the first production facility were flown to Arauca from the U.K. disassembled. Other major equipment for the power system, including transformers, switchgear, and motor control centers, which was susceptible to road transport damage, was also flown in.

National grid connection

By late 1989, the 230 kv national grid connection is scheduled for connection in Caño Limón. The connection will be made at the central plant and operated in parallel with the Caño Limón power system.

Distribution to the main townships of the region via some 120 km of aerial lines is presently under construction by the utility company. Initially, the towns will be supplied power from the Caño Limón system to provide a reliable 24 hr source of power.

The Caño Limón system will continue to provide backup power to the townships after the grid connection is available.

CUBA

CAPITAL: Havana
MONETARY UNIT: Pesos
REFINING CAPACITY: NA
PRODUCTION: NA
RESERVES: NA

CUBA'S NATURAL GAS FLOW, STILL VERY small, climbed to a record.

Flow quintupled in 1987 compared to 1986. Production, which exceeded 1 bcf for the first time, comes almost entirely from north coast fields east of Havana.

Much of the production is so sour it can't be used without processing plants not now available. Hydrogen sulfide content of associated gas in Varadero field is as much as 18.14%. Because Cuba still hasn't determined its gas resources, Havana officials haven't decided whether it is worthwhile to build a gas processing plant.

New data on Varadero field show that the well with 18.14% hydrogen sulfide content in its associated gas flowed 1,329 b/d of oil in its third year of production. That's 6-7% of total Cuban 1987 crude production. Another Varadero well on production for 11 years is producing 331 b/d of oil. Varadero field is believed to extend offshore.

ECUADOR

CAPITAL: Quito
MONETARY UNIT: Sucres
REFINING CAPACITY: 123,300 b/cd
PRODUCTION: 310.1 Mb/d
RESERVES: 1,350,000 Mbbl

WORK WAS HITTING ITS STRIDE AT YEAREND 1988 under exploration contracts international companies signed with Ecuador's Corp. Estatal Petrolera Ecuatoriana (CEPE) since enactment of the current petroleum law in 1982.

Units of British Petroleum Co. plc, Conoco Inc., and Occidental Petroleum Corp. and partners found oil on their blocks, all in the Oriente region. Other contractors had begun or were to soon begin exploration programs.

Most advanced is BP's campaign for its Block 7 discovery. The company asked CEPE to approve unitization and commerciality of its Payamino find, which apparently links with the state company's Paraiso and Pucuna discoveries to the north.

Earlier in 1988, CEPE 3 Pucuna flowed a combined 5,748 b/d of 27-35° gravity oil.

Other discoveries

On Block 15, which in 1985 became the first acreage awarded under the current law, Oxy gauged 1,200 b/d of 23° gravity oil at 1 Indillana, its fourth and last wildcat planned for the block.

One of the earlier tests, 1 Limoncocha, flowed 2,104 b/d of 28° gravity oil from an interval productive in CEPE-Texaco Inc.'s giant Shushufindi field on a separate structure 5 miles north. Oxy was evaluating the discoveries and might request a 2 year extension of the Block 15 exploration program.

Conoco was drilling another wildcat, 1 Ginta, on promising Block 16. Earlier, Conoco 2 Daimi confirmed the 1 Daimi discovery. A previous test, 2 Amo, was dry. In addition to the Daimi discoveries, Conoco was evaluating strikes at 1 Amo and 1 Bogi.

The Exxon Corp.-Hispanoil combine, was drilling 1 Tzapinom—its second and last wildcat on Block 8. Exxon 1 Dayuno flowed 710 b/d of 13° gravity oil in 1987.

Other exploration

Elsewhere in the Oriente region, Tenneco was drilling 1 Tigrillo, first of four wildcats planned under its contract for Block 12. Target depth was 12,850 ft.

A group led by Ste. Nationale Elf Aquitaine planned to spud its first wildcat, 1 Sunka, on Block 14. The contract calls for three more wildcats.

Petro-Canada planned to drill its first

wildcat on Block 9 early in 198. And groups involving Unocal Corp. on Block 13 and ARCO on Block 10 were starting geophysical work.

On the coast, the Texaco-Pecten Ecuador Co. combine spudded its third and last wildcat, 1 Calceta, on Block 6. The earlier wells, 1 Ricuarte and 1 Chone, were dry.

CEPE takeover

Ecuador Oil Minister Diego Tamariz outlined to the country's congress how CEPE would take over certain operations previously carried out by private companies.

CEPE was to assume operation of the Lago Agrio-Balao crude oil pipeline July 1, 1989, and the 28,500 b/d Anglo refinery at La Libertad the following Nov. 30.

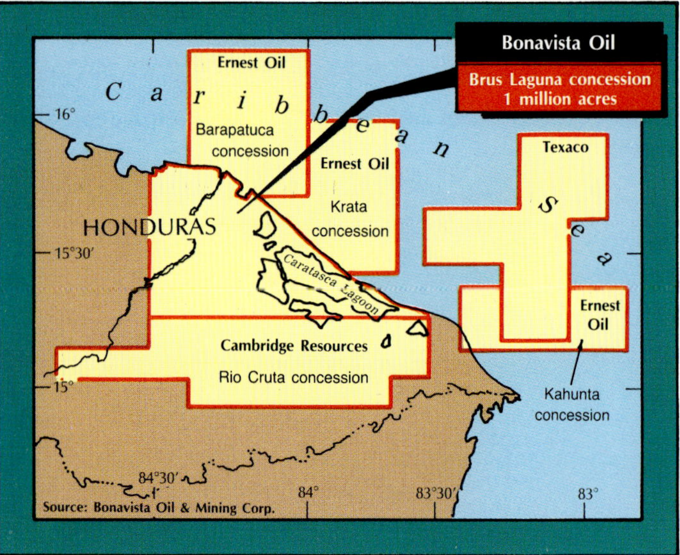

In 1990 it will assume operation of the CEPE-Texaco joint venture's Oriente fields July 1—although the contract with Texaco will continue until June 1992—and operation of Repetrol's 8,500 b/d La Libertad refinery Aug. 19.

CEPE plans to continue producing at the present level of about 310,000 b/d but reduce output 10,000 b/d from CEPE-Texaco's Shushufindi field to boost flow from CEPE's own fields by a similar volume.

GUYANA

CAPITAL: Georgetown
MONETARY UNIT: Dollars
REFINING CAPACITY: NA
PRODUCTION: NA
RESERVES: NA

EXPLORATION WAS SET TO RESUME OFF GUYANA.

The government and its Guyana Natural Resources Agency (GNRA) awarded two offshore exploration licenses.

One went to a combine of Guyana Exploration Ltd., an Isle of Man company owned by U.S. interests, and Petrel Petroleum Corp., Houston. GNRA awarded the other license to a combine of Lasmo Oil (Guyana) Ltd. and BHP Petroleum (Guyana) Inc.

Guyana has seen no exploration since 1983, when Home Oil Co. Ltd., Calgary, shut down operations after testing oil in noncommercial volumes in an onshore wildcat about 200 miles south of Georgetown.

Offshore, a Royal Dutch/Shell Group affiliate had noncommercial hydrocarbon shows in the 1 Abary wildcat.

Word of pending offshore contracts caused a political stir in neighboring Venezuela, which long has disputed its border with Guyana, claiming territory as far east as the Essequibo River. Venezuela's Foreign Ministry later said the contract areas are outside the disputed region.

The Essequibo territory covers 53,000 sq miles between Guyana's Essequibo River and Venezuela's eastern border. The jurisdictional question has been before the United Nations since 1982.

Past Guyanan efforts to explore the region have sparked strong protest from Venezuela and nationalist groups in the territory. Relations between the countries were tense during the administration of the late Guyanan President Forbes Burnham but improved in the administration of Desmond Hoyte.

Contracts similar

The new licenses are similar, both based on production sharing agreements in the event of discoveries.

Lasmo, with a 60% interest, and BHP, 40%, was to begin shooting 2,000 line km of seismic survey on the 11,418 sq km Satira block. The license has an initial term of 3 years with options for two successive 3 year periods.

Lasmo-BHP has an option to drill, depending on seismic results. If drilling is successful the partners can apply for a 20 year production license.

Guyana Exploration and Petrel also agreed to conduct 2,000 line km of seismic survey. The Western Anchorage seismic vessel began operations in the summer of 1988.

The combine planned to begin drilling early in 1989 in an exploration period lasting as long as as 10 years. The combine's block covers 12,450 sq km.

Petrel says main objectives are mid-Cretaceous carbonates in structures or well defined reefal trends. Oil source in the Guyana basin is Cretaceous La Luna shale.

Affiliates of Conoco Inc., Tenneco Inc., Royal Dutch/Shell Group, and Deminex operated in the region earlier.

At least one well had hydrocarbon shows.

HONDURAS

CAPITAL: Tegucigalpa
MONETARY UNIT: Lempiras
REFINING CAPACITY: 14,000 b/cd
PRODUCTION: 0
RESERVES: 0

EARLY EXPLORATION RESULTS ON THE BRUS Laguna concession in Honduras led to seismic surveys on the block.

Bonavista Oil & Mining Corp., Houston, started up 350 line miles of seismic surveys on the 1 million acre concession in the Mosquitia basin.

The work follows geological, geochemical, gravity, and aeromagnetic surveys the company undertook in 1985-87.

The Honduran government extended the concession, scheduled to expire in September 1988, to allow Bonavista and its joint venturers to do more work.

A 1985-86 collation of exploration data on Honduras by Litton Resources Group, Denver, established the presence of large structures and the potential for major oil reserves, Bonavista said.

The Mosquitia basin has many oil seeps. Based on plate tectonic reconstructions, structural and sedimentary correlations with southern Mexico and Guatemala suggest that Mesozoic sourcing may be present in marine and nonmarine facies, Bonavista said. Furthermore, coal measures of Triassic and Jurassic age may underlie the Cretaceous and Cenozoic sections, providing effective oil and gas sources.

PERU

CAPITAL: Lima
MONETARY UNIT: Intis
REFINING CAPACITY: 172,438 b/cd
PRODUCTION: 141.7 Mb/d
RESERVES: 456,806 Mbbl

PERU'S EFFORT TO PROMOTE OIL EXPLORATION AND development by international operators has become law.

President Alan Garcia signed a package of petroleum

incentives late in 1987 based on the government's earlier proposal to Congress.

Companies signing contracts in Peru now may acquire four blocks vs. three earlier. They will have 7 years for exploration instead of 6 and be able to recover costs in 5 years. In addition, Peru's central reserve bank is to guarantee an automatic mechanism for free direct and immediate availability of part of companies' foreign exchange generated by oil exports or production. Foreign exchange delays have been a problem for some operators.

Also under the new law, state owned Petroleos del Peru SA (Petroperu) is to receive a greater share of revenues from fuel oil sales to the state. The government also will encourage additions of catalytic cracking capacity or the equivalent at the La Pampailla and Iquitos refineries, which handle low quality jungle oil.

Petroperu is to give priority to development of natural gas and condensate at Aguaytia in the central jungle.

Peru, which rescinded contracts of three key foreign operators in 1985, imported 2 million bbl of crude oil in 1987 and expected to import 6 million bbl more in 1988.

The country's production fell to 147,500 b/d in November 1987, while refinery runs remained at 162,000 b/d. Petroleum products demand was about 130,000 b/d. Peru exports the balance, mostly residual oil.

Petroperu expected Peruvian crude oil production to rebound in 1989. It forecast an average of 147,300 b/d for the year, compared with 142,000 b/d in 1988. Output averaged 178,800 b/d in 1986 and 163,600 b/d in 1987.

Talara secondary recovery project

The Occidental Petroleum Corp. of Peru-Bridas Exploracion y Produccion SA combine early in 1989 was to start rehabilitating its secondary recovery project on Peru's northern coast.

Oxy-Bridas will drill six wells and work over 100 wells at Talara under a new service contract fee agreement with Petroperu. Talara production averaged about 7,200 b/d in 1987, compared with nearly 9,000 b/d the year before. It fell further in 1988. The state owned company expected production to increase by 3,000 b/d as a result of the new agreement with Oxy-Bridas. Work slumped during a dispute between Oxy-Bridas and Petroperu over payment of fees and other problems. Oxy-Bridas claimed the government owed it about $22 million in fees. The dispute extended to fields and exploratory acreage in Oxy's jungle areas.

Revival of development drilling

Oxy reached an outline agreement with the government to revive development drilling stalled by a foreign exchange dispute. And Shell Exploradora y Productora del Peru BV denied that it withdrew from talks with the government over development of its giant San Martin/Cashiriari gas discoveries in the central southern jungle. Shell said the government suspended talks.

Peru has been importing 22,000 b/d of crude oil to offset slumping production. Part of the problem had been Oxy's inability to buy foreign exchange to pay for maintenance and drilling in the northern jungle. Production there fell to 50,000 b/d from 80,000 b/d 2 years ago.

Under the outline agreement, Peru's central bank will sell Oxy $7-9 million/month in foreign exchange. Oxy will proceed with a suspended 24 well development drilling program estimated to cost $60 million. It also will return 30 wells to production out of 90 shut in for lack of spare parts and supplies. Oxy expects to boost production to 70,000 b/d by early in 1989.

Oxy also was preparing to drill its first wildcat on Block 36 in the central jungle and had asked for a neighboring block. Both blocks are near Shell's giant gas discovery.

In Lima, Shell said it was seeking extension of a negotiating deadline when the government suspended talks on the $1.3 billion gas development project. Shell was having trouble raising financing for state run Petroperu.

In a preliminary agreement early in 1988, Shell agreed to cover cost of developing the fields, estimated at $410 million, plus one fourth of the cost of the planned pipeline, estimated to total $700 million. It agreed to help Petroperu find financing for its part of the project.

Liquids production from the field is expected to reach a maximum stable rate of 40,000-45,000 b/d, which would be carried to the coast. Initial markets for the gas would be in the center of the country along the pipeline route and later to the north and south of Lima. Some dry gas would be injected into the reservoir. Liquids would be separated into LPG and

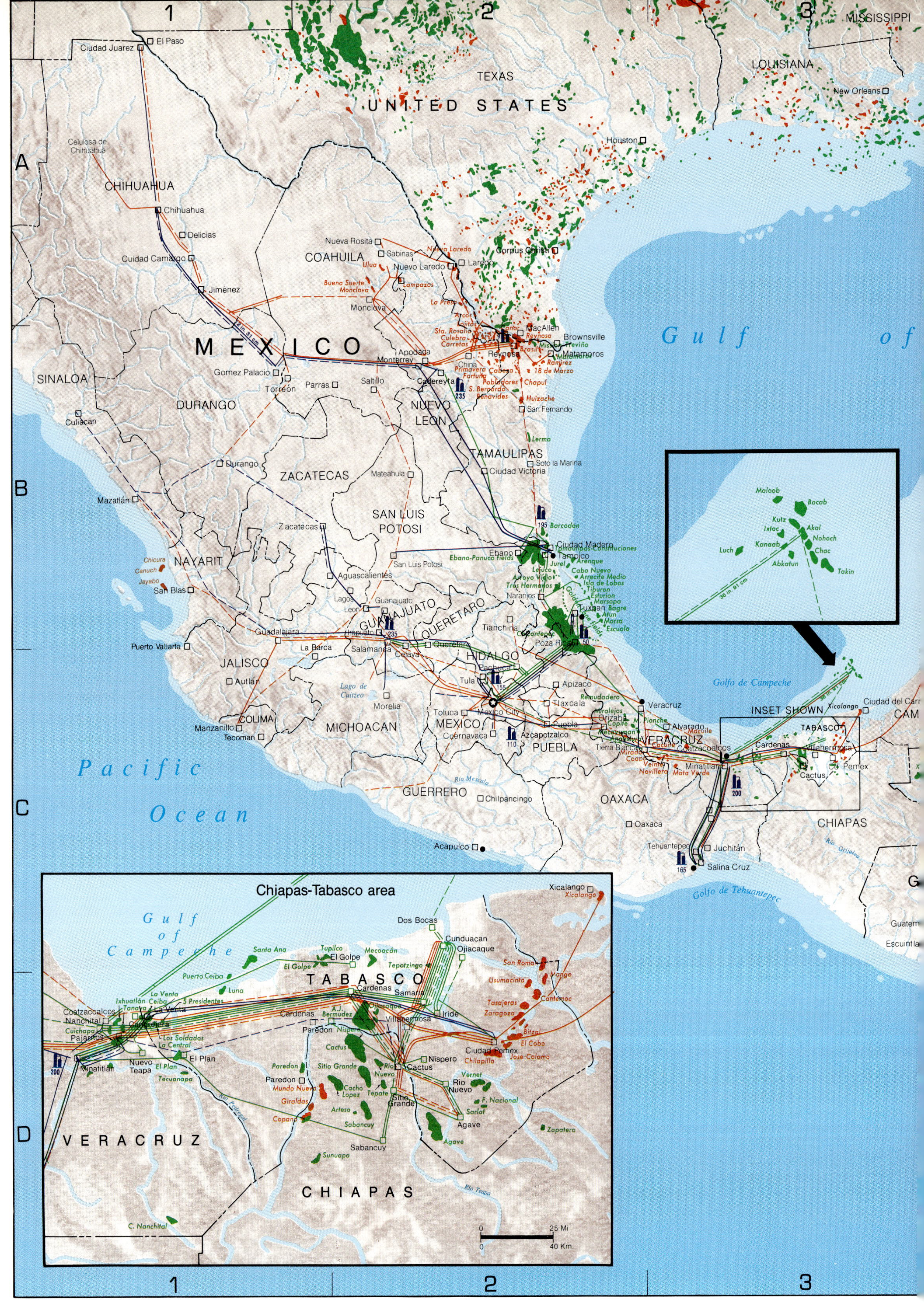

UNITED STATES
TEXAS
LOUISIANA
MISSISSIPPI
New Orleans
Houston
El Paso
Ciudad Juarez
Celulosa de Chihuahua
CHIHUAHUA
Chihuahua
Delicias
Ciudad Camargo
Jimenez
Nueva Rosita
COAHUILA
Sabinas
Ciudad Camargo
Ulua
Buena Suerte
Monclova
Monclova
Nuevo Laredo
Laredo
Corpus Christi
MEXICO
Apodaca
Monterrey
Gomez Palacio
Torreon
Parras
Saltillo
Cadereyta
China
Primavera Cobeza
Parfum
Reynosa
McAllen
Brownsville
Matamoros
Ramirez
18 de Marzo
Pobladores
S. Bernardo
Benavides
Chapul
Huizache
San Fernando
SINALOA
DURANGO
Durango
Culiacan
ZACATECAS
Zacatecas
Mateahula
Lerma
Ciudad Victoria
Soto la Marina
TAMAULIPAS
NUEVO LEON
SAN LUIS POTOSI
Barcodon
Ciudad Madero
Tampico
Ebano-Panuco fields
San Luis Potosi
Naranjos
Aguascalientes
NAYARIT
Chicura
Canuch
Jayabo
San Blas
Lagos
Leon
GUANAJUATO
Irapuato
Salamanca
Celaya
Queretaro
QUERETARO
Tianchinol
Poza Rica
Tuxpan
JALISCO
Guadalajara
La Barca
Puerto Vallarta
Autlan
Lago de Chalco
Morelia
Tula
HIDALGO
Apizaco
COLIMA
Manzanillo
Tecoman
MICHOACAN
Toluca
Mexico City
MEXICO
Cuernavaca
Azcapotzalco
Tlaxcala
PUEBLA
Puebla
Veracruz
VERACRUZ
Coatzacoalcos
Minatitlan
Tierra Blanca
Mata Verde
GUERRERO
Chilpancingo
Acapulco
OAXACA
Oaxaca
Tehuantepec
Juchitan
Salina Cruz
Pacific Ocean
Gulf of
Golfo de Campeche
Golfo de Tehuantepec
TABASCO
Cardenas
Villahermosa
Pemex
Cactus
CHIAPAS
Ciudad del Carmen
CAM
Xicalango
INSET SHOWN
Maloob
Bacab
Kutz
Ixtoc
Akal
Nohoch
Kanaab
Chac
Luch
Abkatun
Takin
96 in 91 cm
Chiapas-Tabasco area
Gulf of Campeche
Xicalango
Dos Bocas
Santa Ana
Tupilco
Mecoacan
Cunduacan
Ojiacaque
El Golfe
Tepetzingo
San Roman
Mango
TABASCO
Cardenas
Samaria
Usumacinto
Cantemoc
Puerto Ceiba
Luna
La Venta
5 Presidentes
Ixhuatlan Ceiba
Tasajeras
Zaragoza
Coatzacoalcos
Nanchital
Tanatu
Bermudez
Villahermosa
Iride
Bitzal
El Cobo
Jose Colomo
Nuevo Teapa
El Plan
Cardenas
Paredon
Cactus
Nispero
Chilapilla
Ciudad Pemex
Vernet
Minatitlan
Paredon
Sitio Grande
Rio Nuevo
Cactus
Rio Nuevo
F. Nacional
Paredon
Mundo Nuevo
Cacho
Tepate
Sto. Grande
Sarlat
Capana
Giraldos
Arteso
Agave
Agave
Zapatera
Sabancuy
Sunupa
Sabancuy
VERACRUZ
CHIAPAS
Rio Teapa
C. Nanchital
0 25 Mi
0 40 Km

heavier fractions near Petroperu's La Pampilla refinery outside Lima. Additional facilities are planned for export of products and subsequent fractionation into gasoline, kerosine, and other products to be marketed independently.

Major natural gas project

Work on Peru's first major natural gas project, which Petroperu views as the forerunner to a bigger gas venture farther south, was to have begun in 1988.

Work was due to begin when Petroperu signed a financing contract with Mexico's Petroleos Mexicanos (Pemex), leader of a group that will perform much of the work.

Other group members are Condux CP, also of Mexico, and Bruse SA and Construcciones Upaca SA, both of Peru.

The group in July 1988 signed a $30.7 million contract with Petroperu for construction of gas processing facilities, a 15 MMcfd pipeline from Aguaytia gas field to Pucallpa, and associated liquids lines.

Petroperu is spending $15 million to complete development of the Block 31 gas field, estimated to hold reserves of 255 bcf of gas and 21.3 million bbl of condensate.

Mexico will use a government to government credit line to finance goods and services for a total of about $20.4 million, 90% of the project's

MEXICO-CENTRAL AMERICA

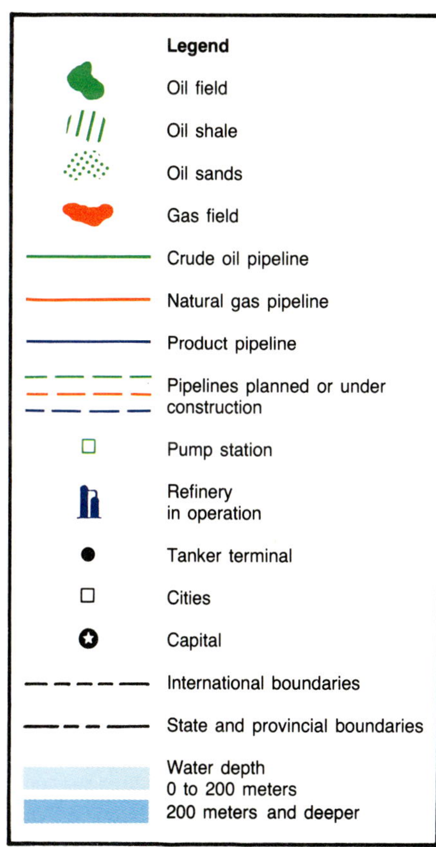

The main components in Petroperu's Aguaytia gas project

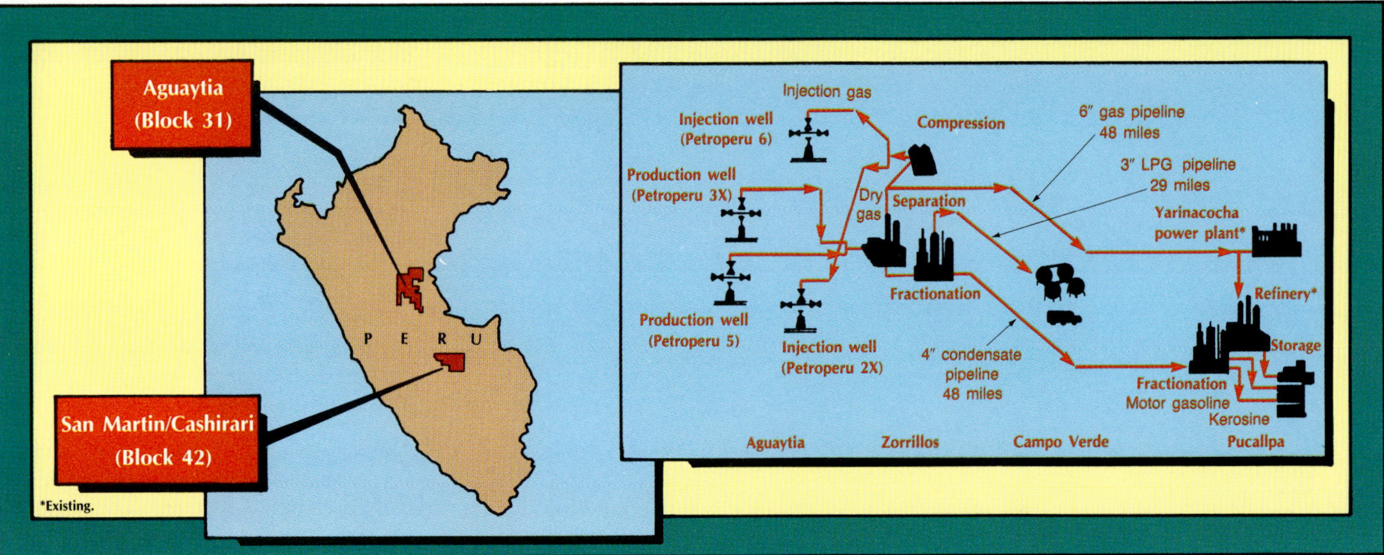

estimated foreign currency costs. It also will help Petroperu arrange the rest of the equipment purchases on a supplier financed basis.

Cofide SA, Peru's state development finance corporation, has committed to provide the equivalent of $6.75 million on $8 million worth of local costs. The corporation might provide more funding.

Petroperu is advancing the remote project with an eye on Shell Exploradora y Productora del Peru BV's giant San Martin and Cashiriari gas and condensate discoveries on Block 42, about 300 miles southeast. Reserves are estimated at 12 tcf of gas and 650 million bbl of condensate. Petroperu is developing the Aguaytia reservoir at first with two injection wells and two producers.

In 1987, it worked over the 2X well, drilled by Mobil Oil Corp. after it discovered the field in 1962. The well is one of the initial injection wells but may be converted to production later. Mobil halted its exploration program in the area in 1968 after determining that development was uneconomic due to location.

Petroperu plans gas compression, separation, and fractionation units at Zorrillos. From there a 29 mile, 3 in. pipeline will carry LPG to a terminal at Campo Verde. A 48 mile, 6 in. pipeline will carry dry gas not reinjected to Pucallpa. The Yarinacocha electrical power plant and the 2,400 b/d Pucallpa refinery will use 12 MMcfd, replacing about 1,000 of residual fuel oil.

The backed out resid will be barged to San Jose de Saramuro for blending in the North Peruvian Pipeline with production from Petroperu's northern jungle fields.

Aguaytia project units will be able to produce 650 b/d of natural gasoline, 750 b/d of kerosene, and 700 b/d of LPG.

Field production will be 30 MMcfd, with reinjection of 16-20 MMcfd, varying with demand. Pemex hopes to complete its part of the project in 1 year. It will use high rivers during the rainy season to move heavy equipment.

Petromar adds rigs

Petromar SA, offshore subsidiary of Petroperu, was to add two rigs to the five working off the country's northern coast.

Serpetro SA, Petroperu's technical services subsidiary, was expected to receive a 3 year contract for the new rigs in association with a unit of Saipem SpA.

Petromar also invited offers for two rigs to directionally drill two to eight offshore wells from onshore. The rigs are to be capable of drilling as much as 12,000 ft measured depth at angles to 60°. The company estimates cost of service at about $5 million at the free exchange rate. It hoped to increase production to 30,000 b/d by yearend 1988 from about 26,000 b/d.

Foreign exploration encouraged

Peru, encouraging exploration by foreign operators, faced growing economic pressure to reverse a 6 year oil production slide.

The country was expected to spend more on petroleum imports than it makes on oil exports during 1988, said Jaysuno Abramovich, president of Petroperu.

Peru began importing light crude in 1987 to sustain refinery runs and compensate for decreasing gravities and volumes of domestically produced crude.

According to unofficial estimates, Peru was to import 4.4 million bbl of crude and 7.1 million bbl of products in 1988. Exports, mainly of heavy fuel oil, could total 14.2 million bbl.

He expected Petroperu to spend $260 million for imported crude and products in 1988 and to earn $174 million on oil exports. Imports of crude would cost $74 million, diesel $84 million, kerosine $89 million, and other products $13 million.

In 1987, Petroperu's import costs totaled $157 million vs. export revenues of $291 million. Abramovich projected average crude oil production at 150,000 b/d for 1988. Refinery runs were expected to average about 165,000 b/d, Peruvian oil demand 143,000 b/d. In 1987, demand averaged 131,000 b/d.

Drilling plans

Petroperu, which lost $404 million in 1987 due to low petroleum prices and high taxes, was to spend about $30 million on exploration in 1988. That includes drilling by Petromar, the subsidiary formed to operate nationalized offshore fields of Belco Petroleum Co. of Peru.

The state company planned to spud a wildcat at Chambira on the southern end of northern jungle Block 8. The 14,000 ft test will cost an estimated $8 million.

Petroperu planned five wildcats on the northern coast. Petromar slated six offshore wildcats. Petroperu also planned 60 development wells and 14 secondary recovery wells on the northwest coast and four development wells in the central jungle.

Oxy of Peru was to spud its first wildcat on Block 36. Oxy spudded a wildcat on the Jibaro extension of northern Block

1AB after drilling a dry hole at Matusari.

Oxy offered to drill 24 infill wells on its northern jungle block.

Petroperu also was discussing letters of intent for exploration with Mobil Oil Corp. and Unocal Corp.

Mobil is interested in Huallaga basin Block 53, formed in 1987 in the combination of Block 29 with parts of Blocks 28 and 30. Investment could be about $100 million.

Unocal expressed interest in Block 39, north of Shell Exploradora y Productora del Peru BV's big gas discovery in the central southern jungle. The Total-Deminex-Hispanoil combine conducted seismic surveys in the area during the mid-1970s.

VENEZUELA

CAPITAL: Caracas
MONETARY UNIT: Bolivares
REFINING CAPACITY: 1,201,100 b/cd
PRODUCTION: 1,658 Mb/d
RESERVES: 58,083,900 Mbbl

WORLD CLASS RESULTS OF A 12 YEAR REVIVAL OF exploration in Venezuela have enabled the country—one of two Western Hemisphere members of the Organization of Petroleum Exporting Countries—to upgrade its crude oil production.

Following major discoveries of light and medium crude in southern Lake Maracaibo, along the border with Colombia, and in the Eastern Venezuela basin, state owned Petroleos de Venezuela SA (Pdvsa) no longer must rely for future production on heavy and extra-heavy crudes from the Orinoco oil belt.

Pdvsa has no intention of abandoning heavy oil technological development. The Orinoco belt alone holds reserves of 26 billion bbl of extra-heavy oil and bitumen and resources totaling an additional 134.2 billion bbl.

So experiments with steamfloods and gas lift continue. Furthermore, Pdvsa affiliate Maraven SA is placing on stream the country's first pipeline using Core Flow technology, in which extra-heavy crude travels in an annulus of water. And the country expects to be exporting 50,000 b/d of a combustible heavy crude-water mixture called Orimulsion by the end of 1989.

But the exploration/production emphasis is shifting to light and medium crude in the near and medium terms. An aggressive exploration program concentrates on light and medium oil. With 100,000 sq km of sedimentary basin on Venezuela's continental platform and 350,000 sq km onshore, and with only 30,000 sq km assigned for exploration at present, opportunities abound.

Pdvsa plans to drill 138 exploratory wells during 1988-93, hoping to add 14.4 billion bbl of light and medium crude plus condensate to its 58 billion bbl of current reserves.

Of that, 2.7 billion bbl is expected to come from Lake Maracaibo and Falcon state, 400 million bbl alone from the southern part of the lake. Extensions of old fields in Barinas state, in conjunction with further discoveries in Apure near new fields along the Colombian border, will add an expected 400 million bbl of light and medium crude reserves. Most—330 million bbl—will come from the Guafita area offsetting Colombia's giant Cano Limon field in Apure. But the biggest additions of light and medium crude oil reserves will occur in eastern Venezuela, where Pdvsa affiliate Lagoven SA is developing giant El Furrial field and Corpoven SA is developing its adjacent Musipan discovery.

Pdvsa expects these and surrounding discoveries in northern Monagas state to yield another 8.6 billion bbl. About 1.8 billion bbl will come from elsewhere in the eastern basin.

Production targets

Pdvsa's aim is to boost productive capacity to about 3 million b/d within the next few years from a steady 2.6 million b/d. This represents a rebuilding of productive capacity, which reached nearly 4 million b/d in 1970 but slumped after international oil companies, foreseeing nationalizations of the mid-1970s, quit exploring.

Capacity fell to less than 2.5 million b/d in 1976-81 as the national oil company absorbed the former producers and regenerated an exploration program. It rebounded to its present level in 1982.

Pdvsa decided to emphasize light and medium crudes in 1986, when the potential to do so became apparent following the Guafita, northern Monagas, and southern Lake Maracaibo discoveries. By 1987 it had squeezed production of heavy and extra-heavy crude to about 500,000 b/d out of peak production of 1.8 million b/d. By comparison, in 1984, when total production averaged about the same, heavy and extra-heavy oil accounted for about 750,000 b/d.

Assuming a rise in Venezuela's OPEC quota, Pdvsa hopes by 1992-93 to be producing 2.2 million b/d of crude, only 300,000 b/d of it heavy and extra-heavy.

More light oil sought

Pdvsa divides its exploration plans into two "strategic programs"—one involving moderately risky wells in high potential areas, the other oriented more to rank wildcat drilling.

The less risky strategy concentrates on the light oil areas of Lake Maracaibo and prospects surrounding recent light and medium crude oil discoveries. Biggest drilling focus will be in the northern Monagas area of the El Furrial and Musipan giants.

In the riskier work, Pdvsa officials are discouraged by drilling so far on the El Tablazo prospect north of production in Lake Maracaibo.

They express greater hope for the Norandino Front area to the south. At Guarumen, north of Barinas, geophysical work has identifed a "huge trap" that might contain 400 million bbl of oil.

And in the Pantano Delta area of Delta Amacuro state, Pdvsa has no current drilling plans but has conducted preliminary geophysical work and will shoot 4,000 line km of

Venezuela's 1988-93 exploration plans

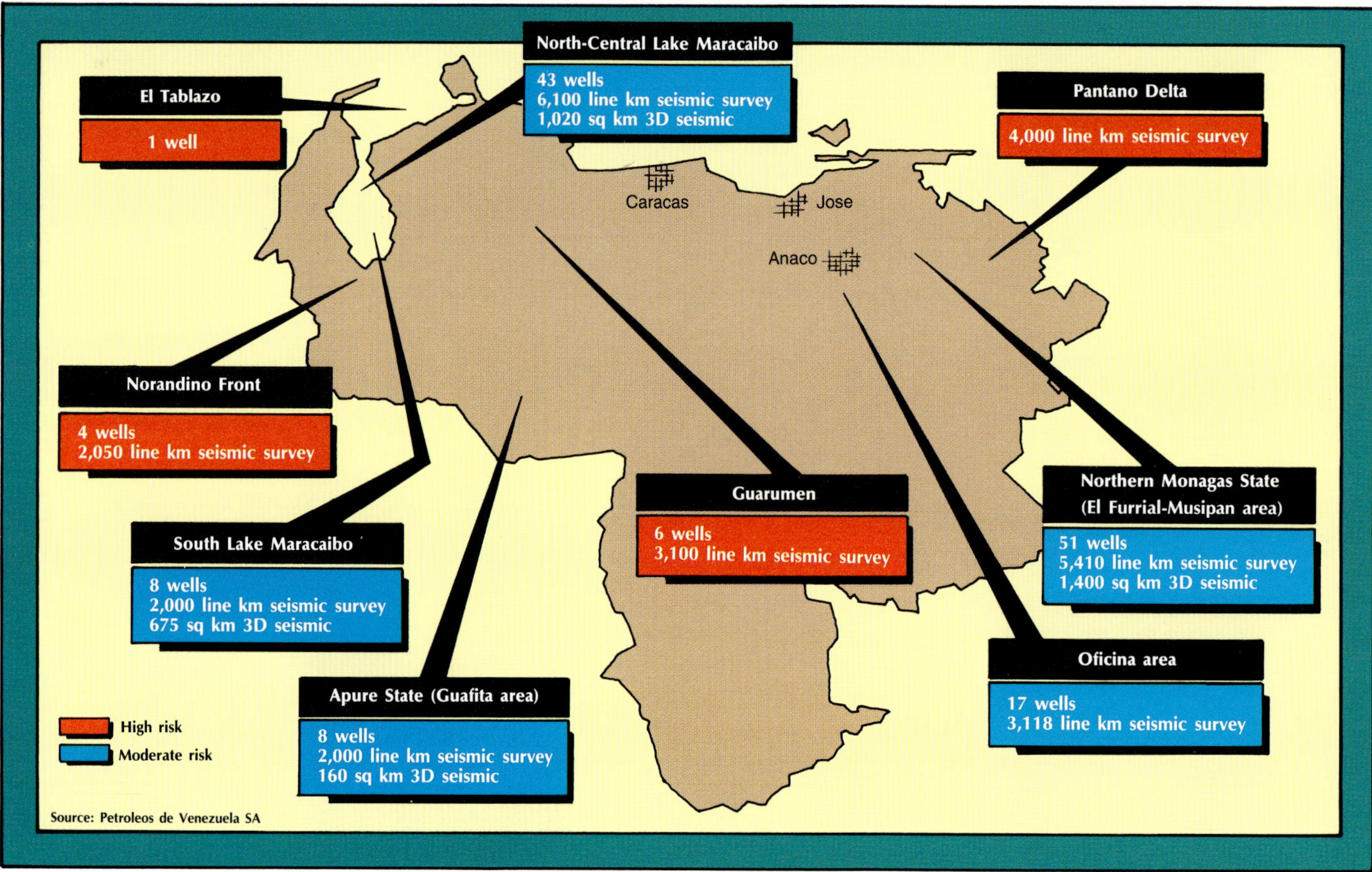

seismic surveys.

Eastern Venezuela focus

Pdvsa officials expect the northern Monagas discoveries to provide all new productive capacity necessary to reach the 3 million b/d target.

At El Furrial, about 18 miles west of Maturin, operator Lagoven SA has identified four structures. It completed two wells in 1986 and three wildcats and four development wells in 1987. It planned five wildcats, seven development wells, and one outpost well in 1988. Six El Furrial wells were producing about 30,000 b/d at midyear 1988. Lagoven expects productive capacity to reach 100,000 b/d by 1990. The company had proved 832 million bbl of reserves by yearend 1987 and expects this to climb to 1.35 billion bbl.

At Musipan, Corpoven SA has drilled two wells and plans a total of 27. Mid-1988 production was 10,000 b/d. A 16 in. pipeline carries El Furrial and Musipan production about 4 miles to old Jusepin oil field, where Lagoven is adding compression to handle associated gas.

It moved four compressor stations from its southern district to Jusepin and planned to be handling 20 MMscfd by midyear. A 30 MMscfd compression unit was to have been, and two centrifugal compressors with nominal capacity totaling about 90 MMscfd are to start up in late 1990. The gas eventually will enter the 900 MMscfd cryogenic complex in Anzoategui to the west. Corpoven sells about 12 MMscfd on the local market, and 20-22 MMscfd is injected at Jusepin.

New life for old fields

Pressure response to the injected gas has revived hope for Jusepin field. Discovered in 1938 with reserves originally estimated at more than 900 million bbl of 36° gravity crude, Jusepin was overproduced during World War II. Flow reached 60,000 b/d, and reservoir pressure plummeted from its original level of about 2,500 psi. Production had been around 300-500 b/d, reservoir pressure 500 psi, with 750,000 bbl of reserves remaining. After injection of about 7 bcf of rich gas from the new fields, however, Jusepin reservoir pressure jumped to 650 psi. Pdvsa officials think the increase may signal potential for improved Jusepin oil recovery. The reservoir has porosity of 36% and permeability of 5-80 md, 40 md in the injection area.

Company affiliates are studying other ways to improve recovery from old eastern Venezuelan light and medium oil fields. One possibility is injection of carbon dioxide extracted from gas handled by the Anaco-Jose cryogenic complex. CO_2 content of the current stream is 6-8%.

Corpoven and Intevep SA, Pdvsa's research affiliate, are performing reservoir simulation studies and laboratory miscibility tests and plan pilot projects in eastern Venezuela in 1989. Pilot injection rates will be about 4.5 MMscfd of CO_2, probably from field extraction units.

Pdvsa envisions commercial CO_2 injection projects beginning in 1994 with injection rates of 10-40 MMscfd.

Submersible pumps also are improving oil recovery. Corpoven has had good results in an old field in Guarico—Budare, discovered in 1958, with about 20 million bbl of reserves remaining from 200 million bbl originally in place.

After working over 14 wells and installing submersible pumps, the company boosted field production to an average 5,500 b/d of 23° gravity crude for the past half year from less than 3,500 b/d before.

Action in the west

Submersible pumps are boosting production in one of the

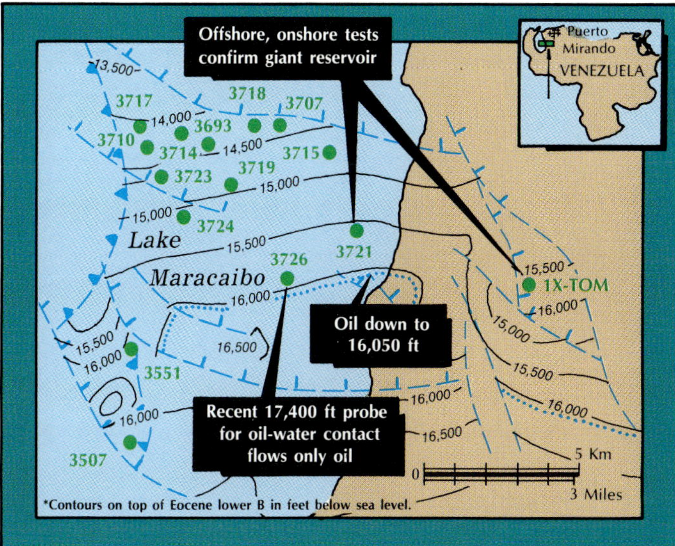

new fields in the west.

Corpoven installed electrical centrifigual pumps in two of 20 wells in big Guafita field near Colombia, boosting production to 10,000-11,000 b/d/well.

The company was producing 40,000-50,000 b/d in the area, piping production to older fields at Barinas for eventual transportation to the 105,000 b/d refinery at El Palito on the coast.

The low-GOR field's productive capacity is to reach 65,000 by 1993, but water encroachment may be a problem.

In Lake Maracaibo, Maraven is drilling 70 wells to develop its 800 million bbl Ceuta South-Southeast light and medium crude discovery. Drilling and production schedules haven't been determined, but the Pdvsa subsidiary says the field will sustain production of 120,000 b/d for 10 years.

Maraven estimates reserves at 800 million st-tk bbl and anticipates waterflood of the productive Eocene Misoa reservoir. The discovery well, Maraven 3715X VLG, drilled to 17,686 ft in September 1985, flowed 4,600 b/d of oil through a ¾ in. choke. An onshore wildcat, Maraven 1X TOM, drilled to 16,524 ft, flowed an average 2,000 b/d. Two rigs recently have been drilling in the area. The Misoa formation has produced in the main Ceuta field since 1959. Flow from 101 wells averages 54,741 b/d of oil. The recent discoveries are in three Misoa sandstones designated upper and lower B and Misoa C.

Of the 70 wells planned, about 30% will be onshore. Water depths of the offshore portion of the field range from 50 ft to 100 ft. Maraven reports no drilling problems. The 26°-36° gravity crude is sweet, with small amounts of paraffin that probably will require treatment. Maraven can use existing pipelines to handle production through Puerto Mirando. And in the South Lake area, where Lagoven has discovered 120 million bbl of light oil and condensate, France's Ste. Nationale Elf Aquitaine will help test horizontal drilling of the Cretaceous at target depths of about 14,000 ft. The project, involving horizontal sections of 1,000-2,000 ft, will start in the early 1990s.

Lagoven, which accounts for about half the lake's production, is letting Maracaibo potential slip to make room for flow from eastern Venezuela. Its western division produces oil from about 5,000 lake wells—87% on gas lift, 11% flowing, and 2% on mechanical pump. The company's lake productive potential dropped to 907,000 b/d at yearend 1987 from 917,000 b/d a year earlier. Natural declines would have pulled it to 750,000 b/d at the end of 1987 without Lagoven's

1987 program of 93 workovers and drilling of 32 production wells, 16 steam injectors, 12 water injectors, and four other wells.

Lagoven's lake production at midyear 1988 averaged 780,000-785,000 b/d, with heavy oil accounting for most of the increment below capacity. Of the total, 520,000 b/d is medium gravity crude, 125,000 b/d light crude, and the rest heavy.

The company boosted light and medium crude production by about 25,000 b/d by moving two 75 MMscfd gas compression modules from the Urdaneta heavy oil area in the western part of the lake to the Lagunilla producing region in the east, joining them with an identical pair of modules. The move took 10 months.

Upgrading potential

Discoveries of light and medium crude have reshaped more than Venezuela's exploration and production plans.

Discoveries in the west, for example, are producing enough gas to keep that part of the country self-sufficient beyond expectations of just a few years ago. Plans, therefore, have changed for the 500 mile Nurgas network, which will carry eastern Venezuelan gas westward. With western Venezuela now expected to remain in surplus, the Nurgas system will end at Moron, site of a petrochemical complex. The leg westward from there to Rio Seco won't be built.

More important internationally, a lighter average barrel helps Venezuela sell crude abroad and generates maximum revenues from exports constrained by an OPEC quota.

At the same time, Venezuela is finding ways around the OPEC limit. Like some other OPEC members, it is expand-

JOSE EXTRACTION PLANT, part of the Eastern Venezuela Cryogenic Complex, has been operated during capacity tests at inputs of 75,000 b/d of propane and heavier liquids, 5,000 b/d more than design capacity. The facility's loading jetty (shown in photo) has ship mooring capacity of 11,000-75,000 cu m.

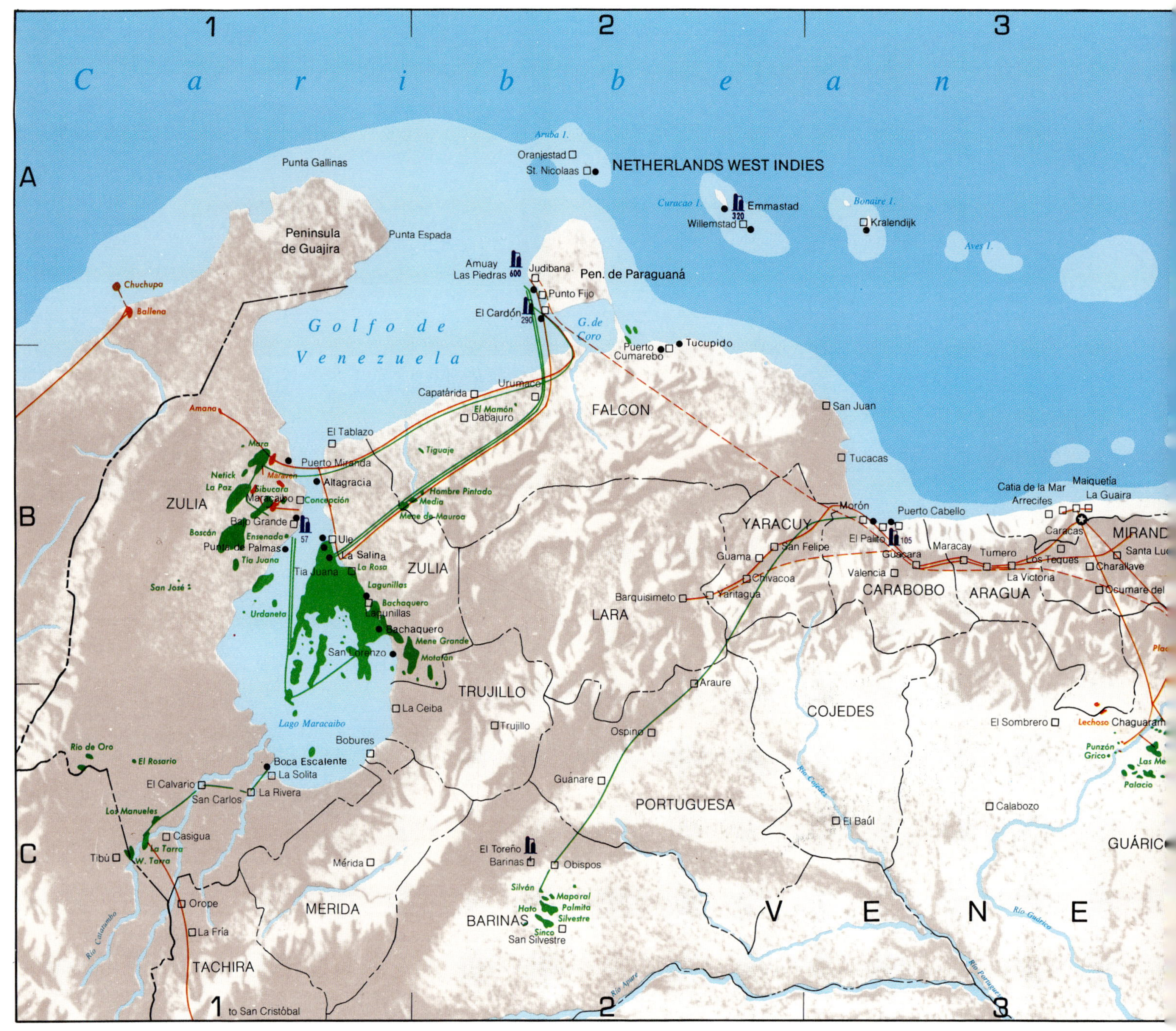

ing its petrochemical industry partly to use surplus gas—which isn't subject to quota—to boost exports.

And in a tactic new to OPEC and perhaps vexing to some of its members, Venezuela is calling Orimulsion a nonconventional hydrocarbon exempt from OPEC ceilings. The combustible emulsion includes 65-80 vol % Orinoco crude suspended as tiny droplets in water. A surfactant prevents coalescence of the oil droplets.

Because water is the continuous phase, Orimulsion viscosity is much less than that of the 7-10° gravity Orinoco crude, making it much easier to transport. The emulsion can replace coal or fuel oil in some burners. Officials claim sales potentials of 600,000 b/d by the mid-1990s and 1 million b/d by 2000. Atop quota-limited production growing ever lighter and more valuable, this novel use of abundant crude might not only increase export revenues but also serve another key Venezuelan objective—to be viewed in the U.S. and Europe as a long term source of petroleum in a secure part of the world.

Processing activity

Pdvsa and Unocal Corp. tentatively agreed to form a joint venture through which each company will indirectly hold 50% interest in Unocal's 147,000 b/d refinery at Lemont, Ill.

The transaction also includes 14 terminals, more than 100 service stations, a lube oil blending and packaging plant, and contracts with a network of 190 marketers in 12 Midwest states. Unocal expects to realize more than $500 million from the deal. Pdvsa would receive half interest in the assets plus a 135,000 b/d long term crude supply contract.

A condition is the joint venture's securing nonrecourse financing of at least $400 million under terms both parties find acceptable. The venture would make and sell Unocal 76 branded products. The preliminary agreement doesn't cover Unocal's truckstops or chemical distribution operations.

Pdvsa wants to acquire foreign downstream interests representing outlets for 700,000 b/d of its crude oil production. In the U.S., it now owns half interest in Citgo Petroleum

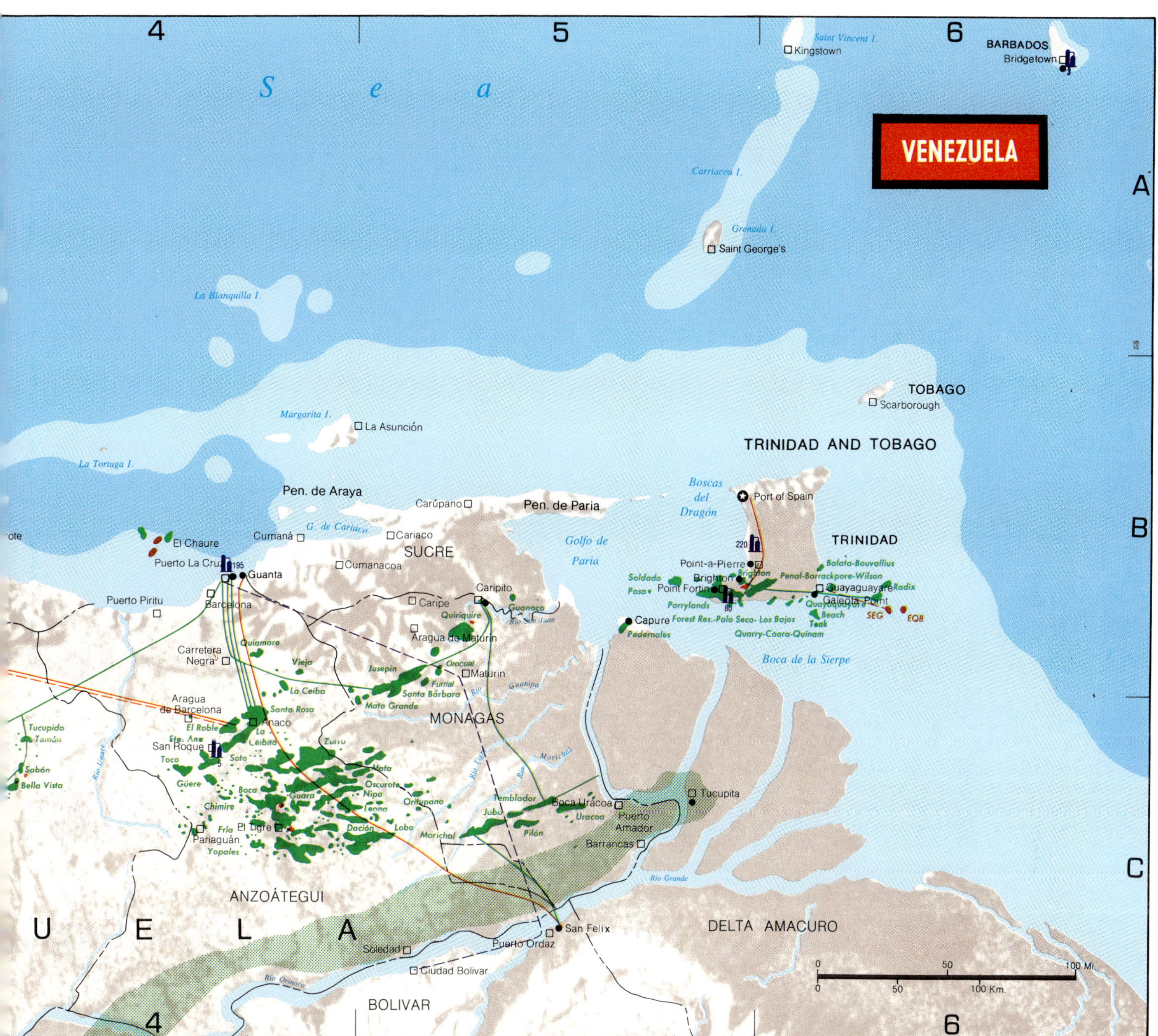

Corp. and Champlin Refining Corp. and planned to close purchase the remaining interest in Champlin. It also holds downstream interests in West Germany and Sweden.

But Carlos Andres Perez, elected Venezuelan president Dec. 4, 1988, said during his campaign that he would halt Pdvsa's aggressive foreign investment program. Perez was president during nationalization of the Venezuelan petroleum industry in 1976. Completion of the Unocal joint venture would guarantee Pdvsa sales of at least 585,000 b/d through foreign downstream interests. The company has 1.23 million b/d of refining capacity in Venezuela.

State owned Pequiven SA of Venezuela and Veba Oel AG of West Germany will build a $276 million propylene and polypropylene plant at the petrochemical complex under development in eastern Venezuela. The plant, near Jose, Anzoategui state, will have design capacities of 250,000 metric tons/year of propylene and 70,000 tons/year of polypropylene. Pequiven expects to export 70% of the production. The plant will go on stream in 1992.

Expansion is in prospect for the Eastern Venezuela Cryogenic Complex in Anzoategui state.

Corpoven SA, operator of the two site natural gas processing complex, has begun preliminary engineering studies for a 50% increase in design capacity, now 800 MMscfd. And capacity may be doubled, officials say.

Pequiven launched a major petrochemicals expansion program when it signed agreements for construction of several large plants.

Projects include expansions of the state owned company's two main complexes at Moron, Carabobo, and El Tablazo Bay, Zulia, as well as grassroots construction near Jose, Anzoategui. Planned investments, some in conjunction with joint venture partners and other investors, total $2.23 billion during 1988-92. Objectives are to increase use of Venezuelan natural gas, produce domestically some products now imported, and boost exports. The grassroots complex will be near a natural gas processing plant operated by Pdvsa.

IPE

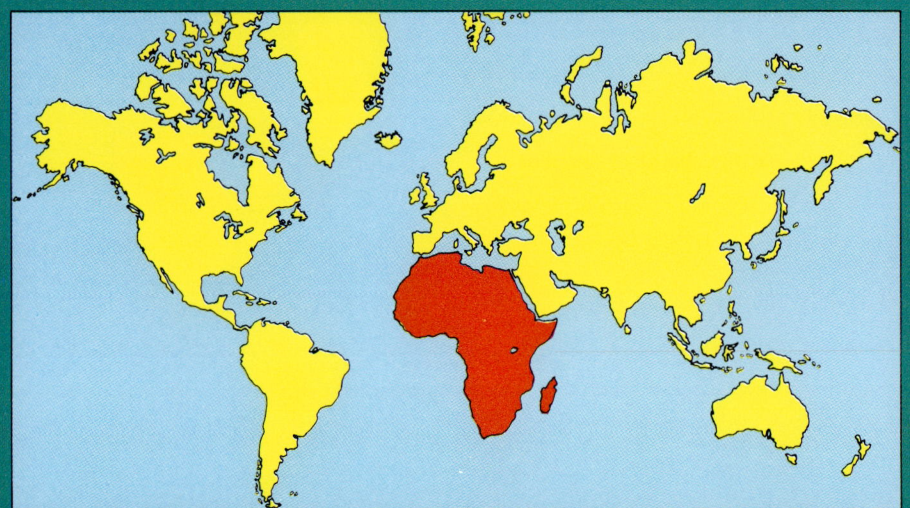

ALGERIA

CAPITAL: Algiers
MONETARY UNIT: Dinars
REFINING CAPACITY: 464,700 b/cd
PRODUCTION: 666.8 Mb/d
RESERVES: 8,400,000

A NEW COMMERCIAL OUTLOOK ON THE OIL AND GAS business was starting to emanate from Algeria in 1988.

Foreign companies were being lured back into the exploration business with new production-sharing contracts.

And in the LNG business, where Algeria is one of the major producers, exports to the U.S. resumed at world market prices. Deliveries to Britain were due to resume late in 1988 and new contracts were signed for deliveries to Turkey and Greece, all at competitive market prices.

Excluded from this turnaround in attitudes were Algeria's traditional customers for LNG in Europe. Sonatrach, the Algerian state energy company, was still insisting on prices that make imported LNG up to 30% more expensive than gas from other sources.

As a result LNG liftings declined and gas companies in France, Belgium, and Spain were in dispute with Sonatrach over prices.

Algerian exploration

The first agreement under Algeria's new regulations governing production-sharing agreements was signed between Agip SpA and Sonatrach at the end of 1987. A similar production-sharing agreement covering the Dzioua block in the Sahara was signed between Sonatrach and CEPSA of Spain.

A number of other companies were talking to Sonatrach and further new production-sharing agreements were expected.

Terms for the Agip and CEPSA agreements are similar. If commercial oil or gas is found, development costs will be shared 50-50. Oil will be shared 35/65 in Sonatrach's favor for production up to 15,000 b/d, rising to 12.5-87.5 over 75,000 b/d.

The Agip agreement covers the Rom discovery made by the company's Algerian affiliate in Block 403 at the end of 1987. The acreage lies 440 miles south of the Mediterranean coast and 125 miles west of the Tunisian border.

Agip entered Algeria in 1980 and started exploration work on permits 403 and 407 covering a total of 10,000 sq km in the Great Eastern Erg of the Sahara Desert. Agip Algerie ran 2,670 miles of seismic and completed the Rom 1 well in 1986. The well tested 5,000 b/d of 37° gravity crude from a zone at 10,825 ft. First appraisal of the find started in March 1988, and Agip was running further seismic in the area. Development work will include the construction of a pipeline from the field to the coast. Agip said it expects to start production at the end of 1990. Output will build to 30,000 b/d.

Additional work was scheduled for the two permits. Agip was committed to drill an additional wildcat well and run a further 745 miles of seismic. The company then enters a second exploration phase lasting 2 years during which it must drill another well. If it drills a second

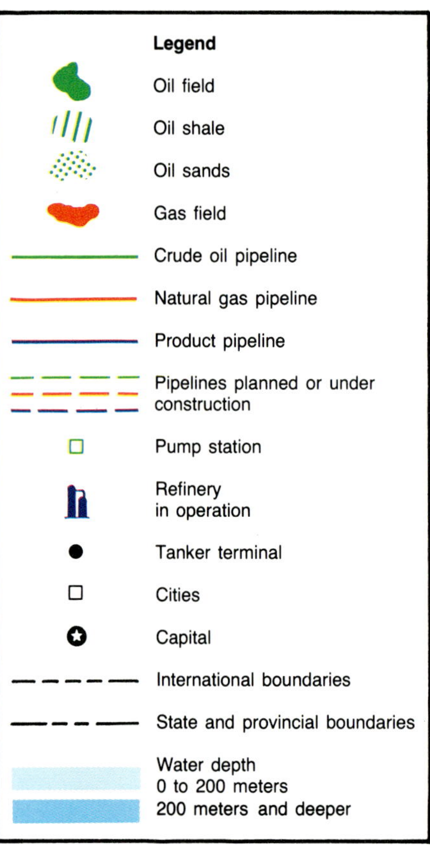

AFRICA

FRANCE
ITALY
YUGOSLAVIA
ROMANIA
BULGARIA
GREECE
ALB.
Black Sea
Caspian Sea
S.
S. R.
Lisbon
PORTUGAL
SPAIN
Madrid
Rome
Genoa
Istanbul
Nicosia
TURKEY
Tehran
IRAN
Strait of Gibraltar
Ceuta
Maison Carree
Skikda
Bizerte
Athens
40
CYPRUS
SYRIA
Damascus
Baghdad
IRAQ
Rabat
Melilla
Oran
Algiers
323
120
Tunis
MALTA
LEBANON
Damascus
70 110
ISRAEL
JORDAN
Kuwait
Mohammedia
Casablanca
Sidi Kacem
Arzew
La Skhira
Djebel-Abderrahmane
Mediterranean Sea
Alexandria
Cairo
Suez
53 128
100
Safi
MOROCCO
TUNISIA
42 in 107 cm
40 in 101 cm
Hassi R'Mel
Bir-Tlacin
23
Benghazi
Tobruk
Imbarka
Meleiha
Razzak
Gharadig
Belayim
Essaouira
Djebel-Jeer
Sidi Dhan
Hassi Messaoud
2/16 in 41 cm
Sider
Ras Lanuf
Dahra
Amal
Zelten
Alamein
Ras Gharib
Geisum
Tenerife
130
Las Palmas
El Aaiún
Canary I.
ALGERIA
LIBYA
Beda
Samah
Gialo
Sorir
EGYPT
Riyadh
SAUDI ARABIA
Villa Cisneros
MAURITANIA
MALI
NIGER
CHAD
SUDAN
Port Sudan
21
Red Sea
YEMEN
10
DEMOCRATIC YEMEN
Nouakchott
AFRICA
Khartoum
Kosti
18
Assab
170
DJIBOUTI
Djibouti
SENEGAL
Dakar
Srbikotane
Diam Niade
Bamako
Ougadougou
Niaméy
Lake Chad
N'Djamena
Muglad
Shair
Tabaldi
Abu Gabra
Unity
Melut
Bentiu
ETHIOPIA
Addis Ababa
GAMBIA
GUINEA BISSAU
Bissau
Conakry
GUINEA
BURKINA FASO
Kano
Kaduna
90
CENTRAL AFRICAN EMPIRE
SOMALIA
SIERRA LEONE
Freetown
10
IVORY COAST
GHANA
TOGO
NIGERIA
100
Makurdi
Bangui
UGANDA
Kampala
Lake Rudolf
KENYA
Nairobi
Mogadishu
18
Monrovia
LIBERIA
13
Abidjan
50
Accra
Lome
20
Porto Novo
Lagos
Warri
Port Harcourt
CAMEROON
Yaoundé
Lake Victoria
Mombasa
Espoir
Belier
Tano
10
Salt Pond
26
Oloibin
Bonny
60
Ekumbo
Manbo
Kwa Kwa
RWANDA
Kigali
BURUNDI
Bujumbura
90
Zanzibar
Gulf of Guinea
Sovellabb
Bata
EQUAT. GUINEA
CONGO
ZAIRE
TANZANIA
Dar es Salaam
Sao Tom
Libreville
GABON
Cape Lopez
Port Gentil
23
Tchibanga
Emeraude
Anguille
Grondin
Torpille
Brazzaville
Kinshasa
Lake Tanganyika
6 in 15 cm
Sango-Sango
South
Pointe Indienne
Pointe Noire
Yonga
Senda
CABINDA
Takula
Emergente
Malonga
Mavanga
Sulele
32
Ango-Ango
Malongo North
Moroni
Comoro I.
Atlantic
Luanda
Benfica
Tobias
Luanda
Mbini
ANGOLA
Lobito
MALAWI
Lilongwe
MADAGASCAR
Tamatave
16
Tananarive
Ocean
Ndola
24
ZAMBIA
Lusaka
Lake Kariba
Zomba
PEOPLE'S REPUBLIC OF MOZAMBIQUE
Mozambique Channel
NAMIBIA
ZIMBABWE
Salisbury
Umtali
Beira
Windhoek
Walvis Bay
BOTSWANA
Thohoyandou
Panda
Matola
Mmabatho
Pretoria
Maputo
SWAZILAND
Indian
Johannesburg
16
Durban
65
Sasolburg
78
LESOTHO
REPUBLIC OF SOUTH AFRICA
Umtata
Ocean
Cape Town
90
Cape of Good Hope
Mossel Bay

0   300   600 Mi.
0 200 400 600 Km.

Oil discovery

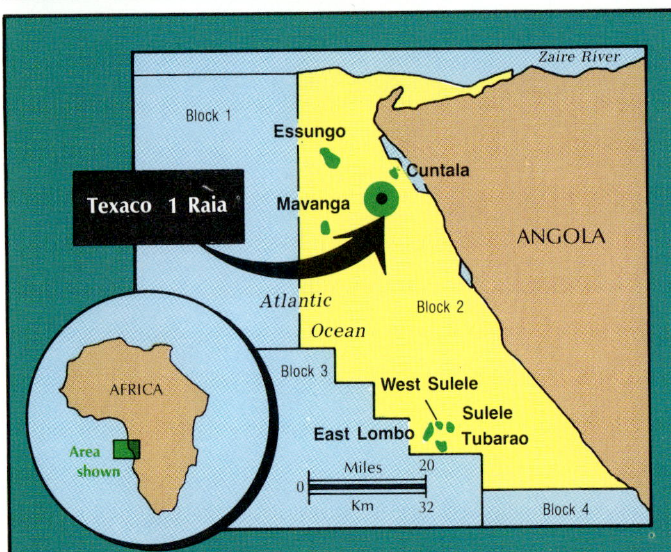

Offshore Cabinda

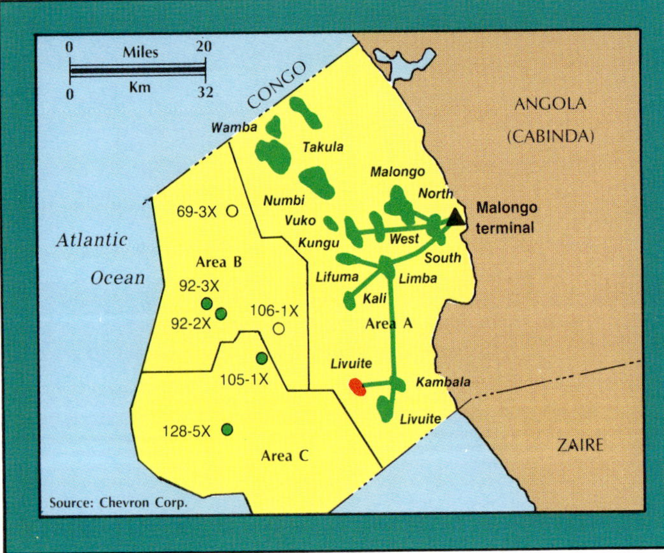

well, it will reduce relinquishment at the end of this period from 40% to 25%.

A third phase covering 3 years also has the requirement for a single well with the ability to avoid further relinquishment by drilling a second well.

Agip says the agreement with Sonatrach will enable the company to recover all exploration expenses from production revenues.

Investment in the field will be funded equally by Agip and Sonatrach.

Sonatrach has continued to undertake exploration for itself. In the Erg area it made an interesting new discovery with the Rhourde Er-Rouni wildcat about 210 miles east of Hassi Messaoud. The well flowed high-quality crude in significant quantities. Sonatrach was planning an appraisal of the find.

Total CFP, working outside the terms of the new exploration deals, contracted to undertake an 18-month seismic survey for Sonatrach.

The French company was to shoot two surveys in the Hassi Messaoud area, one in the northern Sahara, and a fourth in the south.

LNG deals

Algerian crude-oil production averaged a steady 650,000-700,000 b/d. Throughout the years of national ceilings on output agreed by the Organization of Petroleum Exporting Countries, Algeria has been one of the few members to observe its quota.

Industry sources say Algeria would be hard pressed to increase its output significantly. However, it does have the capacity to make substantial increases in the LNG trade.

Algeria's insistence in the early 1980s on linking gas export prices to crude prices put an end to its U.S. business and forced European customers to make drastic cuts in deliveries. Three U.S. customers were contracted to take over 0.5 tcf of gas annually, and the loss hit the Algerians hard. In an attempt to retrieve part of the lost LNG trade with the U.S., new agreements were reached with Panhandle and Distrigas of Boston.

Sonatrach caused a stir in European gas circles by entering a joint venture with Panhandle to deliver 158 bcf/year over a 3-year period. The deal will give 63% of the revenue to the Algerians and 37% to Panhandle.

At the end of 1987 Sonatrach delivered the first of 17 cargoes to Distrigas of Boston. The price of $2.60/MMBTU, related to the price of competing products, was followed by a series of other competitively priced deals.

A contract with Turkey for 70 bcf/year of gas over a 20-year period was priced at around $2/MMBTU. BOTAS, the Turkish state pipeline company, will build a reception terminal and regasification plant at Eregli on the Sea of Marmara. Deliveries are scheduled to start in 1992.

In a second agreement Algeria will sell 424 bcf of LNG to Greece under a 20-year contract due to start in 1991. The Greeks were paying $2.25/MMBTU in a contract which gives the buyer considerable flexibility of liftings.

This new approach to pricing enabled Sonatrach to win back one of its oldest customers, British Gas plc. Before major reserves were found in the North Sea, British Gas helped Sonatrach to pioneer the LNG trade. British Gas has an option to take 12 cargoes a year for 3 years. If it took all 36 cargoes, Algerian imports would total 21 bcf. The LNG will form part of the British Gas storage and peak-shaving operations. The price was thought to be $2.15/MMBTU tied half to the price of heavy fuel oil and half to heating oil.

British Gas was able to take advantage of the new competitively priced supplies on offer from Sonatrach as the regasification plant at Canvey Island on the Thames estuary had been kept in working order since the original Algerian supply contract expired in 1981.

Europeans'
turn

The flurry of competitively priced, flexible deals left the established European customers wondering when their turn would come to renegotiate contracts on more commercial terms.

At midyear 1988 customers in France, Belgium, and Spain had been disappointed. Algeria showed no inclination to spread commercial contracts any further.

Gaz de France, the biggest of the European customers, had been trying to renegotiate its three supply contracts without success while Enagas of Spain and Distrigaz of Belgium went to arbitration over pricing and liftings.

Under three contracts signed in 1965, 1972, and 1982, Gaz de France was committed to lifting of 323 bcf/year, but there has been substantial underlifting because of the slower-than-expected growth in the French gas market.

Gas prices were originally calculated using a formula linked to the official prices of eight OPEC crudes. When crude prices went into freefall, an interim measure was introduced based on an adjusted netback value of crudes.

Efforts to turn the interim agreement into a longer-term pricing accord have failed.

At the beginning of 1987 Gaz de France reverted to the old basket of crude prices formula and Sonatrach continued to bill its customer using the interim agreement.

Political attempts to solve this impasse have failed, and neither side has shown much real inclination to come to grips with the pricing and liftings problem.

Now that the French presidential election is over, Algeria looks more intent on going over the head of Gaz de France and trying for a political settlement with the French Government. State-owned Gaz de France points out that the last time Algeria and France reached a political agreement to a gas pricing dispute in 1982, it was left paying 27% more for Algerian gas than for supplies from other sources.

Based on the 1982 formula Gaz de France was paying $1.97/MMBTU, roughly the same as other supplies. The Algerians are continuing to bill the French gas company on the basis of the 1986 temporary formula at $2.79/MMBTU.

Since the beginning of 1987 when the billing discrepancies started, the cumulative difference between the two systems has built to $450 million.

Cooperative
agreements

Algeria, Libya, and Tunisia moved nearer to linking the three countries with a gas pipeline by signing a protocol in Tunis early in 1988. The new 250-mile gas line will run from the junction of the Transmed line at the Tunisian border through Tunisia to Zuwara on the Libyan coast.

Construction cost is estimated at $400 million.

The Algerian gas will fuel an 800,000-kw power station at Zuwara and may also provide the basis for new petrochemical and fertilizer plants. The agreement in Tunis creates a joint Libyan-Algerian company, owned by Libya's National Oil Corp. and Sonatrach of Algeria, which will finance the project.

A second company has also emerged from the protocol. The Maghreb Arab Co. will be owned by all three countries and will be responsible for planning and later building the pipeline.

Earlier Libya and Algeria agreed to set up the Libyan-Algerian Oil Co., jointly owned by Sonatrach and NOC. It will search for and develop hydrocarbons in both countries and may later expand into foreign exploration projects. Headquarters of the new company will be Algiers. There will be a subsidiary office in Libya. The company has a budget of $60 million to cover the first 3 years of operations. At the same time Algeria and Libya agreed to establish a joint geophysical company.

The Libyan Arab-Algerian Geophysical Co., with a budget of $19 million, will provide seismic and magnetic surveys to companies working in Algeria and Libya. NOC of Libya and Enterprise Nationale de Geophysique of Algeria have set the new company the target of operating three seismic crews within 2 years. Two will operate in Libya and one in Algeria. The company will set up headquarters in Tripoli with associate offices in Algeria.

ANGOLA-CABINDA

CAPITAL: Luanda
MONETARY UNIT: Kwanza
REFINING CAPACITY: 32,100 b/cd
PRODUCTION: 449.3 Mb/d
RESERVES: 2,024,000 Mbbl

CABINDA GULF OIL CO. LTD. (CABGOC) ADVANCED plans to boost production off Cabinda with platform installations costing about $100 million.

Development and other activity in Takula, Wamba, and Numbi fields off the Angolan enclave will account for more than 75% of Cabgoc's producing budget and 75% of its production during the next 5 years. Angola's production is expected to peak in 1990 at 410,000-513,000 b/d.

Cabgoc production off Cabinda reached 275,000 b/d at yearend 1987. Cabgoc output from 14 of 15 fields producing in Concession Area A averaged 231,800 b/d in 1987, up 22% from 1986.

Meantime, Cabgoc was evaluating several recent oil discoveries for possible development in Concession areas B and C off Cabinda. Chevron Corp. subsidiary Cabgoc is operator of the 2,700 sq mile concession off Cabinda in association with Soc. Nacional de Combustiveis de Angola (Sonangol).

Cabgoc planned to install the production platforms in Wamba and Numbi fields to double their combined production to more than 140,000 b/d. 1988 figures showed Numbi at 44,000 b/d of oil and Wamba at 26,000 b/d. Schedule calls for installation of the first platform in 1989 and the second in 1990.

Wamba and Numbi output was limited in 1988 because it was processed on Takula platform. In addition, Cabgoc will install a waterflood platform in Takula field to extend production of about 110,000 b/d. Water injection of 350,000 b/d is to begin by yearend 1989.

Cabgoc is having fabricated a 90 MMcfd gas compression platform for installation in Takula that will be used for gas lift, LPG extraction, and dry gas reinjection in Malongo fields. LPG extraction and gas reinjection in the Malongo complex presently entail three compression platforms with total 63,000 hp of compression and a large tanker for LPG storage and export of about 6,000 b/d. Nine production gathering stations produce to the 1.7 million bbl capacity Malongo terminal. Crude is treated, stored, and exported through the Malongo buoy 8 miles offshore.

The two Takula platforms produce oil directly into the Chevron Shipping Co. storage tanker Afran Ocean, where it is stored and lightered to export tankers.

Cabgoc kept three rigs active in 1988. It launched an aggressive exploration program in areas B and C in late 1986, maintaining two rigs active at eight well sites through 1987. Its 92-2X discovery tested oil at 2,382 b/d and follow-up 92-3X flowed 3,820 b/d. Cabgoc evaluated the strikes with a 5,800 km 3-D seismic survey. The 105-1X flowed hydrocarbons from eight zones covering a vertical interval of 2,625 ft. The 128-5X flowed oil at a rate of 3,122 b/d.

Texaco discovery

A combine led by a subsidiary of Texaco Inc. tested its eighth discovery on a 1 million acre tract off Angola. The group's 1 Raia wildcat, 6½ miles offshore on Block 2, flowed an average 2,880 b/d of 35.6° gravity oil through a ½ in. choke from Cretaceous dolomite pay below 8,494 ft. It was drilled to 11,250 ft in 66 ft of water.

Site is 12 miles southeast of Essungo oil field, a November 1975 discovery, and 4 miles southwest of Cuntala oil field, a September 1978 discovery. The area lies about 200 miles north of the Angolan capital, Luanda.

The Texaco group acquired Block 2 under a production sharing agreement with state owned Soc. Nacional de Combustiveis de Angola (Sonangol) in 1980.

Since then the combine has been producing Essungo and Cuntala fields in the north part of the block. Oil production began late in 1987 in East Lombo, Tubarao, and West Sulele fields in the south part of the block.

CONGO

CAPITAL: Brazzaville
MONETARY UNIT: CFA francs
REFINING CAPACITY: 21,000 b/cd
PRODUCTION: 134.8 Mb/d
RESERVES: 710,000 Mbbl

BP PETROLEUM DEVELOPMENT (OVERSEAS) STARTED SURVEYS in a lightly explored onshore area of Congo. It's the company's first acreage in that country.

A BP led group planned to explore for oil on the 502 sq mile Kayes Block B concession. Interests are BP 30%, BHP Petroleum (Pty.) Ltd. 20%, and Hydrocongo, the state oil corporation, 50%.

The group will spend about $25 million during the 4 year period of the license. The program will consist of at least 217 line miles of seismic surveys, along with the drilling of several wells. Significant discoveries have been made in the Atlantic Ocean off Kayes Block B.

Coastal Corp. has conducted exploration on Block B and drilled a small gas discovery on Kayes Block A to the north. The Congolese government licensed Block A to a unit of Chevron Corp. and Kayes Block C to the south to Conoco Inc. BP, Chevron, and Conoco were jointly running an aerial survey to map the area. Because of continual cloud cover there is almost no satellite information on Congo.

EGYPT

CAPITAL: Cairo
MONETARY UNIT: Pounds—livres
REFINING CAPACITY: 489,203 b/cd
PRODUCTION: 851.3 Mb/d
RESERVES: 4,300,000 Mbbl

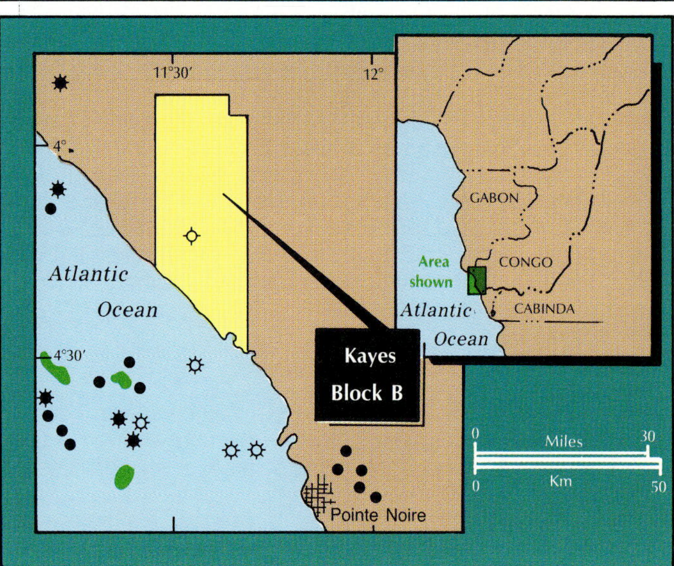

EGYPT'S WESTERN DESERT IS STILL THE MOST prospective region, easily accessible to foreign oil companies, along the southern rim of the Mediterranean Sea.

International companies and a band of independents have put the vast Desert region high on their list of priorities following a series of small but profitable oil finds in the area. A network of crude pipelines linking the largest discoveries has made development of new finds easier. The next stage of the hydrocarbon infrastructure, a major gas pipeline into Alexandria, has been started.

Output from the Western Desert was running at around 54,600 b/d, and production from present developed fields was set to rise to a peak of nearly 70,000 b/d. But the small structures will start to decline quickly, and by 1991 production from the fields already under development will fall back to around 50,000 b/d.

The Egyptian government wants to maintain its crude oil output at around 880,000 b/d throughout this period. Achieving this target will require an additional 240,000 b/d of production to come on stream over the next 4 to 5 years. Again, the bulk of this new production will probably come from the Gulf of Suez. But the state planners are hoping that new discoveries in the Western Desert will boost production levels from the area to around 100,000 b/d.

According to the government planners, the completion of the new Shell Winning gas pipeline to Alexandria and other possible gas pipeline projects will enable production to rise from fields where flaring restrictions are holding down output. The 167 mile pipeline from the Badr el-Din concession to Amiriyah, west of Alexandria, was made possible by the negotiation of a gas clause into the company's production-sharing contract. This gave oil companies rights to develop and sell any gas found on new concessions or on acreage where agreement had been reached to include the gas clause retroactively.

Foreign companies generally agreed that the Shell-style contract makes exploration in gas-prone areas an economic proposition. Shell says that now the gas clause has been introduced into new concessions, exploration may take a different direction. Exploration in the gas-prone areas is now viable, it says. Licensing awards brought a number of new operators in the Western Desert. Britoil plc, absorbed into British Petroleum, was awarded the 1,110-sq-mile West Natrun concession. Under the terms of the 7 year agreement, the operator must drill five wells and spend $19.5 million.

WEST AFRICA
GHANA
TOGO
BENIN
NIGERIA
CAMEROON
Kaduna
Bida
Baro
Ilorin
Ibadan
Makurdi
Donga
Yola
Lome
Porto Nova
Lagos
Seme
Tema
Salt Pond
Tano
Lake Volta
Meren
Okan
Forcados
Ea
Pennington
Warri
Okrika
Oguta
M'Bede
Kolo Creek
Idoho
Ekulama
Okubie
Port Harcourt
Alesa-Eleme
Pointe Limboh
Douala
Victoria
Malabo
Douala
Yaounde
Ebolowa
EQUATORIAL GUINEA
Sanaga Sud
Kribi
Bata
RIO MUNI
Oyem
Atlantic Ocean
Sao Tome and Principe
Libreville
Booue
Lambarene
GABON
Mossaka
Konzi
Lopez
Clairette
Tchengue
Anguille
Cape Lopez
Port Gentil
Odombo
Gentil-O.
Ozouri
Animba
Torpille
Ikenge
CONGO
Gamba-Ivinga
Brazzaville
Kinshasa
ZAIRE
Zaire River
Yanga-Sendji
Loango
Pointe Indienne
Sendji
Emeraude
Malongo
W. Malongo
Kunguio
Takula
Pointe Noire
CABINDA
Cabinda
Mibale
Moanda
Essungo
GCO
Guntala
Mavanga
N'Zombo
Quinguila
Cabeca Da Cobra
Maleva North
Sulele
E. Lombo
Ambriz
Luanda
Cauaco
Mulenvoss
Benfica
Uacongo
Barra do Cuanza
Galinda
Tobias
Nova Gaia
Bengo
Cuanza
ANGOLA
Lobito
Lobito
0   100   200   300 Mi.
0 100 200 300 400 Km.

GABON
Cape Lopez
Clairette
Port Gentil
Port Gentil-Ocean
Ocean
Anguille
Tchengue
Ozouri
Doree
Merou
Girelle
Torpille
Atlantic Ocean
Pelican
Baudroie
Marin
Baliste
Grondin
Gonelle
Barbier
Mandaros
Breme
Olende
Batanga
Breme
PLW
Ongindjo
Loanda
Animba
Bilape
Gongoue
N. Tumbenyoni
Pointe Weze
Mporaloko
Chikawar
Rembo-Katto
Odombo
Assewe
Ikenge
Omboue
0   25 Mi.
0   40 Km.

CONSIDERABLE EXPLORATION and production activities are being carried out by Shell Winning in the Western Desert of Egypt (Shell photo).

SPECIALLY TRACKED VEHICLES are used by Conoco in the rugged terrain of Egypt's Western Desert (Conoco photo).

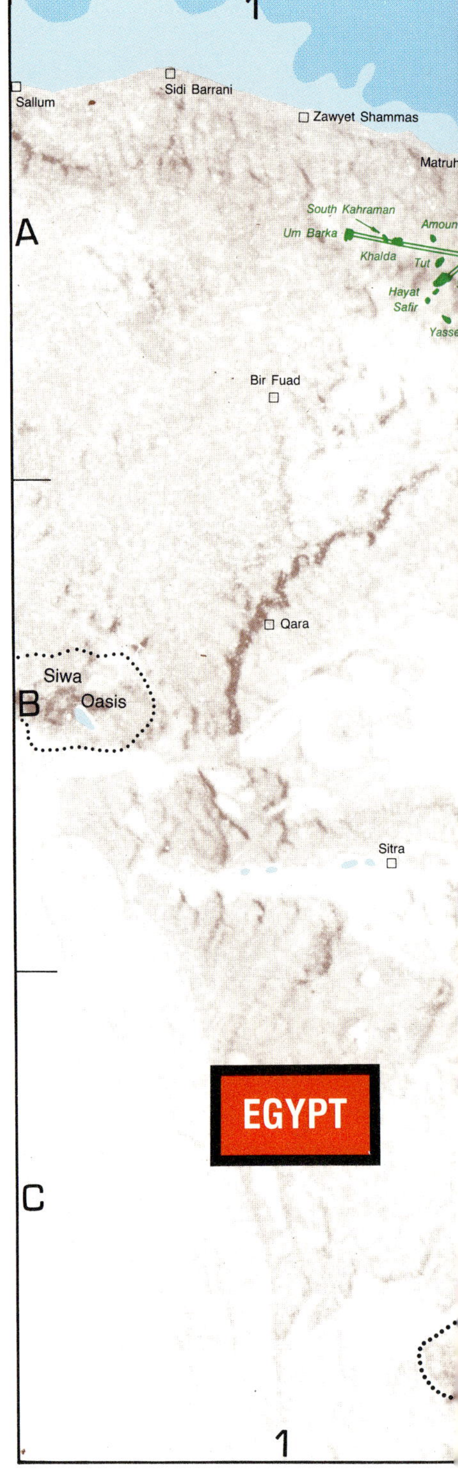

Gas demand

Demand for gas in Egypt will be at a premium in the short term as the authorities try to cope with the shortfall in electricity supplies from the Aswan High Dam on the Nile. Over the longer term gas will be used to phase out liquids in the domestic and industrial sectors as crude-oil production falls in the mid-1990s and the availability of crude for export becomes tight.

The 1,750,000 kw of capacity from the Aswan dam supplies about 22% of Egypt's electricity requirements. Years of drought in the upper reaches of the river have reduced the flow to the point where the hydropower output has been cut by 50%. New generating capacity of 315,000 kw was due to be commissioned in the fall of 1988, and the electricity authorities reckon they have a further 1,000,000 kw in reserve. To supplement this, a crash program of power-station expansion has started in Cairo and the Delta.

Replacing hydropower with spare capacity in existing thermal stations has already taken about 40,000 b/d of fuel oils and has reduced the availability of export crudes to about 400,000 b/d. As more oil-burning units are activated, the pressure on Egyptian crude exports will grow.

Meeting the longer-term objective of expanding the industrial and domestic markets for gas will require large-scale investments in distribution networks.

Completion of the national gas grid will allow more gas from the IEOC/Marathon/BP Abu Madi gas field near the mouth of the Delta to find its way into power stations and

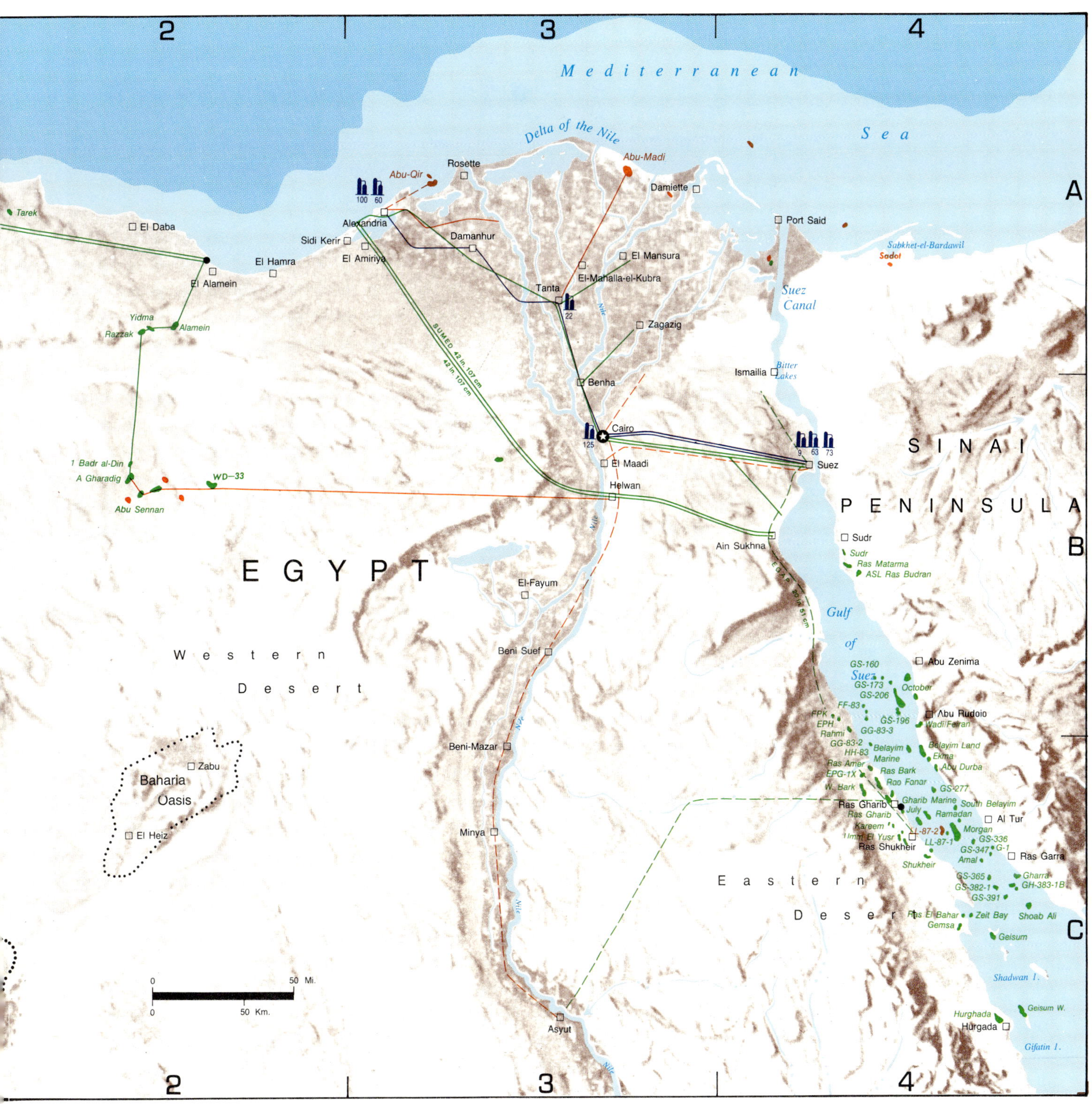

industry in the Delta and the Cairo area.

The government was also investigating the possibility of using associated gas from the Conoco-operated Khalda concession in the Western Desert to fuel a small power station and desalination plant at Marsa Matruh on the Mediterranean coast to the northwest of the field. Further developments in petrochemicals are also on the agenda using gas from the Abu Qir offshore gas field.

The Phillips-operated WEPCO venture, working as a contractor to EGPC, is expanding the Abu Qir offshore facilities to handle output of 350 MMcfd. Two small drilling platforms will be installed offshore, linked to another process train at the onshore production unit which has a capacity of 240 MMcfd.

Shell pioneers

Pioneering work by Shell Winning on the gas clause enabled the Royal Dutch/Shell subsidiary to become the first company to take advantage of the new operating conditions.

The company is laying the 167 mile pipeline of 20 and 24-in. diameter from its Western Desert fields to Alexandria, where gas will be fed into the grid that covers the Delta area. Total cost of the project is $350-400 million. Production is scheduled to start up in mid-1990. The flow of gas will build to a plateau of 150 MMcfd. Provision of a gas transportation system will open the way for new oil and gas developments in the area around the Shell fields.

The company, working through the Bapetco joint venture operating company, has been producing up to 12,000 b/d

Where the new concessions are in Egypt

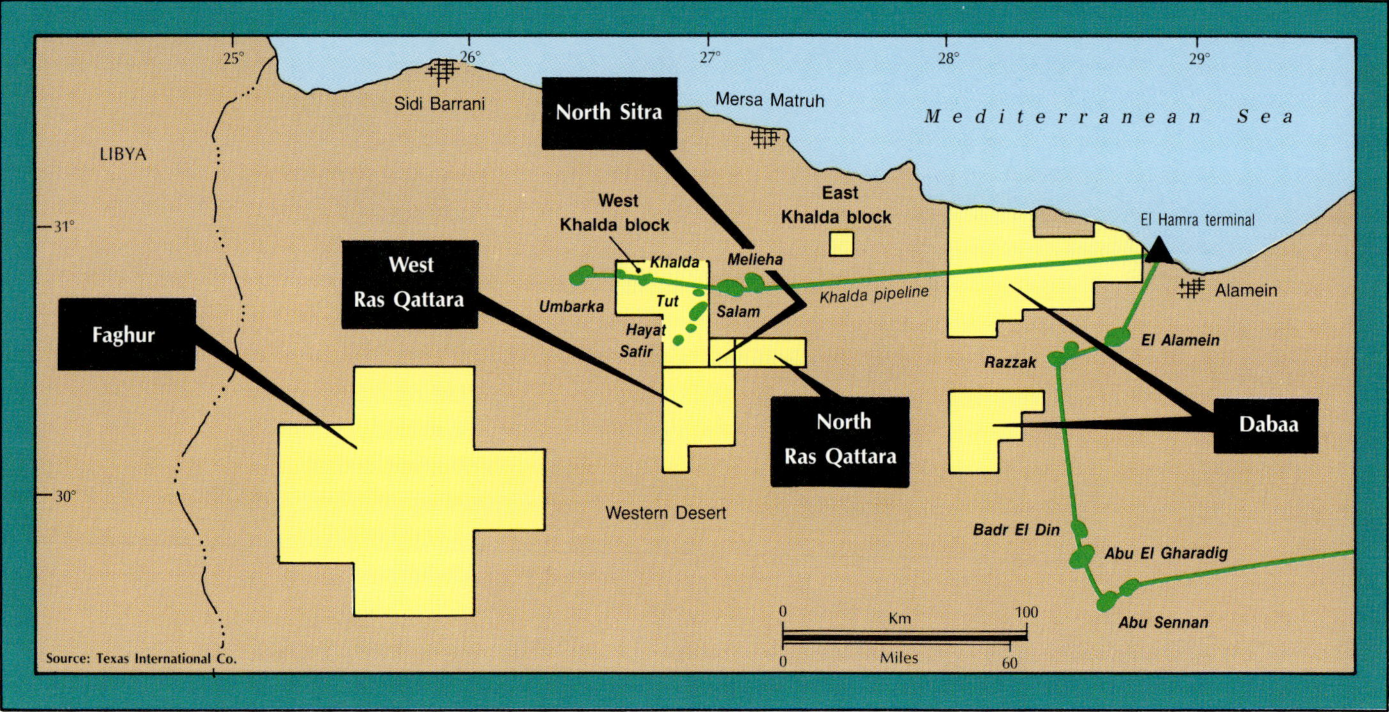

from the Badr el-Din field. Oil is transported through a 12-in. spur line to the el-Hamra terminal on the Mediterranean coast. Gupco is also using the spur line to evacuate crude from the production facilities in the Abu el Gharadiq area.

The new gas line to Alexandria is part of Bapetco's overall development plan for this part of the Western Desert. In addition to increasing output from BED-3, Shell will lay a 43-mile, 10-in. oil line from the Sitra field into the main production system.

Shell expects to have two drilling rigs at work through to 1991. The company drilled two wells in the Qarun concession, which lies between Amoco acreage and the Nile.

Much of the area to the east of Bapetco is held by Gupco, a joint Amoco Egypt Oil Co. and EGPC company, which is the major producer of oil and gas in Egypt.

The bulk of its production comes from the Gulf of Suez, but at an early stage the company ventured into the Western Desert, where it found the Abu Gharadiq field and a number of small associated fields. Crude oil is exported to the coast through the WEPCO facilities on the Alamein area fields, and a gas pipeline was built to supply gas to the Helwan industrial area of Cairo.

The introduction of the gas clause into new and some existing contracts is likely to activate more gas development in the Western Desert. Amoco was already working on a project to lower the suction pressure on its desert gas operations and increase daily output.

Gupco operates a group of five fields in the Western Desert of which Abu Gharadiq and Razzak are the largest. Liquids production was averaging more than 20,000 b/d. The gas supply from the area is 125 MMcfd and is the fourth largest source of gas entering the Egyptian distribution network after Abu Madi, Abu Qir, and EGPC Gulf of Suez.

Overall, Amoco feels the prospects for finding more small or medium sized fields in the Western Desert are good. It was negotiating to acquire new concessions. The West Qarum area, which lies immediately to the east of the Abu Gharadiq area, was the latest concession acquired in 1985.

During 1987 Gupco production averaged 493,000 b/d. The bulk of the output comes from the established October, Morgan, July, and Ramadan fields.

Conoco operations

Potentially the biggest of the Western Desert developments is operated by Khalda Petroleum Co, a venture by Conoco Inc., Texas International Co., and EGPC. Production from the Khalda concession increased to about 20,000 b/d from the Salam, Khalda, Tut, Hayat, and Safir fields on the Khalda concession, which was pioneered by Phoenix Resource Co., an affiliate of Texas International.

Since Conoco farmed into the concession and took over the operatorship, a 40,000-b/d capacity production facility has been built on the Salam field. The satellite fields are linked into this unit. Crude oil is exported though a 16-in., 10 mile spur line into the Meleiha field's pipeline system. Salam will act as the main processing point for other developments by the group in the area.

Conoco acquired five new concessions in the Western Desert covering about 3 million acres. This raises the company's gross exploration acreage to 4.84 million, making the company one of the biggest permit holders in the area.

As a result of the new concessions, the groups led by Conoco have made a major commitment to increased exploration over the next 3 years. They will spend $27 million and drill a minimum of 11 wells. The new exploration permits, Dabaa (1,540 sq miles), North Ras Qattara (118 sq miles), North Sitra (39.5 sq miles), West Ras Qattara (395 sq miles), and Faghur (2,720 sq miles) are east, south, and southwest of the producing areas on the Khalda permit.

The new licenses will also bring three new companies into the Western Desert—units of Norsk Hydro, Oranje Nassau, and Norpetrol A.S. Since these concessions were confirmed, Nippon Mining of Japan has acquired an interest through a deal with Conoco Inc. under which the Japanese will provide $40 million towards exploration on the Egyptian concessions and Conoco acreage in Ireland, France, Tunisia, Somalia and Indonesia. Conoco has also farmed into Phillips Petroleum's 100%-owned South Umbarka concession. It now has a 35% holding.

Oil from the Khalda concession is linked to the pipeline to the El Hamra terminal built and operated by Agiba Petroleum, a joint venture by EGPC, International Egyptian Oil Co (IEOC), an affiliate of Agip SpA, International Finance Corp., and Denison Mines Ltd. Output from the Meleiha concession was running at around 15,000 b/d and was projected to rise to around 20,000 b/d. Exploration was continuing, and the group made a new discovery on the permit. The Falak 2 wildcat was drilled to a depth of 13,000 ft and tested 3,000 b/d of 43° gravity crude and 14 MMcfd of gas from the Jurassic and Lower Cretaceous.

IEOC was also awarded the 1,025-sq-mile Kanayes concession. Over a period of 7 years the company is obliged to drill 12 wells and spend $42 million. The company also acquired a 1,192-sq-mile concession in the Nile Delta area. This is a short-term permit running for only 2 years and requires only two wells.

IEOC is also a major producer in the Gulf of Suez and is backing its contention that the Egyptian Mediterranean off northern Sinai may also produce commercial oil and gas. After a series of disappointing results, IEOC's partners have withdrawn from the offshore permits, leaving the Italian company as the sole foreign shareholder.

Phillips Petroleum Co. was one of the pioneers in the Western Desert. In the 1960s it found and developed Alamein, Yidmas, and Umbarka fields. Subsequent exploration failed to make significant new finds, which put the program into low gear until the Khalda, Meleiha, and Badr El-Din discoveries were made.

Operating through WEPCO, Phillips was awarded the 2,317-sq-mile South Umbarka permit and shot an extensive seismic program. The permit adjoins the Umbarka permit where WEPCO operates a small field that has been tied into the Meleiha pipeline system. WEPCO operates the El Hamra terminal on the Mediterranean coast which handles the crude-oil output from the entire area. The terminal was expanded to meet increased throughput and can handle tankers up to 200,000 dwt.

GABON

CAPITAL: Libreville
MONETARY UNIT: CFA francs
REFINING CAPACITY: 24,000 b/cd
PRODUCTION: 175 Mb/d
RESERVES: 720,000 Mbbl

GABON WAS TO PLACE RABI-KOUNGA FIELD, ITS biggest oil reservoir, on stream early in 1989, several months ahead of its planned start-up date.

Shell Gabon, jointly owned by Royal Dutch/Shell Group and the Gabonese government, planned to commission temporary production facilities.

The facilities were to provide a low level of production until midyear 1989, when production was expected to rise to about 30,000 b/d. Start of production originally was targeted for midyear.

By the end of 1989, Rabi-Kounga flow is expected to reach 80,000 b/d.

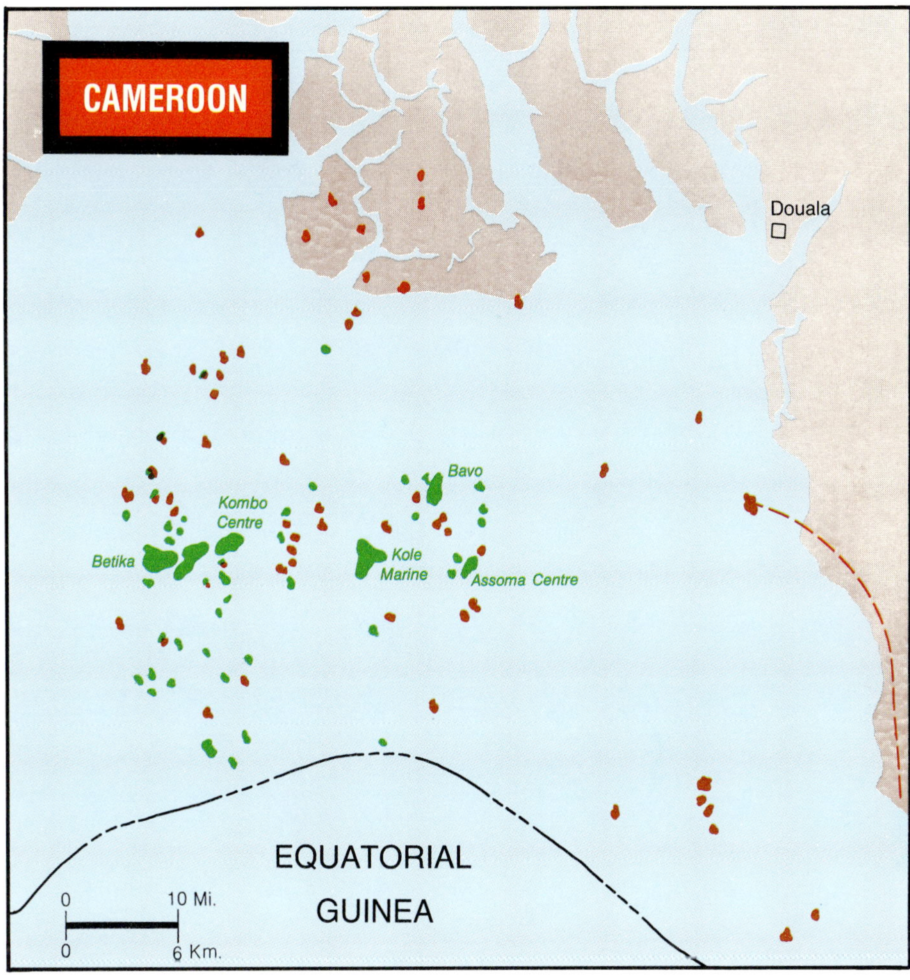

Peak production of 120,000 b/d is scheduled for the end of 1990. The $500 million onshore development project will transform Gabon's production prospects. Gabon produces about 200,000 b/d of oil, mainly from of a series of small offshore fields. The scope of the Rabi-Kounga project is increasing. Reserves estimate has increased to 428 million bbl from 370 million bbl since development drilling began. Shell Gabon also expanded the drilling program. It has decided to drill 80 wells instead of 53. It expected to complete 31 wells by yearend 1988. The remote jungle field is 31 miles from the coast and 62 miles from Shell Gabon's export terminal at Gamba. An 84 mile, 18 in. pipeline was being laid to Gamba to handle production. A longer line to Elf Gabon's export terminal at Cap Lopez will be in operation when peak production is reached.

Oil from the field comes from the reservoir sands of the Gamba and Dentale formations, which form part of a rift basin of Early Cretaceous age, the Dianonga basin, according to geologist M.G. Boeuf of Shell Gabon. The cap rocks are late Aptian evaporitic deposits (Ezanga formation). Boeuf says that trapping in the field is provided by structural closure at the base of the Ezanga salt, where Gamba and Dentale reservoirs are in fluid communication.

The field was discovered on the basis of reinterpretation of seismic data acquired between 1970 and 1977 and the integration of 1981-82 data that showed a marked improvement in quality and resolution.

Boeuf reflects that more than 20 years have elapsed between the two most important discoveries ever made onshore Gabon: Gamba-Ivinga in 1963-67 and Rabi-Kounga in 1984-85.

During this time, he says, efforts were made to improve structural control over an area that remains difficult from both a seismic and a logistic point of view.

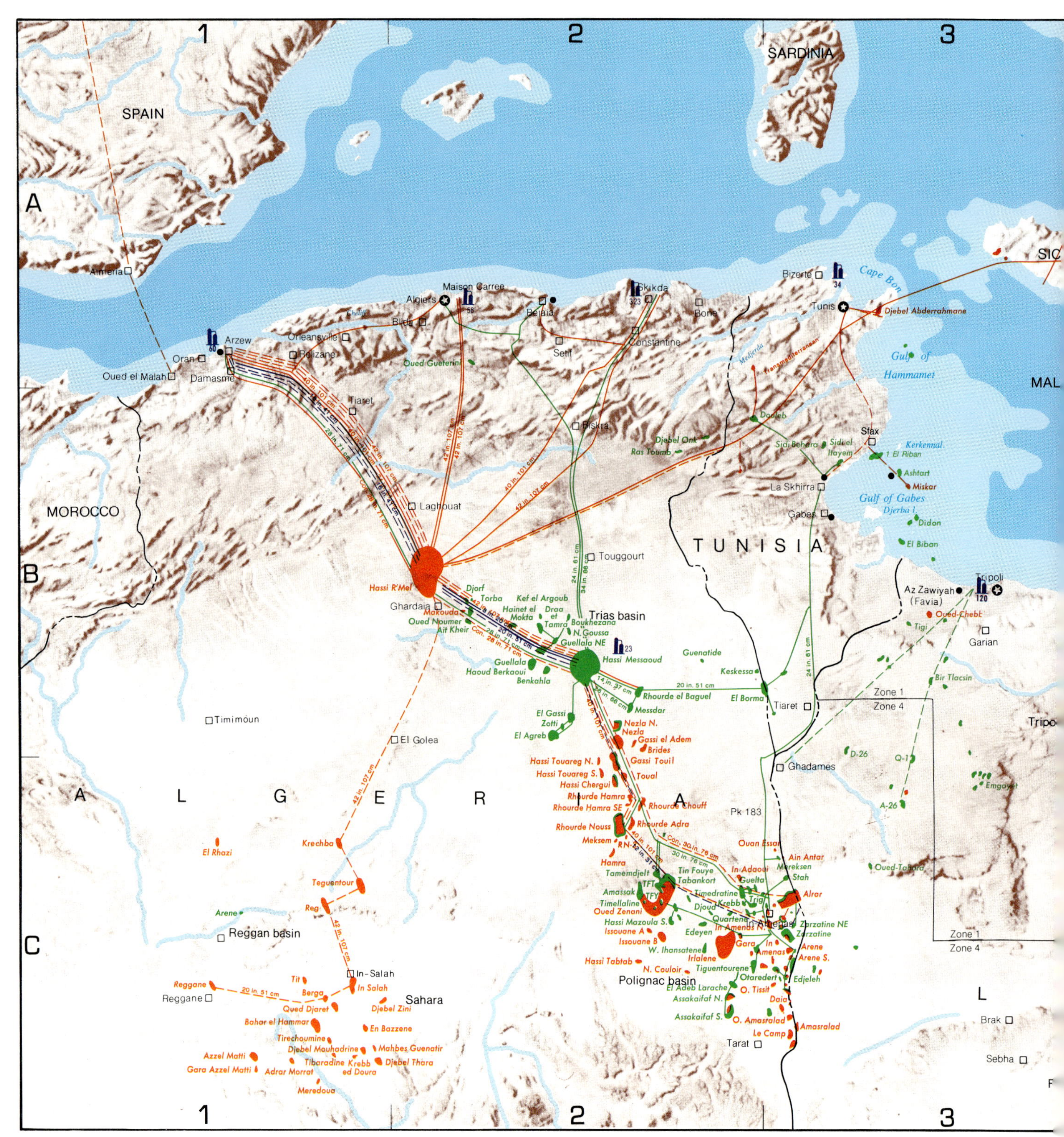

New concession

A group led by Petrofina Exploratie Gabon BV acquired an exploration concession in Gabon, about 38 miles north of the large Rabi discoveries.

The Alombie block, covering 914 sq miles, was awarded as part of Gabon's third licensing round. Petrofina and partners were obligated to begin exploration by midyear 1988. The initial contract covers 4 years and includes seismic and drilling commitments.

GHANA

CAPITAL: Accra
MONETARY UNIT: Cedis
REFINING CAPACITY: 26,600 b/cc
PRODUCTION: 0
RESERVES: 1,000 Mbbl

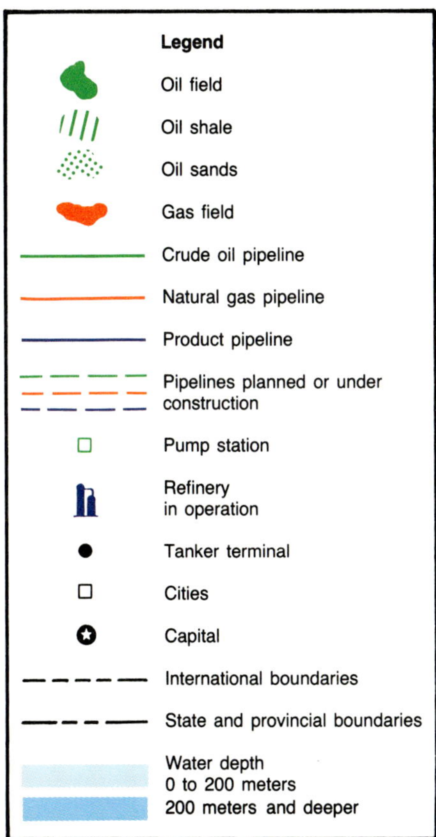

PETRO-CANADA INTERNATIONAL ASSISTANCE CORP.

(Pciac), Ottawa, agreed to assist Ghana in an exploration program to be conducted by a group led by ARCO.

Pciac will contribute as much as $10 million (Canadian) in Canadian goods and services to help Ghana National Petroleum Corp. earn an interest in an undisclosed offshore block to be acquired by the ARCO combine, in which units of Unocal Corp. and the Royal Dutch/Shell Group are partners.

A supplemental agreement calls for Pciac to work with ARCO to stimulate more purchases of Canadian goods and services by the operating group.

In addition, the Dutch government will finance the cost of a drillship for a multiwell program.

Pciac has been helping Ghana for several years, beginning with seismic data reprocessing in 1983.

LIBYA

CAPITAL: Tripoli
MONETARY UNIT: Dinars
REFINING CAPACITY: 329,400 b/cd
PRODUCTION: 1,012.5 Mb/d
RESERVES: 22,000,000 Mbbl

LIBYA HAS SEEN A STEADY EXODUS of foreign companies from the exploration and production sector.

U.S. companies have been forbidden by Washington to operate, and many foreign companies have had reservations about operating in Libya.

Libya's National Oil Co. NOC was making an effort to lure foreign expertise back into the Libyan industry. New terms for production-sharing contracts have been drawn up and new acreage was being offered both onshore and offshore.

There is only one significant foreign producer, a subsidiary of Agip SpA, which operates Bouri, Libya's first offshore field to be put into production.

Bouri field contains about 5 billion bbl of oil and more than 2.5 tcf of gas in place, NOC reported in an article in OPEC Bulletin, a monthly publication of the Organization of Petroleum Exporting Countries.

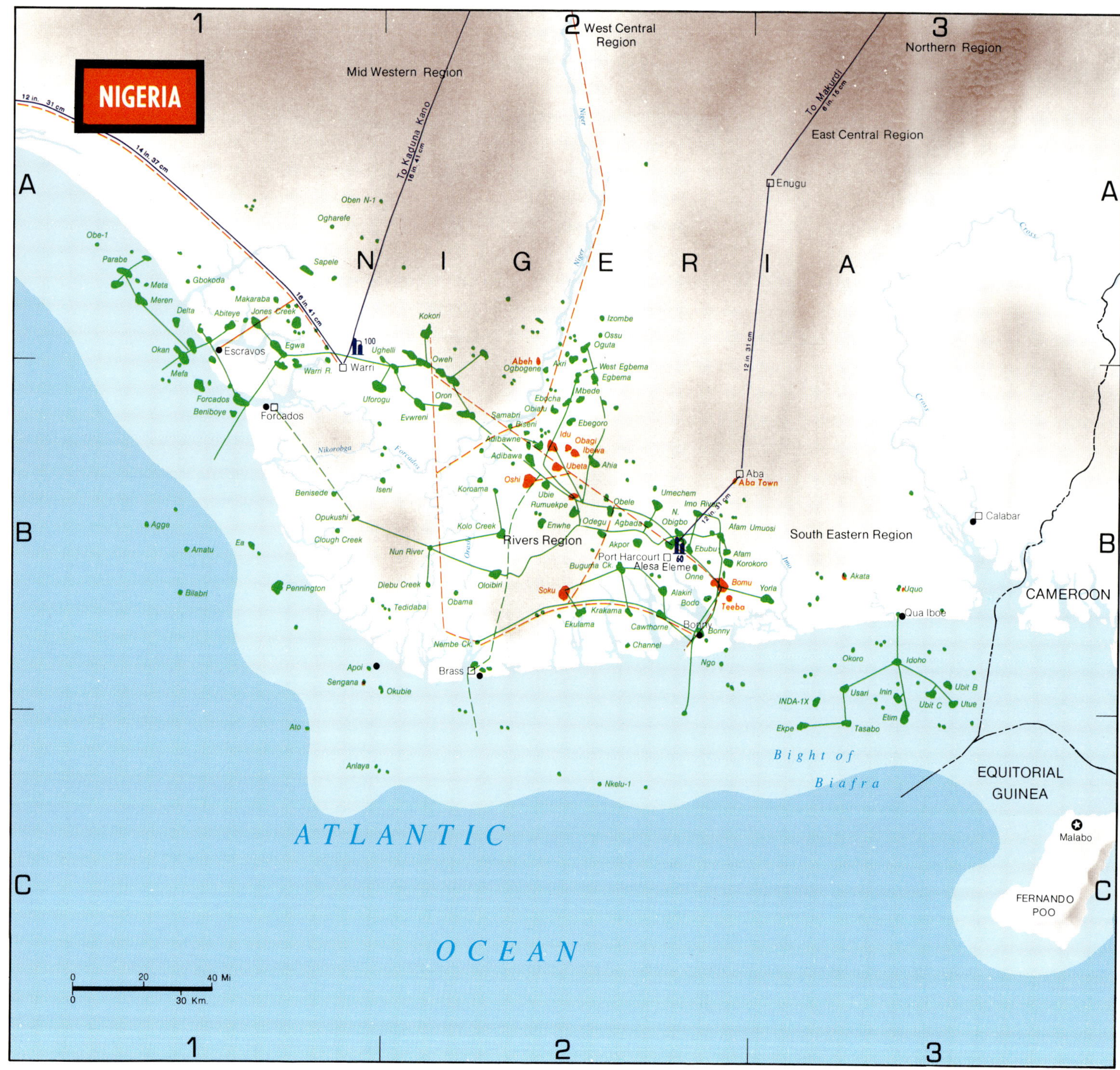

NOC said it may be possible to recover as much as 75% of the 25-27° gravity oil in place, the first 50% of it without use of enhanced recovery techniques.

Libya sees the field, its first offshore, as a symbol of its independence from what it calls the U.S. dominated international oil industry.

Libya has assured OPEC it is trimming light crude production from onshore fields to avoid overproducing its quota as Bouri field flow builds. OPEC raised Libya's production quota 4.1% to 1.037 million b/d effective Jan. 1, 1989.

NOC and Agip-North Africa Middle East are developing Bouri field, discovered in 1976 on a 14,500 sq mile block in the Gulf of Gabes.

"When we have developed all our offshore oil and gas resources, a new North Sea will emerge on this side of the Mediterranean," said Hamid Al-Hourim, manager of the Libyan branch of the Libyan-Italian joint venture developing Bouri.

The field, 3 miles wide and 20 miles long, lies in 558 ft of water 93 miles northwest of Tripoli.

Under the $1.5 billion first phase of Bouri development, NOC installed the DP-4 production platform in September 1986 and the DP-3 production platform in June 1987.

Commercial production, which began in August 1988 at 8,000 b/d, was slated to reach 50,000 b/d by yearend 1988.

DP-4 is a 66 slot platform and, at 67,000 tons, is one of the biggest, heaviest offshore structures in the world. DP-3 weighs 35,000 tons.

The initial plan calls for drilling of 30 wells from DP-4 and 20 wells from DP-3.

With eventual drilling of as many as 105 wells during Phase 1 of development, production could reach 150,000 b/d of oil by 1990 and continue at that level for 35 years without enhanced recovery.

Libya plans to install three more production platforms—DP-1, DP-2, and DP-5—in a second phase after it has fully evaluated Phase 1 performance.

Libya and Tunisia in 1988 agreed to jointly undertake development of a 1,158 sq mile zone in the northwest part of Bouri straddling the demarcation line between the two countries' territorial waters.

Tunisia will get 10% of the revenues from this part of the field for investment in joint economic projects to be managed by a Tunisian-Libyan company set up for that purpose.

According to industry sources Libya would like to increase offshore as well as onshore activity. This objective could be helped by the end of a protracted territorial dispute between Libya and Malta. At the end of 1987 the two countries formally ratified an agreement on the maritime borders according to a ruling from the International Court of Justice. Industry sources say that in the short term this is more likely to stimulate drilling in Maltese waters.

Libya, like several other OPEC producers, was planning to expand its foreign refining and marketing activities. An NOC subsidiary has been established to invest in downstream activities outside Libya.

Libya established the holding company Oil Investments International Co., capitalized at $450 million. The company is owned equally by NOC, Libyan Arab Foreign Investment Co., and the Libyan Arab Foreign Bank.

OII's mandate is to oversee Libyan foreign energy investments. Its first investments were the state's 70% interest in the Tamoil operation in Italy, originally made through the bank, and its 50% stake in Coastal Corp.'s 80,000 b/d Hamburg, West Germany, refinery.

At home, Libya has a refining capacity of 380,000 b/d. Local consumption was running at 90,000-100,000 b/d. Bulk of the processing was at the 220,000-b/d Ras Lunuf unit, which was running at under half capacity.

Libya was also planning a major expansion of petrochemical facilities at Ras Lunuf, which has a capacity to produce 360,000 tons/year of ethylene and 220,000 tons/year of propylene. The Ras Lunuf Oil & Gas Processing Co. said it planned to move downstream to enable the basic production plants to reach full capacity. The new units include an 80,000-ton/year low and high-density polyethylene plant and 80,000-ton/year polypropylene plant. The project also involves a gas extraction plant link to a 60,000-ton/year methyl tertiary butyl ether (MTBE) plant and a unit to produce a similar quantity of benzene. There will also be facilities for limited output of LPG.

Production sharing agreements

Royal Dutch/Shell Group signed the first production sharing agreement under Libya's revised terms for exploration and production.

The agreement with NOC covers the 3,735 sq mile Block 147 in northwestern Libya's Ghadames basin, which extends into Algeria and Tunisia. The block was first licensed to Shell under more onerous terms imposed in 1980.

NOC also signed a new production sharing agreement with Soc. Energia Montedison, Milan, for a block immediately northeast of Shell's acreage.

Libya's more flexible attitude has attracted other foreign companies into negotiating for acreage, including five with existing Libyan interests: Veba Oel AG, Wintershall AG, Agip SpA, Austria's OMV, and Bulgaria's state oil company. Other companies negotiating with NOC are Turkish Petroleum Co., Cie. Francaise des Petroles, Braspetro, INA-Naftaplin, Repsol SA, International Petroleum Corp., and Ranger Oil (U.K.) Ltd.

Areas offered include much of the Gulf of Sirte and the Ghadames, Sirte, and Cyrenaica onshore basins.

U.S. companies were forbidden from participating by the 1986 presidential executive order that prohibited American companies from involvement in Libya.

President Reagan, on his last working day in office, modified those economic sanctions to allow U.S. oil companies to resume operating in Libya. The companies still can't ship Libyan oil to the U.S., send U.S. goods to Libya, or employ U.S. nationals there. Other U.S. policy toward Libya is unchanged.

The changes were designed to protect U.S. interests, eliminate a Libyan windfall selling the U.S. firms' share of produced oil, and protect the companies—Conoco, Marathon, Amerada Hess, Oxy, and W.R. Grace—from a potential breach of contract claim by the Libyan government.

Terms under the 1980 production sharing contracts reflected the high oil prices and expectations at that time. Drilling under these contracts found mainly small accumulations that could not be developed commercially.

MADAGASCAR

CAPITAL: Antananarivo
MONETARY UNIT: Francs
REFINING CAPACITY: 16,350 b/cd
PRODUCTION: 0
RESERVES: 0

MADAGASCAR YIELDED ITS FIRST SIGNIFICANT flows of light hydrocarbons.

Petro-Canada International Assistance Corp. (Pciac) 1 West Manambolo on the country's west coast flowed gas and an undisclosed volume of condensate from two Cretaceous sand intervals. A drillstem test at 5,824-37 ft flowed at the rate of 6.3 MMcfd for about 3 hr with no production decline. Drillstem test at 5,399-5,408 ft flowed 9.5 MMcfd for 8 hr with no decline.

Pciac drilled the Morondava basin well, 200 miles west of Antananarivo, to 8,530 ft with a Peter Bawden Drilling Ltd. rig under an exploration aid agreement with Madagascar's Office Militaire National pour les Industries Strategiques.

Pciac called 1 West Manambolo the first potentially commercial discovery among about 65 wildcats drilled in the Morondava and adjacent basins.

Most exploration has concentrated on deeper Carboniferous-Jurassic Karroo sediments. The discovery in overlying Cretaceous sediments presents a new exploration opportunity in an area of about 38,600 sq miles onshore, Pciac said.

NIGERIA

CAPITAL: Lagos
MONETARY UNIT: Naira
REFINING CAPACITY: 414,500 b/cd
PRODUCTION: 1,358.3 Mb/d
RESERVES: 16,000,000 Mbbl

NIGERIA WAS TAKING STEPS TO PRIVATIZE ITS state owned oil company. Nigerian National Petroleum Corp. became independent of the state and likely will form separate companies for exploration/production, refining, and marketing. Supplying state industries, including the beleaguered electricity authority, will henceforth be strictly for cash. NNPC now can fix its own prices for oil and products.

New gas pipeline

NNPC commissioned a 211 mile, 36 in. natural gas transmission line from Warri to an electric power plant at Egbin, near Lagos. The pipeline is the longest segment of Nigeria's Escravos-Lagos system.

Tullow concession

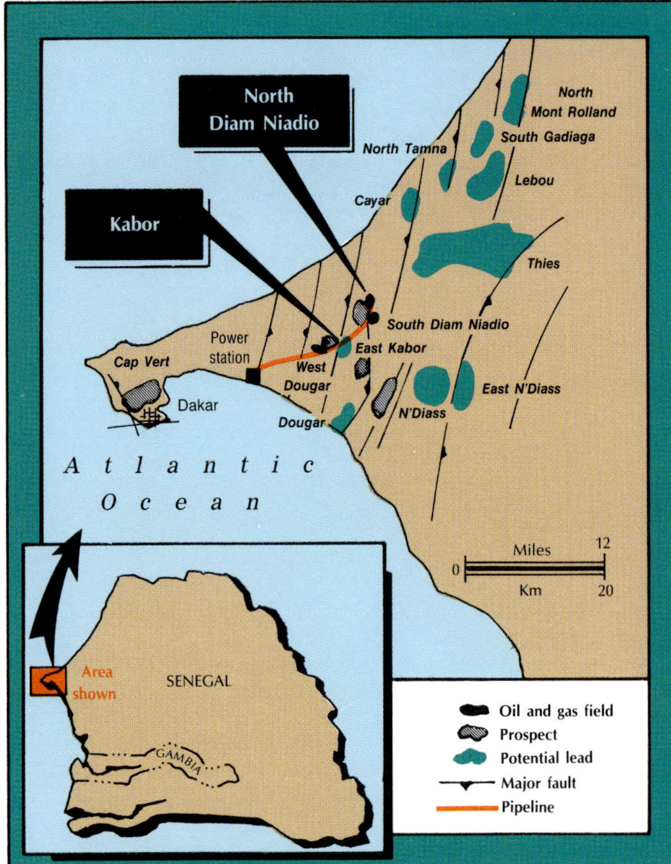

It can carry 450 MMscfd of gas, but the plant, operated by Lagos (Egbin) National Electric Power Authority, uses only about 200 MMscfd at peak generation. Most of the natural gas comes from Ughelli and Ogoturu area fields. The gas is treated at Warri, where NNPC started up a petrochemical complex. A Saipem SpA/Snamprogetti SpA joint venture laid the transmission line between Warri and Egbin.

Mannesmann Anlagenbau AG of West Germany handled construction of the Escravos-Lagos system's western gas gathering system, including facilities at Warri. It involves 65 miles of 8-24 in. pipeline, mostly in swampy areas of Bendel state. Under a separate contract, Mannesmann provided supervisory control and a data acquisition system for the project.

Entrepose International SA of Paris built the eastern gathering system, involving 20 miles of 30 in. pipeline from Otorogu to Warri.

Edop field production

Oil production from Edop field off Nigeria ultimately may reach 150,000 b/d.

The field, operated by Mobil Producing Nigeria in a joint venture with NNPC, had three wells capable of producing a total of 20,000 b/d via the Edop A platform. Plans call for five additional wellhead platforms, 30-40 wells, and a production platform with capacity of 150,000 b/d.

Edop A, in 105 ft of water about 28 miles south of Mobil's Qua Iboe terminal, is tied back to Mobil's Inim field production platform by an 18 in. pipeline. The 1981 discovery came on stream late in 1987.

To the west, Agip Nigeria began development of Agbara oil field 26.7 miles off Brass, Rivers State. A platform jacket was under construction in Italy and Port Harcourt. Crude will

New Nigerian gas transmission line

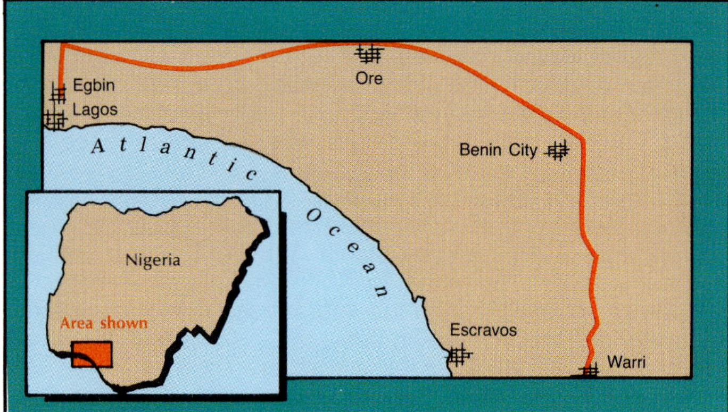

be carried by pipeline to the Brass terminal.

New fertilizer complex

Nigeria commissioned its first world scale fertilizer complex.

Start-up in 1988 of the 1,000 metric ton/day nitrogen-phosphorus-potash mixed fertilizer plant at Onne, Rivers State, near Port Harcourt, marked full operation of an $800 million project intended to serve domestic and export markets.

In 1987, Nigeria made its first export of ammonia from the Onne complex's 1,000 metric ton/day ammonia plant and started up the 1,500 metric ton/day urea plant at Onne.

The project is a joint venture of National Fertilizer Co. of Nigeria Ltd. (Nafcon) and Dresser Industries Inc. unit M.W. Kellogg Co.

The ammonia plant was operating at 115% of design capacity, the urea plant at 107%, and the mixed fertilizer plant at design capacity.

The ammonia plant uses Kellogg's process, the urea plant the Stamicarbon process, and the mixed fertilizer plant Jacobs Engineering Group technology.

Natural gas feedstock comes from Alakiri field 8.7 miles away.

Adjoining the fertilizer complex are refining and petrochemical plants.

Completion marks years of delay on the Onne project. Nigeria and its consultants Scientific Design Co. in 1979 sought bids on the project after completing 3 years of feasibility studies. A group led by Kellogg won the contract to build the project, with initial plans for mechanical completion in spring 1983.

Protracted negotiations saw the Kellogg contract signing delayed until mid-1980.

Meantime, Kellogg acquired an equity interest in the project in mid-1981, about the time financing agreements were signed between Nigerian and export-import banks in Japan and the U.S. Further delays totaling 21 months in releasing letters of credit to free financing for the project postponed work until after the first completion date. The venture then set the new mechanical completion date for July 1987.

In 1986, a group of Nigerian banks provided loans to supplement the export-import bank loans.

The complex will meet all of Nigeria's granular urea demand and about 60% of the country's mixed fertilizer needs.

That will save Nigeria about $100 million/year in fertilizer imports. Nigeria says the project is economic because its internal rate of return is estimated at 16% based on current world prices.

The Nafcon plant has shipped products to Spain, France, West Germany, Belgium, Ivory Coast, Cameroon, the U.K., and the U.S. Initial domestic orders total 75,000 metric tons of urea and more than 200,000 metric tons of mixed fertilizers.

Kellogg as minority partner in the venture with Nigeria's Federal Ministry of Industries will sell its interest to Nigeria during a fixed term when Kellogg will oversee operations and maintenance while training Nigerians to take over operations.

That's expected to take 4 years.

SENEGAL

CAPITAL: Dakar
MONETARY UNIT: CFA francs
REFINING CAPACITY: 29,800 b/cd
PRODUCTION: 0
RESERVES: 0

IRELAND'S TULLOW OIL PLC STARTED A FOUR WELL drilling program in Senegal.

The aim is to boost natural gas production near the capital, Dakar.

Tullow, in partnership with Senegal's state oil company and a local concern, refurbished a gas system near Dakar with the aid of World Bank financing.

The system moves 500 Mcfd of gas from Kabor and North Diam Niadio fields to a coastal power station, backing out diesel fuel.

The fields were placed on stream by a unit of Ste. Nationale Elf Aquitaine.

Production stopped in 1980 when Elf dropped the concession.

Tullow planned to increase production to 1 MMcfd during the second half of 1988 and believed the drilling program could push production to 3-5 MMcfd.

The additional produced gas also will be burned in the power station.

The takeover of Elf's 3,470 sq km former concession gave Tullow access to liquids production coming from two wells.

Production from a third well began late in 1987 when a liquids separation plant became available. Production for 1987 totaled 11,000 bbl.

The new drilling program was preceded by a high resolution seismic survey over the Kabor and North Diam Niadio structures to select sites for two wells.

Further seismic was shot in a 315 km survey on other prospects on the concession.

Completed early in 1987, it will form the basis of siting the two other wells in the program.

Drilling contractor will be Puma Drilling (U.K.) Ltd., a Tullow subsidiary.

Norway's Smedvig Drilling acquired a 10% interest in the rig and will operate it.

SOUTH AFRICA

CAPITAL: Johannesburg
MONETARY UNIT: Rand
REFINING CAPACITY: 433,500 b/cd
PRODUCTION: 0
RESERVES: 0

SOUTH AFRICA'S STATE OWNED SOEKOR TESTED ITS biggest oil and gas discovery.

The E-AR1 well, in Mossel Bay off the country's southern coast, flowed a combined 10,200 b/d of oil and 22 MMcfd of gas from two zones. Site is 75 miles southwest of the town of Mossel Bay.

South African Minister of Economic Affairs Danie Steyn called the well the most promising result of a lengthy offshore drilling program.

The program has produced a commercial offshore gas-condensate discovery that is being placed on production to provide feedstock for a plant to produce transport fuels and middle distillates.

The latest well is Soekor's fourth offshore oil discovery in the area.

The four productive structures, about 4-6 miles apart, have not generated enough reserves to warrant development.

Other structures lie nearby, and the company hopes the drilling program will uncover enough reserves to support installation of a floating production platform.

The latest well, which was drilled by the Actinia semisubmersible, flowed 5,400 b/d of oil from one zone and 4,800 b/d of oil from a second.

At midyear 1988, the E-AD1 well flowed as much as 7,000 b/d of oil.

Water is about 500 ft deep in the area.

Soekor said the area's geology is very complex, and four discoveries the past 2 years show the company is "on the right track."

Soekor had the Omega semisubmersible working in Mossel Bay and the Nymphea semisubmersible drilling off South Africa's west coast, so far without success.

The Mossel Bay project was on schedule to begin gas production by June 1991. Condensate production will start in first quarter 1992.

When the offshore field is in full production, the onshore refinery under construction will be producing about 4 million l./day of liquids.

All the gas will be used in the process for manufacturing liquids.

No gas will be used outside the plant. The newly privatized owners have a license from Sasol Ltd., Johannesburg, to use the Fischer Tropsch process for producing liquids from coal.

The project is 50% owned by the South African government through the Central Energy Fund and 30% by Gencor, which is managing the project.

The other 20% interest, which is owned by an undisclosed shareholder, may provide the basis for further privatization.

Work has started on the refinery and the offshore drilling/production program.

SUDAN

CAPITAL: Khartoum
MONETARY UNIT: Pounds—livres
REFINING CAPACITY: 21,440 b/cd
PRODUCTION: 0
RESERVES: 300,000 Mbbl

A JOINT VENTURE OF UNITED STATES OIL CO. AND Panoco Sudan Ltd. was to begin development in late 1988 of a 12 year old Chevron group gas/condensate discovery in the Red Sea off Sudan.

Reserves estimates of Suakin field, in 600 ft of water 30 miles off Port Sudan, are 864 bcf of gas and 123 million bbl of condensate.

Plans called for reinjection of gas until the market for it improves.

Usoco plans an ammonia/urea fertilizer plant in Sudan within 3 years to use the gas.

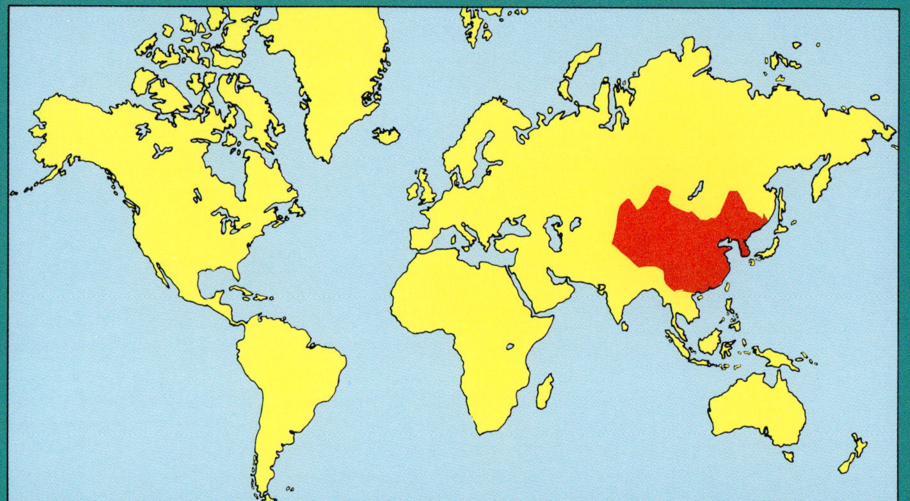

CHINA

CHINA

CAPITAL: Beijing
MONETARY UNIT: Yuan renminbi
REFINING CAPACITY: 2,200,000 b/cd
PRODUCTION: 2,690.3 Mb/d
RESERVES: 23,550,000 Mbbl

CHINA AND FOREIGN OPERATORS WERE PRESSING exploration in several areas of the country at yearend 1988.

China National Petroleum Corp., formerly the Ministry of Petroleum Industry, made what it calls a "breakthrough discovery" in the Lunnan area of the northern Tarim basin in central Asia's Xinjiang Uygur Autonomous Region.

CNPC's 2 Lunnan flowed 4,289 b/d of high quality, light crude through a 38/64-in. choke from 30 1/2 ft of pay in the Triassic. The well is 224 miles southwest of Urumqi, the regional capital of Zinjiang. Total depth was 17,129 ft. Reports on the well's gas flow conflicted but ranged from 1.9 MMcfd to 3.9 MMcfd. Higher oil and gas flows were expected when more Triassic zones were perforated.

Bohai Oil Corp. was drilling two appraisal wells in Jinzhou 9-3 field in Liaodong Bay, a northeast arm of Bohai Sea. Bohai Oil calls the strike its third major discovery in the bay, after Suizhong 36-1 oil field and Jinzhou 20-2 gas/condensate field. Bohai Oil said Jinzhou 9-3 field is likely to merit development.

A group led by Sun Orient Exploration Co. was starting exploration in Contract Area 09/36 in Bohai Bay.

And China National Offshore Oil Corp. (Cnooc) and Total Chine of France extended marginal Wei 10-3 oil field in Beibu Gulf with a high flowing appraisal well. The two companies must decide whether to develop the new area of the field before Total Chine's exploration contract expires in August 1989.

Tarim exploration

Encouraged by the 2 Lunnan discovery, CNPC decided to deploy more seismic crews and drilling rigs in the Tarim basin to speed exploration in 1989.

CNPC was to begin appraising the Lunnan area and exploring other promising structures. Qi Daqing of China Features, Beijing, reported the 2 Lunnan discovery found oil in 11 formations with a total thickness of 198 ft and oil and water in six other formations with an additional 95 ft of pay.

The oil and gas flows from Triassic are considered a breakthrough, CNPC said, because the formation is prevalent in the basin. The other potential producing intervals are in Upper Tertiary, Lower Cretaceous, Jurassic, and Ordovician. The well's high yield also shows that porosity and permeability are adequate at as deep as 13,000 ft.

CNPC and Geophysical Service Co. of the U.S. have explored the basin since 1982 under a contract extended in 1986 through 1989. The work has shown the basin has the potential of large or even extraordinarily large fields, Beijing Review said. It said the first 5 years of prospecting yielded 165 comprehensive reports that identified 10 prospects.

Central Tarim basin drilling in the Taklimakan Desert costs at least $823/ft, or about $16 million/well based on required depths, according to Beijing Review. Total productive capacity of 400,000 b/d is required to justify development of the region.

Field conditions made geophysical surveys difficult. Temperatures range as high as 126° F., and sand dunes are as high as 600 ft. Diesel fuel, explosives, grain, and water were supplied from the desert's border, a 435-500 mile round trip. Water underlies a large area of the basin at 6-30 ft, but it's mostly alkaline.

Offshore action

Exploration or appraisal drilling was under way or planned in Bohai Sea and the South China Sea's Beibu Gulf.

Bohai Oil's 9-3-1 Jinzhou, second of two exploratory wells, flowed a combined 1,157 b/d of oil and 219 Mcfd from a main pay zone at 5,459-5,577 ft. The other exploratory well also flowed oil at commercial volumes. The field is 34 miles southeast of the city of Jinzhou in Liaoning Province.

Bohai Oil said commerciality is likely because of the field's relative closeness to shore and high quality of its crude. Data were too scarce to accurately estimate reserves, the company said.

China Features said the Sun Orient group decided to go ahead with exploratory drilling on Beibu Gulf Contract Area 09/36 because seismic data Sun Orient collected and interpreted under a March 1987 geophysical survey contract with Cnooc showed the area "potentially oil rich." The 3 year agreement obligates the group to drill at least one wildcat.

DRILLING PROCEEDS offshore as China moves to tap more energy supplies. Onshore, a gas processing plant operates in Sichuan Province, the country's biggest gas producing area.

Cnooc-Total's Wei 10-3-13 well, in Beibu Gulf's Wei 10-3 oil field, flowed 4,692 b/d of oil and 2.98 MMcfd of gas. Total depth was 7,005 ft. The field, discovered by Total Chine late in 1982, is comprised of northern, central, and southern fault belts. Most exploration has been conducted on the central fault belt, where trial production began in late 1986 from an eight well platform. Three of the wells are still producing. The central belt's reserves seem to have been overestimated or are harder to produce than was anticipated, China Features said.

The new well, on the northern fault belt, "effectively enlarged the field's oil rich acreage." Wei 10-3 field produced more than 2.19 million bbl of oil in 1987 and more than 1.825 million bbl in the first 10 months of 1988. All was sold for foreign currencies, either at home or abroad.

Seismic data gathering, interpretation

China was stepping up a program designed to upgrade its seismic data gathering and interpretation technology.

Two orders for geophysical equipment and technology from U.S. companies totaled more than $27 million, with still larger orders in the offing.

China's Ministry of Petroleum Industry let contract to Applied Automation Inc. (AAI), Bartlesville, Okla., for $14.9 million in seismic systems to be used in frontier areas of China's western basins. AAI calls it one of the largest single source instrumentation purchases by the Chinese.

In addition, the ministry and Cnooc let contracts to GeoQuest Systems Inc. and Digital Equipment Corp., both of Houston, totaling more than $12.6 million. GeoQuest says the order is the petroleum industry's largest procurement of interactive geophysical technology.

AAI was to ship the seismic systems through first half 1989. Involved were seven large and two very large Opseis systems that include nine central recording stations and 660 remote seismic units. Chinese officials said the size and timing of the purchase were tied to the country's need for continuing oil exploration, especially in the western regions.

The GeoQuest/Digicon order resulted from a 1 year solicitation by the Chinese that was the most comprehensive competition for such technology seen in the industry, said GeoQuest Vice Pres. Mike Hare.

Although the Chinese have been upgrading their seismic technology for the past 5 years, the recent orders show an accelerated push midway through the government's current 5 year plan for its petroleum industry. The order represents the first phase of GeoQuest's efforts to target the China market for interactive geophysical technology.

"We expect Phase 2 to be at least half again as big—perhaps 50 systems," Hare said. A recent influx of World Bank funding has contributed to the burst of Chinese equipment/technology acquisition.

GeoQuest's Interactive Exploration Systems (IES) will be installed at major oil fields and geophysical research centers in China. The Chinese will use Digicon MicroVAX 3000 microcomputers for GeoQuest workstations.

In addition, GeoQuest and Chinese geoscientists will cooperate in a comprehensive, long term technology exchange program at Dagang oil field center to develop advanced interpretation techniques and capabilities tailored to China's geophysical and geological needs.

The IES systems will be installed in the cities of Jilin, Xian, Zhuo Xian, Guang Zhou, Zhang Jiang, Gao Bei Dan, and Shanghai, as well as Dagang, Shengli, Daqing, Liao He, Zhong Yaun, Chong Qing, Dian Guia Qi, and Bohai oil fields, and Qing Hai Province, Hua Dong University, and Beijing Geophysical Research Institute.

Hainan Island discovery

CSR Orient Oil Pte. Ltd. may be the first non-Chinese company to discover oil onshore China.

The company, operator for an Australian group, recovered oil on two cased hole drillstem tests of the 1 South Jinfeng wildcat on Hainan Island. On an 18 hr drillstem test at 7,259-68 ft, oil flow declined from an initial rate of 1,650 b/d to 226 b/d of oil at the end of the test. Gas flow increased from a rate of 1.75 MMcfd to 3.84 MMcfd after 18 hr.

An earlier drillstem test at 7,347-7,413 ft flowed formation water at a rate of about 4,000 b/d during 2 hr. The test was designed to examine the reservoir characteristics of a lower water sand believed to contain hydrocarbons updip from 1 South Jinfeng, CSR said. On Sept. 26, 1988, the well flowed at a rate of 285 b/d of oil and 430 Mcfd of gas at the end of a 9 hr test at 7,283-7,309 ft. Total depth is 7,776 ft. The group is considering appraisal drilling.

CSR, with a 30% interest, operates the exploration program in cooperation with Hainan Provincial Petroleum Development Corp. Other interest owners are Bond Petroleum International Inc. 30%, BHP Petroleum (China) Inc. 20%, and Barcoo Petroleum NL and National Mutual Life Association of Australasia each 10%.

South China Sea contract

China awarded affiliates of Texaco Inc. and two Japanese companies a contract area in the South China Sea and cut

Breakout of China's energy supply/demand

	Million tons coal equivalent	Coal	Oil	Gas	Hydro-power
			Percent share		
PRODUCTION					
1970	309.9	81.6	14.1	1.2	3.1
1975	487.5	70.6	22.6	2.4	4.1
1980	637.4	69.5	23.7	3.0	3.8
1985	855.5	72.8	20.9	2.0	4.3
1987	910.0	72.2	21.0	2.2	4.6
CONSUMPTION					
1970	292.9	80.9	14.7	0.9	3.5
1975	454.3	71.9	21.1	2.5	4.6
1980	602.8	72.2	20.8	3.1	4.0
1985	770.2	75.2	17.0	2.2	4.8
1987	845.0	75.1	17.6	2.3	5.0

Source: Petroleum Industry Research Foundation Inc.

China's oil situation

	Consumption (Million b/d)	Percent growth	Production (Million b/d)	Percent growth
CONSUMPTION, PRODUCTION				
1975	1.34	—	1.54	—
1980	1.75	5.5	2.12	6.5
1985	1.83	0.9	2.49	3.3
1987	2.08	6.6	2.68	3.7
1990*	2.39	4.8	2.89	2.5

	Crude	Products	Total
		1,000 b/d	
EXPORTS			
1983	297	108	405
1984	440	125	565
1985	600	133	733
1986	570	70	640
1987	545	59	604
1988†	530	48	578

*Government forecast. †Estimated.

Source: Petroleum Industry Research Foundation Inc.

Some of China's main oil pipelines

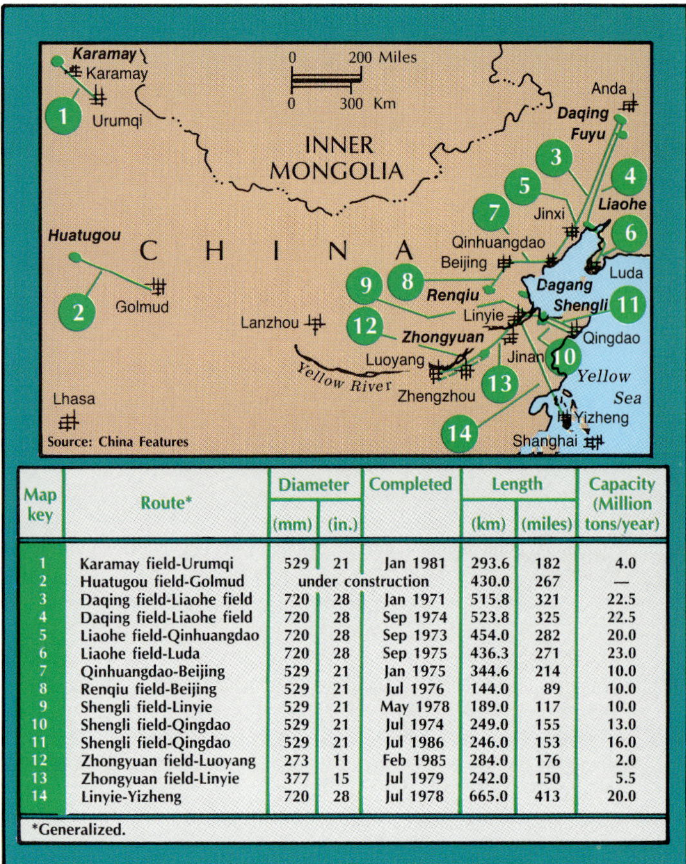

Map key	Route*	Diameter (mm)	(in.)	Completed	Length (km)	(miles)	Capacity (Million tons/year)
1	Karamay field-Urumqi	529	21	Jan 1981	293.6	182	4.0
2	Huatugou field-Golmud	under construction			430.0	267	—
3	Daqing field-Liaohe field	720	28	Jan 1971	515.8	321	22.5
4	Daqing field-Liaohe field	720	28	Sep 1974	523.8	325	22.5
5	Liaohe field-Qinhuangdao	720	28	Sep 1973	454.0	282	20.0
6	Liaohe field-Luda	720	28	Sep 1975	436.3	271	23.0
7	Qinhuangdao-Beijing	529	21	Jan 1975	344.6	214	10.0
8	Renqiu field-Beijing	529	21	Jul 1976	144.0	89	10.0
9	Shengli field-Linyie	529	21	May 1978	189.0	117	10.0
10	Shengli field-Qingdao	529	21	Jul 1974	249.0	155	13.0
11	Shengli field-Qingdao	529	21	Jul 1986	246.0	153	16.0
12	Zhongyuan field-Luoyang	273	11	Feb 1985	284.0	176	2.0
13	Zhongyuan field-Linyie	377	15	Jul 1979	242.0	150	5.5
14	Linyie-Yizheng	720	28	Jul 1978	665.0	413	20.0

*Generalized.

the size of an offsetting contract area.

Awarded Contract Area 15/31 consists of all or part of five blocks covering almost 316,000 acres in the Pearl River Mouth basin 95 miles south-southeast of Hong Kong. Exploratory drilling was to begin early in 1989.

The area consists of S 1/2 of 15/31, all of 15/32, 28/01, and 28/02, and N 1/2 of 28/08.

Contract Area 15/33, awarded in 1983, was amended to cover three blocks totaling more than 237,000 acres. Seismic surveys were planned for 1989. As awarded in 1983, the area consisted of four blocks including Block 15/33, and covered about 316,000 acres in 230-330 ft of water. It now consists of Blocks 15/34, 28/03, and 28/04.

Interests in both contract areas are Texaco's Getty Oil International (Orient) Inc. 48.4%, Japex Nanhai (Pearl River) Ltd. 31.3%, and Huanan Oil Development Ltd. 20.3%.

The companies jointly own Pearl River Oil Operating Co. Pearl River is working with Nanhai East Oil Corp., which is responsible for implementing the contracts for China National Offshore Oil Corp.

In 1986 Pearl River's XJ 34-3-1 wildcat flowed 37° gravity oil on two tests at a combined rate of 1,874 b/d with minor gas volumes through a 1 in. choke. The discovery is near a wildcat drilled by the Chinese in 1979. Sun Orient Exploration Co. and Texas Eastern Orient Inc. relinquished their interests in Contract Area 15/33 in 1987.

Gas/condensate discovery

Esso China Ltd. and Shell Exploration (China) Ltd. tested a gas/condensate discovery east of Hainan Island in the South China Sea.

The companies tested four intervals in the Wenchang 9-2-1 discovery well. Combined flow rate was 25.4 MMcfd of gas and 2,160 b/d of condensate. The best flow from a single interval was 9.4 MMcfd of gas and 1,450 b/d of condensate. More evaluation was needed to determine the accumulation's size, Esso China said. The discovery is in the Contract Area 39/11 extension, 99 miles east of Hainan Island in 387 ft of water. It is the pair's second discovery in the area.

Esso-Shell's Wenchang 19-1-2, formerly 19-1-S, flowed 3,200 b/d of 35° gravity oil in mid-1984. That discovery is in Contract Area 40/01, an east offset to 39/11.

Esso China operates both areas and participates 50-50 with Shell Exploration in partnership with Nanhai West Oil Corp., a unit of China National Offshore Oil Corp.

Another South China Sea strike

The Agip-Chevron-Texaco (ACT) group was to delineate a South China Sea oil strike, which yielded the biggest flow rate disclosed off China.

ACT's Huizhou 26-1-1 wildcat flowed oil from seven zones at a combined rate of more than 26,000 b/d. Individual flow rates ranged from 741 to 5,925 b/d. Oil gravities were 27-40°. The combined flow tops by more than 10,000 b/d the biggest flowing well reported off China, said Kim Woodard, president of Washington, D.C., consultants China Energy Ventures Inc. He estimated water depths in the discovery area at 300-350 ft.

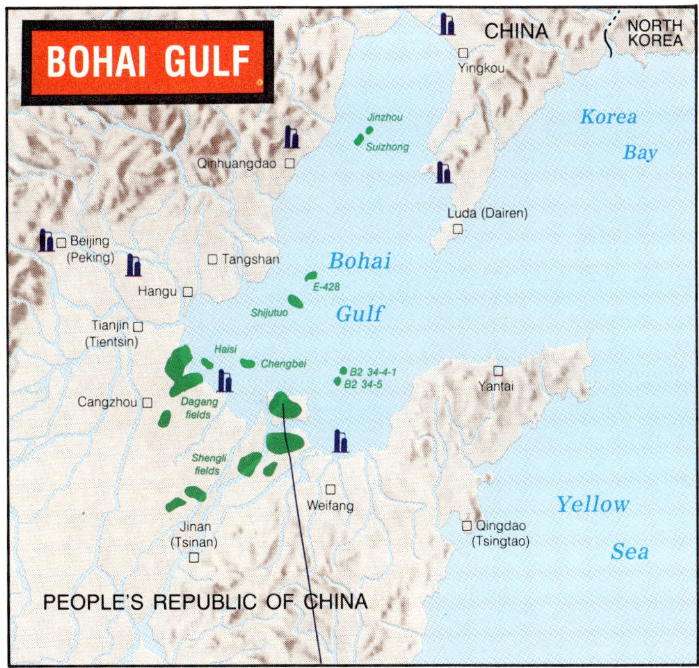

The discovery is in Contract Area 16/08 about 15 miles southwest of Huizhou 21-1 oil field, which ACT group plans to develop in the 1990s.

Huizhou 26-1-1 is about 170 miles southeast of Guangzhou in the Pearl River Mouth basin. ACT group moved the rig that drilled the discovery well to Contract Area 16/04. The Nanhai V, under contract from China Nanhai West Drilling Co., was to have spudded Huizhou 10-1-1 at a site 50 miles northeast of the discovery. ACT group expects production to start in 1990 from Huizhou 21-1, also in 16/08. Oil flow is to peak at 23,000 b/d about 1 year later. Each ACT partner holds a one third interest in the 600,000 acre Contract Area 16/08 and the 800,000 acre Contract Area 16/04. The group is conducting exploration in cooperation with Nanhai East Oil Corp., a branch of Cnooc.

Partners in ACT group are Agip (Overseas) Ltd., Chevron Overseas Petroleum Ltd., and Texaco Petroleum Mij. (Nederland) BV.

Gas discovery, development drive

China was mounting a drive to find and develop more gas, an energy source it has long neglected. The onshore and offshore push is designed to help fuel the country's drive toward modernization, says China Features.

Gas supplied only 3% of the country's total energy consumption in 1986, compared with a 70% share for coal and 20% for oil, China's two biggest energy sources.

Gas production in 1986 amounted to 13.4 billion cu m (473.2 bcf), including associated gas. About 60% of production went to customers outside of oil and gas fields. The rest, mainly associated gas, was consumed in the fields.

The Ministry of Petroleum Industry aims to increase the role gas plays in meeting increasing energy demand. Oil can't do the job alone. Some of the country's biggest oil fields have reached maturity. And oil is a valuable commodity for export.

Coal is the country's biggest energy resource. Reserves estimated at 782 billion tons appear sufficient for a century of production. But coal consumption is hampered by a lack of adequate railways.

Exploitable hydropower resources are estimated at 380 million kw, enough to boost electrical power production to 20 times the 1985 level. But as western observers have pointed out, development of those resources, which lie in the remote Southwest, would flood hundreds of thousands of hectares of arable land and must await China's use of high voltage, long distance, transmission technology. That leaves gas as the most likely candidate to supply more energy.

Gas exploration goal

China's goal is to find at least two gas fields with reserves of more than 100 billion cu m (3.531 tcf) each or three or four fields with 50 billion cu m (1.766 tcf) each, Energy Minister Wang Tao told China Features writer Qi Daqing. "If such a goal can be fulfilled," Wang Tao said, "the industry will be on its feet soon."

A scarcity of big gas discoveries does not mean big gas fields do not exist, Wang Tao said. "They are there. The problem is we haven't found them. Modern geological theorists believe structures should be studied in a bigger context, such as the formation and evolution of a whole basin. Previous exploration in China was much too confined to single structures, making big finds unlikely."

Now, to boost gas reserves, top priority will go to exploration. Wang Tao said, "From now to the end of 1990, gas related efforts will be centered on prospecting."

Results will decide whether the gas industry can boost its production considerably in the next 5 year plan (1991-95).

Chosen for "immediate, concerted" exploration are four onshore and two offshore areas: the Sichuan, Shanganning, Dongpu, and Songliao onshore basins, Liaodong Bay, and the Yinggehai basin of the South China Sea.

The Sichuan basin in Sichuan Province, China's No. 1 gas producing area, is the main hope for more major discoveries and a substantial production increase in the short term.

Recent exploratory drilling found two promising structures in the eastern part of the province. A well drilled on the larger of the pair, the 100 sq km (36.61 sq mile) Qilixia structure, flowed 334,000 cu m/day (11.79 MMcfd) of gas.

In the south central part of the province, a wildcat drilled on the Moxi structure of similar size flowed 119,000 cu m/day (4.2 MMcfd).

Also deemed "promising" is the Shanganning basin, which extends through Shanxi and Gansu provinces and the Ningxia region of Northwest China. China Features said all wells drilled in a 1,000 sq km (386.1 sq mile) area of the basin have found commercial volumes of gas or at least gas shows. One wildcat flowed more than 160,000 cu m/day (5.65 MMcfd) of gas from Ordovician limestone.

The Dongpu basin, on the border area of Henan and Shandong provinces, holds the Zhongyuan complex of oil fields. Zhongyuan gas reserves, including associated gas, stood at 115 billion cu m (4.06 tcf) at the beginning of 1987. Gas reserves are increasing at a rate of more than 10 billion cu m (353.1 bcf)/year.

Gas exploration in the Songliao basin of Northeast China centers in Daqing and Jilin fields. The Daqing field administration is exploring for gas in a 10,000 sq km (3,861 sq mile) outlying area in the northern part of the basin.

Results of wells drilled in the eastern, northern, and southern parts of the basin have been "positive," China Features said. A wildcat drilled about 150 km (93 miles) northeast of Harbin, capital of Heilongjiang Province, flowed more than 130,000 cu m/day (4.59 MMcfd) of gas from Jurassic pay at about 1,000 m (3,280 ft).

Jilin's administration is stepping up gas exploration in the Dehui-Yulin area, not far from Changchun, capital of Jilin Province. A wildcat drilled on the 39 sq km (15.06 sq mile) Sijiazhi structure flowed more than 100,000 cu m/day (3.531 MMcfd) of gas.

In Liaodong Bay, the northeast arm of Bohai Sea, development of JZ20-2 gas field was to have begun in 1988. Productive capacity will be 1.37 million cu m/day (48.38 MMcfd), with flow set to begin late in 1990.

"Indications are that gas reserves are not limited to JZ20-2, and further search for gas is going on," China Features

Production from major oil fields

Field	1987 production (1,000 tons)	Percent change from 1986
Daqing	55,550	—
Shengli	31,600	+ 7.1
Liaohe	11,350	+ 13.4
Huabei	7,950	− 20.5
Zhongyuan	6,800	+ 7.9
Xinjiang	5,750	+ 4.5
Dagang	4,170	+ 6.6
Jilin	2,860	+ 20.9
Henan	2,520	+ 0.8
Changqing	1,420	− 1.6
Jianghan	1,010	− 1.9
Jiangsu	625	+ 11.8
Yumen	534	− 2.5
Chengbei	362	+ 62.0
Wei-10-3	353	+126.2

Source: Ministry of Petroleum Industry

Addition of productive capacity

Field	Capacity buildup — 1,000 tons —	Volume in new zones
Daqing	4,090	—
Shengli	5,440	3,780
Liaohe	2,500	2,500
Huabei	409	409
Zhongyuan	1,500	1,270
Xinjiang	931	616
Dagang	524	373
Jilin	419	318

Source: Ministry of Petroleum Industry

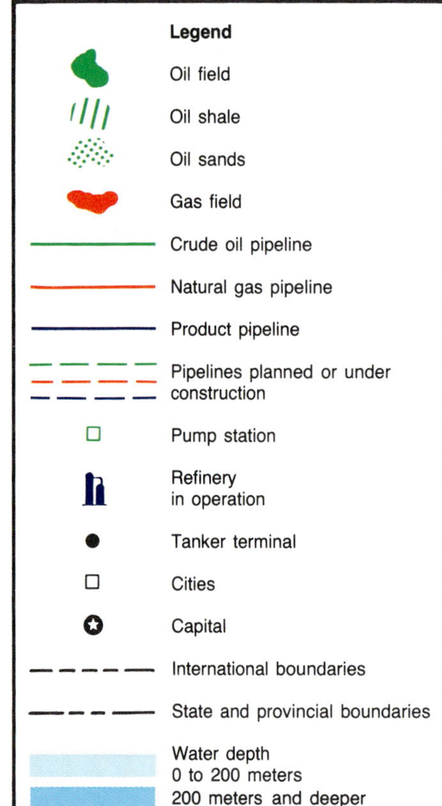

JACK UP works in the Bohai Sea, an area in which China is boosting offshore productive capacity.

said.

Most of China's gas production comes from two major areas: Sichuan Province of Southwest China and Daqing oil field of Northeast China. Sichuan is by far the largest gas production base in China. About 70% of the country's gas fields are there. In 1986 Sichuan produced 5.8 billion cu m (204.8 bcf) or a little more than 40% of the national total. The province supplied 5.2 billion cu m (183.6 bcf) for consumption outside of field use.

Daqing, China's biggest oil field, produced 2.3 billion cu m (81.2 bcf) of associated gas in 1986. Other big associated gas producers are Liaohe, Dagang, Zhongyuan, and Shengli oil fields in that order.

Oil fields provided 2.8 billion cu m (98.9 bcf) to outside consumers.

Figures supplied to China Features by the Ministry of Petroleum Industry show that 66% of the gas supplied for consumption outside of field use went to the petrochemical industry as feedstock. Of the balance, 8.6% served as fuel for metallurgy, 6.6% for residential use, 3% for electrical power generation, and 15.8% for other purposes.

Lack of an adequate pipeline system makes residential gas service rare outside of Sichuan Province and oil fields. But the gas pipeline system is growing. A line has linked Beijing with Huabei oil field in Hebei Province, supplying about 50,000 cu m/day (1.766 MMcfd) of gas to the capital.

Expansion of line capacity to 150,000 cu m/day (5.297 MMcfd) was under way.

Another pipeline has enabled Tianjin to obtain gas from Dagang oil field.

Those systems and a few others move gas to about 2 million residential customers in seven large cities. Another 18 million burn LPG, produced mainly by petrochemical plants, as residential fuel.

Most of the gas supplied as petrochemical feedstock is used in the manufacture of fertilizers. China has eight large chemical fertilizer plants, each rated at 300,000 tons/year of ammonia. In 1986 they consumed about 3 billion cu m (105.9 bcf) of gas. More than 1,000 small to medium size fertilizer plants consumed an undisclosed volume of gas.

China produced about 68 million tons of chemical fertilizers in 1986 and imported more than 10 million tons. Consumption is expected to double by 2000.

Striving for self-sufficiency in chemical fertilizers, China is building three ammonia plants, each rated at 300,000 tons/ year. Feasibility studies were under way for three more.

By the end of 1995, officials estimate, the gas industry will be called upon to supply gas feedstock for about 20 big chemical fertilizer plants. They will consume about 8 billion cu m (282.5 bcf)/year, equal to the total volume supplied for consumption outside of field use in 1986.

Oil production recovery

China's oil production had recovered strongly at yearend 1988 from oil production setbacks caused during the summer.

But in spite of rising production, exports were expected to decline because of China's burgeoning domestic demand and continued support for the Organization of Petroleum Exporting Countries.

Floods shut in 300 wells in the nation's biggest field— Daqing—reducing flow there by more than 70,000 b/d to slightly less than 1 million b/d. An official Beijing report said Chinese October 1988 crude output averaged a record 2.788 million b/d, compared with 2.695 million b/d in July 1988. Flow during the first 10 months of 1988 averaged 2.710 million b/d, up 2.07% from the same period in 1987 year. Chinese oil production during first half 1988 averaged 2.687 million b/d, up only 1.5% from the first 6 months of 1987. Beijing hoped increasing offshore flow would help boost crude production to 3 million b/d in 1990. However, production gains will have to accelerate sharply during the next 2 years if that target is to be achieved.

Chinese gas production for the first 10 months of 1988 was about the same as in the same 1987 period.

Beijing officials say that despite anticipated higher domestic production in 1989, Chinese crude and products exports will be reduced, not increased.

Zheng Dunxun, president of China National Chemical Import and Export Corp. (Sinochem), said his company supports OPEC's pricing and production agreement aimed at reducing output to 18.5 million b/d and returning prices to $18/bbl in 1989.

Sinochem is China's only oil export/import firm. Zheng said reducing oil exports would not hurt Chinese producers because domestic energy shortages accompanying China's rapid economic development "will not be overcome in the near future." He added that it has long been China's policy to support OPEC's actions to stabilize the world's oil market.

Zheng estimated that Sinochem would earn $700 million less in 1988 than initially expected because of sagging world oil prices. Reduced overseas shipments and unstable world market prices caused oil to lose its former status as China's top foreign exchange earner to textiles. Zheng was optimistic about the world oil market in 1989. He said that even with reduced oil exports, China could earn as much as in 1988 if prices rise to OPEC's $18/bbl target.

INFILL DRILLING under way in central Daqing field of China's Heilong-jiang Province.

China's oil exports peaked at 600,000 b/d in 1985. Japan has been by far the largest buyer of Chinese oil during recent years.

Gas field development in South China Sea

ARCO China Inc. and Kuwait Foreign Petroleum Exploration Co. (Kufpec) signed an agreement with Cnooc to proceed with the first gas field development in the South China Sea.

The signing formalizes a draft agreement covering development of Yacheng 13-1 gas field 65 miles south of Hainan Island. Cnooc holds a 51% interest in the project, operator ARCO 34%, and Kufpec 15% through its Santa Fe Minerals (Asia) Inc. unit. Project cost is not disclosed.

Start-up of the field, which ARCO-Santa Fe discovered in 1983, is scheduled for 1993. Peak productive capacity is estimated at 500 MMcfd.

The field at first will provide via subsea pipeline 125-135 MMcfd of gas for new industrial development on Hainan Island, where China has established a special economic zone. More gas could move later through an extension of the pipeline north to Guangdong Province, Hong Kong, and other provinces on the mainland.

The partners also have begun a joint study with Japan's Nissho Iwai Corp. to determine feasibility of an LNG export project for some of the Yacheng gas.

Bohai Sea development plans

China was moving ahead with preliminary plans for a development program that could cost as much as $1.35 billion in giant Suizhong 36-1 heavy oil field in Liaodong Bay of Bohai Sea.

Bohai Oil Corp., a subsidiary of Cnooc, tentatively estimated 1.2 billion bbl of heavy oil in place and possible recovery of as much as two thirds of that. The Chinese believe the field is the largest found to date off their shores and count it among the world's 10 largest offshore oil deposits. Estimating reserves will be one of the first tasks leading to possible development, which could involve steam-

flooding.

Bohai Oil let a 6-9 month contract to Gustavson Associates Inc., Boulder, Colo., to conduct an in depth geological and reservoir engineering study of the field. The U.S. State Department Trade and Development Program (TDP) has approved a $645,000 grant to Gustavson for the contract. TDP expected to let a $530,000 follow-up contract of similar duration in 1989. Work under that contract will involve detailed project engineering design, including determination of the number of platforms and type of crude transportation facilities required, capital requirements, and markets for the crude.

Bohai Oil hopes to be able to produce about 20,000 b/d of oil and 50 MMcfd of gas by 1990, says Xinhua News Agency.

The company is working to open Bozhong 28-1 and Bozhong 34-2/4 fields in cooperation with Japanese companies and is independently working to open two other fields.

Discovered in 1986, Suizhong 36-1 field lies in nearly 100 ft of water 31 miles off Suizhong, Liaoning Province. Suizhong followed the discovery of Jinzhou 202-1 gas field to the northeast in 1984. The Suizhong discovery well, SZ36-1-2D, cut several oil bearing sandstones in the Paleogene below 4,774 ft in the central Liaoxi uplift. Eight appraisal wells have been drilled. At least one well drilled below the heavy oil found shows of light oil and associated gas in another formation.

A large normal fault with northwest dip seals the Suizhong reservoir, whose pay sands are of fluvial deltaic nature. Areal distribution is unpredictable and difficult to trace on seismic sections, especially where beds are thin. High resolution processing is required to reprocess several key seismic lines across the field. The most recent seismic work includes a 165 ft spaced, three dimensional survey shot across the entire field in 1987.

The development project will provide a good vehicle to funnel more U.S. equipment and technology into China, said the National Council for U.S.-China Trade, Washington, D.C. The Chinese probably will spend $289.56 million, about 21% of total expenses, on equipment and expenses from foreign countries. The U.S. could be expected to export as much as 65% of that, the council said.

Occidental Petroleum Corp. contacted the Chinese about the possibility of becoming involved in the Suizhong project, a source in Beijing said. Suizhong oil is 14-17° gravity, very viscous, with a gas-oil ratio of about 25:1. Reservoir sand is loose, and poorly defined oil-water contacts may cause production problems. Temperatures in the Liaodong Bay area range from 0 to 100° F., and half the bay freezes for about 30 days/winter, with ice 1.5-6 in. thick.

The U.S.-China council estimated that a maximum of 300 wells might be required to produce the field, but the Gustavson study will examine the number of wells needed. Based on a Cnooc estimate, approximate costs based on 80 wells would be $361.74 million. Breakout is drilling 80 production wells $102.4 million, 80 completions $25.6 million, production facilities $188.03 million, production operation preparation $3.5 million, training $500,000, additional development related costs of $4.5 million, administration $9.83 million, and contingencies $27.38 million.

Introduction of a relatively small amount of energy into the reservoir may be enough to produce the oil, said John Gustavson, president.

China has studied development of Jinzhou 202-1 gas field to help supply Northeast China's gas distribution grid, but the field's gas might be used to fuel steam generators for recovery of oil from Suizhong 36-1 field, Gustavson said. His company plans to study production technologies used in Central California heavy oil fields near Bakersfield.

Also in the Bohai Sea, Bohai & MHI Platform Engineering Co. Ltd., Tanggu, Tianjin, China, let contract to Mitsubishi Heavy Industries Ltd., Tokyo, for a 15,000 b/d crude oil separation and utility facility for use in BZ34-2/4E oil field. The unit will replace a conventional production and processing facility presently in use on the field's production platform. The new facility will be part of a system in which oil produced at two 10,000 b/d capacity wellhead platforms will be transferred through a 1.2 mile pipeline to a floating production and storage unit (FPSU) at single point mooring. In the facility, to be installed on the FPSU, oil will be processed for degassing and dehydration and then stored. FPSU storage capacity is 377,388 bbl of oil. Processed oil will be exported by tanker to Japan.

Mitsubishi planned to deliver the unit in April 1989 to Shanghai for installation on the deck of the FPSU, which is to start operating in September 1989.

And China plans to boost oil productive capacity in the Bohai Sea.

Bohai's earliest producing field, Chengbei, was flowing about 8,000 b/d of oil from 50 wells and has yielded 2.38 million bbl since start-up June 30, 1987.

BOC and Japan China Oil Development Corp. (Jcodc) signed a contract to develop two smaller fields northwest of Longkou.

BZ28-1 field, expected to produce 7,400 b/d of oil, was to go on stream in 1988. East BZ34-2/4, which will produce 8,260 b/d, is to go on production at the end of 1989.

Two structures discovered independently by BOC since 1986 in Liaodong Bay have merited particular attention, says Zhao Qinghua, writing for China Features. BOC, state owned, plans to start production in 1990 from the SZ36-1 structure in 98 ft of water 31 miles off Jinxi, where eight wells have been drilled. Unconsolidated sands with 230-436 ft of net pay at 4,921-5,905 ft hold reserves of 879 million bbl.

BOC was to drill several wells in 1988 in JZ20-2 gas/condensate field, which it plans to place on stream by 1991. Reserves are estimated at 706 bcf of gas, 22 million bbl of condensate, and 73 million bbl of crude. A 520,000 ton/year urea plant and a 300,000 ton/year ammonia plant will be built in Jinxi to use the gas.

In Liaodong Bay, independently explored by BOC with a $30 million World Bank loan and a $1.175 billion grant from the U.S. Trade Development Program, many foreign oil companies have expressed interest in financial participation, Zhao reported. More exploration is under way or planned. Laizhouwan Oil Development Co., a group made up of Japan Petroleum Resources and five other nongovernment companies, began exploring Laizhou Bay early in 1988. The

China's petroleum exports and trade balance*

	1982	1983	1984	1985	1986	1987	1988	1989	1990	Growth rate 1985-90 (%)
PETROLEUM EXPORT VOLUME (1,000 b/d)										
Crude	293.6	296.6	440.2	600.0	570.0	544.0	552.0	557.0	560.0	−1.4
Daqing	—	—	396.2	480.0	427.5	374.0	345.0	313.3	280.0	−10.8
Shengli	—	—	44.0	120.0	142.5	170.0	207.0	243.7	280.0	+16.9
Products	107.6	107.6	124.9	136.1	109.2	111.0	99.0	83.0	68.0	−13.9
Total	**401.2**	**404.2**	**565.1**	**736.1**	**679.2**	**655.0**	**651.0**	**640.0**	**628.0**	**−3.2**
AVERAGE PRICE ($/bbl)										
Crude	30.30	26.72	25.08	23.97	11.43	16.38	18.25	20.13	21.00	
Daqing	—	—	—	—	12.00	17.00	19.00	21.00	22.00	
Shengli	—	—	—	—	11.00	15.00	17.00	19.00	20.00	
Products	35.42	33.53	29.68	29.15	18.86	22.00	25.00	26.00	27.00	
PETROLEUM EXPORT VALUE (billion $)†										
Crude	3.25	2.89	4.03	5.25	2.38	3.25	3.68	4.09	4.29	−4.0
Products	1.39	1.32	1.35	1.45	0.75	0.89	0.90	0.79	0.67	−15.4
Total	**4.64**	**4.21**	**5.38**	**6.70**	**3.13**	**4.14**	**4.58**	**4.88**	**4.96**	**−6.0**
FOREIGN TRADE (billion $)†										
Nonoil exports§	17.30	17.95	19.65	20.66	27.80	30.58	33.64	37.00	40.70	+13.6
Total exports§	21.94	22.16	25.03	27.36	30.93	34.72	38.22	41.88	45.66	+10.2
Imports¶	18.94	21.32	26.75	42.26	42.90	44.19	45.51	46.88	48.28	+2.7
Net balance	+3.00	+0.84	−1.72	−14.90	−11.97	−9.47	−7.29	−5.00	−2.62	

*Base case. †Current dollars. §fob. ¶cif.

Source: China Energy Ventures Inc. from data supplied by General Administration for Customs of the International Monetary Fund, People's Republic of China State Statistical Bureau, and the National Council for U.S.-China Trade

group was expected to drill its first well early in 1989.

Amoco Orient Petroleum Co. was awarded a 2.19 million acre contract area south of Qinhuangdao in 1987.

Bohai history

BOC technicians used bamboo poles and sent divers to fathom the seabed when the company began operations in 1965. The first encouraging prospects for oil in the Bohai seabed came in 1979 following a geophysical survey of the continental shelf. The state decided to allow foreign participation to gain foreign funds and technology.

BOC signed a contract in May 1980 with Japan National Oil Corp. for joint oil and gas exploration and development in a 9,846 sq mile area of the Bohai Sea.

Before the contract ended in 1987, the venture found oil in 11 of the 16 structures it explored. Of the 35 wells completed by May 1987, 25 are high flowing gas and oil producers.

The contract stipulated that Jnodc would relinquish all contract areas to BOC by June 1987 except 985 sq miles, which it would retain until 1990 for further exploration and appraisal. Work was under way on equipment to develop BZ28-1 and East BZ34-2/4 fields under the September 1987 agreement. BOC has the construction contract for BZ28-1 field equipment, to consist of a $36 million floating production and storage unit.

Bohai-Mitsubishi Corp., another joint venture, was awarded the main construction contract for East BZ34-2/4 field facilities.

Another contract went to SBM, Monaco.

Ste. Nationale Elf Aquitaine drilled three dry holes between 1980 and 1983 in a 3,629 sq mile area and dropped the acreage. Since then it has provided $50 million in loans toward exploitation of West 428 field. Repayment will be in oil.

BOC's growth

BOC, in cooperation with foreign companies the past 9 years, has found 10 fields with a combined productive area of about 30 sq miles.

Reserves of more than 876 million bbl and 212 bcf of gas have been verified. The work is the result of $820 million in foreign investment and brought BOC an additional $620 million in revenue from subcontracting and provision of labor and technical services.

BOC controls 12 service companies that undertake geophysical surveys, offshore drilling, geological services, offshore engineering projects, ship repair, and communications and provide transportation and storage. The companies employ nearly 10,000 persons.

BOC also has established six joint companies with partners from Japan, France, U.K., and Norway. They include China-France Bohai Geoservices Co. Ltd., a joint venture with Geoservice Co. of France; China-GECO Geophysical Co., a joint venture with GECO of Norway; and China Bohai Racal Positioning & Survey Co., a joint venture with Racal Survey of the U.K.

With foreign exchange, BOC has purchased a pipelaying ship, several supply boats, and an oil tanker. It has brought in foreign technology which it is able to offer elsewhere. For example, in Xinjiang Province of Northwest China, the company has a contract to conduct cementing and mud logging for 16,400 ft wells. Overseas, BOC has provided 55 welders to help build offshore platforms in Abu Dhabi.

Pipeline activity

Pipeline construction is growing in China after a lull in the early 1980s.

Work began on a 267 mile crude oil pipeline in western

Qinghai Province to connect Huatugou field to the town of Golmud, says a report in China Features, Beijing.

Golmud is the starting point of a 621 mile products pipeline to Lhasa, capital of Tibet Autonomous Region.

China is building a refinery, set for completion in 1991, at Golmud to process Huatugou crude.

Pipelines rank fourth among China's major transportation means. By the end of 1987, China had built 57 cross country oil pipelines totaling about 7,450 miles in length. The pipelines transported more than 584 million bbl of oil in 1987, 62% of total shipments.

China will boost oil production to 4 million b/d by 2000. About 75% of the oil will come from Northeast and East China, where trunk lines connect six large oil fields. The rest of the production will come from Northwest China. Pipeline construction will surge in the 1990s and beyond in that area, where conditions are extremely harsh.

With China's overall economic situation greatly improved under new open policies, construction of several new trunk lines is expected to start, Zhao said. These include a 236 mile line in Inner Mongolia, a 174 mile coal gas pipeline to Harbin in Northeast China, a second line between Zhongyuan oil field and Luoyang, and a 168 mile line from Dongying to transport gas to Jinan.

Several major pipeline networks have been built in China through the years.

Dual 325 mile pipelines in Northeast China link Daqing oil field, the country's biggest, and Liaohe, its third biggest. The two fields produce more than 60% of China's crude oil.

The pipelines extend south from Daqing and fork at Liaohe field, with one extension to Luda, another to Qinhuangdao. Large terminals have been built in both port cities.

Qinhuangdao's terminal has two 630,000 bbl floating roof tanks. The tanks, 262 ft in diameter and 72 ft high, are the largest in China.

Crude reaching the ports is shipped to foreign buyers, Shanghai, and other China coastal areas for domestic consumption.

Part of the oil moves through another pipeline from Qinhuangdao to Beijing for processing in a large refinery.

Another large East China pipeline serves four large oil fields: Dagang near Tianjin, Huabei south of Beijing, Zhongyuan in central Henan Province, and Shengli at the mouth of the Yellow River.

These trunk lines include dual 329 mile pipelines that extend southeast from Shengli field to the port of Qingdao and a 546 mile pipeline south from Linyie to a terminal at Yizheng on the Yangtze River. The latter transports crude to petrochemical plants and other industries in Jiangsu Province, Shanghai, and other cities along the Yangtze.

A 779 mile network transports crude from Karamay field in Xinjiang Province.

Three other small pipeline networks serve Yumen oil field in Gansu Province, Jianghan field in Hubei Province, and other fields in Guangdong Province.

Construction resumes

The pace of pipelaying slowed in the early 1980s, when existing oil fields had been hooked up. Then a 176 mile line from Zhongyuan oil field to Luoyang was built in 1983-85.

In July 1986 a second line from Shengli oil field, China's second largest, to Qingdao began operating. The line allowed Shengli field to boost production to 600,000 b/d in 1987 from 528,000 b/d in 1986.

The pipeline will become China's most modern by mid-1989. China Petroleum Engineering Construction Corp.'s Pipeline Survey and Design Institute and Nova Corp., Calgary, plan to equip it with nine microwave stations and computers to control flow.

The line will be heated to transport high viscosity crude,

and 1.88 million bbl in storage was to be built at Qingdao terminal by yearend 1988.

Gas pipeline construction also has been proceeding.

China has more than 3,500 miles of gas lines to transport production of 484 bcf/year. China figures that about 11 bcf/year is transported over long distances. China will discover an additional 28.6 tcf of gas during 1986-90, more than double the addition made in the previous 30 years, according to a plan by the Ministry of Petroleum Industry.

The country will raise production to 530 bcf/year by 1990.

About 70% of China's gas fields are in Sichuan Province, which produced 571 MMcfd of gas in 1987, nearly half the country's total. The province has 1,780 miles of pipelines moving gas to industrial and home users.

Three gas pipelines with a total length of 385 miles have been built in eastern China to supply gas from Zhongyuan oil field to the cities of Kaifeng, Anyang, and Cangzhou.

Early technology

Trenches were hand dug and pipe was transported by horse carts and 5 ton lorries when China started building pipelines in the Northeast in the early 1970s.

A coat or two of asphalt was the only anticorrosion measure, brick tanks and prestressed concrete tanks with vaulted roofs were built in deep mountains, and explosives were used to blast trenches across rivers.

Large amounts of pipeline construction equipment have been imported since 1975. It includes tractors and trailers from Daimler-Benz AG of West Germany and Caterpillar Tractor Co. of the U.S. and 8-16 ton cranes from Kato Works in Japan.

A $4 million horizontal drilling rig imported from the U.S. in 1977 drilled holes for several pipelines.

X ray, ultrasonic magnetic particles, and fluorescent penetration have replaced ultrasonic detectors to check weld integrity. Epoxy resin, epoxy agent with asphalt, and "yellow jacket" plastic foam coating are used in corrosion prevention.

Downstream operations

China is pursuing joint ventures and other deals covering downstream operations in western nations.

A Chinese state owned petroleum company has agreed to acquire a 50% interest in Coastal Corp.'s U.S. West Coast refining/marketing business.

China and Guyana signed several agreements covering interest free loans for private sector downstream projects in Guyana.

Sinochem signed a memorandum of agreement with Coastal to acquire for an undisclosed amount of cash half interest in a joint venture to own and operate Coastal's West Coast downstream operations. Coastal operations involved are its Hercules, Calif., refinery, fuel oil terminals in Los Angeles and Coos Bay, Ore., and marketing operations in San Francisco, San Pedro, and Long Beach, Calif., and Portland, Ore. Coastal subsidiary Pacific Refining Co. operates the Hercules refinery northeast of San Francisco. Refinery capacity is 55,000 b/d throughput and 2.2 million bbl storage. Another Coastal unit, Western Fuel Oil Co., operates the fuel oil terminals, which have combined products storage capacity of 1.4 million bbl. In addition, Pacific Refining and Western Fuel Oil have extensive marketing and export operations serving the West Coast and Pacific Rim countries with a full range of transportation and residual fuels.

The memorandum of agreement calls for capital outlays to upgrade the Hercules refinery to boost production of higher octane gasolines and other transportation fuels. In addition, Sinochem and the joint venture company will negotiate a crude supply agreement.

The West Coast is perhaps the most competitive U.S. refining/marketing region. Imports of Chinese gasoline spurred controversy within the downstream industry in California during the early 1980s.

Relations between Coastal and Sinochem began in 1978, when Coastal became the first U.S. company to buy and import crude from China.

China and Guyana's agreements cover interest free loans of $5.5 million to finance private sector projects. One of the projects entails construction of a factory to produce 5 million high density polyethylene bags/year, mainly for farm products. Another agreement calls for a plant to be built by Demerara Distilleries Ltd. to recover and utilize methane from its distillation operations.

Petrochemicals expansion at home

China is expanding its petrochemical industry to become a major exporter of petrochemicals in the 1990s.

Keys to the effort are aggressive imports of western technology, broad reorganization of the industry, and sweeping reform of its management system, writes Liu Jianjun in the official weekly Beijing Review.

China's petrochemical industry blossomed in the 1970s with large scale expansion of its oil production capacity. The industry consists of more than 40 large petrochemical complexes and many smaller enterprises employing more than 1 million persons. Total sales are more than 40 billion yuan/year ($10.748 billion U.S.).

The Chinese petrochemical industry's oil processing capacity is 110 million metric tons/year, with a yield of more than 2,000 petrochemical products, almost all for domestic use.

In 1987, petrochemical production included 1.63 million metric tons of ethylene, more than 1.1 million metric tons of synthetic fibers, 140,000 metric tons of synthetic rubber, more than 800,000 metric tons of plastics, and 17.03 million metric tons of chemical fertilizers.

In the early 1980s, China introduced centralized management and unified planning for its plants, setting up several integrated petrochemical complexes.

In 1983, it formed the China National Petrochemical Industrial Corp. (Cnpic), an economic entity with ministerial status. Cnpic oversees 40 large and medium sized refining, petrochemical, chemical fiber, and chemical fertilizer businesses linked to about a dozen ministries or commissions, as well as related research institutes and institutions of higher learning.

Cnpic is responsible for drafting general programs for reform and development that are then worked into specific annual quotas under contracts with branch companies, factories, and individuals. At the same time, it gives the enterprises full autonomy and responsibility for profits.

During 1983-87, the value of Cnpic's total industrial production jumped by an average 8.8%/year. Production of major products increased by 30-80%, with some doubling or tripling. In the same period, the corporation completed more than 10 grassroots petrochemical complexes at a total investment of almost 30 billion yuan, or 10% more than the combined total investment for the previous 30 years.

A special focus for growth has been ethylene development. China's seven major ethylene plants in 1987 produced more than twice the volume of 1986.

Although Cnpic has been importing some advanced equipment, such as catalytic cracking units for the Liaoyang and Beijing Yanshan plants, its ethylene industry sorely needs technological upgrading and debottlenecking. Current ethylene production falls short of domestic demand. Supplies of polyethylene and polypropylene are tight.

Through capacity additions and imports of ethylene plants, China plans to boost ethylene capacity to more than 2 million metric tons/year.

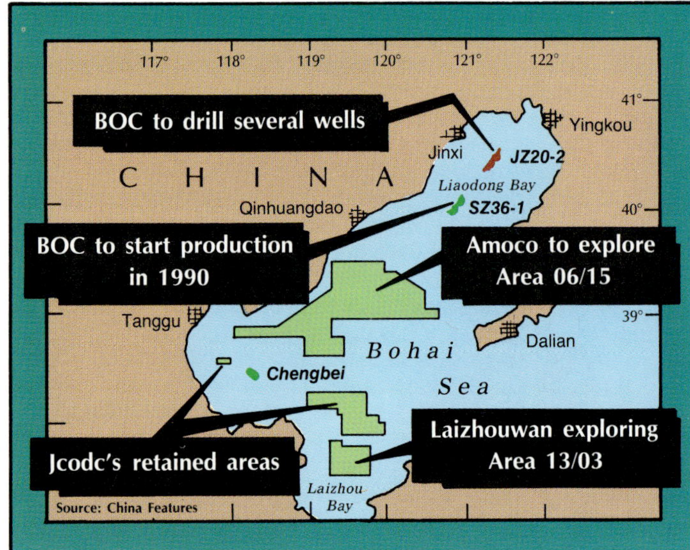

China's production of chemical fibers also has grown by an average 20-30%/year to rank fourth in the world after the U.S., Japan, and U.S.S.R. in 1987, but it still is low on a per capita basis.

Chemical fiber supply/demand is roughly in balance, but demand growth will require a doubling of production to more than 2 million metric tons/year.

China's petrochemical industry in 1987 imported $660 million in technology, equipment, and capital. Advanced technology and key equipment accounted for $300 million of that total, machinery, accessories, and spare parts $110 million, and chemical feedstocks, steel products, and cement about $100 million.

China's imports of technology and equipment since the 1970s cover more than 240 contracts totaling more than $6 billion. Ancillary equipment imports total more than $1 billion.

Cnpic gave priority to importing equipment and funds for key projects such as the few 300,000 metric ton/year capacity ethylene plants. Foreign loans for ethylene plants that size came from the Export-Import Bank of Japan at Daqing and C. Itoh & Co. Bank at Qilu in Shandong Province.

Since 1984, China's petrochemical industry has signed loan agreements or contracts covering $1 billion in foreign funds with banks in 30 countries.

Aside from grassroots construction projects, Cnpic imported plants to produce derivatives, including a 60,000 metric ton/year low density linear polyethylene plant at Qilu, a 70,000 metric ton/year polypropylene unit from Union Carbide at Qilu, and a 70,000 metric ton/year polypropylene plant at Shanghai from Italy.

In 1987, Cnpic's petrochemical exports were valued at $900 million.

During 1983-87, China exported $6.1 billion worth of petrochemicals and refined products to more than 60 countries.

Cnpic's push into the world market is progressing with exports of its technology, labor and training services, and bids for 12 international projects.

The corporation has signed three contracts to date for foreign projects, including an oil shale plant in Jordan and a joint venture with a U.S. petrochemical company to produce furnaces for a Chinese business.

Reform measures enable Cnpic to export 16 petrochemicals directly through a multichannel network instead of through foreign trade departments.

Selected areas and fields

The China Petrochemical International Undertaking Co. under Cnpic plans to set up trade offices in the U.S., Japan, and Hong Kong.

In recent years, China's petrochemical industry has signed agreements for 18 joint ventures with companies in the U.S., Japan, Canada, Singapore, and Hong Kong.

Technology exchanges and related cooperative agreements have been completed involving more than 20 projects with 15 companies in the U.S., Canada, France, Italy, and Japan.

Such efforts are intended to upgrade the level of technological expertise in China's petrochemical industry. During 1983-87, Cnpic invested several billion yuan in completing 413 technology upgrading projects. Those efforts attracted $300 million in foreign investment. Of 138 projects planned to expand productive capacity or trim energy consumption, 60 have been completed.

A notable example was the Hubei chemical fertilizer plant, which had a cooperative agreement, including purchase of technology and equipment, with a Danish company. Because of the cooperation, the Hubei plant was able to boost its capacity of ammonia to 10 million metric tons/year from 8.5 million metric tons/year while cutting its energy use by 10%, and the Danish company did $1 million worth of business.

China's efforts to upgrade its petrochemical industry have spurred cooperative ventures with more than a dozen countries, including the U.S., Japan, Netherlands, Denmark, Switzerland, and West Germany.

Shanghai project

A unit of Occidental Petroleum Corp. will license technology to a Chinese state owned company for construction of a major polyvinyl chloride paste resin plant in Shanghai. It will be the first project by a U.S. chemical company in Shanghai and the biggest PVC paste resin plant in China.

Plans call for the 20,000 metric ton/year plant to start up at yearend 1989 in the WuJing chlor-alkali complex.

Occidental Chemical Corp. will provide its paste resin technology to Shanghai Sitco International Trading Co.

OxyChem also will buy certain equipment for the plant, provide onsite technical supervision for construction and start-up, and in the U.S. supervise the technical training of plant operators and supervisors. Bechtel Inc. is engineering contractor on the project.

Coal dominating energy sector

China's biggest business problem is to keep growth from outstripping economically available goods and services.

The problem applies very much to China's energy sector, which will continue to be dominated by coal, the country's most plentiful fossil fuel, says John H. Lichtblau, Petroleum Industry Research Foundation Inc., New York.

China produced 660 million tons of coal equivalent and consumed 635 million tons in 1987, placing it among the world's three largest coal nations with the U.S. and U.S.S.R.

Coal accounted for at least 72% of China's total energy production and 75% of primary energy consumption last year. Its dominance is not expected to decline the rest of the century.

Oil supply/demand also will grow, but demand is expected to outpace production, said Lichtblau, who attended a Chinese-American energy symposium in Nanjing sponsored by the U.S. Department of Energy and China's State Planning Commission. Lichtblau held other discussions in Beijing with government officials and representatives of foreign oil companies.

The most likely way to balance domestic oil supply/demand will be a combination of shortages, higher domestic prices, and lower exports, Lichtblau said.

China's economic growth in the first 2 years of its current 5 year plan (1986-90) has substantially exceeded the targeted rate of 6.7%/year.

China forecasts primary energy production growth of 3.5%/year through 2000 and coal production growth of 4-4.5%/year.

Some of the increase may go into export, which in 1987 reached a record 11.5 million tons.

Coal's use creates air pollution, and so far little has been done to contain it.

Most raw coal is burned without processing or cleaning, and electrical power stations do not have scrubbers.

Power demand is rising rapidly. China's first two nuclear plants are due to start up by the early 1990s, but nuclear power is projected to account for only 2.5-3% of electricity generation by 2000.

Despite 6-7%/year growth in power generation, supply of about 500 billion kw-hr is not enough to meet demand. Power shortages idled nearly 20% of Chinese industrial capacity in 1987.

"Thus, if China's economy is to grow at anywhere near the foreseen rates, electric power generation will have to keep pace and with it the growth in coal burning for this purpose," Lichtblau said.

Hydropower is expected to continue to provide 20% of China's electricity through 2000.

Natural gas production may rise to 1.5-1.8 tcf/year from the present 500 bcf/year, displacing only about 50 million tons/year of coal out of perhaps 1 billion tons/year burned by then.

Because coal is a major energy source for every sector of the Chinese economy—even providing 80% of railroad locomotive fuel—increased air pollution may be part of the price the country has to pay for the remarkable speed of industrialization and rise in living standards, Lichtblau said. ▣

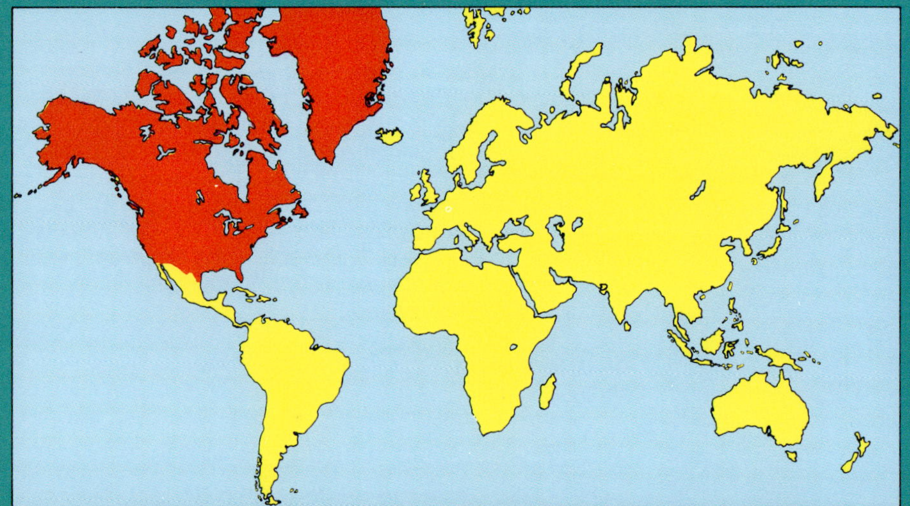

NORTH AMERICA

UNITED STATES

CAPITAL: Washington, D.C.
MONETARY UNIT: Dollar
REFINING CAPACITY: 15,556,700 b/cd
PRODUCTION: 8.206 million b/d
RESERVES: 26.5 billion bbl

THE U.S. PETROLEUM INDUSTRY BEGAN 1989 WITH oil production lagging the 1987 volume and drilling activity down from that year.

Crude oil prices had rebounded from depressed levels earlier in 1988, but spot gas prices showed little change from early 1988. Although the Baker Hughes Inc. average weekly count of active rotary rigs at 936 for the year was off only two from the 1987 average, the tally of 940 for the last week of 1988 was off 205 from yearend 1987.

The best news for the industry lay downstream, where profits from sales of petroleum products and petrochemicals were at record highs for many companies. However, refiners worried that gasoline, their big ticket item, was fair game for increased federal taxes when the incoming President, Republican George Bush, cast about for ways to reduce a huge budget deficit inherited from President Ronald Reagan.

Another nagging worry for industry executives was a rising volume of oil imports, signaling too much dependence on supplies that could be cut off quickly by a war or embargo.

To buoy U.S. activity and slow the tide of imports, industry urged things like tax incentives for exploration/development, along with increased access to federal acreage offshore and in northern Alaska. There was no assurance it would get either one from Bush, whom many oilmen had supported in his 1988 campaign for the White House.

Downstream profits buoy OGJ group

Downstream operations lifted profits for most of the 23 U.S. oil companies whose performance was tracked quarterly by Oil & Gas Journal.

During the first 3 quarters of 1988, the group's net income jumped 74.4% from a year earlier to $16.647 billion. Group third quarter net profits climbed 34.1% from the same quarter of 1987 to $4.552 billion. Profits totaled $6.039 billion in the second quarter and $6.027 billion in the first.

Companies in the group ranged in size from Quaker State Corp. with assets of $612.6 million at the end of 1987 to Exxon Corp. with assets of $74.042 billion. Most were integrated operations.

Weak crude oil prices for the first 9 months of 1988 helped cut feedstock costs and boosted refining and petrochemical earnings. Companies scored downstream earnings gains in the U.S. and abroad. Refining margins widened as product prices held steady on the strength of rising demand, which also lifted volumes.

Energy Information Administration (EIA) data showed that U.S. demand for petroleum products for the first 3 quarters of 1988 averaged 16.884 million b/d, up 303,000 b/d from the same period of 1987. The International Energy Agency (IEA) estimated non-Communist product demand at 49.23 million b/d, up 2% from the first 3 quarters of 1987.

U.S. refiners' acquisition cost for U.S. and imported crude oil fell to an average $13.94/bbl from $15.92 during the period, and average domestic field prices slipped to $11.61/bbl from $13.64.

Meanwhile, the U.S. average pump price for all types of motor gasoline climbed to 96.24¢/gal from 94.83¢/gal a year earlier. The average price of residential heating oil moved up to 80.27¢/gal from 78.37¢/gal in 1987.

Postings for the benchmark West Texas intermediate crude oil provided an example of price volatility during 1988.

The year opened with postings at $16.50/bbl, which rose to a high for the year at $17.50 late in April, fell to $12.75 for part of October-November and stood at $15.50 at yearend. One large buyer made 20 changes in postings for West Texas intermediate during the year.

An increase in the average refinery utilization rate to 84.8% from 83.2% helped reduce operating costs and further improved margins.

Here is a sample of company refining and marketing results for the first 9 months of the year:

• Exxon reported profits of $414 million in the U.S., compared with a $50 million loss a year earlier, and $835 million outside the U.S., up from $185 million in 1987.

• Mobil Oil Corp. had income of $333 million in the U.S., up $263 million, and $368 million outside the U.S., up $197 million.

• Chevron Corp. earned $411 million in the U.S. vs. $2 million a year earlier and $154 million outside the U.S. vs. $26 million in 1987.

• Texaco Inc. earnings totaled $402 million in the U.S., compared with a loss of $63 million in 1987. Outside the U.S. earnings were $291 million, up from $118 million.

• Shell Oil Co.'s oil products segment earned $480 million, up from $272 million.

• Amoco Corp.'s earnings from worldwide refining, marketing, and transportation climbed to $631 million from $147 million in 1987.

Lower crude costs also boosted earnings from chemical operations.

These are examples of 9 month chemical profits: Phillips Petroleum Co. $690 million vs. $333 million for the same period of 1987; Mobil $482 million, more than twice the year-earlier net; ARCO $670 million vs. $280 million; Amoco $539 million, up 77%; Shell $480 million vs. $272 million; Texaco $135 million vs. $50 million; Chevron $283 million, up 56.4%; and Exxon $966 million, up $403 million.

Almost all OGJ group companies had lower exploration/production earnings. An exception was Phillips, which posted a gain to $443 million from $234 million a year earlier.

Thirteen of the companies in the group reported capital and exploration expenditures of $19.824 billion for the first 9 months of 1988, up 29.7% from the same 1987 period.

U.S. independent operators, with assets concentrated in oil and gas production, turned in much weaker profit performance during 1988.

In the first half, combined earnings of 138 independents amounted to $2.059 billion, up only 8.7% from first half 1987. By contrast, virtually the same group of producers reported combined profits up 301% for first half 1987.

Industry spending
seen as inadequate

The U.S. petroleum industry should have no difficulty funding from internal cash flow a 4 year domestic spending program estimated at $183 billion by Salomon Bros. (Fig. 1).

But it's questionable whether that will be enough to ensure acceptable rates of return or significantly improve U.S. energy security, the New York investment firm said.

Capital spending by oil and gas companies in the U.S. will rise by about 4.8%/year in 1989-90, Salomon Bros. told a December 1988 Arthur Andersen & Co. symposium in Houston. And the rate of U.S. petroleum capital spending growth will more than double to 10.9%/year in 1991.

Although U.S. upstream spending will grow by about 17.9% in 1988-91, it will be outpaced during the period by capital spending for refining, marketing, and pipelines at 22% and for chemicals and other businesses at 40%.

Further, U.S. finding costs won't remain at the low level of 1988 but aren't likely to reach the high levels of the early 1980s. And operators on average probably will replace only about half of their U.S. production in 1989. That will send more exploration dollars abroad.

Salomon Bros. estimated the 8 year average petroleum finding cost in the U.S. since 1980 at $9.80/bbl of oil equivalent (BOE). The marginal cost of developing a barrel of oil in the U.S. from lease purchase through well casing ranged in that

period from a high of $13/BOE to slightly more than $6/BOE.

"We observe that finding costs are elastic and depend to a great extent on oil and gas prices," Salomon Bros. said. "The more cash the industry has to reinvest, the higher finding costs tend to run, and vice versa.

"Thus we do not believe that U.S. finding costs are likely to jump back to the $12-13/BOE of the early 1980s because the industry's current level of diminished cash flow cannot support that much waste."

Revisions, or adding reserves via accounting adjustments, contributed 2.5 bbl of oil reserves for each barrel added through drilling in 1987.

Salomon Bros. estimated the structural, or embedded, finding cost in the U.S. at 1988 drilling activity levels at about $8/BOE.

The company estimated that operators will spend $28 billion to acquire leases and develop oil and gas resources in the U.S. in 1989. At $8/BOE, that would yield reserves of

U.S. petroleum industry spending

	1988	1989	1990	1991	4 Years
			Billion $		
Upstream related	28	28	30	33	119
Downstream related	9	10	10	11	40
Chemicals and other	5	6	6	7	24
Total	**42**	**44**	**46**	**51**	**183**

Source: Salomon Bros. Inc.

ARTIST'S CONCEPT of the Placid Oil Co. group's Green Canyon 29 system in the Gulf of Mexico shows floating drilling/production platform connected to subsea drilling/production template via rigid riser on Green Canyon Block 29, subsea satellite wells on Ewing Bank Block 999 and Green Canyon Block 31, shallow water processing platform on Ship Shoal Block 207, and flow line and export pipeline bundles. Drawing courtesy of Enserch Exploration Partners Ltd. (Fig. 2).

UNITED STATES

State	No. Refineries
Alabama	2
Alaska	6
Arizona	1
Arkansas	3
California	28
Colorado	2
Delaware	1
Georgia	2
Hawaii	2
Illinois	6
Indiana	4
Kansas	7
Kentucky	2
Louisiana	17
Michigan	4
Minnesota	2
Mississippi	5
Montana	4
Nevada	1
New Jersey	6
New Mexico	3
North Dakota	1
Ohio	4
Oklahoma	6
Oregon	1
Pennsylvania	8
Tennessee	1
Texas	31
Utah	6
Virginia	1
Washington	7
West Virginia	1
Wisconsin	1
Wyoming	6
TOTAL	182

CANADA

Province	No. Refineries
Alberta	5
British Columbia	6
New Brunswick	1
Northwest Territories	1
Nova Scotia	2
Ontario	6
Quebec	3
Saskatchewan	2
Newfoundland	1
TOTAL	26

For individual locations and capacities, see detailed maps on following pages.

ALASKA
YUKON
NORTHWEST TERRITORIES
VICTORIA ISLAND
Amundsen Gulf
DISTRICT OF FRANKLIN
Fairbanks
Willow
Kenai
Anchorage
Valdez
Drift River
Gravina Point
KODIAK
Fort Good Hope
Norman Wells
Great Bear Lake
Arctic Circle
DISTRICT OF MACKENZIE
DISTRICT OF KEEWATIN
Inuvik
Kinsin Point
Whitehorse
Haines
Skagway
Juneau
Yellowknife
Great Slave Lake
Dubawnt Lake
TERRITO
Hudson Ba
Pacific Ocean
VANCOUVER
BRITISH COLUMBIA
Prince Rupert
Kitimat
Prince George
ALBERTA
Edmonton
Calgary
Medicine Hat
SASKATCHEWAN
Saskatoon
Regina
Pembina
Trans Canada
Lake Athabasca
Reindeer Lake
CANADA
MANITOBA
The Pas
Lake Winnipeg
Winnipeg
ONT
Lake Nipigon
Victoria
Vancouver
Olympia
Seattle
Tacoma
Spokane
WASHINGTON
Portland
Salem
Eugene
OREGON
Boise
IDAHO
Helena
Billings
MONTANA
Clearbrook
Duluth
NORTH DAKOTA
Bismarck
MINNESOTA
Minneapolis
St. Paul
WISC
SOUTH DAKOTA
Pierre
WYOMING
Casper
Cheyenne
NEBRASKA
Omaha
Lincoln
IOWA
Des Moines
Reno
Carson City
NEVADA
Great Salt Lake
Salt Lake City
UTAH
Denver
COLORADO
U
S
Kansas City
Topeka
KANSAS
Wichita
MISSO
Sacramento
Oakland
San Francisco
CALIFORNIA
Fresno
Coalinga
Las Vegas
Estero Bay
Bakersfield
Santa Barbara
Ventura
Los Angeles
San Diego
Tijuana
Rosarito
Mexicali
Four Corners
El Paso
ARIZONA
Phoenix
Tucson
Nogales
Nogales
Santa Fe
Albuquerque
NEW MEXICO
Lubbock
Amarillo
Panhandle
OKLAHOMA
Little Rock
ARKANSAS
LOUISIA
TEXAS
Austin
San Antonio
Laredo
Nuevo Laredo
Corpus Christi
Port Isabel
Hermosillo
Guaymas
Cd. Obregon
Navojoa
Bombas Cantina
MEXICO
Ciudad Juarez
Baton R
KAUAI
NIIHAU
OAHU
Honolulu
Barber's Point
Ewa Beach
MOLOKAI
Wailuku
LANAI
MAUI
KAHOOLAWE
HAWAII
Hilo

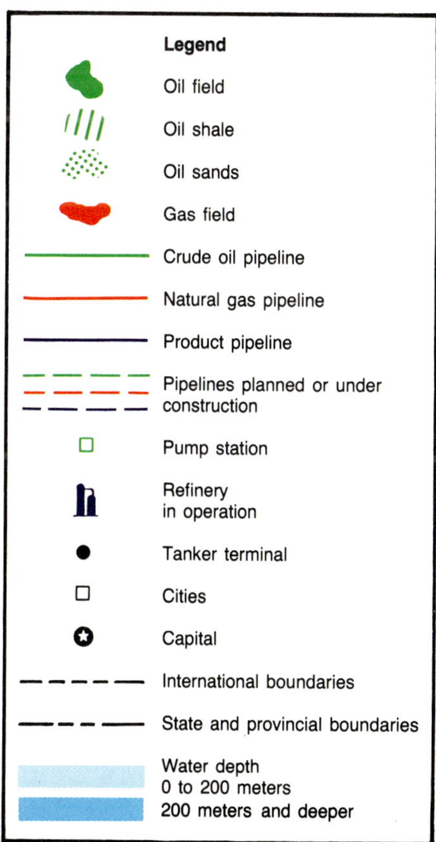

3.5 billion BOE, only 55% of the volume U.S. oil and gas fields produced in 1988.

Boosting the U.S. replacement rate to 80% would force finding costs up to $10/BOE, and increasing the replacement rate to 100% would hike finding costs to $14/BOE or more.

Thus to replace 80% of U.S. production, the industry would have to spend $50 billion/year, and to replace 100% of production it would have to spend $88 billion/year—more than three times the level Salomon Bros. predicted for 1989.

Salomon Bros. said a larger than normal portion of downstream spending in 1989 will be earmarked for "soft assets," including convenience stores, oil change facilities, and image building items such as service station lighting, signs, and pumps.

As gasoline demand continues to grow, however, more outlays will be needed for new conversion equipment such as cokers, crackers, and reformers. No grassroots refineries are likely to be built or needed in the U.S. the next 5 years, the comapny said.

To keep pace with product demand, industry will have to spend $2-3 billion/year on new or upgraded conversion capacity, Salomon Bros. estimated. Some of that money will come from redirecting soft asset expenditures, some from upstream budgets, and some from nontraditional suppliers of capital such as crude exporting countries' joint ventures in U.S. refining assets.

"Current high profit margins in refining virtually guarantee the availability of capital for this sector," Salomon Bros. said.

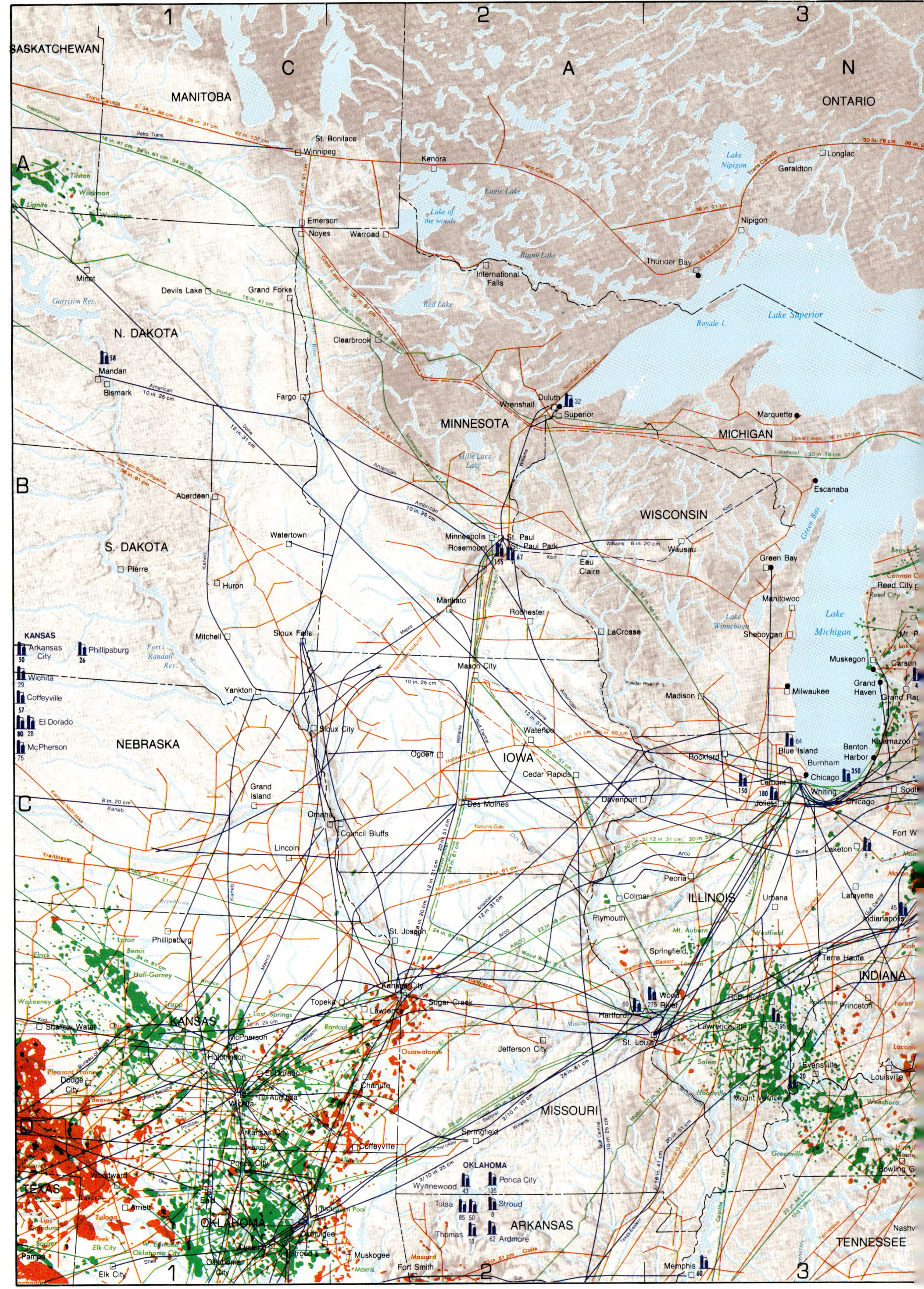

SASKATCHEWAN
MANITOBA
ONTARIO
C A N A
N
St. Boniface
Winnipeg
Kenora
Lake Nipigon
Longlac
Geraldton
Tilston
Workman
Lignite
Westhope
Emerson
Noyes
Warroad
Eagle Lake
Lake of the woods
Rainy Lake
Nipigon
Thunder Bay
Minot
Devils Lake
Grand Forks
Red Lake
International Falls
Lake Superior
Royale I.
Garrison Res.
N. DAKOTA
Clearbrook
Marquette
Mandan
Bismark
Fargo
Wrenshall
Duluth
Superior
MICHIGAN
MINNESOTA
Escanaba
Aberdeen
Mille Lacs Lake
WISCONSIN
Green Bay
S. DAKOTA
Watertown
Minneapolis
Rosemount
St. Paul
St. Paul Park
Eau Claire
Wausau
Manitowoc
Pierre
Huron
Mankato
Rochester
Sheboygan
Lake Michigan
Reed City
KANSAS
Mitchell
Sioux Falls
LaCrosse
Madison
Milwaukee
Muskegon
Grand Haven
Carson
Arkansas City
Phillipsburg
Mason City
Lake Winnebago
Wichita
Fort Randall Res.
Yankton
NEBRASKA
Sioux City
Waterloo
Rockford
Blue Island
Benton Harbor
Kalamazoo
Coffeyville
El Dorado
McPherson
Ogden
IOWA
Cedar Rapids
Burnham
Chicago
Whiting
Grand Island
Des Moines
Davenport
Lemont
Joliet
Omaha
Council Bluffs
Fort W
Lincoln
Peoria
Laketon
St. Joseph
Colmar
ILLINOIS
Urbana
Lafayette
Indianapolis
Plymouth
Phillipsburg
Mt. Auburn
Westfield
Terre Haute
INDIANA
St. Joseph
Springfield
Kansas City
KANSAS
Topeka
Lawrence
Sugar Creek
Wood River
Hartford
St. Louis
Lawrenceville
Princeton
Stratton Water
McPherson
Evansville
Louisville
Hutchinson
El Dorado
Jefferson City
Salem
Mount Vernon
Dodge City
Augusta
Chanute
MISSOURI
Wichita
Arkansas City
Springfield
Greenville
Ponca City
Coffeyville
Ponca City
OKLAHOMA
Wynnewood
Tulsa
Stroud
ARKANSAS
Ardmore
Thomas
Muskogee
TEXAS
OKLAHOMA
Oklahoma City
Fort Smith
Memphis
TENNESSEE
Elk City
Pampa

NORTHEASTERN U.S.

QUEBEC
Gouin Res.
Lac Au Goeland
Abitibu Lake
Smooth Rock Falls
Hearst
Timmins
Noranda
Rouyn
Sudbury
North Bay
Lake Nipissing
Parry Sound
Chalk River
Ottawa
Massena
Trans Canada
Lake St. John
Saguenay River
Baie-St. Paul
Quebec
St. Romuald
100
Shawinigan Falls
Montreal
120 87
Montpelier
Champlain Lake
VERMONT
NEW HAMPSHIRE
Concord
MAINE
Moosehead Lake
Bangor
Blue Hill
Augusta
St. John
Weldon
238
NEW BRUNSWICK
Chaleur Bay
St. Lawrence River
Fundy Bay
Portland
Portsmouth
Newington
MASSACHUSETTS
Boston
Massachusetts Bay
Cape Cod
Worcester
Springfield
Hartford
Providence
CONNECTICUT
RHODE ISLAND
Rhode Island Sound
Nantucket I.
New Haven
Bridgeport
Long Island Sound
Albany
Utica
Syracuse
Rochester
Oswego
Pulaski
Oneida Lake
NEW YORK
Binghamton
Elmira
Scranton
Linden
Perth Amboy
New York
55
Port-Reading
MICHIGAN
Lake Huron
Saginaw Bay
West Branch
Owen Sound
Georgian Bay
Manitoulin I.
Bay City
Saginaw
123 70 71
Flint
Corunna
London
Sarnia
Detroit
De Clute
Trenton
Tilbury
127 125
Toledo
Lake Erie
Cleveland
Youngstown
Akron
OHIO
Springfield
Columbus
Lakeview
St. Mary's
Zanesville
Heath
Nanticoke
Toronto
Port Credit
Oakville
80
Clarkson
Hamilton
Brantford
Buffalo
Tonawanda
95
60
Haldimand
Norfolk
Dunkirk
Bayham
Erie
60
Warren
Bradford
Farmers Valley
Smethport
Housville
Emlenton
Freedom
Bloomfield
PENNSYLVANIA
Harrisburg
Philadelphia
NEW JERSEY
Thorofare
Westville
125
Paulsboro
145
Wilmington
140
Delaware City
Dover
DELAWARE
Delaware Bay
Newark
Steubenville
Parkersburg
Charleston
Catlettsburg
Newell
213
Falling Rock
WEST VIRGINIA
Pineville
Betsy Layne
Hazard
Welch
Nora
KENTUCKY
Knoxville
Lexington
Frankfort
Cincinnati
Calves
Altoona
Columbus
MARYLAND
Baltimore
Leesburg
Washington
VIRGINIA
Cove Point
Chesapeake Bay
Richmond
51
Yorktown
Hampton Roads
Goodwin Neck
Norfolk
Portsmouth
Roanoke
Greensboro
Winston-Salem
Raleigh
Tarboro
NORTH CAROLINA
Charlotte
Pamlico Sound
Cape Hatteras
Atlantic Ocean

0 50 100 200
0 100 200 300 Km.

4 5 6
A D A
B
C
D

Breakout of capacities — Table 1

ETHYLENE	1986 – Million lb/year –	1988	POLYETHYLENE	1986 – Million lb/year –	1988
Amoco	2,100	2,280	Allied	900	1,150
Aristech	500	—	Amoco	350	—
BASF/UTP/Borg-Warner	750	780	Cain*	940	950
Cain*	—	2,700	Chevron	1,480	1,495
Chevron	2,215	2,265	Dow	2,455	2,520
CCPC†	1,400	—	DuPont	740	740
Dow	3,700	4,000	Eastman	620	620
DuPont	1,925	1,000	El Paso	405	505
Eastman	1,300	1,400	Exxon	1,190	1,315
El Paso	520	520	Hoechst Celanese	280	310
Exxon	3,012	3,200	Mobil	950	1,000
BF Goodrich	350	350	Phillips	1,380	1,430
Lyondell	2,800	2,800	Quantum	3,245	3,975
Mobil	1,050	1,563	Soltex	1,100	1,200
Occidental	275	550	UCC	2,030	2,130
Olin	64	20	Westlake	60	460
Phillips	2,154	2,300			
Quantum	2,070	2,175	**Total**	**18,125**	**19,800**
Shell	4,040	4,040			
Sun	225	225			
Texaco	1,050	1,185			
UCC	2,810	2,810			
Vista	650	670			
Unannounced and estimated	—	100			
Total	**34,960**	**36,933**			

*Acquired May 2 by Occidental. †Foreunner of Cain.

Source: Bonner & Moore Market Consultants

Partial breakout of U.S. imports* — Table 2

Leading suppliers	Volume (1,000 b/d)	% of total imports	% of products suplied
Saudi Arabia	1,022	14.6	6.0
Canada	979	14.0	5.8
Venezuela	797	11.4	4.7
Mexico	736	10.5	4.3
Nigeria	554	7.9	3.3
Iraq	324	4.6	1.9
U.K.	304	4.3	1.8
Algeria	285	4.1	1.7
Virgin Islands†	222	3.2	1.3
Angola	200	2.9	1.2
OPEC members	**3,358**	**47.9**	**19.8**
Persian Gulf countries	**1,491**	**21.3**	**8.8**

*Crude oil and products, January-September 1988. †Products only.

Source: Department of Energy

Who got what in Tenneco sale — Table 3

Buyer	Asset	Price (Million $)
Amoco	Rocky Mountain division	900
ARCO	Pacific Coast division	670-700
British Gas	Tenneco Oil International*	194.5
Chevron	Offshore division	2,600
Conoco	Tenneco Oil Norway	115
Fina	Gulf Coast, Southwest divisions	600
J. Philip Handy	Retail marketing division	NR
Mesa	Midcontinent division	715
Mobil	Processing, wholesale marketing	560
Shell	Tenneco Oil Colombia	500
Seagull	Houston Oil & Minerals	56.2
Star Gas	Blue Flame, Greene Propane	NR

*Did not include Norwegian and Colombia exploration/production assets, which were acquired by other companies.

Point Barrow
Barrow
Smith Bay
Simpson
Arctic Plains
Harrison Bay
Beaufort Sea
Tuktoyaktuk Pen.
Koakoek
Kopanoar
Issungnak
Amauligak
Atkinson
Itiyok
Prudhoe Bay field unit
Prudhoe Bay
Endicott
Kuparuk
Point Thomson
Flaxman Is.
Pitsiulak
Ivik
Tarsut
Mayogiok
Tuk
Richard I.
Adgo
Malik
Taglu
Ya Ya
Niglintgak
Kumak
Titalik
Kugpik
Reindeer
Parsons Lake
Umiat
Gubik
E. Umiat
Kavik
Arctic Wildlife Reserve
Mackenzie Bay
Camden Bay
Inuvik
Naval Petroleum Reserve No. 4
Brooks Range
Fort Good Hope
Norman Wells
Arctic Circle
Fort Yukon
Circle
YUKON
NORTHWEST TERRITORIES
Tanana
Fairbanks
North Pole
Delta Jch
Dawson
ALASKA
Alaska Range
Willow
Palmer
Anchorage
Valdez
Gravina Point
Cordova
Chugach Mountains
C A N A D A
Whitehorse
Watson Lake
Haines Jct.
Beluga River
Ivan River
Granite Point
North Cook Inlet
Trading Bay
Birch Hills
McArthur River
Swanson River
Drift River
MGS
Nikiski
Kenai
Kenai
Kenai Pen.
Cook Inlet
Montague I.
Yakataga
Yakutat
Skagway
Haines
Juneau
Alaska Peninsula
Kodiak
Kodiak I.
Sitka
Wrangell
Ketchikan
G u l f
o f
A l a s k a
ALASKA

Localized service through an international network of convenient locations

You're on the go throughout the world. And we're staying right up with you. That's why wherever you operate you'll find trained Halliburton men nearby with service, tools and materials selected and proven for each locale.

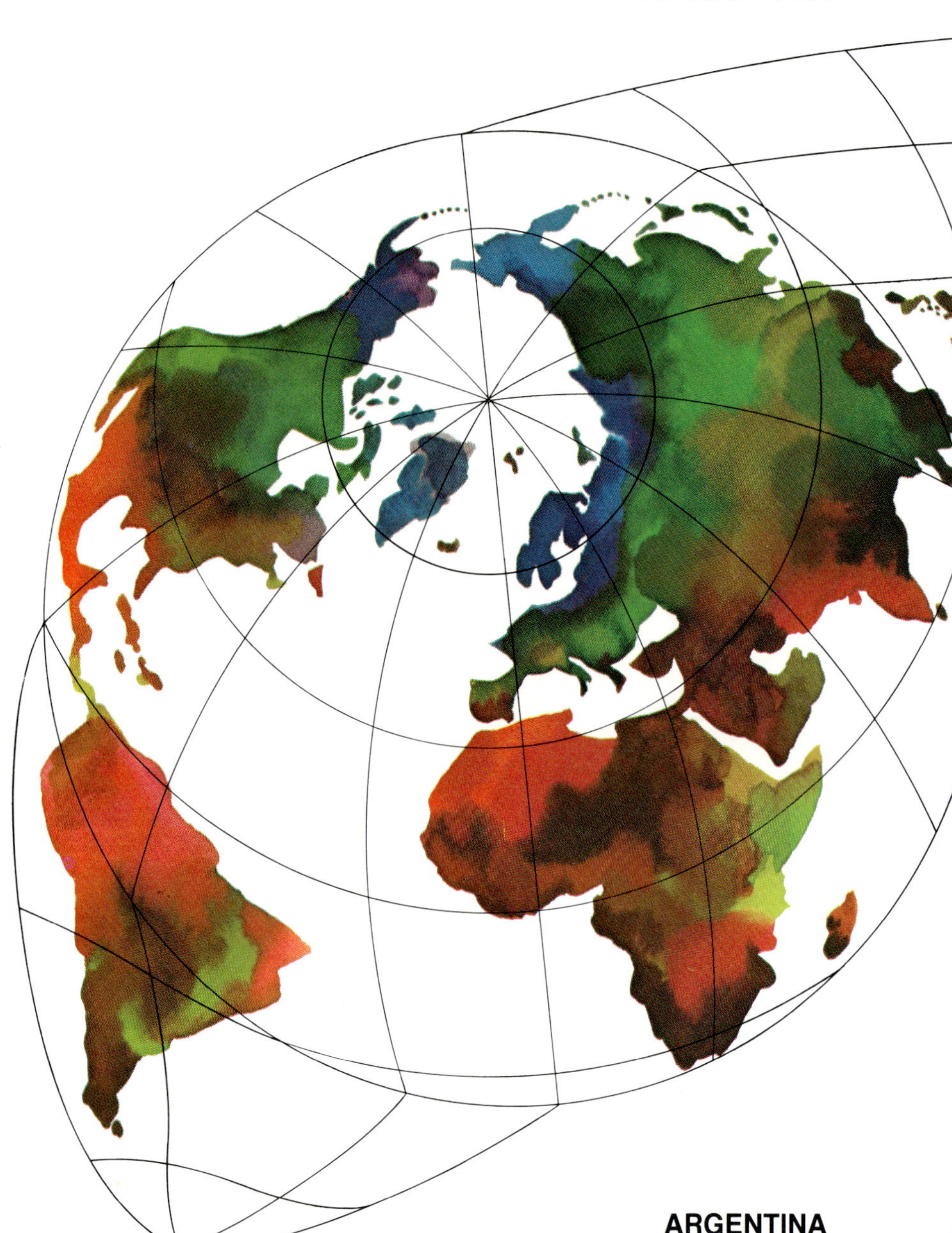

ALGERIA

Halliburton Limited
ALGIERS
Villa Touati
Djenane Achabou-Zouaoua
Dely-Ibrahim/Cheraga
Ph: 213 + 648428

ANGOLA

Halliburton Overseas Ltd.
LUANDA
Av. 4 de Fevereiro No. 42
Edificio Secil, Andar 11
Ph: 373758/373459

ARGENTINA

Halliburton Argentina S.A.
(Division Office)
1340 BUENOS AIRES
Maipu 942, Piso 14
Phone: 541 + 312-8413
8300 NEUQUEN
Casilla de Correo 166
Phone: 54 + 943-23470
9900 COMODORO
 RIVADAVIA, Chubut
Casilla 177,
Barrio Industrial
Ph: 54 + 967-24919
5507 LUJAN de CUYO, Prov.
 de Mendoza
Casilla de Correo 32
Ph: 54 + 61 + 981869

AUSTRALIA

**Halliburton Australia
Pty. Ltd.**
DRY CREEK, South
 Australia 5094
44 Churchill Rd. (Extension)
Phone: 618 + 349 + 4588
PERTH, West
 Australia 6109
20 Malcolm Road
Maddington 6109
Ph: 61 + 9 + 459-6444
SALE, Victoria 3850
304 Raglan
Phone: 61 + 51 + 443484

AUSTRIA

**Halliburton Company
Austria GmbH**
A-1213 VIENNA
Postfach 107
Ph: 43 + 2246-4333

BAHRAIN

Halliburton Limited
MANAMA (Division Office)
Box 515
Phone: 973 + 258866

BOLIVIA

Halliburton Company
SANTA CRUZ
Casilla 482
Phone: 591 + 33 + 44117

BRAZIL

**Halliburton — IMCO do
 Brasil-Servicos
 Comercio e Industria,
 Ltda.**
20.000 RIO DE JANEIRO, RJ
 (Division Office)
Caixa Postal 2038
Ph: 55 + 21 + 262-7210
40000 SALVADOR BAHIA
Caixa Postal 939
Ph: 55 + 71 + 246-2433

BRUNEI

Halliburton Limited
KUALA BELAIT
P.O. Box 393
Phone: 6733 + 22156

CANADA

**Halliburton Services
Limited**
CALGARY, Alberta T2P OS2
 (Division Office)
1100 Calgary House
550-6th Ave. S.W.
Phone: (403) 269-6141

Technical Centre
CALGARY, Alberta T2V 6V1
5140 Skyline Way N.E.
Phone: (403) 269-6141

CHILE

**Halliburton-IMCO
 Servicios Petroleros
 (Chile) Ltda.**
PUNTA ARENAS,
 Magallanes
Casilla de Correos 1307
Phone: 5661 + 223953

**CHINA, PEOPLE'S
REPUBLIC OF**

**Halliburton Overseas
Limited**
SHEKOU, Shenzhen
Room 316
China Merchants Bldg.
Shekou Industrial Zone
Shenzhen, P.R.C.
Ph: 86 + 755-92470
BEIJING, P.R.C.
No. 508, 5th Floor
Block A-2
Lido Commercial Bldg.
Lido Center
Jichang Road

COLOMBIA

Cia. Halliburton de Cementacion y Fomento
BOGOTA
Apartado Aereo 4549
Phone: 571 + 292-0211

DENMARK

Halliburton Company Germany GmbH
DK-6705 ESBJERG
P.O. Box 2066
Phone: 45 + 5 + 145444

ECUADOR

Cia. Halliburton de Cementacion y Fomento
QUITO
Apartado 83 A
Phone: 593 + 2 + 520079

EGYPT, A.R.E.

Halliburton Limited
CAIRO
P.O. Box 1227
Ph: 20 + 2 + 853001

FRANCE

Halliburton Company
(Division Office)
LE FLORESTAN
4 Ieme Etage
2 Boulevard Vauban
78180 St. Quentin En
   Yvelines
64142 BILLERE CEDEX
B.P. 209, Lons
Phone: 33 + 59 + 321446
91801 BRUNOY CEDEX
B.P. 18, Rue de La Foret
Zone Industrielle
Epinay Sous Senart
Epinay, France
Phone: 331 + 60465550

GABON

Halliburton — IMCO Gabon S.A.R.L.
PORT GENTIL
B.P. 507
Ph: 241 + 752196

WEST GERMANY

Halliburton Company Germany GmbH
D-3100 CELLE
Bruchkampweg 7
Postfach 380
Phone: 49 + 5141 + 88560

GREECE

Halliburton Ltd.
65410 KAVALA
P.O. Box 1525
Phone: 30 + 51 + 831842

GUATEMALA

Halliburton Company
GUATEMALA CITY
2DA. Calle 15-96
Zona 13
Phone: 502 + 2 + 318185

HOLLAND

Halliburton Company Germany GmbH
2595 ES THE HAGUE
(Division Office)
Emmapark 3
Phone: 31 + 70 + 475211
1976 BE IJMUIDEN
Holland Branch
Amperestraat 1/G
Ph: 31 + 2550 + 34924

INDIA

Halliburton Offshore Services, Inc.
NEW DELHI
B-7/110A Safdarjang Enclave
Phone: 91 + 11 + 601-052

INDONESIA

P.T. Halliburton Indonesia
JAKARTA (Division Office)
Jln. Kemang BangKa 1/28
Phone: 62 + 21 + 798-0568
JAKARTA (District Office)
Wisma Pe De, 4th Floor
Jln. M. T. Haryono Kav. 17
Phone: 6221 + 829-7108
BALIKPAPAN
Kampung Damai
RT. XIII-A/58
Sepinggan Bypass
Phone: 62 + 542 + 23278

ITALY

Halliburton Italiana S.P.A.
MILANO
Via L. Tolstoi 86
20098 Zivido
(San Giuliano Milanese)
Phone: 392 + 98491451

JAPAN

Halliburton Overseas Ltd.
TOKYO 106
5th Floor, Maruyama Bldg.
3-8, 2-Chome Azabudai
Minato-Ku
Phone: 81 + 3 + 586-9271

KENYA

Halliburton Ltd.
NAIROBI
P.O. Box 62216
Phone: 254 + 2 + 762-191

KUWAIT

Halliburton Limited
AHMADI
P.O. Box 9022
Phone: 965 + 3984801

EAST MALAYSIA

Far East Oilwell Services Sdn. Bhd.
MIRI, SARAWAK 98007
P.O. Box 583
Phone: 60 + 85 + 651043

WEST MALAYSIA

Far East Oilwell Services Sdn. Bhd.
KUALA LUMPUR 50250
Suite 708, 7th Floor
Peinas Intl.
Letter Box #35
Jalan Sultan Ismail
Phone: 60 + 3 + 2619244

MEXICO

Halliburton de Mexico S.A. de C.V.
MEXICO 06600, D.F.
(Division Office)
Paseo Reforma No. 76-1203
Phone: 905-566-8088
CUIDAD, del Carmen
Calle 31 #254
Avenida Aviacion
Phone: 52938 + 213-71
REFORMA, Chis.
Km. 10.4 Carretera a
   Reforma
Phone: 52932 + 80111
TAMPICO, Tamps
Apartado Postal 1164
Phone: 52 + 121 + 381-20
VILLAHERMOSA, Tabasco
Cesar Sandina No. 63
Esq. A. Quevedo
Phone: 52 + 931 + 30347
REYNOSA, Tamps
Km. 103.5 Carretera
Monterrey Reynosa
Phone: 62 + 892 + 300-86

NEW ZEALAND

Halliburton Overseas Ltd.
NEW PLYMOUTH
P.O. Box 7160
Phone: 64 + 67 + 72405

NIGERIA W.A.

Halliburton Nigeria Limited
LAGOS
P.O. Box 3694
Phone: 234 + 1 + 615444
WARRI
P.O. Box 359
Phone: 234 + 53 + 231300
PORT HARCOURT
P.O. Box 462
Phone: 234 + 84 + 335619

NORWAY

Halliburton Overseas Ltd.
4056 TANANGER
P.O. Box 67
Phone: 47 + 4 + 696733

SULTANATE OF OMAN

Halliburton Limited
MUSCAT
Box 9081
Phone: 968 + 603246

PAKISTAN

Halliburton Limited
ISLAMABAD
P.O. Box 1136
Phone: 92 + 51 + 823924

PERU

Empresa de Servicios Tecnicos Petroleros S.A. (ESTEPSA)
TALARA
Apartado 7A
Phone: 51 + 74 + 334-668
LIMA
Apartado 4988
Phone: 51 + 14 + 420258
IQUITOS
Apartado 563
Phone: 51 + 94 + 236426

PHILIPPINES

Halliburton Services
MAKATI, METRO MANILA
Ground Floor Makati
   Tuscany
6751 Ayala Ave.
Phone: 63 + 2 + 871 + 892

QATAR

Halliburton Limited
DOHA
P.O. Box 3036
Phone: 974 + 671111

SAUDI ARABIA

Halliburton Company
DHAHRAN
Box 657
Phone: 966 + 3 + 856-1616

SINGAPORE

Halliburton Limited
SINGAPORE 9123 (Division Office)
P.O. Box 156
Orchard Point P.O.
Phone: 65 + 3377741

THAILAND

Halliburton Company
BANGKOK, 10110
137/3 Phayathai X-Ray
Computer Center
Asoki Road
Soi 21 Sakhumvit
Phone: 662 + 258-3431

TRINIDAD, W.I.

Halliburton Trinidad Limited
SAN FERNANDO
P.O. Box 57
Phone: 809 + 6579181

TUNISIA

Halliburton Limited
SFAX
Zone Industriella
Plage Poudriere
Route Sidi Mansour Km 3,5
Phone: 216 + 4 + 27487
TUNIS
32 Bis Avenue
Chedly Kallala
Phone: 2161 + 288-558

TURKEY

Halliburton Company
ANKARA
Tunus Caddesi 50-6
Kavaklidere
Phone: 90 + 41 + 282921
DIYARBAKIR
P.K. 187
Phone: 90 + 831 + 26020

UNITED ARAB EMIRATES

Halliburton Limited
ABU DHABI
P.O. Box 57
Phone: 971 + 2 + 553000
DUBAI
P.O. Box 3111
Phone: 971 + 4 + 341588

UNITED KINGDOM

Halliburton Manufacturing & Services Ltd.
LONDON W1R 0EL
(Regional Offices)
17 Hanover Square
Phone: 44 + 1 + 629-7611
GREAT YARMOUTH,
   Norfolk NR3O 1QF
South Denes Road
Ph: 44 + 493 + 30300
ABERDEEN AB2 OES,
   Scotland
Howe Moss Crescent
Kirkhill Industrial Estate
   Dyce
Phone: 44 + 224 + 771991
ARBROATH, Angus DD11,
   2NF, Scotland
P.O. Box 9
Phone: 44 + 241 + 77333

VENEZUELA

Cia. Halliburton de Cementacion y Fomento
CARACAS 106
(Region Office)
Apartado 61229 Chacao
Phone: 58 + 2 + 751-7265
EL TIGRE, Edo. Anzoategui
Apartado 221
Phone: 58 + 83 + 351114
MARACAIBO
(Division Office)
Apartado 698
Phone: 58 + 61 + 920540
LAS MOROCHAS
Ave. Intercomunal
Phone: 5865 + 74-277

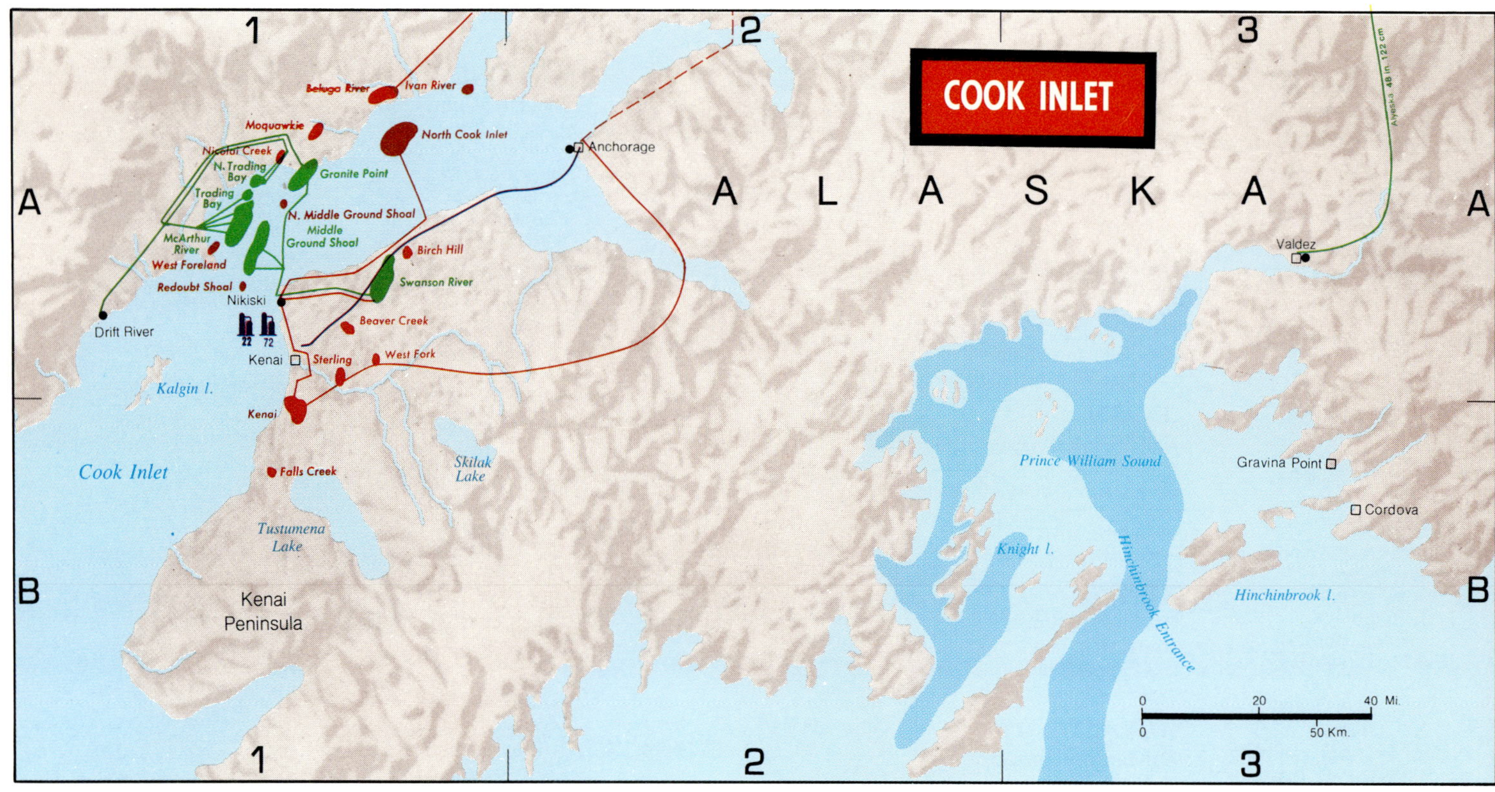

Its projections excluded nonrecurring outlays incurred to acquire reserves or other firms' assets.

Salomon Bros. predicted that U.S. companies' domestically generated cash flow will reach $250 billion during 1988-91, compared with a total capital spending need in the period of $183 billion.

"This apparent financial surplus of $67 billion is more illusion than reality," the company said, "because those funds will be invested in the search for and development of petroleum reserves overseas rather than in the U.S."

Because marginal returns on incremental upstream spending fall very sharply at higher spending levels, redirecting that $67 billion to domestic projects would not be enough to close the U.S. production replacement gap, Salomon Bros. said. It concluded that 100% production replacement in the U.S. would call for added outlays of $233 billion during the next 4 years.

Even if those added funds were available, the investment would produce poor financial returns, said Salomon Bros.

In the early 1980s, when the U.S. replacement rate was much better, marginal exploration and production investments achieved returns of about 3%. Marginal exploration and production returns in 1988, with spending and replacement rates sagging, averaged about 8%.

"The industry is investing its domestic cash flow surplus in other countries where geology and tax systems are often more favorable than they are in the U.S.," Salomon Bros. said.

"During the past 8 years, the 30 largest private sector oil firms in the world achieved foreign finding costs of $5.51/BOE vs. $9.81/BOE in the U.S.

"Similarly, the foreign reserve replacement rate of those firms in 1980-87 was 100% vs. 74% in the U.S."

Offshore leasing plan clears a hurdle

As 1989 began, a federal appeals court upheld the Interior Department's 5 year offshore leasing plan but ordered further study of how drilling off Alaska and the Lower 48 West Coast could affect migratory wildlife.

Environmental groups led by the National Resources Defense Council and several states—California, Florida, Massachusetts, Oregon, and Washington—filed the lawsuit in a drive to overturn Interior Sec. Don Hodel's leasing plan for the Outer Continental Shelf.

Platform Bullwinkle

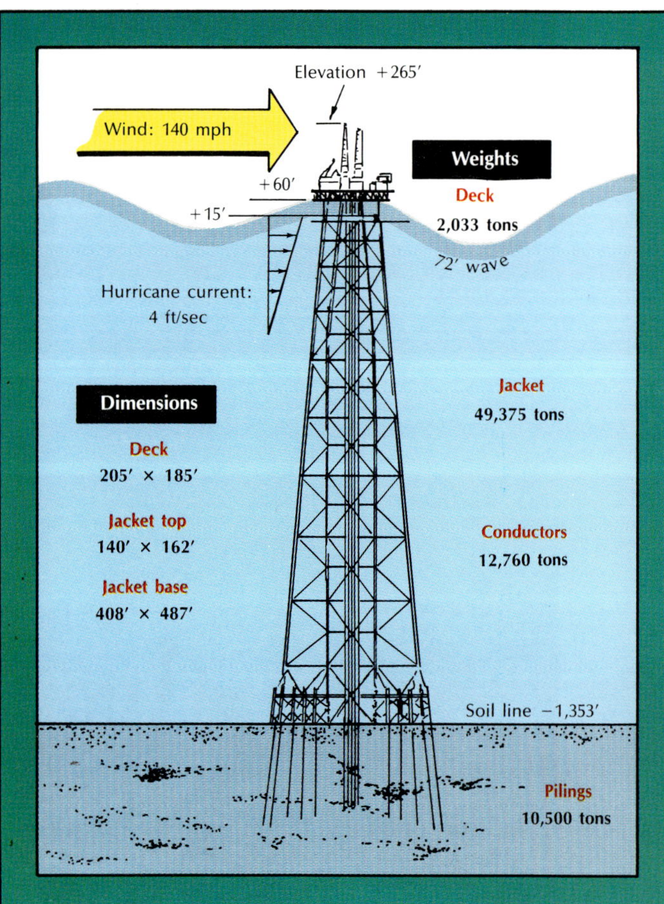

CALIFORNIA

Pacific Ocean

OREGON
IDAHO
NEVADA
ARIZONA
MEXICO

Grants Pass
Medford
Talent
Ashland
Klamath Falls
Rome
Triangle
Fields
Eureka
Tompkins Hill
Grizzly Bluff
Petrolia
Shasta Lake
Alturas
Goose Lake
Humboldt
Elko
Corning
Susanville
Black Butte
Greenwood
Rice Creek
Rancho Capay
Willow
Beehive Bend
Durham
Sycamore
Grimes
Arbuckle
Buckeye
Kirk
Wild Goose
Dunnigan Hills
Sparks
Reno
Virginia City
Carson City
Blackburn
Santa Rosa
Conway Ranch
Edhunters Lake
Sacramento
Miller
Maine Prairie
River Island
Petaluma
The Geysers (steam)
Benicia
Hercules
Martinez 225
Pinole Pt.
Benicia
Martinez
Isleton
Currant
Trap Spring
Eagle Springs
Soda Springs
Kate Springs
Bacon Flat
Grant Canyon
140 126
Richmond
San Francisco
Rodeo
Oakland
Hayward
Livermore
Roberts Is.
350
118
Half Moon Bay
Redwood City
LaHonda
Oil Creek
Moody Gulch
San Jose
McMullin Ranch
Vernalis
Modesto
Fresno
Tonopah
Warm Springs
Mono Lake
Santa Cruz
Sargent
Hollister
Monterey Bay
Salinas
Chowchilla
Mollat Ranch
CALIFORNIA
Bishop
Raisin City
Gill Ranch
Bitterwater
Vallecitos
Helm
Monroe Swell
East Coalinga
Extension
Coalinga
Riverdale
King City
Quinado Canyon
Paris Valley
McCool Ranch
San Ardo
North Dome
Kettleman
Hanford
Pyramid Hills
Harvester
Trico
Shale Point
Lost Hills
Semitropic
McKittrick
Oildale
San Luis Obispo
South Belridge
Cymric
McKittrick
Elk Hills
Mountain Poso
Kern Front
Kern River
Round Mountain
Bakersfield
Palomo
10
10
18 41 20
Guadalupe
Santa Maria
Santa Maria Valley
Midway Sunset
Russell
Ranch
South
Cuyama
Yowlumne
Tejon
Lompoc
Zaca
Indian Springs
Overton
Point Arguello
Molino
Point Conception
Sacata
Pescado
Goleta
Santa Barbara Hills
Hondo
Dos Cuadras
Pitas Point
Santa Clara
Island
Newhall
Aliso Canyon
Ventura
Oxnard
Sockeye
Las Vegas
Henderson
Boulder City
Hoover Dam
Nelson
Davis Dam
Kingman
Black Mesa
Barstow
Needles
Yucca
Santa Barbara Channel
San Miguel Island
Santa Rosa Island
Santa Cruz Island
Inglewood
Playa Del Rey
Montebello
Brea Olinda
Santa Fe Springs
Los Angeles
Torrance
Long Beach
Wilmington Trend
San Pedro Channel
Santa Barbara Island
Huntington Beach
Newport West
San Bernardino
Santa Catalina Island
San Clemente Island
Gulf of Santa Catalina
Carlsbad
Four Corners
Colorado River
San Diego
San Diego Bay
Tijuana
El Centro
Mexicali
Yuma

State   No. Refineries
McKittrick........1
Carson..........2
Hanford.........1
Wilmington.....4
Bakersfield......3
El Segundo.....1
Richmond.......1
Santa Maria.....1
Long Beach.....1
Oildale..........1
Benicia..........2
Santa Fe Springs....1
Torrance........1
Newhall.........1
Oxnard..........1
Hercules........1
Martinez........1
Los Angeles....2
Rodeo...........1
South Gate.......1

Carson
30 215
Wilmington
65 5 75 130
Signal Hill
8 15
Torrance
23
Santa Fe Springs
46 40
El Segundo
485

Oxnard
4
Newhall
21
Los Angeles
108
Long Beach
41

0   50   100 Mi.
0   50   100 Km.

SANTA MARIA BASIN

Florida withdrew from the suit after reaching a compromise with Interior. The American Petroleum Institute (API) intervened on behalf of the government.

Hodel's plan called for 37 OCS lease sales during January 1987-June 1992 covering acreage in the Gulf of Mexico and off Alaska, California, Washington-Oregon, and the East Coast.

The District of Columbia Circuit Court of Appeals upheld Interior in all challenges the lawsuit made to the leasing schedule. The court's three judges ruled that Hodel's schedule was reasonable, and he adequately explained his criteria for excluding certain areas from the schedule.

The challengers also had charged that the sale plan was based on unreasonably high oil prices, along with undervalued environmental costs, and failed to ensure that the public would receive fair market value for oil leases.

In each of those allegations, the court found that Hodel had explained his methodology well and had reached reasonable decisions based on facts.

The petitioners also claimed Interior should have considered various conservation policies, such as increasing auto fuel economy standards, as a partial alternative to the 5 year program.

Hodel had argued he did not have to consider conservation as a partial alternative to OCS leasing because full OCS production and conservation combined cannot meet the nation's energy needs.

The court disagreed, saying the National Environmental Policy Act requires Hodel to consider alternatives even if they do not reduce the need for OCS leasing.

But it added, "The continuing need for development of the OCS, however, reduces the scope of the required consideration. We hold that the secretary's accounting for conservation alternatives was adequate."

The court sustained a complaint that the environmental impact statement (EIS) for the 5 year plan inadequately considered the cumulative effects of simultaneous OCS development in the Pacific and Alaskan regions on migratory species, particularly whales and salmon.

The petitioners contended the cumulative effect of simultaneous development will be greater than the sum of development in each area considered separately because migratory species will have to swim through each area.

They said Hodel should have considered that and canceled or deferred some of the proposed lease sales in the two regions.

The court agreed that the EIS did not contain enough analysis of the issue.

It said, "In each place in which the EIS even mentions interregional impacts of OCS development, it merely announces that migratory species may be exposed to risks of oil spills and other 'impacts' throughout their routes.

"These perfunctory references do not constitute analysis useful to a decisionmaker in deciding whether or how to alter the program to lessen cumulative environmental impacts.

"Therefore we remand this matter to the secretary for further consideration and any revisions in the 5 year program that consideration may warrant."

The court suggested that Interior examine cumulative effects of simultaneous interregional OCS development in "a single, coherent section rather than fragment his analysis by area," identifying the species and OCS activity along their migratory routes and examining whether staggered lease sales would lessen the risk.

More deepwater tracts leased in gulf

The U.S. offshore exploration industry, as expected, in 1988 took another step into ultradeep water of the Gulf of Mexico. The move took place in the eastern Gulf of Mexico by way of Sale 116—the third area-wide lease sale of the 1980s in the eastern gulf and the first since 1985.

Most previous ultradeepwater leasing activity had been in the central and western gulf regions.

Of the 115 blocks that received bids in Sale 116, 70% were in water deeper than 3,000 ft, including a group of four in the Elbow region off Florida in a world record 11,000 ft of water. The 15 companies that participated in the sale, held Nov. 16 in New Orleans, made 135 bids totaling $47,559,746. Total high bids amounted to $41,582,298—a record low in area-wide leasing history.

At the last previous eastern gulf area-wide sale, held prior to the 1986 oil price collapse, 22 companies offered total high bids of $124 million for 82 blocks. Besides the influence of lower prices, existence of only two eastern gulf discoveries hurt interest in Sale 116.

Mobil Exploration & Producing U.S. Inc. was the big player in Sale 116 in terms of net high bid exposure. The company offered high bids of $11.3 million net for interests in 32 blocks.

Chevron U.S.A. Inc. was second, offering high bids of $5.3 million net for interests in 28 blocks.

The biggest players in terms of the number of tracts acquired were Odeco Oil & Gas Co. and Murphy Oil U.S.A. Inc. The two companies each acquired interests in 38 tracts with a net high bid exposure of $2.2 million each.

Conoco Inc. was second, acquiring interests in 36 blocks for a net $4.3 million.

Willbros has built pipelines or related facilities in 48 nations

And we have constructed more pipelines than any other company in the world, working in every type terrain and climate — including the frozen Arctic, the swampy jungles of Africa, the deserts of the Middle East and the mighty Andes Mountains of South America. Willbros has more than 80 years of successful experience, turnkey capabilities and unsurpassed record of completing jobs on schedule and on budget. Let Willbros construct your next energy project.

For more information about Willbros and its services, call or write: L. R. Hamilton, Willbros USA, Inc., 2431 East 61st Street, Suite 700, Tulsa, Oklahoma 74136. Phone: 918-748-7000.

WILLBROS International, Inc.
WILLBROS USA, Inc.
WILLBROS Energy Services Company
WILLBROS Colombia, S.A.
WILLBROS (Overseas) Limited
WILLBROS Drilling, Inc.
WILLBROS West Africa, Inc.
WILLBROS Middle East, Inc.
WILLBROS (Nigeria) Limited
WILLBROS Butler Engineers, Inc.
The Oman Construction Company

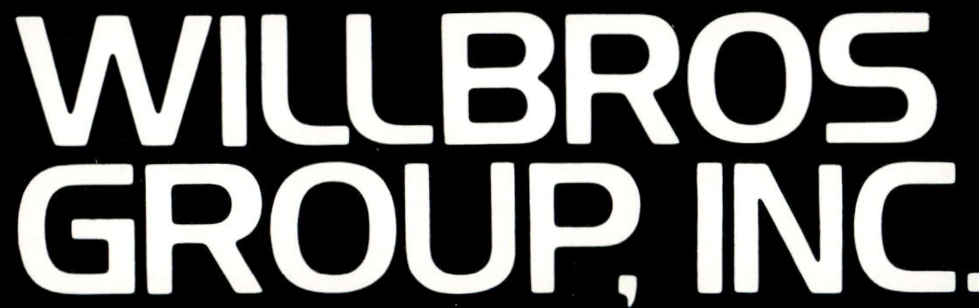

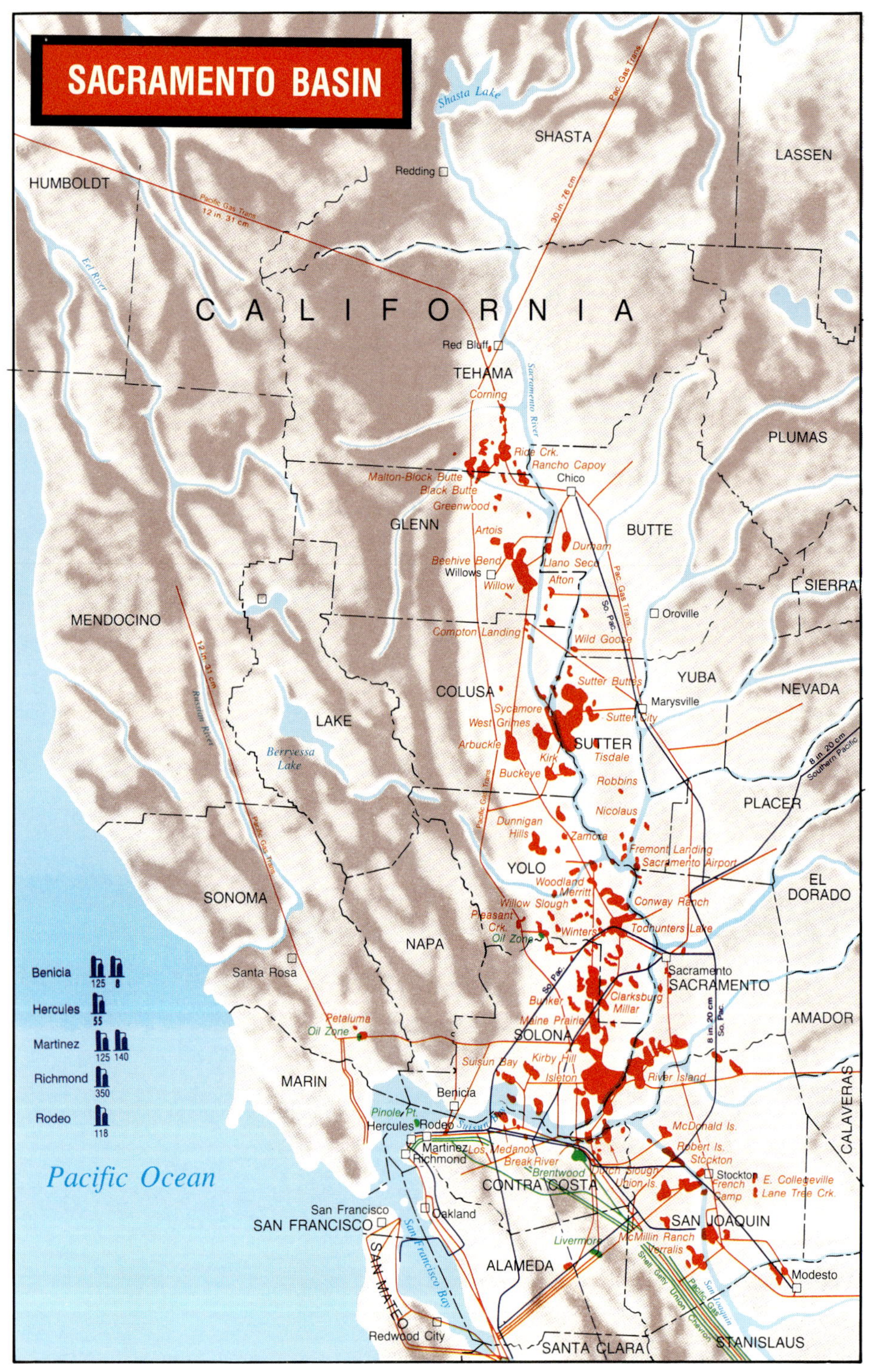

The previous record was 1,613 ft, held by Brazil's Petroleo Brasileiro SA's 3-RJS-376-D subsea completion in Marimba field off Rio de Janeiro.

Enserch Exploration Inc., Dallas, said Green Canyon 31-4 went on stream at 6:30 a.m. Nov. 14, flowing 2.5 MMcfd of gas. Enserch was a partner in the operation.

The flow rate was gradually increased to 8.15 MMcfd and 1,066 b/d of condensate Nov. 16 through a $^{21}/_{64}$ in. choke with 4,800 psi flowing tubing pressure. The flow was to be stabilized for a few days before the choke was opened further.

Green Canyon 31-4 was the second well on stream in Placid's floating/subsea production system. The first well, Ewing Bank Block 999-1, went on stream Oct. 13. It was flowing 5.15 MMcfd of gas and 1,363 b/d of condensate Nov. 16.

Gas and condensate production was being commingled and shipped to the system's shallow water processing platform through a 16 in. export line, planned for later use exclusively for gas transportation. A 14 in. export line was to be placed on stream when warranted by added production.

Placid hoped to have four wells on stream by yearend 1988, including another satellite subsea completion on Green Canyon Block 31 and a template well on Green Canyon Block 29, where the floating portion of the production system was anchored.

The Placid group began a new era of ultradeepwater development in the gulf when it started producing 5 MMcfd of gas and 1,550 b/d of condensate Oct. 13, 1988, into the Green Canyon Block 29 floating/subsea production system from the Ewing Bank Block 999-1 subsea satellite well. It was the gulf's first floating production system (Fig. 2).

The project overcame numerous delays stemming from Placid's debt problems, a pipeline bundle towing accident, and poor weather.

Total cost of the Green Canyon 29 project was estimated at $350 million. Cost of the floating/subsea production system was about $150 million, and the pipeline system and shallow water processing platform cost $50 million.

The project was still marginally economic but not as economic as first envisioned in 1983 because oil and gas prices had dropped more than 50%.

The project was planned to eventually have a mix of 12 template and eight satellite wells on production. The mix might change, however, depending on future drilling results.

Ewing Bank 999-1 is in 1,462 ft of water 4,375 ft north of the Green Canyon 29 subsea drilling/production template.

The satellite well established a Gulf of Mexico water depth production record, eclipsing the 1,025 ft mark held by Shell Offshore Inc.'s Platform Cognac. Ewing Bank 999-1's record fell when the Green Canyon 31-4 well went on production.

Meanwhile, Penrod 72, the floating portion of Green Canyon 29's drilling/production system, was anchored

High bid of the sale was $3,142,000 offered by Amoco Production Co. for De Soto Canyon Block 133 in about 6,500 ft of water.

Amoco beat out a $289,000 bid from Shell Offshore Inc.—Shell's only bid in the sale—and a $1,427,600 bid from a combine of Texaco U.S.A., Union Exploration Partners Ltd., Conoco Inc., Odeco, and Murphy.

Gulf operations
set deepwater marks

Operations in the Gulf of Mexico during 1988 shattered the world water depth production record.

Placid Oil Co., Dallas, and partners set the new mark late in the year when they started production from a Green Canyon Block 31-4 subsea satellite well in 2,243 ft of water in the Gulf of Mexico.

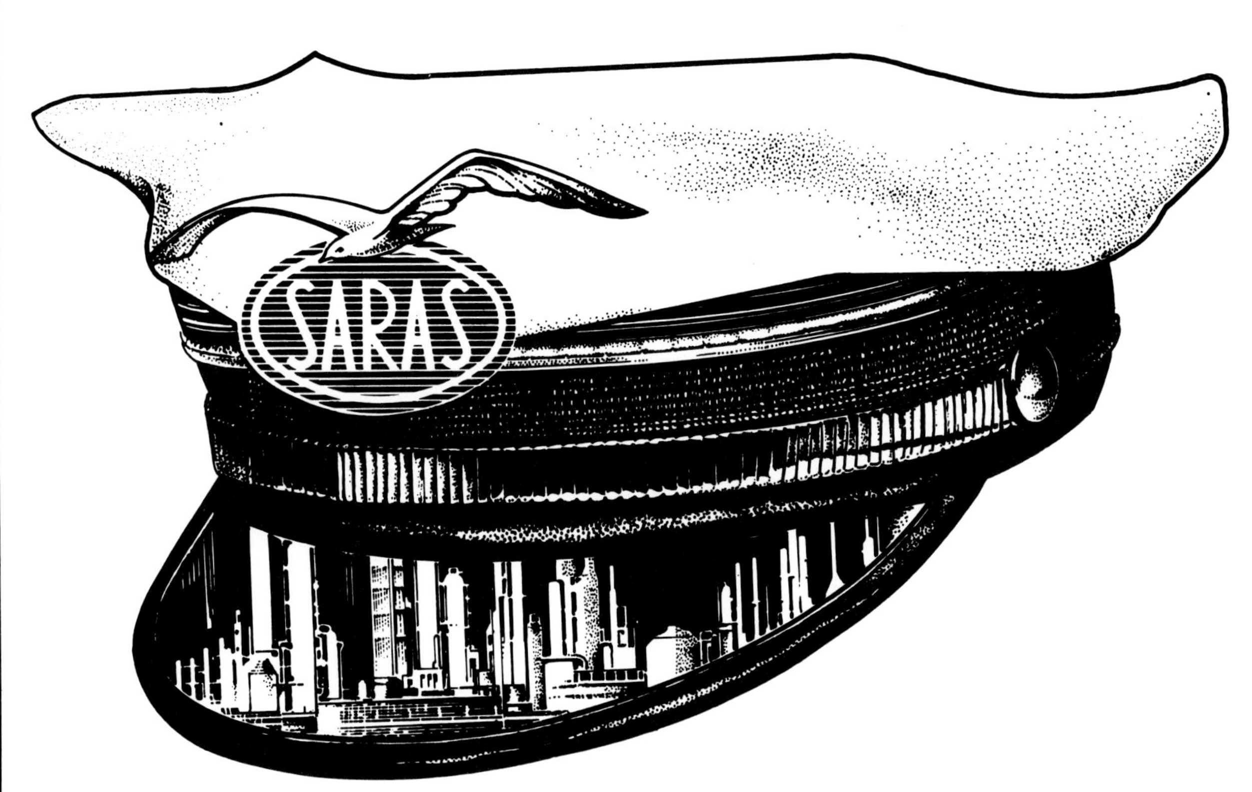

The largest oil refinery in the Mediterranean at your service

SARAS has always advanced technologically to process crude oil on behalf of third parties with the highest added value.

With a processing capacity of 18 million metric tons per year (360,000 BSD) and 5 million tons per year of conversion capacity, SARAS refinery can receive crude oil from tankers of up to 260,000 DWT, store into its huge tank farm (4 million cu.mt.) and deliver the entire range of refined products according to customer requirements.

SARAS S.p.A. RAFFINERIE SARDE

HEAD OFFICE - 20122 MILAN - GALLERIA DE CRISTOFORIS, 8 - TEL. (02) 77371 - TELEX 311273 - FAX (02) 790640
REFINERY - 09018 SARROCH (CAGLIARI) - S.S. SULCITANA KM. 19 - TEL. (070) 90911 - FAX (070) 900209
BRANCH OFFICE - 00187 ROME - SALITA S. NICOLA DA TOLENTINO, 1-B - TEL. (06) 4742701 - FAX (06) 4742701

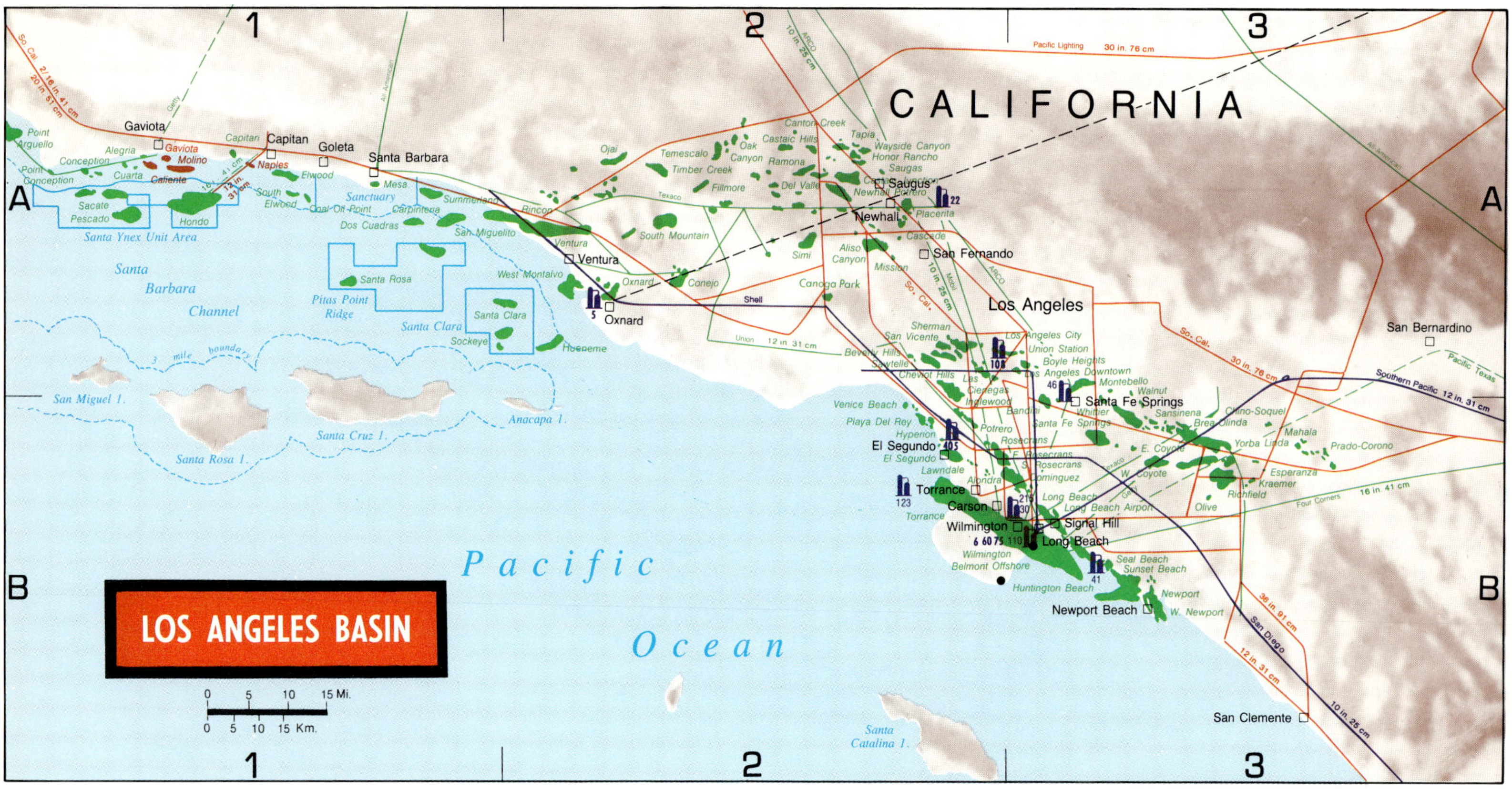

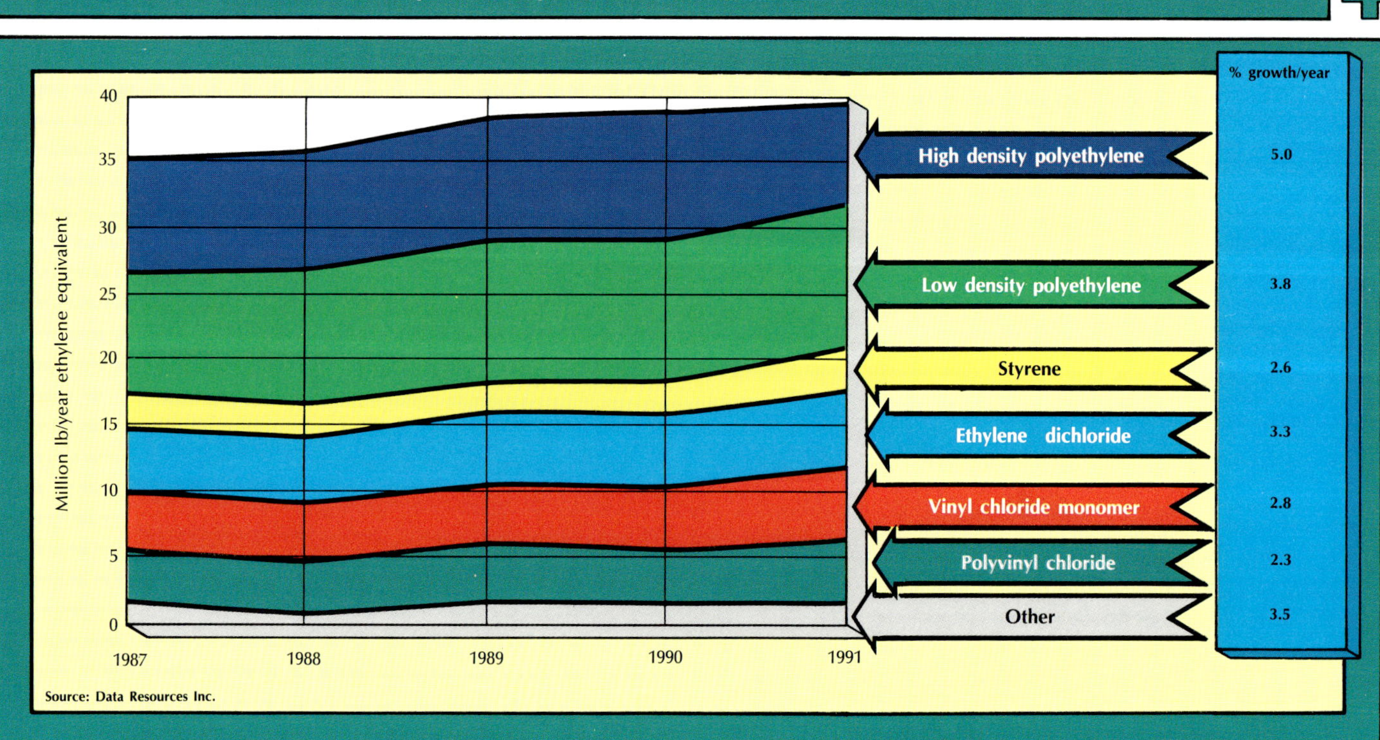

above the subsea template in 1,522 ft of water.

It set the gulf's water depth production record for a surface platform.

Oil and gas production underwent first cut processing on Penrod 72, which was capable of processing 120 MMcfd of gas, 45,000 b/d of liquid hydrocarbons, and 22,500 b/d of water.

Space was allotted on the semisubmersible for future separation, compression, dehydration, and water treatment equipment.

A 50 mile export pipeline bundle carried oil and gas to a final cut processing platform set in 101 ft of water in Ship Shoal Block 207. Gas was then sold to Pontchartrain Natural Gas System and the oil to JM Petroleum Corp. under 6 month contracts.

In addition to Enserch, Placid's partners were Louisiana-Hunt Petroleum Corp., Opubco Resources Inc., HI Production Co., Penrod Drilling Co., Petro Hunt Corp., and Prosper Energy Corp.

Exxon took a farmout on all but two of Placid Oil Co.'s

Demand outlook for ethylene by derivative **4**

Source: Data Resources Inc.

Outlook for U.S. ethylene, polyethylene capacity 

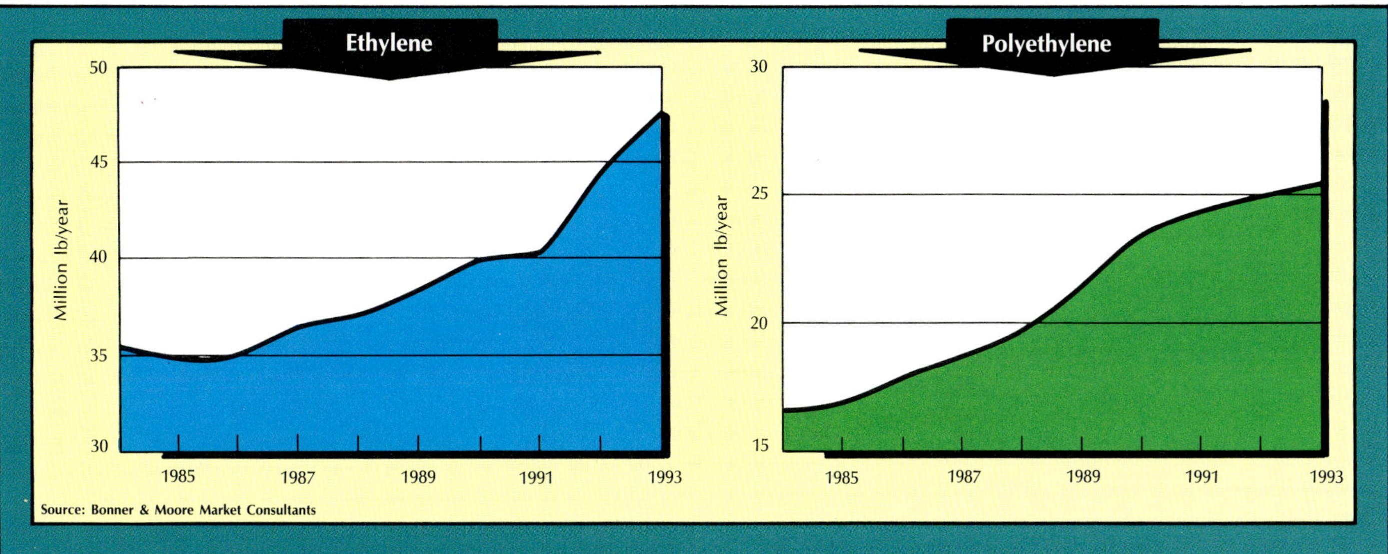

Green Canyon blocks and all of Placid's Mississippi Canyon and Garden Banks blocks in the Gulf of Mexico. The farmout, completion of which awaited Placid's reorganization out of bankruptcy, involved 38 Green Canyon blocks, 14 Garden Banks blocks, and six Mississippi Canyon blocks.

All of Placid's partners in the blocks, with the exception of Enserch, also farmed out their interests to Exxon. Placid and its partners maintained their working interests in Green Canyon Blocks 29 and 31, as well as an overriding interest in the blocks farmed out to Exxon.

The deal was the largest farmout of deepwater blocks in the gulf. Water depths on the blocks range from about 600 ft to about 3,800 ft.

With the farmout, Exxon became operator of about 170 Gulf of Mexico tracts in water deeper than 1,000 ft, Minerals Management Service figures showed. Shell Offshore Inc. remained the biggest ultradeepwater lessee with 365 gulf blocks in more than 1,000 ft of water.

Other milestones
in offshore operations

Among industry's other milestones in the Gulf of Mexico, Shell Offshore Inc. in mid-1988 set its massive Platform Bullwinkle jacket in 1,353 ft of water on Green Canyon Block 65, breaking Platform Cognac's world water depth record for conventional platform installation.

In addition, the gulf's first application of tension leg platform technology moved closer to operations. Conoco Inc. and partners in September 1988 landed the 1,400 ton foundation template that will anchor its Jolliet field tension leg well platform (TLWP) in 1,760 ft of water on Green Canyon Block 184.

Off Alabama, Mobil Exploration & Producing U.S. Inc. in April 1988 began industry's first production in Mobile Bay when it placed on stream Mary Ann gas field. That capped a 20 year, $400 million campaign to tap the world's first offshore Jurassic Norphlet reservoir.

Heerema Marine Contractors launched the 1,365 ft tall, 49,375 ton jacket for Platform Bullwinkle from its H-851-S barge shortly after 6 p.m. May 31. The jacket was lowered onto the seafloor at 9:30 a.m. June 1 at a site about 150 miles west-southwest of New Orleans.

Operations to drive the first of four critical piles began early June 2. It was expected to take about 5 days to drive all four piles, after which the jacket could withstand the full force of a hurricane.

Oil production was scheduled to begin in first quarter 1989.

With two drilling rigs in place, Platform Bullwinkle will stand 1,615 ft high—161 ft higher than the world's tallest building, Sears Tower in Chicago. Total structural weight will be more than 77,000 tons, most of it in the jacket (Fig. 3).

The jacket was designed and engineered by Shell, which let contract in 1985 to Gulf Marine Fabricators Inc. to fabricate, transport, and install the jacket. Its construction took 3 years at Gulf Marine's Ingleside Point, Tex., yard near Corpus Christi and involved 14 major subassemblies of four flat panels in the upper jackets, six blocks in the core, and four stiffened panels.

Installation culminated the jacket's 332 nautical mile voyage that began with passage through the Corpus Christi Ship Channel. It was the biggest single structure to transit the channel.

In all, 28 piles were scheduled to secure the jacket. The 84 in. diameter piles, fabricated by McDermott Marine Construction at its Morgan City, La., yard, were to be driven to penetrations of as much as 437 ft with underwater hammers.

Platform Bullwinkle will produce oil through a 7½ mile pipeline to Shell's Platform Boxer on Green Canyon Block 19. A gas pipeline will be installed later, but a route had not yet been determined.

Shell next was scheduled to install its Shell 12 and 21 platform rigs to begin directional drilling of Bullwinkle development wells. Drilling was expected to continue for 3-4 years.

Once the wells have been drilled, rigs will be removed and permanent production facilities installed.

Jolliet's single piece, 190 ft long, 190 ft wide foundation template, built by Gulf Island Fabrication Inc., was stabbed over the 24 slot drilling template during a 2 week operation using Micoperi SpA's 7000 crane barge.

Jolliet's TLWP was scheduled to be transported to Green Canyon 118 and its steel tethers attached and anchored to the foundation template in mid-1989.

Eleven of the field's proposed 20 development wells had been drilled.

Jolliet's 312 ton drilling template, also manufactured by Gulf Island Fabrication, was landed in mid-1987.

Conoco is operator of and holds a one-third interest in the $400 million project. Partners, each with a one-third interest, are OXY U.S.A. Inc. and Texaco Producing Inc.

Productive capacity of the TLWP will be 35,000 b/d of oil and 50 MMcfd of gas.

How U.S. gasoline consumption relates to income, price*

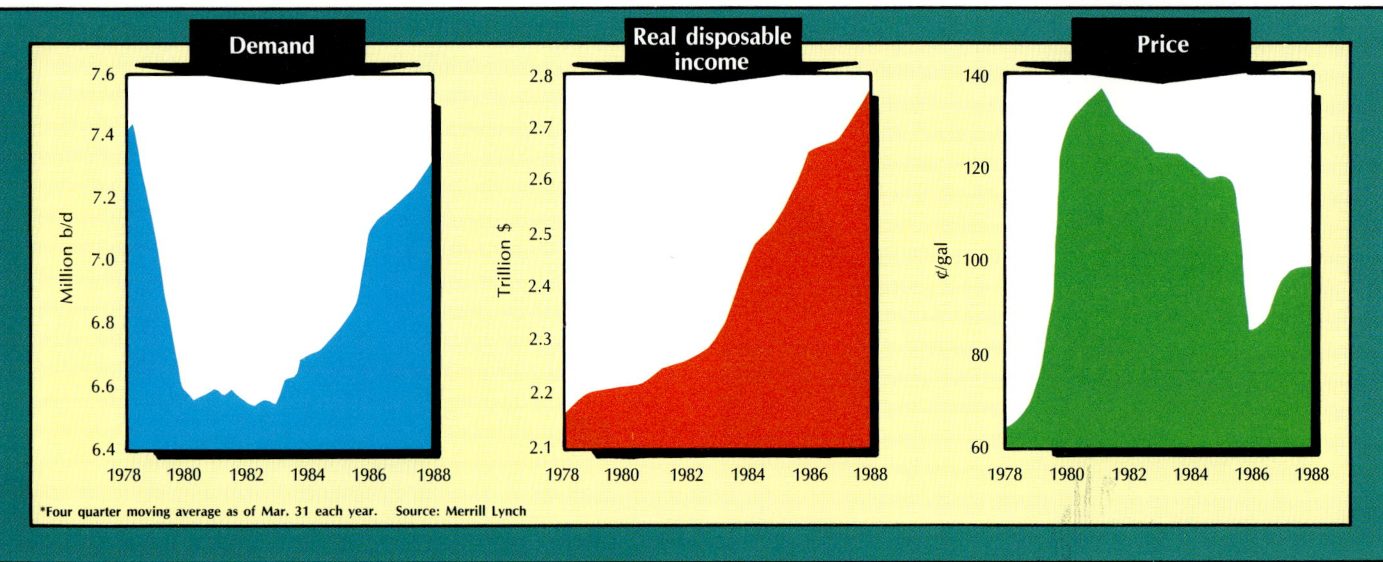

Mobil began Mary Ann production at a rate of 30 MMcfd from two wells tapping a reservoir at more than 20,000 ft under state Blocks 76, 77, 94, and 95 between Dauphin Island and Fort Morgan Peninsula.

Platform equipment strips liquids from the gas, which then moves through a 14½ mile pipeline to Mobil's treatment plant near Coden, Ala. The plant removes hydrogen sulfide, carbon dioxide, and other impurities and delivers the treated gas to the Transco pipeline system.

Petrochemicals face slightly leaner times

U.S. petrochemical producers are likely to face slightly leaner times in the next few years beyond 1989 but are not likely to relive the very depressed conditions of the late 1970s and early 1980s.

Strong demand, soaring prices, and tight supplies of most key petrochemicals sparked a flurry of expansions of olefins capacity in the U.S. Those expansions involved actions ranging from minor debottleneckings to construction of world class grassroots ethylene plants.

Along the Texas Gulf Coast alone, operators announced more than 150 petrochemical plant expansions totaling $5 billion in work, the Houston Economic Development Council (HEDC) reported. That region provided two thirds of major petrochemicals produced in the U.S.

HEDC produced its estimates before late 1988 announcement of plans by Formosa Plastics to build a $1.3 billion ethylene plant at Port Lavaca, Tex., and by Amoco Chemical Co. to build a 1.5 billion lb/year ethylene plant on the Gulf Coast. If all the announced expansions proceed, industry officials said, the U.S. petrochemical industry is likely to find itself with surplus capacity by the mid-1990s—perhaps earlier. The extent of that surplus largely will determine the economic health of the U.S. petrochemical industry during the rest of the 1990s.

A forecast by Data Resources Inc. said high density polyethylene will pace petrochemical demand growth through 1991, rising by 5%/year as U.S. demand for leading petrochemicals nudges 40 million lb/year of ethylene equivalent (Fig. 4). Announced expansion plans buttressed a Bonner & Moore Market Consultants' (B&M) forecast, unveiled in September 1988, that a jump of about 29% in U.S. ethylene capacity in the ensuing 5 years could trigger a slide in ethylene prices.

B&M based its analysis on a survey of U.S. ethylene producers' plans for adding capacity In addition to announced and unannounced plans, B&M incorporated its estimates for added capacity (Fig. 5 and Table 1). B&M estimated U.S. ethylene producers will hike capacity to 47.245 billion lb/year in 1993 from 36.933 billion lb/year in mid-1988. The biggest year to year jumps, totaling more than 3 billion lb/year each, will occur in 1991 and 1993.

"Reported plans for ethylene expansion are larger than most of the industry expects," said Bill Urquhart, B&M manager of chemical planning.

Of the total expansion, about 4.5 billion lb/year comes from announced new plants by Quantum, Phillips, and Dow. The remaining 6 billion lb/year of capacity additions in the next 5 years consists of 5.2 billion lb/year in expansions and 800 million lb/year in restarts of idle capacity.

All U.S. ethylene producers except three will add capacity through expansions, rerates, or restarts, mostly during 1991-93, Urquhart said. B&M's analysis made no provisions for idling of older, smaller plants.

Another factor in the outlook for a surplus was the potential for more Canadian supplies moving into the U.S. because of announced major ethylene expansions by Dow Chemical Canada Inc. and Nova Corp., coupled with the incentive of the U.S.-Canada Free Trade Agreement.

Concern mounts over gasoline tax

With U.S. gasoline demand nudging the record volume of about 7.4 million b/d reached in 1978, analysts, consumers, and industry spokesmen early in 1989 were sounding warnings about the effects of possible increases in the federal tax on the refining industry's No. 1 product.

Merrill Lynch, New York financial firm, said a sharp increase in the tax, coupled with an economic slowdown, could slice gasoline demand and squeeze refining/marketing profits. Such an increase could threaten state highway improvement programs, said the Highway Users Federation. And American Petroleum Institute Pres. Charles DiBona warned, "The severe damage that would rapidly result from such a discriminatory, regressive tax increase—including higher inflation, slower growth, and increased unemployment—would outweigh the environmental and energy security benefits of the induced conservation.

"And since a significant fraction of the revenues generated by a gasoline tax increase would be offset by declines in other sources of government income or absorbed in new expenditures needed to cope with the damage caused by the tax, those revenues cannot realistically be expected to

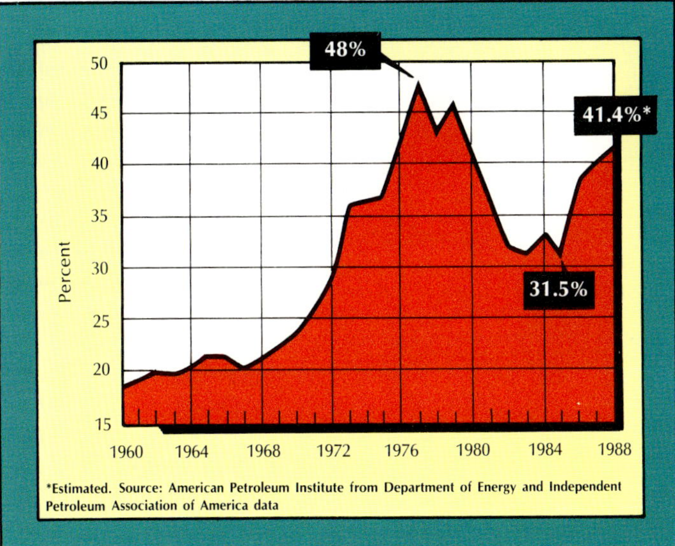

Imports' share of U.S. oil consumption 7

*Estimated. Source: American Petroleum Institute from Department of Energy and Independent Petroleum Association of America data

vanquish the deficit."

DiBona said the federal budget deficit, caused by excessive growth in federal spending and not too low taxes, should be cut by slowing growth in government outlays. If increased revenues still were needed, the fairest choice would be a broad based value added tax, he said.

Merrill Lynch sought to assess the effect of hiking the federal gasoline tax combined with a slowdown in the U.S. economy. The company developed a model that tracked the effects of both events on gasoline demand (Fig. 6).

The model showed that if gasoline prices rose 25% and real disposable income climbed about 2.6%, after rising 3.8% in 1988, U.S. gasoline demand could fall by about 2.5%, or 200,000 b/d in 1989.

Merrill Lynch figured the 25% increase as a 20%/gal jump in the federal gasoline tax together with a 4%/gal rise stemming from an expected improvement in oil market fundamentals. The federal excise at the time was 9.1¢/gal for gasoline and 15¢/gal for diesel fuel.

A decline in demand would lessen the pressure on refining conversion capacity, alleviating the supply tightness felt in 1988 and thereby putting pressure on gasoline prices at the refinery gate, Merrill Lynch said. With U.S. oil production on the decline, refiners were increasingly looking to imports to fill a widening supply/demand gap. Data compiled by API showed that U.S. dependence on imported oil stood at 41.4% of 17.743 million b/d of oil demand in 1988, the highest since 1980 and not too much shy of the 48% record of 1977 (Fig. 7). The Department of Energy reported that the biggest sources of U.S. imports in 1988 were Saudi Arabia and Canada, each of which accounted for about 14% of total U.S. oil imports (Table 2).

Oil reserves up, gas reserves down

U.S. oil reserves took a surprising jump in 1987, mainly because of revisions for fields in Alaska. Gas reserves declined again. The Energy Information Administration, in a summary of its annual reserves report, said crude oil reserves were 27.256 billion bbl at yearend 1987, up a net 367 million bbl from yearend 1986 after production of 2.873 billion bbl. The 1.4% increase was mainly due to the largest gross positive revision EIA has recorded: 3.687 billion bbl. Most of that was due to large increases of 1.121 billion bbl from infill drilling, enhanced recovery, and reassessment of Alaskan field performance.

EIA also listed gross increases in the Lower 48, mainly

because of expansions in thermal recovery projects for heavy oil in California (505 million bbl) and enhanced recovery projects in West Texas (477 million bbl).

The revisions were supported by improved economics as the average U.S. crude oil price increased to $15.41/bbl in 1987 from $12.51/bbl in 1986. Crude reserves adjustments and revisions added 2.549 billion bbl gross in 1987, nearly three times the 1986 volume and roughly twice the 1977-86 average of 1.426 billion bbl/year. Crude oil extensions, new reservoirs in old fields, and new field discoveries added 691 million bbl in 1987, up from 1986 but still 22% less than the 891 million bbl/year average in 1977-86.

New field discoveries doubled to 96 million bbl in 1987 but were still only 59% of the 1977-86 average of 162 million bbl. More than half of the 1987 discoveries, 58 million bbl, came from Niakuk field in the Beaufort Sea off Alaska's North Slope. Reserves added by extensions, 484 million bbl, were 20% higher than in 1986. New reservoirs in old fields, 111 million bbl, were up 37% from 1986. However, both were substantially below the 1977-86 yearly averages.

Indicated additional reserves, defined as crude oil volumes that may become economically recoverable from known reserves through application of improved recovery methods using current technology, totaled 3.649 billion bbl.

"The presence of large indicated additional reserves in the North Slope of Alaska, California, West Texas, and New Mexico implies that significant upward revisions to proved reserves will continue in the future," EIA said.

Dry gas reserves slipped to 187.211 tcf at the end of 1987 from 191.586 tcf at yearend 1986. As a result, U.S. gas reserves had declined for 6 consecutive years, and the latest 2.3% drop was the largest of that period.

The latest decline resulted from a combination of the lowest total discoveries (7.175 tcf) recorded by EIA, somewhat higher production (16.114 tcf) than in 1986, and a net increase in the 1987 reserve base through revisions and adjustments (4.564 tcf) that was a little lower than in 1986 although still nearly three times the 1977-86 average.

EIA's reserves figure includes 24.6 tcf on the Alaskan North Slope that do not have a market at present. Although producers have removed the reserves from the proved category because they cannot be sold, EIA noted the produced gas created a positive net cash flow from stripped liquids and from use as a fuel in field and pipeline operations. Given the economic production of North Slope gas, EIA chose to retain it as proved reserves.

Total discoveries of dry natural gas were down 20% to 7.175 tcf in 1987, the lowest discovery level during 1977-87. New field discoveries were roughly flat at 1.089 tcf but still 57% lower than the 1977-86 average of 2.539 tcf.

Extensions (4.587 tcf) and new reservoirs discovered in old fields (1.499 tcf) were substantially down in 1987.

"These discoveries reflect exploratory gas drilling that dropped to even lower levels than in 1986," EIA said. "In fact, they were the lowest since 1973."

Reserves of natural gas liquids were 8.147 billion bbl, essentially unchanged from the 8.165 billion bbl at the end of 1986. The first full year of operation for large gas processing plants in New Mexico and on the Alaskan North Slope helped sustain NGL reserves.

Gas markets spark pipeline plans

Growing demand for gas in the U.S. Northeast, Florida, and California spawned a flurry of construction proposals by gas transmission lines in 1988. At yearend, the Federal Energy Regulatory Commission had not sorted out enough proposals in its approval process to set off a wave of pipeline construction.

Construction of crude oil and petroleum products pipelines was barely noticeable when compared with gas transmission activity.

WESTERN U.S.
BRITISH COLUMBIA
ALBERTA
SASK
WASHINGTON
OREGON
IDAHO
MONTANA
NEVADA
UTAH
WYOMING
CALIFORNIA
ARIZONA
Pacific Ocean
Vancouver
Nelson
Trail
Cherry Point
Ferndale
Port Angeles
Anacortes
Seattle
Tacoma
Olympia
Aberdeen
Longview
Vancouver
Portland
Salem
Corvallis
Eugene
Roseburg
Medford
Klamath Falls
Yakima
Richland
Pasco
Walla Walla
Pendleton
The Dalles
Bend
Baker
Ontario
Caldwell
Boise
Mountain Home
Twin Falls
Pocatello
Moses Lake
Spokane
Coeur d'Alene
Moscow
Lewiston
Missoula
Anaconda
Butte
Dillon
Helena
Great Falls
Babb
Kevin
Cut Bank
Mosby
Bascom
Hardin
Laurel
Billings
Hardin
Cody
Greybull
Thermopolis
Fort Washakie
Pinedale
La Barge
Rock Springs
Hiawatha
Logan
Brigham City
Ogden
Salt Lake City
Woods Cross
Provo
Roosevelt
Vernal
Rangely
Craig
Grand Junction
Montrose
Durango
Farmington
Bloomfield
Gallup
Flagstaff
Winslow
Prescott
Phoenix
Tucson
Ajo
Nogales
Douglas
Eureka
Reno
Donner Pass
Carson City
Yerington
Fallon
Elko
Ely
Blackburn
Currant
Tonopah
Fredonia
Sacramento
Stockton
Hayward
Modesto
Hercules
Richmond
San Francisco
Oakland
Redwood City
San Jose
Santa Clara
Santa Cruz
Salinas
Santa Rosa
Arbuckle
Rio Vista
Chowchilla
Fresno
Hanford
Coalinga
Oildale
Bakersfield
Las Vegas
Boulder City
Hoover Dam
Kingman
Estero Bay
San Luis Obispo
Santa Maria
Mc Kittrick
Santa Barbara
Ventura
Oxnard
Newhall
Los Angeles
Long Beach
Santa Fe Springs
San Bernardino
South Gate
Carson
El Segundo
Torrance
Wilmington
Signal Hill
Huntington Beach
Carlsbad
San Diego
Tijuana
Mexicali
Yuma
El Centro
Rosarito
0 100 200 Mi.
0 100 200 Km.

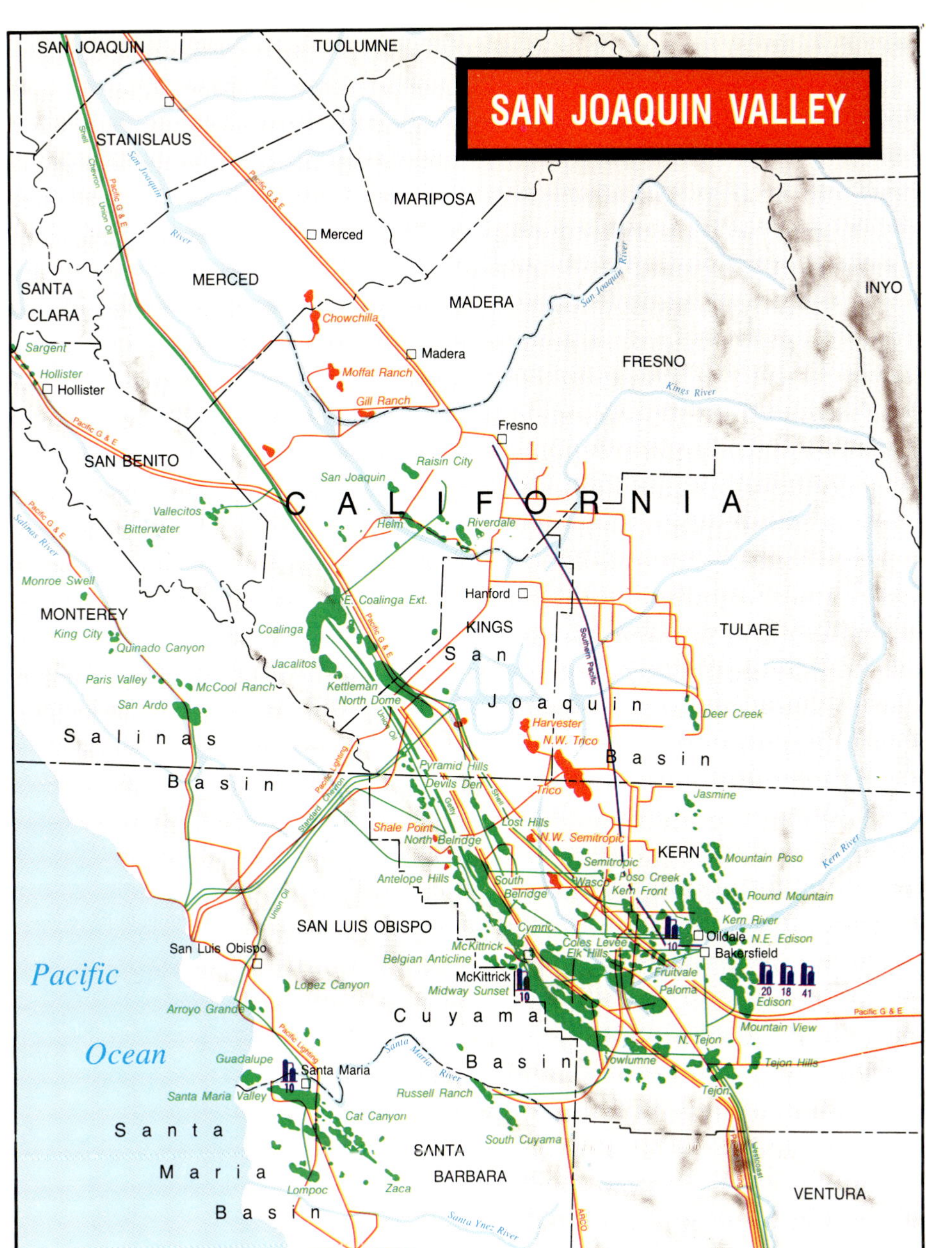

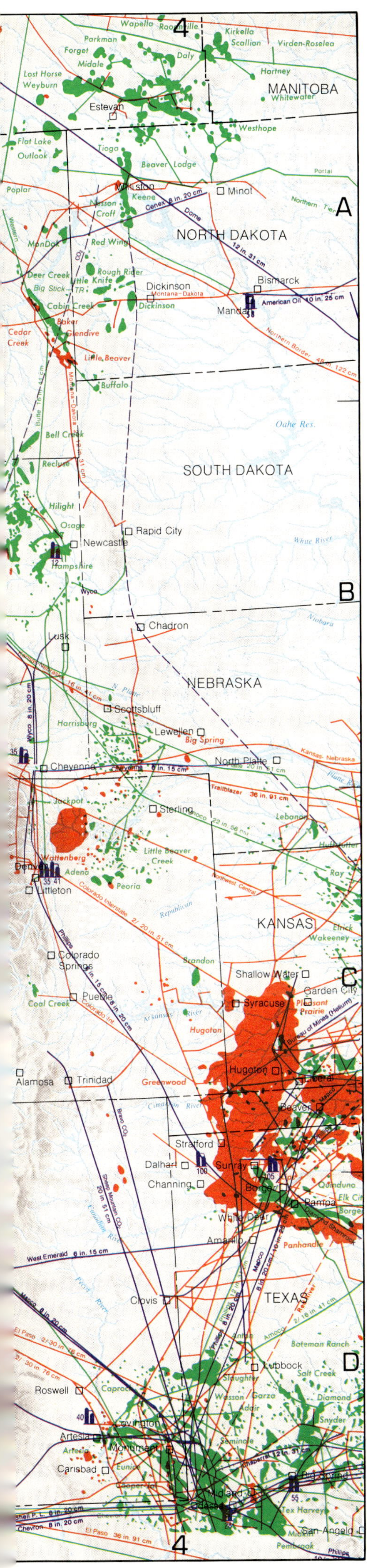

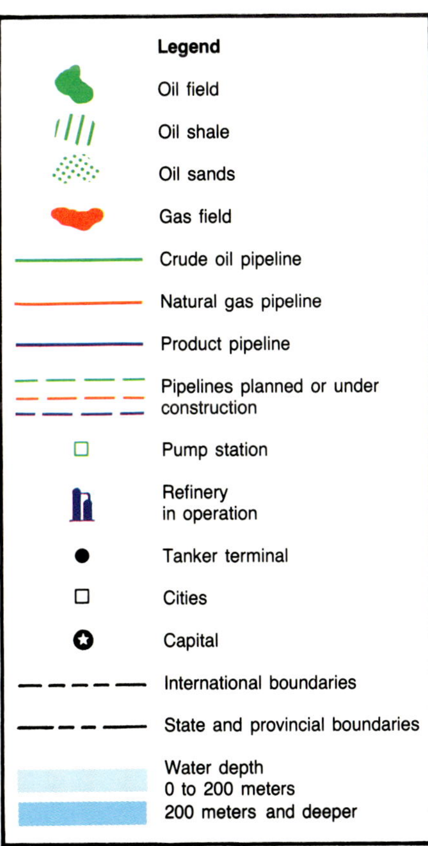

1989 was barely under way when it became apparent that construction on California's first interstate gas pipeline could start before the year was over. FERC issued a final certificate authorizing Coastal Corp.'s Wyoming-California Pipeline Co. (WyCal) to build a $665 million, 1,000 mile, 24-36 in. system from Southwest Wyoming to Kern County, Calif., to supply fuel for cogeneration and thermal enhanced oil recovery projects.

WyCal, still negotiating contracts with shippers, said it expected construction to begin in second half 1989 and deliveries to start by early 1991. No other line proposed to serve California's big EOR/cogeneration market had received a FERC certificate, although three competing proposals were still alive. Transmission plans were flourishing despite a lingering surplus of gas supply. Pipeliners, convinced the surplus could not last too much longer, were preparing for the long term market.

The American Gas Association at midyear 1988 predicted that the U.S. gas "bubble" would shrink from a high

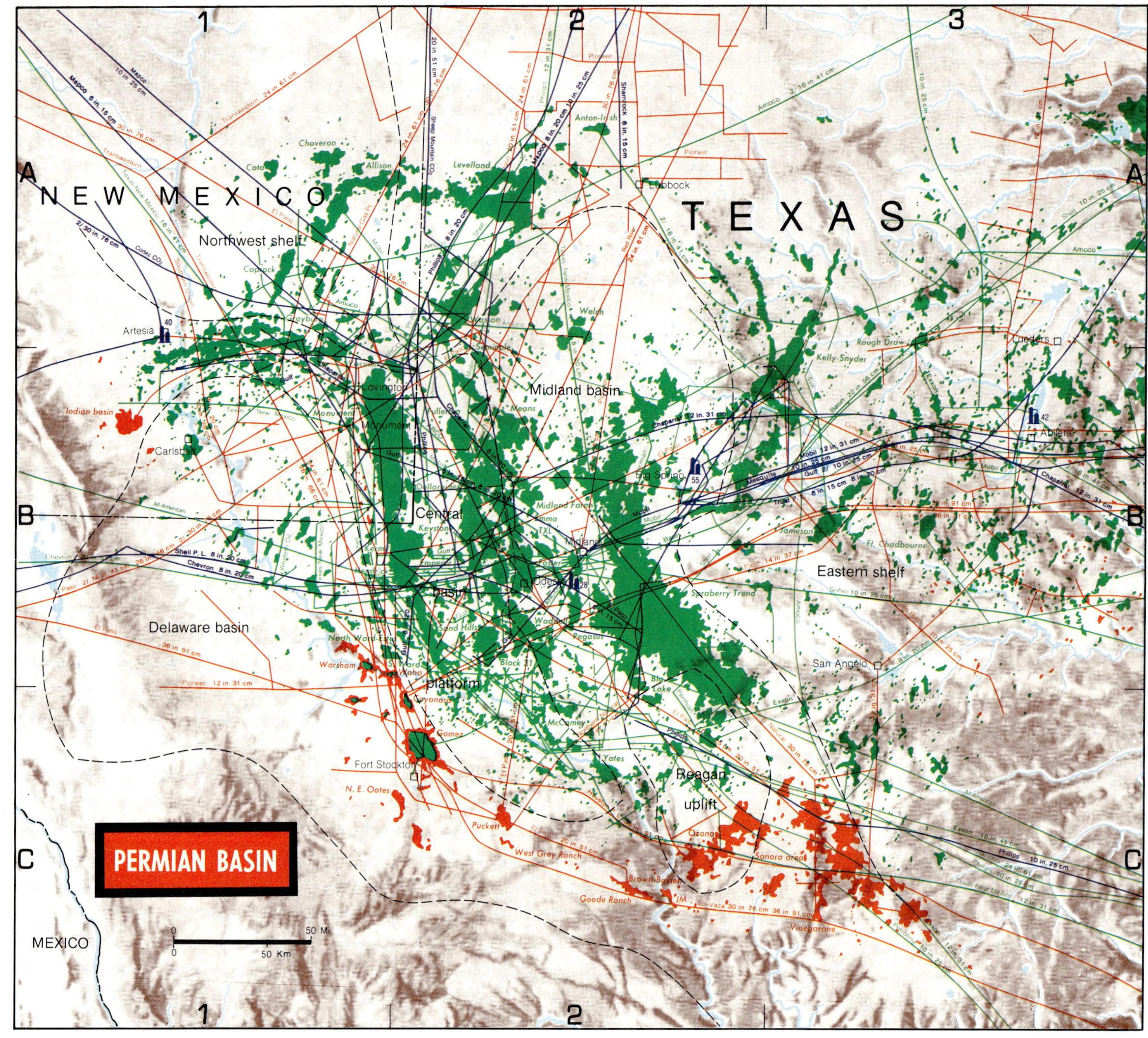

of 4.6 tcf in 1983 to only 1.5 tcf in 1988 and disappear by 1990 as supply and demand come into approximate balance.

The outlook reflected rising gas demand and declining productive capacity. AGA expected gas demand to grow 6% in 1988, the biggest annual increase of the previous 2 decades. It pointed out that the current slack in the gas transmission network could vanish during periods of peak demand. So interruptible and "best efforts" customers could experience supply cutoffs during winter months.

The wrenching reorganization drive set off by the 1986 depression in the petroleum industry took a concentrated form among U.S. gas transmission companies.

Many had their exploration and production assets on the auction block as a result of heavy leveraging and soaring commitments to take or pay settlements with producers whose gas sales and prices fell short of contract terms.

Tenneco Inc. was the pacesetter among gas pipelines in the sale of nontransmission assets. Put up for sale in September 1988, its exploration/production and refining/marketing assets in the U.S. and abroad had drawn more than $7.3 billion in high bids from industry by mid-October. Sales in the biggest transaction of its kind in history were being concluded as 1989 approached (Table 3).

As a group, gas transmission companies with upstream affiliates were among the most highly leveraged in the U.S. petroleum industry. They report an average long term debt that represents 132% of stockholder equity in fiscal year 1987, twice the industry average of 60.22%. The proceeds from most asset sales were used to reduce debt, thereby boosting stock prices. However, there were other reasons to sell upstream assets. Some managers believed they could achieve a better rate of return by investing in other operations. Others said a more focused company was easier to manage and provided investors a clearer window with which to view an asset's true value. Not all gas pipelines were trying to sell their upstream assets. Some were seeking to buy some of the properties put into play by their competition.

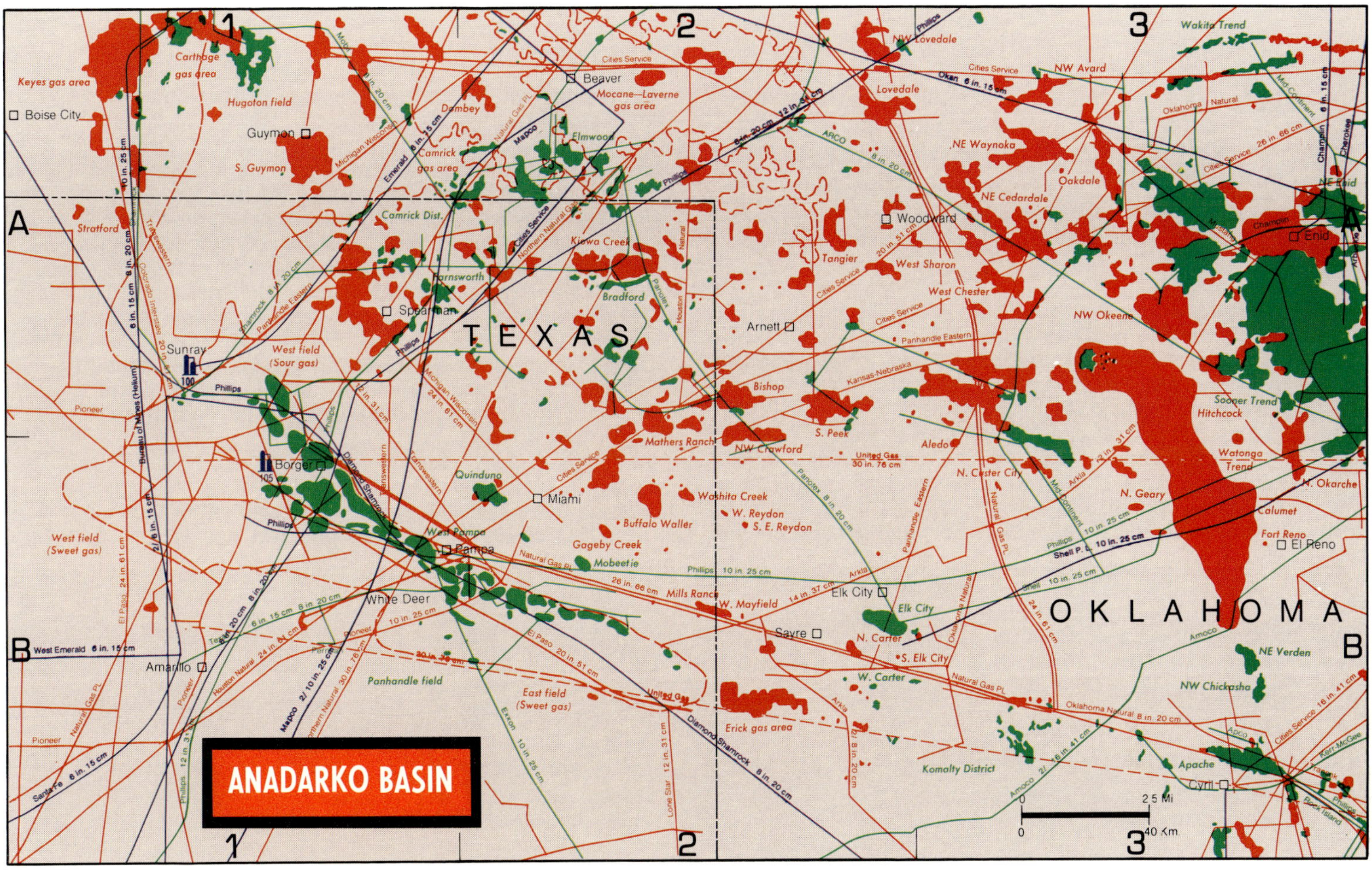

ANADARKO BASIN
TEXAS
OKLAHOMA
Keyes gas area
Boise City
Carthage gas area
Hugoton field
Guymon
S. Guymon
Dumbey
Beaver
Mocane—Laverne gas area
Elmwood
NW Lovedale
Lovedale
NW Avard
Camrick gas area
Camrick Dist.
Stratford
Sunray
West field (Sour gas)
Kiowa Creek
Farnsworth
Spearman
Bradford
Arnett
Woodward
Tangier
West Sharon
West Chester
NW Okeene
NE Waynoka
NE Cedardale
Oakdale
Enid
Borger
Quinduno
Miami
Buffalo Waller
Washita Creek
Gageby Creek
Mobeetie
Mathers Ranch
NW Crawford
S. Peek
Bishop
Aledo
Sooner Trend
Hitchcock
Watonga Trend
N. Geary
N. Okarche
Calumet
Fort Reno
El Reno
West field (Sweet gas)
West Pampa
Pampa
White Deer
Mills Ranch
W. Mayfield
Elk City
Elk City
N. Custer City
N. Geary
Amarillo
Panhandle field
East field (Sweet gas)
Erick gas area
Sayre
N. Carter
S. Elk City
W. Carter
S. Cester City
NE Verden
NW Chickasha
Pioneer
Komolty District
Apache
Cyril
2.5 Mi
40 Km

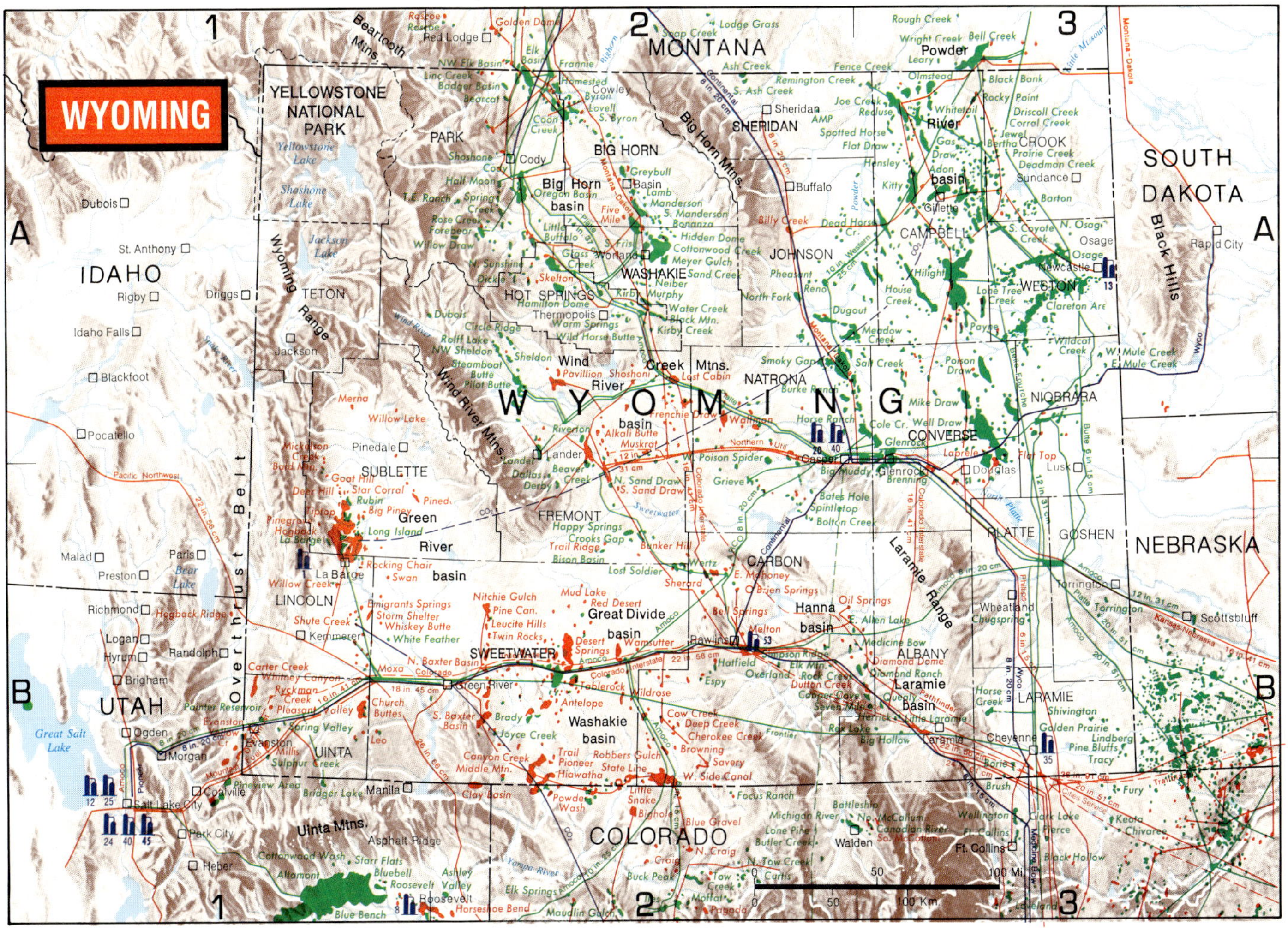

WYOMING
YELLOWSTONE NATIONAL PARK
MONTANA
SOUTH DAKOTA
IDAHO
WYOMING
NEBRASKA
UTAH
COLORADO
Dubois
St. Anthony
Rigby
Idaho Falls
Driggs
Blackfoot
Pocatello
Malad
Preston
Richmond
Logan
Hyrum
Brigham
Paris
Bear Lake
Randolph
Ogden
Morgan
Salt Lake
Great Salt Lake
Park City
Heber
TETON
Jackson
SUBLETTE
Pinedale
LINCOLN
Kemmerer
UINTA
PARK
Cody
Big Horn
BIG HORN
Greybull
Basin
WASHAKIE
HOT SPRINGS
Thermopolis
FREMONT
Lander
Riverton
HOT SPRINGS
Wind River
NATRONA
Casper
CONVERSE
Douglas
Glenrock
CARBON
Rawlins
SWEETWATER
Green River
Rock Springs
Great Divide basin
Washakie basin
Hanna basin
ALBANY
Laramie
Medicine Bow
SHERIDAN
Sheridan
Buffalo
JOHNSON
CAMPBELL
Gillette
CROOK
Sundance
WESTON
Newcastle
NIOBRARA
Lusk
PLATTE
GOSHEN
Torrington
Scottsbluff
Wheatland
Chugwater
LARAMIE
Cheyenne
Ft. Collins
Walden
Craig
COLORADO
Rapid City
Black Hills
Powder River
Big Horn Mtns
Wind River Mtns
Laramie Range
Overthrust Belt
Uinta Mtns
50 Mi
100 Km

CANADA

CAPITAL: Toronto
MONETARY UNIT: Dollar
REFINING CAPACITY: 1,855,800 b/cd
PRODUCTION: 1.594 million b/d
RESERVES: 6,785,514,000 bbl

THE CANADIAN PETROLEUM INDUSTRY ENTERED 1989 after taking two historic steps.

As a result of action in 1988, the industry was poised to hike oil and gas exports to the U.S. It also made another move toward offshore field development.

Canadian approval of a free trade agreement (FTA) with the U.S., approved earlier by the American government, was destined to add a dimension to burgeoning energy trade between the countries.

It assured producers and consumers on both sides of the border that, except in emergencies, neither country will impose new trade restrictions.

The assurance may have no immediately discernible effect on two way energy trade—at more than $10 billion (U.S.) in 1987, the world's largest—because growth already was so strong. But it did provide an important planning tool.

In September 1988, Mobil Oil Canada Ltd. let the first engineering contract related to development of Hibernia oil field off Newfoundland to Newfoundland Geosciences Ltd., St. John's, Newf., for comprehensive engineering of the Hibernia site.

A British drillship was scheduled to begin a geotechnical drilling program, and a remotely operated vehicle was to perform a photo survey of the Atlantic Ocean floor at the site on the Grand Banks.

Canadian governments and a group led by Mobil had agreed in principle the preceding July on a financial aid package that paved the way for development of the field.

The $8.5 billion (Canadian) project was designed to yield

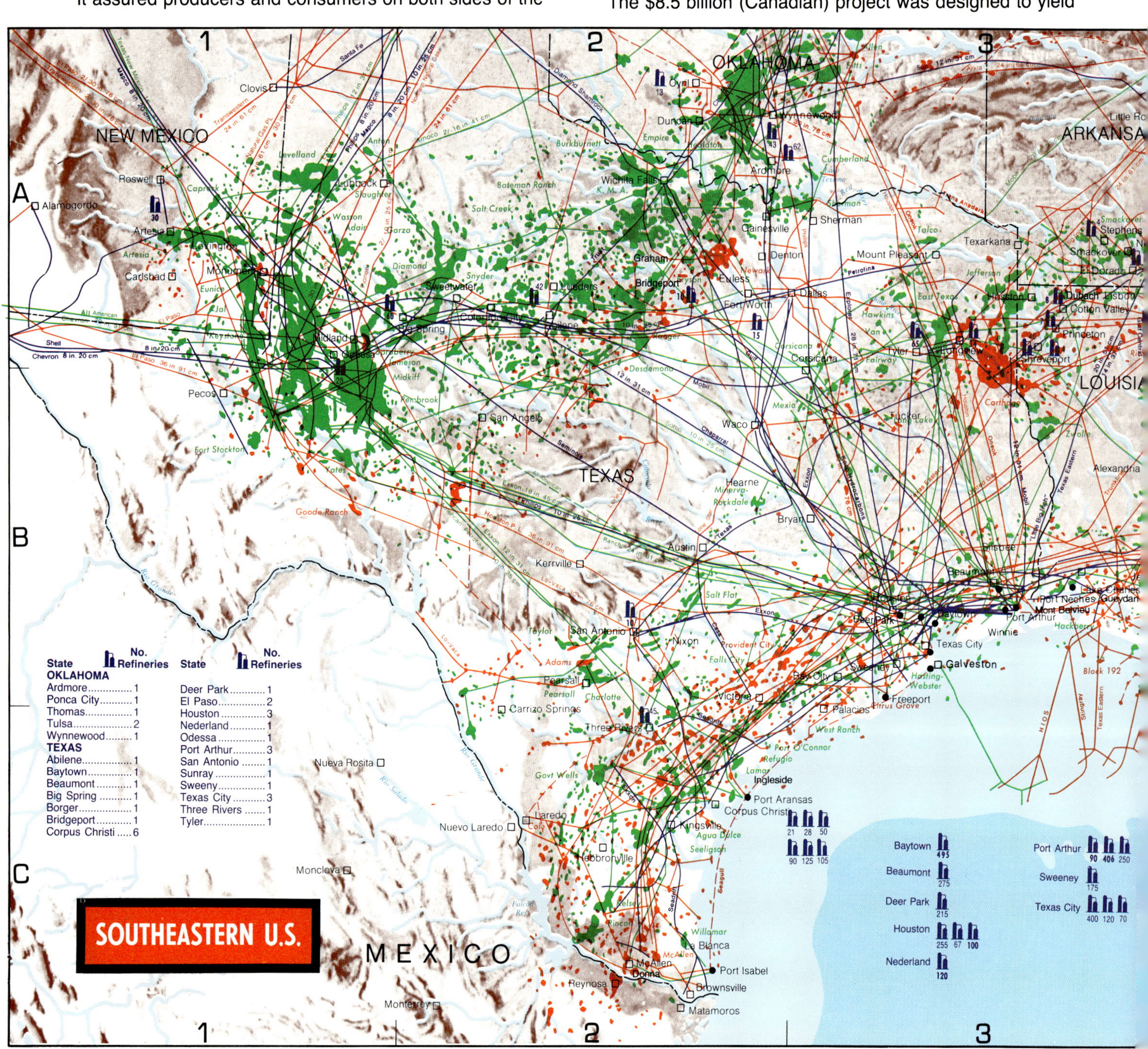

State	No. Refineries	State	No. Refineries
OKLAHOMA		Deer Park	1
Ardmore	1	El Paso	2
Ponca City	1	Houston	3
Thomas	1	Nederland	1
Tulsa	2	Odessa	1
Wynnewood	1	Port Arthur	3
TEXAS		San Antonio	1
Abilene	1	Sunray	1
Baytown	1	Sweeny	1
Beaumont	1	Texas City	3
Big Spring	1	Three Rivers	1
Borger	1	Tyler	1
Bridgeport	1		
Corpus Christi	6		

Location	Capacities
Baytown	495
Beaumont	275
Deer Park	215
Houston	255, 67, 100
Nederland	120
Port Arthur	90, 406, 250
Sweeney	175
Texas City	400, 120, 70

Canada's first commercial volumes of offshore hydrocarbons.

Free trade agreement

Canadians ensured approval of the FTA when they reelected the Progressive Conservative party Nov. 21, 1988. U.S. President Ronald Reagan signed American ratification of the pact the preceding Sept. 28.

Canadian Prime Minister Brian Mulroney called Parliament into session the week of Dec. 12 to complete implementing legislation for the FTA before the Jan. 1 deadline.

Free trade was the major issue of the Canadian election campaign. In energy, trade is mostly from Canada to the U.S. Canada exported a net 992 bcf of gas to the U.S. during 1987, 34% of marketed production and 5.77% of U.S. gas demand (Fig. 1).

It shipped a net 592,000 b/d of crude to the U.S. in 1987, 47.8% of production and about 3.5% of U.S. demand.

Canada provided 800,000 b/d of crude, products, and liquid petroleum gases, or 13% of U.S. imports.

U.S. demand for Canadian oil has grown as its oil production declined. One large producer voiced the possibility for more growth in U.S. demand for Canadian oil, mainly in the Midwest.

U.S. gas producers had a different problem. FTA approval by the U.S. occurred during a lengthy gas deliverability surplus.

U.S. gas producing and consuming interests disagreed about whether FTA discriminated in favor of Canadian gas producers but seemed to agree that it sank chances for an oil import fee.

What was certain was that U.S. gas producers, unlike their Canadian counterparts, languished in a maze of federal regulations, unresolved take or pay disputes, and related limitations imposed by Congress, Federal Energy Regulatory Commission, and courts.

The U.S. is Canada's only customer for oil and gas

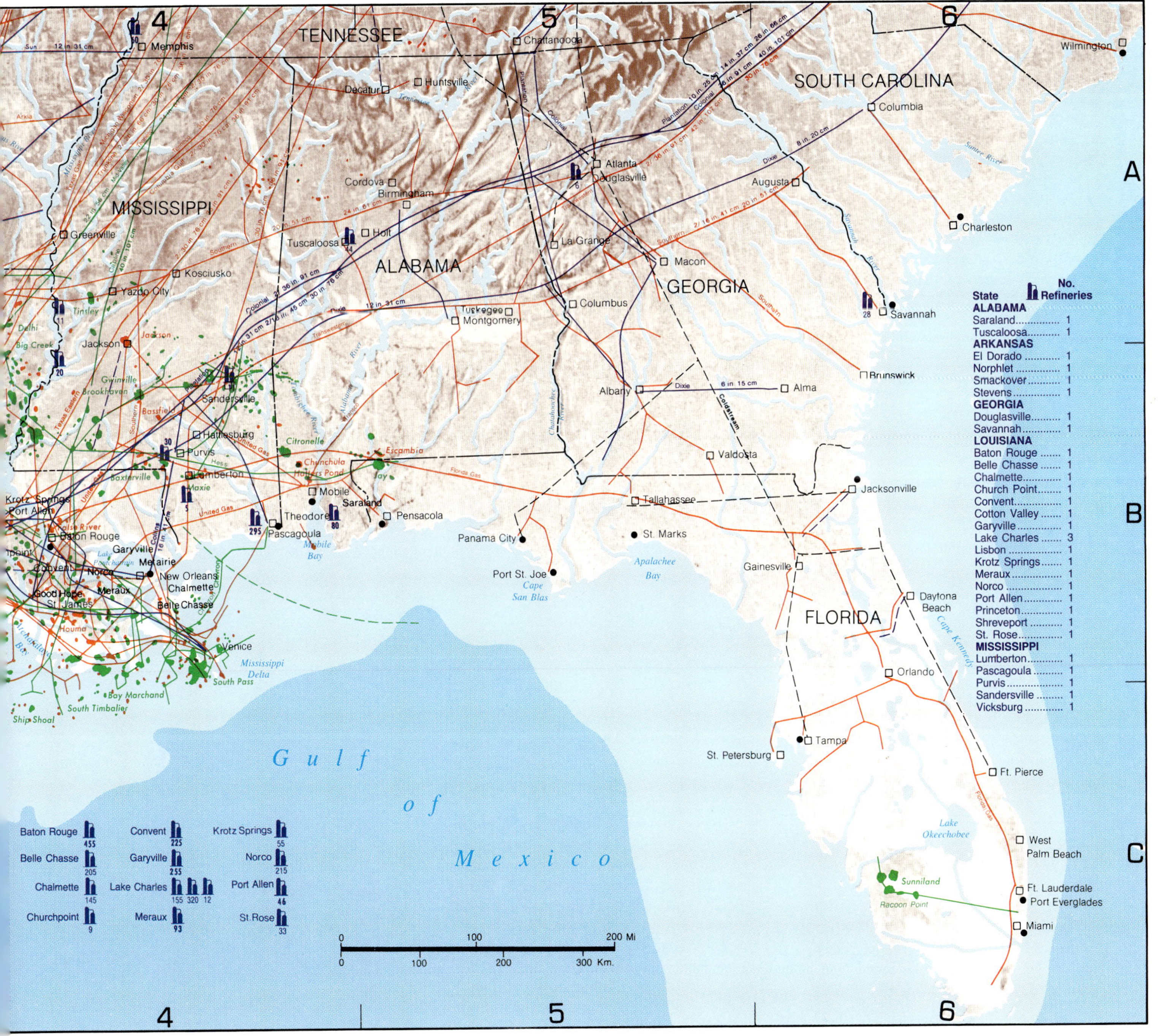

State	No. Refineries
ALABAMA	
Saraland	1
Tuscaloosa	1
ARKANSAS	
El Dorado	1
Norphlet	1
Smackover	1
Stevens	1
GEORGIA	
Douglasville	1
Savannah	1
LOUISIANA	
Baton Rouge	1
Belle Chasse	1
Chalmette	1
Church Point	1
Convent	1
Cotton Valley	1
Garyville	1
Lake Charles	3
Lisbon	1
Krotz Springs	1
Meraux	1
Norco	1
Port Allen	1
Princeton	1
Shreveport	1
St. Rose	1
MISSISSIPPI	
Lumberton	1
Pascagoula	1
Purvis	1
Sandersville	1
Vicksburg	1

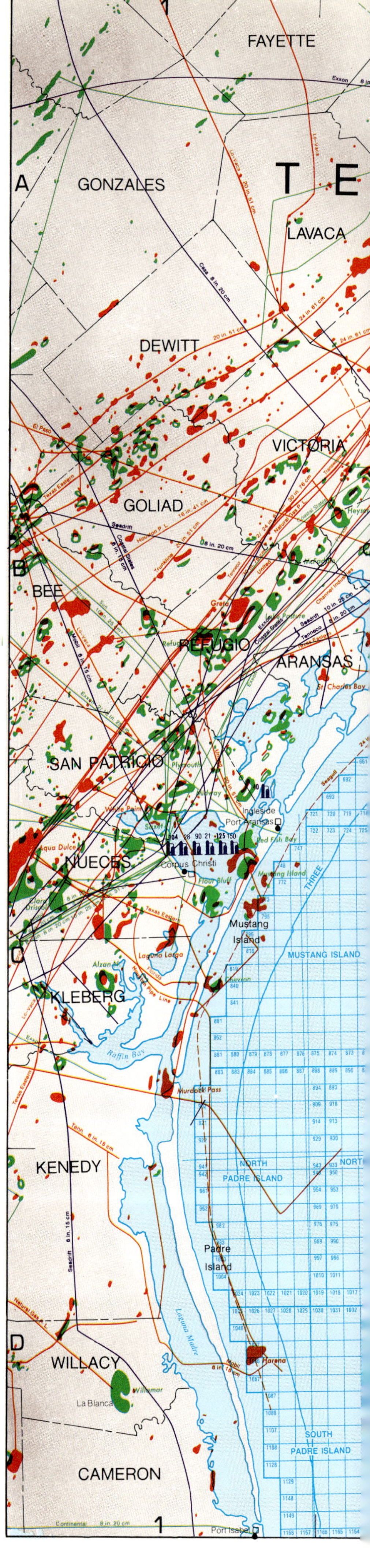

exports. Most of Canada's oil exports go to the U.S. Midwest, while most of its gas exports go to the U.S. West Coast (Fig. 2).

In other 1987 energy trade, Canada bought 15.8 million tons of U.S. coal, 20% of U.S. exports; Canada sent the U.S. about 43,350 gigawatt-hr of electricity, about 1.8% of U.S. demand; and Canada supplied about 30% of the western world's uranium.

With two exceptions, the FTA droppd restrictions on energy shipments, such as quotas, export or import taxes, and minimum import or export price requirements. It also barred new restrictions.

The exceptions involved national security and shortages. National security changes were limited to cases concerning nuclear nonproliferation or military threats.

Another clause called for equal sharing of cuts in the event of a shortage. Canada can impose export quotas if it cuts production and must provide the U.S. with proportionate access to supplies.

"Proportionate" meant that Canada must allow U.S. customers to bid for the same average share of Canadian oil production that the U.S. took in the previous 36 months. Canada has to declare the shortage.

All trade between the two countries is valued at about $150 billion/year, the world's largest trade partnership.

How Canada, U.S. viewed the FTA

Canada's energy industry favored the FTA and welcomed the Conservative victory.

Canadian Petroleum Association (CPA) Pres. Ian Smythe summarized the benefits FTA provided the energy sector.

FTA did not displace federal agreements to provide financial support for megaprojects that include development of Hibernia field, an Alberta-Saskatchewan heavy oil upgrader, the OSLO oil sands project in northern Alberta, and a gas pipeline from the mainland to Vancouver Island.

FTA's stemming of the threat of revived federal interference benefited industry, Smythe said.

George Lawrence, president of the American Gas Association, said his group was pleased with the outcome of the Canadian national election.

"The consensus seems clear that this is going to be very, very good for both countries and should serve as a candle for the rest of the countries of the world on trade agreements."

Lawrence said stabilized gas trade between Canada and the U.S. will serve as a supply umbrella for the U.S. which will permit pursuit of new markets. American producers will participate in 90-95% of those new markets, so the FTA should be good for them as well as for Canadian exporters.

U.S. producers charged that Canadian gas producers gained market advantage with the FTA.

Mitchell Energy & Development Corp. Chairman George P. Mitchell pointed out that Canadian federal and provincial governments provide exploration incentives not available in the U.S. Therefore, Mitchell said, FTA will encourage consumption of Canadian oil and gas at the expense of U.S. oil and gas.

But Martha Musgrove, director general of the Canadian Department of Energy, Mines, and Resources' natural gas branch, said FTA merely legitimized what already existed between Canada and the U.S. It also added stability and predictability to the market, Musgrove said.

Attempts by Canadian exporters to fulfill or increase their authorized levels of exports until the mid-1990s will

TEXAS GULF COAST
Gulf of Mexico
AUSTIN
HARRIS
FORT BEND
CHAMBERS
JEFFERSON
GALVESTON
Galveston Bay
WHARTON
BRAZORIA
MATAGORDA
Matagorda Bay
Matagorda Peninsula
Matagorda Island
Freeport
Sweeny
Wadsworth
Collegeport
Citrus Grove
High Island
High Island East Addition
High Island South Addition
High Island East Addition South Extension
West Cameron
West Cameron West Addition
West Cameron South Addition
Sabine Pass
GALVESTON
GALVESTON SOUTH ADDITION
BRAZOS
BRAZOS SOUTH ADDITION
LEAGUE LINE
MUSTANG ISLAND EAST ADDITION
CORPUS CHRISTI NG 14-3
PORT ISABEL NG 14-6
EAST BREAKS NG 15-1
ALIMINOS CANYON NG 15-14
SOUTH PADRE ISLAND EAST ADDITION
Blk 405 L
Blk 140 L
Blk 485 L
Blk 538L
Blk 541L
Blk 722S
Blk A-76
Blk 71 L
Blk 140 L
Blk 160L
Blk 228S
(Blk 52L)
0 10 20 30 40 50 Mi
80 Km
ATTENTION USER: This special OCS environmental map (visual)
has been prepared from existing sources to accompany an
Environmental Statement. THE BUREAU OF LAND MANAGEMENT,
USDI, does not guarantee the accuracy to the extent of
responsibility of liability for reference therein.

This OCS visual is not to be used for
navigational purposes, nor is it a legal document
for federal leasing purposes. Prepared by the
New Orleans Outer Continental Shelf Office,
New Orleans, La.

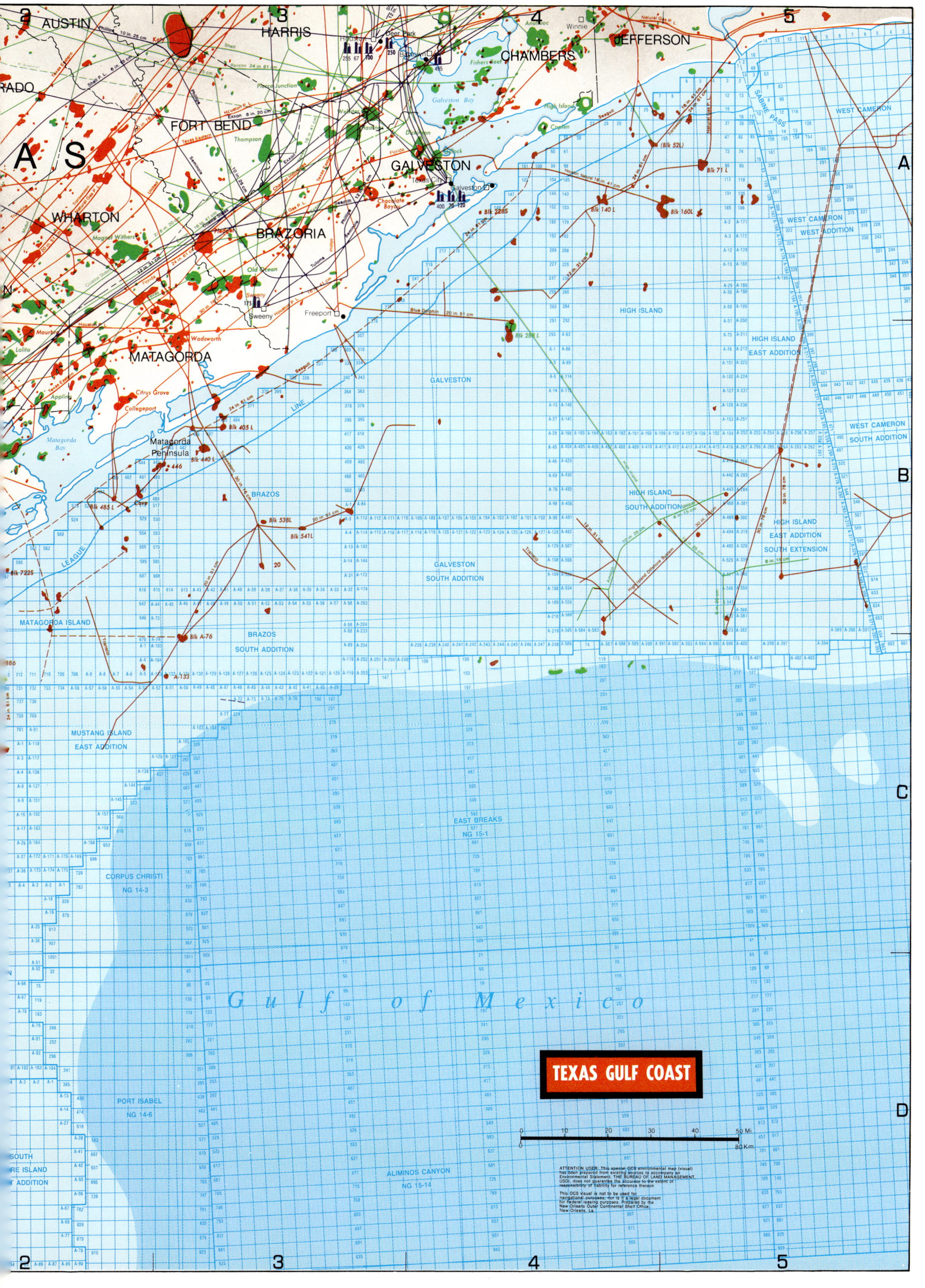

121

TEXAS
LOUISIANA
BEAUREGARD
ALLEN
ST. LANDRY
JEFFERSON DAVIS
ACADIA
CALCASIEU
Lake Charles
Jennings
Churchpoint
Raynes
Lafayette
ST. MARTIN
IBERVILLE
ASCE
LAFAYETTE
Maurice
Port Neches
Port Arthur
Sabine Lake
Hackberry
CAMERON
Calcasieu Lake
Mud Lake
Cameron
Holly Beach
Big Lake
Grand Lake
White Lake
Lac Hope
VERMILION
IBERIA
New Iberia
Avery Island
Erath
Abbeville
ASSUMPTIO
ST. MARTIN
Charenton
ST. MARY
Weeks Island
Vermilion Bay
Cote Blanche
West Cote Blanche Bay
Garden City
Morgan City
Bayou Sherman
Pecan Island
WEST CAMERON
Blk 45
West Cameron 65
THREE
LINE
Tiger Shoal
Robin Island
Atchafalaya
Bayou Penchant
HIGH ISLAND
Sabine Pass
Blk 149
Blk 180
Blk 63
Blk 192
Blk 89
Vermilion 84
Blk 76
Blk 71
235
SOUTH MARSH ISLAND NORTH ADDITION
S. Marsh Is. 257
S. Marsh Is. 265
Blk 1001
Blk 32
Blk 28
EAST CAMERON
Blk 131
WEST CAMERON WEST ADDITION
West Cameron 264
Blk 260
Blk 104
VERMILION
Blk 227
136
EUGENE ISLAND
Blk 175
Blk 188
SHIP
HIGH ISLAND EAST ADDITION
SOUTH MARSH ISLAND
Blk 88
Blk 45
Blk 198
Blk 205
Blk 66
Blk 464
West Cameron 475
WEST CAMERON SOUTH ADDITION
Tennessee Line
VERMILION SOUTH ADDITION
Blk 276
HIGH ISLAND EAST ADDITION SOUTH EXTENSION
West Cameron 551
EAST CAMERON SOUTH ADDITION
VERMILION SOUTH ADDITION
SOUTH MARSH ISLAND SOUTH ADDITION
EUGENE ISLAND SOUTH ADDITION
Blk 330
Ship Shoal
SHIP SOUTH A
Terpon
East Cameron 359
Vermilion 380
Vermilion 397
GARDEN BANKS NG 15-2

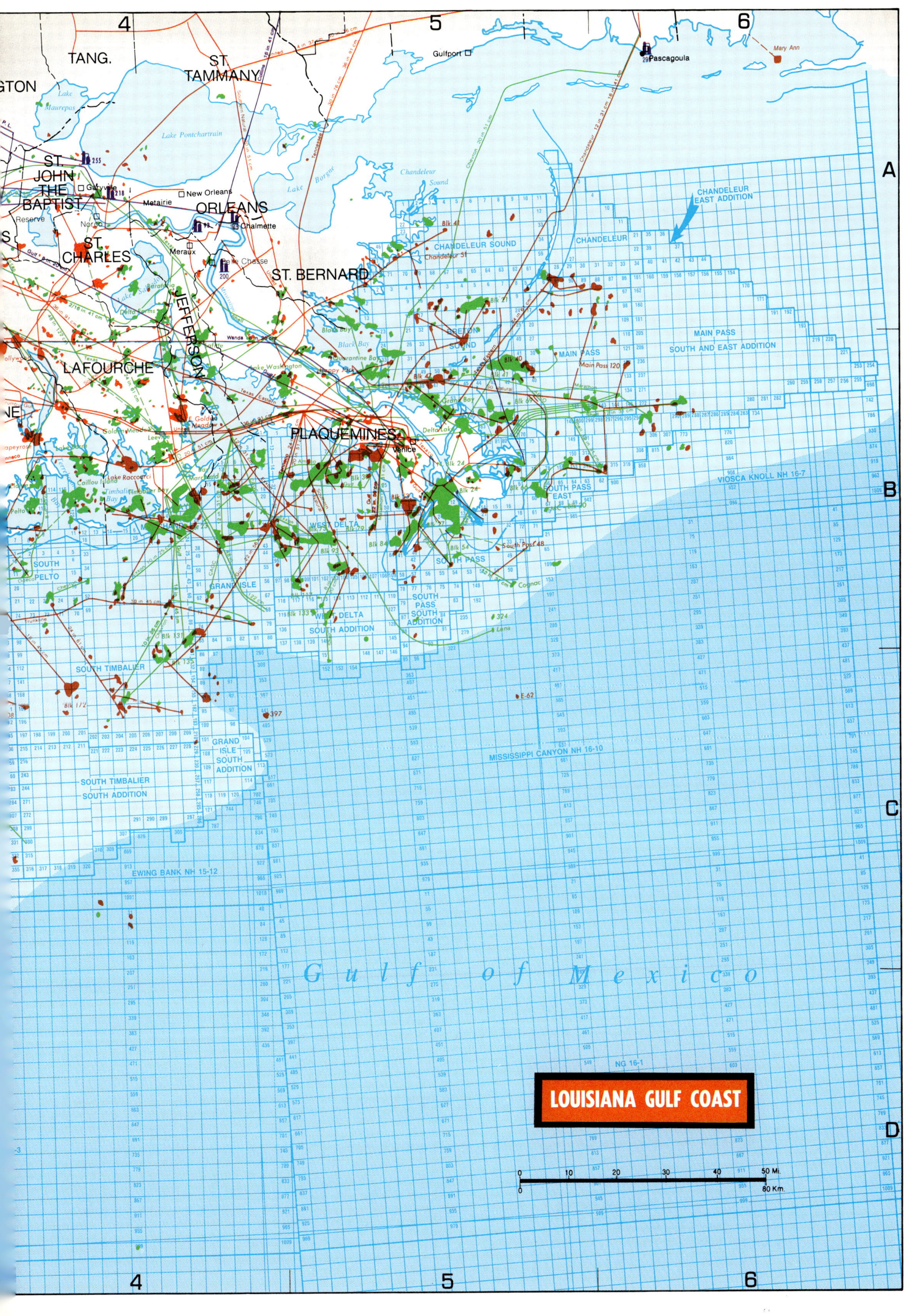
LOUISIANA GULF COAST
Gulf of Mexico
TANG.
ST TAMMANY
Gulfport
Pascaguola
Mary Ann
Lake Maurepas
Lake Pontchartrain
Lake Borgne
Chandeleur Sound
CHANDELEUR EAST ADDITION
ST JOHN THE BAPTIST
Garyville
Reserve
Norco
Metairie
New Orleans
ORLEANS
Chalmette
Meraux
Belle Chasse
ST. BERNARD
ST CHARLES
CHANDELEUR SOUND
Chandeleur
Chandeleur St
MAIN PASS
MAIN PASS SOUTH AND EAST ADDITION
Main Pass 120
JEFFERSON
Wanda
Black Bay
Quarantine Bay
Lake Washington
Texas Eastin
LAFOURCHE
Grand Bay
Delta Lake
Venice
PLAQUEMINES
Golden Meadow
Lake Raccourci
Timbalier Bay
Caillou Island
SOUTH PASS EAST
VIOSCA KNOLL NH 16-7
SOUTH PASS
South Pass 48
Cognac
SOUTH PELTO
GRAND ISLE
WEST DELTA
WEST DELTA SOUTH ADDITION
SOUTH PASS SOUTH ADDITION
Lena
SOUTH TIMBALIER
E-62
Trunkline
GRAND ISLE SOUTH ADDITION
SOUTH TIMBALIER SOUTH ADDITION
MISSISSIPPI CANYON NH 16-10
EWING BANK NH 15-12
Gulf of Mexico
NG 16-1
LOUISIANA GULF COAST
0 10 20 30 40 50 Mi.
0 80 Km.

encounter and promote increased price competition, said a study conducted by Gas Research Institute (GRI), Chicago.

In later years, however, U.S. demand for Canadian gas may not be fully satisfied unless contracts provide for further development. The FTA may facilitate emergence of such long term arrangements, GRI said.

The Canadian Gas Association predicted that Canadian domestic gas demand will increase steadily, rising 25% to 2.6 tcf/year by 2000.

Pressure was increasing on Alberta's Energy Resources Conservation Board (ERCB) to change its policy that limited gas export permits to 15 years. Some companies said the policy is not consistent with deregulation of the gas industry, and companies were continuing to sign export sales deals for terms beyond 15 years.

U.S. market
for Canadian crude

Refiners beyond Chicago in the U.S. Midwest may provide a market for 50,000-100,000 b/d of additional Canadian heavy crude and bitumen, said Jack Montgomery, Amoco Corp. supply vice-president.

Canada shipped the U.S. 261,000 b/d of those hydrocarbons in 1985, 329,000 b/d in 1986, 350,000 b/d in 1987, and 413,000 b/d in the first 6 months of 1988. The U.S. industry for some years has viewed 400,000-450,000 b/d as the maximum U.S. capacity for Canadian conventional heavy crudes and bitumen without major hardware changes.

The U.S. West Coast oil market was saturated, Gulf and East Coast markets had ample supplies of better quality and more competitive crudes available from other sources, and price and quality competition was fierce in Far East markets, Montgomery said.

Canadian producers could reach Wood River, Ill., or Catlettsburg, Ky., refineries by barging or reversing either of two northbound pipelines.

The price of Canadian heavy oil/bitumen was established by approximate market parity with Mexican Maya crude at the point of competition in the Chicago area.

Competing with Maya at Wood River instead of Chicago resulted in 45¢/bbl less producer netback at Edmonton, Alta., the product of 20¢/bbl less pipeline tariff on Maya and an assumed 25¢/bbl more on Canadian heavy.

"To the extent that pricing is on an Edmonton-plus transportation basis, the 45¢/bbl lower netback could be incurred on sales at Twin Cities and Chicago as well as Wood River," Montgomery said.

Cutting Canadian prices to fight for markets south of Chicago would force Gulf Coast suppliers to decide whether to fight for northern markets.

Many would not. Venezuela was considering buying a one half interest in the Unocal Corp. refinery near Chicago as a secure market for Venezuelan crude. The plant was running nearly 50,000 b/d of heavy Canadian crude, Montgomery noted.

Another element of the agreement, elimination of U.S. duties on products imported from Canada, put offshore refiners, which still must pay the duties, at a competitive disadvantage.

Canadian gas exports hit record volume

Canada was selling record volumes of gas to the U.S. as 1989 approached.

Volume for the 1987-88 contract year ended Oct. 31 was estimated at 1.3 tcf, up from 921 bcf in contract year 1986-87, and sales were forecast to increase.

Export prices and netbacks to producers also were rising, Canada's National Energy Board (NEB) reported.

The export price averaged $2.28 (Canadian)/Mcf in September 1988, up from $2.26 in the preceding August and $2.21 in July. Producer netback averaged $2/Mcf, up from $1.97 in August and $1.91 in July.

September sales to the U.S. totaled 93.89 bcf, up 29% from September 1987 sales. Sales had averaged 94.6 bcf/month since the previous April.

U.S. industry executives forecast a growing export market for Canadian gas at an October 1988 energy conference in Calgary.

Separately, GRI predicted late in 1988 that the volume of Canadian gas available to U.S. markets will increase until the mid-1990s and then begin declining (Fig. 3).

GRI noted Canada had exported large volumes of gas to the U.S. since the late 1950s, ranging from 25% to 40% of annual Canadian production. But Canadian gas satisfied only a small share of U.S. gas consumption, and Ottawa regulated the volume and price of exports.

However, Canada relaxed controls, replacing the protection formula with a more market-oriented policy. Export prices also were decontrolled, except that they were prohibited from being lower than comparable prices to Canadian consumers.

"Expectations of the availability of Canadian gas supplies in U.S. forecasts tend to show Canadian natural gas exports gradually growing through the projections' horizons before tapering off after 2000," GRI said.

"Canadian forecasts tend to show the opposite trend, such that exports grow rapidly between now and the early to mid-1990s and then decline."

NEB in December 1988 approved a $576 million expansion of the TransCanada PipeLines Ltd. system. The project will enable Alberta gas producers to move another 114 bcf of gas to Canada and U.S. markets during the 1989-90 contract year, the company said.

Most of the gas was scheduled to begin flowing by Nov. 1, 1989. About 60% of the added gas was earmarked for U.S. Midwest and Northeast markets. The rest was destined for Canadian consumers, mainly in southern Ontario.

TransCanada planned to apply to the NEB early in 1989 for an even bigger expansion of its system, scheduled to carry a price tag of $1 billion plus.

The approved expansion was to be the largest on TransCanada's system since 1982. The project called for laying 208 miles of pipeline at various locations in each province along its system and addition of eight compression units with a combined capacity of 230,000 hp.

Alberta producers can expect to receive about $250 million in additional net revenue during 1989-90 based on current price forecasts, TransCanada said. Revenues will increase thereafter as gas prices strengthen.

NEB had warned earlier that lack of pipeline capacity could cause interruptions in gas deliveries during peak demand by industrial users.

NEB predicted that utilization of the Nova Corp. pipeline system in Alberta will reach 93% in 1989. The central section of TransCanada's transcontinental system was headed for 99% utilization during 1988 and 1989.

Hibernia development gets government boost

The Hibernia field development agreement among the federal and Newfoundland governments and Mobil's group called for an aid package worth about $3.3 billion in grants, interest rate relief, and loan guarantees for the project.

One of Mobil's partners, Chevron Canada Resources Ltd., boosted its interest in the block that holds Hibernia and several other fields on the Grand Banks (Fig. 4).

Hibernia lies in 262 ft of water, 194 miles east of St. John's.

Signing of the development agreement followed 2½ years of negotiations on the project that had resumed after collapsing with oil prices in 1986.

Canadian Minister of Energy, Mines, and Resources Marcel Masse said the project's green light will stimulate other activity off Canada, adding that the governments' investments and shares of risk and revenues are "fair and reasonable."

Total costs for Hibernia during an estimated 18-25 year life were estimated at more than $20 billion. Operating costs during a 20 year period were pegged at $11.5 billion.

Added to the first phase cost of $5.2 billion for Hibernia were outlays for a bottom-anchored concrete production platform and $700 million for oil tankers and loading facilities (Fig. 5).

The first phase cost estimate was up from an earlier estimate of $4.2 billion.

Government aid will entail a direct outlay for 25% of the preproduction capital costs, up to a ceiling of $1 billion, a ceiling of $1.6 billion in loan guarantees covering preproduction capital costs, and $300 million in interest rate relief. Newfoundland agreed' to abolish the retail sales tax on project capital items, cut the retail sales tax on operating costs to 4%, move to a "price sensitive" royalty, and provide $11 million for specific engineering work in the province.

The project's construction phase will include a dry dock for the platform's concrete base and an assembly site at Come-by-Chance, Newf., for the production facilities' main support frame. The dock, supported by a $95 million contribution from the Canadian joint offshore development fund, will be available for other offshore developments.

The agreement called for setting up contractor packages as a way to encourage Canadian participation in engineering, fabrication, construction, operation, and related supplies and services.

Assuming an oil price of $20-25/bbl in 1987 U.S. dollars, Canada will receive $2.4-4.7 billion in revenues from the project. The government of Newfoundland and Labrador will earn net revenues of $2.6-5.1 billion from the project under the same price scenario.

The agreement also called for governments to share the revenue gain if oil prices surge or reserves prove larger than expected. However, if oil prices drop to less than $12.50 (U.S.)/bbl, the federal and Newfoundland governments will get no revenue from the project, including repayment of government aid.

The governments do not expect a return on their investment until prices reach $24(U.S.)/bbl. Mobil estimated Hibernia's breakeven price at $22 (U.S.)/bbl.

The Hibernia group has the option to withdraw from a commitment to spend at least $1 billion on the project if first phase costs exceed $5.6 billion.

The 1979 discovery is to start production of about 110,000 b/d in 1995, 30 years after Mobil acquired its exploration permit (Table 1).

Production is to peak in 1996 at 150,000 b/d from reserves estimated at 525-650 million bbl.

<table>
<tr><td colspan="2">Hibernia chronology   Table 1</td></tr>
</table>

1965—Exploration permit issued to Mobil.

1966—Exploration begins.

1972-76—First evidence of oil potential, six wildcats drilled, $36 million spent.

1978—Renewal permits issued to Mobil, Gulf, and Petro-Canada.

1979—Chevron and Columbia Gas join the group, Chevron drills and tests Hibernia P-15 discovery well.

1980—Hibernia discovery announced.

1980-84—Field delineation phase, nine wells drilled, eight successful, $465 million spent.

1985—Mobil submits development plan, group makes first proposal for government fiscal regime.

1986—Government makes its first proposal for fiscal regime, group makes second fiscal proposal.

1987—Second government fiscal offer, group's third proposal.

1988—Third government offer Mar. 25, group increases cost estimate to $5.2 billion May 20, intensive negotiations begin in Toronto July 4, terms of final offer agreed on July 9, statement of principles signed July 18.

1989—Definitive agreements in place.

1989-95—Construction phase.

1995-2012—Production phase.

Mobil holds a 28.1% interest in the field. Its partners are Gulf Canada Resources Ltd. 25%, Petro-Canada 25%, and Chevron 21.9%.

Chevron in September 1988 agreed to pay $82.75 million for Columbia Gas System Inc.'s Canadian Atlantic assets. Columbia had asked for bids the preceding May.

Chevron's purchase included a 5.5% interest in the Hibernia joint venture block, in which Chevron previously owned a 16.4% interest.

The block held five major oil and gas discovery areas—Hibernia, Hebron, Nautilus, Mara, and South Mara—and a number of undrilled prospects, all in the Jeanne D'Arc subbasin. Columbia also owned acreage off Prince Edward Island and on the Labrador shelf.

The total deal involved almost 21,000 net acres in about 30,000 gross acres.

What's planned
for Hibernia hardware

The Mobil combine spelled out plans for a concrete, gravity based structure (GBS) supporting production topsides to develop Hibernia, with as many as 64 wells drilled from the structure.

An undetermined number of subsea wells beyond the practical range of directional drilling from the GBS will be needed to complete development. Subsea wells will be connected via flow lines to either production manifolds, production headers, or the GBS.

The system will be capable of treating and storing produced crude and removing water and gas.

Two articulated loading platforms (ALPs) will stand 2-3 miles off the structure to load three shuttle tankers. Each ALP will have three major components: a base, column, and loading head.

The column will attach to the base by an articulated joint that will allow the column to incline in any direction, with the buoyancy for the structure providing upright restoring force.

The loading head will be mounted on a large bearing assembly to permit rotation. The base design is either a gravity or a piled steel structure.

Articulated columns of similar design have been used in the North Sea since 1975.

Three 1,200 dwt tankers with ice strengthened hulls will transport crude to shore. Special strengthening measures were being developed to prevent damage from collision with pieces of glacial ice too small to be reliably detected.

One multipurpose support vessel, a state of the art ice clearing vessel, will be built especially for Hibernia.

The GBS will consist of a concrete base slab, a caisson reaching from the seabed to about 16 ft above mean water level, and four shafts extending above the caisson.

The entire structure will be constructed from high strength, reinforced concrete. Water and iron or other loose, highly dense material will provide ballast.

Four shafts will support the topsides facilities, including two 32 well drilling shafts, a riser shaft, and a utility shaft.

The GBS will contain complete pumping systems and will support risers for intrafield pipelines.

Topsides production facilities will consist of production, utility, drilling, electrical, and personnel support systems on three levels: the cellar deck, module deck, and weather deck.

The cellar deck will consist of a main support frame (MSF) accommodating utility and process equipment. The MSF will be an open steel truss structure, supporting all facilities above the concrete columns.

The module deck will contain process, utility, and drilling facilities in modules supported by the MSF. About 16 modules ranging in size from 39 ft by 85 ft by 92 ft high to 59 ft by 112 ft by 69 ft high will occupy the module deck. The modules will contain gas compression, gas injection, wellheads, drilling fluids, water injection, power generation, and accommodation facilities.

A 300 bed accommodation module will be located at one end of the topside facilities.

The top of the modules on the module deck will serve as the weather deck. Two derricks/substructures, drillpipe racks, and drilling office modules will be located on the weather deck.

Other activity
on the Grand Banks

Among other action on the Grand Banks, a structure in the Whiterose complex contains a significant oil accumulation with possible reserves of 200-300 million bbl, Husky Oil Operations Ltd. disclosed late in 1988.

In addition, Petro-Canada took another step toward development of Terra Nova oil field after reporting "very encouraging" results from the latest delineation test. The company and its partners awarded a preliminary engineering contract to evaluate development options.

Husky's Whiterose E-09 step-out cut a thick, continuous oil reservoir within a single formation. The company based the reserve estimate on the drilled thickness at the step-out and the areal extent indicated from preliminary geophysical mapping.

Relatively compact field geometry and water depth favor future development using a floating production system, Husky said.

Whiterose E-09 was drilled in 407 ft of water on a separate structure 4.3 miles south of Husky's Whiterose N-22 discovery well and 4.1 miles east of the Whiterose J-49 delineation well.

Husky said more delineation drilling was needed to confirm the reservoir's extent. Further delineation and development hinge on suitable economics, including favorable oil prices, it said.

Husky released Bow Drill 3 semisubmersible from the E-09 location July 2 and held well information confidential until

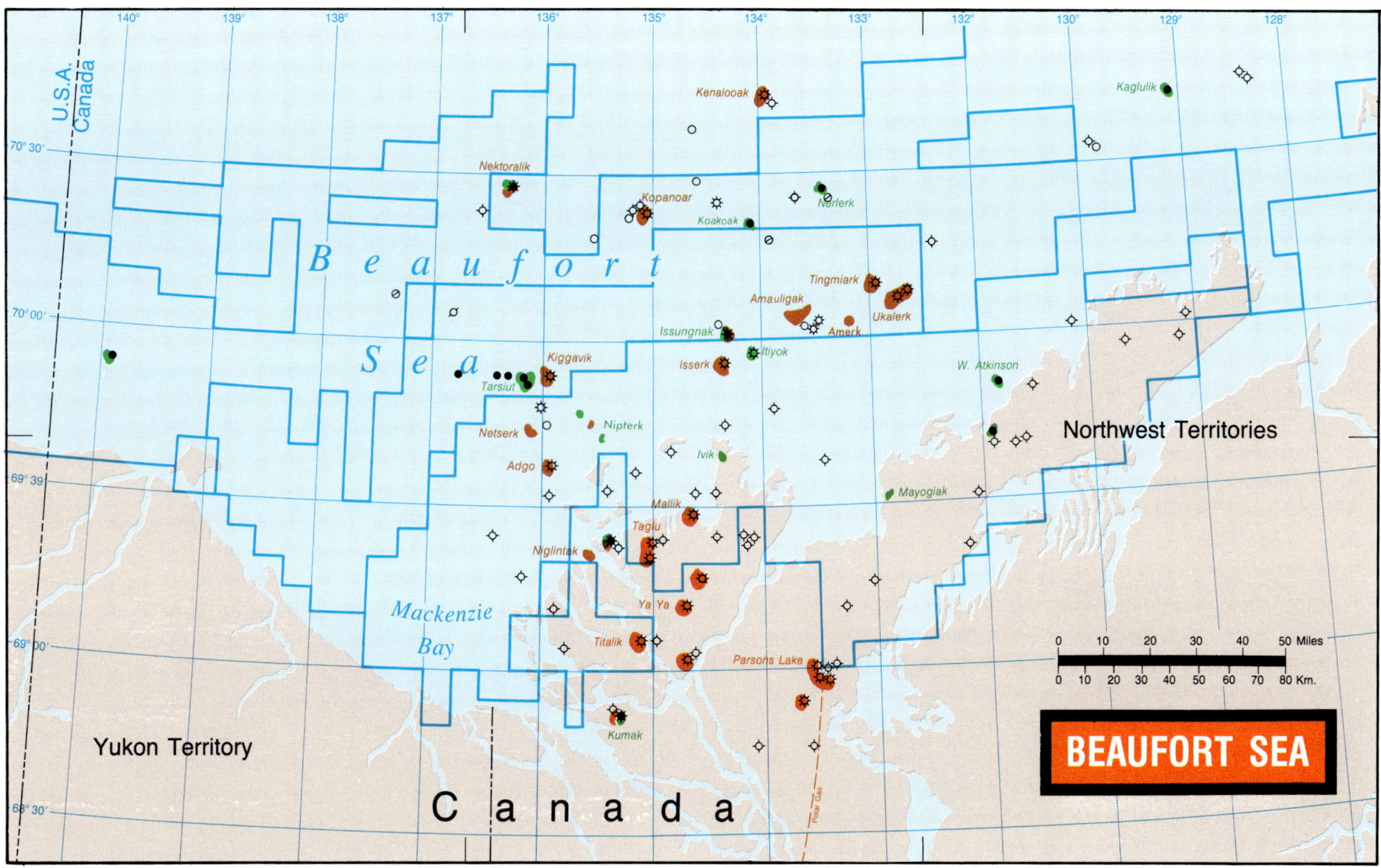

closing Sept. 28 of a Newfoundland land sale.

Husky and its partners were trying to resolve a disagreement among some of the original working interest owners in acreage surrounding E-09 as to the final interests in the well after rig release.

Cost participants in the zones tested were Husky 45.3125%, Bow Valley Industries Ltd. 38.28125%, Gulf Canada 9.375%, and Parex, a general partnership, 7.03125%.

Husky tested several Cretaceous zones separately and together at Whiterose E-09.

It flowed a combined 5,219 b/d of oil and 3.02 MMcfd of gas through a 1 in. choke from 9,705-96 ft and 9,813-72 ft.

Flow was 1,447 b/d of oil and 820 Mcfd of gas through a $^{28}/_{64}$ in. choke from 9,631-64 ft and 430 b/d of oil and 300 Mcfd of gas through an $^{18}/_{64}$ in. choke from 9,565-91 ft.

Those four intervals, tested together, flowed 4,036 b/d of oil and 2.28 MMcfd of gas through a $^{48}/_{64}$ in. choke.

The well also flowed a combined 4,511 b/d of oil and 2.96 MMcfd of gas through a 1¼ in. choke from 9,526-9,798 ft and 9,812-72 ft. It flowed 488 b/d of oil and 1.03 MMcfd of gas through a $^{24}/_{64}$ in. choke from a Jurassic zone at 12,011-086 ft.

Petro-Canada cut 226 feet of net oil pay in its Terra Nova C-09 well, significantly exceeding what had been expected. The well, drilled to 11,942 feet in 305 feet of water by the Sedco 710 rig, flowed at a combined rate of 16,000 b/d.

Two intervals in a lower zone and one interval in an upper zone were tested at the C-09.

Flow rates on the tests were 8,301 b/d, 4,309 b/d and 3,354 b/d.

C-09 was the last in a series of delineation tests planned on the structure, a 1984 discovery. Interests in C-09 were Petro-Canada 75%, Canterra Energy Ltd. 8.3%, ICG Resources Ltd. 6.7%, Parex 5%, and Mossbacher Operating Ltd. 5%.

Petro-Canada Chairman Bill Hopper said Terra Nova field could produce 40,000 to 50,000 b/d as early as 1993.

Development eyed in Beaufort Sea

Elsewhere in Canadian frontier regions, a group led by Gulf Canada planned in summer 1988 to begin conceptual engineering aimed at field development and pipeline construction for its Amauligak oil field in the Beaufort Sea.

That step marked the end of a delineation drilling program for the 1984 Amauligak J-44 discovery. Four of the delineation wells drilled on the east side of the structure flowed oil and gas (Fig. 6).

Later, Gulf Canada gauged as much as 6,660 b/d of oil on four tests in the Amauligak O-86 delineation well.

Gulf Vice Pres. Glen Russell said no decision had been made to proceed with development, although a tentative plan unveiled early in 1987 envisioned about 2.5 million bbl/year of water borne production. Next would come construction of a pipeline up the Mackenzie Valley to markets.

With "excellent producibility," Amauligak is in a class with Hibernia oil field, Russell said.

Amauligak 2F-24B, the latest delineation at midsummer 1988, flowed oil on four drillstem tests. Three of the intervals flowed 5,975 b/d of oil, the limit of test equipment, helping to confirm reserves of 500 million bbl.

It was the fourth hole drilled from the berm location on the eastern portion of the structure. The Amauligak I-65A discovery found oil in the West Amauligak structure, across a fault from East Amauligak.

The Molikpaq rig that drilled Amauligak 2F-24B, 46½ miles northwest of Tuktoyaktuk, N.W.T., was to be stacked for the winter in a nearby harbor.

Amauligak 2F-24B interests were Gulf Canada 50.25%, Husky Oil Operations Ltd. 17.5%, Norcen Energy Resources Ltd. 15%, ATS Exploration Ltd. 10.1625%, Texaco Canada Resources Ltd. 3.2875%, and Atcor Ltd. 3.8%.

Site for Amauligak O-86 was 2 miles west of Gulf's Amauligak I-65A oil discovery, earlier reported as a delineation well. Gulf said reserves in Amauligak O-86 and I-65A, both across a fault west of the East Amauligak reservoir, were in addition to about 500 million bbl credited to East Amauligak. Gulf held a 50.25% interest in the O-86 well. Its partners were BP Petroleum Development Ltd. 28.9625%, Husky Oil Operations Ltd. 17.5%, and Texaco Canada Resources Ltd. 3.2875%.

Industry loses ground gained in 1987

Net income of Canada's petroleum industry fell sharply in first half 1988, losing some ground gained in 1987 from the hard times of 1986.

Low oil prices and a stronger Canadian dollar sapped 1988 earnings, a federal agency said. Higher operating costs and increased federal sales and excise taxes also helped cause the drop.

The Petroleum Monitoring Agency said the Canadian industry's first half 1988 net income was $1.3 billion, compared with $1.8 billion for the first 6 months of 1987.

Canadian companies posted net income of $3.4 billion in 1987, compared with a loss of $2.7 billion in 1986.

The collapse in international oil prices during 1986 and asset writedowns of about $3 billion contributed to the weakest financial performance in the Canadian industry's recent history, PMA said.

No further writedowns were required in 1987. Other factors in the improved 1987 performance were reduced noncash charges, elimination of the petroleum and gas revenue tax (PGRT), reduced royalty rates and royalty holidays, lower interest charges, and lower operating costs.

First half 1988 figures were based on data from 127 companies that represented about 90% of Canadian industry revenues. Figures for 1987 and 1986 were based on data from 135 companies that accounted for about 90% of revenues.

PMA said the Canadian industry's return on equity fell to 6.5% in first half 1988 from 9.9% in the same period of 1987.

Despite lower revenues, spending on oil and gas development increased 47% to $4.4 billion, for a reinvestment rate of 93%. The industry committed to new capital projects after cash buildups in 1987, but oil and gas prices fell shortly after the plans were made.

A bright spot was downstream activity. Downstream income increased 26% to $590 million in first half 1988, due largely to increased petrochemical sales. Profit from petrochemicals increased 73% to $264 million.

Upstream operations earned $2.2 billion in 1987 after losing $1.8 billion in 1986 (Fig. 7).

Downstream net income rose 7% to $900 million. And other Canadian and foreign operations turned a $310 million profit after a $1.8 billion loss in 1986.

Net capital spending rose slightly, but gross spending fell 6% to $7.5 billion.

Upstream spending was a record low $5.3 billion, down 12%, as all categories of upstream spending fell. Record high upstream spending was $9.3 billion in 1985.

The average price received for conventional crude during 1987 was $21.61/bbl, 19% higher than in 1986. Gas prices fell 25% to $1.67/Mcf.

The industry achieved a 9.1% return on shareholder equity in 1987, compared with a negative 7.1% in 1986. The industry's 1987 figure was 2.4 percentage points below the 12.1% realized by other nonfinancial industries.

Amoco makes biggest transaction

Industry reorganization saw Amoco Canada Ltd. complete Canada's biggest business transaction, the $5.5 billion (Canadian) purchase of Dome Petroleum Ltd., Calgary. That resulted in Canada's largest gas producer and second largest oil producer.

The deal was final Sept. 1, 1988, the same day the Supreme Court of Canada rejected a bid by Toronto busi-

nessman Abdul Rehman Premji for an injunction to block it. It was the last court battle involving shareholders and creditors holding Dome's $6 billion in debt since the companies announced a sales agreement 16 months previously.

The agreement had been approved by the courts as fair to all parties. Dome's domestic and international creditors had settled earlier in the year.

Amoco Canada Pres. Don Stacy said his company planned heavy oil development and renewal of exploration in the Beaufort Sea, where Dome was a pioneer operator. Amoco guaranteed to spend $2.5 billion on exploration and project development in Canada during the ensuing 5 years. It also was committed to a public share issue in Canada within 5 years.

Plans included an outlay of more than $100 million for heavy oil development in the Cold Lake region of Alberta, where it obtained Dome properties on the Primrose block.

Upstream activity shows increase

Canadian operators stepped up their pace of drilling in 1988, mainly on the strength of oil development in Alberta, by far Canada's biggest oil and gas producing province.

An unofficial tally near yearend showed that industry had drilled 7,640 wells during January-October 1988, 56% ahead of the count in the same period of 1987. Alberta accounted for 71% of 1988 activity, Saskatchewan almost 25%. Another breakout of activity showed that about half of 1988 wells were oil completions, about 25% gas, and 25% dry.

CPA's official count for 1987 listed 7,078 wells drilled, an increase of 18% from depressed activity in 1986. There also was a 14.9% gain in footage drilled—5,373 miles of hole in 1987, compared with 4,678 miles in 1986.

The 1987 success rate for all drilling, 74.8%, was essentially flat with 1986. New field wildcat success rate for 1987 stood at 28.4%—64 discoveries out of 225 wells drilled—down 2.5 percentage points from 1986. Canada's average production of 1.594 million b/d of oil during the first 9 months of 1988 was up 5.1% from the volume of 1987. Gas production during January-August 1988 amounted to 2.894 tcf, up 30% from the same period in 1987.

NEB's latest official estimate, issued in 1988, pegged Canada's reserves of conventional crude oil at 730.8 million cu m as of Dec. 31, 1986, down 6.6 million cu m from yearend 1985. Initial—before production—reserves of conventional crude oil stood at 2,698.6 million cu m, up 62.4 million cu m by means of discoveries, extensions, and revisions.

Marketable gas reserves in the conventional producing areas of the country amounted to 75.3 exajoules, down 2 exajoules.

Initial reserves were 136 exajoules, up 1.3 exajoules.

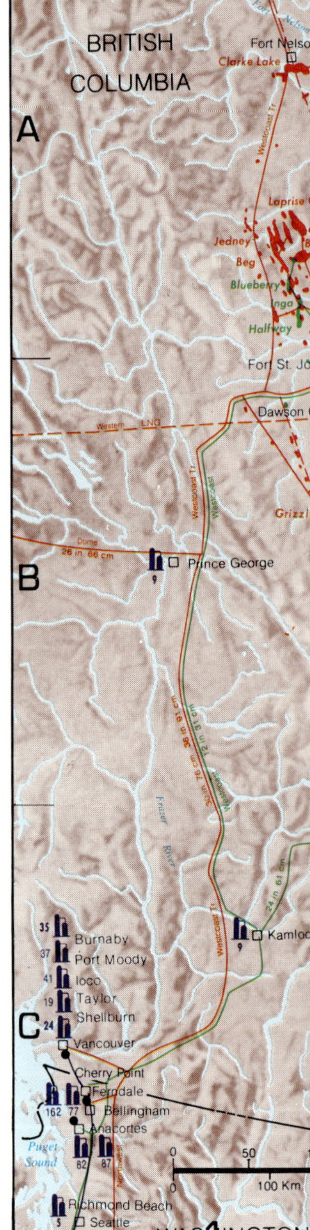

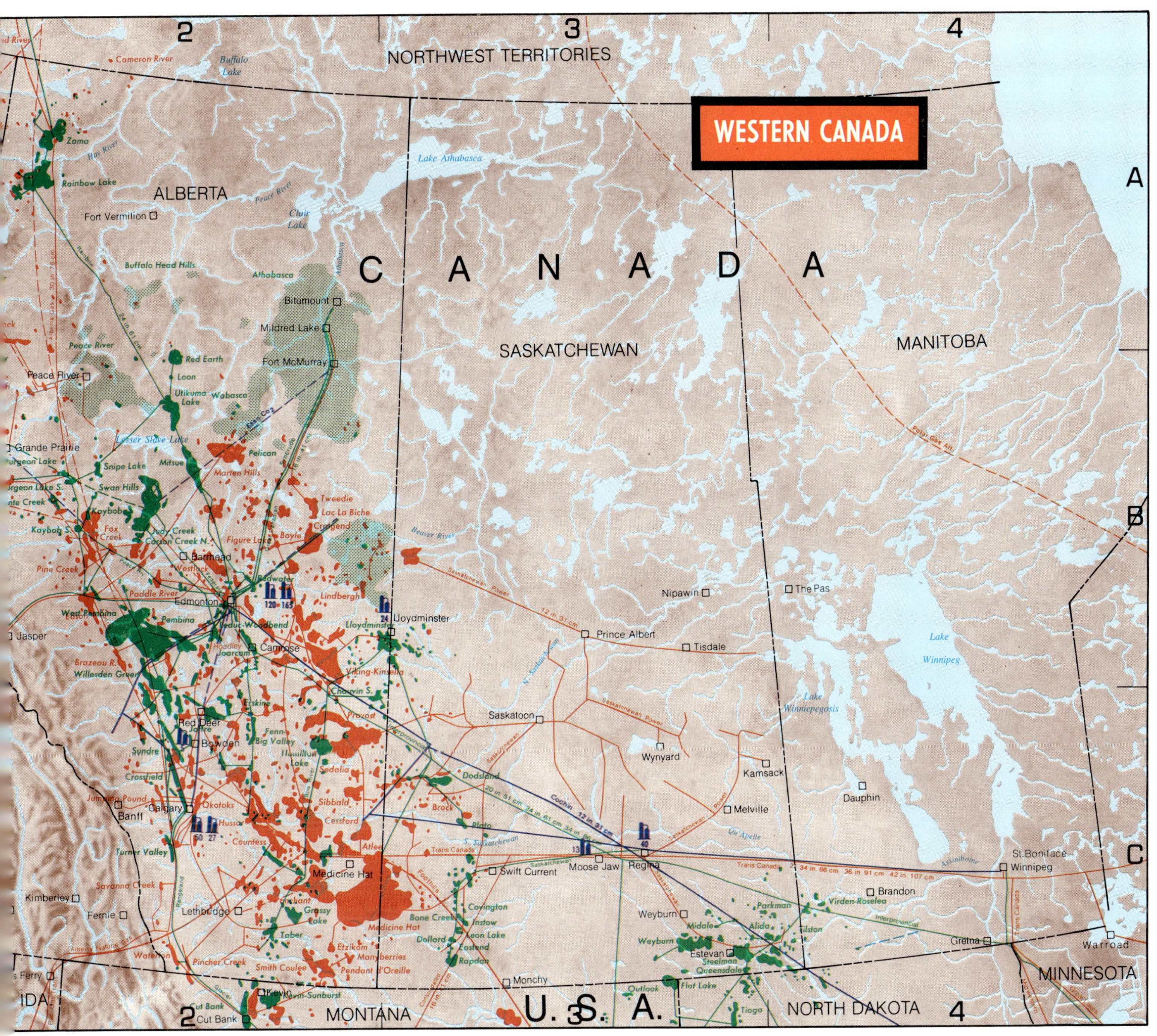

NEB's outlook to 2005

The NEB late in 1988 predicted rising prices for Canadian gas and an end to the country's oil self-sufficiency.

NEB's biennial report outlined a number of scenarios and emphasized that price is the critical supply factor in the period to 2005. It also said its estimates could be changed by environmental concerns, such as limits on the burning of fossil fuels.

The board forecast that oil imports will reach 956,000 b/d, about 60% of domestic consumption, by 2005 if prices stay at $15 (U.S.)/bbl and domestic development is limited. At $30/bbl, Canada would be self-sufficient in oil with a surplus of 327,000 b/d available for export.

The board said it could not forecast prices because markets are glutted, but it foresaw several scenarios based on high and low paths ending in 2005 at a price of $20-30 (U.S.)/bbl.

It said Canada could lose its status as a net exporter within 7 years if prices stay low. Persistent low prices will lead to a shortfall of 188,700 b/d by 1995. If prices rise, an oil surplus of exports over imports could rise to 371,000 b/d by 1995.

NEB established reserves of conventional crude at about 4.59 billion bbl at the end of 1986. It forecast 3.48-4.12 billion bbl of light and heavy crude will be found by 2005.

The board expected the price of gas to increase to make high cost arctic production economic within 10 years. It expected gas exports to increase to 1.5 tcf/year in 1992.

Field prices for Canadian gas will at least double to $3 (Canadian)/Mcf and reach as much as $4/Mcf by 2005. Price gains will be so strong that even under a "low case" outlook for prices, Mackenzie Delta gas will enter production by 1999.

Export demand will reduce gas surpluses in Alberta and deplete remaining low cost reserves. The Canadian industry will then develop higher cost reserves, including those in frontiers.

PEC Decoking System

As its name indicates the Pig Decoking Method consists of using pigs in order to remove carbon or iron sulfides coat stuck to steel pipes of fired heater.   Our Pig Decoking Method is of course covered by patents in the USA, United Kingdom, West Germany, France, and Japan, and has proved its effectiveness through our long experience including many different types of heating furnace.

The method of decoking using pigs is extremely simple, and the process itself is safe and sure.   Furthermore the total working cost is much lower than any other conventional methods, and the annual "heat consumption" for operating fired heater undergone our decoking method can recover the same level as new tubes since the heat insulating layer by coking is totally removed thanks to our decoking process resulting in tens of millions of yen worth of energy saving.

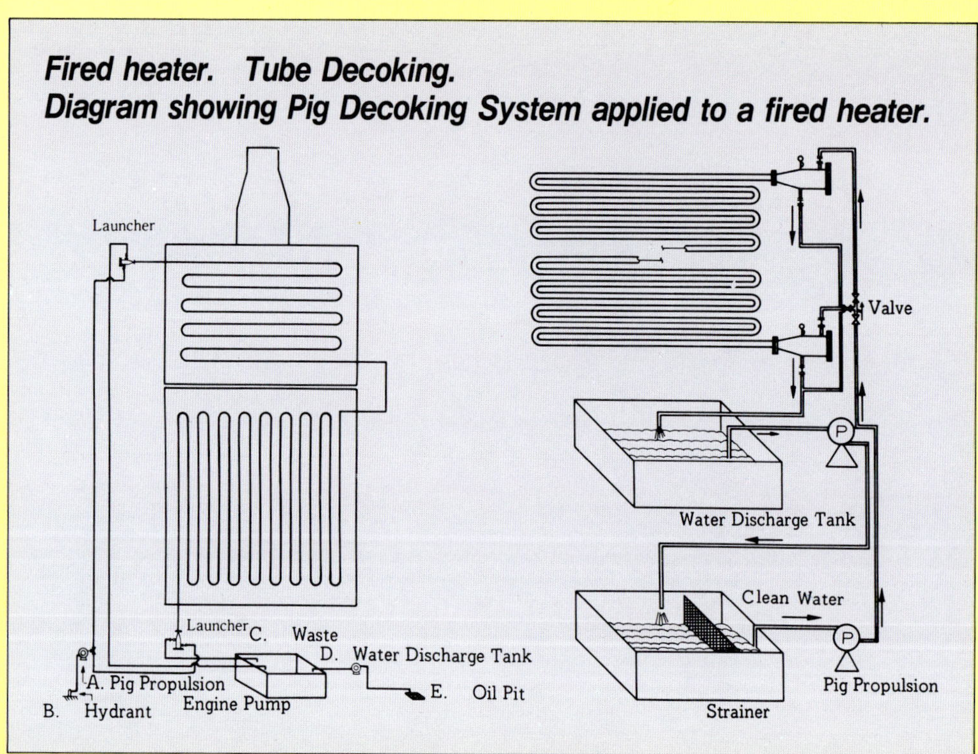

PEC Pig Systems Provide :

- Complete tube interior face cleaning.
- 180 degree "U Bends" for all Pigs.
- Water pressure driven Pigs system.
- Multiple runs per Pig.
- Multiple cutting surfaces per Pig.
- Long runs are a standard procedure.
- Operate up to 170 degrees F.
- Returns interior face to new condition.
- Pipe size from 2" thru 120" standard.

[PEC] POLY-PIG ENGINEERING CO.,LTD.
1-12-8, Minamikugahara Ohta, Tokyo, Japan
Telephone : 03-756-4411   Facsimile : 03-756-2188

PEC INTERNATIONAL (USA) INC.
16637 West Hordy, Rd., Houston, TX77060 U.S.A.
Telephone : 713-873-6622   Facsimile : 713-873-6662

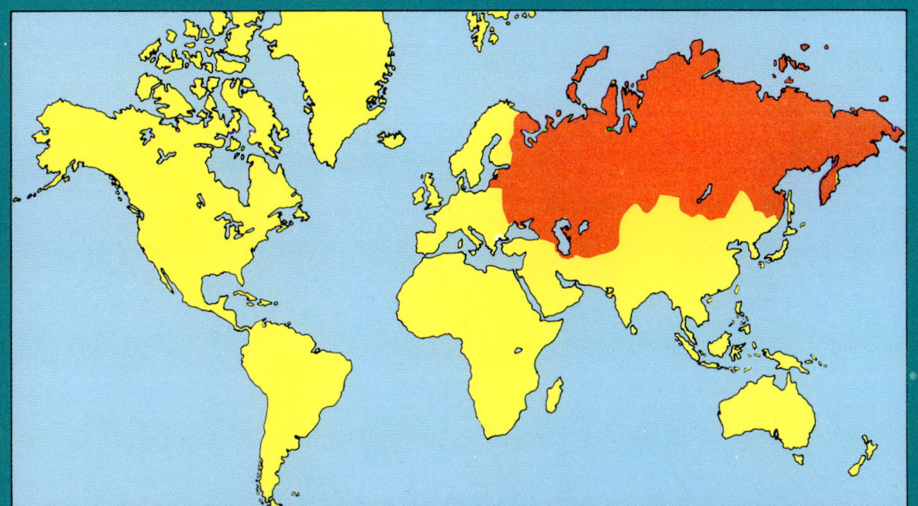

RUSSIA

CAPITAL: Moscow
MONETARY UNIT: Ruble
REFINING CAPACITY: 12,300,000 b/cd
PRODUCTION: 12,477 M b/d
RESERVES: 58,500,000 M bbl

THE SOVIET UNION HAS ORDERED INCREASED OIL and gas production for 1989.

The goal is to prevent further erosion in revenues from exports of crude oil, refined products, and gas.

If the new official targets—augmented by pledges from the nation's oil and gas associations to exceed them—are achieved, the U.S.S.R. will be well above the lower limit of its original 1990 goal for crude and condensate flow.

The objectives also indicate that in 1990 Soviet gas production may top the maximum volume forecast for that year when the current 5 year plan was announed in 1986.

The 5 year plan's initial 1989 crude and condensate target was 12.472 million b/d. The new 1989 goal, including the regional oil production associations' promises to exceed the official target, is 12.632 million b/d, which is 155,000 b/d above the production level as 1989 opened.

The oil production associations pledged to achieve 1988 flow of 12.516 million b/d. A slump in summer production cut crude and condensate flow during the first 9 months of 1988 to 12.469 million b/d, down 19,000 b/d from the same 1987 period.

The U.S.S.R.'s officially reported oil production was 12.48 million b/d in 1987, 12.3 million b/d in 1986, and 11.9 million b/d in 1985.

Moscow originally set the 1990 crude/condensate production goal at 12.5-12.8 million b/d.

While the U.S.S.R.'s 1989 oil production is at best targeted to grow about 1%, gas flow is expected to increase 6-7% compared with 1988.

The Soviet 5 year plan originally called for gas production to reach 28.24 tcf in 1989. The new target this year, including the production associations' pledge to exceed the initial objective, is 28.95 tcf.

The gas production associations set a 1988 goal of 27.08 tcf. Officially reported Soviet gas production was 25.66 tcf in 1987, 24.22 tcf in 1986, and 22.70 tcf in 1985.

The U.S.S.R.'s original gas production target for 1990 was 29.48-30.01 tcf, with most of the increase over 1985 coming from supergiant Yamburg field north of the Arctic Circle in western Siberia, Sovetabad in Turkmenia near the Iranian border, and Karachaganak and Astrakhanskoye in the pre-Caspian depression.

Yamburg's produciton capacity is now believed to exceed 3.5 tcf/year. That compares with about 13 tcf/year for Urengoi, the world's largest gas field. It lies astride the Arctic Circle southeast of Yamburg.

On the basis of energy (calorie) equivalence, the Soviet Union's gas flow was on line to surpass its crude/condensate production for the first time during 1988.

Official data showed that natural gas would provide about 39.4% of the U.S.S.R.'s production in terms of "standard fuel," containing 7,000 cal/kg, in 1988. That compares with 37.7% in 1987, 36.6% in 1986, 35.8% in 1985, and 27.1% in 1980, 19.1% in 1970 and 7.9% in 1960.

By contrast, the proportion of crude and condensate in overall Soviet fuel production was about 38.9% during 1988 vs. 40% in 1987, 40.6% in 1986, 41.1% in 1985, 45.5% in 1980, and an all time high of 45.7% in 1978.

Coal represented about 20.2% of the U.S.S.R.'s fuel production in 1988 against 20.6% in 1987, 21% in 1986, 21.2% in 1985, 25.2% in 1980, 35.4% in 1970, and 53.9% in 1960.

Other fuels—peat, shale, and wood—accounted for only about 1.5% of Soviet fuel production in 1988 in terms of energy equivalence vs. more than 1.7% in 1987, 1.8% in 1986, 1.9% in 1985, 2.2% in 1980, 4.4% in 1970, and 7.7% in 1960.

Low prices slice
Soviet oil fortunes

Depressed oil prices, combined with fast rising oil and gas production costs, have played key roles in placing the Soviet Union in its worst financial bind since World War II.

The world's No. 1 oil producer is for the first time in the postwar period openly predicting a budget deficit, estimating that it will be 36.3 billion rubles in the red during 1989. That is equal to about $58 billion (U.S.) at Moscow's current arbitrary currency exchange rate of $1.60/ruble.

What's more, latest official 6 month figures show that 1988 may be the first year since 1976 that the U.S.S.R.'s total imports exceed exports.

From January through June 1988 the Soviets had total exports of less than 33 billion rubles vs. imports of nearly 33.4 billion rubles.

The Soviets said the sharp drop in world oil price since

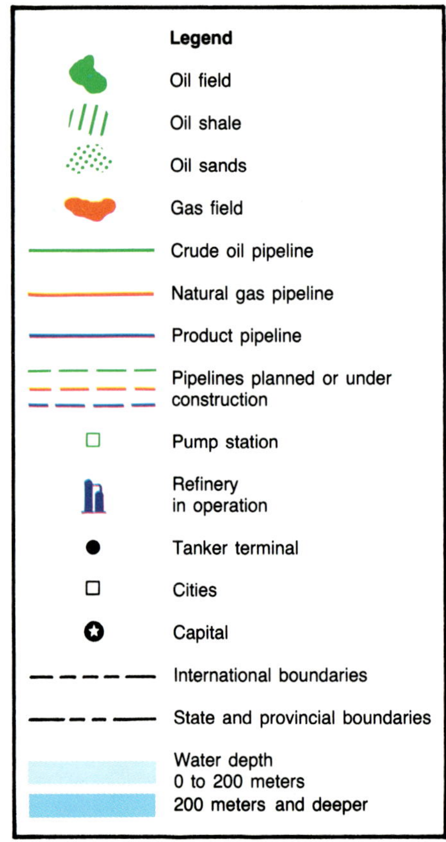

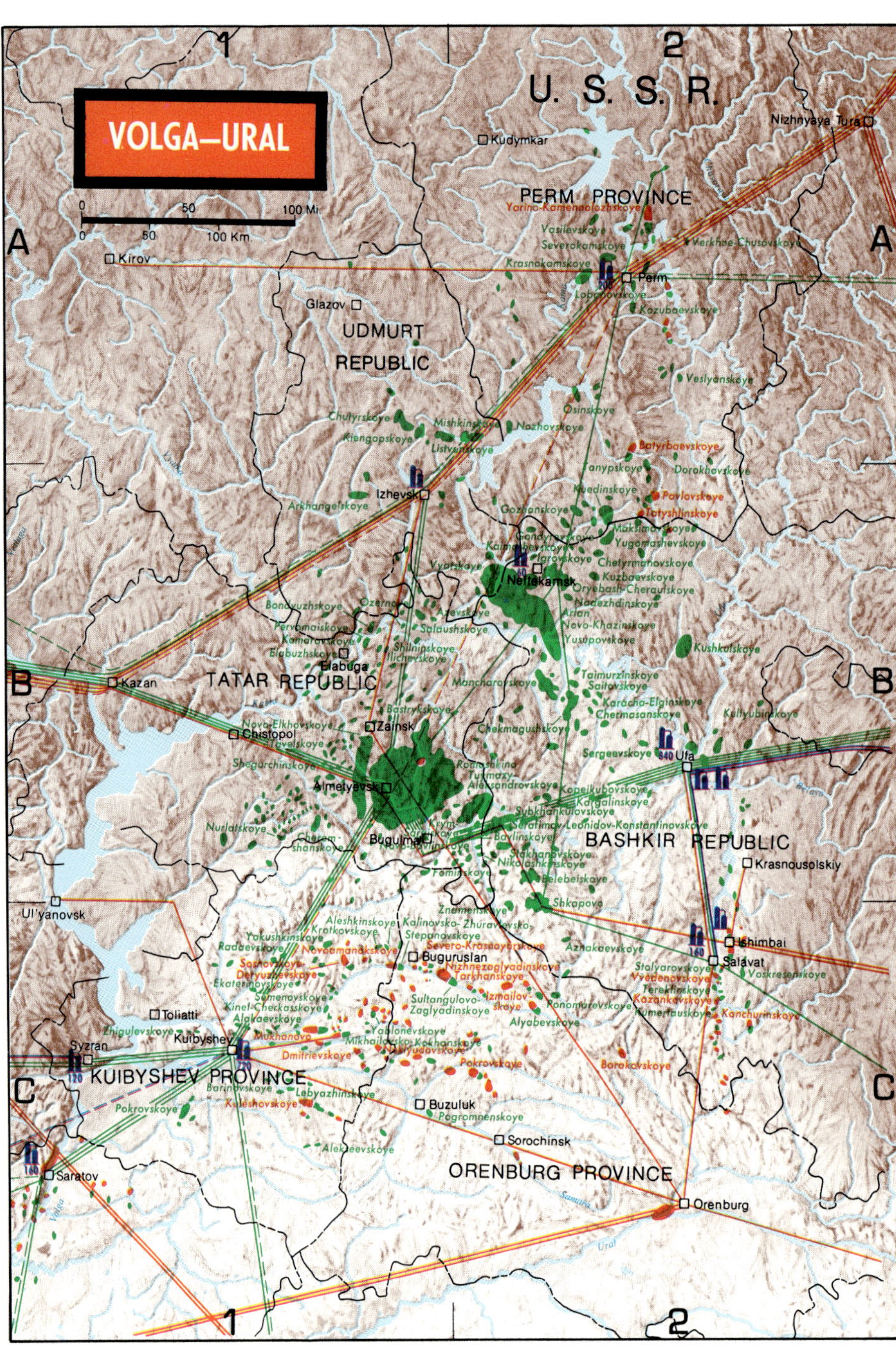

1985 has cost it about 40 billion rubles ($64 billion). This occurred even though the Soviet Union sold a record 3.915 million b/d of crude and products worth more than $36.5 billion in 1987 vs. 3.725 million b/d valued at $35.9 billion in 1986.

During 1985, the U.S.S.R.'s crude and products exports totaled only 3.334 million b/d. But revenues from foreign oil sales amounted to more than $45 billion.

Peak revenues from Soviet oil exports were $49.5 billion in 1984.

The U.S.S.R. has kept its budget in the black and maintained a favorable foreign trade balance most of the time since World War II largely because of its fast growing oil production and exports.

In 1987, crude and products sales to other countries accounted for 33.5% of total exports. Gas exports contributed another 9.4%.

Oil's importance in Soviet exports to hard currency customers is especially great.

Crude and product exports to Italy represented nearly 65% of total U.S.S.R. sales to that country in 1987, and gas deliveries accounted for 18.5%.

Crude and products in 1987 made up more than 45% of Soviet exports to West Germany. Gas deliveries brought in another 35%.

More than 58% of France's 1987 imports from the U.S.S.R. was provided by oil and 23.3% by natural gas.

The Soviet Union maintained record or near record oil exports to its leading western European trading partners during first half 1988.

Early last year, the price of U.S.S.R. export blend 32° gravity crude was nearly $16/bbl. By October 1988, the price had dropped to $12.45/bbl.

Tengiz oil field called supergiant

The Soviet Union has revealed data indicating that its Tengiz oil field near the Caspian Sea's northeast coast may rank among the world's largest supergiants.

Discovered in 1979, Tengiz hasn't been fully evaluated. But the Soviets say the field's "explored" (proved plus probable) reserves exceed 2.5 billion metric tons (18.25 billion bbl).

The field has been evaluated to a depth of 5,100 m (16,732 ft), according to Moscow reports in 1988. The Soviets believe that oil may be found as deep as 6,000 m (19,685 ft).

Tengiz reportedly has only one huge oil reservoir in subsalt Carboniferous limestone. Top of the pay zone is at 13,271 ft. The field's area is said to cover "hundreds of square kilometers."

If the estimate that Tengiz has more than 18 billion bbl of reserves is reasonably accurate, Tengiz is in the upper echelon of the world's supergiant fields.

The Soviet Foreign Trade Association, V/O Sojuznefteexport, exports these fine products:

- Soviet export blend
- High-octane motor gasoline
- Low-sulfur diesel fuel
- Fuel oils
- Straight-run gasolines
- Wide range of lubricating oils
- Paraffin
- Petroleum coke
- Solvents and other petroleum products

SOJUZNEFTEEXPORT

32/34 Smolenskaya pl.; 121200 Moscow, USSR / Tel.: 253-94-88 or 253-94-89 / Telefax: 244-22-91 / Telex: 411148 A

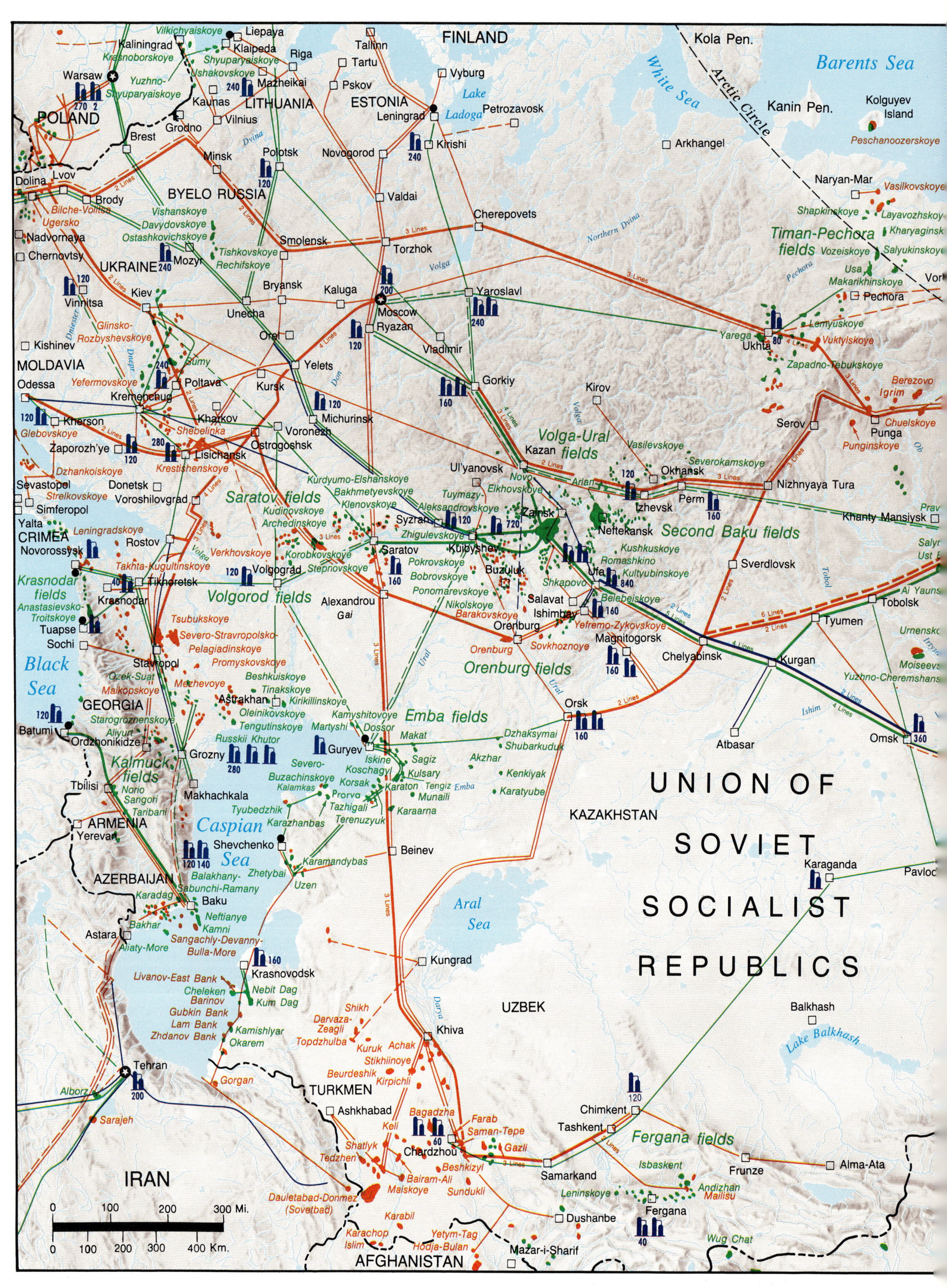

UNION OF SOVIET SOCIALIST REPUBLICS
FINLAND
POLAND
Warsaw
Kaliningrad
Klaipeda
Liepaya
Riga
Tallinn
Tartu
Vyburg
Kola Pen.
Barents Sea
Kanin Pen.
Kolguyev Island
Arkhangel
Naryan-Mar
White Sea
Arctic Circle
Vilkichyaiskoye
Krasnoborskoye
Yuzhno-Shyuparyaiskoye
Shyuparyaiskoye
Nshakovskoye
Mazheikai
LITHUANIA
Kaunas
Vilnius
Grodno
Brest
Pskov
ESTONIA
Leningrad
Lake Ladoga
Petrozavosk
Kirishi
Peschanoozerskoye
Vasilkovskoye
Shapkinskoye
Layavozhskoye
Kharyaginsk
Timan-Pechora fields
Vozeiskoye
Salyukinskoye
Usa
Makarikhinskoye
Pechora
POLAND
Dolina
Lvov
Brody
Minsk
Poloptsk
Novogorod
Valdai
Cherepovets
Northern Dvina
Dvina
BYELO RUSSIA
Bilche-Volitsa
Ugersko
Nadvornaya
Chernovtsy
UKRAINE
Mozyr
Vishanskoye
Davydovskoye
Ostashkovichskoye
Tishkovskoye
Rechifskoye
Smolensk
Torzhok
Volga
Yaroslavl
Kirov
Serov
Punga
Chuelskoye
Punginskoye
Ob
Zapadno-Tebukskoye
Yarega
Ukhta
Lemyuskoye
Vuktyiskoye
Berezovo
Igrim
Vor
Vinnitsa
Kiev
Bryansk
Kaluga
Moscow
Ryazan
Vladimir
Gorkiy
Kishinev
MOLDAVIA
Odessa
Yefermovskoye
Kremenchug
Glinsko-Rozbyshevskoye
Sumy
Poltava
Orel
Yelets
Kazan
Volga-Ural fields
Vasilevskoye
Okhansk
Izhevsk
Perm
Nizhnyaya Tura
Severokamskoye
Khanty Mansiysk
Kherson
Glebovskoye
Zaporozh'ye
Dzhankoiskoye
Kharkov
Shebelinka
Voronezh
Michurinsk
Ostrogozhsk
Saratov fields
Kurdyumo-Elshanskoye
Bakhmetyevskoye
Klenovskoye
Syzran
Zhigulevskoye
Ul'yanovsk
Novo-Elkhovskoye
Tuymazy-Aleksandrovskoye
Arlan
Zainsk
Neftekansk
Ufa
Kushkuskoye
Romashkino
Kultyubinskoye
Second Baku fields
Sverdlovsk
Tobolsk
Tyumen
Sevastopol
Strelkovskoye
Simferopol
Yalta
CRIMEA
Novorossysk
Leningradskoye
Rostov
Donetsk
Voroshilovgrad
Lisichansk
Krestishenskoye
Kudinovskoye
Archedinskoye
Korobkovskoye
Stepnovskoye
Pokrovskoye
Bobrovskoye
Ponomarevskoye
Saratov
Kuibyshev
Buzuluk
Shkapovo
Belabeiskoye
Yefremo-Zykovskoye
Salavat
Orenburg
Sovkhoznoye
Magnitogorsk
Chelyabinsk
Kurgan
Krasnodar fields
Anastasievsko-Troitskoye
Tuapse
Sochi
Takhta-Kugultinskoye
Tikhoretsk
Krasnodar
Verkhovskoye
Volgograd
Volgorod fields
Alexandrou Gai
Nikolskoye
Barakovskoye
Orenburg fields
Orsk
Atbasar
Omsk
Black Sea
Stavropol
Tsubukskoye
Severo-Stravropolsko-Pelagiadinskoye
Promyskovskoye
Beshkuiskoye
Tinakskoye
Kirikillinskoye
Oleinikovskoye
Tengutinskoye
Astrakhan
Kamyshitovoye
Martyshi
Dossor
Makat
Emba fields
Dzhaksymai
Shubarkuduk
GEORGIA
Batumi
Ordzhonikidze
Starogroznenskoye
Aliyurt
Russkii Khutor
Grozny
Kalmuck fields
Norio
Sangori
Taribani
Makhachkala
Tyubedzhik
Ozek-Suat
Maikopskoye
Mezhevoye
Guryev
Iskine
Sagiz
Koschagyl
Korsak
Prorva
Karaton
Tengiz
Munaili
Severo-Buzachinskoye
Kalamkas
Tazhigali
Karazhanbas
Terenuzyuk
Kulsary
Akzhar
Kenkiyak
Karaarna
Karatyube
KAZAKHSTAN
Karaganda
Pavlod
Tbilisi
ARMENIA
Yerevan
AZERBAIJAN
Astara
Bakhar
Karadag
Baku
Neftianye Kamni
Sangachly-Devanny-Bulla-More
Aliaty-More
Balakhany-Sabunchi-Ramany
Shevchenko
Caspian Sea
Zhetybai
Uzen
Karamandybas
Beinev
Aral Sea
Balkhash
Lake Balkhash
Tehran
Alborz
Sarajeh
Gorgan
Livanov-East Bank
Cheleken
Barinov
Gubkin Bank
Lam Bank
Zhdanov Bank
Nebit Dag
Kum Dag
Kamishlyar
Okarem
Krasnovodsk
Kungrad
Khiva
Darya
UZBEK
Chimkent
Tashkent
Karaganda
IRAN
TURKMEN
Ashkhabad
Keli
Bagadzha
Farab
Saman-Tepe
Gazli
Samarkand
Leninskoye
Fergana fields
Isbasket
Frunze
Alma-Ata
Andizhan
Mailisu
Fergana
Wug Chat
Dushanbe
Shikh
Darvaza-Zeagli
Topdzhulba
Kuruk
Achak
Stikhiinoye
Beurdeshik
Kirpichli
Shatlyk
Tedzhen
Bairam-Ali
Maiskoye
Sundukli
Chardzhou
Beshkizyl
Karabil
Yetym-Tag
Hodja-Bulan
Dauletabad-Donmez (Sovetbad)
Karachop
Islim
Mazar-i-Sharif
AFGHANISTAN
0   100   200   300 Mi.
0   100 200 300   400 Km.

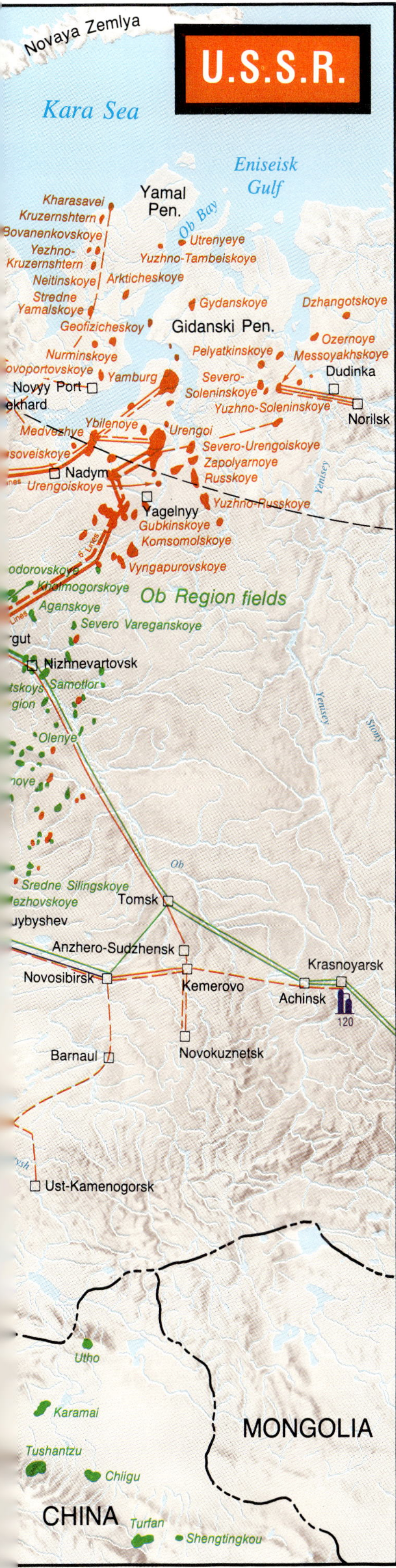

The Congressional Research Service, U.S. Library of Congress, reported there are 37 supergiant oil fields in the world, each with recoverable reserves greater than 5 billion bbl. Those fields contained an estimated 51% of all oil discovered to date. Tengiz wasn't listed among the supergiants in the Library of Congress study. Moscow said in 1987 an accurate estimate of Tengiz oil reserves would not be available until 1995.

The nation expects oil flow from the field to average 60,000 b/d during 1989. Crude production was to have started in first half 1988, but field work was behind schedule. Drilling rigs bought from Romania were found unsuitable for use in Tengiz because of extremely high downhole pressures and severe H_2S corrosion.

Soviet manufacturers in 1988 were unable to meet delivery schedules for the advanced equipment required at Tengiz, forcing Moscow to buy more than it hoped from the West.

Soviet geologists several years ago favorably compared Tengiz reserves with those of the U.S.S.R.'s biggest oil field, Samotlor in western Siberia. It was also said that Tengiz has good prospects for raising maximum flow to 1 million b/d, a figure that seems modest in view of the new estimate of field reserves.

Tengiz was rarely mentioned during the early 1980s. When it was cited, emphasis was placed on the enormous difficulties facing the field's developers rather than on the reservoir's size. Moscow began to cite the importance of Tengiz at the same time Soviet officials were promoting a huge joint venture petrochemical complex at the field with western partners.

Plan to hike "oil mine" production

The U.S.S.R. reports that heavy crude production from its Yarega "oil mine" in the Komi Autonomous Republic was a record 10,000 b/d in 1987, with much higher production planned in the 1990s.

Construction of four new shafts and rebuilding of two old ones are expected to hike Yarega production to 32,000 b/d. Located in the far northeastern part of European Russia, the Yarega facility combines steam injection from the surface with the drilling of closely spaced wells from chambers about 660 ft below the surface.

Oil flows by gravity into tunneled areas beneath the pay zone.

Previous Soviet plans for increasing Yarega's production to more than 30,000 b/d were set back by a serious fire and methane explosion in one of the underground chambers in 1986.

The U.S.S.R. began commercial production from Yarega in 1939. But use of conventional methods, with holes drilled from the surface, permitted recovery of only 1.5-2% of oil in place.

Directional drilling of small diameter holes from the underground chambers boosted recovery to 4-4.5%. Maximum production by this method reached 7,800 b/d in 1952 but declined to 172 b/d in 1975.

Later use of steam injection has increased Yarega's crude recovery to "more than 50% of oil in place," the Soviets report. From 1975 through 1987, Yarega's total production was nearly 41 million bbl.

Experience obtained from employment of Yarega's "thermomining" method, along with data on foreign heavy oil operations, has resulted in proposals to use similar techniques in other old petroleum regions of the U.S.S.R. where production of remaining crude by conventional means would be "very difficult or impossible."

In other Russian drilling-production activity:

● The Kasimov machine building plant in Baku delivered workover rigs for use in marshy western Siberian oil fields. The rigs, mounted on large, all-terrain vehicles, are more maneuverable than any previously manufactured in the country. The Soviets ordered 10 self-propelled units and related tools and equipment from IRI International of the U.S. to work in Tengiz oil field and Karachaganak gas/condensate field.

Value of the IRI contract is $30-35 million. IRI constructs the rigs at Pampa, Tex., shipping one/month beginning in November 1988. The rigs have depth capacities of 15,000 ft and working pressures of 10,000 psi, able to operate in temperatures to −49° F.

Ancillary equipment includes substructures, mud systems, and hydrogen sulfide detection and alarm systems. IRI is owned 39% each by Dresser Industries Inc. and Ingersoll Rand Co. and 22% by IRI management.

● Promsyrioimport ordered 80,000 metric tons of seamless tubular products from USX Corp.'s U.S. Steel International Inc. The steel is produced at Fairfield, Ala., and exported through Mobile, Ala. Deliveries were scheduled for October 1988 through March 1989.

● The U.S.S.R. has started development of another giant western Siberia gas field, Komsomolskoye, about 112 miles south of supergiant Urengoi gas field. Proved plus proba-

TOBOLSK petrochemical complex on the Irtysh River in western Siberia recently increased butadiene production to rated capacity. Two U.S. companies — Combustion Engineering Inc. and McDermott International Inc. — have signed a letter of intent for a joint petrochemical venture with the Soviets at Tobolsk.

ble reserves are estimated at about 16 tcf. Discovered in 1966, the field produces from a single Cenomanian zone at 2,979-3,346 ft. Wells will be drilled in clusters, with rigs moved on rails to new sites.

● The Soviets spudded a well targeted to 20,341 ft in giant Karachaganak. Also it will spud a well projected to 22,966 ft. Sour gas production from Karachaganak, discovered in 1978, moves by pipeline to supergiant Orenburg field for processing. The field has Carboniferous and Permian pay at 12,254-17,831 ft.

● The country ordered $7.8 million (U.S.) in drilling equipment from Dreco Energy Services Ltd., Edmonton, Alta. The sale to Chimmashexport covers degassing systems and downhole tools for hydrogen sulfide service.

● It let a 2 year, multimillion dollar contract to Travis Chemicals Inc., Calgary, Alta., for chemicals for Tengiz field.

● The U.S.S.R. retired its first mobile offshore rig, the Apsheron jack up, which began work in the Caspian Sea in 1966. The Soviets will use the rig, renamed Siazan, for experimental work and as an "historical example of offshore drilling equipment." It was rated to 6,000 ft in 50 ft of water.

● The Soviet Union last year prepared to drill the first development well in supergiant Bovanenkovskoye gas field more than 250 miles north of the Arctic Circle on western Siberia's Yamal Peninsula. Moscow plans for Bovanenkovskoye to begin commercial production in 1991. Discovered in 1971, the field is believed to hold total reserves of 146.5 tcf, third largest in the U.S.S.R.

A 250 metric ton rig on location in the field is equipped for use in permafrost conditions. All working areas are heated, and the rig is designed to operate in the area's coldest weather.

● The Soviets spudded western Siberia's first superdeep hole south of Urengoi gas/condensate field and 106 miles south of the Arctic Circle. Drilling, by the Uralmash rig in an area free of permafrost, will take at least 5 years. Target depth is 26,246 ft.

Exploration activity

The U.S.S.R. has started exploratory drilling in the Baltic Sea in an area well north of the D-6 field off Kaliningrad Province.

Work was suspended off Kaliningrad late in 1986 following protests by conservationists and regional officials who contended that D-6 development was an unacceptable threat to a coastal region containing rare plants and animals.

Soviet reports say the new drillsite is off the coast of the Latvian Republic, about 31 miles form the port of Liepaya. The fixed D-6 platform was about 100 miles farther south in 98 ft of water, 14 miles off the Kaliningrad Province sector of the Kurshskaya Spit.

Drilling off Liepaya is being carried out from the "stationary" platform Shelf-8, according to Moscow. However, all previous Soviet drilling rigs in the Shelf series have been semisubmersibles. Before construction of fixed platforms off Kaliningrad Province, the Communist exploration group Petrobaltic, including the Soviet Union, Poland, and East Germany, operated a single mobile drilling rig, a jack up, in the Baltic Sea.

Drilling from Shelf-8 is under the direction of the Arctic Offshore Oil and Gas Exploration Association (Aoogea), Murmansk. It's the organization responsible for the country's exploration program in the Barents Sea.

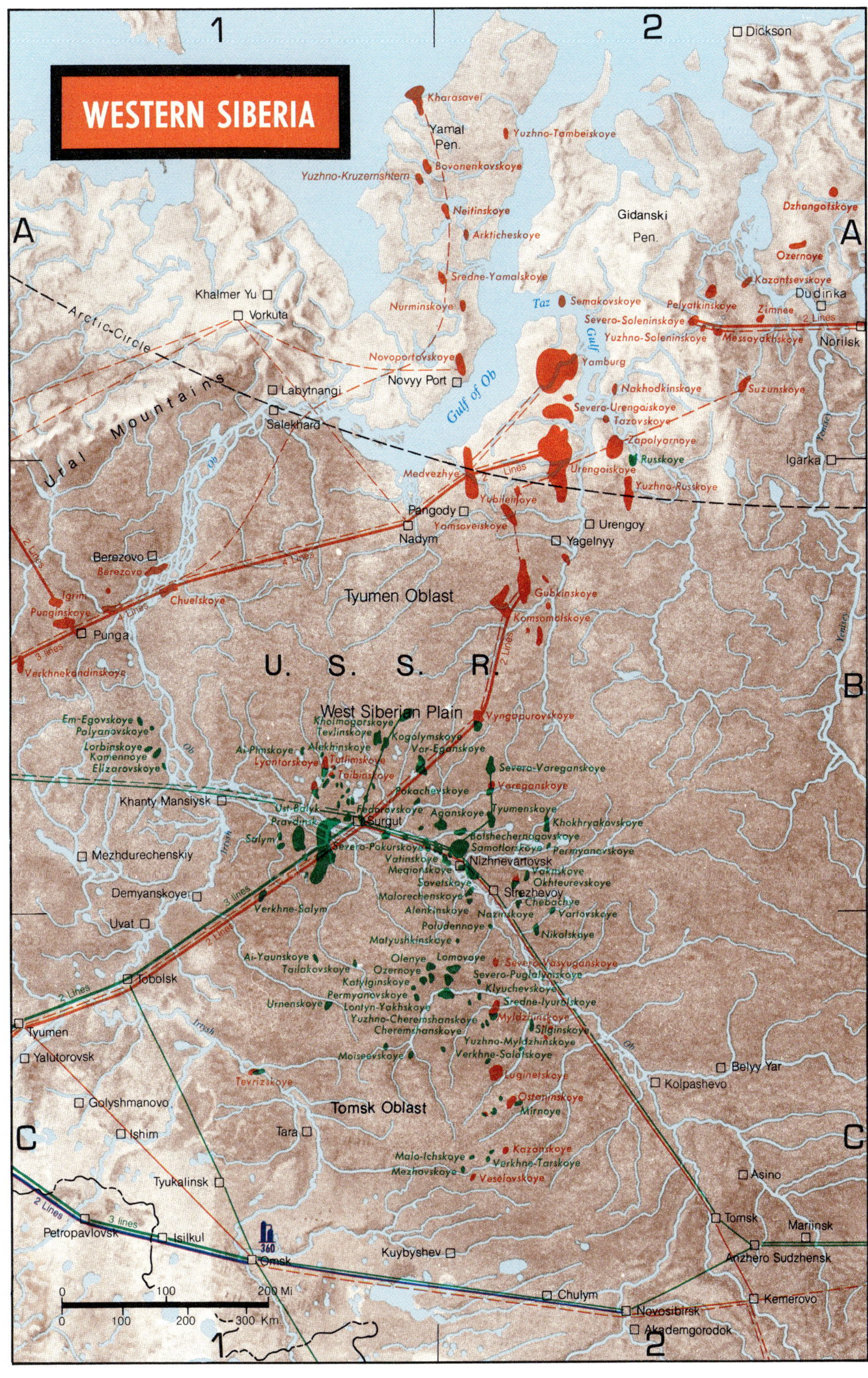

Meanwhile, a storm of high level controversy has enveloped Soviet claims late in 1987 of substantial success in Barents Sea exploration.

The spark for the dispute came from a veteran Soviet petroleum engineer who resigned from the Communist party to protest mismanagement and suppression of information on the state of the U.S.S.R.'s offshore arctic campaign.

The new, highly pessimistic assessment of results achieved by Aoogea differs sharply from previous glowing reports on Soviet accomplishments in the Barents Sea. Those reports appeared in October 1987 during a visit to Murmansk by Communist party chief Mikhail Gorbachev.

At that time, the Moscow press announced one oil and three gas strikes in the sea north of European Russia.

The optimistic statements were completely at odds with earlier reports emphasizing Aoogea's failure to meet drilling goals and its ineptness in handling the entire arctic offshore exploration program from the time of the association's 1981 establishment through 1986.

According to the Communist party newspaper Pravda, the continued sad state of Barents Sea petroleum exploration and attempts by Aoogea's management to conceal its errors illustrate the difficulties confronting Gorbachev's campaign for glasnost (openness) in the U.S.S.R. and perestroika (restructuring) of Soviet industry in the face of strong opposition from long entrenched bureaucracy.

Pravda, which devoted most of an entire page to an investigation of criticism leveled against Aoogea, noted that a 1985 study by a Murmansk newspaper and a provincial watchdog agency found that the association had fulfilled its offshore arctic exploration program by only 54.1% during 1981-85.

In mid-1987, Pravda reported, another check on Aoogea's work showed that serious shortcomings persist.

"Nothing has changed. The deficiencies remain the same," Pravda said.

Moreover, Aoogea's general director privately warned the Pravda correspondent not to write anything about the association's affairs. Earlier, Aoogea officials had told editors of the Murmansk daily Polyarnaya Pravda (Polar Truth) that if they continued to criticize the association their articles would either be ignored or refuted.

In other exploration highlights:

1. Drilling has resumed in the world's deepest hole, SG-3 on the Kola Peninsula west of Murmansk. Originally targeted to 15,000 m (49,912 ft), SG-3 reached 39,586 ft in 1984, remaining there until recently. The well is being deepened by 10-12 m/day. The rock structure of a core recovered from 12,000 m differs hardly at all from those obtained from 8,000 m and 6,000 m, say the Soviets.

2. Drilling of the U.S.S.R.'s second deepest well—Saatly SG-1, in Azerbaijan's Kura River basin southwest of Baku—has again been stalled by downhole problems.

Moscow reports that SG-1 has gone below 8,300 m. But it has been deepened by only 200 m during the past 6 years.

The hole has cost about 25 million rubles (close to $40 million). That was the projected cost for taking the hole to 11,000 m (36,089 ft). Soviet officials say that even if, as speculated, oil is found at 9,000-11,000 m it could not be produced because the U.S.S.R. doesn't have the necessary equipment.

3. A team of Soviet geologists from the Academy of Sciences is studying the advisability of drilling one or more superdeep wells in eastern Siberia to obtain data on possible commercial recovery of hydrogen from great depths.

Purpose of the tests would be to find data supporting theories advanced regarding formation of the solar system, the earth, and the earth's core. The long term economic payoff from the drilling, according to the geologists, could be discovery of a new, inexhaustible source of ecologically clean energy: hydrogen.

The first Soviet well aimed at finding streams of hydrogen would be drilled to 6-8 km (19,685-26,246 ft) in the Tunkinskaya Depression west of Lake Baikal and southwest of the city of Irkutsk. Cost of the hole would be about 50 million rubles.

A Soviet Uralmash BU-15000 rig, rated to 15,000 m, may be used to drill the Tunkinskaya Depression hole if the project is approved.

4. The Soviets spudded a stratigraphic test in the Volga-Ural area's Tatar Republic targeted to an area record 22,966 ft. Drilling is expected to take 7 years.

5. The U.S.S.R. last year began exploratory drilling in the Caspian Sea with a new semisubmersible, Shelf-5. It joined four other semis at work in the Caspian Sea. Shelf semis are rated to 19,685 ft.

Crude line to Turkmenia refinery

The Soviet Union has completed construction of a crude pipeline from oil fields in western Siberia's northern Tyumen Province to a refinery under construction at Neftezavodsk near Chardzhou in Southeast Turkmenia.

Located near the Amu-Darya River north of the Afghanistan border, the Neftezavodsk refinery was originally slated

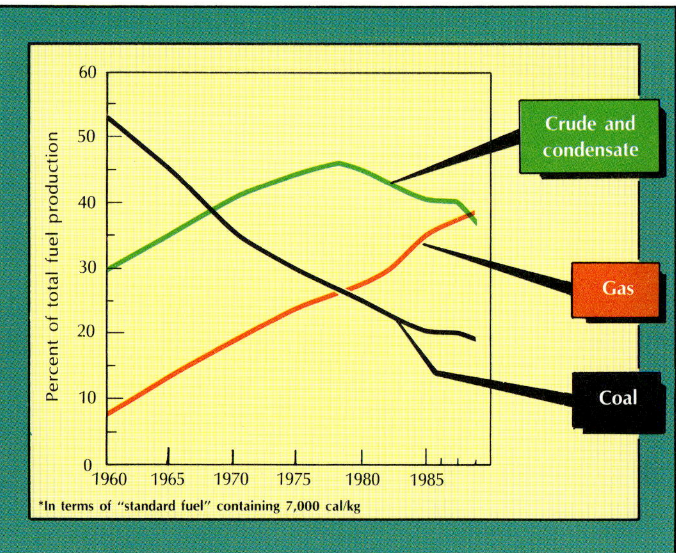

to go on stream in the late 1970s. Scheduled completion was delayed until 1980, then yearend 1985.

The Soviets still haven't announced that the Neftezavodsk facility is on commercial production. First phase of the refinery is believed to include a 60,000 b/d crude unit.

The new crude pipeline from western Siberian fields to Neftezavodsk is a 400 mile extension of the 1,020 mile link completed in 1983 from Pavlodar in northeastern Kazakhstan to the 120,000 b/d Chimkent refinery in southern Kazakhstan north of Tashkent. Originating at Surgut on the Ob River in Tyumen Province, the pipeline runs south to Omsk, then southeast to Pavlodar.

Elsewhere in the area, the Russians began testing a large diameter gas pipeline from Tyumen to Omsk. Laterals are being laid to district centers near Omsk, which has a large refinery and petrochemical complex. Gas comes from fields in northern Tyumen Province.

In far eastern Russia, plans have been delayed to lay a second natural gas pipeline from Okha on northern Sakhalin Island across the Tatarsky Strait to Komsomolsk-on-Amur. Russia cited slow progress in development of oil and gas fields on the northeastern shelf of the island. Also delayed was a proposed gas pipeline from Okha more than 500 miles southwest of Khabarovsk.

Striving to hike gas exports

Moscow believes continued growth in western European gas consumption will create demand so large by the end of the century it can be met only by greater dependence on the Soviet Union.

Soviet leaders persist in that belief even though gas exports to western Europe during the 1980s are not yielding the huge economic windfall Moscow expected.

Neither are those exports providing the extensive Communist penetration of energy markets in North Atlantic Treaty Organization countries that some U.S. strategists feared.

The U.S.S.R., world leader in gas reserves, production, and exports, will encounter increasingly stiff competition for gas sales in western Europe through the mid-1990s, at least. Norway, for example, has its sights set on North Sea gas sales to Sweden and other Scandinavian countries.

It's clear the Soviet Union has gas to spare.

Industry trade publication Oil & Gas Journal figures show that Russia at yearend 1987 held proved gas reserves of 1.45 quadrillion cu ft—38% of world reserves. No other

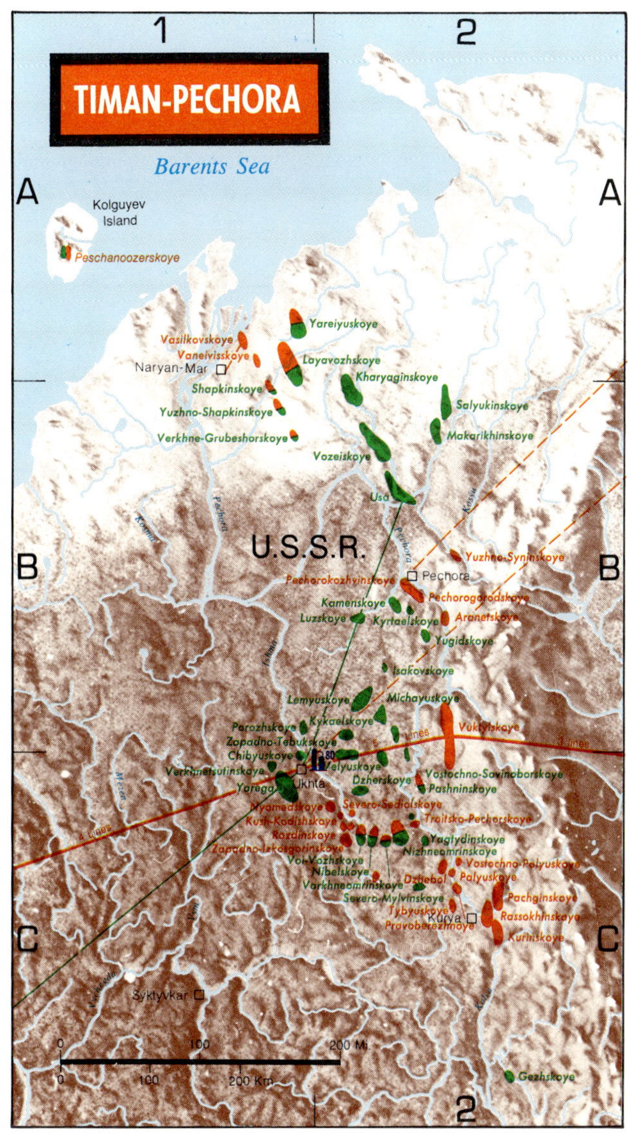

country even approaches that volume. Iran, with 489.4 tcf, is a far distant second.

What's more, Soviet authorities claim the U.S.S.R. has 40-50% of the world's gas resources, dwarfing volumes likely to be found in the North Sea.

North Sea production is seen as the main impediment to substantially increased Soviet sales to western Europe during the 1990s.

Moscow estimates that a single Soviet gas field, supergiant Urengoi, had initial explored reserves more than double those now proved for the entire North Sea.

Western Siberia's second largest field, Yamburg, where reserves have barely been tapped, also is believed to have greater production potential than all North Sea fields found to date.

Urengoi gas has been shipped to western Europe for more than 5 years.

Yamburg deliveries are slated to begin this year.

The Soviets emphasize that their potential for increasing gas exports without causing domestic shortages is vast.

The U.S.S.R. in 1987 exported only 11.6% of its gas flow vs. 31.4% of its oil production.

In the future, the U.S.S.R. will push still harder to substitute gas for oil domestically and in eastern Europe. The goal is to maximize the volume of crude and products that can be sold worldwide for desperately needed hard currencies.

The Soviets report they are trying to open new gas markets in "several" western European countries in the near term.

But even if those efforts are successful, they will have comparatively little effect on total Soviet gas exports outside the Communist bloc during the next 4 years.

Moscow is well aware of its limited potential for increasing near term gas sales to western Europe. That's why the Soviets have been in no hurry to complete a 56 in., 535 mile transit pipeline that will carry more Soviet gas across Czechoslovakia to western Europe.

The U.S.S.R. still has surplus capacity in the three pipelines that link the Soviet Union and western Europe via Czechoslovakia.

The new Czech line has been under construction with Soviet assistance since early 1983, when Moscow predicted that it would be operating at full capacity of 25 billion cu m/year by Jan. 1, 1988.

If, as now expected, the new Czech gas line is completed this year, it will have taken 6 years to build. The Soviets have repeatedly demonstrated their capability of laying 56 in. gas pipelines four times as long as the Czech project in less than 3 years.

Sweden is a prime target for Soviet gas sales in the near term.

Soviet newspaper Izvestia says the U.S.S.R. hopes to export 1-1.5 billion cu m/year to Sweden "by the beginning of the 1990s."

Moscow has proposed extending the growing gas pipeline system serving Finland westward across the southern Gulf of Bothnia to Sweden. Such a project couldn't be completed until the 1990s.

In the interm, the U.S.S.R. could sell small volumes of liquefied natural gas to Sweden. The Soviets in 1987 delivered 34,500 metric tons of LNG to Norway, up from 12,600 tons in 1986.

The U.S.S.R. is expected to maintain its monopoly of the Finnish natural gas market during the foreseeable future. Sales to Finland climbed to a record 1.64 billion cu m in 1987 from 1.24 billion cu m in 1986.

The volume of Soviet gas exports to Finland will continue to increase in the long term but is unlikely to exceed 3 billion cu m/year by 2000. Slumping prices paid by Finland for crude and products bought from the U.S.S.R. during the past several years have created a big trade deficit for the Soviets, resulting in heavy pressure on Helsinki to buy as much Soviet gas as possible.

The U.S.S.R. expected to add another western European country, Switzerland, to its list of natural gas customers in 1988. Sale volumes are likely to be small.

Spain remains a possible buyer of Soviet gas.

Belgium, which during the early 1980s considered importing gas from the U.S.S.R., still hasn't worked out a contract with Moscow.

With only two exceptions—1980 and 1982—the U.S.S.R. has increased gas exports to western Europe every year since the first deliveries were made to Austria in 1968.

But revenues from gas sales to western Europe plummeted in 1987 as prices fell for the second straight year.

The U.S.S.R. received just a little less than 1.8 billion rubles for 40.23 billion cu m of gas sold in 1987 to Austria, Italy, West Germany, Finland, and France. In 1986 the Soviets were paid more than 2.6 billion rubles for 37.87 billion cu m of gas exported to the same countries, and in 1985 the U.S.S.R. earned a record 3.26 billion rubles on sales of only 34.2 billion cu m.

The U.S. Central Intelligence Agency, which substantially underestimated Soviet gas production during the 1980s, in 1978 projected the U.S.S.R.'s total gas exports would reach 77.8 billion cu m in 1985. Total delivered that year to Communist and non-Communist countries was 75.8 billion cu m.

CIA believed Soviet gas exports to western European countries would be 34.7 billion cu m in 1985 and was exactly

SOVIET oil wells produce from the pre-Caspian depression's Karaton field (photo at left), just north of supergiant Tengiz. Tengiz is in a flat, barren area subject to flooding during Caspian Sea storms. Oil production continues to increase in western Siberia as the U.S.S.R. develops growing numbers of small fields (photo at right). Western Siberia's Tyumen province, by far the U.S.S.R.'s largest producing area, accounted for half of the aboveplan 1988 Soviet crude and condensate flow through mid-April that year.

on target.

But some western analysts who forecast that rapidly rising revenues from Soviet gas exports would overtake fast falling receipts from foreign oil sales by the late 1980s were mistaken.

The U.S.S.R. in 1987 sold a record 3.915 million b/d of crude and products worldwide valued at more than 22.8 billion rubles vs. 84.4 billion cu m of gas worth less than 6.4 billion rubles.

Oil represented 33.5% of the total value of Soviet exports in 1987, up from 32.9% in 1986. Gas accounted for 9.4% of the country's total 1987 export revenues, down from 10.8% in 1986.

Joint enterprises, petrochemical output

A Mitsubishi Corp. group will study the feasibility of building a $5 billion petrochemical complex in western Siberia under a contract with the Soviet government.

The study will cover construction in the Nizhnevartovsk area in middle western Siberia of 15 units to produce intermediate polymers and specialty plastics for export and domestic use.

The Soviets expect to export at least 40% of the products.

Feedstock for the complex would be ethylene supplied by a network of gas processing plants the Soviet Ministry of Oil Refining and Petrochemical Industry plans to build at a cost of about $10 billion in the Surgut and Tobolsk areas west of Nizhnevartovsk.

The Soviet government wants to start construction of both complexes within the next few years and complete them in the latter 1990s. The projects eventually will be integrated.

Feasibility studies on the gas processing plant projects are under way by a combine of Combustion Engineering Inc., Mitsubishi, and Mitsui & Co.

Joining Mitsubishi in the petrochemical project study are Mitsui and Chiyoda Corp. Press reports from Tokyo said C-E also is a participant, but the company wouldn't confirm it.

The Soviets are pursuing a number of joint ventures with foreign companies in petroleum ventures in the U.S.S.R.

Soviet plans to hike petrochemical production and exports to offset sharply reduced revenues from foreign oil and gas sales have fallen short of Moscow's expectations during the current 5 year plan.

The U.S.S.R. designated 1986-90 for unprecedented expansion of its petrochemical industry.

It may still achieve that general goal. But failure to boost key sectors of petrochemical production at the targeted high rates, mainly by their efforts alone, has forced the Soviets to turn to the West for massive assistance through joint ventures.

Perestroika of the petrochemical industry has lagged government goals in part because Soviet manufacturers of processing equipment failed to deliver advanced technology on schedule.

Short of hard currencies to place orders in the West, Moscow now counts on joint enterprises with capitalists to provide improved equipment and more efficient management methods required to inject some vigor into its petrochemical development program.

There is no doubt the U.S.S.R. has the potential to become a petrochemical giant—and even to rank first in the world, matching its performance in oil and gas production. It recently has been raising hydrocarbon production faster than contemplated by the official plan.

During first half 1988, the U.S.S.R.'s crude and condensate flow averaged 12.514 million b/d. That's higher than for any 6 month period in history and compares with 12.462 million b/d in the same period of 1987.

Natural gas production continues to show big gains. Flow

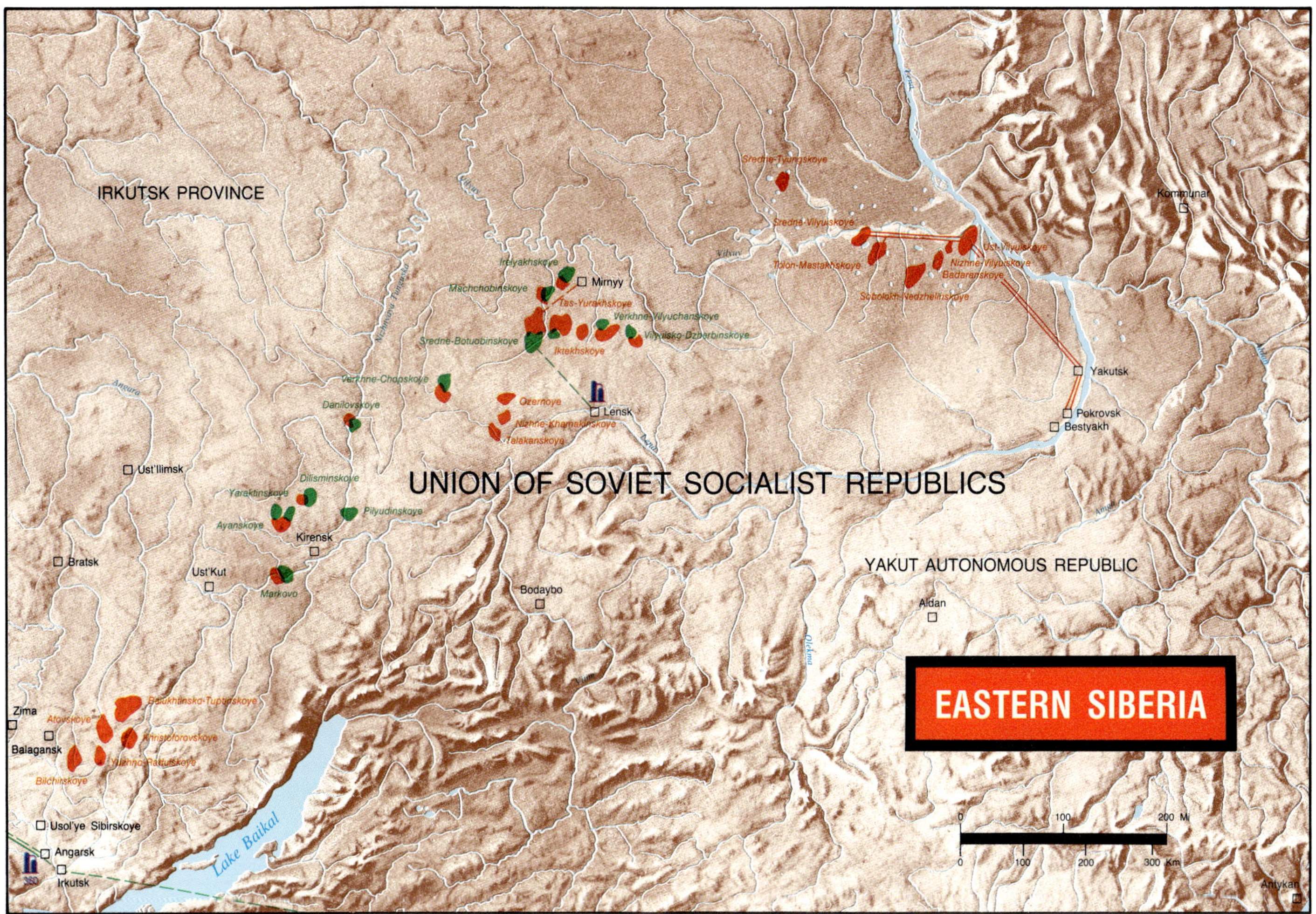

was 13.52 tcf during first half 1988, up from 12.67 tcf during January-June 1987.

Moscow still expects gas production to exceed 35 tcf/year during the next decade.

The U.S.S.R.'s next 5 year economic plan (1991-95) will again give high priority to increased petrochemical production. But the Soviets are unlikely to become a major factor in high quality petrochemical exports before the mid-1990s, at a time when the currently strong global petrochemical market may weaken because of large additions to capacity in many parts of the world.

Moreover, the U.S.S.R. must cope with widespread domestic petrochemical shortages in such areas as plastics before it can begin larger scale exports of such products.

Even before perestroika, Soviet economists complained about severe imbalances in domestic industrial output, distribution breakdowns, and the preponderance of raw materials in the U.S.S.R.'s foreign trade.

Production and consumption of oil, gas, iron, steel, cement, agricultural machinery, and natural fibers continue to grow as other industrial nations increasingly switched emphasis to hydrocarbon fuel conservation and manufacture of synthetic materials and electronic equipment.

While Soviet refinery crude runs rose, domestic airline flights were canceled because of localized jet fuel shortages. In some areas of the U.S.S.R., agriculture and transportation still suffer from tight supplies of diesel fuel and gasoline.

With by far the world's biggest natural gas production and consumption rates, spot shortages of this fuel persist during the winter.

During 1987, Soviet production of chemical equipment and spare parts fell 3% from 1986 and was only 85% of plan. During first half 1988, chemical equipment and spare parts manufacture jumped 11% from the comparable 1987 period but was still below plan.

New capacity for production of 450,000 metric tons/year of ammonia, 51,000 tons/year of synthetic resins and plastics, and 22,000 tons/year of chemical fibers was provided in 1987. But production targets for ammonia and synthetic resins and plastics weren't achieved.

During first half 1988, mineral fertilizer production grew 4%, chemical fiber manufacture rose 3% from the same 1987 period, and both production goals were exceeded. But the Soviets also reported shortfalls in production of ammonia, polyethylene, polystyrene, and polyvinyl chloride.

Russia is by far the world's largest mineral fertilizer producer. But in 1986 it manufactured only 4.4 million metric tons of synthetic resins and plastics, compared with 22 million tons for the U.S., 8.8 million tons for Japan, and 7.6 million tons for West Germany in 1985.

Included in the U.S.S.R.'s 1986 synthetic resins and plastics production were only 1.18 million tons of polyethylene, 104,000 tons of polypropylene, 525,000 tons of polyvinyl chloride resin and copolymers of vinyl chloride, and 449,000 tons of polystyrene and copolymers of styrene.

The 1987 plan called for 1.25 million tons of polyethylene, 130,000 tons of polypropylene, 600,000 tons of polyvinyl chloride resin and copolymers of vinyl chloride, and 500,000 tons of polystyrene and copolymers of styrene.

Soviet synthetic fiber production was slated to increase to 922,000 tons in 1987 from 856,000 tons in 1986.

Russia is doing well as a methanol producer. Production jumped to 3.2 million tons in 1986 from 1.9 million metric tons in 1980. The 1987 target was 3.33 million tons.

Ethylene production increased to only 2.8 million tons in 1986 from 2.7 million tons in 1985. The 1987 goal was less than 3 million tons.

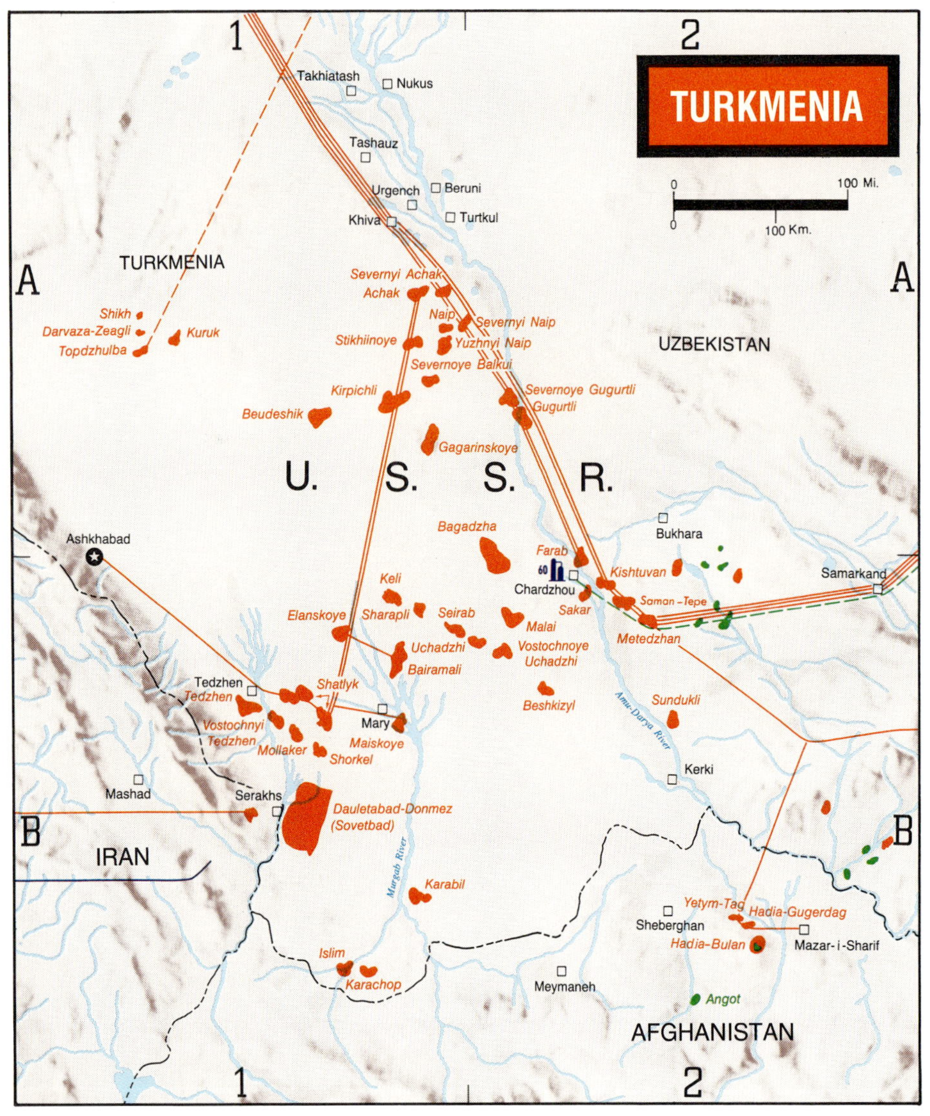

Propylene manufacture grew modestly to nearly 1.3 million tons in 1986 from 1.2 million tons in 1985. Target for 1987 was about 1.4 million tons.

Phenol production rose to 515,000 tons in 1986 from 502,000 tons in 1985. The 1987 target was 521,000 tons.

Official Soviet production figures, especially when presented in percentages and growth rates, are highly deceptive. Thus the U.S.S.R. claims that its chemical and petrochemical industry grew 10-fold from 1960 to 1986, compared with only 4.8-fold for the U.S., for example.

However, the same official Soviet source shows that the U.S.S.R.'s production of synthetic resins and plastics was 10% of U.S. production in 1960, 17% in 1970, 17% in 1980, and 19% in 1985.

From 1961 through 1986, Moscow reports that its synthetic resins and plastics production increased 11%/year, compared with 8.4%/year for the U.S. between 1961 and 1985. However, in terms of volume, U.S. growth was 746,000 tons/year, and the U.S.S.R.'s was 157,000 tons/year.

The Soviets point out that between 1961 and 1986 chemical fiber production in the U.S.S.R. rose 7.8%/year vs. 6.3%/year for the U.S. But during the same period, the U.S. hiked production of chemical fiber by 117,000 tons/year, compared with 49,000 tons/year for the Soviet Union.

Similar discrepancies are evident in the U.S.S.R.'s volumetric vs. percentage growth rates in production of such petrochemicals as ethylene, propylene, and polystyrene.

Soviet petrochemical production shows up even more unfavorably when calculated on a per capita basis.

Thus, in 1986, the U.S.S.R. manufactured 35.2 lb per capita of synthetic resins and plastics vs. 198 lb in the U.S., 282 lb in West Germany, 161 lb in Japan and Czechoslovakia, 139 lb in East Germany, 134 lb in France, 123 lb in Italy, 99 lb in Bulgaria, 88 lb in Hungary, 77 lb in the U.K., 62 lb in Romania and Yugoslavia, and 40 lb in Poland.

The 1986 situation was comparable in chemical fiber production, with Soviet per capita production less than 12 lb vs. 40 lb in East Germany, more than 35 lb in the U.S. and West Germany, 33 lb in Japan, 29 lb in Romania, 26 lb in Italy and Czechoslovakia, and 24 lb in Bulgaria.

The sharply falling value of Soviet exports, especially sales of oil and gas to the West, has deprived the U.S.S.R. of critically needed hard currency to buy petrochemical and chemical manufacturing equipment from the West.

Two years ago, 33.5% of Soviet export revenues came from foreign sales of crude oil and petroleum products and another 9.4% from gas sales.

Revenues from foreign oil deliveries plummeted from a record 30.9 billion rubles in 1984 to 28.2 billion rubles in 1985, 22.5 billion rubles in 1986, and 22.8 billion rubles in 1987 despite large gains in export volumes.

Value of gas sales to other countries dropped from a peak of 7.7 billion rubles in 1985 to 7.4 billion rubles in 1986, and to 6.4 billion rubles in 1987—again despite higher overall delivery volumes.

With hydrocarbons leading the decline in total Soviet foreign sales from a high of 74.4 billion rubles in 1984 to 68.2 billion rubles in 1987, Moscow was forced to cut purchases of machinery and equipment from the West. The U.S.S.R.'s total imports of equipment for its petrochemical and chemical industry fell from 1.18 billion rubles in 1984 to 1.04 billion rubles in 1985, 865 million rubles in 1986, and 701 million rubles the following year. The U.S.S.R. has failed in its efforts to boost petrochemical and chemical exports. During 1984-87 Soviet foreign sales of such products dropped from 1.28 billion rubles to 1.05 billion rubles.

Foreign agreement

In a foreign joint enterprise, the first Soviet-British engineering and trade venture has taken shape.

The joint venture, named Asetco Ltd., will modernize and expand Soviet ethylene and polyethylene plants at a cost of $151-189 million. The announcement of a Soviet enterprise with a foreign partner no longer surprises anyone, but Asetco goes beyond conventional limits, the publication Sotsialisticheskaya Industriya reported. Forming the joint stock company were the Soviet production association Stavropolpolimer in Stavropol and Orgsintez in Kazan, the U.S.S.R. Ministry of the Chemical Industry's Giproplast Institute (State Institute for Designing Enterprises for Production Plastics and Semifinished Materials), the British engineering firm John Brown Engineers & Constructors Ltd., the British bank firm Morgan Grenfell Group, and the Moscow Narodyni Bank.

Union Carbide Corp., Danbury, Conn., will be responsible for product sales outside the Soviet bloc. Asetco will use

Union Carbide's proprietary Unipol process to upgrade the plants. Linde AG of West Germany agreed to participate in Asetco as a subcontractor.

Other Soviet enterprises, foreign firms, and banks also may become stockholders, the Soviet publication said.

Asetco will increase modernized plants' capacities by one-third and modify them to manufacture premium products for sales on domestic and foreign markets. All outlays for modernization are to be recouped through revenues from polyethylene exports.

A venture with a foreign banking group as an equal partner is the main difference between Asetco and other Soviet joint ventures. The Soviet side, thereby, is freed from direct currency outlay, the publication said.

Participation by the Soviet design institute will permit a sharp cut in the time required for preparation of technical documentation.

Construction was due to begin in third quarter 1988.

About 40-50% of the plants' output will be exported.

Capacity of a plant at Kazan on the Volga River will be doubled to 400,000 metric tons/year of polyethylene, and capacity of a plant at Budyennovsk in the Caucasus region will be increased 50% to 300,000 metric tons/year.

Asetco will be registered on the island of Jersey in the English Channel. Management will be conducted by a board of directors.

Upstream deals

Moscow is pressing upstream joint ventures in and outside Russia with British, French, Japanese, and U.S. companies.

International production joint ventures represent one of three efforts that an official says may make more Soviet oil available for export.

The others are conservation of energy and resources—part of economic restructuring—and investment in refining technology more sophisticated than that in use at present, said Anatoly Tischenko, director, All Union Science & Research Institute for Organization, Management, and Economics of the Oil & Gas Industry of the U.S.S.R.

The U.S.S.R. recently entered a joint venture for offshore development design and is negotiating several others for production work.

The design venture, reported in a Tass dispatch from London, involves the U.S.S.R.'s Industrial Construction Bank and Kuibyshev Construction-Engineering Institute and the British firm J.P. Kenny.

Called Intershelf, the venture will provide engineering and economic assistance to Soviet agencies and enterprises and to firms in other countries.

Involved in negotiations for production ventures are Occidental Petroleum Corp., Chevron Corp., and another U.S. company that hasn't been identified.

Work by the venture involving the unidentified firm would take place on Sakhalin Island, Tischenko told a London conference organized by the Institute of Petroleum. Oil and other hydrocarbons produced by joint ventures would be exempt from state balance accounting, he said. They would be available for export.

Tischenko said the Soviet oil industry is in a transition to sharply higher production costs. During the next three 5 year plan periods, investment in drilling will increase to 38.3 billion, 42.8 billion, and 48.6 billion rubles. Total industry investment, 83 billion rubles in the current 5 year plan, will increase in the next three periods to 117 billion, 130 billion, and 138 billion rubles. Quality of current reserves is declining, Tischenko said, and the number of fields producing sour crude is increasing. New discoveries tend to be in remote areas with poor oil industry infrastructure. □ IPE

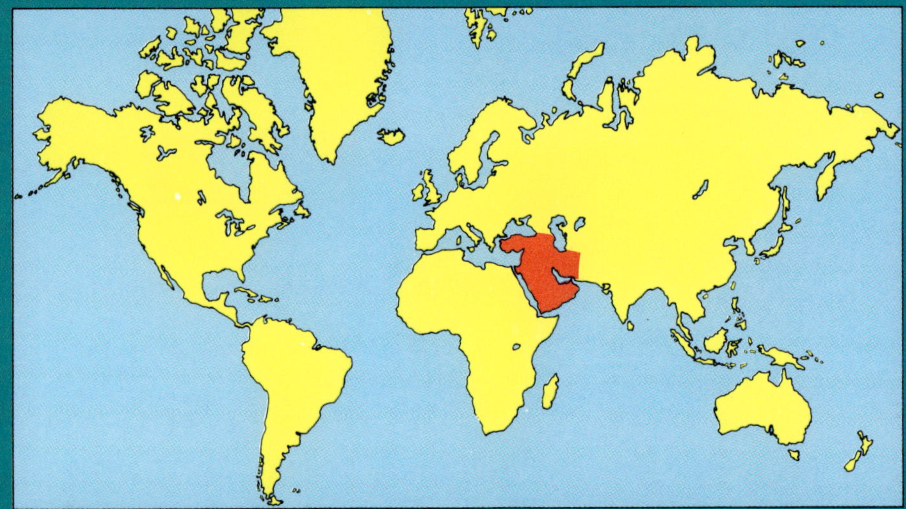

MIDDLE EAST

IRAN

CAPITAL: Tehran
MONETARY UNIT: Rial
REFINING CAPACITY: 530,000 b/cd
PRODUCTION: 2,207.5 M b/d
RESERVES: 92,850,000 M bbl

AN AUG. 20, 1988, CEASE-FIRE IN THE 8 YEAR OLD war between Iran and Iraq spawned elaborate plans by Iran to rebuild oil productive capacity, crude export facilities, and refinery capacity.

Iran plans to raise oil production capacity to 3.6 million b/d in 1993 from 3.1 million b/d in 1989 and almost double natural gas output to 8.95 MMcfd in 1998 from 4.6 MMcfd.

Middle East Economic Survey, quoting an oil ministry publication, said Iran will boost refining capacity to 1.3 million b/d in 1993 from 743,000 b/d in 1989. National Iranian Oil Co. will purchase 11 drilling rigs and spend $450 million on exploration between 1989-93. The company hopes to find at least 5 billion bbl of crude.

The official Islamic Republic News Agency (IRNA) quoted Iranian Oil Minister Golamreza Aghazadeh as saying Iran will boost exports and refining capacity as part of the postwar reconstruction of the country.

Aghazadeh said the new loading terminal under construction at Ganeveh in the northern Persian Gulf began operating shortly after the cease-fire took effect. NIOC has built four jetties at Ganaveh with a capacity of 2 million b/d.

Aghazadeh said in a separate statement before the cease-fire Iran will proceed with a 236 mile pipeline from Gurreh pump station on the Iranian mainland opposite Kharg Island to Bandar Taheri outside the Strait of Hormuz.

Saipem SpA of Italy and two Iranian contractors will build the 600,000 b/d line. Iran let contract to Smit International Marine Services BV, Rotterdam, to supply a 1 mile subsea pipeline and a single buoy mooring at Bandar Taheri. Iran plans to expand capacity to 1 million b/d later.

Aghazadeh said there will be a major refinery expansion program in an attempt to phase out product imports. The Abadan refinery, destroyed early in the war, will be rebuilt, but Aghazadeh didn't disclose the size of the replacement unit.

Investment will also be made in a 220,000 b/d refinery and petrochemical complex at Arak in western Iran. Iran previously considered building a 135,000 b/d refinery at Arak.

Iran has taken steps to complete the 880 mile, 48-56 in. IGAT 2 natural gas pipeline from Kangan gas field on the Persian Gulf to the Soviet Union.

It let contract to Saipem to lay about 20% of the pipeline not finished when the project ceased during the 1978-79 Iranian revolution. Meanwhile, Iran and the Soviet Union have signed a gas sales agreement, says IRNA.

Volumes will average 290 MMcfd, but IRNA gave no details on a start-up date for the deal or indication of prices.

It said, however, that the agreement makes provision for volumes to be increased at a later date.

At a Moscow meeting, Iranian and Soviet officials also discussed economic and trade agreements covering water, electricity, and steel projects.

Started in 1976, the pipeline was to be part of a deal to deliver 1.6 bcfd of gas to the southern U.S.S.R., which would supply equal volumes to Iranian customers in Europe.

The Iranian news agency said corroded

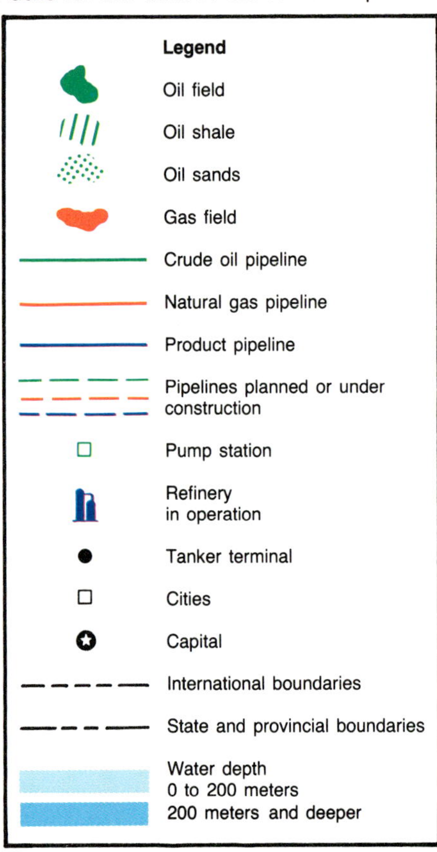

MIDDLE EAST
Black Sea
SOVIET UNION
TURKEY
SYRIA
LEBANON
ISRAEL
JORDAN
IRAQ
IRAN
TURKMENISTAN
Caspian Sea
SAUDI ARABIA
KUWAIT
BAHRAIN
QATAR
UNITED ARAB EMIRATES
OMAN
Gulf of Oman
Persian Gulf
Strait of Hormuz
Arabian Sea
SUDAN
ETHIOPIA
Red Sea
NORTH YEMEN
SOUTH YEMEN
Rub al-Khali
Batumi
Tbilisi
Yerevan
Nakhichevan
Evlakh
Baku
Derbent
Krasnovodsk
Cheleken
Gorgan
Neka
Bandar Shah
Ashkhabad
Mashhad
Sarakhs
Torbat-e Heydariyeh
Shahrud
Semnan
Tehran
Astara
Tabriz
Reza'iyeh
Rasht
Qazvin
Qum
Kashan
Arak
Azna
Daran
Isfahan
Shahreza
Yazd
Kerman
Sarcheshmen
Bandar Abbas
Jask
Muscat
Mina al Fahal
Sohar
Sur
Salalah
Mukalla
Sainut
Aden
Taizz
Mocha
Hodeida
Sana
Marib
Salit
Massawa
Port Sudan
Jidda
Mecca
Medina
Yanbu
Habigh
Ziba
Al Wajh
Aqaba
Ma'an
Jerusalem
Amman
Zarqa
Damascus
Beirut
Tripoli
Tartus
Banias
Latakia
Tel Aviv
Jaffa
Haifa
Homs
Hama
Deir al-Zor
Haleb
Dortyol
Yumurtalik
Mersin
Diyarbakir
Elazig
Malatya
Gaziantep
Nineveh
Mosul
Kirkuk
Baghdad
Basrah
Khanaqin
Kermanshah
Ahwaz
Abadan
Fao
Kuwait
Ahmadi
Ras Khafji
Qaisumah
Burayadah
Riyadh
Hofuf
Dhahran
Dammam
Manama
Sitra
Doha
Abu Dhabi
Dubai
Sharjah
Buraydah
Rafha
Badanah
Jalamid
Turaif
Qaryatain
Uwaiqilah
Shubah
Nakhichevan

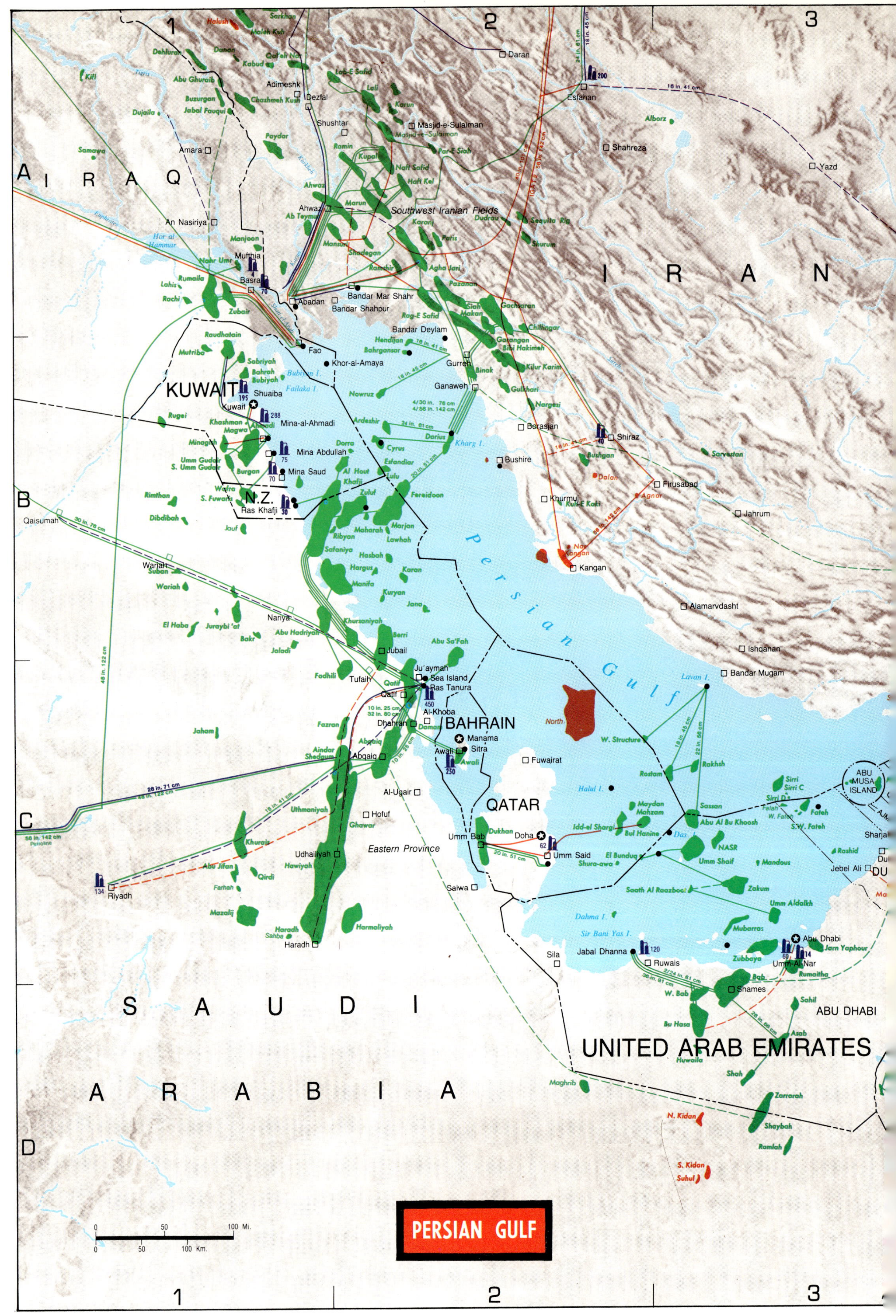
IRAQ
IRAN
KUWAIT
N.Z.
SAUDI ARABIA
BAHRAIN
QATAR
UNITED ARAB EMIRATES
ABU DHABI
Persian Gulf
PERSIAN GULF
Eastern Province
Southwest Iranian Fields
ABU MUSA ISLAND
Holush
Sorkhan
Maleh Kuh
Daran
Dehlura
Danan
Kabud
Gol eh Nar
Kilf
Abu Ghuraib
Lop-E Safid
Leli
Esfahan
200
Adimeshk
Dezful
Karun
Alborz
Shahreza
Buzurgan
Jabal Fauqui
Chashmeh Kush
Shushtar
Masjid-e-Sulaiman
Masjid-e-Sulaiman
Dujaila
Paydar
Romin
Kupali
Naft Safid
Haft Kel
Par-E Siah
Yazd
Amara
Samawa
Ahwaz
Marun
Karanj
Pazis
Dudrau
Southwest Iranian Fields
Seeuita Rig
An Nasiriya
Ab Teymur
Shurum
Hor al Hammar
Shadegan
Manjoon
Muftnia
Ramhir
Agha Jari
Pazanan
Gachsaran
Nahr Um
Basra
70
Abadan
Bandar Mar Shahr
Slah Makan
Chillingar
Rumaila
Lahis
Rachi
Zubair
Fao
Bandar Shahpur
Rag-E Safid
Gatangan
Bibi Hakimeh
Raudhatain
Bandar Deylam
Kilur Karim
Norgesi
Mutriba
Hendijan
Bohrgansar
Gurren
Binak
Gulfhari
Sabriyah
Bahrah
Bubiyah
Bubiyan I.
Ganaweh
Borasjan
Shiraz
KUWAIT
Shuaiba
Failaka I.
Nowruz
Sarvestan
195
Kuwait
288
Khashman
Mogwa
Ardeshir
4/30 in. 76 cm
4/56 in. 142 cm
Bushgan
40
Rugei
Mina-al-Ahmadi
Dorra
24 in. 61 cm
Darius
Kharg I.
Dalan
Firusabad
Minagish
Mina Abdullah
Cyrus
Esfandiar
Bushire
Khurmuj
Agar
Jahrum
Umm Gudar
S. Umm Gudar
75
Burgan
70
Mina Saud
Al Hout
Khafji
Lulu
Fereidoon
20 in. 51 cm
Kuh-E Kaki
Wafra
S. Fuwaris
Zuluf
Rimthan
Dibdibah
Ras Khafji
38
No Ruznan
Qaisumah
30 in. 76 cm
Jouf
Ribyan
Maharah
Marjan
Lowhah
Kangan
Alamarvdasht
Warjah
Suban
Wariah
Safaniya
Hasbah
Hargus
Karan
Ishqanan
El Haba
Juraybi'at
Bakr
Nariya
Khursaniyah
Berri
Manifa
Kuryan
Jana
Joladi
Abu Hadriyah
Abu Sa'fah
Lavan I.
Bandar Mugam
Jubail
Fadhili
Ju'aymah
Sea Island
Ras Tanura
Rakhsh
Sirri
Sirri C
Tufaih
Qatif
450
North
W. Structure
Rostam
Sirri D's
Fateh
W. Fateh
S.W. Fateh
Qatif
10 in. 25 cm
32 in. 80 cm
Al-Khoba
BAHRAIN
Manama
Sharjah
Jaham
Fazran
Dhahran
Damam
Sitra
Fuwairat
Halul I.
Sassan
Rashid
Jebel Ali
DU
Aindar
Shedgum
Abqaiq
Awali
Awali
250
Maydan
Mahzam
Abu Al Bu Khoosh
Das I.
NASR
Mandous
Khurais
28 in. 71 cm
48 in. 122 cm
Al-Ugair
QATAR
Idd-el-Shargi
Bul Hanine
Umm Shaif
Zakum
Umm Aldalkh
Ma
Abu Jifan
56 in. 142 cm
Petroline
Uthmaniyah
18 in. 41 cm
Hofuf
Ghawar
Umm Bab
Dukhan
Doha
20 in. 51 cm
Umm Said
El Bunduq
Shura-awa
Saath Al Raazboo
134
Riyadh
Qirdi
Farhan
Hawiyah
Udhailiyah
Salwa
Dahma I.
Mubarras
Mazalij
Haradh
Sahba
Harmoliyah
Haradh
Sir Bani Yas I.
Sila
Jabal Dhanna
120
Ruwais
Abu Dhabi
69
Jarn Yaphour
114
Umm Al-Nar
Rumaitha
Shames
W. Bab
Sahil
ABU DHABI
Bu Hasa
Asab
UNITED ARAB EMIRATES
Huwaila
Shah
Maghrib
Zarrarah
N. Kidan
Shaybah
Ramlah
S. Kidan
Suhul
50
100 Mi.
50
100 Km.

segments of the existing pipeline have been repaired.

NIOC has been working on a $1 billion project for gas treatment plants for Kangan and nearby Nar gas fields. The plants are to be operational within the year.

Iran commissioned a 25 mile pipeline to carry oil products from the Soviet Union inland from the Caspian Sea port of Anzali. Construction, costing $35 million, took place during the first 9 months of 1988.

The Soviet Union has an agreement to swap products for 100,000 b/d of Iranian crude.

In other transportation activity, Iran and Turkey will make a final effort to establish feasibility and financing of a $4 billion crude pipeline from Iran's Ahwaz fields to the Dortyol export terminal on the Turkish Mediterranean coast. In 1988, they announced plans for a 1,150 mile, 1 million b/d line but stumbled over financing.

A joint committee will review project economics, and if agreement isn't forthcoming, the project will be dropped.

Industry sources say Iran will pay a little more than two thirds of the construction bill. Turkey will pay the balance and probably will be responsible for much of the construction work.

Turkey pursued a policy of strict neutrality throughout the Iran-Iraq war. In addition, Turkey buys a sizable part of its crude oil requirements from Iran and Iraq and has a sizable countertrade business with the two nations.

Turkey also provides the largest single pipeline for Iraq's crude, handling about 1.5 million b/d through a terminal on the Mediterranean. From the moment the Iraqi line was conceived, Turkey made it clear that an Iranian pipeline alternative to gulf tanker routes also would be welcome.

However, bringing the Iranian project to this advanced stage has been difficult. Early in 1985 Iran and Turkey signed an accord covering a crude oil line and a gas line to serve the Turkish market and form the first stage of a gas export route to Europe.

The Turkish pipeline company, Botas, undertook two feasibility studies for the oil project. In 1987 Botas's boss, Nazihi Berkam, disclosed that Iran had rejected the feasibility studies. The project then seemed to slip into limbo.

One of the reasons behind the long gestation period for the pipeline has been Iran's reluctance to accept Turkish proposals for the route to the Mediterranean.

Iran's preference was for an export terminal at Trabzon on Turkey's Black Sea coast less than 100 miles from the border with the Soviet Union. Iran argued that the Black Sea route is shorter, cheaper, and more secure.

Turkish authorities want the pipeline to feed into a terminal at Dortyol, favoring that site because shipping and storage facilities are in place to serve a Turkish pipeline from the inland refinery at Batman. This infrastructure could be expanded to handle increased shipments of Iranian crude.

Dortyol is within sight of the Yumurtalik terminal, which in 1988 handled about 1.5 million b/d of crude from Iraq.

In an exploration highlight, the Iranian Oil Ministry in 1988 reported a significant gas discovery about 45 miles southeast of the Kangan gas fields.

It said the new field can produce about 920 MMcfd from two pays.

NIOC last year let a $1.25 million contract to Safety Boss Ltd., Calgary, to control a blowout that burned for 2 months, destroying a service rig.

The well produces 70,000 b/d of oil and 150 MMcfd of natural gas.

Location wasn't disclosed. Safety Boss estimated that it would take 2 months to control the well and restore normal production.

Meanwhile, Iran is developing central and southern area oil fields that will boost production by 300,000 b/d by 1990, Middle East Economic Survey reported.

The Soviet Union's Technoexport is developing the Mishrif reservoir in West Qurna field. Expected production is 200,000 b/d.

Rawati field is due second phase development, boosting total capacity to 600,000 b/d.

Contracts are to be signed for development of Saddam, Khabbaz, and Khormah fields, each with capacity of 20,000-30,000 b/d.

In the country's processing activity, Iran started construction of a 250,000 metric ton/year petrochemical plant in the northwest part of the nation, near the Russian border. The plant, whose product slate isn't disclosed, will cost about $600 million.

Elsewhere, the giant petrochemical complex at Bandar Khomeini, about 90% complete at the September 1980 outset of the Iran-Iraq war, is a total loss, say Japanese officials.

The 13 plant, $5.5 billion joint venture of a Mitsui-led group and Iranian National Petrochemical Co. was repeatedly bombed by Iraq.

Mitsui group wants the project declared a writeoff so it can claim about $1.3 billion from insurance policies, but Iran wants it completed.

To help boost refinery capacity, Iran plans to rebuild part of the Abadan plant and order a new unit at Bandar Abbas.

As 1989 opened, Chiyoda Corp., Tokyo, and Snamprogetti SpA, Milan, were set to win a $2 billion contract to build a 230,000 b/d oil refinery for NIOC at Bandar Abbas. It is scheduled to be operational by 1992.

The project is the first major new refining project in Iran since the start of the Iran-Iraq war.

To help finance the project Japan's four major trading houses, Mitsui, Mitsubishi, Marubeni, and Sumitomo, will purchase a volume of Iranian crude equivalent in value to half the total construction costs.

Iran's deputy prime minister Hamid Mirzadeh said $70 million will be spent on repairs to the 630,000 b/d Abadan unit, which was badly damaged during the war.

He said NIOC hoped to have 130,000 b/d capacity available from Abadan by April this year.

Mirzadeh, in charge of postwar reconstruction, said the Kangan gas treatment plant was scheduled to start up in April.

Gas liquids from the plant will be exported, and the dry gas will be piped into the domestic distribution system.

Iraqi jets considerably damaged the plant in a rocket attack in summer 1988 when the 1.2 bcfd plant was 85% complete.

Mirzadeh said Iran will rebuild the Kharg Island export terminal, a petrochemical plant near Bandar Khomeini, and the Ahwaz steel plant.

The Iranian parliament made $735 million available for the first phase of the reconstruction program, and a further $344 million is available in foreign currency, he said.

In an imports-exports deal in 1988, Iran offered to sell France $500 million worth of crude oil, with about 70% of the cost offset against deliveries of French agricultural products and fertilizers.

The deal, involving about 100,000 b/d, would require France to lift a ban imposed in 1987 on Iranian oil imports.

Iran expects oil exports to earn $9 billion in the Iranian year that began Mar. 21, 1988, compared with $8.66 billion the previous year, according to Persian news reports.

IRAQ

CAPITAL: Baghdad
MONETARY UNIT: Dinar
REFINING CAPACITY: 318,500 b/cd
PRODUCTION: 2,679.2 M b/d
RESERVES: 100,000,000 M bbl

FOLLOWING THE CEASE-FIRE IN THE IRAN-IRAQ WAR, Iraq has been in the midst of a program to hike productive capacity to 4 million b/d. And it has started preliminary work to reactivate Persian Gulf export facilities put out of action early in the war.

Neither country has revealed how it will pay for the work or service the huge debts built up during the fighting. The countries want to boost oil exports and increase foreign exchange earnings quickly.

Iraq has exported token volumes of oil through the Persian Gulf and plans to restore tanker loading capacity at the head of the gulf.

Terminals at Khor al Amaya and Fao were put out of action early in the war. Iraq has two temporary single buoy moorings in storage ready to resume exports.

Surveys have started on the subsea line that linked the mainland to the Khor al Amaya terminal.

If it is usable, the temporary loading buoys could be activated quickly.

Whether to repair the main Khor al Amaya facilities hasn't been decided. Much will depend on the state of the subsea feeder lines. A decision also is needed on the onshore tanker terminal at Fao. All units, including storage, were destroyed.

Iraqi exports through the gulf would prove popular with customers as long as peace persists. However, Iraqi investment in pipelines to the Mediterranean Sea and the Red Sea and overland transport through Jordan have ensured that Iraqi production of as much as 2.6 million b/d has continued unhampered.

Work has started on expansion to 1.65 million b/d of the pipeline through Saudi Arabia to the Red Sea. Throughput of the Turkish pipelines also would be increased so Iraq's total export capacity would rise to about 4 million b/d.

Iraq, also spending for new production facilities, awarded a contract to Technip Geoproduction in connection with development of Khabaz oil field in Central Iraq. Technip in turn let contract to Thermodyn, part of the French Framatome group, to supply compressor trains to handle sour gas from the field.

The unit will move gas to a treatment plant supplied by Technip.

Khabaz is one of three fields in the area with a potential productive capacity of 20,000-30,000 b/d that has been earmarked for early development. The major increase in capacity will come from development of the Mishrif reservoir in the West Qurna structure in southern Iraq by the Soviet company Technoexport.

Mishrif production will provide another 200,000 b/d in 1990.

That will be followed by the development of the Ratawai reservoir, which will boost total production from the field in the early 1990s to 200,000 b/d.

Iraq also let contract to Bechtel's U.K. affiliate for consultancy services for the proposed $2 billion Petrochemical Complex No. 2, which will be built in Central Iraq south of Baghdad.

The complex will have a 450,000 ton/year ethylene and derivatives plant and other downstream units.

Iraq let contract to Mannesmann of West Germany for development of Saddam oil field, expected to produce 45,000 b/d of 36° gravity crude and 300 MMcfd of gas. The contract covers oil and gas processing plants, gas gathering and degassing facilities, and pipelines.

Another contract is to be let for 60,000 b/d Khormal oil field.

Elsewhere, India agreed to import 2.5 million metric tons of crude from Iraq during fiscal 1988-1989 in a deal worth $140 million.

Much of the payment will take the form of defrayments of Iraqi debts to Indian companies for various construction projects during the past 9 years.

On the export front, Iraq has ceased oil exports through Aqaba, Jordan.

Under a contract with Jordan Ports Corp. that expired in April this year, Iraq stopped the export of 7,000 metric tons/day of crude through Aqaba. It since has stuck the Iraqi ports for oil exports. In 1987, Iraqi exports via Aqaba totaled $200 million.

Iraq reports that its 112% increase in crude oil reserves in 1987 capped 20 years of what the country's oil minister calls "a fairly heavy exploration and delineation program."

Iraq estimates reserves at 100 billion bbl.

Minister of Oil Issam Al-Chalabi said about 175,000 line km of seismic survey has been shot in Iraq since 1968. The work identified "hundreds of structural anomalies" and provided more data on previously known structures.

During the same period, the country drilled 350 exploration and appraisal wells on 72 structures, 54 of which had never been drilled.

The drilling campaign yielded 45 new field oil and gas discoveries, "some of them supergiant and giant fields."

That boosted Iraq's field total to 64 oil fields, 13 of them

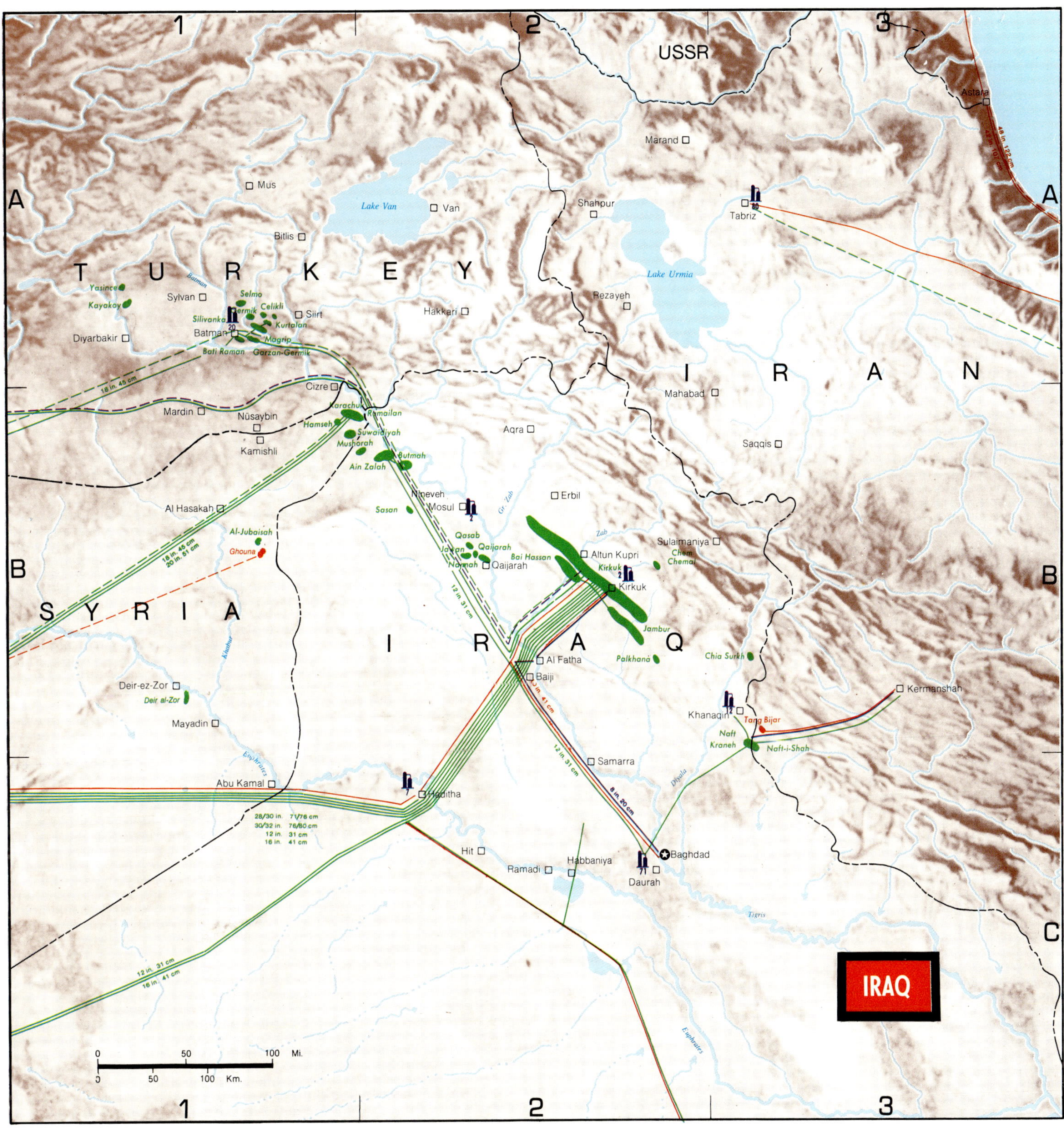

with free gas, and 7 gas fields.

Al-Chalabi said Iraq has six supergiant fields and 22 giants.

And Iraq is adding pipeline capacity for flexibility to move more crude exports to the Red Sea through Saudi Arabia. It expects to complete in about another year a second pipeline linking northern oil fields near Kirkuk to southern fields near the Persian Gulf.

The 400 mile, 42 in. line will have a capacity of 900,000 b/d, says Middle East Economic Survey.

Iraq's twin pipelines through Turkey to the Mediterranean Sea have 1.5 million b/d combined capacity.

KUWAIT

AS THE NEW YEAR ENTERED, THE BRITISH government extended for 2 years the Kuwait Investment Office's deadline for reducing its stake in British Petroleum

KUWAIT CITY SKY-LINE along Fahd al Salem thoroughfare, formerly one of Kuwait's main commercial shopping districts. (Photo courtesy Kuwait Oil Co.)

plc to 9.9% from 21.6%. In a separate move, BP expected to sign an agreement to sell its world mineral business to RTZ Group, London. The sale, which could raise $3.5-4 billion, triggered reports that BP might use the proceeds to repurchase a block of 700 million of its shares from KIO.

Britain originally gave KIO 12 months to reduce its interest in BP after the Monopolies and Mergers Commission reported the holding might not be in the public interest.

KIO, owned by the Kuwaiti government, asked Britain for as long as 5 years to reduce its holding to minimize its loss on the sale, taking place in a poor market.

A spokesman for KIO said the extension to 3 years was satisfactory even though it fell short of the requested timetable for selling BP shares.

Industry sources said KIO may have been given more time to sell its BP shares in return for a promise that the 11.7% shareholding would not be sold as a block to a single buyer. British Secretary of State for Trade and Industry Lord Young accepted the unanimous MMC conclusion that it would not be in Britain's interest for BP to have a major shareholder who also is a member of OPEC.

BP Chairman Sir Peter Walters, who earlier expressed unease at the size of the Kuwaiti holding, welcomed the government's decision.

The main issue now is orderly disposal of the shares, he said. KIO said it was extremely unhappy with the report's outcome and the way its conclusions were reached.

Meanwhile, UOP, Des Plaines, Ill., signed a letter of intent to furnish Kuwait National Petroleum Co. a fluid catalytic cracking unit, power recovery system, and two Merox process units for the MTBE/alkylation/FCC revamp project at KNPC's 288,000 b/cd Mina Al-Ahmadi refinery. UOP also will design a new LPG Merox process unit and LPG fractionation facilities.

OMAN

CAPITAL: Muscat
MONETARY UNIT: Omani Riyal
REFINING CAPACITY: 76,932 b/cd
PRODUCTION: 596,700 b/d
RESERVES: 4,071,160 M bbl

OMAN HAS AWARDED EXPLORATION ACREAGE TO units of Amoco Corp. and Wintershall AG.

Amoco will drill six wildcats and spend $14 million during the first 3 years of the license covering the 15,000 sq km Affar block in the Wahiba Sands area of eastern Oman. It will spend another $7 million and $6 million in two later 2 year periods. Any oil production would be split 80:20 in favor of Oman, gas 70:30.

Wintershall has similar terms for its 27,000 sq km Saiwan area of southern Oman. It will spend $4.5 million in the first 3 years and $3 million in each of four 2 year renewal periods. International Petroleum Ltd., Vancouver, B.C., spudded 1 Diba in 330 ft of water on the Batinah concession in the northern Gulf of Oman outside the Strait of Hormuz. Drilling

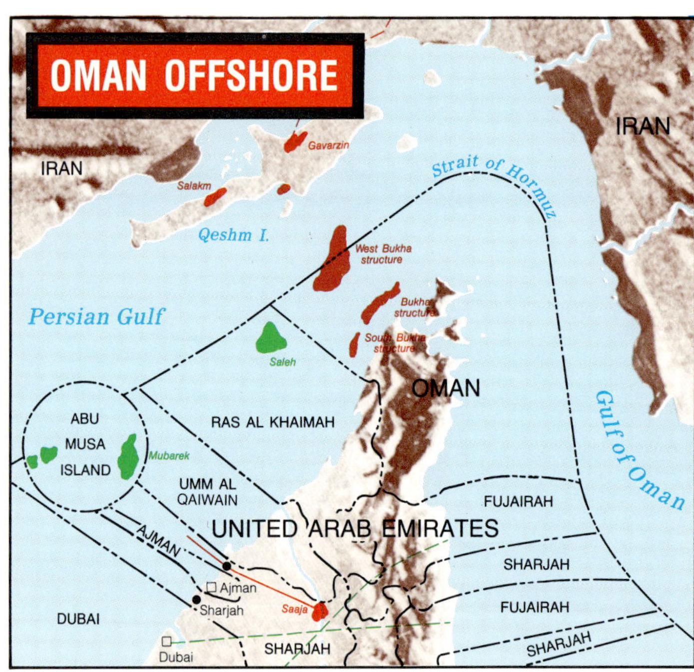

to Cretaceous at 13,100 ft on the 66 sq mile structure expected to take 90 days.

The well is the first for the Sonat George Richardson drilling rig, thought to be the first semisubmersible to operate in the Persian Gulf region. Petroleum Development Oman found light oil fields in southern Oman with its Rajaa 2 and Zummurud wildcats. Rajaa 2 is about 4 miles from Rajaa 1, which started producing in 1988. Zummurud is about 5 miles southwest of Rahab.

Oman will have a floating methanol plant off its coast starting up in 1990. Operator Ocean Phoenix Ltd. affiliate Offshore Gas Developments Ltd., Grand Cayman, signed agreements with Oman for gas supply and with Hoescht Celanese Chemical Corp. to buy its 500,000 ton/year Clear Lake, Tex., methanol plant to be upgraded and installed on a converted VLCC moored off Oman.

Suneinah concession
of northern Oman

Occidental of Oman Inc. in partnership with Chevron, is operator of the 3 million acre Suneinah concession in northern Oman. To date, discoveries have been made in Shams and Wadi Rafash gas/condensate fields and Safah oil field. Only Safah is being developed currently. The others await further delineation.

Fifty development wells were drilled in Safah during 1987-88: 28 in 1988 and 22 in 1987. Production is from the Shuaiba limestone formation. By the end of October 1988, Oxy Oman reports, the field was producing more than 15,000 st-tk bo/d. The oil is 43° gravity, and reserves are estimated at 60-80 million bbl by primary depletion, the company says. After treatment at the Safah production facilities, oil is exported via Oxy Oman pipeline to Lekhwair, where it joins the PDO pipeline system for transportation to the storage and export facilities at Mina al Fahal.

Future development plans for the field include more in-fill drilling and a possible gas reinjection secondary recovery scheme. The estimated peak production level of 20,000+ st-tk bo/d is anticipated to be achieved during 1989, Oxy Oman says. Field life is expected to be 20-25 years.

The company says it plans an exploration program this year in the Sharqi area, which is in the concession's southern area, and an 18,000 ft test in the Mountain Front area in the northern sector of the concession. Further delineation wells at Shams and Wadi Rafash will follow in 1990, the company notes.

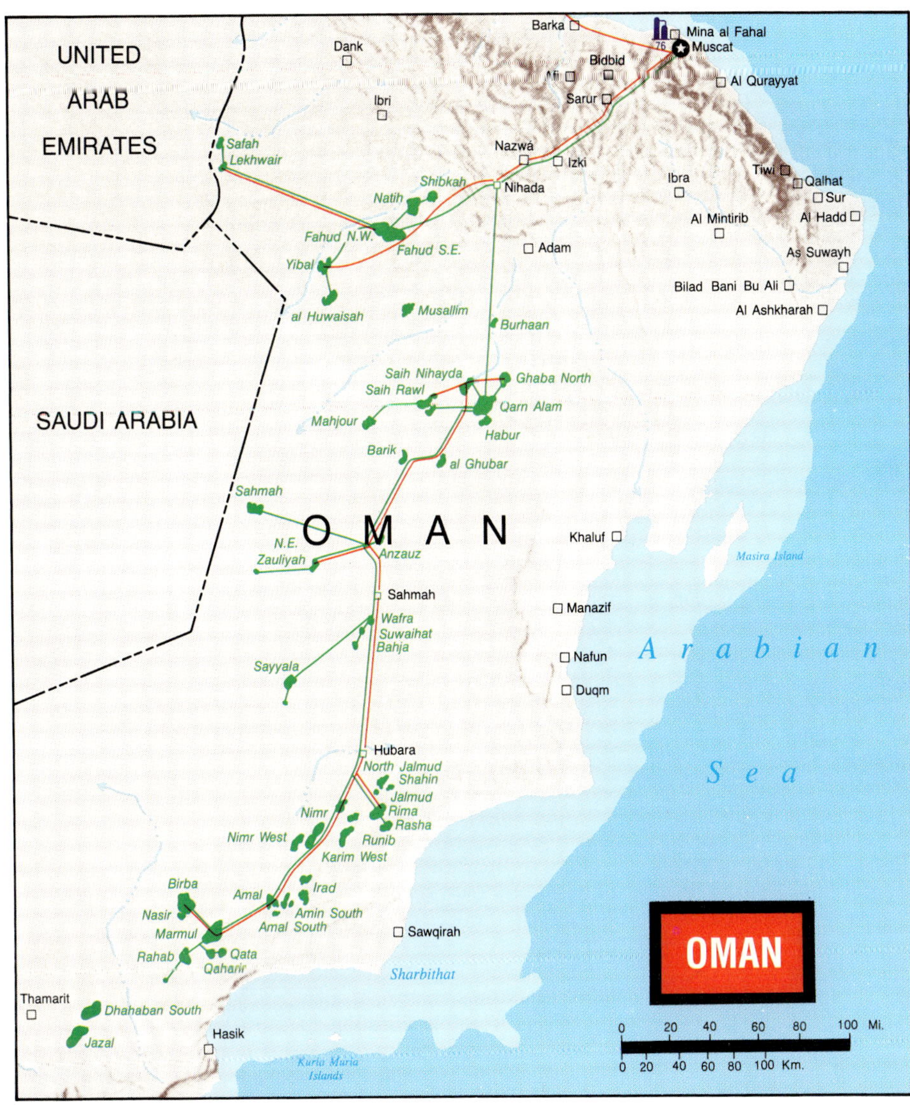

holds a concession covering most of the reservoir. However, the southern edge extends onto an offshore block held by a five company group led by Wintershall AG of West Germany. The government and the combine of Wintershall, International Ocean Resources Inc., Veba Oel AG, Deutsche Schachtbau, and Gulfstream Resources Canada, have been involved in a long contractual wrangle that only last year was resolved.

Problems arose when Qatar declined to participate in a gas development project based on a Wintershall gas discovery made on the fringe of the reservoir in 1980. The government says it suggested the group should develop the discovery on its own as allowed by terms of its production sharing contract.

At the same time, the Wintershall group was involved in separate discussions about carrying out the main North field development project. Failure to agree led to arbitration initiated by the group, which alleged breach of contract.

Arbitration seems to have taken a middle course that has left neither party disappointed. The government is happy because arbitrators ruled that the production sharing contract is still valid and has not been wrongfully terminated. They did not award the group damages.

Wintershall says the judgment confirmed and extended certain contractural rights. And it also laid down that there should be opportunity to agree on relinquishment of all or part of the group's rights. The value of those rights would be determined by further arbitration.

Elsewhere, Qatar Fertilizer Co. will add a 1,500 metric ton/day ammonia unit to its fertilizer complex at Umm Said, reports Norsk Hydro, a 25% partner with QGPC. Current capacity is 1,800 tons/day. Feedstock will come from North field.

QATAR

CAPITAL: Doha
MONETARY UNIT: Riyal
REFINING CAPACITY: 62,000 b/cd
PRODUCTION: 349,200 b/d
RESERVES: 3,150,000 M bbl

QATAR GENERAL PETROLEUM CORP. LET CONTRACT to Neyrfor-Weir and joint venture partner Nasir Bin Khalid Al Thani & Sons to provide turbo and directional drilling services for first phase development of offshore North field. Neyrfor-Weir will drill 16 wells with two rigs from two wellhead platforms.

The Maersk Victory jack up spudded the first well in 170 ft of water about 50 miles offshore. North, one of the few big development projects under way in the Middle East, holds an estimated 150 tcf of reserves. First phase development will provide gas only for domestic use.

Longer term, Qatar sees North gas feeding LNG export chains to the Far East, supplying its Persian Gulf neighbors, and even finding a market in western Europe through pipeline and LNG links.

Two wellhead platforms are installed in the field. Offshore processing facilities will be added later.

The $950 million project is scheduled to start gas deliveries to local consumers toward the end of 1991.

Capacity will be about 800 MMcfd.

The project is under control of state owned QGPC, which

SYRIA

CAPITAL: Damascus
MONETARY UNIT: Syrian Pound
REFINING CAPACITY: 243,744 b/cd
PRODUCTION: 273,300 b/d
RESERVES: 1,730,000 M bbl

IN NORTHERN SYRIA, BP PETROLEUM DEVELOPMENT Ltd. acquired 3,300 sq km Block 2, adjacent to the main oil producing province.

As sole licensee, BP is required to drill at least three wildcats during the license's initial 3 year term. BP has an option to extend the term for another 5 years.

Damascus Petroleum Ltd., wholly owned subsidiary of Occidental Petroleum Corp., plans exploration on its 2.8 million acre Bosra block in southwestern Syria.

It will shoot 2,200 line km of seismic surveys and start drilling in the last half of 1989. Damascus and state owned Syrian Petroleum Co. signed a production sharing agreement in March 1988. Oxy holds 40% interest.

Partners are Canadian Offshore International Ltd., Hydroil AS, and Wilhemsen Oil & Gas KS.

In field development, Al-Furat Petroleum Co. let contract to Technoexport/Salzgitter Lummus group as design and procurement subcontractor for production and pipeline facilities for Omar oil field near Deir-Es-Zor and other facilities in Thayyem field.

Technoexport is responsible for a 56 mile, 24 in. pipeline, gathering lines, other supplies, and all construction. Start-up on work at Omar began at yearend 1988, and April this year at Thayyem.

TURKEY

CAPITAL: Ankara
MONETARY UNIT: Lira
REFINING CAPACITY: 724,654 b/cd
PRODUCTION: 50,000 b/d
RESERVES: 381,000 M bbl

TURKEY'S STATE OWNED TURKIYE PETROLLERI Anonim Ortakligi (TPAO) and Chevron International Ltd. agreed to form a 50-50 joint venture to explore a 500,000 acre onshore tract in Turkey.

Chevron, operator, will conduct a 30 km experimental seismic program to determine best technology. It then will shoot about 150 line km of seismic survey.

The agreement calls for drilling of a wildcat within 4 years.

In another exploration deal, Turkey permitted Finland's Neste Oy to assign a 40% interest in six exploration licenses in southeastern Turkey to Idemitsu Turkey Oil Exploration Co. Ltd. Neste will remain operator and hold 50% interest in the licenses, which cover almost 1,000 sq miles near Cizre and Mardin.

The remaining 10% is held by Aladdin Middle East Ltd. Last year, Turkey began laying distribution lines in Ankara in preparation for natural gas deliveries from the U.S.S.R. About 40,000 buildings are to have gas service during winter 1989.

The Soviet Union began gas deliveries to European areas of Turkey in 1987.

Elsewhere in Turkey's transportation industry, Oiltanking G.m.b.H., Hamburg, and Turkpetrol, a Turkish independent oil trader, signed a letter of intent to build a tank terminal near Iskenderun in southern Turkey.

Initial storage capacity will be 300,000 bbl. The terminal will handle tankers as large as 50,000 dwt.

In an LNG deal, Turkey, Algeria, and Libya negotiated for possible imports of LNG to reduce Turkey's dependence on natural gas from Russia.

Turkey wants to import a total of 56.5 bcf/year from the North African countries.

Turkish imports of Soviet gas are to reach 211 bcf/year.

A pipeline from Bulgaria into western Turkey, commissioned in June 1987, was to be extended to Ankara in 1988.

UNITED ARAB EMIRATES

CAPITAL: Abu Dhabi Town
MONETARY UNIT: Dirham
REFINING CAPACITY: 180,000 b/cd
PRODUCTION: 1,442.9 M b/d
RESERVES: 98,105,000 M bbl

ABU DHABI HAS REVAMPED THE ADMINISTRATION OF its state controlled oil and gas sector.

The biggest and most powerful of the United Arab Emirates created a Supreme Petroleum Council to take over the role of the Abu Dhabi Department of Petroleum and the Board of Directors of Abu Dhabi National Oil Co. (Adnoc).

The structure of the states' oil and gas industry in the Middle East has been changing since crude oil prices went into a nosedive and the seemingly ever expanding state oil companies came under pressure to generate better returns through greater operating efficiency.

Creation of the council concentrates oil power into the hands of Khalifah bin Zayid al Nuhayyan, the crown prince and president of the national executive council.

Under him, executive power is held by Suhail al-Mazroui as secretary to the Supreme Council and general manager of Adnoc.

Two other key members of the old oil regime—Adnoc Chairman Shaikh Tahnoun bin Muhammad Al Nuhayyan and Abu Dhabi Oil Minister Mana Said al-Otaiba—get seats on the Supreme Council.

Otaiba continues as U.A.E. oil minister and will retain his responsibilities for OPEC.

The next step is likely to be merging of the staff and functions of the petroleum department with Adnoc.

Meantime, Abu Dhabi's International Petroleum Investment Co. received Spanish government approval to raise its 10% holding in Cia. Espanola de Petroleos SA (Cepsa) to 15%.

IPIC wanted to increase its holding in Spain's leading private refiner to 20%.

Cepsa operates a 130,000 b/d refinery at Tenerife in the Canary Islands and a 160,000 b/d plant near Cadiz. The deal includes a 60,000 b/d contract for Abu Dhabi crude.

IPIC paid about $520 million for the 10% stake in Cepsa, and it gained two seats on the board of directors.

Adnoc, in another move, last year postponed a maintenance shutdown of the 114,000 b/d Ruwais refinery. And it restarted 70,000 Umm al Nar refinery following a month long overhaul.

Elsewhere, Technip Geoproduction, Paris, received a telex of intent from Abu Dhabi's National Petroleum Construction Co. for relocation of the ADMA-OPCO accommodation platform on the Zakum Central complex in the Persian Gulf. The platform will be moved as it stands. A barge will be positioned between the platform legs and attached to a second barge for stability.

The company has a method of cutting the platform legs for which a patent is pending. After legs are cut, the barge will lift topsides clear, using a controlled ballasting system.

The unit will be moved to its new location slightly more than 6 miles away. Work was scheduled to begin last year.

In Dubai, ARCO Dubai was scheduled to test a new geological concept with an exploratory drilling campaign near its Margham gas/condensate field by yearend 1988.

ARCO shot new seismic, reprocessed data, and traded data with neighboring operators.

ARCO has been producing Margham condensate in partnership with Britoil plc since 1984. Output in 1987 averaged about 20,000 b/d. In Ras al Khaimah, Phillips Petroleum International Corp. Mid East took a 50% stake in International Petroleum Ltd.'s (IPL) onshore concession. Swedish Exploration Consortium AB took a 7% stake.

Seismic programs of 186 km and 38 km were shot in 1988. Also IPL was last year to start Al Khatt-IX wildcat, targeted to 6,500 ft to test Thammama carbonates.

Ras al Khaimah Offshore Petroleum Ltd. (Rakopet) completed the 7 Saleh development well in Saleh oil field, directionally drilled from the 6 Saleh wellhead platform. The new well produces 1,750 b/d and 14.8 MMcfd of gas.

Rakopet is owned by Wintershall AG, International Petroleum Corp., Overseas Petroleum & Investment Corp., and Neste Oy.

Off Sharjah, Crescent Petroleum Co.'s Mubarek oil field, damaged by an Iranian gunboat attack in April 1988, was out of commission for 2 months. A survey showed substantial

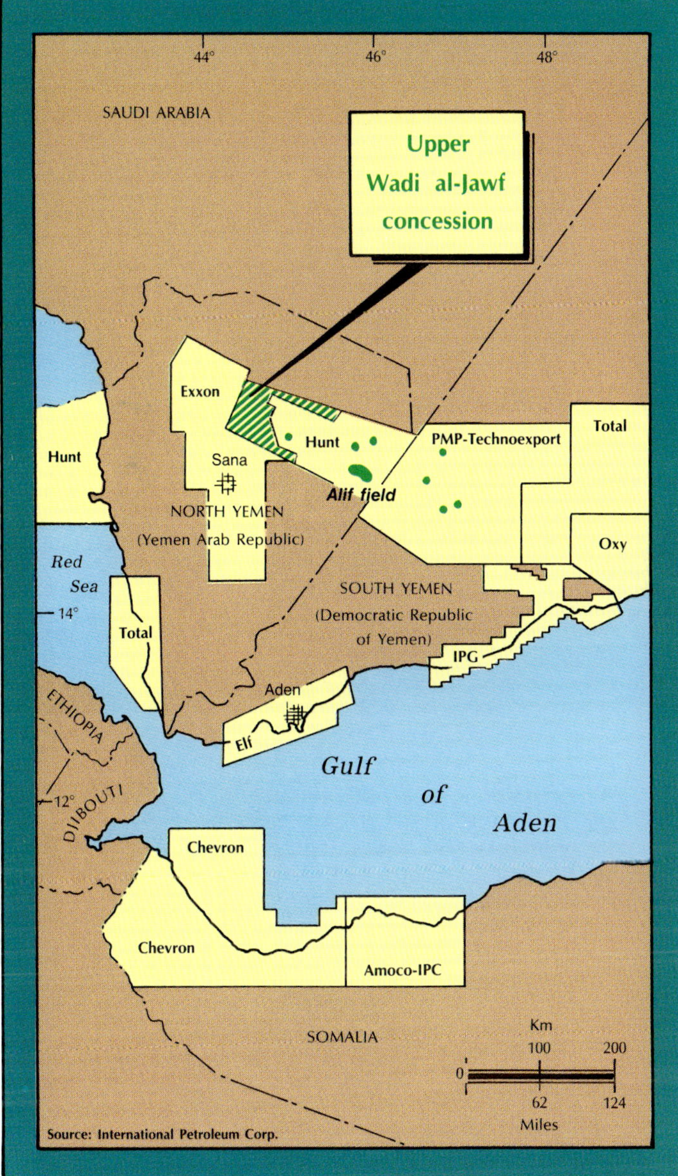

damage to the 112,744 dwt York Marine storage vessel. Before the attack production averaged 8,000-9,000 b/d.

NORTH YEMEN

CAPITAL: Sana
MONETARY UNIT: Rial
REFINING CAPACITY: 10,000 b/cd
PRODUCTION: 159,900 b/d
RESERVES: 1,000,000 M bbl

INTERNATIONAL PETROLEUM CORP. OF DUBAI WILL explore a newly acquired 1,659 sq mile concession in North Yemen. The acreage in the Marib al-Jawf basin was relinquished by Yemen Hunt Oil Co., which pioneered exploration in the area and placed Alif field on production at the end of 1987.

The Upper Wadi al-Jawf acreage covers the western end of Yemen Hunt's original concession.

IPC plans a 250-370 line mile seismic survey once the exploration agreement has been formally ratified by the People's Assembly. The seismic program will be followed by a drilling program. Yemen Hunt concentrated its exploration and development efforts in and around Alif field and associated discoveries.

The area of IPC's concession has only been lightly explored. The concession agreement was signed by IPC wholly owned subsidiary International Petroleum (Bermuda) Ltd. and North Yemen's Ministry of Oil and Mineral Resources. Production from the Yemen Hunt area started at 135,000 b/d and built toward 200,000 b/d by yearend 1988. Oil moves through a 260 mile pipeline to the coast for export.

As 1989 neared, Yemen Hunt announced another significant discovery in the concession. It's delineating a reservoir yielding 44° gravity oil from eight wells on Asad al-Kamil field about 10 miles north of Azal field.

In other action involving Yemen Hunt, the company and its affiliates won summary judgment in a district court in Texas in litigation brought by Arabian Shield Development Co. and Dorchester Master Limited Partnership, Dallas. The plaintiffs claimed the Hunt interests cheated them out of a concession in North Yemen. Arabian Shield plans to appeal.

Meanwhile, North Yemen and South Yemen formed a joint exploration and development venture, Yemen Oil Co. for Investments & Mineral Resources, to operate in a disputed border area. The area is between Yemen Hunt's Marib al-Jawf concession and the Shabwa area of South Yemen, where oil has been found by a Soviet contractor.

Elsewhere, Texaco Exploration Yemen Inc. took a farmout from Total Cie. Francaise des Petroles on the 2.3 million acre onshore and offshore Al Mukha concession in North Yemen. The companies will jointly conduct a 2 year second exploration phase involving seismic surveys, geologic field studies, and drilling of a wildcat.

SOUTH YEMEN

CAPITAL: Aden
MONETARY UNIT: Rial
REFINING CAPACITY: 161,500 b/cd
PRODUCTION: 12,500 b/d
RESERVES: 3,380,000 M bbl

STE. NATIONALE ELF AQUITAINE SIGNED AN exploration and development agreement for the 18,390 sq mile Sirr Hazar permit in northern South Yemen. The 6 year pact calls for shooting of 6,000 line km of seismic surveys and drilling of four wells during the first 4 years.

Elf Petroland BV, Elf's Dutch subsidiary, will be operator with 100% interest. The country is pressing plans to begin exports of oil and natural gas. With Soviet help, South Yemen has found oil reserves it claims total 3.38 billion bbl.

Work has begun on a 140 mile pipeline from new oil fields in the Shabwa area to a coastal export terminal. The $135 million project, involving three pump stations, will start up at a capacity of 30,000 b/d.

Ali Salim al Baidah, secretary general of the Socialist party, said capacity will increase to 120,000 b/d when four more oil wells in the Soviet operated fields are brought on stream in 1991-92. Britoil plc and Lasmo Oil (Aden) Ltd. will take farmouts on the Aden Abyan block, awarded to a unit of Ste. Nationale Elf Aquitaine in 1987. Britoil will take a 30% interest and Lasmo 25% in the 19,000 sq km concession centered on the city of Aden. Aeromagnetic surveys will be followed by seismic surveys and drilling. After a late 1988 visit to Oman by South Yemen Pres. Haidan Abu Bakr al Attas, a communique said most of the elements needed to settle a 20 year border dispute between the two countries are in place. They also signed a broad cooperation agreement that includes oil and mineral exploration. IPE

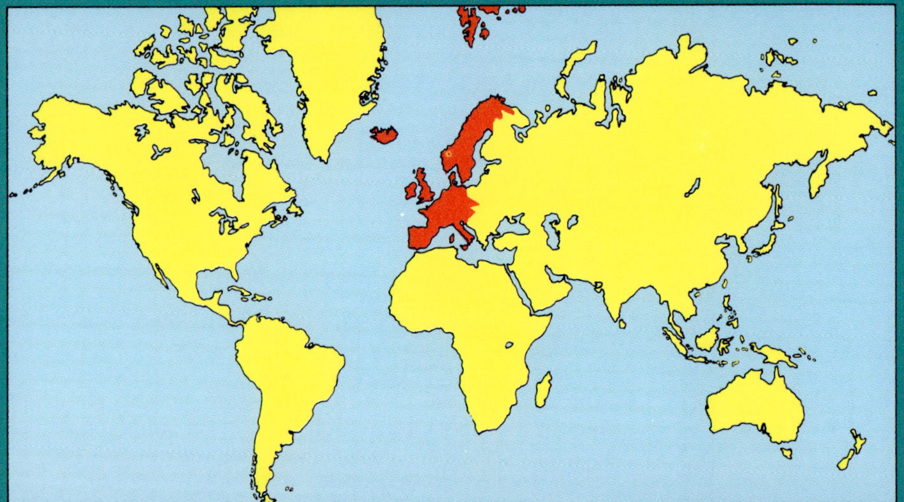

AUSTRIA

CAPITAL: Vienna
MONETARY UNIT: Schilling
REFINING CAPACITY: 204,000 b/cd
PRODUCTION: 23.2 Mb/d
RESERVES: 100,000 Mbbl

OMV AG, the Austrian state owned company, reached agreement with the Norwegian state oil company, Statoil, for a long term supply of natural gas.

Under the agreement, Statoil will sell the gas to Ruhrgas AG at the latter's terminal on Germany's northern coast. Ruhrgas will move the methane to the Austrian border, where Statoil will buy it back for resale to Austria.

Under the $3.5 billion deal, Statoil will supply OMV with 35 billion cu ft/year of gas for 20 years, starting in 1993.

Negotiations on the contract began in 1986, after Statoil signed contracts to supply Germany, France, Belgium, and the Netherlands with gas from Sleipner and Troll gas fields in the Norwegian North Sea over a 20-year period.

The Austrian deal was delayed because Ruhrgas declined to transport gas that it did not own, and would not allow Statoil to use its system, although for a fee, to transport the gas.

Ruhrgas said such an arrangement could tempt its German customers to buy gas directly from foreign suppliers and demand that Ruhrgas transport it for a fee.

The contract primarily is for gas from Troll offshore field, which is not due to go on production before 1996. So the Norwegians will make initial deliveries from other Norwegian North Sea fields.

Exploration outlook

Rohoel-Aufsuchungs Gessellschaft 1 Weizberg, a wildcat in western Austria near the German border, flowed gas at a rate of 7 MMcfd.

The operator is a 50-50 venture of Royal Dutch/Shell Group and Mobil Corp.

The operators also had an oil and gas discovery at the 1 Muhlreith near Salzburg. The hole flowed 2,500 b/d of oil from one zone and 3.4 MMcfd of gas from a shallower sand.

OEMV Aktiengesellschaft, Austria's national oil company, gauged 2.8 MMcfd of sweet gas from the Triassic at 11,000 ft at a wildcat east of Salzburg. The 1 Molln was planned for 17,400 ft.

It is 15 miles northeast of OEMV's 1 Gruenau oil discovery, drilled in 1987, which found 39° gravity oil in Upper Cretaceous sands at 16,000 ft.

BELGIUM

CAPITAL: Brussels
MONETARY UNIT: Franc
REFINING CAPACITY: 630,500 b/cd
PRODUCTION: 0
RESERVES: 0

The Belgian chemical industry continued to expand in 1988 at a healthy rate.

Finland's Neste Oy was doubling capacity of its Beringen polypropylene plant to 240,000 metric tons/year at a cost of $100 million.

The addition will go on stream in the late 1990s using Neste's technology. Meanwhile, the state company said it would shift its plastics marketing operations to Brussels.

Exxon Chemical plans a $21 million expansion of its Antwerp hydrocarbon solvents plant. The 80,000 metric tons/year addition will boost capacity to 680,000 tons.

The project will increase desulfurization, hydrogenation, and fractionation capacities and add storage tanks and loading facilities.

Exxon also was expanding by 50% the capacity of its 245,000 metric ton/year polyethylene plant at Meerhout. The expansion will cost $30 million.

Pertrofina SA agreed to buy British Petroleum Co. plc's 50% share in the 306,000 b/d SIBP SA refinery at Antwerp. Petrofina now holds the other 50%.

Norway's Statoil and Himont Inc. revived plans for a joint project to build a 180,000 metric ton/year propylene plant at Antwerp in 1990.

The $190 million project will use propane feedstock supplied by Statoil. And the deal gives Statoil access to Himont technology for expansion and debottlenecking of its polypropylene plant at Bamble in southern Norway from 70,000 tons/year to 90,000 tons/year,

DENMARK

CAPITAL: Copenhagen
MONETARY UNIT: Krone
REFINING CAPACITY: 176,500 b/cd
PRODUCTION: 95.5 Mb/d
RESERVES: 861,000 Mbbl

Denmark planned to abolish royalties and reduce state company participation requirements on tracts it will offer in an onshore and offshore licensing round in 1990.

The third Danish North Sea licensing round will offer all unallocated acreage, including for the first time some blocks relinquished by Dansk Undergrunds Consortium.

Offshore discoveries in 1987 nearly doubled Denmark's proved oil reserves, from 440 million bbl to 861 million bbl.

FINLAND

CAPITAL: Helsinki
MONETARY UNIT: Markka
REFINING CAPACITY: 241,000 b/cd
PRODUCTION: 0
RESERVES: 0

Finland plans to ease foreign access to its $2.3 billion petroleum markets to avoid future reprisals from the European Community.

Presently, only state owned Neste Oy and Suomen Petrooli, the Finnish unit of Soviet Sojuznefteexport, handle oil imports for the country. Under the liberalization plan, foreign firms may have direct access to a portion of Finnish oil imports contracted from the U.S.S.R.

Meanwhile, Neste Oy wants to expand its market share in domestic and regional fuels marketing through acquisition or joint venture.

The state company had difficulty meeting its 1988 commitment to import 2 million metric tons of Soviet crude, since it could not use all the oil. Under a trade protocol, Neste buys excess crude to balance accounts with its Soviet barter trade partner, then resells it to third parties.

The two countries finally agreed to use international market trends to determine the value of Soviet crude bartered for Finnish goods.

In the past, annual protocols have fixed prices, the 1988 price being $18/bbl. In the future, the value of trade will be based on reviews of oil prices in the international market.

In 1987 Neste Oy started up Scandinavia's first alkylation unit, a 170,000 metric ton/year, $42 million facility at its Porvoo refinery.

Neste Oy also plans to become Europe's biggest producer of polyethylene and polypropylene, with the goal of polyolefins capacity of 1.2 million metric tons/year by the early 1990s.

In 1987 Neste stated up a new $100 million polypropylene unit at Porvoo, which together with a new plant due to go on stream in 1990 at Beringen, Belgium, will bring the company's total production to nearly 400 million metric tons/year.

Neste Oy bought 8.5 million shares of Sovereign Oil & Gas plc, London, making it the company's largest shareholder with a 29.9% interest.

The state oil company also is building a $10 million, 26 mile extension of its southern Finland natural gas pipeline into greater Helsinki. Main customers will be an electric power plant, district heating plants, and major industries.

Neste Oy is expanding its 60,000 b/d Naantali refinery to produce jet fuel, some of which will be exported. The refinery will produce up to 30,000 metric tons/year of jet fuel.

Petrofina SA and Neste Oy planned to invest nearly $620 million in a new company, Finaneste, which will build a steam cracker capable of producing 450,000 metric tons/year of ethylene at Antwerp. Petrofina holds 65% of the new venture and Neste Oy the balance.

Finaneste has taken over two steam crackers and auxiliary plants of Petrofina's NV Petrochim SA. Debottlenecking of those plants and the new cracker will boost Finaneste's ethylene capacity to 1 million tons/year by early 1991.

FRANCE

CAPITAL: Paris
MONETARY UNIT: Franc
REFINING CAPACITY: 1,875,970 b/cd
PRODUCTION: 68.7 Mb/d
RESERVES: 206,030 Mbbl

France's oil industry plans to double research spending in the next 5 yeas to stay at the forefront of technology and maintain its rank as the world's third largest supplier of oil services and equipment.

An R&D program worked out jointly by Elf Aquitaine, Total CFP, Institut Francais du Petrole, and France's oil service/supply industry has earmarked $850 million to $1.02 billion during 1989-93, vs. $425 million in 1984-88.

The 5-year program will concentrate on cutting E&P costs, but longer term the goal is to improve arctic petroleum technology.

The French government lifted its ban on Iranian oil imports, which had been imposed in July 1987 after the two countries broke diplomatic relations after an Iranian embassy official refused to be questioned about 1986 terrorist bombings in France.

The government's first step in 1987 was to allow oil companies to exchange oil for agricultural products case by case in barter deals, but arrangements were so complicated few companies did so. It later totally lifted the ban.

Before the ban, Iran was France's largest oil supplier, providing 719,000 tons of oil in June 1987, or 6.6% of French oil imports.

Exploration action

In an extension of the Paris basin play, Elf Aquitaine found oil in a Paris suburb. A hole drilled in the middle of Ivry-sur-Seine recovered oil form a reservoir at 6,500 ft.

Total CFP and Coparex completed a seismic survey on the Permis de La Manche, west of the British Channel Islands in French waters. Exploration in the area during the early 1980s was unsuccessful.

Premier Consolidated Oilfields plc received a permit to explore the Perigord concession in the Dordogne region of Southwest France. Premier will spend at least $175,000 for geophysical work in the first 2 years of the 5 year permit. Total exploration spending will be $1.5 million.

Conoco Inc.'s French unit asked France for permission to

Location — No. Refineries

United Kingdom
South Killingholme 1
Fawley 1
South Humberside 1
Coryton 1
Port Clarence 1
Eastham, Cheshire ...1
Shell Haven 1
Stanlow 1
Scotland
Grangemouth 1
Dundee 1
Wales
Milford Haven 2
Llandarcy 1
Pembroke 1
Dyfed
Belgium
Antwerp 4
France
L'Orcher 1
La Mede 1
Mardyck 1
Reichstett-Vandenheim 1
Donges 1
Feyzin 1
Grandpuits 1
Fos sur Mer 1
Port Jerome 1
Notre Dame de Gravenchon 1
Berre l'Etang 1
Petit Couronne 1
Dunkirk 1
Lavera 1
Netherlands
Rotterdam 3
Pernis 2
Amsterdam 1
Vlissingen 1
West Germany
Vohburg 1
Burghausen 1
Godorf 1
Harburg 1
Heide 1
Duisburg 1
Neustadt 1
Neustadt-Donau 1
Ingolstadt 1
Karlsruhe 2
Worth 1
Gelsenkirchen 1
Wesseling 1
Lingen 1
Mannheim 1
Salzbergen 1

Location — No. Refineries

Italy
Milan 1
Pavia 1
Taranto 1
Gela 1
Marittima 1
San Quirico 1
Siracusa 1
Mantova 1
Busalla 1
Porto Marghera 1
Melilli 1
Milazzo 1
Naples 1
Rome 1
Sarroch 1
Novara 1
Priolo 1
Livorno 1
Cremona 1

NORWAY
Mongstad
Bergen
Karmoy
Stavanger
Oslo
SWEDEN
Lysekil
Goteborg
Malmo
DENMARK
Fredericia
Kalundborg
Copenhagen
Skaelskor
SCOTLAND
Flotta
Orkney I.
Tiffany
Fraserburgh
St. Fergus
Forties
Grangemouth
Glasgow
Ekofisk
UNITED KINGDOM
North Tees
North Sea
Rostock
Stettin
Hamburg
Brunsbuttel
Wilhelmshaven
Bremen
GERMAN DEMOCRATIC REPUBLIC
Berlin
Dresden
Prague
N. IRELAND
Belfast
I. of Man
Liverpool
Manchester
Leeds
IRELAND
Dublin
Whitegate
Bantry Bay
WALES
Birmingham
Bracton
ENGLAND
Milford Haven
Fishguard
London
Crawley
Wytch Farm
English Channel
Dunkerque
NETHERLANDS
Groningen
Amsterdam
Rotterdam
Antwerp
Brussels
BELGIUM
Bonn
Bremen
FEDERAL REPUBLIC OF GERMANY
Frankfurt-am-Main
Karlsruhe
Baden-Wurttemberg
Ingolstadt
Vohburg
Munich
Linz
AUSTRIA
Graz
Le Havre
Rouen
Gargenville
Vernon
LUXEMBOURG
Nancy
LeMans
Paris
Coulommiers
Grandpuits
Donges
Nantes
Tours
Blois
Orleans
FRANCE
Vierzon
Roussines
Vichy
Lyon
Grenoble
Chazelles
Ambes
Bordeaux
Basel
Bern
Zurich
SWITZERLAND
Geneva
Turin
Milan
Porto Marghera
Ravenna
Trieste
Koper
Genoa
Savona
Ligurian Sea
CORSICA
Rome
Bay of Biscay
Parentis
Gaviota
La Coruna
Vizcaya
Orpina
Bilbao
San Sebastian
Vitoria
Logrono
Burgos
Valladolid
Oporto
PORTUGAL
Bayonne
Lacq
El Serrablo
Lourdes
Toulouse
ANDORRA
Gulf of Lions
Marseille
SPAIN
Zaragoza
Madrid
Barcelona
Tarragona
Dorede
Angula
Salmoneta
Tarraco
Montananzo
Casablanca
Castellon
Ampoto
Castellon de la Plana
Valencia
Puertollano
Cordoba
Huelva
Rota
Cadiz
Algeciras
Cabo Ruivo
Lisbon
Sines
Gulf of Cadiz
Strait of Gibraltar
Almeria
Cartagena
Minorca I.
Majorca I.
Iviza I.
Mediterranean Sea
Porto Torres
SARDINIA
Cagliari
Italy
Milan 80
Sannazzaro, Pavia 210
Taranto 90
Porto Marghera 80
Gela 80
Falconara, Marittima 78
Genoa, San Quirico 130
Augusta, Siracusa
Mantova 55
Busalla 46
Naples 100
Rome 80
Meilli 220
Milazzo 160
Sarroch 285
Novara 240
Palermo
SICILY
Rabat
MOROCCO
Oran
Arzew
Algiers
ALGERIA
Skikda
Tunis
TUNISIA
Pantelleria I.
MALTA
North Atlantic Ocean

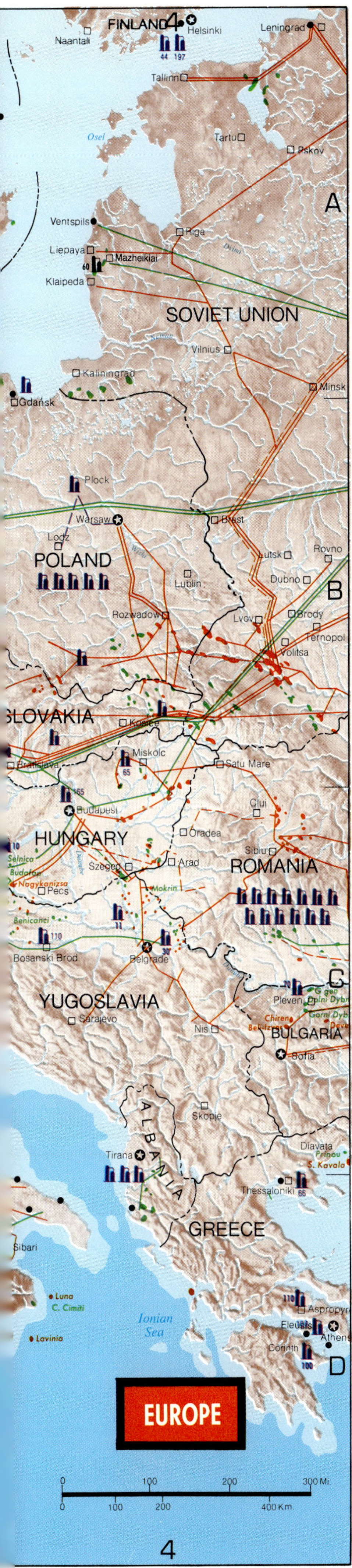

farm out to Nippon Mining Co. Ltd. of Japan its Permis d'Auch acreage in Southwest France.

It would be the first Japanese involvement in French exploration and production. It follows Nippon Mining's agreement to invest $40 million in 10 Conoco exploration projects.

Triton France SA tested 212 b/d of oil at a Paris basin wildcat. The Triton 1 Maincy, 1.2 miles south of St. Germain oil field on the Melun Permit, flowed from a Triassic sand at 7,367-431 ft. Triton holds a 50% interest in the permit and Total Exploration SA the rest.

Refining outlook

Elf France will cut distillation capacity at its Feysin refinery to 94,000 b/d from 154,000 b/d as part of a restructuring program.

It also will spend $340 million during 3 years to improve refinery productivity and upgrade its marketing system. And it plans to reduce its workforce 1,400 positions, or 30%.

Petroleos del Norte SA of Spain will establish a network of gasoline stations in Southwest France as part of its expansion abroad.

It will be the first Spanish company to enter a foreign gasoline retail market in preparations for Spain's entry into the European Community. Petronor has a 240,000 b/d refinery at Bilbao in northern Spain, near the border with France.

Solvay Group of Brussels was expanding capacity of its Sarralbe high density polyethylene plant to 120,000 metric tons/year.

ARCO Chemical Europe has started up a $340 million propylene oxide-/tertiary butyl alcohol plant at Fos-Sur-Mer.

The plant can produce 330 million lb/year of PO, 790 million lb/year of TBA, and 110 million lb/year of propylene glycol. A 10,800 b/d methyl tertiary butyl ether unit was nearing completion.

Mobil Chemical Co. was building a plant at Notre Dame de Gravenchon (Seine-Maritime), France, that will expand its polyalpha olefin base stock capacity for synthetic lubricants by 60%.

Ste. Francaise Exxon Chemical plans to boost capacities of its olefins plants in France. It will increase capacity at its Notre Dame de Gravenchon (Seine-Maritime) plant to 320,000 metric tons/year from 300,000 tons/year, and capacity may be increased to 400,000 tons/year later.

Shell Francaise completed a $27 million revamp of its Etang de Berre refinery. The work was needed to accommodate a fluid catalytic cracking unit Shell moved to Berre from Pauillac.

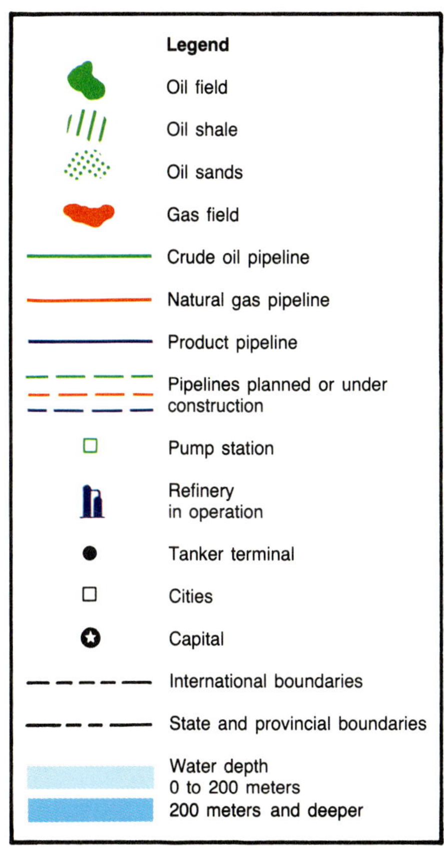

The existing distillation unit at Berre needed upgrading to bring it up to the operational level of the relocated FCC unit.

The 130,000 b/d refinery, 8 miles west of Marseilles, was shut down 10 weeks for the work.

GREECE

CAPITAL: Athens
MONETARY UNIT: Drachma
REF. CAPACITY: 384,500 b/cd
PRODUCTION: 21.8 Mb/d
RESERVES: 20,000 Mbbl

Greece and Turkey eased their tensions in 1987, lending hope that oil exploration might resume in the Aegean Sea.

The two countries have been at odds over Aegean drilling rights and territorial limits. Greece claims a 10-mile limit off its coasts while Turkey respects only a 6-mile limit.

Leaders of the two countries agreed to hold annual summit meetings, establish joint committees to negotiate solutions, and otherwise improve communications.

In 1987, Greece's parliament ratified an agreement ending the threat of expropriation of foreign operators' Aegean Sea oil and gas interests.

The government gave up its claim to acquire as much as a 51% interest in Prinos and South Kavala fields.

Greece had threatened to nationize offshore oil and gas fields operated by North Agean Petroleum Co. But under an agreement with NAPC leader Denison Mines Ltd., Toronto, the government will relinquish a claim to acquire up to 51% of the fields.

NAPC must seek government approval before launching further exploration in Greek waters and the government's stake in nonproducing areas around the fields will increase to 15% from 10%.

Public Power Corp. let a $1 million contract to Xcel Group Inc. for supply of research and development equipment in a program aimed at gasification of low BTU lignite.

Hellenic Aspropyrgos Refinery SA started up a $15 million, 63,000 metric ton/year MTBE plant and associated propane and butane units at its 110,000 b/d Aspropyrgos refinery near Athens.

Petrola Hellas SA has let contract for a 17,500 b/d hydrodesulfurization unit, an amine unit, sour water stripper, and associated facilities at its 108,000 b/d refinery at Elefsis. The new units are needed to meet European Community emissions standards.

ITALY

CAPITAL: Rome
MONETARY UNIT: Lira
REFINING CAPACITY: 2,450,200 b/cd
PRODUCTION: 92.9 Mb/d
RESERVES: 739,000 Mbbl

The advent of Vega field nearly doubled Italy's domestic production in 1988, to 92,900 b/d from 50,000 b/d the year before.

Motedison SpA and partners started production in 1987 at Vega field, Italy's largest oil field.

Vega field reserves are estimated at 250-300 million bbl. It is expected to produce 15° gravity crude for 22 years from the Lower Jurassic Inici carbonate, which holds as much as 300 ft of gross pay.

Vega field was opened in 1981 in 440 ft of water in the Mediterranean Sea off Marina di Ragusa, Sicily.

Vega's oil is stored in a converted 250,000 ton tanker, with a capacity of 1 million bbl, and is shipped to Montedison's 50,000 b/d Priolo refinery in Siracusa and to other Mediterranean refineries.

Italy announced a new national energy plan that calls for a 15-year investment of more than $60 billion to trim dependence on foreign energy supplies to 75% in 2000, compared with 81% in 1988.

The plan calls for increased conservation and development of alternate fuels to reduce acid rain emissions. The plan also calls for large reductions in emissions of lead, carbon monoxide, sulfur dioxide, and nitrogen oxide, but does not call for increases in electricity production from nuclear plants.

Italy was negotiating with the Soviet Union to buy about 15 billion cu m/year, up from 9 billion.

Italy's ENI and Montedison merged their chemical activities in 1988 into an independent corporation in which they will equally share a controlling interest.

ENI and Montedison each held a 40% interest in the joint venture, named Enimont, and the balance was offered to foreign and domestic investors.

Folded into the venture were ENI's Enichem SpA affiliate and Montedison's Priolo refinery, base chemicals, pesticides, and fiber producing affiliates.

Enimont immediately became one of the world's 10 largest chemical companies with total assets of $6.8 billion, 50,000 employees, and plants in 70 countries.

Meanwhile, ENI was negotiating with unnamed oil producing countries about possible joint ventures in its refining and marketing business.

ENI's Agip Petroli unit handles about 700,000 b/d of products in Europe, mostly in Italy. It also has a small refining/marketing business in North Africa.

Elf Italiana, a wholly owned subsidiary of Elf Aquitaine Group, has acquired Italrex, the Italian subsidiary of the French group Coparex.

Italrex has interests in about 20 onshore and offshore licenses in Italy, mostly in the Adriatic area, and operates about half of them. Most are exploration permits.

Italy's Agip Raffinazione plans to revamp the catalytic reforming unit at its Porto Marghera refinery near Venice. The project will add continuous catalyst regeneration.

IRELAND

CAPITAL: Dublin
MONETARY UNIT: Irish Pound
REFINING CAPACITY: 56,000 b/cd
PRODUCTION: 0
RESERVES: 0

The quest for oil production continued off Ireland in 1988, but with no positive results.

The country needs only 100,000 b/d of oil production to become self sufficient, and the presence of hydrocarbons has been shown in offshore wildcats time and again.

The only production off Ireland is gas. Marathon Petroleum Ireland Ltd. operates Kinsale Head gas field off the south coast city of Cork.

To encourage oil exploration, the Irish government drastically revised the tax structure for new discoveries, offering oil companies the second most advantageous terms in Europe (after the Netherlands)

Exploration in Irish waters has been depressed several years, despite fine tuning of the tax system in 1987. Only three holes were drilled in 1987, none a significant discovery.

The government abolished the 8% to 16% royalty and offered free depreciation on exploration and development

FRANCE
English Channel
Bay of the Seine
Gulf of St Malo
Bay of Biscay
Mediterranean Sea
Gulf of Lions
BELGIUM
GERMANY
LUXEMBOURG
Luxembourg
SWITZERLAND
ITALY
SPAIN
ANDORRA
FRANCE
Paris Basin
Aquitaine Basin
Cherbourg
Le Havre
Port Jerome
N.D de Gravenchon
L'Orcher
Caen
Petit-Couronne
Rouen
Beauvias
Gargenville
Vernon
Lisieux
Bernay
Brest
Lorient
Vannes
Rennes
Vern-sur-Seiche
Laval
Le Mans
Chateaubriant
Angers
Nantes
Donges
Cholet
Les Sables d'Olonne
Parthenay
Fontenay-le-Comte
Poiters
Chauvigny
Chatellerault
Chateauroux
Niort
La Rochelle
Sugeres
Saintes
Royan
Le Verdon
Pauillac
Chazelles
Periqueux
Tulle
Brive
Ambes
Bordeaux
Bergerac
Lavergne
Cazaux
Parentis
Lugos
Mothes
Cabeil-Ychoux
Lucats
N. Mimizan
Mont-de-Marsan
Auros
Marmande
Agen
Villefranche
Decazeville
Rodez
Montauban
Tartas
Dax
Orthez
Lussagnet
uch
Castres
Lacq
Bayonne
Pont d'As
St. Faust
St. Marcel
Proupiary
Oloron
Rousse
Meillon
Mozeres
Lourdes
Tarbes
Pierrefitte
Pamiers
Carcassonne
Narbonne
Besiers
St. Girons
Toulouse
Vizcaya
Bilbao
San Sebastián
Vitoria
Logroño
Huesca
Barbastro
Monzon
Balaguer
Manresa
Gerona
Palamos
Zargoza
Lerida
Fraga
Fayón
Tarrega
Igualada
Tarrasa
Mataró
Barcelona
Alcaniz
Teruel
Vinaroz
Tarragona
Angula
Salmonete
Casablanca
Tarragona
Castellon
Amposta
Dorado
Montanazo
Castellon de la Plana
Valencia
Perpignan
La Junquera
Blois
Tours
Verzon
Bourges
Nevers
Issoudun
Chateauroux
Briare
Auxerre
Montbard
Dijon
Besancon
Beaune
Autun
Dole
Chalon
Lons-le-Saunier
Cressier
Moulins
Vichy
Roanne
Vindecy
Macon
Bourg
Lyon
Feyzin
Vienne
St. Etienne
Valence
Annemasse
Lake Geneva
Geneva
Annecy
Chambery
Grenoble
L'Isle
Avignon
Montpellier
Frontignan
Sete
Fos-sur-Mer
Lavera
Berre l'Etang
La Mede
Marseille
Toulon
Manosque
Revigny
Gallician
Valenciennes
Mons
Charleroi
Maubeuge
Charlesville
Reims
Metz
Bellange
Nancy
Epinal
Paris
Coulommes
Chaony
Grandpuits
Valence
Brie-Chailly-Chartettes
Villemer
Gisy-les-Nobles
St. Martin-de-Bossenay
Troyes
St. Firmin-des-Bois
Chueltes
Chateaurenard
Clermont-Ferrand
Les Ancizes
Issoire
Roussiness
Moulins
Antar 12 in. 31 cm
12 in. 31 cm
10 in. 25 cm
18 in. 45 cm
24 in. 61 cm
20 in. 61 cm
16 in. 41 cm
8 in. 20 cm
0   50   100 Mi.
0   50   100 Km.
MINORCA
MAJORCA
IBIZA

expenditures. The entire expenditure could be written off against a corporation's tax in the first year of production or carried forward.

The tax changes also removed the state's right to participation, which in 1986 was reduced from 50% to a maximum of 25%.

That action encouraged British Petroleum and Amoco Ireland to drill deep prospects in the Porcupine basin in the summer of 1988.

And Marathon announced a $50 million, 4 year explora-

Repsol. The leading industrial company of Spain.

Repsol explores and produces oil and gas in more than 15 countries. Refines almost 50 % of the crude oil processed in Spain, transforming it into gasolines, lubricants, and other oil products.

Repsol is leader in research, production and export of petrochemical products and plastics. And it is also the leading company of LPG in Europe.

A great company. The leading corporation of Spain and the seventh European oil company.

Repsol, S. A. Paseo de la Castellana, 89
Madrid 28046. España
Tel. (91) 348 81 00. Telex 48162 RESOL E
Fax (91) 455 76 71

Repsol. Times move with us.

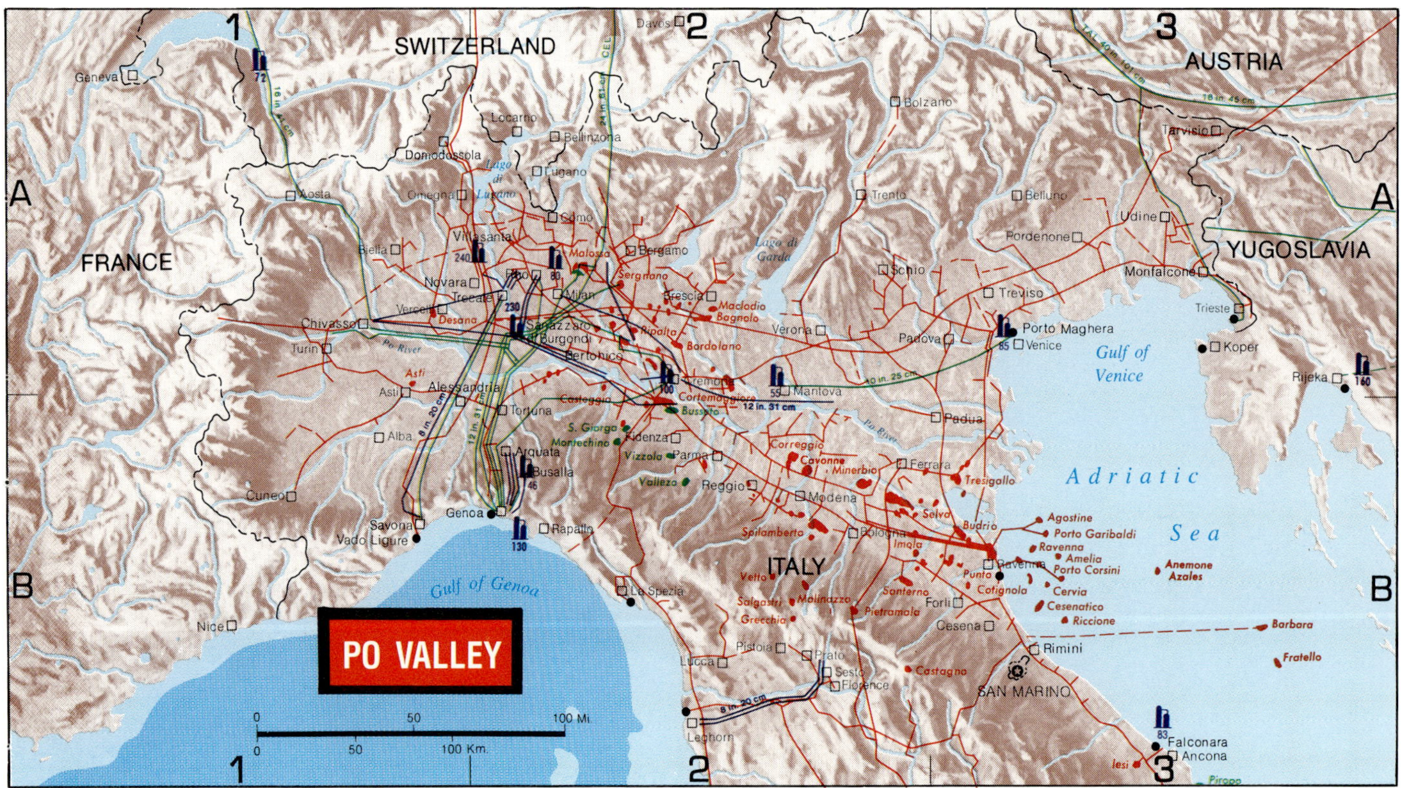

tion program that will include 6,000 km of seismic surveys and up to 10 wildcats.

NETHERLANDS

CAPITAL: Amsterdam
MONETARY UNIT: Guilder
REFINING CAPACITY: 1,380,700 b/cd
PRODUCTION: 83.7 Mb/d
RESERVES: 205,590

Demand for Dutch natural gas in foreign and domestic markets dropped about 12% in 1988.

Norway agreed in 1988 to supply the state owned Dutch Electricity Generation Group 2 billion cu m/year of gas for 20 years.

The state-owned Statoil, Saga Petroleum, and Norsk Hydro will sell the gas, which will constitute 10% of the entire energy consumption of the Netherlands.

The deal marked the first time that Norwegian companies have sold gas which will be used directly in the generation of electricity. The gas will be delivered from several Norwegian fields, to be determined later.

The first phase of the agreement calls for the sale of 1 billion cu m/year beginning in 1995 and jumping to 2 billion the following year. The Dutch electricity group plans to bring two gas-fired generation plants on stream in the northern part of the country in 1995 and 1996.

Gaz de France has negotiated a small reduction in the price of gas it buys from Nederlandse Gasunie of Holland.

Elf Aquitaine Holding BV and Wintershall Nederland BV restructured their interests in Dutch affiliates they held equally.

Elf kept all of Delfzee Petroleum BV, which acquired Amax Petroleum Corp. in 1987. Wintershall kept all of Delfzee Oil & Gas BV, which bought Pennzoil Nederland BV in 1988.

Delfzee Oil & Gas acquired Pennzoil's interest in two producing blocks, a block under development, ten explora-

tion blocks, and interests in related pipelines and gas plants.

Englehard Corp. was building a $50 million research and manufacturing plant at Terneuzen to produce fluid cracking catalysts for refiners.

NORWAY

CAPITAL: Oslo
MONETARY UNIT: Krone
REFINING CAPACITY: 239,400 b/cd
PRODUCTION: 1,069.1 Mb/d
RESERVES: 10,435,000 Mbbl

Norway parted company with the Organization of Petroleum Exporting Countries late in 1988, substantially increasing oil production at one of its largest new fields.

OPEC had been seeking to sustain world oil prices at $18/bbl by reducing production, an effort Norway had supported by cutting oil output 7.5% for more than a year.

But Norway decided to let its production, about 1 million b/d, jump to 1.24 million when the Osedberg field came on stream.

One factor behind the government's decision was the fact that slumping oil prices threatened to force drastic cutbacks in the national budget.

Statoil changes

The Norwegian state oil company, Den Norske stats oljeselskap AS (Statoil), saw a change in leadership in 1988.

Harald Norvik was selected to succeed Arve Johnsen as Statoil's head. Johnsen resigned due to huge cost overruns at the Mongstad refinery and terminal expansion project.

Norvik, 41, was managing director of the Aker-Norcem owned construction firm Astrup-Hoyer. He also was secretary to former Norwegian Prime Minister Oddvar Nordli and state secretary in the Energy Ministry.

Statoil now says the Mongstad refinery never will be profitable.

Cost overruns are estimated at about $1.3 billion of the final cost of $2.2 billion for the refinery expansion and related terminal work.

To be completed in mid 1989, the project will hike capacity to about 133,000 b/d from about 82,000 b/d. The refinery won't just process Norwegian crude but the cheapest available oil.

Under Norvik, Statoil was restructured into three divisions from six: exploration/production, refining/marketing, and petrochemicals.

And the Norwegian parliament (Storting) authorized the sale of the government's interest in oil and gas fields as long as the government finds such sales are in the national interest.

The first such sale is likely to be the government's 31% interest in Snorre oil field.

Norway owns more than 50% of most of the oil and gas fields found on the Norwegian Continental Shelf since January 1985. The state's shares were split into two parts, with Statoil receiving 20% and the government holding 31% directly.

The government explained that due to uncertainty over future oil prices, it is considering selling some of its properties.

It explained that in some cases, the state could collect as much from the taxation of oil companies as from owning and selling the oil and gas, and at the same time, reduce the risk of a downturn in oil prices.

Barents dispute

Norway's hopes for settling its dispute with the Soviet Union over the Barents Sea international boundary were discouraged in 1988.

The energy ministry also was disappointed by dry holes in the Norwegian sector of the Barents. Giant structures mapped by seismic survey offered promise of large oil fields, but five initial wildcats found no oil.

The Jurassic targets were dry and only small volumes of gas were found in shallower Triassic zones.

Barents exploration will continue, however. Norway planned to offer 11 key blocks for exploration as part of the second phase of its 12th licensing round.

And it planned to offer 15-25 blocks, a relatively large number, in 1990.

The Norwegian government had hoped that the Soviet Union was ready to reach an accord on the boundary dispute that blocked exploration on 155,000 sq km, about 11% of the Barents Sea.

Norway was encouraged by the fact that the Soviets had settled a long standing disputed with Sweden regarding a Baltic Sea boundary question.

But in the Barents Sea, the Soviets surprised Norway by proposing a "zone of confidence" covering the disputed zone and an area of undisputed territory on either side.

The Soviets proposed that the two countries control the zone equally, allowing operators to explore and develop any discoveries as long as profits were shared equally.

Norway rejected that proposal, saying a clear demarcation line is needed to avoid potential disputes over oil, gas, and other natural resources in the Barents.

Company activity

Saga Petroleum AS was the target of two merger attempts in 1988.

Saga rejected a merger proposal from France's Ste. Nat. Elf Aquitaine because Elf, as it had in a 1986 offer, wanted to control the surviving company.

Then Total CFP, another large French oil company, tried to boost its 5% interest in Saga by purchasing the Aker Group's 20% interest and Den norske Creditbank's 5%

interest.

Saga's directors recommended that stockholders reject rule changes which would have allowed the Total offer to proceed.

Meanwhile, Saga also offered to buy a 20% interest in Norwegian Oil Consortium from the Aker Group and a 10% interest from two Norwegian shipping groups.

Norsk Hydro SA and Statoil joined forces with Enron Gas Supply Co. to jointly market and sell Norwegian liquefied natural gas in the U.S.

Plans call for initial volumes of 250 MMcfd to be delivered by 1993.

That could increase in the late 1990s, the Norwegian partners said.

The joint venture will seek to develop markets and terminal capacity in the U.S. for the LNG. There are three LNG terminals in the U.S. and no liquefaction facilities in Norway.

PORTUGAL

CAPITAL: Lisbon
MONETARY UNIT: Escudo
REFINING CAPACITY: 313,300 b/cd
PRODUCTION: 0
RESERVES: 0

The Portuguese government has identified a number of likely candidates for its privatizion of state-owned companies.

It is considering selling part of Petroleos de Portugal EP (Petrogal), which operates Portugal's two refineries, and Cia. Nacional Petroquimica (CNP), the country's largest petrochemical producer.

The possibility of injecting private capital into the money losing CNP operation persuaded the government to cancel plans to shut down the company at the end of 1987.

Spain's Repsol Quimica SA was negotiating to buy a 10% interest in CNP, which operates a petrochemical complex at Sines, south of Lisbon, based on a 300,000 metric ton/year ethylene plant.

The ruling party sees privatization as the way to liberalize the Portuguese economy by improving efficiency and reducing subsidies required to keep money losing state companies in business.

It plans to prefer sales to small investors and company employees over sales to foreign investors. Petrogal operates two refineries: a 200,000 b/d unit at Sines and a 90,000 b/d plant at Leca de Palmeira, near Porto.

The government had planned to close the CNP complex at Sines due to mounting losses, about $1.2 billion during 1980-85.

Officials blamed CNP's problems on too much capacity and undercapitalization.

The Sines complex, completed in 1977, was too large for the Portuguese market but was blocked from exports due to a world petrochemical glut.

The government owns several other companies operating petrochemical plants.

Empresa de Polimeros de Sines operates a 120,000 ton/year low density polyethylene unit, a 60,000 ton/year high density polyethylene plant, and a 50,000 ton/year polypropylene plant at Sines.

Petroquimica e Gas de Portugal has plants at Cabo Ruivo, where it can produce up to 220,000 tons/year of ammonia, 15,000 tons/year of phthalic anhydride, and 20,000 tons/year of plasticizer.

And Quimica de Portugal has plants at Barreiro with capacity of 270,000 tons/year of ammonia and 90,000 tons/year of urea.

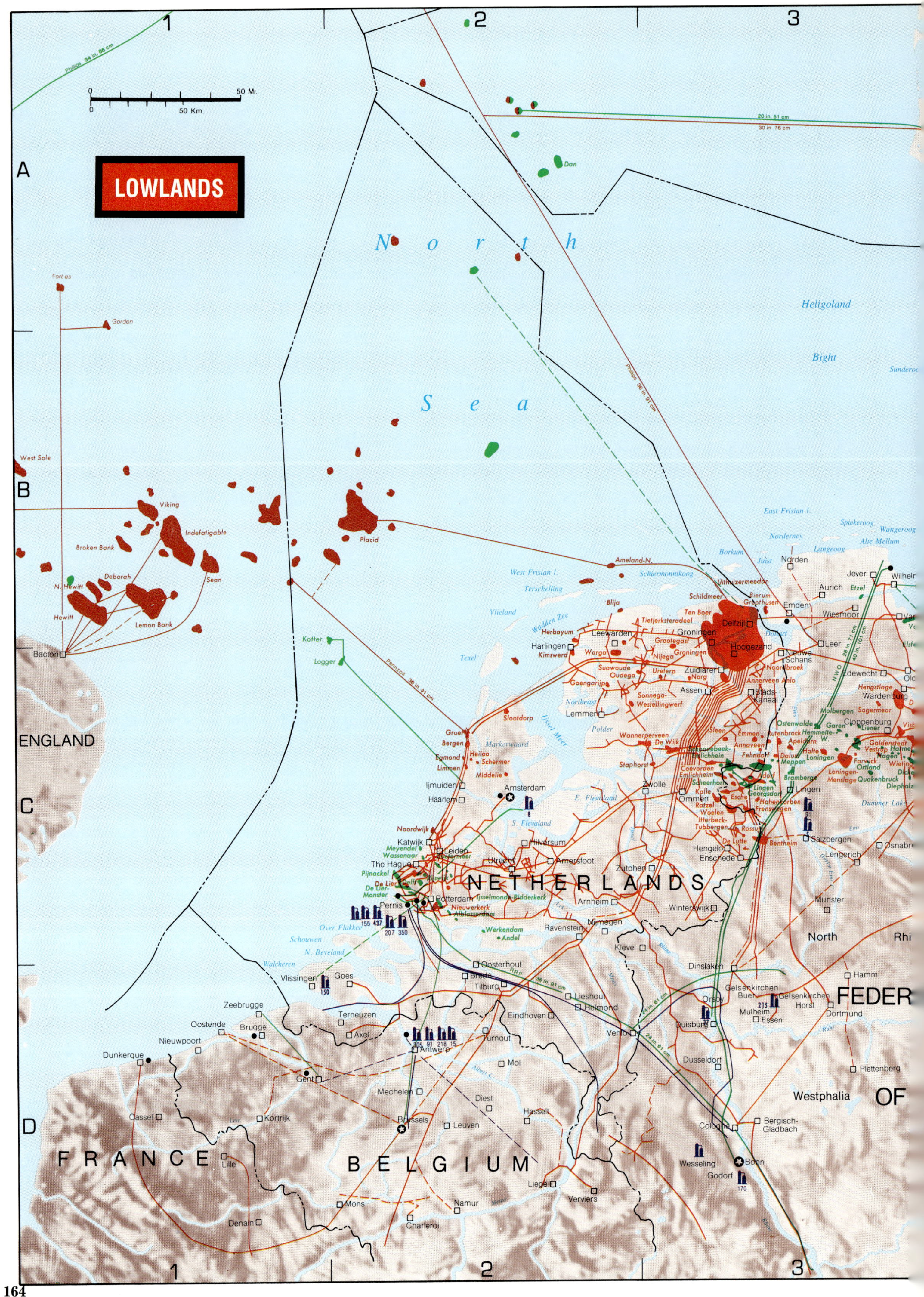

LOWLANDS
North Sea
Heligoland Bight
Sunderoo
ENGLAND
NETHERLANDS
BELGIUM
FRANCE
North Rhi
FEDER
Westphalia OF
50 Mi.
50 Km.
Forties
Gordon
Fort es
West Sole
Viking
Indefatigable
Placid
Broken Bank
Deborah
Sean
N. Hewitt
Hewitt
Lemon Bank
Bacton
Dan
Kotter
Logger
Ameland-N.
West Frisian I.
Schiermonnikoog
Terschelling
Vlieland
Wadden Zee
Texel
Borkum
Juist
Norden
East Frisian I.
Spiekeroog
Wangeroog
Norderney
Langeoog
Alte Mellum
Jever
Wilhel
Etzel
Aurich
Wiesmoor
Var
Ve
Emden
Leer
Elsf
Olc
Nieuwe Schans
Edewecht
Uithuizermeeden
Bierum
Groothusen
Schildmeer
Ten Boer
Delfzijl
Blija
Tietjerksteradeel
Leewarden
Groningen
Hoogezand
Dotart
Wardenburg
Harlingen
Herbayum
Kimswerd
Warga
Grootegast
Groningen
Zuidlaren
Norg
Hengstlage
Sagermoor
Cloppenburg
Vist
Suawoude
Oudega
Ureterp
Annerveen Anlo
Stads Kanaal
Molbergen
Liener
Northeast
Geengarijpe
Sonnega
Westellingwerf
Assen
Sleen
Emmen
Ostenwalde
Vestrop
Hotmer
Apeldom W.
Goldenstedt
Wietjin
Lemmen
Sonnega
Dalum
Forwck
Hagen
Polder
Wannerperveen
De Wijk
Fehndoff
Holte
Loningen
Ortland
Wiefjin
Slootdorp
Staphorst
Meppen
Loningen-
Menslage
Quakenbruck
Diepholz
Gruet
Bergen
Zwolle
Kalle
Esche
Georgsdorf
Lingen
Dummer Lake
Egmond
Heiloo
Schermer
Scheerhorn
Hohenkorben
Ombren
Ratzel
Woelen
Itterbeck-
Tubbergen
Salzbergen
Lengerich
Øsnabre
Limmen
Middelie
Coevorden
Emlichheim
Ijmuiden
Markerwaard
De Lutte
Bentheim
Haarlem
E. Flevaland
Hengelo
Enschede
Amsterdam
S. Flevaland
Munster
Noordwijk
Hilversum
Katwijk
Meyendel
Wassenaar
Leiden
Solfermeer
Amersfoort
Utrecht
Zutphen
The Hague
WinterWijk
Pijnackel
De Lier
Monster
Rotterdam
IJsselmonde
Ridderkerk
Arnhem
Pernis
Nieuwerkerk
Alblasserdam
Nijmegen
Over Flakkee
Werkendam
Andel
Ravenstein
155 437
Schouwen
207 350
Kleve
N. Beveland
Walcheren
Dinslaken
Hamm
Oosterhout
Breda
Gelsenkirchen
Vlissingen
Goes
Buer
Gelsenkirchen
Tilburg
Liesbout
Horst
Dortmund
150
Zeebrugge
Terneuzen
Helmond
Mulheim
Essen
Oostende
Brugge
Axel
Eindhoven
Duisburg
Nieuwpoort
Orsoy
215
Dunkerque
Gent
Turnout
Venlo
Dusseldorf
Plettenberg
165 91 218 15
Antwerp
Mol
Ruhr
Cassel
Mechelen
Diest
Bergisch-
Gladbach
Kortrijk
Brussels
Leuven
Hasselt
Cologn
Lille
Wesseling
Bonn
Mons
Namur
Liege
Verviers
Godorf
170
Denain
Charleroi

SWEDEN
Copenhagen
Kalundborg
Malmo
DENMARK
Arhus
Fredericia
Korser
Nyborg
Skaelskor
Baltic
Sea
Flensburg
Kiel Bay
Schwedeneck
Kiel
Preetz
Fehmarn
Stralsund
Kolberg
Koslin
Rostock
Reinkenhagen
Kolberg
Schleswig
Warnau
Pion
Boostedt
Heide
Heide
Wolin
Holstein
Brunsbuttelkoog
Bramstedt
Lubeck
Wismar
Molln
Schwerin
Stettin
Hamburg
Harburg
Potrau
Walez
Sinstorf
Reitbrook
Dibbersen
Mechelfeld-S.
Volkensen
Taaken
Achim
Schwedt
Lower   Saxony   Basin
Wittenberge
GERMAN
DEMOCRATIC
REPUBLIC
POLAND
Landsberg
Barrien
Verden
Bomlitz
Nienburg
Eystrup
Ahrensheide
Lube
Hademstorf
Steimbke
Elze
Orrel Hankensbuttel
chhorst-
Suderbruch
Thoren
Wietze
Eldingen
Wittingen
renburg
Siedenburg
Fuherberg
Hohne
Kneseback
Berlin
Staffhorst
Thonse
Nienhagen
Weesendorf-S.
Vorhop
Rudersdorf
Siedenburg
Tambuhren
Ehra
Frankfurt-an-
Vogt
Landesbergen
Misburg
Hardesse
Gifhorn
Lachendorf
der Oder
Steinhuder
Lehrte
Lieferde
Ahrenborstel
Bokeloh
Lake
Hanigsen
Roxforrde
Brandenburg
Potsdam
Beckedorf
Hannover
Lehrte
Eddesse
Meerdf.
Wollsburg
Rybaki
Buckeburg
Krensberg-
Olheim
Ruhme
Golzow
Potreniten
Molme
Meerdf.
Rautheim
Guben
Oberg
Braunschweig
Sarstedt
Broistedt
Holmstedt
Staakow
Kemme
Magdeburg
Hohenassel
Salzgitter
Coltbus
ghausen
Bad
Zary
Zagan
efeld
Salzdetfurth
Fallstein
aderborn
EPUBLIC
Kassel
Halle
Schwarzheide
ANY
Volkenroda
Leipzig
Muhlh
Leuna
Lutzen
Bohlen
Gorlitz
Langensalza
Espenhein
Bad
Zeitz
Roslitz
Dresden
Hersfeld
Erfurt
Zittau
Gera
Karl Marx Stadt
Hesse
Zwickau
Zaluzi
Plauen
CZECHOSLOVAKIA
Kralupy
Prague
Karlovy Vary

SPAIN

CAPITAL: Madrid
MONETARY UNIT: Pesceta
REFINING CAPACITY: 1,285,000 b/cd
PRODUCTION: 30.4 Mb/d
RESERVES: 25,955 Mbbl

Spain continued to be the hotspot of company restructuring and of new refining and petrochemical projects in Europe, as the country prepared for entry into the Economic Community.

The government launched an alternative fuels development program with state and private investments expected to total $1.8 billion by 1995.

The goal is to boost production and use of hydro, geothermal, solar, and wind energies as well as fuels from biomass and solid waste. Production from those sources is expected to reach 30,400 b/d of oil equivalent by 1995, compared with 6,000 b/d last year.

The government approved a request by International Petroleum Investment Co. of Abu Dhabi to increase its 10% holding in Cia. Espanola de Petroleos SA (Cepsa), Spain's largest private refiner, to 15%. IPIC had wanted to increase the holding to 20%. IPIC was supplying Cepsa with 60,000 b/d of crude.

Cepsa operates a 130,000 b/d refinery at Tenerife in the Canary Islands and a 160,000 b/d plant near Cadiz.

Spain's state owned Enagas agreed to buy 35 billion cu ft of gas over 30 years form Norway's Statoil.

Deliveries are due to start from Troll and Sleipner fields in the North Sea no later than 1996. France must also approve the deal, because the gas must traverse that country.

Meanwhile, Algeria and Morocco have signed an agreement that would allow construction of a natural gas pipeline from Algeria through Morocco and under the Strait of Gibraltar to Spain. Morocco would receive some of the gas.

Production outlook

Chevron Oil Co. of Spain and state owned Empresa Nacional del Gas SA (Enagas) have agreed to develop Marismas gas field in Guadalquivir Valley, southern Spain.

The field has reserves of about 25 billion cu ft. It is jointly owned by Chevron and Instituto Nacional de Hidrocarburos, Enagas's parent.

Spain was considering whether to develop a 1978 Gulf of Cadiz gas discovery.

Repson Exploracion SA, formerly Hispanoil, was appraising the field, which lies in 195-460 ft of water.

In 1983 the field's reserves were estimated at 100-125 billion cu ft of gas and development costs at $200-225 million. In 1982 Chevron Oil Co. of Spain said the gas price would have to rise to about $7-7.50/MMBTU or reserves to 200-250 billion cu ft to make the field commercial.

Shell Espana NV conducted geophysical surveys in the Els Ports area of Castellon Province in eastern Spain, in preparation for drilling a wildcat.

Marketing changes

Spain was opening its petroleum market to foreign companies in 1988, in order to qualify for entry into the European Community in 1992.

It opened gasoline marketing to foreign competition, with the goal of cutting the monopoly held by Cia. Arrendataris del Monopolio de Petroleos SA (Campsa), the national products distribution company.

Another goal was to increase the number of retail outlets in the country. In early 1987 Spain had only 415 gasoline stations per million vehicles, about a third of the average in the European Community.

Campsa was partially privatized in 1985, with four private companies now holding 41.9% of its equity.

Texaco was one of the companies planning to establish a products marketing network in Spain by acquiring several medium sized distribution companies.

BP Espana SA and Petroleos del Mediterraneo SA (Petromed) agreed to form a joint venture, BPMed, to take over the companies' Spanish oil marketing activities.

The agreement doesn't involve operations covered by monopoly regulations or Petromed's 112,000 b/d refinery near Valencia. The new company's initial sales are expected to total 500,000 metric tons/year of fuel and lubricants, mainly aviation and marine fuels.

BPMed will invest $200 million during the next 5 years in a network of 150 gasoline retail stations and associated facilities in Spain.

Campsa also planned a major investment program to expand and improve gasoline outlets, in preparation for liberalization of the market.

Refining projects

Cia. Espanola de Petroleos SA increased visbreaking capacity to 40,000 b/d from 35,000 b/d at its 160,000 b/d refinery at Santa Cruz de Tenerife in the Canary Islands. It also improved the refinery's pollution control systems.

The same company started up a 45,000 ton/year methyl tertiary butyl ether plant at its 160,000 b/d Algeciras refinery.

Oliofiat Iberica SA, part of the Fiat group, plans to build an $11 million lube oil plant at Sagunto by 1990. Most of the production will be exported.

Petromed plans to increase the catalytic cracking capacity of its 120,000 b/d Castellion refinery to 25,000 b/d from 15,000 b/d. The $15 million project is due completion by early 1990.

Mexico's Petroleos Mexicanos was negotiating to acquire 10% interest in Repsol SA, Spain's state owned oil group. Pemex is Spain's largest crude supplier and owns a 34.4% interest in Petroleos del Norte, operator of the 240,000 b/d Bilbao refinery.

Petroleo Repsol SA planned to expand its Tarragona ethylene cracker from 350,000 metric tons/year to 385,000 metric tons/year. It also will install a $5 million advanced process control system at the Corruna refinery.

Empresa Nacional del Gas SA started up the $62 million gas terminal and regasification facility in 1988. It handles Algerian LNG, distributing up to 106 MMcfd to nearby communities. Enagas operates another LNG terminal at Barcelona and is building one at Cartagena.

The natural gas is from a new LNG terminal at Palos de la Frontera, Huelva, which receives Algerian shipments.

Petrochemicals outlook

Atochem Espana SA, a subsidiary of France's Atochem SA, plans to build a 120,000 ton/year polyvinyl chloride plant at its Prat de Llobregat, Barcelona, complex.

The Spanish government must approve the $60 million project, which is due completion by early 1991. A substantial part of the plant's production will be exported.

Atochem, a unit of France's Atochem SA, also planned to increase capacity at its Tarragona polystyrene plant to 75,000 metric tons/year from 50,000 tons/year. Atochem acquired the plant from ARCO Chemical International in April 1987.

Tarragona Quimica SA, a Spanish unit of West Germany's Hoechst AG, plans to invest $65 million to increase high density polyethylene capacity of its Tarragona petrochemical complex to 125,000 metric tons/year from 100,000. It will also resume production from a 75,000 ton/year vinyl acetate plant closed in 1983.

Using ore and coal from its own mines, Algoma produces molten iron which is converted into steel in its basic oxygen steelmaking plant. Seamless tubes are then produced from continuously cast steel. Algoma's automated heat treating process, including inside-outside quenching, builds in the high strength properties of seamless casing, pipe and tubing.

Performance-proven seamless tubulars

Whatever the requirements, you can rely on Algoma seamless oil country tubulars. "Soo" casing is performance-proven in oil and gas regions around the world, from the Tuscaloosa Trend to the North Sea. The special grades of seamless Soo have been developed to meet the energy industry's need for casing that withstands deep-hole pressures and other hostile conditions. This high strength casing line is manufactured to rigid international quality standards in one of North America's most modern tubular products mills…offering rail and highway connections with all North American points, and direct seaway access for trans-oceanic shipments.

Algoma offers assured supplies of seamless Soo casing today…and a greater range of world class tubular products including casing, line and standard pipe, mechanical tubing, couplings and coupling stock. The size ranges are as follows:

Casing	114.3 mm to 298.4 mm / 4½″ to 11¾″
Tubing	60.3 mm to 114.3 mm / 2⅜″ to 4½″
Line and Standard Pipe	60.3 mm to 323.8 mm / 2⅜″ to 12¾″
Mechanical Tubing	60.3 mm to 323.8 mm / 2⅜″ to 12¾″
Couplings and Coupling Stock	

SOO casing	**90***	**95**	**125**	**140**	**155**
YIELD STRENGTH min. psi	90000	95000	125000	140000	155000
max. psi	105000	110000	150000	165000	180000
TENSILE STRENGTH min. psi	105000	110000	135000	150000	165000
ELONGATION min. % in 2″	26	20	18	17	15
HARDNESS ROCKWELL C max.	19 to 25	27	35	—	—

FOR A FREE PRODUCT BROCHURE AND FURTHER INFORMATION – WRITE:

API Grades also available including: H-40, K-55, C-75,* L-80,* Modified N-80,* N-80, Q-125 and P-110 grades.

*Grades suitable for sour gas service.

ALGOMA SEAMLESS

Our customer commitment is continuous.

THE ALGOMA STEEL CORPORATION, LIMITED
Sault Ste. Marie, Ontario, Canada P6A 5P2
CALGARY—Phone:(403) 263-8990, FAX: (403)233-0636
TORONTO—Phone:(416) 276-1400, FAX: (416)276-1452

ALGOMA TUBE CORPORATION
HOUSTON—Phone:(713) 465-8998, TWX:910-881-1573

Union Explosivos Rio Tinto has converted its 300,000 metric ton/year petrochemical plant at Huelva to natural gas from naphtha feedstock, saving the company about 15% in feedstock costs.

Industrias Quimicas Asociadas SA, a Royal Dutch/Shell subsidiary, was resuming production at its 70,000 metric ton/year ethylene cracker at Tarragona. It was closed in 1982 due to poor market conditions. The firm plans to increase production by 20,000 tons/year at a cost of $8 million.

Repsol Quimica SA paid $22 million for a 20% stake in Aiscondel SA, a Spanish plastics producer, giving the state petrochemical company its first access to polyvinyl chloride production.

Repsol also plans to spend $10 million to double the 20,000 metric ton/year capacity of butadiene/styrene based thermoplastic elastomers at Santander.

Aiscondel SA will raise its polyvinyl chloride capacity to 130,000 metric tons/year from 100,000 tons/year. The project is part of a $45 million, 3 year program to expand and modernize the company's units at Tarragona and Huesca.

Empresa Nacional de Fertlizantes SA will invest about $190 million during 1988-92 to modernize its production facilities. Most of the improvements will be at ammonia and urea plants at Puertollano and Cartagena.

The company started up Spain's largest, most modern complex for production of ammonia based fertilizers, a $150 million complex at Sagunto, near Valencia. It can produce 1,100 tons/day of ammonium nitrate, 1,400 tons/day of calcium ammonium nitrate, and 750 tons/day of nitric acid.

The project was part of the restructuring of the Spanish chemical industry, which calls for closure of obsolete plants, construction of new plants to make higher value products, and a greater concentration of companies through mergers and takeovers.

SWEDEN

CAPITAL: Stockholm
MONETARY UNIT: Krona
REFINING CAPACITY: 426,500 b/cd
PRODUCTION: 0
RESERVES: 0

Sweden's decision to phase out nuclear power by 2010 has made it an attractive market for the gas exporting counties of Europe.

If Sweden adheres to its plans to phase out its extensive nuclear power system, its gas demand could rise from the current 62 MMcfd to 500 MMcfd by 2010.

Norway wants to ship gas to Sweden from the North Sea or possibly from the Haltenbanken area of the Norwegian Sea. And the Soviet Union is offering supplies through Finland or via the extensive northern European supply network.

The OK Petroleum cooperative and the Alex Johnsen trading group have signed an initial agreement with the Soviets to buy 14.1 billion cu ft/year using pipelines though West Germany and Denmark.

Also, Finland's state owned Neste Oy has offered to build a pipeline across the Baltic Sea to Sweden to move about 53 billion cu ft/year of Soviet gas.

Meanwhile, Norway's Statoil has four options to supply Sweden: a pipeline link from Ekofisk field to the Danish offshore gas gathering system and then into Sweden through existing lines; a 372 mile line from Sleipner field through the Skagerrakat to Gothenburg; a line across Norway from the Karsto Statpipe terminal; or another line from the Haltenbanken across central Norway into southern Sweden.

The Swedes are interested in purchasing both Soviet and Norwegian gas. Swedegas AB was considering whether to accept the trans-Finland option. It would involve an extension of the trunk line through which Finland imports Soviet gas.

The Finnish system, with a capacity of 272 billion cu ft/year, runs through Kouvala and ends at Tampere, with a spur line to Helsinki.

Two extension routes are under consideration: one would run from Tampere to the Finnish west cast and from there across the Gulf of Bothnia to Gavle, Sweden. A more southerly route would run to Hanko on Finland's southern coast and there across the gulf to Norrtalje northeast of Stockholm.

Swedish officials are more interested in the southern route, since Stockholm would be a major market for the gas.

Sweden granted a 3,860 sq mile concession in the Baltic Sea to Norway's Statoil, Svnska Shell AB, and Oljeprospektering (Opab), Sweden's state exploration company.

In the first 3 years, Shell and Statoil will conduct seismic surveys and reprocess existing data. The group can extend the term to 5 years by agreeing to drill. Statoil and Shell are joint operators for exploration.

And Statoil bought an underground naphtha storage unit at Stenungsund in southern Sweden from the government and will convert it into Europe's largest underground propane storage unit.

The project will cost $20.3 million and have a capacity of 265,000 tons.

SWITZERLAND

CAPITAL: Bern
MONETARY UNIT: Franc
REFINING CAPACITY: 132,000 b/cd
PRODUCTION: 0
RESERVES: 0

Operations were shut in at Raffinerie du Sud-Ouest SA in Collombrey due to a crude transportation cost dispute with Snamprogetti, which operates the pipeline servicing the refinery.

The plant runs about bout 1.2 million tons/year of crude, and has been operating at only 40-50% of capacity.

Earlier, BP (Schweiz) AG sold its 25.5% share of the 70,000 b/d refinery to Gatoil (Suisse) SA/Gatoil SRT SA. The sale raises the Gatoil companies' share to 72.1%. The other shareholder is Agip Suisse SA.

Skandinaviska Raffinaderi AB is revamping its 210,000 b/d refinery to change the VGO-hydrodesulfurization unit into a mild hydrocracking unit.

Gas de France and Switzerland's Gasnet will lay a 67 mile pipeline linking a storage unit at Etrez to the Swiss gas network.

Gas de France will store gas for the Swiss system under a recent agreement. Completion of the pipeline is due by 1990.

UNITED KINGDOM

CAPITAL: London
MONETARY UNIT: Pound
REFINING CAPACITY: 1,803,100 b/cd
PRODUCTION: 2,376.3 Mb/d
RESERVES: 5,175,000 Mbbl

The Energy Department changed the offshore leasing terms for the 11th round of licensing to ensure that companies do not leave acreage undeveloped.

Formerly, under standard U.K. leasing terms, companies had 6 years to complete work programs before relinquishing 50% of a license. The remaining 50% could be held for up to 30 years.

Under the new terms, if after 12 years of the 30 year term, field development has not been approved, the government can require the licensee to surrender the lease.

The government also was offering extended terms of 48 years for frontier acreage. The 50% relinquishment will be imposed after 8 years. After that, companies have another 16 years to submit a field development plan or possibly lose the lease.

The government stopped taking North Sea royalties in oil at the beginning of the year, and ended its participation agreements.

Royalty oil was handled by the Oil and Pipelines Agency, which no longer is involved in oil trading.

The government introduced participation agreements in the mid 1970s, entitling it to acquire up to 51% of North Sea oil at market price.

Since British National Oil Corp. was abolished in 1985 no participation oil had been taken, but the state had retained the rights to the oil though OPA.

BP-Kuwait plans

British Petroleum Co. plc planned to buy back 790 million shares of its stock from the Kuwait Investment Office for $4.32 billion. The deal, will allow Kuwait to reduce its BP holding from 21.6% to 9.9%, and still make a $684 million on the transaction.

The deal coincided with BP's proposed $4.32 billion sale of its worldwide minerals business to RTZ Corp. plc. BP Minerals owns interests in Canada, the U.S., Mexico, Brazil, Norway, Zimbabwe, South Africa, Australia, Papua New Guinea, and Indonesia.

Earlier in the year RTZ Corp. sold its oil and gas division to a U.K. subsidiary of Ste. Nationale Elf Aquitaine for $575.9 million.

Britain's Monopolies and Mergers Commission had ruled that it was not in the public interest for a government — Kuwait — that was a member of the Organization of Petroleum Exporting Countries to be a major shareholder in a British oil company.

Kuwait had offered not to increase its share and to guarantee that it wouldn't use its 21% interest to further its commercial or political goals.

Kuwait bought the BP shares after the British government's attempted sale of its 32% stake in BP flopped in the wake of the October 1987 stock market collapse.

Wytch Farm

BP Development Ltd. was building a $28.5 million pipeline project to serve Wytch Farm oil field in southern England.

The project includes a 56 mile, 16 in. oil line to the Southampton area and a 31 mile, 8 in. gas line to Sopley, Hampshire.

The project will involve 26 river crossings, 68 road crossings, and four railway crossings.

The pipeline will allow the field's production to increase, feeding the crude to Esso Petroleum's 300,000 b/d Fawley refinery on England's south coast.

Wytch Farm field was producing 6,000 b/d of oil from the Bridport reservoir but BP plans to drill more wells to tap the larger Sherwood reservoir and increase production to 60,000 b/d in 1989. The pipelines also could be used to produce Wareham field, a Bridport and Sherwood field 2.5 miles offshore from the Wytch farm area.

BP drilled an extension well which extended Wytch Farm field offshore, cutting a 249 ft hydrocarbon column in the Triassic Sherwood and flowing 1,000 b/d of oil.

Enterprise Oil plc reported an oil discovery at West Frisby, Lincolnshire.

The well, drilled to 6,000 ft about 8 miles from Welton oil field, flowed 745 b/d and 119 b/d of oil on two drillstem tests.

BP tested a gas discovery on Block 110/3b just east of British Gas plc's Morecambe gas field in the Irish Sea.

The hole flowed at rates as high as 30 MMcfd. It was drilled as part of a farmout in which BP and Acre Petroleum Ltd., a British Gas subsidiary, earned an interest in the license.

Processing outlook

BP Chemicals International Ltd. has let a $60 million contract to recommission and expand mothballed capacity at its Grangemouth, Scotland, petrochemicals plant.

Capacity for high density polyethylene will be modernized, recommissioned, and expanded by 150,000 metric tons /year to 250,000.

Linear low density polyethylene capacity will increase to 125,000 tons/year from 95,000. BP also was considering a 270,000 metric ton/year expansion of ethylene capacity at its Grangemouth plant, doubling production. If it proceeds, the addition could be put on stream in 1992.

And BP let a $1.45 million contract for overhaul of the hydrocracker at its 178,000 b/d refinery at Grangemouth. The plant was damaged by fire in 1987.

Shell Lubricants U.K. dedicated a $120 million, fully computerized lubricating oils formulating, blending, and distribution center at Ellesmere Port, Cheshire.

The center, next to Shell's 262,000 b/d Stanlow refinery, makes extensive use of robots and automated vehicles. It replaces a production center at Shell Haven on the Thames and two older plants at Stanlow and Barton.

Shell U.K. Ltd. began work on a $112.8 million alkylation plant as part of an upgrade of its Stanlow refinery. Commissioning is due in mid-1989. Work is underway on a $357 million catalytic cracker and a $169 million gas separation complex, both due completion in 1989.

Mobil Oil Corp. was spending $110 million to install a continuous catalytic reformer at its 145,000 b/d Coryton refinery in Sussex. The 33,000 b/d unit will replace two platinum reformers. Petrofina (U.K.) Ltd. was planning a 139 mile products pipeline from Immingham, South Humberside, to a depot at Buncefield, Herfordshire. Construction will involve four major main river crossings.

WEST GERMANY

CAPITAL: Bonn
MONETARY UNIT: Deutsche Mark
REFINING CAPACITY: 1,518,200 b/cd
PRODUCTION: 77.9 Mb/d
RESERVES: 408,000 Mbbl

Texaco Inc. sold its 99.12% interest in Deutsche Texaco AG for more than $1.2 billion to Reinisch-Westfalisches Elektrizitatswerk AG (RWE) of Essen, West Germany's eighth largest corporation.

The country's Federal Cartel Office raised no objections to the sale. Deutsche Texaco, acquired by Texaco in the mid-1960s, had net production of 13,000 b/d of oil and 40 MMcfd of gas in West Germany, and an interest in Dubai operations that netted 39,000 b/d of oil.

The company owned interests in two refineries with combined capacity of more than 140,000 b/d and had about 1,900 branded retail stations throughout West Germany, accounting for 10% of the country's gasoline sales.

Following the sale, RWE was negotiating with Statoil for

supply of 80,000 b/d of feedstock.

RWE also said it would convert two furfural refining units at its Grasbrook lubricants plant to use Texaco Development Corp.'s MP Refining Process.

Mobil Erdgas-Erdol GmbH sold its three drilling rigs to ITAG Harmann von Rautenkranz of Celle.

It said its future German drilling will be on a contract basis.

Restructuring, forced by increasing operating losses and overcapacity, continued in the West German refining industry in 1988. Eedoel-Raffinerie Duisburg planned to close its 40,000 b/d Duisberg plant.

Wintershall AG announced plans to shut down its 70,000 b/d Mannheim refinery, and started talks on a refining/marketing joint venture with Saarberg Oel & Handel GmbH, the Saarbergwerke AG unit that has a 25% stake in the 144,000 b/d Erdoel Raffinerie refinery at Neustadt.

Libya's state oil company got a half interest in a Coastal Corp.'s 90,000 b/d refinery at Hamburg in exchange for a firm crude supply agreement.

Coastal acquired the Holburn Europe Raffinerie GmbH refinery from Esso AG in 1987. Libya had been supplying the refinery about 75,000 b/d at a netback price. The deal also satisfied another Coastal unit's $45 million obligation on a Libyan exploration concession. 🄸🄿🄴

NORTH SEA

THE PIPER ALPHA PLATFORM DISASTER AFFECTED almost all aspects of operations off Northwest Europe during the second half of 1988. The July 6 explosions and fire that claimed 167 lives aboard Piper Alpha in the U.K. North Sea spurred calls for tighter safety regulations and government oversight of offshore operations throughout Offshore Northwest Europe (see accompanying article).

Operators in all sectors of the region have undertaken safety reviews. Off the U.K., companies face new rules requiring installation of subsea emergency shutoff valves (ESVs), among other measures.

Concerns that the disaster might crimp U.K. offshore development evaporated as the review process for development proposals continued unabated. Development activity also was brisk off Norway, despite volatile oil prices.

Drilling overall remains strong off Northwest Europe, buoyed largely by the U.K. North Sea gas basin action and activity in the Dutch sector. Much Offshore Northwest Europe work during the weather window was tied to plans for new gas gathering facilities. The Piper Alpha disaster squeezed U.K. oil production in 1988. Piper is one of six fields that had been producing in all about 300,000 b/d into the Flotta trunk oil line to the Orkney Islands.

Piper Alpha operator Occidental Petroleum (Caledonia) Ltd. was undertaking an offshore pipeline diversion and safety valve installation program that was expected to enable the trunk line to reopen before yearend 1988. That would restore 165,000 b/d of oil production.

The tragedy occurred when U.K. North Sea oil flow already was depressed by a heavy summertime maintenance program. Closure of the Flotta line ensured that production did not make its usual brisk late summer recovery.

U.K. offshore output was about 2.2 million b/d and was likely to climb to almost 2.4 million b/d by yearend 1988.

Norwegian oil production was likely to jump significantly near yearend 1988 or early 1989 when Norsk Hydro AS was to bring Oseberg field on stream. Later in 1989 Den norske Stats oljeselskap AS (Statoil) is to start up Veslefrikk field.

By yearend 1989 Norwegian production could be 1.25 million b/d. Oil production from the other two main sectors off Northwest Europe was steady with Danish output averaging 90,000-100,000 b/d and Dutch production 55,000-60,000 b/d.

Development plans continue

Immediately after Piper Alpha, it seemed that safety concerns might slow down processing of offshore development applications by the U.K. Department of Energy.

DOE continued to process applications but told field operators it expected changes in the siting of ESVs. Offshore safety committees will become a statutory requirement, as will other regulations stemming from preliminary findings of a technical inquiry into cause of the Piper Alpha disaster.

U.K. offshore operators continued to submit development applications during a weak oil market, designing stand-alone and satellite developments to be economic at current oil prices.

Drilling buoyant

Despite the tragedy, Northwest Europe's 1988 summer drilling recovery continued into fall.

The region's total rig fleet jumped to 129 from 124 at the first of 1988, and the number of rigs stacked fell to 24 from 40.

The gas drilling boom in the southern U.K. North Sea and consistent drilling in Dutch waters sharply hiked jack up utilization and charter rates.

Only two jack ups were stacked vs. 12 at the first of the year. Jack up rates are about $30,000/day, almost twice those for semisubmersibles.

County Natwest Woodmac, Edinburgh, reports more operators using semis in jack up territory—in less than 330 ft of water.

Active rigs in U.K. waters totaled 54—33 in exploration and appraisal and 21 in development.

Elsewhere off Northwest Europe, the Netherlands at yeared 1988 had six rigs in exploration and eight in development, Norway nine in exploration and appraisal and two in development, and Denmark four in development.

Southern basin start-ups

Two new gas production systems in the U.K. North Sea's southern gas basin started up.

Phillips Petroleum Co. U.K. Ltd. started deliveries from Audrey field through the Loggs pipeline system to Theddlethorpe.

BP Petroleum Development Ltd. started up Cleeton and South Ravenspurn fields through a pipeline to a new reception terminal at Dimlington in North Humberside, U.K.

Audrey can produce 450 MMcfd. Phillips installed unmanned wellhead platforms for six wells and a nearby subsea well in Audrey in Blocks 49/11a and 48/15a.

The platform is linked by a 10.5 mile, 20 in. line to Conoco U.K. Ltd.'s V-fields complex and the Loggs pipeline. Audrey is also controlled from the Conoco unit.

Phillips and partners Conoco, Fina Exploration Ltd., Britoil plc, Agip (U.K.) Ltd., Lasmo North Sea plc, Acre Petroleum Ltd., and Elf Oil & Gas Ltd. were discussing a second development phase to be implemented in 1989-90.

Flow from the two BP fields, known as the Villages development, will peak at 220-270 MMcfd in 1989.

The Villages field complex, 100% BP owned, entails a central production complex and satellite drilling platform in Cleeton and three wellhead platforms in South Ravenspurn. Project was completed £60 million ($105 million) under budget of £760 million ($1.33 billion).

Cleeton and South Ravenspurn lie mainly in Blocks 42/29 and 42/30. Hamilton Brothers Oil & Gas Ltd. is developing North Ravenspurn field in Blocks 42/30 and 43/26 for startup in 1990.

The Hamilton Bros. project, using a gravity based concrete platform, will ship gas through the Cleeton pipeline.

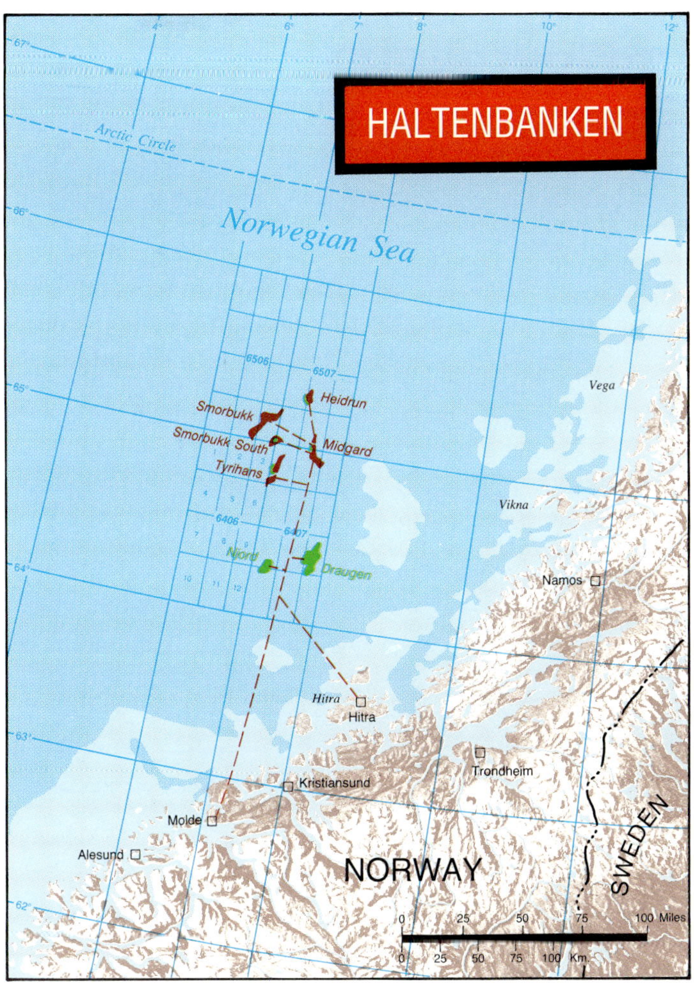

Hamilton Bros. let a £20 million ($35 million) contract for fabrication and offshore commissioning of an integrated deck for North Ravenspurn to Redpath Offshore, part of the offshore and structural division of Trafalgar House Group.

Mobil North Sea Ltd. claimed an industry record by signing a gas sales contract and a transportation agreement for its 100% owned Camelot field a little more than a year after drilling the successful appraisal well that triggered the development decision.

Camelot, in southern North Sea Blocks 53/1a and 53/2, has reserves of 250 bcf. It was scheduled to start production in third quarter 1989 and peak at 70 MMcfd.

Mobil was installing a small wellhead platform that will be tied back to and controlled from the East Leman complex operated by Amoco (U.K.) Exploration Co.

The gas sales agreement was signed with British Gas plc (BG), and the transportation agreement with Amoco's East Leman group will cover all transportation, processing, and measurement services.

Mobil signed an interim agreement with Amoco for maintenance services in the run-up to the installation program. Under a long term agreement Amoco will also provide operating services for Camelot.

U.K. gas gathering

The most important advance in the U.K. North Sea in 1988 was consolidation of plans for gas gathering systems in the central sector.

BP, seeking a market for sour gas from its Miller field, broke new ground by selling the entire 570 bcf of reserves to the North of Scotland Hydro-Electric Board as generating fuel for a power station at Peterhead north of Aberdeen.

BP will lay a 150 mile, 30 in. line from Miller to the Scottish mainland. Initially it will be dedicated to Miller sour gas but also will have capacity of more than 1 bcfd for third party gas.

Miller is the first U.K. offshore project where 100% of gas reserves have not been sold to BG. That will become more common in the wake of the Monopolies and Mergers Commission recommendation that BG should be limited to buying 90% of production from a field.

The Miller project started as one leg of a large gas gathering system covering the whole central North Sea in competition with several other proposed gas systems.

The direct competitor in the Miller area is Marathon Oil U.K. Ltd.'s proposed line to link its Brae area into St. Fergus. Marathon was in detailed negotiations with Mobil to extend the line to include Beryl field's untapped gas reserves.

Consolidation also marked gas gathering in the southern part of the central U.K. North Sea. Amoco and Texas Eastern North Sea Inc. have joined their proposals for gas gathering lines into a joint venture known as Central Area Transmission systems (CATS).

BG and Amerada Hess Ltd. also are partners in CATS, which was conducting a detailed design of a 36 in. line. A decision on the project was scheduled in first quarter 1989. If the project proceeds, Amoco's Everest and Lomond fields will be first to tie into the system.

Further competition could come from Shell/Esso group's spare capacity in the Fulmar gas line expected in the 1990s.

U.K. development

The southern gas province captured the U.K. offshore development spotlight in 1988.

In late summer Conoco started up the V-fields project and related Loggs gas gathering pipeline. Pipeline throughput will build to about 800 MMcfd the next 3 years.

Among other southern basin gas projects getting a green light during the weather window are Britoil plc's Amethyst development, Shell/Esso's Barque and Clipper fields, and Phillips's Della field.

Off the U.K., installation work during the weather window focused east of the Shetlands, where Shell U.K. Exploration & Production positioned the steel jackets and topsides for Tern and Eider platforms.

Work was also in progress on Amerada Hess's Ivanhoe/Rob Roy project using a converted semisubmersible, Shell/Esso's Kittiwake field, Amoco's Arbroath field—to be developed as a satellite of Montrose field, and BP's Swops tanker concept for Cyrus field.

Marathon has permission for a subsea development in Central Brae field and was expected to submit development plans for East Frigg reservoir in 1989.

Also on tap for possible development were Chevron's Alba oil field, Enterprise Oil's Nelson oil field, and Amerada Hess's Waverley/Brunel project.

Amerada Hess let contract to Global Engineering Ltd. for development studies on the Waverley structure in Block 15/21.

First phase will entail comprehensive review of development options, and second phase will cover further technical evaluation and conceptual engineering design. Contract was to be complete in early 1989 to prepare for development proposal to DOE.

Waverley extends into an adjoining block held by Amoco, where it is known as Brunel. Reserves are estimated at 250 million bbl.

First horizontal well from semisubmersible

BP Petroleum Development Ltd. completed the first horizontal well to be drilled and tested from a semisubmersible rig in the U.K. North Sea.

BP drilled 16/28c-11, tapping the 10 million bbl Cyrus

Operators to feel fallout from Piper Alpha

Fallout from the Piper Alpha disaster in the U.K. North Sea is likely to be felt in the worldwide offshore community for a long time to come.

In October 1988, the director of safety of the U.K. Department of Energy, issued a report placing the explosion in the compressor unit. Director Jim Petrie, in his interim report, said the release of condensate vapor from a section of pipework in the module was the most probable cause for the initial explosion.

A public inquiry into the July 6 tragedy was set to begin on Jan. 19, 1989, and afterwards, recommendations will be made for every operator working in the North Sea. The inquiry is likely to last 6-8 months.

The world's worst offshore accident occurred when the gas leak apparently set off a string of ferocious explosions that annihilated Occidental's 13 year old platform. A total of 167 men died, and as of late December 1988, 32 crewmen remained missing. Ironically, the explosion occurred only 2 weeks after the annual inspection by the U.K. Department of Energy.

More restrictions seen. Once the investigation is complete, operators most definitely will face more stringent restrictions in the North Sea, particularly the U.K. sector. More frequent inspections and an overall review of existing platforms are only two options being mentioned frequently.

Shock waves from the disaster were felt immediately afterwards throughout the producing sector.

In Australia, for instance, the government ordered 19 platforms to be inspected only 1 week after the Piper Alpha disaster. All the installations were given a clean bill of health.

Meanwhile, most industry experts feel it will be years before the ill-fated Piper field is returned to production. London analyst James Capel & Co. speculated the field may not return until at least 1992.

"The government will almost certainly be unwilling to consider a new development plan until the full inquiry into the Piper Alpha accident has been completed—possibly in 12-18 months time," Martin Lovegrove, manager of Capel's Petroleum Service Dept. said in late August.

Piper Alpha was producing 125,000 b/d of oil, but when combined with satellite fields, which also were lost, the total production was about 300,000 b/d.

Lovegrove said the remaining 142 million bbl of oil at Piper may be produced through a smaller structure, or a floating production system. The re-development probably will require only 12-18 wells, as opposed to the 36 wells being produced before the casualty. Redevelopment cost is estimated at $225-350 million.

Lovegrove, as do others, said the industry will be facing operational and financial challenges as a result of the disaster. He said the government most definitely will order operators to inspect their equipment more regularly and meticulously. The government itself probably will step up its installation

THE JULY 6 DESTRUCTION of the Occidental Piper Alpha platform in the U.K. North Sea was the world's worst offshore accident.

inspections, he said, adding operators will be forced to upgrade their smoke, fire, and gas detection equipment.

Black boxes? One recommendation has the government considering every platform be fitted with a black box that would continually monitor and tape key events in the daily life of a platform. Similar to those placed aboard airplanes, the so-called black boxes would be designed to head off potential problems, but Lovegrove said they also would prove invaluable in determining the cause of an accident should one occur.

"All of this, of course, will increase operating costs, although, the scale would probably not be frightening—5%, or so, on existing fixed operating costs," he predicted.

However, if the investigation suspects design flaws may have contributed to the massive loss of life, Lovegrove said extensive alterations of existing and new structures may be demanded.

"This, in turn, would undoubtedly be expensive and could cause the temporary restriction, or even the cessation of production, while remedial work is in progress," he said. "The vital question facing field operators and partners would then be: Are the changes worthwhile given the current low oil prices."

The biggest problem facing investigators is the lack of material evidence, since the control room and most of the topsides were destroyed. **IPE**

---

reservoir in Block 16/28, in 360 ft of water with the Sedco 700 semisubmersible. The rig was specially modified and fitted with a top drive system for the well.

It was completed in 41 days, 10 days ahead of schedule and tested about 6,000 b/d of oil through 1 in. choke. Productivity was significantly better than would be expected from a conventional well in the same area, BP said.

BP was preparing Cyrus for development with its custom built Swops production vessel. After testing, the horizontal well was suspended for further evaluation and possible completion as a Swops producer in first quarter 1989.

Target was Paleocene Andrew at 8,265 ft true vertical depth subsea (Tvdss).

Final horizontal hole section was parallel to and 52.5 ft vertically above oil/water contact at 8,390 ft Tvdss. Vertical tolerance on the horizontal well path was ±16.4 ft, and minimum length expected for the horizontal section was 1,312 ft.

BP used a combination of electronic measurement-while-drilling tools and steerable downhole motor throughout drilling. It used a drill string simulator to determine bottomhole assemblies required to minimize torque and drag, thus allowing the bit to be controlled and steered to the target.

The simulator predicted that heavy downhole pipe strings

BERGROHR HERNE

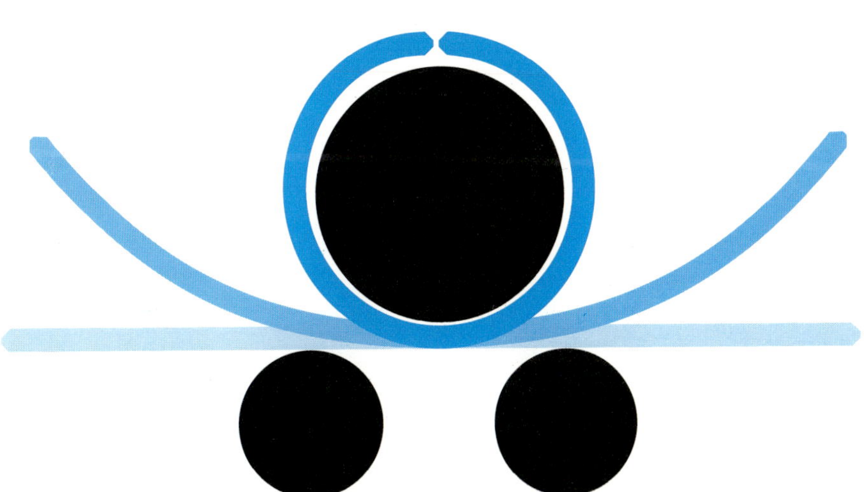

NORTH SEA
Atlantic Ocean
SCOTLAND
N. IRELAND
IRELAND
ENGLAND
WALES
Atlantic
Ocean
Outer Hebrides
Inner Hebrides
North Channel
Irish Sea
Celtic Sea
English Channel
St George's Channel
Firth of Clyde
Solway Firth
Morecambe Bay
Liverpool Bay
Cardigan Bay
Bristol Bay
Firth of Tay
Firth of Forth
Moray Firth
Cromarty Firth
Bantry Bay
Galway Bay
Shannon
Lough Neagh
Isle of Man
Kirkwall
Orkney Is.
Lerwick
Shetland
Flotta
Wick
Beatrice
Nigg Bay
Cromarty Firth
Fraserburgh
St. Fergus
Peterhead
Cruden Bay
Inverness
Aberdeen
Dundee
Glenrothes
Finnart
Ardyne Point
Grangemouth
Hunterston
Edinburgh
Glasgow
Couseland
Ardrossan
Newcastle
Carlisle
Port Clarence
Darlington
Middlesbrough
Lancaster
Heysham
Bradford
Manchester
Liverpool
Sheffield
Eastham
Stanlow
Ellesmere Port
Derby
Formby
Birmingham
Londonderry
Belfast
Dublin
Limerick
Wexford
Bantry
Cork
Whitegate
Kinsale Head
Milford Haven
Angle Bay
Neath
Llandarcy
Pembroke
Cardiff
Gloucester
Oxford
Swindon
Bristol
Bath
Reading
Exeter
Plymouth
Weymouth
Bournemouth
Fawley
Kimmeridge
Wytch Farm
Honey Grove
0   50   100 Mi
0   50   100 Km.
Complete offshore supply and service base
Offshore supply and service base
Specialized service base

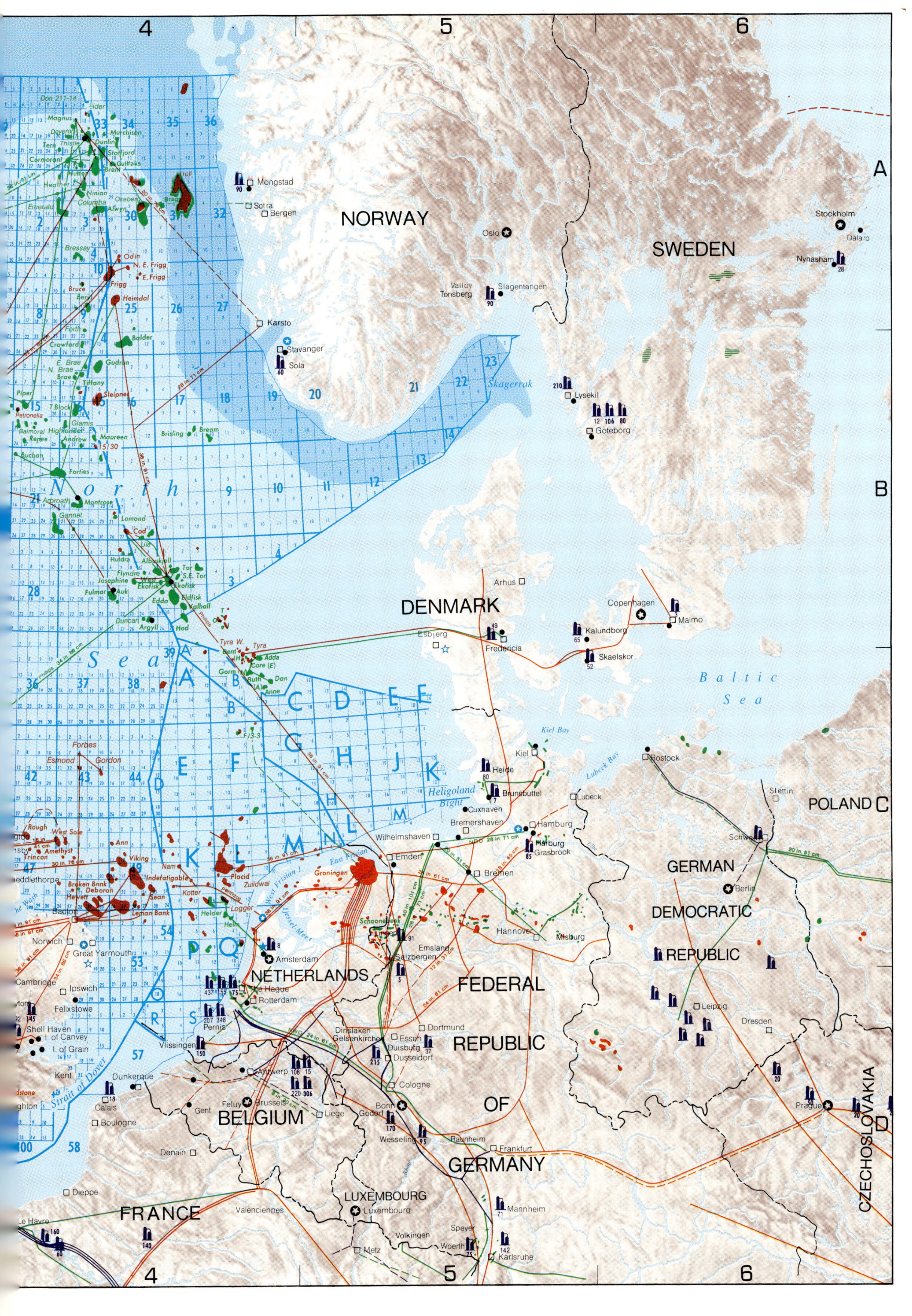

NORWAY
SWEDEN
Stockholm
Dalaro
Nynasham
Mongstad
Sotra
Bergen
Oslo
Valloy
Tonsberg
Slagentangen
Karsto
Stavanger
Sola
Skagerrak
Lysekil
Goteborg
DENMARK
Arhus
Copenhagen
Malmo
Esbjerg
Fredericia
Kalundborg
Skaelskor
Baltic Sea
Kiel Bay
Lubeck Bay
Kiel
Heide
Brunsbuttel
Lubeck
Rostock
Stettin
POLAND
Cuxhaven
Bremershaven
Wilhelmshaven
Hamburg
Harburg
Grasbrook
Bremen
Emden
GERMAN
DEMOCRATIC
REPUBLIC
Berlin
Leipzig
Dresden
Groningen
Hannover
Misburg
NETHERLANDS
Amsterdam
The Hague
Rotterdam
Pernis
Salzbergen
Emsland
FEDERAL
REPUBLIC
OF
GERMANY
Prague
CZECHOSLOVAKIA
Vlissingen
Dinslaken
Gelsenkirchen
Essen
Duisburg
Dusseldorf
Dortmund
Cologne
Antwerp
Bonn
Godorf
Wesseling
Raunheim
Frankfurt
BELGIUM
Gent
Feluy
Brussels
Liege
Denain
Dunkerque
Calais
Boulogne
Strait of Dover
Mannheim
LUXEMBOURG
Luxembourg
GERMANY
Dieppe
Le Havre
FRANCE
Valenciennes
Volkingen
Metz
Speyer
Woerth
Karlsruhe
North Sea
Heligoland Bight
Great Yarmouth
Norwich
Cambridge
Ipswich
Felixstowe
Shell Haven
I. of Canvey
I. of Grain
Kent
Dieppe

were not required to prevent buckling.

It took 30 days to run and cement 9⅝ in. casing at the top of the reservoir. At this stage, the hole angle had built to 80°.

Upon drilling out of casing shoe, the hole angle built to 90° and continued to total depth of 12,214 ft below rotary table. Total length of hole from casing shoe to bottomhole was 2,525.6 ft with a 1,853 ft horizontal section.

Once the hole angle had built to more than 45°, the drill string consisted almost entirely of standard weight pipe.

Torque and drag were almost exactly as predicted, while weight transmission and bit control were very good, BP said. It ran a 5 in. slotted liner and set it across the reservoir interval before testing.

Norwegian gas

Viability of gas projects in the central North Sea ultimately could depend on negotiations between BG and the Norwegian gas export group led by Statoil that started in late 1988.

Gas will be available from two major developments off Norway in the 1990s. Norske Shell's Troll project is not scheduled to come on stream until 1996, but Statoil's Sleipner field will start up in 1993.

Statoil let contract for Sleipner's gravity base to Norwegian Contractors and will let the fabrication contracts in 1989. The platform will handle 725 MMcfd of gas and 125,000 b/d of liquids.

BG can expect strong opposition to Norwegian gas imports from U.K. operators claiming sufficient domestic reserves to meet the U.K.'s medium term needs, and even stronger opposition from U.K. petroleum services/suppliers fearing loss of development work in the next 5 years.

Early in 1989, Statoil also was to begin organizing the Zeepipe gas pipeline project to deliver gas from Sleipner to Zeebrugge, Belgium, beginning in 1993.

The Zeepipe group will route the pipeline through Norwegian, Danish, West German, Dutch, and Belgian waters. It dropped an alternative route through U.K. waters because of permit difficulties and hiked planned diameter of the 515 mile initial section to 40 in. from 38 in. It will lay pipe in 1991-92.

Norwegian subsea projects

Norway entered the subsea era during the summer of 1988 with start-up of Statoil's Tommeliten field and Elf Aquitaine Norge AS's East Frigg development.

Tommeliten initially will provide gas for injection into Ekofisk field, and East Frigg will move gas into the central Frigg production facilities to offset declining output from the main field.

Norsk Hydro completed the concrete gravity platform for Oseberg field and installed the unit offshore. It was to start up yearend 1988 or early 1989 and produce about 100,000 b/d in 1989. Peak production from the present development will be about 240,000 b/d, but a third platform, approved

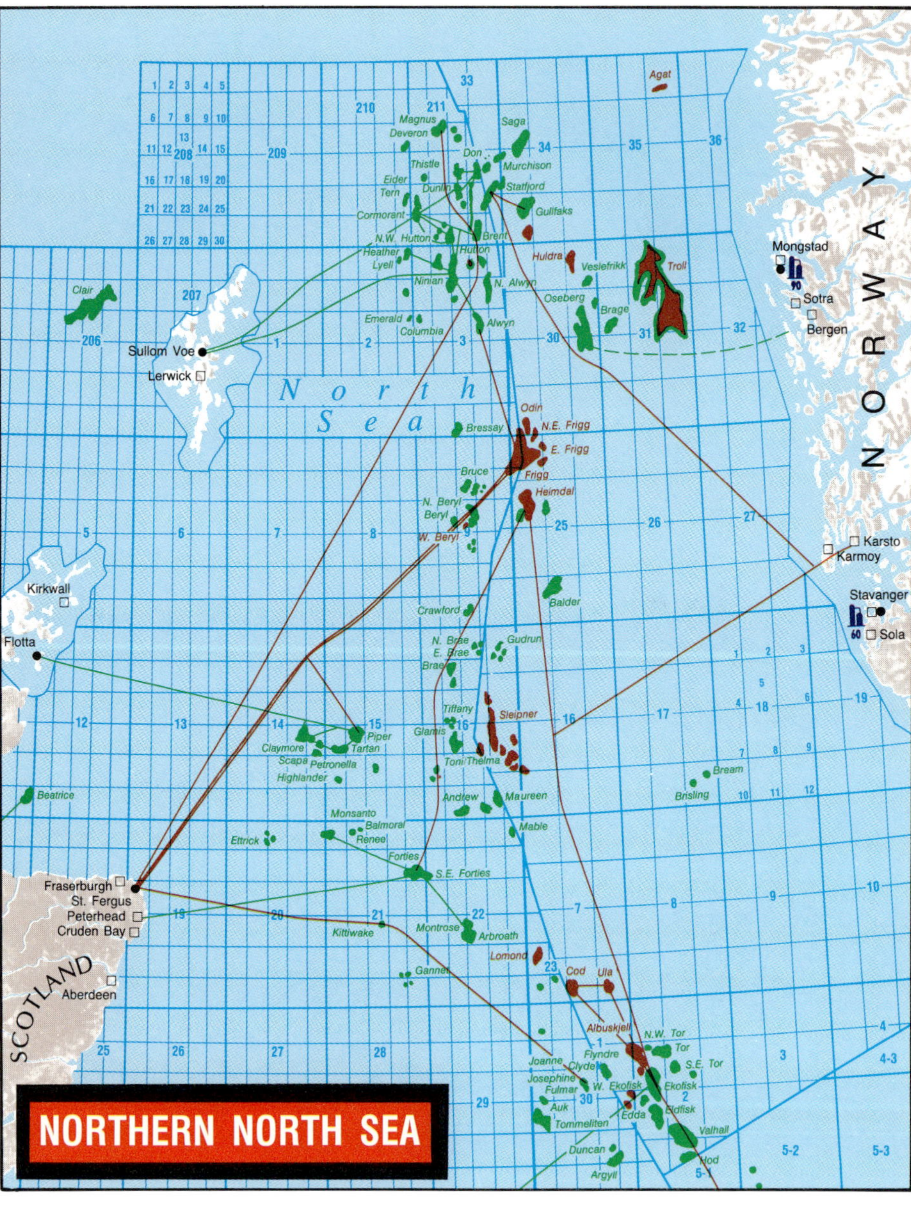

earlier in 1988, will raise output to 300,000 b/d.

Statoil's 65,000 b/d Veslefrikk field, using a floating production unit bridge-linked to a wellhead tower, was on schedule for installation in 1989. BP Petroleum Development Norway's 60,000 b/d Gyda field should start up in 1990.

Throughout summer 1988 Statoil moved further ahead of schedule with its Gullfaks C platform, which will be the biggest production unit in the North Sea.

The most unusual project under construction was the 140 m diameter concrete wall that will enclose the concrete storage tank in the main Ekofisk production complex. The unit, designed to protect the tank from seabed subsidence, will be installed during the summer of 1989.

The Norwegian government early in 1988 set the pace for future development with approvals for Saga Petroleum's 750 million bbl Snorre field using the first tension leg platform in the Norwegian North Sea.

It followed this with approval of the first Haltenbanken development in the Norwegian Sea, Conoco's early production system for Heidrun—which will be followed by permanent oil production facilities based on a concrete TLP.

Amoco Norway AS won approval for a single wellhead platform for Hod field tied back to Amoco's Valhall production facility.

Exploratory drilling in Norwegian waters continued to

THERE'S MORE TO TOTAL THAN MEETS THE EYE.

There is oil from the North Sea to the Middle East, from the Far East to Tierra del Fuego, in oceans and deserts, all over the world.
Total is searching for it.

There is a potential for energy in oil, gas, coal, uranium and photovoltaics as well.
Total is developing it.

There is a demand for crude oil and natural gas conversion into gasoline, fuels, LPG and lubricants.
Total is providing it.

The world is in a need of energy, Total devotes its energy to meeting it.

So there is more to Total than meets the eye, a lot more.

THE ENERGY EXPERTS

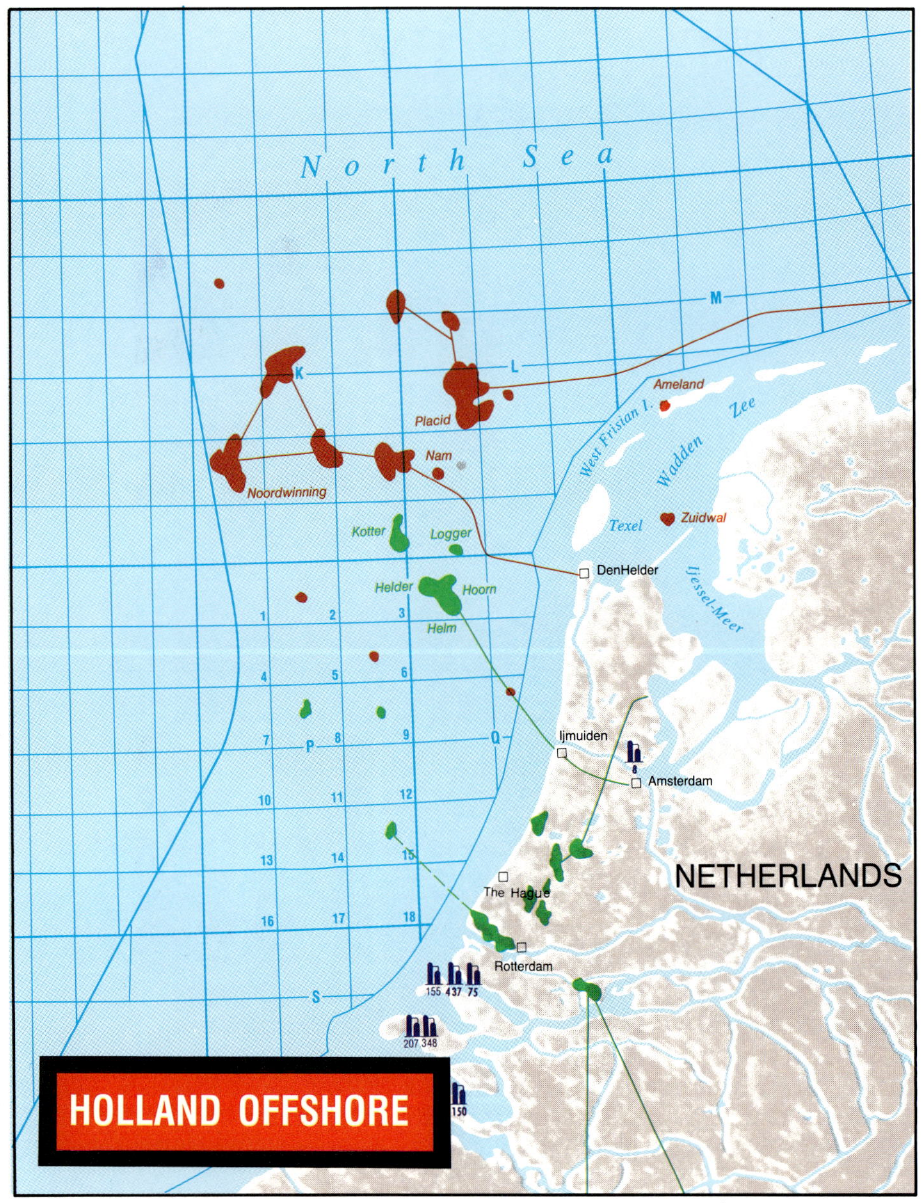

part of the Dutch North Sea to gas development.

Combined cost of the line and the four field projects will be 2.1 billion guilders ($2.1 billion U.S.). Design and construction of the pipeline will claim half of the budget.

The pipeline project will be undertaken in two stages.

In the first, pipe will be laid from the L2 development through L15 to the Dutch coast. This 140 mile section and the first two fields will be operational in 1991.

The second phase, linking F3 and F15 into the system, is to be complete in 1992.

Gas will move from landfall to a Gasunie treatment plant near Den Helder. This unit will be doubled in size to handle the new throughput.

A single drilling/production platform will be installed in each field. A total of 25 wells will be drilled.

NAM originally planned to develop F3 oil and gas reserves with a two phase flow pipeline; that idea vanished when crude oil prices collapsed.

No decision had been made on transport of F3 liquids.

The alternatives of tanker loading or another pipeline were still under consideration.

Haltenbanken activity

Saga Petroleum AS was nearly ready to declare Midgard field commercial.

Midgard is the main pillar in any gas development for the frontier Haltenbanken region in the Norwegian Sea off Norway, says Saga.

The field is expected to play the role of gas guarantor for Haltenbanken. The region's first two fields earmarked for development, Heidrun and Draugen, will produce oil and associated gas. Midgard could produce 70-500 bcf/year of gas.

"It will play an active, flexible role in matching a wide range of possible export scenarios and volumes from Haltenbanken," Saga said.

Saga's discovery of the field in 1981 established the presence of hydrocarbons on Haltenbanken.

The company pegs Midgard reserves at 4.2 tcf of the estimated 5.5 tcf in place, making the field by far the largest gas accumulation found so far on Haltenbanken. Because of excellent reservoir properties, Saga expects excellent, flexible gas deliverabilities. Saga also expects to recover 110 million bbl of condensate.

Saga was searching for a gas buyer and says the field can be on stream about 4 years after authorities and partners approve the project. Expected start-up of production is 1994-95.

Gas would move by pipeline over iceberg scoured seabottom to shore. It would be used domestically and exported via liquefied natural gas tankers and other pipelines.

Saga's partners in Midgard field are Agip SpA, ARCO Norway Inc., Deminex (Norge) AS, Norsk Hydro AS, Norske Shell AS, and Den norske stats oljeselskap AS.

Saga was optimizing three development concepts, two that cover gas production of 70-140 bcf/year and one at 140-

disappoint in 1988, particularly in the Barents Sea.

The most interesting find was in the most southerly part of the Norwegian North Sea where the country's oil development began.

Phillips found oil from Jurassic below the chalk where most Ekofisk oil and gas have been found. Phillips was conducting seismic surveys on the South Eldfisk structure and hoped to win approval for a long term production test.

Dutch gas pipeline

Nederlandse Aardolie Mij. (NAM), jointly owned by Exxon Corp. and Royal Dutch/Shell Group, planned to lay a major gas transmission system in the northern part of the Dutch North Sea.

The 161 mile, 36 in. line will run from NAM's F3 block to a landfall south of Den Helder, Netherlands. Start-up of the 1.7 bcfd capacity Northern Offshore Gas Pipeline is expected in 1991.

The line will run through Petroland's F15 gas field and pick up gas from two proposed NAM gas development projects in Blocks L2 and L15. The four fields have total reserves of 1.765 tcf, all committed for sale to NV Nederlandse Gasunie.

The project, biggest off Netherlands in many years, will provide the infrastructure to open the northern and eastern

MANNESMANN
DEMAG

Mannesmann Demag
Compressors and Pneumatic Equipment
P.O.Box 10 15 07, Wolfgang-Reuter-Platz
D-4100 Duisburg 1, Federal Republic of Germany
Tel.: Germany (2 03) 6 05-1, Fax: (02 03) 6 10 61-63

Compressors for the oil and gas industry

Process compressors are a traditional Mannesmann Demag field of activity. The Compressors and Pneumatic Equipment Division has contributed towards the dynamic development and the economics of chemical and petrochemical processes for a long time now. The product range includes single-stage and multi-stage compressors of centrifugal design, with horizontally or vertically split casings, geared turbocompressors, axial-flow and axial-centrifugal compressors. The degree of standardization we have selected ensures that specific designs meet the appropriate conditions. Ample proof of our international standing is demonstrated by the fact that we have supplied compressor installations all over the world, particularly to the oil and gas industry and the petrochemical industry, ranging from the tapping of energy resources through to the final further-processing stage.

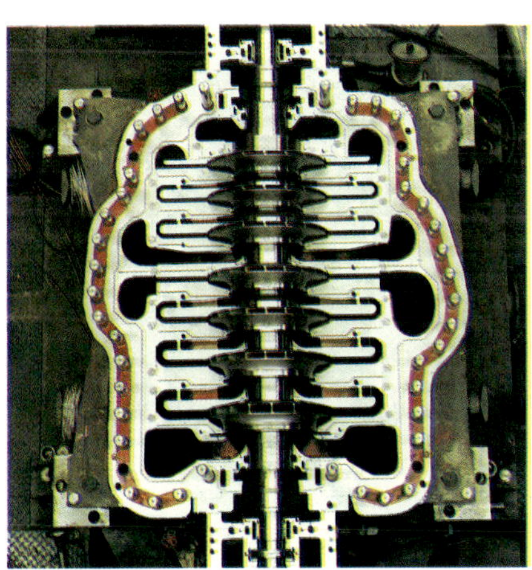

Compressors for gas processing

On offshore production platforms centrifugal compressors, installed as multicasing trains, gather natural gas from several wells and compress the separated product ready for transfer to the mainland. In onshore gas fields, where compressors are only required for a limited period at a particular location, mobile stations are used, i.e. single-stage blowers with complete peripherals or preassembled compact units with multi-stage compressors.

Compressors for gas liquefaction

In natural gas liquefaction plants, centrifugal compressors drive the refrigeration circuits. The compressors are multi-stage and, depending on the process involved, can have interstage sidestreams.

Compressors for pipelines

To compensate for pressure losses in pipelines, centrifugal compressors with fully automatic monitoring systems function with great reliability in booster stations no matter where they are located.

Compressors for petrochemical processing

The availability of an installation depends largely on the in-service behaviour of the compressors installed. The machines we have already supplied to all sectors of the petrochemical industry, in refineries, through to olefin and ethylene plants, demonstrate the reliability and soundness of Mannesmann Demag compressors.

We are represented by:

USA/Canada
Mannesmann Demag Corporation
Compression Equipment Division
1055 Parsippany Boulevard
Parsippany, New Jersey 07054
Tel.: (201) 402-5775
Fax: (201) 402-8452
Tlx.: Western Union 139006
Mannesmann Demag Corporation
1990 Post Oak Blvd., Suite 1800
Houston, Texas 77 056/USA
Tel.: (713) 960-1900
Tlx.: 77-5407

mannesmann *technology*

Mannesmann Rexroth GmbH
Postfach 3 40, Jahnstrasse 3-5
D-8770 Lohr
Phone (93 52) 18-0
Telex 6 89 418

Hydraulics, Gearboxes and Electronics – Composed to Offshore Systems

Having one partner only obviously permits you to take advantage of this well organized and structured product supply program for all requirements of fluid power controls and drives in off/on-shore technology.

Put your trust in these major sources for all your coming projects involving any kind of fluid power controls and drives including electronics and transmissions.

The special combination of product ranges offered by the strong group of Mannesmann companies for hydraulics, electronics and transmissions provides the best laid foundation for your entire projects. Problems caused by product matching do not exist for our organization. From initial contact up to delivery, the coordination of resources for the design of mechanical power systems will be carried out internally by Mannesmann Rexroth.

Let us take care of your complete systems up to and including commissioning.

With our international representation through subsidiary companies and agencies around the world, we ensure the best possible cooperation for handling large-scale international off/on-shore projects.

Use our worldwide systems service operations – for your own sake. Our branches are always around the corner!

Due to our solid experience of many years covering a lot of unusual projects – also for other comparable special applications using fluid power of the same range – we are convinced that we are the supplier for controls and drives you need, no matter how exacting your requirements may be.

Any time you need us for your projects – we will be at your disposal.

High quality components and years of experience in the layout and operation of difficult systems are our contribution to large scale world-wide energy projects.

Hydraulic controls and gear boxes transmitting the power

In the following examples you can see equipment and applications working with components and systems from our five companies.

More and more responsible people involved in off/on-shore technology, have decided to use the best possible products for these applications, i.e. to use products of the Mannesmann Rexroth companies.

When will you make your decision… or have you already done so?

Hydraulic winches anchor the platform

Hydraulic crane for offshore duty

Ekofisk field installation lifted at 6 meters by Mannesmann Rexroth in August 1987

Pipeline Systems

Executed by applying finance resources, production facilities, expertise and experience

International pipeline projects have become highly integrated operations demanding a great degree of responsibility from those involved in carrying them out.

So it is that pipeline contractors find themselves confronted with a comprehensive catalogue of requirements including financing, engineering, procurement, site construction, commissioning, training of clients' personnel, operation and maintenance. These activities, of course, are not restricted to the pipeline itself but apply equally to all related facilities such as pump or compressor stations, storage systems and terminals. Where the whole of these activities is carried out on the same project, so demanding a fully integrated pipeline system, special know-how is called for to achieve the required function and performance.

The pipeline system can be designed and constructed for the transport of all media including oil, oil products, gas, water and slurries.

The satisfactory completion of such complex undertakings can only be achieved by a sophisticated project organization, the efficient coordination of the various activities involved and the strict control of an integrated time schedule.

Mannesmann Anlagenbau's outstanding capabilities in these fields are evidenced by our performance on many such projects – many large, a few gigantic, and all completed with the reliability and efficiency which customers the world over have come to expect from us.

North Yemen's First Oil to reach the Red Sea

The major part of this crude oil pipeline system from the desert town MARIB to the Red Sea port SALIF:
– oil field installations
– central production unit
– 430 km – 24"/26" diameter pipeline
– 3 pump stations
– offshore facilities
was timely completed by Mannesmann Anlagenbau and its joint venture partner CCC within a construction period of 9 months.

MANNESMANN
ANLAGENBAU

Engineering + Construction

Mannesmann Anlagenbau AG
Postfach 30 07 41, Theodorstrasse 90
D-4000 Düsseldorf 30, Fed. Rep. of Germany
Phone (2 11) 6 59-1, Telex 8 586 677
Fax (2 11) 6 59-23 72

General Contracting
Pipeline systems
Plant for the oil and gas industry
Offshore technology
Plant for the chemical and petrochemical industry
Water technology
Environmental and energy technology

Piping Systems
Piping systems for industrial installations, conventional and nuclear power stations
District heating systems and HVAC
Clean room technology

mannesmann *technology* 

MANNESMANN RÖHRENWERKE

Mannesmannröhren-Werke AG
Postfach 11 04, Mannesmannufer 3
D-4000 Düsseldorf 1, Fed. Rep. of Germany
Phone (2 11) 8 75-0, Telefax(2 11) 8 75-3245
Telex 8 581 421

Mannesmann supplies a wide range of tubular products for the exploration, production, transmission, and processing of oil and natural gas, in compliance with all international standards and customers' special requirements.

Oil country tubular goods

Drill pipe, casing and tubing are available in all API sizes and grades as well as in special grades, e.g. arctic grades, special corrosion-resisting grades, or superhigh strength grades.

Drill pipe
in OD's ranging from $2\,^3/_8''$ to $6\,^5/_8''$

With weld-on tool joints according to API specifications or in special design

Casing
in OD's ranging from $4\,^1/_2''$ to 26″

Thread types in accordance with API specifications or special joints, such as BDS, MUST (Mannesmann Ultra Seal Thread), HPC (High Performance Connection), Omega, Mid Omega and Big Omega up to 26″

Tubing
in OD's ranging from $2\,^3/_8''$ to $4\,^1/_2''$

Thread types in accordance with API specifications and special joints, such as gastight TDS, ST/A, ST/C and ST/P

Our representative in the USA:
Mannesmann Oilfield Tubulars Corporation
1990 Post Oak Boulevard, 17th Floor
Houston, Texas 77056
Phone (713) 552-4069, Fax (713) 960-8631
Telex (3112) 27200144

Mannesmann supplies drill pipe, casing and tubing which can withstand the most severe service conditions in oil and gas fields involving combined tension, compression, and bending loads.

Line pipe for oil and natural gas

Seamless line pipe
in OD's up to 26″ in compliance with API Spec 5L, ANSI B 36.10, ASTM, B.S., DIN, as well as other standards and special requirements

HFI-welded line pipe
high-frequency induction welded in OD's from 1″ to 20″ in compliance with API Spec 5L, ANSI B 36.10, ASTM, B.S., DIN, as well as other standards and special requirements

Double submerged-arc welded line pipe (large-diameter pipe)
with longitudinal weld, OD's from 24″ to 64″ in compliance with API Spec 5L, ANSI, ASTM, B.S., DIN, as well as other standards and to customers' requirements

Special steels meeting the most stringent requirements of modern pipeline construction and operation, e.g. under arctic conditions and for offshore applications

All pipe is available PE coated with MAPEC®, the Mannesmann developed 3-layer corrosion protection system. Other types of corrosion protection and coating are also available, e.g. epoxy lining for gas lines.

Custom-bent steel pipe
pipe bends in OD's from $3\,^1/_2''$ to 64″; made from seamless or welded steel pipe on a modern induction heat bending machine; in all customary steel pipe materials matching the high requirements of straight Mannesmann line pipe

Our representative in the USA:
Mannesmann Pipe & Steel Corporation
1990 Post Oak Boulevard, 18th Floor
Houston, Texas 77056
Phone (713) 960-1900, Fax (713) 960-1063
Telex (023) 166545

mannesmann *technology*

280 bcf/year.

A gravity based platform seems most realistic for the highest annual production range, while floating and gravity based concepts were being analyzed for the low range.

Saga had evaluated a concrete monotower with a modularized or integrated deck, as many as 16 well slots, and facilities for tie-in of subsea templates for production of outlying reservoir segments.

The floating production concept centers on a catenary anchored semisubmersible production vessel tied in by flexible risers to two templates, containing at least four subsea production wells and the other two subsea production wells.

Development costs were estimated at $1.1-1.7 billion for the low production range and $1.7-2 billion for the high range and are expected to be reduced with optimization.

Three appraisal wells followed the 1981 discovery. Appraisal drilling ended in early 1987.

Saga expects the field to produce 15-35 years with more subsea wells tied in periodically to maintain production rate.

Midgard is a horst like structure cut into four segments by northeast-southwest trending cross faults. Its reservoirs are in Middle Jurassic Garn and Ile and Lower Jurassic Tilje.

Garn is 180 gross ft thick with 29% porosity, 7,700 md permeability, and 6% water saturation. It contains 3.55 tcf of gas in place. Ile is 250 ft thick with 28% porosity, 4,000 md permeability, 13% water saturation, and 1.6 tcf of gas.

Tilje is 650 ft thick with 24% porosity, 750 md permeability, 29% water saturation, and 250 bcf of gas in place.

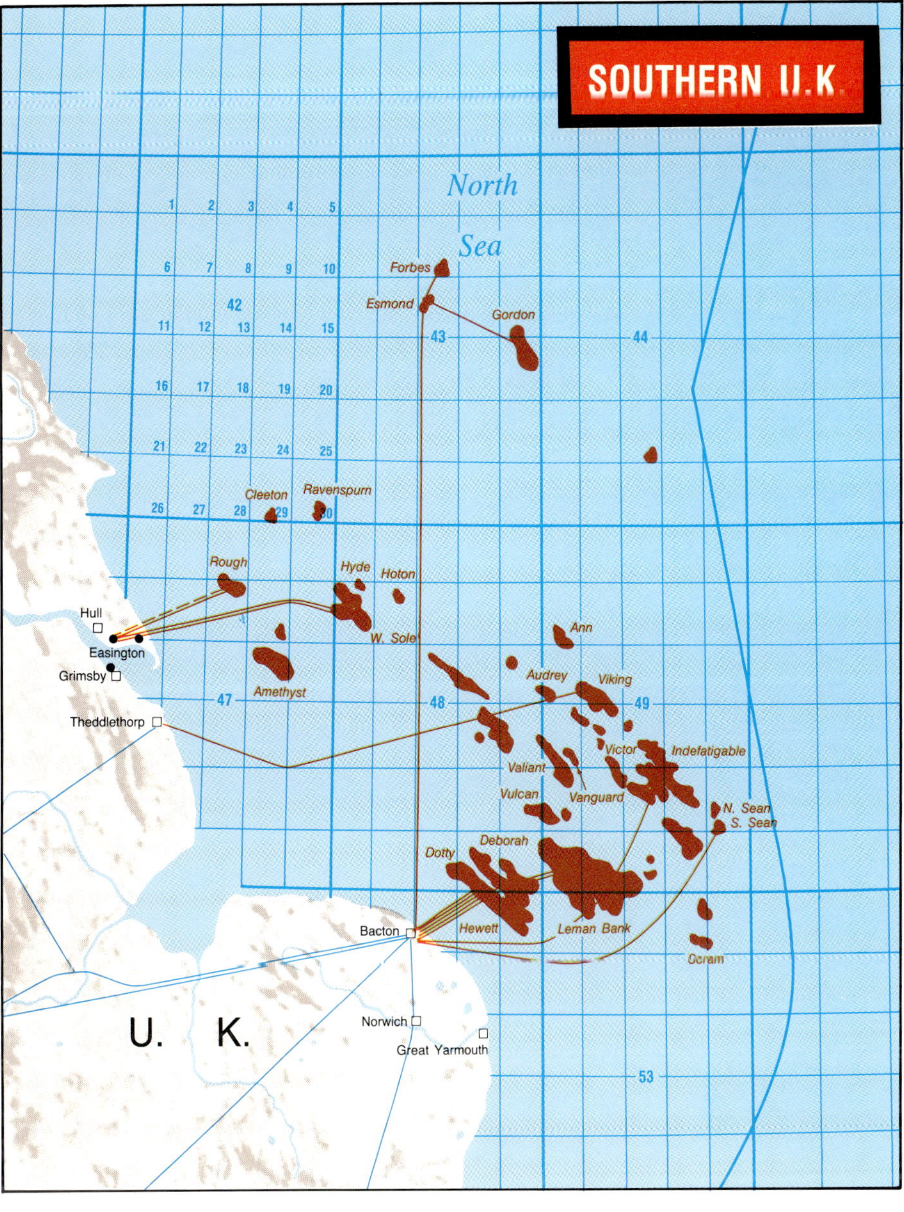

Saga also proved a 125 ft oil zone in one of the segments which may yield "minor reserves" if deemed commercially attractive.

Production rates are expected to be as high as 100 MMcfd/well.

The natural production rate of 10%/year would be close to 390 bcf/year.

The excellent reservoir properties will allow the field to deliver 70-500 bcf/year.

Pressure depletion will be the main production mechanism, but a large aquifer outside the horst is expected to give pressure support.

Midgard gas is 82 mole % methane with about 3 mole % of C5+ and traces of hydrogen sulfide and carbon dioxide. Heating value of processed gas will be 1.15 MBTU/scf.

Norwegian authorities gave the go-ahead for early production from Heidrun and were expected to approve Draugen development late in 1988.

Saga recommended that oil and condensate be buoy loaded on Haltenbanken via two shared buoy loading systems and a common shuttle tanker fleet.

Volumes produced will depend on sequence of field development but likely will be 126,000-157,000 b/d of oil and condensate.

Draugen and Heidrun associated gas may be reinjected for several years while a total gas solution for the mid-Norway offshore region is established.

Saga urged that gas be piped to shore in mid-Norway, probably to landfall west of Molde or Trondheim, to a terminal where NGL is separated and perhaps fractionated.

It estimated the cost of a 280 bcf/year gas trunk line from the offshore fields to an onshore terminal at $415 million.

The terminal will cost $340-850 million, depending on fractionation capacity.

Base gas volumes will depend on which market is being served.

Probable volumes are 150-200 bcf/year, but Haltenbanken fields are easily capable of producing more than 350 bcf/year.

From the mid-Norway terminal, dry gas would be transported to market by one of three methods: to eastern Norway and Sweden via pipeline, to continental Europe and the U.K. through a connecting pipeline to the existing North Sea gas pipeline grid, or to Europe and the U.S. as LNG.

A link to the North Sea grid would require a pipeline of about 250 miles from the terminal and cost $500-600 million, depending on capacity and distance.

With such a connection, Haltenbanken gas could be shipped to Emden, West Germany, Zeebrugge, Belgium, and St. Fergus, Scotland.

Depending on production volume, investment in a gas liquefaction plant would be $690 million to $1.3 billion. IPE

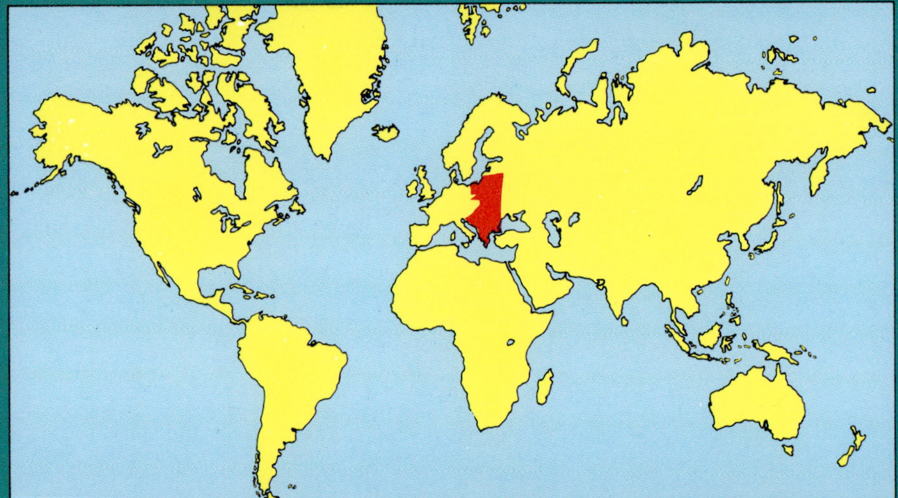

EASTERN EUROPE

EASTERN EUROPE COMMUNIST BLOC OIL PRO-
duction continues to drop.

Yugoslavia's crude flow fell from 83,000 b/d in 1985 and 1986 to slightly more than 77,000 b/d in 1987 and to about 75,000 b/d in first half 1988. Hungary's oil production has slipped from more than 40,000 b/d in 1986 to 38,400 b/d in 1987 and less than 38,000 b/d in the first 6 months of 1988, the latest period for which bloc information is available. Production is likely to continue falling through the 1990s.

Romania still refuses to disclose offical oil production figures. Offshore Black Sea flow began in 1986, but onshore production apparently continues to decline, leaving production in the first 6 months of 1988 close to 215,000 b/d.

Other estimated first half 1988 Communist bloc oil production includes Albania 55,000 b/d, East Germany 10,000 b/d (mostly condensate), Bulgaria 5,000 b/d, Poland 3,000 b/d, and Czechoslovakia 2,000 b/d.

Yugoslavia unlikely to reverse decline

Yugoslavia has reported continued exploration success in Serbia, southeast of Belgrade. But the success most likely will have little effect on the nation's oil production before the 1990s. Prospects are that Yugoslavia will find it very hard to achieve its goal of 165,000 b/d of oil by the end of the century. Gas flow has increased substantially during the 1980s and was expected to reach a peak of more than 3 billion cu m in 1987. However, Yugoslavia's exploration and development programs have been hampered by the nation's financial problems, including runaway inflation, a shortage of hard currency, large trade deficits, mounting foreign debt, and heavy losses incurred by many industrial projects such as refinery expansions during the past decade. Onshore, Yugoslavia continues to explore for hydrocarbons in Croatia and Serbia. At least 10 wells drilled at Stig, near Pozarevac, southeast of Belgrade, have found oil and some gas. Pay depth is 1,700-3,000 m (5,577-9,842 ft). Estimated initial oil yield is 219 b/d/well.

The Yugoslav News newspaper said 1987's planned oil production of 86,000 b/d is only about one fourth of the nation's consumption. More than 200,000 b/d of crude oil and about 20,000 b/d of petroleum products were to be imported. In northeastern Yugoslav's Podravina region, INA-Naftaplin reports finding "rich" gas reserves. Thirteen of 27 wells drilled by mid-1987 were productive. Exploration and development continued in the Podravina area, where preparations are under way to lay gas pipelines and build other facilities.

Yugoslav News said a number of wells drilled on Vis Island in the Adriatic Sea are "very promising."

Geology of the area around the island is believed to be similar to that of Italy's Pescara oil and gas region across the Adriatic. In 1986, it was reported that INA-Naftaplin and other companies planned to drill five wells in the Adriatic off Montenegro.

The wells were to be drilled by the Zegreb-1 semisubmersible, rated to 25,000 ft, before the end of 1990.

About 100 miles southeast of Vis Island, preparations for offshore drilling near the port of Dubrovnik are said to have been completed. Yugoslavia previously discovered commercial gas and small volumes of oil off its northern and central Adriatic coasts. Moscow's Pravda newspaper reported that Adriatic gas will soon move by pipeline to tourist centers on the Istra Peninsula.

Yugoslavia's seven refineries can process about 600,000 b/d of crude, Yugoslav News reported. The Rijeka refinery and several others run crude for foreign customers. Even so, Yugoslavia's refining capacity is badly underutilized.

Unlike most other eastern European Communist countries, Yugoslavia has no near term plans to build more nuclear plants. That's because of growing public opposition. The nation's only nuclear power plant is the 632,000 kw unit placed in operation in 1981 at Krsko. Yugoslavia is counting on the Soviet Union to provide it with increasing volumes of oil and gas. During a November 1987 visit to Belgrade by a Soviet economic delegation, Yugoslav officials were told that they could expect "somewhat higher" crude deliveries from the U.S.S.R. through 1990. Soviet gas exports to Yugoslavia likely will increase faster than oil shipments during the long term. The U.S.S.R. and Yugoslavia late in 1987 discussed

EASTERN EUROPE

Baltic Sea
Copenhagen
Reinkenhagen
Kolberg
Rostock
Hamburg
GERMAN
DEMOCRATIC
REPUBLIC
Roxforde
Brandenburg
Berlin
Rudersdorf
Frankfurt
Gölzau
Rybaki
Fallstein
Dessau
Volkenroda
Muhlh
Schkopau
Lützken
Bohlen
Langensalza
Zeitz
Leipzig
Stettin
Schwedt
Bromberg
Poznan
Grünberg
Krotoszyn
Breslau
FEDERAL
REPUBLIC
OF
GERMANY
Munich
Waidhaus
Pilsen
As
Litvinov
Zaluzi
Kralupy
Prague
Kolin
Pardubice
Brod
Trebic
Brno
CZECHOSLOVAKIA
Neisse
Cosel
Ostrava
Stahava
Pribor
Vacenavice
Hodonin
Malacky
Stefanov
Gbely
Vienna
Schwechat
Bratislava
Vystok
Nitra
Sala
Szony
Komárom
AUSTRIA
Linz
Salzburg
Graz
Maribor
Ljubljana
Zagreb
Sisak
Hijeka
Pula
Koper
Trieste
Padua
Venice
Verona
Corteggio
Minerbio
Bologna
Ravenna
Riccione
Rimmi
Florence
Falconara
Ancona
ITALY
Bellante
Teramo
Cellino
Pescara
San Stefano
Vasto
Valle Cupa
Allano
Civitavecchia
Rome
Fiumicino
Gaeta
Naples
Petroliara-Farnesina
Monte Stillo
Candela
Chieuti
Portocannone
Cupello
S. Salvo
Ripi
Grottole
Ferrandina
Rotondella
Nova Siri
Pisticci
Taranto
Brindisi
Bari
Sibari
San Leonardo
C. Cimiti
Lavinia
Milazzo
Messina
Palermo
Marsela
Mazara del Vallo
Marinella
Lippone
SICILY
San Nicola
Bronte
Gagliana
POLAND
Warsaw
Glinik
Plock
Lodz
Radom
Pulawy
Lublin
Kielce
Sandomierce
Rozwadow
Portynia
Lubaczow
Grobla
Trzebina
Oswiecim
Bochnia
Tarnow
Jaroslaw
Czechowice
Jaslo
Gorlice
Krosno
Przemysl
Smilov
Mikova
Olka
Strázske
Kosice
Zvolen
Sáhy
Nyirbogdány
Miskolc
Leninvaros
Debrecen
Budapest
Százhalombatta
HUNGARY
Nagykanizsa
Lovaszi
Budafap
Zalaegerszeg
Pannonian basin
Pécs
Babocsa
Szeged
Arad
Oradea
Szandaszolos
Pusztaföldvar
Battonya
Hajduszoboszlo
Nadudvar
Biharnagybajom
Demjen
Pedemes
Mezökeresztes
Satu-Mare
Baia-Mare
Transylvania
Cluj
Bieaz
Darmanesti
Sulpacu Barcau
Teleac
Sarmasel
Lucacesti
Miercurea
Singeorgiu
Saros
Noul Sasesc
Sibiu
Gheorghiu-Dej
Bacau
Brasov
Boldest Moreni
Talsani
Cimpina
Frasinu-Mislea
Ploiesti
Pitesti
Gura
Brazi
Bucharest
ROMANIA
Petroseni
Turnu-Severin
Craiova
Oltenia
Giurgiu
Ruse
Arges
Galati
Reni
Izmoil
Braila
Buzau
Faurei
Navodari
Constanta
Lebada
East Swan
Tiulenovo
Kavarna
Varna
Staro Orjachovo
Burgas
Otmanli
Dolni Dybnik
Chiren
Pleven
Gorni Dybnik
Develtek
Beli Izvor
Vraca
BULGARIA
Stara Zagora
Sofia
Plovdiv
Nis
Pristina
Skopje
ALBANIA
Tirana
Cerrik
Stalin
Fier
Divjaka
Ballsh
Patos
Marinza
Vlore
Bubulina
Kastoria
Grevena
Larissa
GREECE
Arta
Delfi
Eleusis
Corinth
Athens
Aspropyrgos
TURKEY
Edirne
Orestias
Philippi
Alexandroupolis
Kavala
Prinau
S. Kavalla
Thessaloniki
Sea of
Marmara
Aegean Sea
Black
Sea
Adriatic
Sea
YUGOSLAVIA
Sarajevo
Dubrovnik
Split
Zadar
Dugi
Novi Sad
Belgrade
Pancevo
Bosanski Brod
Nova Gradiska
Dugo Selo
Sombor
Srbobran
Bizovac
Gospodjinci
Mramorak
Selo
LITHUANIA
Kaunas
Vilnius
Molodechno
Borisov
Minsk
KALININGRAD
Kaliningrad
Krasnoborskoye
Ushakovskoye
Grodno
Baranovichi
Ivatsevichi
Kobrin
Brest
Pinsk
BYELORUSSIA
Pripyat basin
Vishanskoye
Davydovskoye
Ostashkovichskoye
Tishkovskoye
Rechitskoye
Rechitsa
Mozyr
Ovruch
SOVIET UNION
Smolensk
UKRAINE
Lutsk
Rovno
Dubno
Zhitomir
Kiev
Brody
Lvov
Staro-Konstantinov
Ternopol
Vinnitsa
Borislav
Dashava
Kadobnensk-Kalusz-Grabovsk
West Ukrainian fields
Stryy
Dolina
Nadvornaya
Bitkov
Kosmachskoye
Chernovtsy
Kosov
Krasnoilsk
MOLDAVIA
Iasi
Ugeny
Kishinev

laying pipelines to carry Soviet gas to Yugoslavia from existing systems crossing Hungary and Bulgaria. Meantime, Yugoslavia estimates potential oil resources in the southern Adriatic at about 300 million metric tons (2.19 billion bbl).

Belgrade also reports that the first of those five wells to be drilled off Montenegro in extreme southern Yugoslavia is targeted to 5,800 m (19,029 ft) and will cost about $12 million. The Panon jack up—not the Zegreb-1 semi as originally planned—is drilling the first hole near the port of Ulcinj. Southern Adriatic drilling as deep as 7,000 m (22,966 ft) is seen as possible by 1990.

Elsewhere, INA-Naftaplin let a $40 million contract to Pritchard Corp., Overland Park, Kan., for engineering on the 177 MMcfd Molve III gas treatment plant for Molve gas field in northern Yugoslavia.

Start-up is set for 1990. NGL will be transported by pipeline to the Pritchard-designed Ivanicggrad ethane recovery plant for fractionation.

In other Communist bloc action, Chemopetrol Chemical Works, Litvinov, Czechoslovakia, as 1989 neared

Slump in daily oil output by Russia is being offset

THE U.S.S.R.'s AVERAGE DAILY OIL FLOW DURING 1988 slipped for the second time in 4 years, and prospects for a significant rebound in 1989 are bleak.

But the oil output slump by the world's largest producer was more than offset by continued enormous gains in natural gas flow, which has grown steadily since the end of World War II.

Official data show that 1988 Soviet oil production was 624 million metric tons, the same total as reported for 1987. However, with 1988 being a leap year, average production fell to 12.446 million bo/d from 12.480 million bo/d in 1987.

Soviet 1988 crude/condensate production inched up slightly more than 1% over 1983 output of 616 million metric tons (12.320 million b/d). Most of the meager gain during the 5 year period was provided by condensate.

The U.S.S.R.'s oil production suffered especially large declines during the second half of 1988. Crude/condensate flow averaged 12.514 million b/d in the first half of 1988 but only 12.378 million b/d during the final 6 months.

An all time monthly high production averaging 12.650 million b/d was achieved last April. By August, with widespread forest fires hampering work in western Siberia, which produces two thirds of the U.S.S.R.'s crude and condensate, flow skidded to 12.245 million b/d.

Average production for the last 3 months of 1988 recovered to 12.378 million b/d. Even so, the figure was down considerably from the 12.458 million b/d reported during the final quarter of 1987.

Moscow initially set the nation's 1989 crude/condensate target at 12.472 million b/d. But a new goal announced in fall 1988 and including the regional production associations' pledge to exceed the official target, is 12.632 million b/d.

Attainment of this objective is believed highly improbable.

Meanwhile, Soviet natural gas production reached 27.18 tcf last year, another world record and far above the official plan. The 1988 figure is up from 25.66 tcf in 1987 and 24.22 tcf in 1986.

The U.S.S.R.'s current 5 year plan called for 1989 gas flow of 28.24 tcf. But the nation's gas production associations have promised to boost this year's flow to 28.95 tcf.

There's little doubt that the Soviet Union will produce over 30 tcf of natural gas in 1990.

During recent years, a growing number of Soviet authorities have called for an early end to ever higher official annual oil production goals. Now some scientists and economists openly favor a reduction of crude/condensate flow from present levels.

Writing late in 1988 in the official government newspaper Izvestia, Academician Aleksandr Sheindlin declared, "It is now possible to say that it would be economically advisable for our country to lower oil production immediately. The money and resources now expended in efforts to increase or even to maintain the already irrationally high level of oil flow could be much more effectively invested in energy conservation."

Sheindlin emphasized that the U.S.S.R. now uses "2-3 times as much primary energy resources as the United States per unit of gross national product." He said that the United States is by no means the leading nation in energy conservation but even so has increased gross national product 25% since the 1973 oil crisis while cutting primary started up the world's first grassroots commercial Unicracking/feed preparation unit.

The pretreatment unit, engineered by Voest Alpine/Salzgitter Lummus G.m.b.H., produces hydrocracked vacuum gas oil, gasoline fraction, and catalytic reformer feeds.

In Romania, the First of May Machine Building Plant in Ploesti built a drilling rig of its own design rated to 15,000 m (49,212 ft).

The plant also builds rigs rated to 1,200-10,000 m (3,937-32,808 ft). Late in 1988, Romania prepared to tow its newest jack up, Saturn, to its first Black Sea location. Built at Galati on the Danube River, Saturn is the seventh jack up built by the country.

The first six are working in the Black Sea. [IPE]

YUGOSLAVIA'S new bulk polyvinyl chloride plant at Zadar, Croatia, will handle 40,000 tons/year. Photo courtesy Technip.

by enormous gains in natural-gas production

energy consumption per unit of GNP by about one third.

Although Sheindlin's ideas on fuel conservation are gaining support among top level scientists and high government officials who are promoting "perestroika" (economic restructuring), most Moscow planners continue to push for greater production of oil, gas, coal, and nuclear energy. The Ministry of Geology still insists that hydrocarbon production should be increased as fast as possible for the indefinite future.

If Soviet oil production doesn't increase in 1989, the main reason is more likely to be the lingering effects of the ethnic unrest and strikes in the Baku area of Azerbaijan during fourth quarter 1988 than a deliberate government effort to restrict nationwide crude/condensate flow.

Onshore Azerbaijan, where most of the ethnic violence and economic turmoil occurred in fall 1988, is no longer a major Soviet oil producing area. But the Baku district still manufactures up to 60% of the U.S.S.R.'s oil field equipment. An official of Azerbaijan's Petroleum Machine Building Association declared recently that during one period late last year shipments of equipment to oil fields in western Siberia's Tyumen Province and to Tataria, the biggest oil producing area in the Volga-Ural region, were reduced by 50%. Delivery of pumping units and christmas tree equipment was particularly hard hit.

"Each day our association lost more than 2 million rubles of production," the spokesman said.

Strikes in Azerbaijan's petroleum equipment manufacturing plants, together with transportation tieups which prevented materials from reaching the facilities and halted shipments of finished goods, will likely delay placing hundreds of new western Siberian wells on production during 1989. The shortage of equipment made by Baku area factories will also cripple workover activity at oil and gas fields throughout the U.S.S.R.

The U.S.S.R.'s Council of Ministers reported that "enormous damage" was done during fourth quarter 1988 to both the Azerbaijan and national economies.

A state of emergency and curfews were imposed in the Baku area. Special around-the-clock police and military protection was provided for oil and gas fields, pipelines, electric transmission lines, pumping and compressor stations, fuel storage facilities, railroads, and airports.

Authorities said that strike agitators were active in the big Novo-Bakinsky refinery and Azerbaijan's important petrochemical industry. Tens of thousands of ethnic Armenians, including many petroleum industry workers, fled from Azerbaijan; and some were abruptly fired from their jobs as potential troublemakers.

Onshore Azerbaijan, which in 1940 produced 445,000 bo/d—by far the largest volume of any area in the U.S.S.R.—now yields less than 70,000 b/d, mostly from stripper wells. The republic's unrest probably cut this onshore flow by no more than 10,000 b/d.

Offshore Caspian Sea fields, which did not exist before World War II, now account for about 75% (over 200,000 b/d) of Azerbaijan's crude/condensate production and more than 90% of its gas flow. Despite attempts by strikers to prevent movement of men, equipment, and supplies from Baku to Caspian Sea drilling and production platforms, their disruptive activities had comparatively little effect on Azerbaijan's offshore oil and gas flow, according to official reports, although a number of Armenian workers quit their offshore jobs.

Chronology of events for 1988

JANUARY

THE MARKET: Quota busting and discounting doom OPEC's $18/bbl target crude price. Arab light sells for less than $16/bbl on the spot market in New York. One month West Texas intermediate (WTI) futures range between $15.30/bbl and $15.65/bbl.

GOVERNMENT: Australian oil market deregulation takes effect.

LNG: Cabot Corp.'s Distrigas subsidiaries receive their first shipment of Algerian LNG to the U.S. in more than 2 years...President Reagan signs an order allowing export of Alaskan North Slope gas as LNG.

KEY DISCOVERIES, NEW PRODUCTION: CNG Producing reports uncontrolled flows from 1-32 Cottonwood Creek, a deeper zone wildcat in southern Oklahoma's old Hewitt field, of 3,000-4,000 b/d of oil and 2.5 MMcfd of gas...BP Petroleum Development tests 28 MMcfd of gas and 447 b/d of condensate at Hides-1 in Papua New Guinea...Madagascar yields its first significant flows of light hydrocarbons at a west coast well operated by Petro-Canada International Assistance.

MERGERS, RESTRUCTURING: Esso Resources Canada receives approval to buy Sulpetro for $780 million.

LANDMARKS: Petrobras of Brazil installs a subsea tree at 3-RJS-376-D in 1,613 ft of water in the Campos basin, a world water depth production record.

MISHAPS: An Ashland tank collapses during filling at Floreffe, Pa., spilling 750,000 gal of diesel fuel into the Monongahela River.

FEBRUARY

THE MARKET: OPEC output slides as buyers demand discounts and members adopt "flexible pricing." Spot Arab light price drops to $14.45/bbl; WTI futures price reaches $17.66/bbl then falls to $15.75/bbl.

GOVERNMENT: Australia relaxes limits on foreign interests in oil and gas projects.

DISCOVERIES, PRODUCTION: BHP estimates a third Timor Sea discovery, 3 Skua, pushes oil reserves in the Jabiru-Challis area to 100 million bbl.

OPEC MOVES DOWNSTREAM: Abu Dhabi International Petroleum Investment buys 10% of Spain's Cepsa refining group.

TRANSPORTATION: Celeron's All American Pipeline begins continuous deliveries of crude from California to West Texas.

LNG: Distrigas and Sonatrach agree to resume long term shipments of LNG to the U.S.

MERGERS, RESTRUCTURING: Arkla acquires Entex, becoming the third largest U.S. gas distributor...British Petroleum receives U.K. approval to take full control of Britoil...ARCO plans to integrate Tricentrol into its British subsidiary.

LANDMARKS: U.S.S.R. reports world record production for 1987—12.48 million b/d of crude and condensate and 25.66 tcf of gas...Alaska replaces Texas as the No. 1 oil producing state.

MARCH

THE MARKET: Venezuela seeks a meeting of OPEC's price monitoring committee. Arab light prices swing between $13.30 and $14/bbl, WTI futures between $16 and $17/bbl. Spot gas prices plummet 50¢/Mcf in California to $1.65/Mcf.

RESOURCES, RESERVES: USGS trims its estimate of U.S. undiscovered conventional oil resources to 20.2 billion bbl in the Lower 48 and 13.2 billion bbl in Alaska.

NEW FUELS: Venezuela plans commercial production of Orimulsion, a blend of extra-heavy crude, water, and chemicals that can replace coal as a boiler fuel.

MERGERS, RESTRUCTURING: Fletcher Challenge buys New Zealand's 70% interest in Petrocorp in a deal worth $1.175 billion...Kuwait Investment Office boosts its stake in British Petroleum to nearly 22%...Total Minatome announces plans to buy CSX Oil & Gas for $612 million...Texaco wins shareholder, court approval for its plan to emerge from bankruptcy, including a $3 billion payment to Pennzoil to settle the Getty lawsuit.

APRIL

THE MARKET: OPEC's price monitoring committee meets amid reports of price discounting by members; separately, group members meet with non-OPEC exporters. Arab light spot price climbs to as high as $15.45/bbl near month's end from $13.80/bbl at the beginning. WTI futures price peaks at $18.60/bbl.

GOVERNMENT: Nigeria takes steps to privatize Nigerian National Petroleum Corp.

DISCOVERIES, PRODUCTION: ARCO gauges more than 9 MMcfd of sweet gas in a deeper pool wildcat near Wilburton, Okla...U.S.S.R. discloses that Tengiz oil field near the Caspian Sea's northeast coast has proved plus probable reserves exceeding 18.25 billion bbl.

LANDMARKS: OCS Sale 113 attracts bids for a record 684 tracts in the Gulf of Mexico...Amoco and Exxon confirm an oil and gas discovery in a world record 3,560 ft of water on Viosca Knoll Block 956 in the Gulf of Mexico.

PERSIAN GULF WAR: The U.S. destroys four platforms in Iran's Sirri and Sassan oil fields in retaliation for mining of a Navy ship. Iran attacks the Mubarak platform off Sharjah.

MISHAPS: Petrobras' Enchova platform off Brazil catches fire and cuts Campos basin production by 62,000-75,000 b/d...A ruptured storage tank drain line allows 4,200 bbl of heavy crude to spill into marshlands at Shell Oil's Martinez refinery near San Francisco Bay.

MERGERS, RESTRUCTURING: Occidental Petroleum changes the name of its Cities Service subsidiary to OXY Oil & Gas USA.

MAY

THE MARKET: OPEC rejects an offer from six non-OPEC producers to cut output in exchange for lower quotas. Crude price ranges: Spot Arab light $14.40-14.80/bbl, WTI futures $17.14-17.80/bbl.

MERGERS, RESTRUCTURING: TWA Chairman Carl Icahn offers $14.58 billion cash for common shares he doesn't own of Texaco, which rejects the bid...Tenneco offers to sell its nonpipeline petroleum assets.

DISCOVERIES, PRODUCTION: Agip-Chevron-Texaco group gauges 26,000 b/d from seven zones in Huizhou 26-1-1 in the South China Sea...South Africa reports its first commercial oil strike at a Soekor well testing 6,000-7,000 b/d in Mossel Bay.

LANDMARKS: Chukchi Sea Sale 109 nets 653 bids for 351 blocks off Northwest Alaska, a record number of bids for OCS sales outside the Gulf of Mexico.

MISHAPS: A catalytic cracker at Shell Oil's Norco, La., refinery explodes and burns, killing one person and injuring 22.

PERSIAN GULF WAR: Iraq launches its most destructive air raid yet, hitting Iran's Larak Island transshipment facility.

GOVERNMENT: FERC approves the first gas inventory charge for an interstate pipeline...Canada unveils a grant program worth as much as $200 million/year for small oil companies.

RESOURCES, RESERVES: DOE says in a report that 1.059 quadrillion cu ft of gas can be recovered conventionally, mainly through infill drilling, in the U.S. Lower 48.

JUNE

THE MARKET: OPEC ministers extend the first half quota of 15.06 million b/d for members other than Iraq into the second half. Arab light price slides to less than $13/bbl by months' end. WTI sinks to $15.15/bbl.

LANDMARKS: Shell Offshore sets 1,365 ft tall Platform Bullwinkle in 1,353 ft of water in the Gulf of Mexico, setting a world water depth record for platform installation...ARCO unveils the first retail methanol pump in the U.S. in Los Angeles.

MERGERS, RESTRUCTURING: Texaco plans to sell Deutsche Texaco to a West Germany utility company...Nova of Calgary acquires Polysar Energy & Chemical...Texaco defeats Icahn in a proxy fight for board control.

OPEC MOVES DOWNSTREAM: Texaco and Aramco Services plan a joint venture to own 50% of Texaco's three biggest refineries and retail operations in 23 states.

JULY

THE MARKET: The June OPEC accord begins to unravel as key producers ignore quotas. Arab light falls as low as $12.60/bbl.

WTI futures range: $14.50-16.40/bbl.

MERGERS, RESTRUCTURING: Saudi Arabia creates Petroref to operate Petromin's holdings in three export refineries and three domestic plants...Reading & Bates plans to sell its oil and gas assets and move headquarters to Houston from Tulsa...Sun Co. restructures, expanding downstream in the eastern U.S. and spinning off U.S. exploration and production operations.

MISHAPS: A U.S. Navy warship patrolling the Persian Gulf, shoots down an Iranian airliner it mistakes for an attacking warplane, killing all on board...Explosions and fire destroy Occidental (Caledonia)'s Piper Alpha platform in the U.K. North Sea, killing 167 persons. It's the offshore producing industry's worst accident ever.

PERSIAN GULF WAR: Iran says it's willing to accept a cease-fire in its 8 year long war with Iraq.

LANDMARKS: Amoco Norway drills the highest angle well in the Norwegian North Sea, a Valhall hole with an average inclination of 73° and maximum angle of 76.23°...Placid Oil breaks the world water depth record for subsea completions with the 31-4 well in 2,234 ft of water in the Gulf of Mexico.

AUGUST

THE MARKET: OPEC's price monitoring committee meets but takes no action on quota violators. Arab light at times trades between $12.80 and $13.45.bbl. WTI futures price swings between $15 and $16/bbl.

MERGERS, RESTRUCTURING: Elf agrees to buy most of Roy M. Huffington's U.S. oil and gas holdings...Texaco offers to sell its 78% stake in Texaco Canada...Gulf Canada acquires Asamera.

GOVERNMENT: The U.S. windfall profit tax meets its doom when the Senate passes a trade bill approved earlier by the House and supported by President Reagan. The bill has a repeal provision.

PERSIAN GULF WAR: Iran and Iraq agree to a cease-fire.

LNG: Phillips and Marathon win a 15 year extension to their LNG export permit from Alaska's Cook Inlet to Japan.

LANDMARKS: Cape Horn Methanol Ltd. starts up a 750,000 metric ton/year export methanol plant at Cabo Negro, Chile, the world's largest single unit of its kind...BP completes the first horizontal well drilled and tested from a semisubmersible in the U.K. North Sea at 16/28c-11 in Cyrus oil field.

SEPTEMBER

THE MARKET: Quota busting pushes OPEC production to 19.5-2 million b/d. Arab light drops as low as $10.40/bbl, WTI futures $13.40/bbl.

MERGERS, RESTRUCTURING: Amoco Canada completes the $5.5 billion (Canadian) purchase of Dome Petroleum...AOC Acquisition, part of Toronto's Horsham Corp., agrees to buy Clark Oil & Refining from bankrupt Apex Oil for $650 million.

DISCOVERIES, PRODUCTION: Agip SpA and National Oil Co. begin production from Bouri oil field, Libya's first offshore development project...Phillips announces plans to expand the Ekofisk oil field waterflood in the North Sea.

LANDMARKS: Standard Alaska Production completes the world's first long radius horizontal gas injection well. It's in the Eileen West End area of Alaska's Prudhoe Bay oil field.

RESERVES, RESOURCES: EIA says U.S. oil reserves increased 1.4% to 27.256 billion bbl at yearend 1987, mainly due to revisions for Alaska. Gas reserves dropped 2.3% to 187.211 tcf.

GOVERNMENT: A U.S.-Canada free trade agreement clears its last hurdle in the U.S. when approved by the Senate. In Canada, opposition senators balk, leaving the issue up to voters in a general election.

MISHAPS: Odeco's Ocean Odyssey semisubmersible explodes twice and burns while drilling a high pressure gas prospect for ARCO in the U.K. North Sea, killing one person.

OCTOBER

THE MARKET: OPEC's price monitoring committee refuses to recommend a ministerial meeting. OPEC production continues to climb. Arab light spot trades at less than $11/bbl during most of the month, at times dropping below $10/bbl. The WTI futures price drops as low as $12.60/bbl.

TRANSPORTATION: Statistics Canada projects 1988 Canadian natural gas exports to the U.S. of 1.2 tcf, up 20% from 1987.

LANDMARKS: Placid starts the Gulf of Mexico's first floating production system in 1,462 ft of water, Green Canyon Block 29.

GOVERNMENT: The U.K. tells Kuwait Investment Office to cut its interest in British Petroleum to 9.9%.

OPEC MOVES DOWNSTREAM: Venezuela's Pdvsa agrees to purchase the 50% Champlin Refining interest it doesn't already own.

MERGERS, RESTRUCTURING: Tenneco receives more than $7.3 billion in successful bids for its exploration/production and refining/marketing assets...Coastal and China National Chemicals Import/Export Corp. form a joint venture to own and operate Coastal's West Coast refining/marketing business...Burlington Northern plans to spin off Burlington Resources.

DISCOVERIES, PRODUCTION: China sets preliminary plans to develop giant Suizhong 36-1 heavy oil field in Bohai Sea at a cost of as much as $1.35 billion.

NOVEMBER

THE MARKET: OPEC ministers agree to a production quota of 18.5 million b/d, higher than the earlier quota but well below recent production. Arab light spot prices swing between $9.80/bbl and $12.45/bbl. WTI futures price mostly stays below $15/bbl.

LNG: Panhandle Eastern signs a contract to sell Citrus Trading large volumes of Algerian LNG...Shell Oil and Columbia Gas System plan a joint venture to import LNG.

LANDMARKS: Placid breaks the world water depth production record when the Green Canyon Block 31-4 subsea satellite well begins flow in 2,243 ft of water in the Gulf of Mexico.

MERGERS, RESTRUCTURING: Sun Co. completes the $513 million acquisition of Atlantic Petroleum, including a 130,000 b/d refinery at Philadelphia.

NEW FUELS: DOE completes the sale of the Great Plains coal gasification plant in North Dakota to Basin Electric Power Cooperative for a conditional $599 million...Consumers' Cooperative Refineries starts up Canada's first heavy oil upgrader, a $680 million, 50,000 b/d plant at Regina, Sask.

MISHAPS: A fire and explosion kill 22 persons at Bharat Petroleum Corp.'s 140,000 b/cd refinery in Mahul, India.

GOVERNMENT: The U.S. elects George Bush, a former drilling contractor, president...Canada reelects Brian Mulroney prime minister, paving the way for Canadian passage of the free trade agreement with the U.S.

OPEC MOVES DOWNSTREAM: Abu Dhabi receives approval to raise its 10% stake in refiner Cepsa of Spain to 15%.

DECEMBER

THE MARKET: Key OPEC producers indicate they'll cut production to new quota levels. Arab light rises to $13/bbl. The WTI futures price at times exceeds $17/bbl.

MERGERS, RESTRUCTURING: Wintershall plans to shut down its 70,000 b/d Mannheim, W. Germany, refinery and discusses a refining/marketing joint venture with Saarberg Oel & Handel-...Mesa Limited Partnership completes purchase of Tenneco's Midcontinent oil and gas subsidiary for $715 million...Mobil completes its $590 million purchase of Tenneco's 145,000 b/d Chalmette, La., refinery...Sohio Oil buys Mobil's 79,000 b/sd Ferndale, Wash., refinery and interest in Olympic Pipeline for $152.5 million...Gulf Canada agrees to buy the U.S. assets of Home Petroleum for $227 million.

RESERVES, RESOURCES: Yearend worldwide reserves of crude oil and lease condensate increase by 2.3% from yearend 1987 to 907.44 billion bbl, according to Oil & Gas Journal. Gas reserves increase by 4.2% to 3.955 quadrillion cu ft.

OPEC MOVES DOWNSTREAM: Venezuela's Pdvsa agrees with Unocal to form a joint venture that will own Unocal's 147,000 b/d refinery at Lemont, Ill.

DISCOVERIES, PRODUCTION: Amoco 1 Zipperer gauges 48.8 MMcfd of gas in Pittsburg County, Okla., 13 miles southwest of ARCO's Wilburton discovery.

MISHAPS: The Rowan Gorilla I jack up sinks in the North Atlantic while en route from Halifax, N.S., to the U.K. North Sea. No one is injured. IPE

Traveling abroad with common sense

WHEN YOU TRAVEL ABROAD on business, your mind is generally on the work you plan to do when you get to your destination. As a result, some of the routine travel chores you should handle before you go, as you go, and after you get there often get scant attention.

The time to organize all that is before you leave, so you can concentrate on your work after you get there. Don't try to remember all those minor things you have to do that concern traveling away from home and from place to place.

Simply make up a list, check off the things as they are done, and forget about them.

If you don't, you're bound to forget some things. Sure, you remembered to stop the newspaper, and arranged to have someone pick up your mail. But did you check to make sure your insurance policies wouldn't expire while you were gone?

Did you leave a key to your house with a trusted neighbor in case firemen need to get in to fight a fire caused by faulty wiring, or to let the police in to check that report of someone seen prowling around in your vacant house?

Did you arrange for someone to feed your pets, to water your plants, to mow your lawn if you're going to be gone for some time?

Did you check the thermostat just before you left to be sure you weren't spending a lot of money heating or cooling an empty house?

Did you empty the refrigerator to avoid coming back to a smell that will linger for days?

And I'm sure you checked to see if you had your airline ticket, your passport, and your credit cards, but did you make xeroxes of those things to leave with a relative, and take the other copy with you?

Did you bring your address book with phone numbers you might need at home or away (a tip: one of those

"calculator" watches will store up to fifty names and phone numbers right there on your wrist, and you can call them up anywhere in the world at the touch of a button).

You remembered your eyeglasses, of course, but did you tuck away in your suitcase the eyeglass PRESCRIPTION in case you break or lose the originals?

Don't forget a sewing kit, a number of pens, a small magnifying glass to read that tiny lettering on foreign street maps, a convertor to let you use your electric razors and such in strange and varied electric outlets, a list of the prescription drugs you might have to use in case something happens to the supply you brought, and a small electric flashlight for use in blackouts that have been known to occur with great regularity in some countries. Some folks even take along a portable smoke detector to give themselves peace of mind at night.

If you're going to be away for awhile, it might be just as well to make up a little medical package. In it might go aspirin, antidiarrhea medicine, antibiotic ointment for cuts and scratches, calamine lotion for insect bits and sunburn, adhesive tape, gauze bandages, bandaids, Dramamine, sunblock, decongestant, insect repellent, a thermometer, and perhaps even antimalaria pills, halazone for purifying water, and oil of cloves to relieve toothache. Chances are you aren't going to need all that stuff, but if the need does arise, you may be glad you went to the trouble of bringing it along.

Peace
of mind

And speaking of peace of mind, try not to set yourself up as a mark for those larceny-minded folks you'll find in every country. Do a little preplanning. Before you head out to the airport, for instance, put together in one pocket only the things you'll need at the time — your passport, your transporation money, your tips, your airline ticket — and put them someplace handy. DON'T put them in your wallet or purse where you'll have to fumble through everything right out in public to get them.

You should always put your name and address inside your luggage, of course, but DON'T put your address on the outside, only your name and phone number. Too many thieves will be glad to know that the person at the address he reads on the outside of luggage will be conveniently out of town for awhile.

Keep thinking ahead all the time. When reserving seats, try to sit as far forward as you can, so you can beat that mob off the plane.

If you've got a lot of work to do, one of those seats just behind a bulkhead will often give you more room to spread your work out, but you run the risk of having a crying baby for a neighbor, since airlines like to assign those seats to parents with children.

Be realistic — if you hear your flight may be held up, get to a phone and CALL your airline, rather than waiting in that long line at the counter. If you can't get any satisfaction, start calling other airlines.

Using
common sense

Once you arrive at your destination, using common sense will keep you fit and clear-minded for handling your business chores, instead of just trying to survive the trip itself. At first, stick to foods your stomach is used to, until your body gets a little practice at strange cuisine. Remember that alcohol in a drink won't kill the germs in the water, and that salads with raw fruits and vegetables, while good for your figure, may be anathema to your stomach, since food just washed in water doesn't rid itself of bacteria. Eat fruit that you peal YOUR-SELF.

And try to be sensible in other ways. A sign posted in kilometers doesn't require a calculator to figure out. Just multiply it by six and remember that kilometers are only a little more than half a mile long.

A traffic sign posted at 60 km means 60 X 6 = 36 — in other words, about 36 miles.

In a restaurant in which English is not spoken to any great extent, don't use a long sentence which the listener must sort out.

Don't say, "Would you be so kind as to bring me a glass of water, please? I'm thirsty." The waiter may spend considerable time trying to find the operative word in that sentence. Instead, just say "Water", which is much easier to manage for someone not fluent in English.

In fact, if you carry a notebook and pencil, writing the word down (printing it) may be easiest of all.

And it might not be a bad idea to get the waiter to write down the PRICE of the meal to be sure you know what you're getting into.

Sometimes understanding a few foreign phrases can be quite helpful. In Egypt, for instance, IPE mistakenly listed the words for the men's restroom as Hammam Sayyedat, and that for women as Hammam Regal. Eng. Rafiek Nosseir, OPS. Director of Sumed Co., however, immediately pointed out that it is just the opposite — Hammam Regal for men, and Hammam Sayyedat for women. That's the kind of a mistake you wouldn't want to make more than once.

The Samoan Hideaway magazine once came up with Ten Commandments for Travelers. Most of them had to do with traveling for pleasure, but the business traveler should commit two of them to memory.

They are: Blessed is the person who can make change in any country, for lo, that person shall not be cheated.

And most important of all: Remember thy passport and keep it handy — for a person without a passport is a person without a country.

CAMEROON

IF YOU WANTED TO VISIT AFRICA ALL AT ONE TIME, you could just about do so by coming to Cameroon.

It has all the traditional wild animals, for instance, from giraffes to elephants. It has sandy beaches, steaming jungles, mighty rivers, pygmy tribes,and wind-blown grassland areas.

All this with a French flavor, although both English and French are official languages -- try to imagine a language situation much like that of Canada. And, of course, there are some 200 local dialects throughout this diverse country. Tip: If you go to a movie, expect the subtitles to be in French.

More than 8 million persons are spread out over a triangular area about two-thirds the size of Texas, but these people are made up of some 286 different groupings, with a heavy Muslim concentration in the north, and animist and Christian religions tending toward the southern areas.

Next year marks the third decade of independence for Cameroon (carved out of areas colonized by the French and English, which explains the bilingual official languages), and its 18th year as a republic. Developments in both oil production (since 1978), industry, and agriculture makes the economy basically stable, with growing production of lumber, aluminum, textiles, timber, and cotton. Plantations raise bananas, palms, pineapples, tea, and rubber. Cameroon is the fourth-largest producer of cocoa in the world, and the second-largest producer of coffee in Africa. In markets, vendors sell plantains, papayas, yams, sea bass, and chickens. One can even obtain for supper a bit of crocodile meat, or monkey, or porcupine. It's a busy place.

Getting around

It takes about an 11-hour flight to get from New York, say, to the port city of Douala. If you go in July, it'll probably be raining when you get there — it rains 24 days out of the month around that time — but there are plenty of taxis to take you into the city center of Douala, some 8 miles away. To get into Cameroon, you'll need a passport and a ticket out of the country. No visas are needed for short stays. Smallpox and yellow-fever vaccinations are a good idea. Hotels in Douala you might like are Akwa Palace, Box 4007, or Hotel des Cocotiers, Box 310.

Although Douala is the largest city in the country, with more than a million persons, the capital is Yaoundé, farther inland. It is a thoroughly modern city of more than 650,000 persons. There, you might try for accommodations at the Indepencence Hotel, Box 474, or the Sheraton-Mont Febe Palace, Box 4178.

Getting around is not difficult. Taxis are everywhere, but they work on a fixed charge instead of using meters, so be sure you understand the cost to your destination before you start off. There are car rentals available. Roads are not too good — there are only about 1,500 miles of paved roads in the entire country, and the unpaved ones can get pretty tricky, especially during the rains of July and August. The Cameroons are beginning to focus on a big new road-building program, however; lack of roads keeps getting in

the way of development of agriculture and mining, since it's hard to get products to market. Expect to find a lot of road-construction signs if you drive much. An example of the need for an expanded road system is the aluminum smelter at Edéa. It is using IMPORTED alumina, even though more than a million tons of the stuff lies unmined in the northwest part of Camaroon, simply because there's no way to get it to the smelter.

There are also a few railroads for travel, but the best way to get around from city to city is probably Cameroon Airlines, Avenue De Gaulle, Douala.

Business hours

Banks stay open from 7:30 to 11:30 in the morning, then close down to open again at 2:30 for an hour. They close on weekends. Shops operate from 9 to 12:30 each morning, close for 3½ hours, and then open again until 7 at night.

Cameroon is a good place to pick up traditional African items such as carved masks, pottery, wickerwork, and table linens. In Douala, try Ali Baba, Galeria Africa, and Reine Sherman's, as well as the markets at New Bell and Deido.

Speaking of traditional activity, you can go safari hunting by paying a firearm tax and getting a permit from the Secretariat Rural, S/c du Directeur des Eaux et Forêts in Yaoundeé.

If you want someone to handle the whole thing for you, contact Safaris Jacques Guin, BP 9, Maroua. Maroua is near the Waza Wild Life Reserve.

Arangements for traveling about in Cameroon can be made through Camvoyages, 15 Blvd. de la Liberete, Box 4079, in Douala.

For additional information, try the General Delegation for Tourism, Yaoundé, or the Cameroon Embassy, 2349 Massachusetts Ave. NW., Washington D.C. 20008, or at 84 Holland Park, London Wll.

INDIA

THE FIRST THING YOU MAY NOTICE WHEN YOU ARrive in India is that there is a festival going on. That's because there is a festival almost ALWAYS going on — eight major ones in August alone, for example.

India is a religiously oriented country (84% Hindu) and you are reminded of that almost everywhere. You'll be expected to take off your shoes before entering temples and mosques (you can rent some "substitute" shoes most such places, but stockinged feet are considered okay).

You have to keep remembering little things, like not taking items made of leather into such places, and not taking pictures inside temples.

But Indians don't let such traditional beliefs slow their progess in the modern world. Although India's economy rests largely on agriculture, it is also among the top 20 industrial nations of the world.

It doesn't export much, but that's because Indians themselves use up much of what they produce. Its democratic political structure and a large skilled labor force make it attractive to industry.

You'll need a passport to visit India, along with a tourist or transit visa. But be sure to get your visas BEFORE you get to India, because it's hard to get one after you arrive. Get it from Indian embassies or consulates, and try to handle it in person, because it sometimes takes a long time to get it wrapped up by mail.

If you need to extend your stay after you get there, apply to the Foreigners Registration Offices in New Delhi, Bombay, Calcutta, or Madras. If you can't get to any of those places, try the Superintendent of Police in whatever city you are located.

There are a good many places in India — places like the Andaman Islands, or West Bengal, or Nepal or Bhutan — that require special permission to visit.

Unless you're coming in from Africa or South America, you don't need a yellow-fever inoculation, but for your own protection, you'd better be loaded to the gills with shots for smallpox, typhoid, tetanus, hepatitis, and poliomyelitis. You MUST have a certificate of vaccination against cholera or you can't get in.

Stay clear of drinking unboiled water, and that includes using the ice in smaller towns, which is often made with unboiled water.

And there's a couple of things to remember at customs when you enter the country. If you're taking along typewriters, recorders, or such, be sure to declare them upon your arrival, so you can get them out again. You'll also want to make a currency declaration if you're taking in more than $1,000.

All this requires a bit of time, so allow for it in your planning.

Getting around

You'll probably fly in to New Delhi, Bombay, Calcutta, or Madras, since more than 50 airlines serve these facilities.

These airlines make getting around India by air remarkably easy. Indian Airlines has a tremendously comprehensive domestic network, along with a modern reservations system that has speeded up travel considerably. When catching a plane, however, remember that there are still a lot of security checks, which means you probably should get to the airport an hour or two before departure time.

A shuttle bus sent by your hotel is the best transportation from these airports.

You can use taxis, too, but be careful to use only the yellow-topped black taxis, since these are not only licensed, but also charge fixed fares. Just be sure the meter is set at zero before the flag is pulled down. Also, realistically, some taxis charge more than the meter says. But they're not being sneaky — they'll have a sign up in their vehicle showing how much over the meter fare they charge!

Getting around a city by bus isn't generally a good idea, since they are usually quite late and already stuffed with passengers, particularly during rush hours.

You can't rent a car in New Delhi to drive yourself, which is just as well considering the conditions of the road and the traffic problems. But you can rent a taxi for a full 8 hr for about $25. You can even use one of those little metered three-wheelers you'll find almost everywhere, at about half the cost of taxis.

But railroads are the symbol of Indian transportation. It has the second-largest railway system in the world. And if you can keep your head, it is both safe and comfortable.

Reservations are an absolute must. You can get help with reservations and itinerary planning at Tourist Guide Offices at most major stations —In New Delhi, for instance, at Baroda House, Copernicus Road, at Indian Gate.

You might do well to get an Indrail Pass, which must be paid for in non-Indian currency, and which takes you almost anywhere.

Get them through travel agents, or at Railway Central Reservations Offices in New Delhi (on State Entry Road, tel 344-877), Bombay, Calcutta, Madras, Secunderabad, and Hyderabad.

Stick to the air-conditioned first class, the two-tier air-conditioned sleepers, or the air-conditioned chair cars. Anything less than that can put you in dusty, uncomfortable surroundings for the course of your journey.

Use express trains, because most others are remarkably slow. Get to the station at least an hour ahead of time, and scout around for a list that will be posted somewhere which will give your name and the seat/berth number you've been assigned.

On the journey, you can order food from the coach attendant, as well as tea, coffee, and soft drinks.

Things to consider

Although Telex services are quite reliable, using the telephone to conduct long-distance affairs can be nerve-shattering.

It is not unheard of to place a long-distance phone call at one time and have it go directly through, while a day later trying to place the same call might take as long as 24 hours to complete. The best way is to have your hotel operator handle the process, and get in touch with you when she has it set up. There is a service called a "lightning" call that often speeds things up, but it costs eight times as much as a regular call.

Banks are open 4 hr a day during the week, and for 2 hr on Saturday morning, although the State Bank's tourist counter is open from 5 to 7 each evening. You can find that in G Block, Connaught Place, in New Delhi. At the airport a bank branch is open 24 hr a day. Stick to regular banks and hotel currency-changing facilities as a rule.

Not only is it illegal to change money elsewhere, but you've got a good chance of getting stuck with phony Indian money.

You can get plenty of English-language newspapers in the major cities, including the Times of India, the Economic Times, and the Statesman.

About tipping: handle it as you would most places, except that you needn't overtip, and taxi drivers aren't usually tipped at all.

By the way, if you're staying at a private home, be sure to check with your host to see if it's okay before tipping any of the domestics.

Hotels generally do a good job on laundry and dry-cleaning, without costing an arm and a leg.

You might want to get an adapter for your electrical appliances.

Standard in most places is 220 v.

You can't find film too easily, so it's probably a good idea to take a full supply with you when you go. Color processing takes at least a week, but you might be able to speed up things.

In New Delhi, try Mahatta & Co., 59M Block, Connaught Circus, New Delhi (tel 353-0008) for film processing.

If you're looking for the local chapter of Lions, Rotary, or Kiwanis, check the phone book.

A cautionary note when ordering drinks: Although some states are completely dry, and others have "dry days", as a visitor you can get, through the Government Tourist Office, a permit that allows you to order and consume drinks almost everywhere. BUT YOU CANNOT BUY A DRINK FOR A FRIEND WHO DOESN'T HOLD SUCH A PERMIT!

There are, however, a number of hotels licensed by the government in which drinks are properly served. Stick to imported whisky, since local whisky can be pretty bad. Do try a Feni, made in India from cashew juice.

Nightclubs you might want to visit include the Supper Club at the Ashoka Hotel and the Bali Hi at the Maurya Sheraton, both in New Delhi.

When shopping, keep in mind that Government stores sell things generally cheaper than regular stores, but watch out for non-government stores that hang up signs saying vaguely they are "government approved" in the hope you'll think they ARE the government store.

The Kashir Government Emporium on Connaught Circus in New Delhi sells just about everything, and will take foreign currency in payment.

The most fascinating place to shop, probably, is the Chandi Chowk, a local bazaar in Old Delhi. Here, while you're dodging bicycles and beggars, offer the merchants about a tenth of their posted price, and you'll usually get it for about half the original price, after some pleasurable bargaining.

Where to stay; where to eat

India has a wide range of hotels, from five-star luxury affairs to one-star hotels for travelers on a tight budget.

Some are former palaces of maharajas. And in Kashmir, you can even stay on one of those lavish, world-famous house-boats.

In New Delhi, try the Ashok (tel 600-391), Claridge's (tel 301-0211), or the Oberoi (tel 699-571). In Calcutta, you'll like the Hotel Hindustan International on Jagadish Chandra Bose Road, and in Bombay try the Centaur, right there at the Bombay airport.

Where to eat is no problem, but picking from the tremendous variety is a pleasant chore.

In Bombay, try a dish called *Bhelpuri,* made up of puffed rice, flour, potatoes, and onion, seasoned with chutney. *Chapathis* consists of dough rolled thin and cooked like pancakes. For desert, you'll like *paan* — betel leaf with lime paste, stuffed with grated betel nut, aniseed, and cardamom.

You can get North Indian food at the Moti Mahal, located near the Delhi Gate in Old Delhi.

For a really memorable dining experience, however, go to Ahmarabad on the western seacoast, to the Vishalla Restaurant, where you get baked breads, fresh,white butter, jaggery, and yogurt served in large brass dishes to the sound of folk music, while you drink from pots molded by the local potter.

What to see

While at Ahmadabad, by the way, be sure to visit the shaking minarets of Rajpur Bibi's Mosque at nearby Gomtipur.

Pressure exerted inside the inner walls of the minarets sets them vibrating in a remarkable manner.

Certainly you must visit Agra's Taj Mahal, just 125 miles south of New Delhi.

You can make this very special trip on the "Taj Express", which leaves New Delhi at 7 in the morning and has you back by 10 that evening.

In the south of India, at Shravanabelagola, you can hardly avoid seeing the thousand-year-old statue of Lord Bahubalik, since it towers over the countryside and can be seen from 26 km away.

Who to contact

Government of India Tourist Offices can help you along your way before you leave.

There's one at 30 Rockefeller Plaza in New York, one in Dallas, at 13317 Kir Lane,75240 (tel 214-234-2233), one at 21 New Bond Street, London W1, and one in Australia, C/o Air-India, 440 Collins St., Melbourne VIC 300 (tel 03-602-3233).

Others who might help set up travel arrangements include American Express, International Banking Corp., Wengar House, Connaught Place, Box 537 (tel 344-119) in New Delhi, or Kai Travel Private Ltd., Jalan Sanur Beach, Bali Beach Hotel (tel 74-366) in Srigar, Kashmir.

MALAYSIA

SEVENTY-FIVE YEARS AGO, WHEN MALAYSIA FIRST entered upon the oil scene, it faced — and solved — a problem in typically Maylasian manner.

At that time, Miri Well No. 1 struck oil at about 450 ft. Within three years, the area was producing 488,000 bbl/year, but it was being moved out at a snail's pace. Trouble was that a great sandbar across the mouth of the Miri River (which is still there and still causing trouble) prevented tankers from moving in to load the oil. They had to anchor several kilometers out to sea, and the oil was laboriously brought out to them in small boats.

In typical fashion for Maylasia, a simple solution was worked out. A long undersea pipeline was laid out to sea, a flexible hose attached, and the crude was then loaded directly onto the tankers — a state-of-the-art solution for 1914!

They are still doing things in a Malaysian way. With an increased standard of living and industrial growth, city planners are keeping a close eye on the building that has been known to transmute pleasant traditional cities into garish "box" buildings with no character.

City planners now recommend that all buildings reflect Malaysian architecture — tradition-inspired, authentic Malaysian designs suited to a modern city (see an example in the photo at left).

You can get to the capital, Kuala Lumpur, from the U.S. west coast in about 18 hr flying time, and you'll arrive at Skubang Airport, about 14 miles from the city itself. You can catch a cab for about $5 that will take you straight to one of the many fine hotels available — perhaps the Federal, at 35 Jalan Bukit Bintang, or the Holiday Inn, on Jalan Pinang. Be sure you get your reservations well in advance, since they are sometimes hard to come by.

The same goes for business appointments — make them well in advance, and be sure to keep them on time.

If you have a couple of days before your first meeting, get some business cards printed up in both English and Chinese. That's because you more than likely will be dealing with wealthy and influential Chinese businessmen, who play a major role in Malaysia affairs, and they place a great deal of importance on how you respond to social and business contacts.

If you're going to deal with governmental officials, wear a conservative business suit for sure.

Remember that most businessmen take their vacations in December-February, so try to schedule your business meetings at other times.

Banks are open from 10-3 weekdays, and in the mornings on Saturdays.

Things to consider

Try to catch a taxi on the street; if you call for one, the meter starts when they leave to pick you up. You'll see a good many bicycle rickshaws around, but few foreigners use them.

If you decide to rent a car to drive yourself, remember that traffic goes on the left side, British style. Roads in West Malaysia are usually pretty good, except occasionally during the monsoon season, but roads in East Malaysia leave considerable to be desired.

There is a wide mix of races in Malaysia — Chinese, Indians, Kadazans, Muruts, Ibans, Bidayuhs, Europeans, and Malays — and this is reflected in the food they eat. You'll like *satay,* which are pieces of meat barbecued on bamboo skewers and dipped in sweet chili and peanut sauce.

There's also shark's fin soup, steamed dumplings, salted eggs, and a kind of spaghetti called *Meehoon.* Your best bet is to go to the Bintang Restaurant at 44 Jalan Sultan Ismail (tel 425-151) for traditional Malay food. In eating places away from the main cities, you might see if they have any of the freshly brewed coconut beer which some travelers swear by.

And try The Hut, in Petaling Jaya, a restaurant offering shows with folk and classical dances, shadow plays, and Malay music.

For restrooms facilities in hotels and restaurants, men use the one marked Laki Laki, and women use the one marked Perempuan.

For a special treat, you might try the hill resort of Genting Highlands, some 30 miles north of Kuala Lumper, which is perched a mile high up on a hill. It has the only casino in Malaysa, an 18-hole golf course, swimming pools, waterfalls, cave temples, amusement parks, a flower nursery, an indoor gymnasium, and facilities for bowling and horseback riding. It is not inexpensive.

You can get help ahead of time in setting up your trip to Malaysia through the Malaysian Tourist Information Center, Transamerica Pyramid Building, 600 Montgomery Street, San Francisco, Calif. 94111, or at 17 Curzon Street, Mayfair, London W1.

OGJ400 shows 1-year breather for industry

THE U.S. OIL AND GAS PRODUCING INDUSTRY GOT A breather in 1987 at the downstream industry's expense. It was a 1-year trend.

The 1988 Oil & Gas Journal 400—annual snapshot of the biggest companies with significant operations in the U.S. ranked by assets—shows general oil company financial improvement in 1987 from the preceding year, when crude oil prices crashed. In 1987, the Organization of Petroleum Exporting Countries kept a lid on production, crude prices averaged nearly $18/bbl, and refining margins narrowed.

Then came 1988. Crude prices flirted with the $10/bbl level, triggering fears that the year would repeat the disastrous events of 1986. Upstream operations suffered. Refining, with feedstock costs low and products demand strong, soared to record profitability for some companies.

And it all occurred despite expectations throughout most of the industry that 1988 would continue the mild upstream recovery that began the year before.

What happened? OPEC lost—or intentionally relaxed—its grip on production. Key Persian Gulf members deliberately violated their quotas. Group production at times exceeded 20 million b/d—2 million b/d more than the official target. Falling prices devastated the already hard pressed OPEC members adhering to quotas, not to mention nonmember producers everywhere else. By the time ministers met in Vienna during November, there was speculation the group might lose members deciding they had nothing left to gain from staying in the cartel.

The meeting ended with a new ceiling—18.5 million b/d, higher than both the earlier target and expected demand for OPEC crude in the near term. At yearend 1988, there was no certainty that the new accord would remain in place or that it would work, especially if OPEC earned no complementary production restraint from nonmembers.

Once again, industry saw the harsh side of an unpredictable oil market. This time, though, companies—the large ones at any rate—seemed better prepared than they had been in 1986.

Years of capacity cuts and upgrades put refiners in position to profit from 1988's long periods of wide refining margins. During the first 9 months of the year, U.S. refiner crude acquisition costs fell 15.3% to an average of $15.13/bbl. Product prices climbed during the period, with motor gasoline, for example, climbing to 96.24¢/gal from 94.83%/gal in the same period of 1987.

And with demand strong, refineries operated at 84.8% of capacity in the first 9 months of 1988 vs. 83.2% in the same period of 1987.

What this shows is how quickly an oil industry financial snapshot can fade. The year portrayed by the current OGJ400 was poor by comparison for refiners and, if not good for producers, at least an improvement from 1986.

In 1987, average U.S. refiner acquisition cost of crude oil increased to $17.87/bbl. Product prices rose but not as much. Average refiner sales price of finished motor gasoline rose only 11.2% to 59.1¢/gal for the entire year. Average retail pump price of all types of motor gasoline climbed 2.8% to 95.7¢/gal.

Refiners shown on the 1988 OGJ400 to have been squeezed in 1987, therefore, may well have come back strong in the sharply different economic environment of 1988.

For producers, 1987 can hardly be called a recovery year, especially in view of what followed.

Independents appearing on the 1987 list did well just to have survived 1986. Most, in fact, were struggling to some degree, even with the year long respite from falling crude prices.

Mergers and other forms of restructuring continued to churn the list, especially below the 20 biggest companies. The 1988 OGJ400 had 76 companies that didn't appear the previous year. Of those, 39 were limited partnerships. The entire list included 63 limited partnerships and seven royalty trusts.

Only nine companies appearing in the 1987 list fell off in 1988 for not ranking among the top 400 companies in terms of assets. Other reasons for new names included mergers, asset sales, and name changes. One company from 1987 was taken private, three were in Chapter 11 bankruptcy in 1987, and one company was being liquidated. Assets represented by all companies on the 1988 list gained slightly from the 1987 list to $501 billion.

There is some double counting, however. Assets of limited partnerships, for example, may be partly included in assets of progenitor companies retaining significant partnership interests.

Total revenue for the 1988 list increased 5.6% from the prior year's list. And net income leaped 80% to almost $9.4 billion from the OGJ400 of a year earlier. The income increase was moderated by Texaco Inc.'s $4.4 billion loss in 1987, reflecting writedowns associated with a huge legal judgment against it in a case brought by Pennzoil Co.

Exxon Corp. remains the largest company on the OGJ400, with 1987 assets totaling $74.04 billion. The previous year, Exxon's assets totaled $69.48 billion.

Assets total for the smallest company on the 1988 list, Rotan Mosle Energy Fund III Ltd. with $646,000, was quite higher than that of the previous year's list, Imperial Energy Corp. with $149,000.

List-wide profitability indicators improved in 1988: return on equity 5.42%, return on assets 1.87%, and return on revenue 2.1%. But they remain low by historic standards.

Revenues of the 1988 group account for 9.9% of 1987 U.S. gross national product. Companies on the 1988 list represented 70% of U.S. liquids reserves, 62% of U.S. natural gas reserves, 64% of U.S. liquids production, and 67% of U.S. gas production.

Companies on the 1988 list reported capital and exploratory spending of almost $41 billion during 1987 or corresponding fiscal periods, down 5.46% from 1986 spending by companies on the 1987 list. The 1986 spending total was down almost one-third from its predecessor.

Thirty-seven companies on the 1988 OGJ400 met criteria for growth comparisons—stockholder's equity gain during the reporting year, positive net income for the reporting year and its predecessor, and net income growth in the reporting year. Subsidiary companies and newly public companies aren't included.

Only 17 companies qualified for the comparisons on the 1987 list.

Five fastest growers on the current list, and their assets rankings, are Castle Energy Corp. (264), Harken Oil & Gas Inc. (113), Hadson Corp. (97), Freeport McMoRan Inc. (25), and Swift Energy Co. (209). _{IPE}

Rank by total assets 1987	1986	Company	Total assets $000	Rank	Total revenue $000	Rank	Net income $000	Rank	Stockholders' equity $000	Rank	Capital & Expl. spending $000
1	1	Exxon Corp.	74,042,000	1	83,335,000	1	4,840,000	1	33,626,000	1	7,136,000
2	2	Mobil Corp.	41,140,000	2	56,716,000	3	1,258,000	2	16,783,000	3	2,942,000
3	4	Chevron Corp.	34,465,000	4	29,097,000	6	1,007,000	3	15,780,000	4	2,841,000
4	3	Texaco Inc.[1]	34,388,000	3	35,315,000	400	(4,407,000)	6	9,171,000	6	1,990,000
5	5	Shell Oil Co.	26,937,000	6	21,222,000	5	1,230,000	4	14,842,000	5	2,614,000
6	6	Amoco Corp.	24,827,000	5	22,388,000	2	1,360,000	5	12,107,000	2	2,979,000
7	10	BP America[2]	23,287,000	9	15,562,000	7	564,000	8	5,574,000	7	1,549,000
8	7	ARCO (Atlantic Richfield Co.)	22,670,000	8	17,608,000	4	1,244,000	7	5,878,000	9	1,463,000
9	8	Tenneco Inc.	18,503,000	10	14,957,000	398	(218,000)	11	3,773,000	10	1,291,000
10	9	Occidental Petroleum Corp.	16,739,000	7	17,746,000	12	240,000	10	5,164,000	15	757,000
11	13	Sun Co.	12,580,000	14	9,817,000	9	348,000	9	5,229,000	11	1,176,000
12	11	Phillips Petroleum Co.	12,111,000	12	10,917,000	36	35,000	14	1,617,000	16	750,000
13	12	USX (Oil & Gas Segment)[3]	11,586,000	13	10,074,000	8	[4]447,000	—	NA	12	964,000
14	14	Unocal Corp.	10,062,000	15	9,393,000	17	181,000	13	1,766,000	13	877,000
15	16	Enron Corp.	9,528,812	18	5,915,766	379	(29,296)	15	1,613,460	25	225,942
16	15	Conoco Inc.[5]	9,146,000	11	11,097,000	10	[6]277,000	—	NA	8	1,480,000
17	17	Coastal Corp.	7,989,300	16	7,536,000	19	113,100	26	859,800	35	172,300
18	18	Columbia Gas System Inc.	5,426,776	21	2,830,410	20	111,333	17	1,573,704	20	298,774
19	20	Amerada Hess Corp.	5,304,808	19	4,784,589	13	229,860	12	2,158,544	17	454,583
20	21	Transco Energy Co.	4,296,912	20	3,250,237	390	(61,311)	35	650,122	22	284,268
21	23	Ashland Oil Inc.[9]	4,058,025	17	7,311,273	18	133,432	22	1,065,399	18	388,932
22	22	Consolidated Natural Gas Co.	3,946,905	25	2,328,907	16	186,047	16	1,598,184	19	359,699
23	27	Enserch Corp.	3,336,470	22	2,820,167	40	30,485	25	958,156	26	215,381
24	25	Pennzoil Co.	3,304,707	26	1,836,302	31	44,256	36	621,274	30	198,668
25	31	Freeport-McMoRan Inc.	3,155,102	28	1,540,885	11	248,751	23	1,022,235	14	808,926
26	26	Sonat Inc.	3,115,015	30	1,498,684	21	105,032	24	1,002,665	32	189,996
27	28	Kerr-McGee Corp.	3,071,000	23	2,639,000	24	81,000	19	1,363,000	24	235,000
28	29	Mesa L.P.	2,680,960	56	352,930	39	31,907	20	[10]1,295,009	65	57,810
29	32	Arkla Inc.	2,524,402	37	1,191,837	25	78,283	30	750,811	34	176,015
30	30	Meridian Oil Inc.[12]	2,268,495	42	735,912	43	27,370	18	1,431,474	47	102,620
31	33	Mitchell Energy & Development Corp.[13]	2,112,139	49	561,882	73	8,755	39	561,895	45	104,529
32	35	Murphy Oil Corp.	2,066,590	29	1,517,638	384	(43,638)	31	750,066	38	162,932
33	34	Nicor Inc.	1,999,961	31	1,439,002	22	83,735	40	550,151	41	144,251
34	24	Maxus Energy Corp.[14]	1,900,500	43	691,000	399	(539,600)	61	197,000	31	197,500
35	37	Union Pacific Resources Co.[15]	1,883,000	33	1,421,000	15	206,000	33	704,000	29	208,000
36	38	Montana Power Co.	1,862,896	46	629,350	27	62,427	28	800,412	55	79,058
37	39	American Petrofina Inc.	1,835,889	24	2,510,411	23	83,284	32	731,281	27	215,103
38	36	Union Texas Petroleum Holdings Inc.	1,632,773	36	1,257,903	28	55,942	78	132,646	33	188,821
39	42	Louisiana Land & Exploration Co.	1,607,100	40	828,400	48	24,800	38	586,700	28	211,100
40	43	Union Exploration Partners Ltd.[17]	1,554,600	44	684,300	14	207,000	21	[10]1,226,700	36	165,800
41	46	Primark Corp.	1,491,060	34	1,370,673	327	(1,727)	45	426,959	52	81,753
42	45	Cabot Corp.[9]	1,463,132	32	1,436,485	42	28,694	37	590,741	51	94,813
43	48	Anadarko Petroleum Corp.	1,430,486	62	230,746	71	9,084	44	427,123	44	119,634
44	47	Global Marine Inc.	1,386,900	94	94,300	395	(131,500)	381	(49,700)	137	8,600
45	44	BHP Petroleum (Americas) Inc.[18,19]	1,358,004	65	209,402	38	33,332	74	146,322	82	37,475
46	53	Enserch Exploration Partners Ltd.[20]	1,180,733	63	225,082	29	51,779	27	[10]847,655	39	147,551
47	50	National Fuel Gas Co.[9]	1,119,374	41	743,067	32	42,492	47	393,490	43	127,303
48	49	Transco Exploration Partners Ltd.[21]	1,107,308	61	272,948	371	(13,579)	34	[10]673,443	48	101,893
49	62	Freeport-McMoRan Energy Partners Ltd.[22]	1,105,427	78	147,919	67	13,508	29	[10]790,948	61	69,803
50	60	Santa Fe Energy Co.[23]	[24]1,057,500	57	331,200	30	[25]46,500	—	NA	21	[24]293,100
51	52	Questar Corp.	1,046,747	50	497,359	46	25,930	42	472,217	50	97,86

| Worldwide liquids production | | Worldwide natural gas production | | Worldwide liquids reserves | | Worldwide natural gas reserves | | U.S. liquids production | | U.S. natural gas production | | U.S. liquids reserves | | U.S. natural gas reserves | | U.S. net wells drilled | |
Rank	Mill bbl	Rank	Bcf	Rank	Mill bbl	Rank	Bcf	Rank	Mill bbl	Rank	Bcf	Rank	Mill bbl	Rank	Bcf	Rank	Wells
1	637.0	1	1,436.0	1	6,634.0	1	22,365.0	2	276.0	2	735.0	4	2,634.0	2	9,761.0	7	374.0
7	227.0	2	1,429.0	8	2,397.0	2	20,287.0	8	112.0	6	610.0	8	1,024.0	3	7,741.0	3	646.0
3	329.0	5	816.0	2	3,348.0	4	9,444.0	6	185.0	3	708.0	5	1,809.0	4	7,078.0	9	283.0
2	355.0	4	885.0	3	3,089.0	5	7,847.0	5	197.0	1	748.0	7	1,603.0	7	4,783.0	2	699.0
8	214.0	6	665.0	6	2,915.0	6	6,914.0	4	204.0	4	656.0	3	2,782.0	5	6,541.0	1	1,266.0
5	291.0	3	1,022.0	7	2,580.0	3	15,400.0	7	142.0	5	617.0	6	1,670.0	1	10,690.0	4	629.0
4	310.0	20	123.0	5	3,026.0	11	3,890.0	1	309.0	20	105.0	1	2,953.0	17	1,932.0	17	122.0
6	266.0	7	523.0	4	3,041.0	7	6,751.0	3	243.0	7	464.0	2	2,818.0	6	5,854.0	10	272.0
15	55.00	10	372.0	16	407.0	13	2,948.0	13	39.00	8	372.0	14	270.0	11	2,943.0	23	82.90
10	110.0	13	298.0	14	736.0	14	2,932.0	15	26.00	10	282.0	16	222.0	13	2,734.0	13	189.9
14	77.00	14	274.0	13	763.0	16	2,696.0	10	59.00	13	230.0	10	621.0	15	2,232.0	5	453.0
12	91.00	9	407.0	10	909.0	9	5,285.0	12	51.00	11	281.0	12	457.0	9	3,369.0	14	159.0
13	79.68	11	358.8	12	798.0	10	4,746.4	11	57.99	12	274.5	11	620.8	10	3,277.6	6	428.0
11	97.00	8	428.0	11	897.0	8	6,093.0	9	63.00	9	283.0	9	656.0	8	4,399.0	11	266.0
39	4.533	18	129.2	36	42.31	23	1,424.1	37	3.729	17	124.1	30	36.34	20	1,359.9	35	48.30
9	132.0	12	305.0	9	1,031.0	12	3,200.0	14	39.00	14	219.0	13	351.0	16	1,959.0	16	127.7
42	4.011	35	[7]67.63	47	25.96	26	[7]1,068.0	36	4.011	33	[7]67.63	39	25.96	24	[7]1,068.0	55	[8]24.30
59	2.199	33	75.87	61	14.13	32	882.1	60	1.597	31	73.85	76	7.732	29	732.7	—	NA
16	50.00	15	194.0	15	471.0	18	2,018.0	16	25.00	22	103.0	17	201.0	27	771.0	49	30.00
40	4.394	36	67.01	54	18.07	46	399.1	34	4.394	34	67.01	46	18.07	42	399.1	61	22.00
25	13.70	76	14.50	32	54.60	72	198.4	85	0.900	68	14.50	134	2.800	63	198.4	37	43.00
38	4.725	23	109.6	42	30.85	24	1,208.0	35	4.339	19	109.4	36	26.98	22	1,207.0	8	289.9
49	3.099	34	73.47	45	27.16	27	1,054.9	47	2.706	32	73.47	43	23.62	25	1,054.1	20	94.90
27	11.00	24	108.0	20	137.0	30	903.0	22	11.00	23	103.0	19	135.0	26	821.0	26	72.30
30	8.200	31	79.10	29	76.60	28	978.7	24	8.000	30	74.90	25	71.10	31	701.9	38	42.95
56	2.618	49	39.80	85	7.419	70	205.7	49	2.618	46	39.80	79	7.419	61	205.7	66	19.35
26	13.30	29	87.30	25	89.60	35	712.9	25	6.700	29	77.40	32	33.90	36	563.8	50	29.43
31	6.600	26	104.6	26	89.52	19	1,920.6	26	6.600	21	104.6	21	[11]89.52	18	[11]1,920.6	29	67.00
142	0.276	40	55.00	121	4.497	36	704.1	140	0.276	37	55.00	112	4.497	30	704.1	31	55.30
28	9.672	16	150.9	28	84.40	15	2,807.0	23	9.672	15	150.9	23	84.40	12	2,807.0	21	90.50
24	14.10	43	48.00	27	86.70	41	549.5	20	14.10	40	48.00	22	86.70	37	549.5	33	50.22
22	17.90	37	66.50	23	107.0	37	632.7	28	5.800	38	54.80	34	29.80	45	321.1	45	34.50
81	1.100	69	16.60	102	6.200	94	104.9	74	1.100	64	16.60	92	6.200	82	104.9	97	8.500
18	23.24	22	111.6	21	134.2	33	881.9	33	4.587	24	99.73	29	38.45	28	744.2	25	73.90
19	22.30	19	124.9	18	144.6	22	1,474.1	17	21.20	18	120.0	20	130.0	21	1,355.0	27	70.00
110	[16]0.583	70	[16]16.08	119	[16]4.634	48	[16]349.9	129	[16]0.344	93	[16]7.749	158	[16]1.901	72	[16]134.8	64	[16]20.74
34	5.794	59	28.47	41	31.21	56	291.7	29	5.794	56	28.47	33	31.21	48	291.7	43	35.20
17	26.22	21	119.4	22	109.6	21	1,581.7	38	3.634	43	43.86	42	23.71	44	331.7	36	45.83
23	15.80	25	107.7	24	94.10	34	819.3	21	11.30	27	88.00	26	56.10	34	590.2	46	31.00
20	20.00	17	143.0	17	227.0	17	2,524.0	18	20.00	16	143.0	15	227.0	14	2,524.0	34	50.00
193	0.125	127	3.866	382	0.017	162	24.72	191	0.125	124	3.866	380	0.017	157	24.72	—	—
72	1.351	53	35.37	78	8.057	44	494.6	66	1.351	50	35.37	72	8.057	40	494.6	24	78.40
43	3.914	27	95.91	53	19.45	20	1,772.0	40	3.466	25	94.36	50	15.37	19	1,717.1	30	66.80
137	0.322	191	1.300	166	1.789	132	43.50	133	0.322	188	1.300	162	1.789	127	43.50	148	2.660
35	5.101	52	35.51	30	71.22	39	600.7	31	5.101	49	35.51	24	71.22	33	600.7	52	25.00
54	2.668	30	84.06	51	22.70	25	1,200.5	48	2.668	28	84.06	44	22.70	23	1,200.5	19	108.3
154	0.213	79	13.11	71	10.80	82	134.8	151	0.213	71	13.11	64	10.80	71	134.8	22	85.00
33	5.949	28	90.73	49	24.47	42	540.3	27	5.949	26	90.73	41	24.47	38	540.3	62	22.00
51	2.945	41	53.02	34	45.42	40	588.3	45	2.945	39	53.02	28	45.42	35	588.3	41	38.32
21	19.10	58	29.50	19	139.3	66	217.2	19	19.10	54	29.50	18	139.3	57	217.2	15	151.0
48	3.178	57	30.28	57	[16]15.76	43	[16]509.0	43	3.178	53	30.28	48	[16]15.76	39	[16]509.0	67	19.00

Rank by total assets 1987	1986	Company	Total assets $000	Rank	Total revenue $000	Rank	Net income $000	Rank	Stockholders' equity $000	Rank	Capital & Expl. spending Rank	$000
52	57	Total Petroleum (North America) Ltd.	1,014,053	27	1,747,009	378	(26,823)	53	308,296	42	131,941	
53	51	Ocean Drilling & Exploration Co.[26]	997,840	54	390,084	394	(129,325)	41	476,217	73	47,638	
54	54	Equitable Resources Inc.	974,864	53	408,274	33	41,078	43	453,757	56	77,760	
55	61	Hamilton Oil Corp.	955,370	58	320,881	55	21,905	49	349,458	75	41,988	
56	56	Oneok Inc.[27]	919,167	47	587,013	57	19,362	51	323,651	72	49,200	
57	55	Grace Natural Resources Group[28]	[29]917,000	55	388,200	52	[25]22,400	—	NA	96	23,000	
58	64	Diversified Energies Inc.	794,310	45	657,742	44	27,232	58	258,996	60	71,720	
59	59	Zapata Corp.[9]	766,157	76	151,363	396	(155,865)	372	(10,508)	88	31,935	
60	—	Newmont Oil Co.[30]	755,551	88	105,903	35	[25]35,177	—	NA	100	20,541	
61	63	Reading & Bates Corp.[31]	742,407	66	198,442	397	(191,094)	382	(66,525)	120	12,793	
62	69	CENEX (Farmer's Union Central Exchange)[9]	639,130	38	1,138,492	128	1,236	55	281,479	159	5,627	
63	66	Unimar Co.	628,732	70	176,575	383	(40,035)	63	[32]193,166	91	28,197	
64	70	Quaker State Corp.	612,562	39	859,696	387	(48,057)	52	320,330	53	79,396	
65	73	Tesoro Petroleum Corp.[9]	577,628	35	1,320,236	329	(1,776)	65	186,724	110	18,322	
66	74	Helmerich & Payne Inc.[9]	571,348	73	153,915	54	22,016	46	420,833	67	54,170	
67	72	Pogo Producing Co.	565,507	80	141,146	65	16,432	296	1,474	93	26,054	
68	71	Noble Affiliates Inc.	516,793	68	189,570	64	16,447	64	192,403	70	52,607	
69	65	Apache Corp.	504,333	92	102,091	391	(71,035)	79	128,780	87	32,016	
70	96	Triton Energy Corp.[19]	483,761	89	105,865	375	(23,480)	68	176,942	23	242,186	
71	78	Forest Oil Corp.	474,462	84	115,342	80	6,278	100	83,601	71	50,491	
72	68	Apache Petroleum Co. L.P.[33]	473,468	72	155,499	393	(101,181)	66	[10]185,768	89	30,600	
73	81	Adobe Resources Corp.	473,458	91	105,374	45	25,971	48	370,450	81	37,666	
74	90	Pacific Enterprises Oil & Gas Co.[34]	461,356	74	153,820	37	34,815	59	231,105	40	145,164	
75	76	Tidewater Inc.[35]	447,122	71	173,051	389	(56,678)	60	206,150	163	5,376	
76	80	Sunshine Mining Co.	445,481	77	150,249	388	(51,771)	69	176,922	111	18,291	
77	83	Kaneb Services Inc.	432,234	90	105,514	381	(35,500)	89	92,764	119	13,004	
78	88	Sante Fe Energy Partners L.P.[36]	422,500	82	131,200	372	(13,900)	57	[10]262,800	46	103,800	
79	94	NRM Energy Co. L.P.	409,963	104	56,282	385	(44,197)	54	[10]289,387	49	98,755	
80	79	CSX Oil & Gas Corp.[37]	392,130	86	114,273	76	8,126	72	150,508	79	38,141	
81	82	Southern Union Co.	381,764	67	191,580	361	(7,622)	81	111,652	112	18,180	
82	84	Diamond Shamrock Offshore Partners L.P.[38]	365,767	83	119,453	370	(11,782)	50	[10]333,700	66	57,561	
83	115	Ram Industries Inc.[39,35]	338,017	—	—	214	(101)	56	279,560	—	0	
84	92	National Cooperative Refinery Assoc.[40]	335,528	48	572,076	41	28,829	73	[41]148,410	123	12,274	
85	—	Total Minatome Corp.[42]	306,860	98	70,063	369	(11,022)	76	141,625	37	163,554	
86	91	First Mississippi Corp.[40]	296,362	59	295,741	59	18,610	71	158,606	105	19,527	
87	97	Dekalb Energy Cos.[43,27]	294,900	101	66,800	51	[44]23,400	—	NA	76	41,000	
88	95	Seagull Energy Corp.	291,845	75	152,923	75	8,180	84	108,136	115	15,221	
89	93	Sabine Corp.[45]	264,658	100	68,780	360	(7,521)	70	170,864	84	36,407	
90	108	Pelto Oil Co.[46]	263,700	108	44,621	72	8,999	67	181,065	80	38,023	
91	98	Southwestern Energy Co.	262,817	97	74,408	69	9,810	95	87,158	108	18,764	
92	85	Primary Fuels Inc.[47]	256,055	87	113,987	89	[48]4,475	—	NA	68	53,300	
93	101	Quinoco Cos.	245,372	96	80,698	81	6,124	62	194,585	201	2,255	
94	99	NUI Corp.[9]	241,143	60	289,182	74	8,299	114	54,711	97	22,647	
95	100	TGX Corp.	237,065	93	98,594	84	5,744	87	100,273	174	4,618	
96	—	Damson Energy Co. L.P.[49]	236,851	95	85,522	392	(93,599)	83	108,510	185	3,390	
97	111	Hadson Corp.	228,645	51	433,525	83	5,851	96	85,057	102	20,345	
98	104	Allegheny & Western Energy Corp.[40]	205,407	64	223,286	63	17,413	99	84,150	104	19,708	
99	103	NOMECO (Northern Michigan Expl. Co.)[50]	204,163	99	69,250	50	23,516	75	143,265	83	37,378	
100	107	Norcen Energy Resources Ltd.[51]	203,000	149	18,000	308	(1,000)	—	NA	107	19,000	
101	137	Conquest Exploration Co.	199,424	136	22,583	363	(8,381)	80	123,720	127	11,175	
102	106	Prairie Producing Co.[52]	184,040	118	32,822	380	(33,484)	383	(114,232)	98	20,812	

Worldwide liquids production		Worldwide natural gas production		Worldwide liquids reserves		Worldwide natural gas reserves		U.S. liquids production		U.S. natural gas production		U.S. liquids reserves		U.S. natural gas reserves		U.S. net wells drilled	
Rank	Mill bbl	Rank	Bcf	Rank	Mill bbl	Rank	Bcf	Rank	Mill bbl	Rank	Bcf	Rank	Mill bbl	Rank	Bcf	Rank	Wells
36	4.960	66	17.98	37	39.90	52	339.4	50	2.390	80	10.57	68	9.970	101	74.67	99	7.900
29	8.357	45	45.76	38	39.16	57	289.8	41	3.320	42	43.92	55	12.86	56	231.7	74	15.99
82	1.085	48	39.86	77	8.381	38	603.5	75	1.085	45	39.86	70	8.381	32	603.5	18	109.0
57	2.486	42	49.96	31	69.73	31	894.0	131	0.334	109	5.712	130	2.979	143	31.86	234	0.250
107	0.618	77	14.08	133	3.240	92	108.7	100	0.618	69	14.08	125	3.240	80	108.7	71	16.63
76	1.227	60	24.91	87	7.374	73	189.8	70	1.204	57	24.22	81	7.241	64	178.6	86	11.20
115	0.501	63	19.70	139	2.913	84	129.2	110	0.501	60	19.70	132	[11]2.913	73	[11]129.2	91	9.550
161	0.200	84	11.70	157	2.100	83	131.0	206	0.100	82	10.40	191	1.100	83	101.5	—	—
69	1.460	54	32.11	83	7.561	62	250.7	62	1.408	66	15.12	96	5.783	105	62.55	153	2.310
91	0.905	98	8.043	90	6.834	96	98.00	95	0.718	92	8.043	84	6.834	86	98.00	124	4.540
64	1.914	196	1.220	69	11.79	230	7.132	57	1.914	193	1.220	61	11.79	227	7.132	88	10.00
61	2.096	32	79.07	56	16.24	29	967.0	73	1.110	83	9.711	107	4.878	104	64.27	107	6.700
88	0.946	121	4.595	82	7.685	111	60.37	82	0.946	116	4.595	77	7.685	106	60.37	28	67.80
68	1.497	135	3.539	129	3.746	118	52.57	106	0.553	132	3.534	136	2.757	163	23.33	150	2.530
90	0.941	85	11.59	86	7.406	60	271.2	84	0.941	77	11.59	80	7.406	51	271.2	114	5.200
50	3.095	56	30.32	52	20.81	51	344.1	44	3.095	52	30.32	45	20.27	53	254.4	85	11.90
45	3.861	39	58.66	43	30.08	53	331.4	39	3.533	36	57.72	37	26.42	46	316.6	54	24.39
63	2.008	46	40.89	70	11.30	61	270.0	56	2.008	44	40.89	63	11.30	52	270.0	77	14.96
41	4.374	64	19.70	35	43.48	65	219.6	98	0.664	117	4.567	123	3.313	138	35.14	120	4.940
103	0.679	47	40.78	106	5.893	49	345.9	103	0.581	47	39.26	121	3.432	49	272.4	111	5.800
46	3.274	38	64.76	60	14.68	55	312.2	42	3.274	35	64.76	51	14.68	47	312.2	96	8.630
47	3.238	68	17.76	40	33.13	63	248.1	55	2.050	63	17.76	38	26.28	54	248.1	102	7.390
32	6.295	61	23.19	33	52.38	75	167.3	30	5.524	58	21.25	27	[11]52.38	66	[11]167.3	12	206.1
65	1.684	234	0.640	101	6.234	294	2.317	378	0.001	338	0.065	381	0.015	373	0.221	—	NA
79	1.126	74	14.90	96	6.440	79	142.6	76	1.076	72	12.53	99	5.550	89	86.91	121	4.870
95	0.852	86	11.04	125	4.002	109	74.21	89	[11]0.852	78	[11]11.04	116	[11]4.002	102	[11]74.21	72	16.30
37	4.800	55	31.60	44	29.80	67	211.9	32	4.800	51	31.60	35	29.80	58	211.9	32	51.50
85	1.026	65	18.02	67	12.13	78	146.8	79	1.026	61	18.02	59	12.13	68	146.8	—	NA
55	2.625	51	37.95	58	15.68	47	367.5	52	2.153	48	37.95	56	12.75	55	232.5	75	15.03
151	0.227	133	3.658	162	1.955	103	83.99	148	0.227	130	3.658	157	1.955	93	83.99	104	7.000
67	1.553	44	47.18	65	12.48	68	211.4	61	1.553	41	47.18	57	12.48	59	211.4	110	6.200
—	—	—	—	59	15.60	45	460.0	—	—	—	—	49	15.60	41	460.0	—	—
73	1.343	136	3.379	92	6.753	135	41.64	67	1.343	135	3.379	86	6.753	130	41.64	47	31.00
62	2.080	67	17.81	64	12.88	76	157.7	54	2.080	62	17.81	54	12.88	67	157.7	78	14.77
127	0.389	97	8.061	170	1.620	140	37.91	119	0.389	91	8.061	166	1.620	135	37.91	100	7.700
52	2.800	73	15.00	46	26.70	54	322.5	58	1.800	102	6.300	62	11.60	118	49.80	53	25.00
317	0.020	107	6.351	357	0.058	150	29.19	313	0.020	101	6.351	352	0.058	145	29.19	142	3.090
60	2.100	81	12.30	66	12.40	90	113.7	53	2.100	74	12.30	58	12.40	84	101.5	51	27.78
97	0.843	72	15.51	88	7.195	80	137.6	91	0.843	65	15.51	82	7.195	69	137.6	73	16.10
155	0.211	90	9.420	210	0.888	69	210.5	152	0.211	84	9.420	206	0.888	60	210.5	60	22.20
44	3.887	50	39.17	50	23.51	58	281.7	59	1.626	55	29.12	65	10.59	77	121.7	70	17.00
58	2.280	62	20.66	68	11.83	71	201.6	51	2.280	59	20.66	60	11.83	62	201.6	184	1.000
225	0.085	220	0.856	288	0.281	246	4.958	223	0.085	215	0.856	284	0.281	244	4.958	183	1.100
167	0.186	125	4.200	147	2.478	74	169.0	164	0.186	121	4.200	144	2.478	65	169.0	174	1.320
70	1.419	89	9.674	81	7.768	86	124.7	64	1.383	86	9.080	75	7.768	74	124.7	268	0.015
130	0.380	105	6.728	172	1.599	141	37.57	124	0.380	99	6.728	168	1.599	136	37.57	94	8.900
384	0.001	177	1.600	373	0.028	154	27.81	384	0.001	174	1.600	370	0.028	149	27.81	42	36.04
66	1.600	82	12.10	93	6.700	101	85.30	63	1.394	75	12.10	101	5.502	95	81.37	80	14.48
121	0.457	120	4.598	98	6.378	100	85.81	115	0.457	115	4.598	89	6.378	91	85.81	76	15.00
120	0.465	101	6.906	103	6.094	107	76.80	114	0.465	95	6.906	93	6.094	99	76.80	106	6.900
96	0.850	95	8.719	79	7.877	120	51.40	90	0.850	90	8.719	73	7.877	115	51.40	92	9.000

Rank by total assets 1987	1986	Company	Total assets $000	Rank	Total revenue $000	Rank	Net income $000	Rank	Stockholders' equity $000	Rank	Capital & Expl. spending $000
103	102	Home Petroleum Corp.[53,27]	182,517	112	38,418	112	[54]1,847	—	NA	135	9,05
104	125	Snyder Oil Partners L.P.	169,235	106	48,516	85	5,400	77	[10]138,609	57	76,44
105	112	Pyro Energy Corp.	161,050	85	115,247	79	6,396	102	77,861	150	6,01
106	129	Global Natural Resources Inc.	157,334	119	32,505	333	(2,209)	91	89,570	92	27,70
107	117	Statex Petroleum Inc.[55]	147,248	117	33,201	87	[25]5,103	—	NA	99	20,66
108	113	Howell Corp.	146,689	69	183,490	117	1,746	116	54,025	168	5,13
109	75	Moore McCormack Energy Inc.[56]	146,209	122	30,663	107	2,316	128	37,036	77	38,88
110	116	Park-Ohio Industries Inc.	145,241	81	134,586	340	(3,076)	101	80,958	184	3,40
111	119	Avalon Corp.	134,710	132	24,689	183	100	85	102,597	85	34,48
112	122	Wainoco Oil Corp.	126,777	115	33,933	49	24,523	107	65,929	116	14,22
113	121	Harken Oil & Gas Inc.	125,111	52	420,905	95	3,209	105	68,345	128	11,00
114	118	Kaneb Energy Partners Ltd.[57]	121,422	105	52,524	68	11,707	104	[10]69,430	147	6,86
115	109	Damson Oil Corp.[9]	120,554	126	29,037	374	(21,496)	251	3,444	212	1,83
116	164	Presidio Oil Co.	119,097	188	[58]9,054	320	[58](1,466)	133	31,095	161	[58]5,57
117	—	Parker & Parsley Development Partners L.P.	117,266	116	33,296	362	(8,092)	82	[10]110,837	106	19,12
118	126	Sage Energy Co.[40]	115,746	129	27,378	343	(3,232)	109	62,994	126	11,63
119	130	Ensource Inc.	111,583	121	30,701	241	(221)	119	48,288	121	12,63
120	131	Wiser Oil Co.	105,628	120	31,805	97	2,952	90	90,498	114	16,02
121	—	NGC Energy Co.[59]	104,998	156	15,484	350	(4,579)	127	38,403	64	65,20
122	128	Noble Drilling Corp.	103,654	128	28,018	366	(10,148)	94	87,329	140	8,04
123	191	Nerco Oil & Gas Inc.[60]	103,200	159	14,100	358	(6,700)	93	88,000	62	69,30
124	133	San Juan Basin Royalty Trust	102,013	146	18,356	61	17,852	86	[61]101,205	—	
125	—	Prudential-Bache Energy Income L.P. IVP-17	96,998	169	12,201	100	2,691	92	[10]88,104	59	71,85
126	—	Prudential-Bache Energy Income L.P. VP-19	96,463	229	4,614	149	559	88	[10]93,259	54	79,33
127	141	American Exploration Co.[63]	96,094	125	29,687	122	1,491	169	16,253	141	7,83
128	142	Flag Redfern Oil Co.	93,570	123	30,442	93	3,495	106	67,983	109	18,57
129	—	Prudential-Bache Energy Income L.P. IVP-16	93,247	166	13,280	98	2,833	97	[10]84,556	58	71,88
130	139	Integrated Drilling & Exploration Inc.[64]	90,420	139	20,806	220	(127)	98	84,205	153	5,83
131	136	Crystal Oil Co.	89,232	110	40,336	99	2,762	154	21,258	196	2,58
132	138	Consolidated Oil & Gas Inc.[65]	85,953	148	18,026	365	(9,399)	379	(22,620)	228	1,38
133	144	Chieftain Development Co. Ltd.[51,66]	84,218	165	13,365	291	(671)	—	NA	139	8,20
134	120	Kirby Exploration Co. Inc.[67]	82,004	107	47,642	82	6,031	118	49,706	157	5,65
135	165	Plains Resources Inc.	78,856	164	13,514	316	(1,292)	147	24,253	90	[68]29,24
136	145	Tom Brown Inc.[35]	75,253	143	18,959	339	(2,980)	139	28,018	190	3,07
137	—	Prudential-Bache Energy Income L.P. VP-18	74,907	218	5,262	140	860	103	[10]72,098	63	65,58
138	—	Damson Income Energy L.P.[69]	73,716	141	20,405	376	(23,862)	112	[10]58,992	199	2,50
139	—	Berry Petroleum Co.	72,741	103	57,628	58	19,355	113	55,474	160	5,57
140	148	Exploration Co. of Louisiana Inc.	72,435	124	30,190	78	7,570	111	59,001	143	7,49
141	159	Beard Co.[70]	70,219	152	16,462	86	5,336	136	29,401	124	12,17
142	169	Patrick Petroleum Co.	69,162	189	9,009	148	562	130	34,173	117	14,06
143	123	Texas American Energy Corp.	67,022	177	10,364	66	15,704	159	18,252	191	2,98
144	135	MCO Resources Inc.	66,062	133	23,389	377	(25,813)	378	(21,730)	263	72
145	—	Damson Institutional Energy L.P.[71]	65,772	138	20,861	373	(20,084)	122	[10]44,062	259	74
146	—	Hutton/Energy Assets 3rd Energy Ptrsp. A Ltd.	64,519	194	7,913	105	2,390	108	[10]63,581	103	20,13
147	—	Apache Offshore Investment Partnership[72]	63,716	235	4,106	368	(10,785)	232	[10]4,624	136	8,68
148	147	Freeport-McMoRan Oil and Gas Royalty Trust	62,857	145	18,386	62	[73]17,635	110	[61]59,221	—	N
149	150	Pauley Petroleum Inc.[27]	60,222	79	142,042	231	(167)	140	27,688	176	4,22
150	—	Prudential-Bache Energy Income L.P. IIIP-14	59,331	191	8,653	127	1,301	115	[10]54,657	69	53,10
151	153	Equity Oil Co.	56,342	162	13,868	109	2,157	124	40,513	189	3,25
152	140	Unit Corp.	56,066	127	28,888	382	(35,635)	126	40,115	169	5,08
153	151	Houston Oil Trust	53,507	250	3,276	121	[73]1,550	117	[61]53,505	—	N

Worldwide liquids production		Worldwide natural gas production		Worldwide liquids reserves		Worldwide natural gas reserves		U.S. liquids production		U.S. natural gas production		U.S. liquids reserves		U.S. natural gas reserves		U.S. net wells drilled	
Rank	Mill bbl	Rank	Bcf	Rank	Mill bbl	Rank	Bcf	Rank	Mill bbl	Rank	Bcf	Rank	Mill bbl	Rank	Bcf	Rank	Wells
87	0.969	91	9.403	84	7.559	88	122.8	81	0.969	85	9.403	78	7.559	76	122.8	105	7.000
84	1.030	87	10.85	62	13.84	87	124.4	78	1.030	79	10.85	52	13.84	75	124.4	59	22.73
146	0.255	161	2.312	164	1.828	183	17.02	144	0.255	159	2.312	160	1.828	180	17.02	259	0.080
124	0.406	99	7.600	97	6.406	77	153.0	121	[11]0.383	114	[11]5.125	98	[11]5.564	90	[11]86.13	81	14.40
152	0.226	110	6.013	155	2.169	99	87.99	149	0.226	105	6.013	151	2.169	88	87.99	63	21.90
98	0.823	116	5.534	104	5.954	97	89.99	92	0.817	113	5.318	95	5.818	97	80.42	131	4.120
108	0.614	113	5.883	74	10.22	106	79.77	101	0.614	108	5.883	67	10.22	98	79.77	117	5.000
204	0.110	150	2.687	181	1.458	126	45.13	202	0.110	149	2.687	177	1.458	121	45.13	185	1.000
116	0.498	118	5.515	117	4.670	85	128.7	138	0.293	269	0.288	143	2.497	287	2.553	204	0.500
105	0.656	71	15.88	112	5.172	64	247.6	127	0.365	133	3.497	128	3.009	113	53.34	109	6.290
184	0.143	225	0.763	180	1.461	206	12.11	182	0.143	221	0.763	176	1.461	203	12.11	135	3.570
86	1.001	78	13.38	116	4.774	98	89.90	80	1.001	70	13.38	109	4.774	87	89.90	83	13.90
114	0.528	129	3.767	142	2.724	123	49.65	109	[11]0.528	126	[11]3.767	137	[11]2.724	119	[11]49.65	271	0.004
168	[58]0.181	153	[58]2.628	143	2.713	91	109.2	165	[58]0.181	152	[58]2.628	138	2.713	78	109.2	79	[58]14.77
99	0.793	83	11.81	91	6.769	108	75.52	93	0.793	76	11.81	85	6.769	100	75.52	93	9.000
80	1.114	123	4.449	72	10.50	139	37.96	72	1.114	119	4.449	66	10.50	134	37.96	90	9.600
118	0.481	94	8.767	130	3.599	95	99.17	112	0.481	89	8.767	120	3.599	85	99.17	115	5.150
94	0.876	108	6.216	76	9.235	102	84.36	88	0.876	103	6.216	69	[11]9.235	92	[11]84.36	39	41.00
153	0.215	93	9.020	171	1.604	93	108.5	150	0.215	88	9.020	167	1.604	81	108.5	125	4.500
252	0.056	261	0.374	260	0.386	271	3.268	249	0.056	259	0.374	256	0.386	269	3.268	126	4.500
135	0.337	122	4.538	141	2.794	114	58.88	130	0.337	118	4.538	135	2.794	109	58.88	—	NA
248	0.061	80	12.46	195	1.077	59	272.0	245	0.061	73	12.46	193	1.077	50	272.0	197	[62]0.600
139	0.307	147	2.900	109	5.440	124	47.37	136	0.307	146	2.900	102	5.440	120	47.37	—	—
175	0.164	238	0.583	108	5.533	156	27.34	171	0.164	233	0.583	100	5.533	151	27.34	—	—
100	0.739	104	6.743	118	4.658	116	57.16	94	0.739	98	6.743	110	4.658	111	57.16	138	3.420
89	0.946	112	5.964	89	6.976	138	39.05	83	0.946	107	5.964	83	6.976	133	39.05	68	18.70
140	0.306	152	2.670	105	5.898	128	44.42	137	0.306	151	2.670	94	5.898	123	44.42	—	—
188	0.133	145	3.043	193	1.109	130	43.84	185	0.133	143	3.043	190	1.109	125	43.84	89	10.00
71	1.382	88	10.46	107	5.639	127	45.03	65	1.382	81	10.46	97	5.639	122	45.03	163	1.800
122	0.453	114	5.769	111	5.249	89	121.0	117	0.439	112	5.353	104	5.147	79	109.1	166	1.700
111	0.559	260	0.376	110	5.268	195	14.06	104	0.559	258	0.376	103	5.268	191	14.06	216	0.390
—	NA	—	NA	140	2.811	175	20.51	—	NA	—	NA	133	2.811	172	20.51	—	NA
145	0.258	134	3.601	184	1.395	112	60.23	143	0.258	131	3.601	181	1.395	107	60.23	122	4.780
101	0.702	138	3.236	114	5.066	117	53.91	96	0.702	137	3.236	106	5.066	112	53.91	198	0.600
180	0.150	199	1.193	122	4.264	167	22.66	177	0.150	195	1.193	113	4.264	164	22.66	—	—
126	0.390	143	3.068	148	2.440	137	39.76	123	0.382	145	2.933	145	2.440	132	39.76	—	—
53	2.709	233	0.645	39	34.38	211	10.73	46	2.709	229	0.645	31	34.38	208	10.73	224	0.300
183	0.146	131	3.741	300	0.225	178	19.55	180	0.146	128	3.741	296	0.225	175	19.55	155	2.260
109	0.605	158	2.346	134	3.114	193	14.22	102	0.605	157	2.346	126	3.114	189	14.22	48	30.71
162	0.196	198	1.201	200	0.979	196	14.06	158	0.196	194	1.201	198	0.979	192	14.06	157	2.160
131	0.373	172	1.646	136	2.988	192	14.36	125	0.373	170	1.646	129	2.988	188	14.36	—	—
119	0.475	92	9.067	146	2.595	115	58.13	113	0.475	87	9.067	142	[11]2.595	110	[11]58.13	161	1.960
133	0.364	163	2.310	163	1.859	160	26.19	128	0.355	162	2.160	159	1.859	155	26.19	—	—
212	0.097	170	1.883	229	0.617	131	43.70	209	0.097	168	1.883	227	0.617	126	43.70	—	—
342	0.011	165	2.140	236	0.551	146	33.12	338	0.011	163	2.140	234	0.551	141	33.12	—	NA
150	0.236	100	7.428	262	0.379	170	21.92	147	0.236	94	7.428	258	0.379	167	21.92	—	NA
83	1.043	242	0.537	94	6.538	257	4.385	77	1.043	237	0.537	87	6.538	255	4.385	225	0.300
160	0.201	166	2.018	135	3.095	136	40.21	157	0.201	164	2.018	127	3.095	131	40.21	—	—
104	0.658	231	0.659	75	10.13	148	32.67	107	0.543	246	0.443	71	8.224	148	28.04	154	2.310
164	0.191	151	2.682	192	1.136	169	22.40	161	0.191	150	2.682	189	1.136	166	22.40	84	11.91
125	[74]0.391	128	[74]3.789	178	[74]1.507	159	[74]26.61	118	[74]0.391	125	[74]3.789	173	[74]1.507	154	[74]26.61	—	NA

Rank by total assets 1987	1986	Company	Total assets $000	Rank	Total revenue $000	Rank	Net income $000	Rank	Stockholders' equity $000	Rank	Capital & Expl. spending $000
154	187	Calumet Industries Inc.[9]	53,317	111	38,732	336	(2,266)	168	16,555	101	20,437
155	—	Prudential-Bache Energy Income L.P. IIIP-15	51,167	201	7,552	134	1,060	120	[10]47,132	74	43,478
156	179	Bogert Oil Co.[9]	49,394	147	18,183	115	1,754	158	19,390	133	9,725
157	—	Prudential-Bache Energy Income L.P. IIIP-12	48,971	176	10,372	94	3,426	121	[10]44,082	94	25,722
158	170	Thermal Exploration Co.[76,9]	48,842	222	[77]5,119	266	[48](349)	—	NA	129	10,840
159	105	Texas International Co.	48,666	113	35,924	386	(44,222)	384	(143,172)	113	16,159
160	143	Graham-McCormick Oil & Gas Partnership[78]	48,650	155	15,745	353	(5,597)	351	[10](23)	295	379
161	160	Wilshire Oil Co. of Texas	47,503	186	9,198	145	619	181	12,345	183	3,435
162	—	PaineWebber/Geodyne Energy Income L.P. II-A	44,603	260	2,787	151	515	123	[10]43,048	95	25,328
163	—	Clinton Gas Systems Inc.[79]	44,469	109	42,830	177	160	157	19,436	187	3,282
164	171	Fuel Resources Inc.[80,9]	44,369	175	10,414	169	220	167	16,776	131	10,067
165	158	Alamco Inc.	42,734	179	10,184	346	(3,448)	243	3,901	—	NA
166	178	Wolverine Exploration Co.[81]	42,189	182	9,906	26	77,851	162	17,926	78	38,404
167	167	Pend Oreille Oil & Gas Co.[82]	42,105	137	21,023	120	1,607	144	25,521	155	5,708
168	155	OKC L.P.	41,567	114	35,137	77	7,909	138	[10]29,142	—	NA
169	156	Ethyl Corp. (Oil & Gas Division)	41,567	157	14,385	110	[6]1,900	—	NA	118	13,257
170	198	Kelley Oil & Gas Partners Ltd.[83]	41,490	181	10,066	141	839	125	[10]40,376	125	12,091
171	157	General Energy Development Ltd.[84]	41,412	144	18,894	180	107	137	[10]29,381	204	1,979
172	212	Oregon Natural Gas Development Corp.[85]	41,343	198	7,740	244	(240)	141	26,583	145	7,278
173	166	DI Industries Inc.[35]	40,278	212	5,656	111	1,863	179	12,504	297	365
174	168	Fuel Resources Development Co.[86]	39,857	199	7,607	144	640	153	21,689	180	3,975
175	172	Maynard Oil Co.	39,448	154	15,901	136	996	134	29,952	205	1,958
176	180	Samson Energy Co.	38,724	187	9,090	154	472	149	[10]22,971	156	5,69
177	—	Prudential-Bache Energy Income L.P. IIIP-13	38,455	214	5,587	132	1,091	129	[10]35,354	86	34,02
178	—	Callon Consolidated Partners L.P.[87]	37,719	151	16,776	135	1,029	135	[10]29,482	277	57
179	184	Oxoco Inc.	37,499	170	11,996	205	(50)	176	12,760	226	1,50
180	183	Penn Virginia Resources Corp.[88]	36,662	185	9,250	108	[89]2,215	—	NA	132	9,72
181	182	Chapman Energy Inc.[40]	36,571	213	[90]5,624	101	[90]2,667	304	1,301	360	3
182	202	McFarland Energy Inc.	35,934	167	13,103	103	2,553	142	26,113	134	9,48
183	188	Enex Resources Corp.	34,758	174	11,231	153	480	132	31,270	215	1,65
184	177	Devon Resource Investors L.P.	34,616	180	10,072	347	(3,598)	150	[10]22,545	179	4,06
185	190	Plains Petroleum Co.	33,069	131	24,789	70	9,130	146	24,458	172	5,01
186	185	Tipperary Corp.[9]	32,659	142	19,387	60	18,555	374	(11,634)	289	44
187	—	PaineWebber/Geodyne Energy Income L.P. II-B	32,240	363	418	191	34	131	[10]32,058	182	3,62
188	228	Saxon Oil Co.	31,686	168	12,535	129	1,139	160	18,251	193	2,67
189	207	Coho Resources Inc.[91]	31,595	210	5,784	92	4,014	300	1,406	151	5,98
190	206	Fidelity Oil Co.[92]	31,031	204	7,065	88	4,640	155	21,050	149	6,11
191	176	Alexander Energy Corp.[35]	30,647	233	4,244	349	(3,837)	182	12,344	225	1,50
192	175	Walker Energy Partners	29,661	184	9,365	364	(8,840)	237	[10]4,322	249	86
193	189	Deminex U.S. Oil Co.[93]	29,230	153	15,985	314	(1,173)	205	7,208	213	1,76
194	201	Saxon Oil Development Partners L.P.[94]	28,767	171	11,884	123	1,464	148	[10]23,251	197	2,51
195	194	Federated Natural Resources Corp.	28,715	197	7,747	321	(1,473)	246	3,857	192	2,92
196	205	MSR Exploration Ltd.[66]	28,529	240	3,924	242	(221)	152	21,960	214	1,73
197	197	Equitable Petroleum Corp.	28,484	234	4,241	249	(253)	210	6,826	321	15
198	—	Prudential-Bache Energy Income L.P. IIP-8	28,336	195	7,848	119	1,623	145	[10]25,517	138	8,55
199	—	Consolidated Energy Partners L.P.[95,65]	27,942	206	6,696	352	(5,341)	375	[10](19,509)	334	1
200	199	Petroleum Development Corp.	27,870	172	11,807	172	184	189	9,805	244	1,0
201	—	Nucorp Energy Inc.	27,323	158	14,198	173	184	178	12,560	209	1,8
202	186	ConVest Energy Partners Ltd.[96]	26,178	178	10,323	295	(728)	156	[10]20,479	202	2,1
203	192	LL&E Royalty Trust	25,989	134	23,224	53	[73]22,190	143	[61]25,989	—	NA
204	211	Belden & Blake Energy Co.	25,482	193	8,030	296	(747)	164	[10]17,316	148	6,7

Worldwide liquids production		Worldwide natural gas production		Worldwide liquids reserves		Worldwide natural gas reserves		U.S. liquids production		U.S. natural gas production		U.S. liquids reserves		U.S. natural gas reserves		U.S. net wells drilled	
Rank	Mill bbl	Rank	Bcf	Rank	Mill bbl	Rank	Bcf	Rank	Mill bbl	Rank	Bcf	Rank	Mill bbl	Rank	Bcf	Rank	Wells
249	0.059	—	—	279	[75]0.322	—	—	246	0.059	—	—	275	[75]0.322	—	—	—	—
170	0.174	171	1.739	144	2.635	144	34.53	166	0.174	169	1.739	139	2.635	139	34.53	—	—
141	0.278	117	5.519	145	2.598	121	51.27	139	0.278	111	5.519	141	2.598	116	51.27	58	22.90
210	0.099	149	2.804	174	1.576	113	59.24	208	0.099	148	2.804	170	1.576	108	59.24	—	—
182	0.147	190	1.317	202	0.958	180	18.51	179	0.147	187	1.317	200	0.958	177	18.51	—	NA
78	1.174	185	1.379	73	10.48	286	2.578	181	0.145	238	0.535	211	0.839	284	2.578	103	7.230
173	0.165	103	6.793	213	0.860	122	50.80	169	0.165	97	6.793	209	0.860	117	50.80	—	NA
148	0.249	160	2.313	153	2.297	151	29.15	160	0.192	176	1.540	179	1.430	196	13.71	—	NA
279	0.037	215	0.948	165	1.790	185	16.82	274	0.037	210	0.948	161	1.790	182	16.82	—	NA
221	0.087	169	1.920	272	0.331	197	14.03	219	0.087	167	1.920	268	0.331	193	14.03	—	NA
230	0.081	157	2.385	257	0.427	176	20.19	228	0.081	156	2.385	253	0.427	173	20.19	—	NA
274	0.039	155	2.503	271	0.336	164	23.99	270	0.039	154	2.503	267	0.336	160	23.99	239	0.200
149	0.240	154	2.508	355	0.063	—	—	146	0.240	153	2.508	350	[110]0.063	—	—	65	19.40
132	0.369	102	6.903	188	1.310	166	23.71	126	0.369	96	6.903	185	1.310	162	23.71	116	5.100
77	1.192	124	4.262	120	4.516	119	52.05	71	1.192	120	4.262	111	4.516	114	52.05	229	0.280
134	0.357	119	5.306	137	2.985	125	47.25	135	0.309	123	4.084	140	2.624	159	24.15	147	2.700
241	0.069	115	5.691	240	0.533	105	80.42	239	0.069	110	5.691	238	0.533	96	80.42	172	1.370
123	0.440	111	6.003	151	2.308	149	30.99	116	0.440	106	6.003	148	2.308	144	30.99	190	0.900
—	NA	291	0.197	—	NA	337	0.902	—	NA	290	0.197	—	NA	335	0.902	—	—
327	0.016	247	0.476	318	0.168	247	4.928	323	0.016	244	0.476	314	0.168	245	4.928	261	0.070
199	0.121	156	2.443	225	0.649	110	71.64	197	0.121	155	2.443	223	0.649	103	71.64	132	4.000
144	0.264	164	2.280	175	1.567	172	21.29	142	0.264	161	2.280	171	1.567	169	21.29	232	0.266
254	0.054	130	3.754	256	0.438	155	27.38	251	0.054	127	3.754	252	0.438	150	27.38	127	4.420
190	0.129	193	1.285	161	1.964	161	25.77	187	0.129	190	1.285	156	1.964	156	25.77	—	—
106	0.635	148	2.841	123	4.159	199	13.87	99	0.635	147	2.841	114	4.159	195	13.87	214	0.400
178	0.156	140	3.167	198	1.011	181	18.39	175	0.156	139	3.167	196	1.011	178	18.39	213	0.410
318	0.019	167	2.000	283	0.300	129	44.00	314	0.019	165	2.000	279	0.300	124	44.00	—	NA
238	[90]0.075	211	[90]1.033	224	0.652	210	10.78	236	[90]0.075	206	[90]1.033	222	0.652	207	10.78	255	[90]0.100
102	0.700	195	1.268	80	7.828	234	6.182	97	0.700	192	1.268	74	7.828	232	6.182	98	8.300
147	0.255	162	2.312	183	1.418	186	16.82	145	0.255	160	2.312	180	1.418	183	16.82	265	0.028
231	0.081	132	3.702	246	0.509	147	32.78	229	0.081	129	3.702	243	0.509	142	32.78	146	2.775
239	0.072	75	14.62	282	0.317	50	345.5	237	0.072	67	14.62	278	0.317	43	345.5	56	23.00
129	0.383	183	1.445	127	3.804	158	27.13	122	0.383	181	1.445	118	3.804	153	27.13	230	0.270
355	0.008	—	—	215	0.806	—	—	352	0.008	—	—	212	0.806	—	—	—	NA
112	0.555	173	1.641	113	5.096	189	16.22	105	0.555	171	1.641	105	5.096	186	16.22	264	0.040
136	0.327	359	0.040	48	24.81	288	2.554	132	0.327	358	0.040	40	24.81	286	2.554	87	11.13
75	1.304	240	0.570	63	13.00	340	0.800	69	1.289	236	0.542	53	12.90	341	0.700	40	41.00
223	0.086	178	1.551	185	1.364	171	21.79	221	0.086	175	1.551	182	1.364	168	21.79	130	4.195
181	0.148	137	3.303	189	1.215	163	24.43	178	0.148	136	3.303	186	1.215	158	24.43	256	0.100
—	NA	144	3.050	99	6.242	174	20.54	—	NA	142	3.050	90	6.242	171	20.54	129	4.200
113	0.529	176	1.603	115	4.853	190	15.72	108	0.529	173	1.603	108	4.853	187	15.72	245	0.180
174	0.165	192	1.296	247	0.484	209	11.55	170	0.165	189	1.296	244	0.484	206	11.55	209	0.480
176	0.160	315	0.107	132	3.254	255	4.473	172	0.160	314	0.107	124	[113]3.254	253	[114]4.473	—	NA
202	0.111	226	0.760	126	3.869	153	28.40	200	0.111	222	0.760	117	3.869	147	28.40	—	NA
185	0.143	159	2.322	220	0.741	157	27.33	183	0.143	158	2.322	218	0.741	152	27.33	—	NA
201	0.115	141	3.146	176	1.539	104	83.95	199	0.115	140	3.146	172	1.539	94	83.95	243	0.188
319	0.019	229	0.705	307	0.203	168	22.52	315	0.019	225	0.705	303	0.203	165	22.52	133	3.730
117	0.491	168	1.950	124	4.098	194	14.21	111	0.491	166	1.950	115	4.098	190	14.21	180	1.170
138	0.311	175	1.612	149	2.429	203	12.67	134	0.311	172	1.612	146	2.429	200	12.67	211	0.430
92	[97]0.897	142	[97]3.069	128	[97]3.763	198	[97]13.95	86	[97]0.897	141	[97]3.069	119	[97]3.763	194	[97]13.95	226	0.300
143	0.276	180	1.505	138	2.927	184	16.94	141	0.276	178	1.505	131	2.927	181	16.94	167	1.700

Rank by total assets 1987	1986	Company	Total assets $000	Total revenue Rank	$000	Net income Rank	$000	Stockholders' equity Rank	$000	Capital & Expl. spending Rank	$000
205	—	ICG Petroleum Inc.[98]	25,399	251	3,158	150	543	161	18,214	355	43
206	—	Prudential-Bache Energy Income L.P. IIP-10	25,044	215	5,428	114	1,811	151	[10]22,418	122	12,331
207	163	Amax Oil & Gas Inc.	24,364	140	20,557	34	40,797	183	12,013	173	4,923
208	208	Mustang Resources Corp.[99,13]	23,913	205	6,836	306	(986)	195	9,083	210	1,879
209	222	Swift Energy Co.	23,746	173	11,698	91	4,024	184	10,871	220	1,573
210	218	Taurus Exploration Inc.[100,9]	23,205	183	9,536	118	1,725	190	9,650	146	7,088
211	203	KenCope Energy Cos.[40]	22,783	163	13,616	357	(6,502)	177	12,563	261	733
212	225	Barrett Resources Corp.[9]	21,281	238	4,087	290	(656)	171	15,213	171	5,041
213	195	Petroleum Investments Ltd.	21,038	200	7,579	359	(7,150)	193	[109]9,545	282	513
214	209	Baruch-Foster Corp.	20,886	190	8,781	294	(713)	180	12,391	186	3,389
215	—	American National Petroleum Co.	20,276	207	6,406	267	(354)	185	10,686	236	1,131
216	—	Prudential-Bache Energy Income L.P. IIP-11	19,993	237	4,089	125	1,334	163	[10]17,922	130	10,155
217	216	Hershey Oil Corp.	19,408	216	5,402	187	71	174	13,824	195	2,613
218	210	Tucker Drilling Co. Inc.[35]	18,675	192	8,545	318	(1,386)	165	17,233	211	1,875
219	—	Bridge Oil (USA) Inc.[101]	17,761	262	2,700	137	967	200	7,851	227	1,502
220	244	Geodyne Resources Inc.[102]	17,621	160	14,017	90	4,463	207	7,113	262	729
221	—	Columbian Energy Co. L.P.	17,580	268	2,382	334	(2,223)	166	[10]17,024	290	439
222	—	Prudential-Bache Energy Income L.P. IIP-9	17,496	227	4,759	138	965	170	[10]15,790	166	5,177
223	215	Resource Exploration Inc.[9]	16,994	217	5,330	147	589	172	15,160	260	749
224	272	Petromark Resources Co.	15,804	221	5,191	345	(3,297)	356	(624)	293	405
225	—	Garnet Resources Corp.	14,703	309	1,052	228	(155)	175	12,998	206	1,956
226	221	Sabine Royalty Trust	14,593	135	23,058	56	21,589	173	[61]14,161	—	NA
227	—	Daleco Resources Corp.[66]	14,018	336	693	332	(2,119)	337	479	344	75
228	253	Belcor Inc.[103]	13,777	161	13,879	143	713	239	4,128	194	2,658
229	154	Callon Petroleum Co.	13,627	211	5,749	331	(2,113)	194	9,203	270	602
230	223	Adams Resources & Energy Inc.	13,602	102	66,367	106	2,317	365	(4,284)	235	1,153
231	200	Quadra Oil & Gas Inc.[104]	13,446	273	2,228	218	(120)	380	(23,527)	158	5,650
232	—	Red Eagle Resources Corp.[9]	13,231	150	17,568	142	777	260	2,897	162	5,570
233	243	Energy Assets International Corp.[35]	13,204	208	6,262	286	(588)	242	3,936	170	5,043
234	251	Texon Energy Corp.[105]	12,336	281	1,705	356	(6,155)	369	(8,642)	327	136
235	—	Prudential-Bache Energy Income L.P. IIP-7	12,234	230	4,466	131	1,131	186	[10]10,239	280	526
236	230	Michigan Oil Co.[106,40]	12,226	224	4,873	348	[44](3,610)	—	NA	142	7,724
237	—	Prudential-Bache Energy Income L.P. IIP-4	12,150	239	4,006	146	604	188	[109]9,963	—	NA
238	226	Kimbark Oil & Gas Co.	11,899	223	5,055	300	(852)	264	2,517	322	157
239	231	Premier Resources Ltd.[9]	11,788	258	2,856	206	(54)	361	(1,926)	302	299
240	235	Home-Stake Royalty Corp.	11,668	220	5,216	116	1,752	187	10,174	221	1,571
241	—	Prudential-Bache Energy Income L.P. IIP-5	11,315	255	2,998	160	315	191	[109]9,592	352	50
242	239	Energy Management Corp.[27]	11,292	247	[107]3,383	193	[107]29	226	5,071	312	214
243	238	Vermilion Bay Land Co.	11,043	248	3,306	171	187	192	9,561	223	1,537
244	—	Prudential-Bache Energy Income L.P. IIP-6	10,751	228	4,652	133	1,088	197	[108]8,810	375	7
245	245	Dorchester Hugoton Ltd.	10,733	243	3,790	124	1,442	196	[109]9,063	301	314
246	236	Evergreen Resources Inc.[35]	10,710	322	820	328	(1,734)	219	5,616	217	1,628
247	232	Petrominerals Corp.	10,624	202	7,550	293	(704)	206	7,193	246	1,028
248	213	Brock Exploration Corp.[35]	10,427	209	5,975	174	182	266	2,480	203	2,018
249	241	Sage Drilling Co. Inc.[40]	10,419	232	4,331	301	(856)	216	6,004	278	539
250	217	Nugget Oil Corp.[105]	10,281	256	2,982	323	(1,522)	360	(1,557)	268	670
251	237	Seahawk Oil International Inc.	10,198	254	3,017	289	(617)	224	5,086	224	1,534
252	334	Wichita River Oil Co.[108]	10,083	257	2,885	139	917	230	4,788	164	5,282
253	249	Thor Energy Resources Inc.[13]	10,001	196	7,830	196	2	215	6,288	251	847
254	283	Falcon Oil & Gas Co. Inc.[13]	9,901	289	1,520	156	441	220	5,364	178	4,067
255	254	Permian Basin Royalty Trust	9,816	130	26,098	47	25,523	203	[61]7,340	—	0

Worldwide liquids production		Worldwide natural gas production		Worldwide liquids reserves		Worldwide natural gas reserves		U.S. liquids production		U.S. natural gas production		U.S. liquids reserves		U.S. natural gas reserves		U.S. net wells drilled	
Rank	Mill bbl	Rank	Bcf	Rank	Mill bbl	Rank	Bcf	Rank	Mill bbl	Rank	Bcf	Rank	Mill bbl	Rank	Bcf	Rank	Wells
214	0.094	274	0.273	234	0.561	295	2.249	211	0.094	273	0.273	232	0.561	293	2.249	—	NA
260	0.047	182	1.476	203	0.948	152	29.07	257	0.047	180	1.476	201	0.948	146	29.07	—	NA
169	0.176	96	8.181	226	0.642	182	17.28	173	0.159	134	3.386	224	0.642	179	17.28	205	0.500
224	0.086	200	1.188	194	1.093	145	34.31	222	0.086	196	1.188	192	1.093	140	34.31	250	0.130
242	0.067	248	0.472	230	0.597	228	7.229	240	0.067	245	0.472	228	0.597	225	7.229	82	14.00
191	0.128	139	3.207	278	[75]0.323	187	[75]16.50	188	0.128	138	3.207	274	[75]0.323	184	[75]16.50	—	NA
261	0.047	202	1.148	309	0.202	236	5.902	258	0.047	197	1.148	305	0.202	234	5.902	—	—
253	0.056	227	0.732	275	0.326	212	10.54	250	0.056	223	0.732	271	0.326	209	10.54	113	5.560
195	0.123	146	3.005	222	0.684	177	20.08	193	0.123	144	3.005	220	0.684	174	20.08	95	8.680
128	0.388	179	1.530	131	3.317	205	12.36	120	0.388	177	1.530	122	3.317	202	12.36	134	3.590
158	0.206	188	1.332	167	1.754	173	21.28	155	0.206	185	1.332	163	1.754	170	21.28	168	1.680
280	0.037	206	1.124	221	0.728	165	23.82	275	0.037	201	1.124	219	0.728	161	23.82	—	NA
187	0.134	174	1.618	205	0.935	142	37.23	189	0.128	239	0.535	215	0.771	297	2.102	199	0.600
278	0.038	267	0.300	334	0.141	312	1.568	273	0.038	265	0.300	329	0.141	310	1.568	162	1.810
165	0.191	222	0.818	154	2.254	219	9.288	162	0.191	217	0.818	150	2.254	216	9.280	227	0.300
281	0.035	204	1.147	251	0.456	215	9.688	277	0.035	199	1.147	248	0.456	212	9.688	220	0.374
228	0.082	228	0.719	179	1.471	207	12.01	226	0.082	224	0.719	175	1.471	204	12.01	108	6.390
222	0.087	184	1.400	253	0.449	188	16.47	220	0.087	182	1.400	250	0.449	185	16.47	—	NA
245	0.063	186	1.362	299	0.236	216	9.655	242	0.063	183	1.362	295	0.236	213	9.655	112	5.800
156	0.207	210	1.048	173	1.580	224	8.614	153	0.207	205	1.048	169	1.580	220	8.614	252	0.114
262	0.047	322	0.094	241	0.532	373	0.232	324	0.016	321	0.094	359	0.052	371	0.232	—	—
93	0.880	109	6.066	95	6.491	134	41.77	87	0.880	104	6.066	88	6.491	129	41.77	—	NA
328	0.015	256	0.400	212	0.883	143	36.99	325	0.015	254	0.400	208	0.883	137	36.99	206	0.500
250	0.057	366	0.031	290	0.273	367	0.326	247	0.057	365	0.031	286	0.273	365	0.326	—	—
159	0.205	189	1.324	259	0.407	325	1.271	156	0.205	186	1.324	255	0.407	322	1.271	228	0.300
246	0.062	232	0.659	297	0.239	278	2.815	243	0.062	228	0.659	293	0.239	276	2.815	—	—
215	0.092	268	0.299	206	0.923	264	3.684	213	0.092	266	0.299	203	0.923	262	3.684	118	5.000
275	0.039	250	0.434	315	0.170	259	4.311	271	0.039	248	0.434	311	0.170	257	4.311	44	35.00
217	0.090	221	0.820	265	0.354	220	9.250	215	0.090	216	0.820	261	0.354	217	9.250	—	—
291	0.030	230	0.665	316	0.169	239	5.427	287	0.030	227	0.665	312	0.169	237	5.427	—	—
196	0.123	218	0.881	196	1.060	201	12.86	194	0.123	213	0.881	194	1.060	198	12.86	—	NA
211	0.099	201	1.166	177	1.517	221	9.037	212	0.094	219	0.788	174	1.477	229	6.914	137	3.429
166	0.188	244	0.531	168	1.700	244	5.059	163	0.188	241	0.531	164	1.700	242	5.059	—	NA
189	0.133	181	1.496	231	0.589	214	9.873	186	0.133	179	1.496	229	0.589	211	9.873	246	0.180
208	0.101	197	1.205	209	0.890	191	14.92	347	0.009	226	0.672	353	0.058	224	7.334	223	0.320
163	0.195	213	0.978	199	1.000	227	7.426	159	0.195	208	0.978	197	1.000	223	7.426	151	2.460
194	0.125	245	0.497	190	1.193	240	5.208	192	0.125	242	0.497	187	1.193	238	5.208	—	NA
243	[107]0.067	205	[107]1.138	244	0.513	208	11.63	241	[107]0.067	200	[107]1.138	241	0.513	205	11.63	212	[107]0.420
207	0.104	246	0.492	197	1.014	248	4.795	205	0.104	243	0.492	195	1.014	246	4.795	165	1.780
172	0.171	224	0.764	182	1.440	256	4.395	168	0.171	220	0.764	178	1.440	254	4.395	—	NA
—	—	106	6.423	—	—	81	134.8	—	—	100	6.423	—	—	70	134.8	—	—
268	0.044	301	0.145	159	1.994	251	4.544	264	0.044	300	0.145	154	1.994	249	4.544	119	5.000
157	0.207	380	0.010	158	2.054	349	0.541	154	0.207	380	0.010	153	2.054	347	0.541	186	1.000
209	0.100	194	1.279	258	0.410	217	9.592	207	0.100	191	1.279	254	0.410	214	9.592	171	1.450
284	0.034	290	0.215	339	0.118	324	1.307	280	0.034	289	0.215	334	0.118	321	1.307	221	0.351
282	0.035	223	0.811	238	0.544	204	12.57	278	0.035	218	0.811	236	0.544	201	12.57	145	2.817
216	0.091	258	0.381	160	1.965	235	6.173	214	0.091	256	0.381	155	1.965	233	6.173	156	2.241
218	0.090	295	0.170	223	0.666	280	2.744	216	0.090	294	0.170	221	0.666	278	2.744	235	0.250
334	0.013	317	0.104	263	0.378	241	5.158	331	0.013	316	0.104	259	0.378	239	5.158	210	0.470
269	0.044	266	0.303	228	0.622	179	18.72	265	0.044	264	0.303	226	0.622	176	18.72	—	—
74	1.309	126	4.156	55	16.31	133	42.68	68	1.309	122	4.156	47	16.31	128	42.68	—	—

Rank by total assets 1987	1986	Company	Total assets $000	Rank	Total revenue $000	Rank	Net income $000	Rank	Stockholders' equity $000	Rank	Capital & Expl. spending $000
256	227	Trinity Resources Ltd.[109]	9,703	226	4,843	233	(171)	248	3,564	239	1,111
257	242	Summit Energy Inc.[110]	9,673	252	3,124	302	(864)	235	4,459	245	1,029
258	233	Bellwether Exploration Co.[40]	9,637	286	[90]1,577	344	[90](3,238)	355	(450)	304	[90]267
259	250	Cobb Resources Corp.[40]	9,394	277	2,130	245	(245)	212	6,457	250	851
260	265	ZG Energy Corp.[111]	9,357	241	3,908	342	(3,154)	276	2,082	207	1,955
261	260	Com-Tek Resources Inc.[9]	9,335	299	1,282	292	(691)	247	3,738	167	5,172
262	256	Western Energy Development Co. Inc.[112,35]	9,082	294	[113]1,369	351	[113](4,744)	371	(9,813)	315	[113]178
263	—	Xplor Corp.[27]	8,988	253	3,077	204	(49)	199	7,931	343	77
264	281	Castle Energy Corp.[9]	8,940	225	4,870	104	2,502	245	3,890	230	1,356
265	248	Newhall Resources	8,590	219	[114]5,262	126	[114]1,315	204	[107]7,254	333	111
266	259	QED Exploration Inc.[110]	8,578	269	2,331	176	169	213	6,328	232	1,292
267	246	ConVest Energy Corp.	8,565	275	2,224	222	(129)	225	5,075	299	342
268	—	N.Y. Life Oil & Gas Prod. Prop. I-C Ltd.	8,294	263	2,638	237	(198)	201	[107]7,838	144	7,313
269	214	Ratex Resources Inc.[9]	8,180	274	2,227	367	(10,584)	377	(20,749)	329	125
270	—	N.Y. Life Oil & Gas Prod. Prop. II-A LP	8,137	341	660	168	237	202	[107]7,658	152	5,877
271	—	Graham Income Fund 82A L.P.	8,081	270	2,329	164	287	198	[108]8,062	—	NA
272	258	Home-Stake Oil & Gas Co.	7,916	231	4,387	130	1,132	209	6,932	234	1,245
273	278	Nahama & Weagant Energy Co.[13]	7,876	292	1,429	307	(990)	249	3,502	216	1,634
274	220	Lomak Petroleum Inc.	7,687	203	7,475	355	(6,026)	367	(5,925)	175	4,562
275	284	Credo Petroleum Corp.[103]	7,565	327	771	175	182	208	7,054	256	777
276	276	Alta Energy Corp.[40]	7,527	295	1,328	203	(46)	278	2,047	208	1,901
277	—	Search Natural Resources Inc.	7,365	399	34	282	(507)	314	1,042	165	5,182
278	252	Striker Petroleum Corp.	7,322	288	1,548	338	(2,407)	223	5,150	300	325
279	261	Pyramid Oil Co.	7,260	242	3,817	310	(1,034)	214	6,314	248	976
280	286	Whiting Petroleum Corp.	7,169	272	2,246	161	308	218	5,646	241	1,106
281	264	Golden Oil Co.[40]	7,031	285	1,592	270	(389)	222	5,295	319	164
282	271	GeoResources Inc.	6,898	261	2,764	181	105	241	3,980	303	274
283	268	Panhandle Royalty Co.[9]	6,865	259	2,790	152	508	234	4,555	258	765
284	280	Mustang Cos. Inc.	6,638	246	3,500	234	(179)	217	5,769	200	2,300
285	291	Gold King Consolidated Inc.[115]	6,574	291	1,453	257	(305)	228	5,004	219	1,582
286	—	Omni Exploration Inc.[9]	6,516	276	2,172	279	(483)	297	1,471	365	27
287	270	Houston Oil Royalty Trust	6,506	319	830	155	[73]470	211	[61]6,506	—	NA
288	267	Partners Oil Co.	6,431	284	1,594	299	(837)	256	3,217	254	799
289	295	Kelley Oil Corp.	6,333	236	4,103	251	(262)	274	2,118	222	1,557
290	309	Republic Resources Corp.	6,301	303	1,182	258	(310)	309	1,142	373	8
291	285	Prima Energy Corp.[40]	6,287	298	1,298	185	77	221	5,351	288	465
292	257	Basic Earth Science Systems Inc.[35]	6,177	278	1,921	324	(1,534)	270	2,260	271	601
293	—	Eglington Oil & Gas Corp.[116]	6,122	314	915	188	64	370	(9,497)	243	1,078
294	—	Plexus Resources Corp.[40]	6,091	355	510	284	(532)	261	2,822	253	816
295	—	Pinnacle Petroleum Inc.	5,946	310	1,029	298	(786)	244	3,892	296	361
296	273	Petroleum Reserve Corp.[40]	5,888	265	2,522	250	(260)	345	191	356	42
297	296	Burton/Hawks Inc.	5,711	244	3,735	195	16	238	4,136	305	263
298	224	Magic Circle Energy Corp.	5,673	245	3,510	113	1,830	231	4,688	269	668
299	—	Energy Sources Inc.[103]	5,492	308	1,082	325	(1,663)	364	(3,944)	325	13
300	—	N.Y. Life Oil & Gas Prod. Prop. I-A Ltd.	5,396	280	1,894	230	(158)	227	[105]5,060	177	4,174
301	282	Strata Corp.	5,173	266	2,500	326	(1,712)	373	(11,460)	—	NA
302	311	Normandy Oil & Gas Co. Inc.[40]	4,995	324	807	157	417	252	3,381	233	1,274
303	—	N.Y. Life Oil & Gas Prod. Prop. I-D Ltd.	4,992	348	569	184	99	229	[104]4,875	181	3,684
304	277	Altex Industries Inc.[9]	4,928	307	1,083	317	(1,343)	236	4,433	353	4
305	263	Usenco Inc.[103]	4,838	264	2,532	305	(962)	285	1,844	307	25
306	—	Woodbine Petroleum Inc.[9]	4,688	315	914	273	(417)	290	1,686	279	52

Worldwide liquids production		Worldwide natural gas production		Worldwide liquids reserves		Worldwide natural gas reserves		U.S. liquids production		U.S. natural gas production		U.S. liquids reserves		U.S. natural gas reserves		U.S. net wells drilled	
Rank	Mill bbl	Rank	Bcf	Rank	Mill bbl	Rank	Bcf	Rank	Mill bbl	Rank	Bcf	Rank	Mill bbl	Rank	Bcf	Rank	Wells
177	0.159	208	1.101	232	0.585	250	4.549	174	0.159	203	1.101	230	0.585	248	4.549	169	1.660
219	0.090	212	1.009	191	1.161	225	7.591	217	0.090	207	1.009	188	1.161	221	7.591	266	0.020
285	[90]0.033	254	[90]0.407	269	0.343	226	7.511	281	[90]0.033	252	[90]0.407	265	0.343	222	7.511	247	[90]0.170
226	0.085	300	0.146	150	2.354	268	3.355	224	0.085	299	0.146	147	2.354	266	3.355	173	1.340
192	0.127	243	0.532	214	0.852	243	5.104	190	0.127	240	0.532	210	0.852	241	5.104	238	0.209
258	0.050	279	0.243	169	1.637	213	9.945	255	0.050	278	0.243	165	1.637	210	9.945	—	—
203	[113]0.111	328	[113]0.089	187	1.326	285	2.598	201	[113]0.111	327	[113]0.089	184	1.326	283	2.598	—	—
220	0.090	252	0.410	252	0.452	252	4.525	218	0.090	250	0.410	249	0.452	250	4.525	240	0.200
264	0.046	239	0.581	319	0.167	202	12.78	260	0.046	234	0.581	315	0.167	199	12.78	141	3.300
197	0.123	187	1.358	248	0.480	229	7.226	195	0.123	184	1.358	245	0.480	226	7.226	164	1.800
240	0.072	264	0.330	270	0.338	296	2.228	238	0.072	262	0.330	266	0.338	294	2.228	101	7.471
251	0.056	263	0.353	235	0.560	282	2.704	248	0.056	261	0.353	233	0.560	280	2.704	258	0.090
205	0.106	236	0.608	242	0.521	242	5.113	203	0.106	231	0.608	239	0.521	240	5.113	—	NA
304	0.024	209	1.054	281	0.319	249	4.656	300	0.024	204	1.054	277	0.319	247	4.656	—	—
310	0.021	312	0.113	237	0.546	260	4.180	306	0.021	311	0.113	235	0.546	258	4.180	—	NA
206	0.106	216	0.908	243	0.519	245	5.047	204	0.106	211	0.908	240	0.519	243	5.047	—	NA
171	0.172	219	0.872	208	0.902	233	6.733	167	0.172	214	0.872	205	0.902	231	6.733	152	2.350
308	0.023	257	0.386	261	0.384	274	3.082	304	0.023	255	0.386	257	0.384	272	3.082	—	NA
300	0.026	249	0.436	353	0.065	303	1.939	296	0.026	247	0.436	348	0.065	302	1.939	136	3.500
321	0.018	344	0.054	328	0.151	281	2.719	317	0.018	344	0.054	324	0.151	279	2.719	179	1.185
276	0.039	309	0.120	245	0.510	223	8.641	272	0.039	308	0.120	242	0.510	219	8.641	236	0.230
—	NA	—	NA	289	0.278	265	3.569		NA		NA	285	0.278	263	3.569	—	NA
233	0.080	348	0.048	152	2.308	374	0.227	231	0.080	348	0.048	149	2.308	372	0.227	158	2.070
186	0.139	365	0.034	201	0.968	362	0.408	184	0.139	364	0.034	199	0.968	359	0.408	143	3.000
265	0.046	270	0.293	267	0.349	287	2.571	261	0.046	268	0.293	263	0.349	285	2.571	—	—
247	0.062	283	0.239	186	1.348	310	1.629	244	0.062	282	0.239	183	1.348	308	1.629	—	—
227	0.083	379	0.011	211	0.888	375	0.207	225	0.083	379	0.011	207	0.888	374	0.207	—	—
270	0.044	207	1.118	312	[75]0.190	231	[75]7.048	266	0.044	202	1.118	308	[75]0.190	228	[75]7.048	196	0.664
213	0.096	272	0.275	216	0.803	290	2.509	210	0.096	271	0.275	213	0.803	289	2.509	140	3.320
271	0.043	325	0.091	233	0.585	339	0.824	267	0.043	324	0.091	231	0.585	337	0.824	187	1.000
234	0.080	235	0.618	218	0.758	254	4.483	232	0.080	230	0.618	216	0.758	252	4.483	—	—
311	0.021	285	0.234	317	0.169	263	3.801	307	0.021	284	0.234	313	0.169	261	3.801	—	—
343	0.011	334	0.082	368	0.033	358	0.436	339	0.011	333	0.082	365	0.033	355	0.436	218	0.378
356	0.008	237	0.585	359	0.054	222	8.862	353	0.008	232	0.585	355	0.054	218	8.862	128	4.220
372	0.003	375	0.018	384	0.011	383	0.127	369	0.003	375	0.018	383	0.011	382	0.127	—	—
351	0.009	318	0.104	371	0.031	314	1.558	348	0.009	317	0.104	368	0.031	312	1.558	202	0.560
198	0.122	292	0.193	239	0.537	262	3.926	196	0.122	291	0.193	237	0.537	260	3.926	176	1.240
273	0.040	363	0.035	227	0.630	361	0.413	269	0.040	362	0.035	225	0.630	358	0.413	—	NA
348	0.010	326	0.091	347	0.081	277	2.839	344	0.010	325	0.091	342	0.081	275	2.839	231	0.270
294	0.028	305	0.128	325	0.156	329	1.150	290	0.028	304	0.128	321	0.156	327	1.150	195	0.720
200	0.117	214	0.967	207	0.912	253	4.497	198	0.117	209	0.967	204	0.912	251	4.497	188	1.000
255	0.053	336	0.074	219	0.754	292	2.481	252	0.053	335	0.074	217	0.754	291	2.481	244	0.187
379	0.001	323	0.094	389	0.005	342	0.729	379	0.001	322	0.094	389	0.005	339	0.729	—	—
272	0.042	286	0.226	249	0.474	305	1.816	268	0.042	285	0.226	246	0.474	304	1.816	—	—
236	0.079	251	0.414	255	0.441	273	3.193	234	0.079	249	0.414	251	0.441	271	3.193	—	NA
293	0.029	255	0.406	326	0.153	279	2.805	289	0.029	253	0.406	322	0.153	277	2.805	—	—
335	0.013	313	0.113	293	0.263	315	1.530	332	0.013	312	0.113	289	0.263	313	1.530	189	1.000
309	0.022	316	0.106	268	0.347	275	3.078	305	0.022	315	0.106	264	0.347	273	3.078	—	NA
295	0.028	354	0.044	304	0.216	379	0.172	291	0.028	353	0.044	300	0.216	378	0.172	—	—
229	0.082	280	0.240	156	2.114	267	3.359	227	0.082	279	0.240	152	2.114	265	3.359	241	0.200
312	0.021	273	0.275	302	0.220	266	3.543	308	0.021	272	0.275	298	0.220	264	3.543	217	0.380

Rank by total assets 1987	1986	Company	Total assets $000	Rank	Total revenue $000	Rank	Net income $000	Rank	Stockholders' equity $000	Rank	Capital & Expl. spending $000
307	298	Dallas Oil & Minerals Inc.	4,681	297	1,308	238	(207)	233	4,595	231	1,297
308	301	Greenwood Holdings Inc.	4,571	249	[117]3,300	102	[117]2,593	254	3,346	264	[117]713
309	290	Great Eastern Energy & Development Corp.	4,406	283	1,655	243	(235)	240	4,050	240	1,11
310	—	Comstock Resources Inc.	4,381	364	379	277	(462)	283	1,917	198	2,50
311	294	Monogram Oil & Gas Inc.[103]	4,344	300	1,254	219	(124)	366	(5,003)	367	2
312	297	Wexco of Delaware Inc.[118]	4,325	302	1,196	283	(513)	253	3,379	252	82
313	300	Parallel Petroleum Corp.	4,286	329	764	252	(268)	258	3,097	255	78
314	289	Wepco Energy Co.	4,089	311	986	311	(1,066)	272	2,147	311	23
315	304	Energy Capital Development Corp.	3,890	352	529	303	(889)	311	1,101	331	12
316	299	K.R.M. Petroleum Corp.	3,742	279	1,909	315	(1,173)	262	2,727	242	1,08
317	321	Ranger Oil Ltd.[51]	3,656	325	806	167	[44]256	—	NA	—	N.
318	—	Senergy 1986 Ltd.	3,593	333	720	354	(5,887)	250	[103]3,495	154	5,78
319	302	Kimco Energy Corp.[119,40]	3,590	290	1,515	319	(1,454)	295	1,480	238	1,11
320	312	Southern Mineral Corp.	3,547	267	2,389	159	361	259	2,995	218	1,59
321	313	Tyrex Oil Co.[40]	3,488	306	1,122	226	(145)	255	3,277	229	1,38
322	—	Energetics Inc.	3,470	353	529	274	(419)	281	1,997	266	69
323	308	Arch Petroleum Inc.[120,103]	3,372	312	[121]975	162	[121]303	291	1,626	342	[121]7
324	—	N.Y. Life Oil & Gas Prod. Prop. I-B Ltd.	3,344	301	1,227	223	(129)	257	[103]3,103	188	3,25
325	—	International Basic Resources Inc.[40]	3,287	394	80	268	(355)	313	1,057	314	19
326	—	Wright Brothers Energy Inc.[105]	3,223	282	1,699	278	(468)	299	1,431	316	17
327	327	Appalachian Oil & Gas Co. Inc.[40]	3,180	271	2,267	158	417	320	805	272	59
328	293	Valex Petroleum Inc.	3,114	304	1,158	341	(3,142)	331	644	257	77
329	336	Questa Oil & Gas Co.	3,068	313	970	216	(108)	292	1,550	267	67
330	322	Chaparral Resources Inc.[65]	3,055	323	814	271	(391)	275	2,104	286	48
331	345	Impact Energy Inc.	2,974	381	211	269	(374)	306	1,212	298	34
332	305	NP Energy Corp.[35]	2,886	387	173	224	(137)	308	1,162	366	2
333	316	Aspen Exploration Corp.[40]	2,858	356	468	246	(247)	263	2,709	345	7
334	315	Crescent Oil & Gas Corp.[9]	2,631	362	420	297	(767)	271	2,223	338	8
335	360	HarCor Energy Inc.[122]	2,574	296	1,317	198	1	269	2,293	292	41
336	340	Western Energy Resources Inc.	2,552	373	269	285	(532)	265	2,494	237	1,12
337	324	Pantepec International Inc.	2,543	386	183	247	(249)	267	2,457	—	
338	—	Argosy Energy Inc.	2,535	346	603	262	(325)	277	2,074	287	46
339	333	Keldon Oil Co.[110]	2,529	331	736	202	(34)	326	708	376	
340	323	Penn-Pacific Corp.[9]	2,527	367	336	165	259	268	2,391	—	
341	329	Double Eagle Petroleum & Mining Co.[27]	2,321	320	827	221	(128)	273	2,136	281	52
342	328	Intramerican Oil & Minerals Inc.	2,269	332	722	256	(303)	286	1,837	332	1
343	—	Broken Arrow Petroleum Co.[19]	2,259	351	545	182	102	287	1,828	306	2
344	346	Taurus Petroleum Inc.[9]	2,167	316	905	264	(335)	316	1,000	276	5
345	318	Plano Petroleum Corp.[9]	2,150	357	465	287	(604)	279	2,007	285	5
346	310	Roseland Oil & Gas Inc.[9]	2,135	360	[123]429	201	[123](25)	307	1,173	337	[123]
347	332	Nova Natural Resources Corp.[9]	2,122	338	669	239	(207)	280	2,004	341	
348	319	Roberts Oil & Gas Inc.[19]	2,062	354	516	313	(1,112)	347	110	291	4
349	242	Summit Petroleum Corp.[110]	2,042	396	62	261	(323)	346	162	372	
350	341	Petrol Industries Inc.	2,027	287	1,567	166	259	284	1,902	371	
351	337	Albion International Resources Inc.	2,003	369	327	186	75	282	1,948	362	
352	351	Pioneer Oil & Gas Corp.[9]	1,946	321	822	178	160	298	1,459	294	3
353	—	Hailey Energy Corp.	1,933	368	335	335	(2,235)	305	1,243	320	1
354	325	Energy Resources of North Dakota Inc.[35]	1,917	334	704	235	(185)	289	1,712	308	2
355	307	Twin Richfield Oils Inc.[51]	1,888	379	220	275	(455)	350	(10)	—	
356	—	Struthers Oil & Gas Corp.	1,852	335	701	96	2,981	349	37	339	
357	342	General Energy Resources & Technology Corp.[124]	1,846	317	895	211	(70)	321	797	317	1

Worldwide liquids production		Worldwide natural gas production		Worldwide liquids reserves		Worldwide natural gas reserves		U.S. liquids production		U.S. natural gas production		U.S. liquids reserves		U.S. natural gas reserves		U.S. net wells drilled	
Rank	Mill bbl	Rank	Bcf	Rank	Mill bbl	Rank	Bcf	Rank	Mill bbl	Rank	Bcf	Rank	Mill bbl	Rank	Bcf	Rank	Wells
289	0.031	265	0.321	305	0.214	300	1.968	285	0.031	263	0.321	301	0.214	299	1.968	263	0.059
263	[117]0.047	302	[117]0.136	294	0.256	289	2.522	259	[117]0.047	301	[117]0.136	290	0.256	288	2.522	57	[117]23.00
257	0.051	327	0.091	285	[75]0.292	317	[75]1.513	254	0.051	326	0.091	281	[75]0.292	315	[75]1.513	139	3.400
380	0.001	—	—	250	0.457	372	0.250	380	0.001	—	—	247	0.457	370	0.250	—	NA
329	0.015	253	0.410	360	0.054	284	2.684	326	0.015	251	0.410	356	0.054	282	2.684	—	—
286	0.032	275	0.261	332	0.146	326	1.216	282	0.032	274	0.261	328	0.146	324	1.216	181	1.170
320	0.019	288	0.219	276	0.325	298	2.183	316	0.019	287	0.219	272	0.325	296	2.183	177	1.200
296	0.027	294	0.185	286	0.283	302	1.942	292	0.027	293	0.185	282	0.283	301	1.942	267	0.020
367	0.004	335	0.082	273	0.331	218	9.487	364	0.004	334	0.082	269	0.331	215	9.487	—	—
237	0.079	262	0.369	280	0.320	237	5.532	235	0.079	260	0.369	276	0.320	235	5.532	182	1.170
287	0.032	296	0.169	322	0.163	359	0.433	283	0.032	295	0.169	318	0.163	356	0.433	242	0.200
305	0.024	299	0.152	320	0.167	321	1.344	301	0.024	298	0.152	316	0.167	318	1.344	—	NA
235	0.080	355	0.044	274	0.329	380	0.157	233	0.080	354	0.044	270	0.329	379	0.157	192	0.810
297	0.027	217	0.885	346	[75]0.085	238	[75]5.482	293	0.027	212	0.885	341	[75]0.085	236	[75]5.482	149	2.608
283	0.035	362	0.039	296	0.248	357	0.445	279	0.035	361	0.039	292	0.248	354	0.445	—	—
362	0.005	345	0.053	340	0.114	323	1.314	359	0.005	345	0.053	335	0.114	320	1.314	—	NA
301	[121]0.026	304	[121]0.129	217	0.779	313	1.566	297	[121]0.026	303	[121]0.129	214	0.779	311	1.566	257	[121]0.100
259	0.049	271	0.284	303	0.220	297	2.218	256	0.049	270	0.284	299	0.220	295	2.218	—	NA
—	—	360	0.040	—	—	309	1.650	—	—	359	0.040	—	—	307	1.650	—	NA
324	0.017	241	0.545	323	0.159	200	13.36	320	0.017	235	0.545	319	0.159	197	13.36	207	0.500
387	(s)	203	1.148	392	0.001	232	6.840	387	(s)	198	1.148	392	0.001	230	6.840	144	2.835
256	0.053	297	0.168	204	0.942	291	2.494	253	0.053	296	0.168	202	0.942	290	2.494	—	—
344	0.011	281	0.240	356	[75]0.061	272	[75]3.267	340	0.011	280	0.240	351	[75]0.061	270	[75]3.267	170	1.619
352	0.009	303	0.132	327	0.152	270	3.272	349	0.009	302	0.132	323	0.152	268	3.272	—	—
336	0.013	—	—	344	0.098	—	—	333	0.013	—	—	339	0.098	—	—	—	—
385	0.001	307	0.124	338	0.122	258	4.363	385	0.001	306	0.124	333	0.122	256	4.363	—	—
313	0.021	376	0.018	292	0.268	348	0.619	309	0.021	376	0.018	288	0.268	346	0.619	262	0.065
368	0.004	298	0.156	374	0.028	261	4.172	365	0.004	297	0.156	371	0.028	259	4.172	193	0.810
292	0.030	259	0.378	277	0.324	299	2.042	288	0.030	257	0.378	273	0.324	298	2.042	—	—
353	0.009	333	0.084	306	0.206	308	1.741	350	0.009	332	0.084	302	0.206	306	1.741	—	—
381	0.001	329	0.088	385	0.011	306	1.787	381	0.001	328	0.088	384	0.011	323	1.271	191	0.830
244	0.066	—	—	254	0.447	390	0.060	371	0.002	—	—	385	0.010	390	0.060	—	—
314	0.021	321	0.096	264	0.365	330	1.060	310	0.021	320	0.096	260	0.365	328	1.060	—	—
360	0.007	314	0.111	345	0.096	318	1.487	357	0.007	313	0.111	340	0.096	316	1.487	—	—
288	0.032	368	0.028	301	0.221	331	1.058	284	0.032	367	0.028	297	0.221	329	1.058	200	0.570
330	0.014	282	0.240	314	0.179	276	3.035	327	0.014	281	0.240	310	0.179	274	3.035	254	0.108
—	—	384	0.006	—	—	328	1.167	—	—	384	0.006	—	—	326	1.167	—	—
315	0.021	289	0.217	329	0.150	319	1.431	311	0.021	288	0.217	325	0.150	317	1.431	—	—
331	0.014	347	0.052	352	0.070	336	0.911	328	0.014	347	0.052	347	0.070	334	0.911	123	4.670
339	[123]0.012	353	[123]0.045	337	0.130	368	0.314	335	[123]0.012	352	[123]0.045	332	0.130	366	0.314	—	NA
325	0.017	352	0.046	321	0.165	351	0.526	321	0.017	351	0.046	317	0.165	349	0.526	249	0.140
349	0.010	284	0.239	370	0.032	316	1.523	345	0.010	283	0.239	367	0.032	314	1.523	269	0.012
374	0.002	374	0.019	349	0.079	334	0.975	372	0.002	374	0.019	344	0.079	332	0.975	—	NA
232	0.081	372	0.024	100	6.241	363	0.407	230	0.081	372	0.024	91	6.241	360	0.407	—	—
340	0.012	358	0.041	284	0.297	332	1.017	336	0.012	357	0.041	280	0.297	330	1.017	—	—
179	0.151	308	0.124	295	0.252	293	2.377	176	0.151	307	0.124	291	0.252	292	2.377	—	NA
388	(s)	367	0.031	377	0.024	335	0.922	388	(s)	366	0.031	375	0.024	333	0.922	—	NA
303	0.025	388	0.003	324	0.158	394	0.009	299	0.025	388	0.003	320	0.158	394	0.009	233	0.260
—	—	311	0.114	—	—	311	1.613	—	—	310	0.114	—	—	309	1.613	—	—
363	0.005	269	0.296	390	0.004	364	0.384	360	0.005	267	0.296	390	0.004	361	0.384	272	0.001
298	0.027	310	0.119	342	0.107	322	1.320	294	0.027	309	0.119	337	0.107	319	1.320	270	0.010

Rank by total assets 1987	1986	Company	Total assets $000	Rank	Total revenue $000	Rank	Net income $000	Rank	Stockholders' equity $000	Rank	Capital & Expl. spending Rank	$000
358	314	Columbine Exploration Corp.	1,837	293	[125]1,415	322	[125](1,483)	362	(2,683)	349	[125]63
359	339	Castle Royalty L.P.	1,812	374	248	213	(96)	288	[10]1,787	—	0
360	317	Lyric Energy Inc.[105]	1,786	365	370	337	(2,317)	358	(746)	350	55
361	—	Calcasieu Real Estate & Oil Co. Inc.	1,739	340	664	163	298	294	1,499	361	32
362	306	Oxford Consolidated Inc.[35]	1,724	305	1,142	312	(1,090)	335	522	274	583
363	343	Willard Pease Oil & Gas Co.	1,705	345	604	236	(193)	338	451	335	100
364	320	Exploration Co.[27]	1,685	385	194	330	(1,971)	334	552	265	692
365	347	Big Piney Oil & Gas Co.	1,621	382	199	207	(54)	293	1,504	374	8
366	—	Morrison Nuclear Inc.[126]	1,598	328	770	199	0	363	(3,432)	369	19
367	—	Cap Ltd.	1,581	326	778	170	212	301	[10]1,370	324	139
368	—	Gladstone Resources Inc.	1,555	395	73	210	(61)	303	1,330	—	0
369	—	International Management & Research Corp.	1,513	388	159	288	(605)	323	784	347	66
370	355	Hubco Exploration Inc.	1,483	347	602	215	(104)	353	(312)	284	510
371	358	Sheffield Exploration Co. Inc.[40]	1,433	318	892	225	(143)	310	1,111	273	587
372	366	Golden Triangle Royalty & Oil Inc.	1,410	391	111	263	(330)	302	1,362	330	125
373	364	Cheyenne Resources Inc.	1,324	378	227	254	(297)	325	727	310	237
374	348	Reserve Exploration Co.[35]	1,252	358	455	229	(155)	341	378	354	46
375	361	Sharon Energy Ltd.[35]	1,237	337	685	190	35	352	(230)	283	511
376	—	DOL Resources Inc.	1,223	390	132	217	(114)	333	634	—	0
377	359	EMC Energies Inc.[40]	1,206	366	348	227	(152)	336	512	348	65
378	352	Toklan Oil Corp.[27]	1,193	330	761	240	(212)	328	690	323	146
379	349	Love Oil Co. Inc.[19]	1,173	380	212	280	(499)	315	1,011	275	580
380	—	Bogert 1985-III L.P.	1,168	344	616	253	(292)	317	[10]988	—	6
381	369	TPEX Exploration Inc.[40]	1,094	361	423	232	(167)	329	686	318	167
382	356	Wicklund Holding Co.[127]	1,089	372	286	255	(297)	319	825	377	5
383	—	REO Production I-A L.P.	1,083	398	53	197	2	312	[10]1,075	247	1,01[3]
384	350	International Royalty & Oil Co.[9]	1,050	389	135	304	(950)	359	(844)	351	5[1]
385	368	Bolyard Oil & Gas Ltd.[105]	1,019	375	242	212	(85)	339	435	364	28
386	387	Texas Vanguard Oil Co.	1,018	349	567	179	131	330	675	313	198
387	363	Carmel Energy Inc.	1,014	342	656	200	(10)	344	205	—	6
388	354	Eagle Exploration Co.[35]	1,008	350	552	309	(1,000)	368	(6,241)	326	13
389	—	Loch Exploration Inc.	1,002	397	58	281	(505)	357	(720)	—	NA
390	—	National Petroleum Corp. Ltd.[65,66]	1,000	370	311	260	(319)	348	47	370	1
391	—	Minex Resources Inc.[65]	967	392	108	194	28	318	894	363	3
392	—	Texoil Inc.	960	377	228	259	(316)	342	344	346	6
393	365	Bengal Oil & Gas Corp.	936	383	197	272	(393)	354	(367)	336	10
394	370	Tri-Valley Corp.[110]	833	343	617	248	(251)	343	296	309	23
395	—	Winco Petroleum Corp.[9]	801	393	98	208	(59)	322	787	358	3
396	382	Amacan Resources Corp.[105]	780	371	303	189	57	324	769	368	2
397	374	Oil City Petroleum Inc.[27]	765	376	242	265	(348)	340	391	357	4
398	381	Yellowstone Resources Inc.	727	384	197	209	(60)	327	693	328	13
399	375	Hipoint Investments Ltd.	724	359	442	276	(457)	376	(20,069)	340	8
400	—	Rotan Mosle Energy Fund III Ltd.	646	339	666	192	34	332	[10]636	359	3
		Totals	501,191,474		445,743,989		9,371,547		172,805,668		41,630,24[?]

NA = Not Available. (s) indicates less than 500 bbl or 500 mcf. Fiscal yearend is Dec. 31 unless otherwise noted. 1. At fiscal yearend, was operating under Chapter 11. Emerged from Chapter 11 on 4-7-88. 2. Formerly Standard Oil. Subsidiary of British Petroleum Co. plc. 3. Includes 2 operating entities -- Marathon Oil and Texas Oil & Gas. 4. Operating income before foreign income taxes. 5. Subsidiary of E.I. du Pont de Nemours & Co. 6. After tax operating income. 7. Includes production and reserves attributed to both natural gas systems and exploration and production operations. 8. Wells attributable to exploration and production operations only. 9. Fiscal yearend Sep. 30. 10. Partners' capital. 11. May include insignificant amounts outside U.S. 12. Subsidiary of Burlington Northern Inc. 13. Fiscal yearend Jan. 31, 1988. 14. Formerly Diamond Shamrock. Refining operations were spun-off to form Diamond Shamrock R & M Inc. 15. Subsidiary of Union Pacific Corp. Since fiscal yearend has purchased Beard Corp. 16. Includes both cost of service and non-cost of service activities. 17. Unocal owns 86.04% of outstanding partnership units. 18. Subsidiary of Broken Hill Proprietary Co. Ltd. 19. Fiscal yearend May 31. 20. Enserch owns 87% of outstanding partnership units. 21. Transco owns 73% of outstanding partnership units. 22. Freeport-McMoRan owns 59% of outstanding partnership units. 23. Subsidiary of Santa Fe Southern Pacific. 24. Includes small amounts attributable to coal and lumber operations. 25. Operating income. 26. Murphy Oil owns 61% of outstanding stock. 27. Fiscal yearend Aug. 31. 28. Subsidiary of W.R. Grace & Co. 29. Identifiable assets. 30. Subsidiary of Newmont Mini[ng] 31. Since fiscal yearend, company has announced plans to sell all oil and gas reserves. 32. Partn[ers'] capital less demand notes receivable. 33. Apache owns approximately 6.1% of outstand[ing] partnership units. 34. Formerly Pacific Lighting Oil & Gas. Subsidiary of Pacific Enterprises. Si[nce] fiscal yearend has purchased Sabine. 35. Fiscal yearend Mar. 31. 36. Santa Fe Energy owns 82% [of] outstanding partnership units. 37. Subsidiary of CSX Corp. Since fiscal yearend all oil and [gas] operations have been sold to Total Minatome. 38. Maxus owns 81% of outstanding partnership un[its]. 39. Company is an asset based holding company which is 82% owned by Cortez Internatio[nal] 40. Fiscal yearend June 30. 41. Capital and members' equity. 42. Subsidiary of Total Compag[nie] Francaise des Petroles. Since the fiscal yearend has been purchased by Pacific Enterpris[es]. 46. Subsidiary of Southdown Inc. 47. Subsidiary of Houston Industries Inc. 48. Operating inco[me] before corporate overhead and interest. 49. Damson Oil owns 16.39% of outstanding partners[hip] units. 50. Subsidiary of CMS Energy Corp. 51. U.S. operations only. 52. Subsidiary of Placer D[ome] Inc. 53. Subsidiary of Hiram Walker Resources Ltd. 54. Net operating income. 55. Subsidiary of [...] Group Inc. 56. At fiscal yearend was subsidiary of Moore McCormack Resources. Since then has b[een] purchased by Canadian Occidental Petroleum Ltd. 57. Kaneb Services owns 81.3% of outstand[ing] partnership units. 58. Six months ended Dec. 31. 59. Subsidiary of Pacific Gas & Electric

| Worldwide liquids production | | Worldwide natural gas production | | Worldwide liquids reserves | | Worldwide natural gas reserves | | U.S. liquids production | | U.S. natural gas production | | U.S. liquids reserves | | U.S. natural gas reserves | | U.S. net wells drilled | |
Rank	Mill bbl	Rank	Bcf	Rank	Mill bbl	Rank	Bcf	Rank	Mill bbl	Rank	Bcf	Rank	Mill bbl	Rank	Bcf	Rank	Wells
290	[125]0.031	338	[125]0.068	298	0.238	344	0.698	286	[125]0.031	337	[125]0.068	294	0.238	342	0.698	—	—
361	0.006	350	0.047	367	0.035	360	0.431	358	0.006	349	0.047	364	0.035	357	0.431	—	—
364	0.005	339	0.064	330	0.147	269	3.318	361	0.005	339	0.064	326	0.147	267	3.318	219	0.376
369	0.004	369	0.028	388	0.007	382	0.141	366	0.004	368	0.028	388	0.007	381	0.141	248	0.150
266	0.045	293	0.193	310	0.195	392	0.033	262	0.045	292	0.193	306	0.195	392	0.033	215	0.398
306	0.024	320	0.100	291	0.269	301	1.957	302	0.024	319	0.100	287	0.269	300	1.957	—	—
357	0.008	351	0.047	380	0.022	376	0.206	354	0.008	350	0.047	378	0.022	375	0.206	—	—
—	—	324	0.093	—	—	307	[75]1.748	—	—	323	0.093	—	—	305	[75]1.748	—	NA
267	0.045	386	0.004	335	0.140	—	—	263	0.045	386	0.004	330	0.140	—	—	—	—
316	0.021	278	0.246	350	0.075	333	1.004	312	0.021	277	0.246	345	0.075	331	1.004	—	—
373	0.003	382	0.009	362	[75]0.053	283	[75]2.694	370	0.003	382	0.009	358	[75]0.053	281	[75]2.694	—	—
—	—	364	0.035	—	—	352	0.488	—	—	363	0.035	—	—	350	0.488	—	NA
322	0.018	332	0.085	313	0.184	345	0.676	318	0.018	331	0.085	309	0.184	343	0.676	222	0.323
302	0.026	319	0.103	308	0.203	343	0.710	298	0.026	318	0.103	304	0.203	340	0.710	—	—
383	0.001	385	0.006	386	0.008	388	0.065	383	0.001	385	0.006	386	0.008	388	0.065	—	—
354	0.009	370	0.025	343	0.104	365	0.341	351	0.009	370	0.025	338	0.104	362	0.341	194	0.750
299	0.027	373	0.022	311	0.194	384	0.121	295	0.027	373	0.022	307	0.194	383	0.121	208	0.500
326	0.017	342	0.059	341	0.113	355	0.468	322	0.017	342	0.059	336	0.113	352	0.468	175	1.250
370	0.004	371	0.025	364	0.041	378	0.199	367	0.004	371	0.025	361	0.041	377	0.199	—	—
341	0.012	356	0.043	336	0.137	353	0.476	337	0.012	355	0.043	331	0.137	351	0.476	—	—
345	0.011	277	0.248	331	0.147	338	0.899	341	0.011	276	0.248	327	0.147	336	0.899	237	0.224
358	0.008	390	0.001	372	0.030	393	0.011	355	0.008	390	0.001	369	0.030	393	0.011	201	0.570
332	0.014	276	0.258	354	0.064	320	1.349	329	0.014	275	0.258	349	0.064	386	0.085	—	—
323	0.018	330	0.087	361	0.054	356	0.452	319	0.018	329	0.087	357	0.054	353	0.452	—	—
375	0.002	331	0.086	379	0.023	347	0.627	373	0.002	330	0.086	377	0.023	345	0.627	260	0.079
376	0.002	383	0.008	375	0.028	341	0.798	374	0.002	383	0.008	372	0.028	338	0.798	—	—
382	0.001	377	0.015	387	[75]0.008	377	[75]0.200	382	0.001	377	0.015	387	[75]0.008	376	[75]0.200	—	—
377	0.002	341	0.061	383	0.013	304	1.834	375	0.002	341	0.061	382	0.013	303	1.834	—	—
333	0.014	357	0.043	287	0.282	327	1.188	330	0.014	356	0.043	283	0.282	325	1.188	69	18.45
277	0.039	—	—	266	0.350	—	NA	276	0.036	—	—	262	0.350	—	—	—	—
307	0.024	337	0.074	369	0.033	381	0.153	303	0.024	336	0.074	366	0.033	380	0.153	178	1.188
366	0.004	378	0.013	363	0.052	371	0.269	363	0.004	378	0.013	360	0.052	369	0.269	—	—
338	0.012	349	0.048	333	0.144	354	0.469	376	0.002	369	0.026	373	0.028	364	0.336	—	—
378	0.002	391	0.001	348	0.080	386	0.090	377	0.002	391	0.001	343	0.080	385	0.090	—	NA
365	0.005	340	0.064	376	0.027	369	0.313	362	0.005	340	0.064	374	0.027	367	0.313	159	2.000
386	(s)	306	0.128	391	0.003	346	0.667	386	(s)	305	0.128	391	0.003	344	0.667	—	—
389	(s)	343	0.059	—	—	385	0.111	389	(s)	343	0.059	—	—	384	0.111	160	2.000
371	0.004	389	0.001	378	0.023	391	0.052	368	0.004	389	0.001	376	0.023	391	0.052	—	—
350	0.010	361	0.040	358	0.057	370	0.302	346	0.010	360	0.040	354	0.057	368	0.302	253	0.110
346	0.011	387	0.004	351	0.071	387	0.081	342	0.011	387	0.004	346	0.071	387	0.081	—	—
359	0.008	381	0.010	366	0.038	389	0.065	356	0.008	381	0.010	363	0.038	389	0.065	203	0.544
337	0.013	346	0.053	365	0.041	366	0.337	334	0.013	346	0.053	362	0.041	363	0.337	251	0.130
347	0.011	287	0.221	381	[75]0.020	350	[75]0.534	343	0.011	286	0.221	379	[75]0.020	348	[75]0.534	—	NA
	3,743.5		13,449.6		36,237.4		168,880.9		2,343.5		10,226.8		23,182.6		115,239.8		9,543.4

60. Subsidiary of Pacificorp. 61. Trust corpus. 62. Net productive wells only. 63. Since fiscal yearend has purchased all oil and gas operations of Kirby Exploration. 64. Subsidiary of CBI Industries Inc. 65. Fiscal yearend Nov. 30. 66. Financial data converted from Canadian dollars using exchange rates in effect at fiscal yearend. 67. Since fiscal yearend has sold oil and gas operations to American Exploration Co. 68. Includes acquisition of Houston Oil Fields Corp. 69. Damson Oil owns 10.52% of outstanding partnership units. 70. Since fiscal yearend has merged with Union Pacific Resources. 71. Damson Oil owns 6.36% of outstanding partnership units. 72. Apache owns 3.5% of outstanding partnership units. 73. Distributable cash. 74. Year ended Oct. 31. 75. Proved developed only. 76. Subsidiary of Washington Energy. 77. Revenue from unaffiliated customers only. 78. Since fiscal yearend has merged with Snyder Oil Partners. 79. Formerly Leader Development Corp. 80. Subsidiary of Brooklyn Union Gas. 81. Formerly American Quasar Petroleum. 82. Subsidiary of Gulf Resources & Chemical Corp. 83. Kelley Oil owns approximately 3.61% of outstanding partnership units. 84. Formerly Entex Energy Development Ltd. 85. Subsidiary of Northwest Natural Gas Co. 86. Subsidiary of Public Service Co. of Colorado. 87. Callon Petroleum owns 9% of outstanding partnership units. 88. Oil and gas operations only. Subsidiary of Penn Virginia Corp. 89. Operating profit before taxes, interest, depreciation and depletion. 90. Six months ended June 30. 91. U.S. subsidiary of Coho Resources Ltd. 92. Subsidiary of MDU Resources Group. 93. Subsidiary of Deminex Deutsche Erdoelversorgungs GmbH. 94. Saxon Oil owns 67% of outstanding partnership units. 95. Consolidated Oil & Gas owns 61% of outstanding partnership units. 96. ConVest owns 21.5% of outstanding partnership units. 97. Year ended Sep. 1. 98. U.S. operations of Inter-City Gas Corp. — Resources Div. 99. Formerly Zim Energy Corp. 100. Subsidiary of Energen Corp. 101. Subsidiary of Bridge Oil Ltd. 102. Fiscal yearend Feb. 29, 1988. 103. Fiscal yearend Oct. 31. 104. Subsidiary of Canada Northwest Energy Ltd. 105. Fiscal yearend Apr. 30. 106. Subsidiary of Sparton Corp. 107. Eight months ended Aug. 31. 108. Formerly Wichita Industries. 109. Canadian company which operates exclusively in U.S. 110. Fiscal yearend July 31. 111. Formerly Burkhart Petroleum Corp. 112. Since fiscal yearend has merged with Oxford Consolidated. 113. Fifteen months ended Mar. 31. 114. Historical cost basis. 115. Formerly Gerber Energy International. 116. Data includes Osceola Oil & Gas Corp. Both are subsidiaries of Eglington Exploration plc. 117. Fifteen months ended Mar. 31. 118. Subsidiary of WICOR Inc. 119. Formerly Henry Energy Corp. 120. Since fiscal yearend has merged with Usenco. 121. Ten months ended Oct. 31. 122. Formerly Pangea Petroleum Corp. 123. Eight months ended Sep. 30. 124. Formerly Reef Energy Corp. 125. Thirteen months ended Dec. 31. 126. U.S. operating subsidiary of Morrison Petroleums Ltd. 127. Formerly Wicklund Petroleum Corp.

The OGJ400 Company Index

Rank by total assets	Company name	Headquarters city
230	Adams Resources & Energy Inc.	Houston
73	Adobe Resources Corp.	New York
165	Alamco Inc.	Clarksburg, WV
351	Albion International Resources Inc.	Laguna Beach, CA
191	Alexander Energy Corp.	Oklahoma City
98	Allegheny & Western Energy Corp.	Charleston, WV
276	Alta Energy Corp.	Midland
304	Altex Industries Inc.	Denver
396	Amacan Resources Corp.	Salt Lake City
207	Amax Oil & Gas Inc.	Houston
19	Amerada Hess Corp.	New York
127	American Exploration Co.	Houston
215	American National Petroleum Co.	Houston
37	American Petrofina Inc.	Dallas
6	Amoco Corp.	Chicago
43	Anadarko Petroleum Corp.	Houston
69	Apache Corp.	Denver
147	Apache Offshore Investment Partnereship	Denver
72	Apache Petroleum Co. L.P.	Denver
327	Appalachian Oil & Gas Co. Inc.	Chattanooga, TN
323	Arch Petroleum Inc.	Fort Worth
8	ARCO (Atlantic Richfield Corp.)	Los Angeles
338	Argosy Energy Inc.	Salt Lake City
29	Arkla Inc.	Shreveport
21	Ashland Oil Inc.	Ashland, KY
333	Aspen Exploration Corp.	Denver
111	Avalon Corp.	New York
212	Barrett Resources Corp.	Denver
214	Baruch-Foster Corp.	Dallas
292	Basic Earth Science Systems Inc.	Englewood, CO
141	Beard Co.	Oklahoma City
228	Belcor Inc.	Irvine, CA
204	Belden & Blake Energy Co.	North Canton, OH
258	Bellwether Exploration Co.	Houston
393	Bengal Oil & Gas Corp.	London, U.K.
139	Berry Petroleum Co.	Taft, CA
45	BHP Petroleum (Americas) Inc.	Houston
365	Big Piney Oil & Gas Co.	Salt Lake City
380	Bogert 1985-III L.P.	Oklahoma City
156	Bogert Oil Co.	Oklahoma City
385	Bolyard Oil & Gas Ltd.	Denver
7	BP America	Cleveland
219	Bridge Oil (USA) Inc.	San Antonio
248	Brock Exploration Corp.	New Orleans
343	Broken Arrow Petroleum Co.	Scottsdale, AZ
297	Burton/Hawks Inc.	Casper, WY
42	Cabot Corp.	Waltham, MA
361	Calcasieu Real Estate & Oil Co. Inc.	Lake Charles, LA
178	Callon Consolidated Partners L.P.	Natchez, MS
229	Callon Petroleum Co.	Natchez, MS
154	Calumet Industries Inc.	Chicago
367	Cap Ltd.	Oklahoma City
387	Carmel Energy Inc.	Houston
264	Castle Energy Corp.	Blue Bell, PA
359	Castle Royalty L.P.	Kansas City, MO
62	CENEX	St. Paul, MN
330	Chaparral Resources Inc.	Denver
181	Chapman Energy Inc.	Dallas
3	Chevron Corp.	San Francisco
373	Cheyenne Resources Inc.	Cheyenne, WY
133	Chieftain Development Co. Ltd.	Edmonton, Alta., Can.
163	Clinton Gas Systems Inc.	Columbus, OH
17	Coastal Corp.	Houston
259	Cobb Resources Corp.	Albuquerque
189	Coho Resources Inc.	Dallas
18	Columbia Gas System Inc.	Wilmington, DE
221	Columbian Energy Co. L.P.	Topeka, KS
358	Columbine Exploration Corp.	Denver
310	Comstock Resources Inc.	Dallas
261	Com-Tek Resources Inc.	Denver
16	Conoco Inc.	Houston
101	Conquest Exploration Co.	Houston
199	Consolidated Energy Partners L.P.	Denver
22	Consolidated Natural Gas Co.	Pittsburgh
132	Consolidated Oil & Gas Inc.	Denver
267	ConVest Energy Corp.	Houston
202	ConVest Energy Partners Ltd.	Houston
275	Credo Petroleum Corp.	Denver
334	Crescent Oil & Gas Corp.	Denver
131	Crystal Oil Co.	Shreveport
80	CSX Oil & Gas Corp.	Houston
227	Daleco Resources Corp.	Los Angeles
307	Dallas Oil & Minerals Inc.	Fort Worth
96	Damson Energy Co. L.P.	New York
138	Damson Income Energy L.P.	New York
145	Damson Institutional Energy L.P.	New York
115	Damson Oil Corp.	New York
87	Dekalb Energy Cos.	Dekalb, IL
193	Deminex US Oil Co.	Dallas
184	Devon Resource Investors L.P.	Oklahoma City
173	DI Industries Inc.	Houston
82	Diamond Shamrock Offshore Partners	Dallas
58	Diversified Energies Inc.	Minneapolis
376	DOL Resources Inc.	Dallas
245	Dorchester Hugoton Ltd.	Dallas
341	Double Eagle Petroleum & Mining Co.	Casper, WY
388	Eagle Exploration Co.	Denver
293	Eglington Oil & Gas Corp.	Fort Worth
377	EMC Energies Inc.	Casper, WY
322	Energetics Inc.	Englewood, CO
233	Energy Assets International Corp.	Houston
315	Energy Capital Development Corp.	Rosslyn, VA
242	Energy Management Corp.	Denver
354	Energy Resources of North Dakota Inc.	Williston, ND
299	Energy Sources Inc.	Lubbock, TX
183	Enex Resources Corp.	Kingwood, TX
15	Enron Corp.	Houston
23	Ensearch Corp.	Dallas
46	Ensearch Exploration Partners Ltd.	Dallas
119	Ensource Inc.	Houston
197	Equitable Petroleum Corp.	New York
54	Equitable Resources Inc.	Pittsburgh
151	Equity Oil Co.	Salt Lake City
169	Ethyl Corp. (Oil & Gas Division)	Baton Rouge, LA
246	Evergreen Resources Inc.	Denver
364	Exploration Co.	San Antonio
140	Exploration Co. of Louisiana Inc.	Lafayette, LA
1	Exxon Corp.	New York
254	Falcon Oil & Gas Co. Inc.	Odessa, TX
195	Federated Natural Resources Corp.	Traverse City, MI
190	Fidelity Oil Co.	Bismarck, ND
86	First Mississippi Corp.	Jackson, MS
128	Flag Redfern Oil Co.	Midland
71	Forest Oil Corp.	Denver
49	Freeport-McMoRan Energy Part.	New Orleans
25	Freeport-McMoRan Inc.	New Orleans
148	Freeport-McMoRan Oil & Gas Royalty Trust	Houston
174	Fuel Resources Development Co.	Denver
164	Fuel Resources Inc.	Staten Island, NY
225	Garnet Resources Corp.	Houston
171	General Energy Development Ltd.	Houston
357	General Energy Resources & Tech. Corp.	Traverse City, MI
220	Geodyne Resources Inc.	Tulsa
282	GeoResources Inc.	Williston, ND
368	Gladstone Resources Inc.	Dallas
44	Global Marine Inc.	Houston
106	Global Natural Resources Inc.	Houston
285	Gold King Consolidated Inc.	Houston
281	Golden Oil Co.	Denver
372	Golden Triangle Royalty & Oil Inc.	Cisco, TX
57	Grace Natural Resources Group	Dallas
271	Graham Income Fund 82A L.P.	Covington, LA
160	Graham-McCormick Oil & Gas Partnership	Covington, LA
309	Great Eastern Energy & Development Corp.	Englewood, CO
308	Greenwood Holdings Inc.	Englewood, CO
97	Hadson Corp.	Oklahoma City
353	Hailey Energy Corp.	Abilene, TX
55	Hamilton Oil Corp.	Denver
335	HarCor Energy Inc.	Beverly Hills, CA
113	Harken Oil & Gas Inc.	Bedford, TX
66	Helmerich & Payne Inc.	Tulsa
217	Hershey Oil Corp.	Pasadena, CA
399	Hipoint Investments Ltd.	Oklahoma City
103	Home Petroleum Corp.	Denver
272	Home-Stake Oil & Gas Co.	Tulsa
240	Home-Stake Royalty Corp.	Tulsa
287	Houston Oil Royalty Trust	Houston
153	Houston Oil Trust	Houston
108	Howell Corp.	Houston
370	Hubco Exploration Inc.	New Orleans
146	Hutton/Energy Assets 3rd Energy Ptrsp. A Ltd.	Houston
205	ICG Petroleum Inc.	Calgary, Alta., Can.
331	Impact Energy Inc.	Fort Worth
130	Integrated Drilling & Exploration Inc.	Tulsa
325	International Basic Resources Inc.	Tucson
369	International Management & Research Corp.	Philadelphia
384	International Royalty & Oil Co.	Fort Worth
342	Intramerican Oil & Minerals Inc.	Dallas
114	Kaneb Energy Partners Ltd.	Richardson, TX
77	Kaneb Services Inc.	Richardson, TX
339	Keldon Oil Co.	San Clemente, CA
289	Kelley Oil Corp.	Houston
170	Kelley Oil & Gas Partners Ltd.	Houston
211	KenCope Energy Cos.	San Antonio
27	Kerr-McGee Corp.	Oklahoma City
238	Kimbark Oil & Gas Co.	Denver
319	Kimco Energy Corp.	Arlington, TX
134	Kirby Exploration Co. Inc.	Houston
316	K.R.M. Petroleum Corp.	White Plains, NY
203	LL&E Royalty Trust	Houston
389	Loch Exploration Inc.	Gainesville, TX
274	Lomak Petroleum Inc.	Hartville, OH
39	Louisiana Land & Exploration Co.	New Orleans
379	Love Oil Co. Inc.	Denver
360	Lyric Energy Inc.	Amarillo
298	Magic Circle Energy Corp.	Oklahoma City
34	Maxus Energy Corp.	Dallas
175	Maynard Oil Co.	Dallas
182	McFarland Energy Inc.	Santa Fe Springs, CA
144	MCO Resources Inc.	Houston
30	Meridian Oil Inc.	Houston
28	Mesa L.P.	Amarillo
236	Michigan Oil Co.	Jackson, MI
391	Minex Resources Inc.	Riverton, WY
31	Mitchell Energy & Development Corp.	The Woodlands, TX
2	Mobil Corp.	New York
311	Monogram Oil & Gas Inc.	North Canton, OH
36	Montana Power Co.	Butte, MT
109	Moore McCormack Energy Inc.	Dallas
366	Morrison Nuclear Inc.	Calgary, Alta., Can.
196	MSR Exploration Ltd.	Cut Bank, MT
32	Murphy Oil Corp.	El Dorado, AR
284	Mustang Cos. Inc.	Great Bend, KS
208	Mustang Resources Inc.	Houston
273	Nahama & Weagant Energy Co.	Bakersfield, CA
84	National Cooperative Refinery Assoc.	McPherson, KS
47	National Fuel Gas Co.	New York
390	National Petroleum Corp. Ltd.	Dallas
123	Nerco Oil & Gas Inc.	Vancouver, WA

The OGJ400 Company Index

This is the first in a series of histories of the petroleum industry in the major oil and gas producing states.

Alaska is youngest, most significant producer

IN A MODERN CONTEXT, ALASKA'S PETROLEUM IN-dustry is the youngest of any of the major producing states.

Although the first commercial oil was discovered in Alaska at the turn of the century, the Frontier State's modern petroleum era has existed for only a little more than 30 years. It was not until the discovery of Swanson River oil field in 1957 that petroleum made a significant contribution to Alaska's history.

However, petroleum is now the backbone of Alaska's economy, and Alaskan petroleum resources—current and prospective—are the most significant in the U.S.

Alaska accounts for more than 25% of U.S. oil production and about 30% of the country's remaining producible reserves. Its offshore reaches and Coastal Plain of the Arctic National Wildlife Refuge may hold half of the U.S. potential petroleum resource still to be found. The Coastal Plain alone may hold as much as 9 billion bbl of oil, almost equal to the original reserves estimate at Prudhoe. Further, the infrastructure and Trans-Alaska Pipeline System already are in place to exploit that massive North Slope resource. The gas cap at Prudhoe Bay is the biggest single accumulation of natural gas in the Western Hemisphere.

It's that kind of spectacular potential that has lured the petroleum industry to brave the forbidding wilderness and often deadly climate of the 49th state.

Early history

The earliest reports of petroleum in Alaska came from the Inupiat, who told Russian and American whalers and explorers of burning black lakes in the far north.

Seeps along the North Slope shoreline often provided some of the natives a ready fuel by burning oil-soaked tundra.

Ernest Leffingwell mapped the surface geology of part of the land north of the Brooks Range shortly after the turn of the century. He reported oil seeps in the vicinity of Smith Bay in 1914. The pioneering geologist's early efforts led eventually to the discovery of Prudhoe Bay oil field, North America's biggest.

The earliest sustained exploration on the North Slope was conducted mainly by the U.S. Geological Survey, which produced the first report of the region's petroleum geology in 1904. That led to President William G. Harding declaring in 1923 that 37,000 acres of the western North Slope be set aside as a U.S. Naval Petroleum Reserve.

Harding's executive order wiped out what would have been the first oil claims to be staked on the North Slope. The year that Leffingwell reported the Smith Bay seeps, a Wainwright, Alaska, teacher staked a claim on two seeps but did not return. In 1917, gold prospector Sandy Smith found his seep by falling into it. He returned 4 years later to stake a claim for one of the Standard Oil companies, only to come across a competing company attempting the same.

CARIBOU THRIVE within the shadow of oil and gas operations on Alaska's North Slope. Concerns about the caribou almost halted North Slope work in the 1970s, and despite the absence of harm in more than a decade of oil work there, the issue may threaten access to the highly prospective Arctic National Wildlife Refuge. Photo courtesy Alyeska Pipeline Service Co.

USGS conducted geological reconnaissance of Alaska's Naval Petroleum Reserve during 1923-26 at the Navy's behest. Mounting concern over the security of oil supplies led the Navy, again with USGS as prime contractor, to conduct an intensive oil exploration search in and around the reserve. The two began what was to be an 8 year exploration program in the reserve in 1944.

At the program's conclusion in 1953, the Navy had drilled 36 wildcats and 44 coreholes, yielding three gas fields and two oil fields, all too small for anything other than local use. Total price tag: $43 million.

The largest, Umiat, holds reserves of perhaps 30 million bbl. Another small oil field turned up at Cape Simpson, and a gas field at East Umiat yielded flow rates from two wells of 3.4 MMcfd and 8 MMcfd.

The only commercial production resulting from the Navy program occurred south of the village of Barrow. South Barrow gas field and a satellite provide the village with its gas needs and are today the subject of a renewed exploration/development effort to prove more gas reserves.

The daunting terrain, climate, and remoteness of the North Slope left explorationists little opportunity to exploit that region's petroleum potential. Oil would make its first significant commercial contribution to Alaska in the Cook Inlet/Kenai Peninsula region of southern Alaska.

Oil seeps were found in the Cook Inlet area in the mid-19th Century by Russian settlers. A Russian named Paveloff reportedly took the first samples of crude in the Cook Inlet area in 1882.

Another prospector, Edwin Edlemann, staked the first reported claims in the Cook Inlet region in 1892 but later abandoned them.

In 1896, prospector Tom White found oil and gas seeps on the Katalla Slough in the Kenai basin along the Gulf of Alaska. While hunting bear, White fell into a seep. Curious, he returned to set fire to the seep, which burned for a week. After taking samples, White began to promote his discovery. Subsequent cable tool drilling by Chilkat Oil Co. found pay at 336 ft, and an 85 ft fountain of crude ushered in Katalla oil field and commercial oil development in Alaska in 1902.

Earlier wells were drilled on the Kenai Peninsula, without

COOK INLET BASIN launched Alaska's modern petroleum era in the late 1950s and early 1960s. This 1965 photo depicts a jack up drilling a wildcat for Pan American Petroleum Corp. and partners in Cook Inlet.

success, by Alaska Petroleum Co. in 1898.

The lure of more black gold on the Kenai Peninsula even sparked part of the rationale for a group of settlers from Finland the early 1900s. Finnish settlers planning to establish a colony near Aurora on Kachemak Bay filed 65 oil placer claims in 1907-08.

Other exploration for oil ground to a halt in Alaska shortly after the Katalla find because of a controversy between federal agencies over control of Alaskan coal leases. President Theodore Roosevelt ended the controversy by withdrawing all public lands from oil and gas staking, ending placer claim activity. Public oil and gas leasing subsequently was forbidden in Alaska until congressional legislation mandating it passed in 1920.

If a presidential order had not accomplished it, market forces probably would have killed off Alaska's short lived early oil boom. Burgeoning volumes of cheap fuel oil from California were hurting an already marginally economic oil potential in a region whose harshness often defeated the era's drilling technology.

Drilling picked up again the year after Congress passed the Mineral Leasing Act of 1920. Standard Oil Co. of California, Alaska Oil Co., and Associated Oil Co. were

NORTH SLOPE WILDCATTING in the early years was filled with frustration until the discovery at Prudhoe Bay State No. 1. Here, Union Oil Co. of California drills Kookpuk No. 1, later proved dry, near the Colville River delta west of Kuparuk River field.

NORTH SLOPE conditions can be the worst industry encounters. Here, a camp near the Put River, where a rig is being erected, endured a blizzard that brought the wind chill to – 100° F. Photo courtesy BP.

PRUDHOE BAY WELLS were among the most prolific in North American history, with discovery flow rates sometimes topping 20,000 b/d. These early Prudhoe development wells await construction of production facilities and the trans-Alaska pipeline.

active but unsuccessful in the Cold Bay area of Kenai Peninsula. Thereafter, with rare exception, interest by private companies in Alaska's oil potential faded until impending statehood and the postwar boom of the 1950s.

Meanwhile, Katalla produced about 154,000 bbl of oil during 1902-33 from 18 wells drilled to depths of 336-1,810 ft. Chilkat survived the presidential decree because it held patented land at Katalla. It did not survive, however, a fire that destroyed a refinery the company built at Katalla in 1911. The refinery supplied the Cordova area until its destruction on Christmas Day, 1933, marking the end of commercial oil production in Alaska for almost a quarter of a century.

Cook Inlet:
the modern era

As has been the case in other frontier plays, it took many years of study and many dry holes before the Cook Inlet petroleum province was to launch Alaska's modern petroleum era.

At the far eastern corner of the province, Union Oil Co. of California and Ohio Oil Co. began the first private oil drilling in Alaska since the outbreak of World War II, spudding a wildcat near Eureka, about 100 miles northeast of Anchorage.

Exploration soon spread to the western side of Cook Inlet and the Yakataga area of the adjoining Gulf of Alaska province. Phillips Petroleum Co. and Kerr-McGee Oil Industries Inc. drilled two tests to more than 10,000 ft in the Yakataga area. None of the Cook Inlet wells had attained that depth, and by 1957 more than 100 wells had been drilled in the Cook Inlet and Gulf of Alaska provinces without success.

Future efforts in the Gulf of Alaska also would prove fruitless.

Richfield Oil Corp., the predecessor to Atlantic Richfield Co. (ARCO), developed a prospect northeast of the city of Kenai on the Kenai Peninsula in 1955.

Richfield's first instincts were to explore structures under Cook Inlet waters, and it dispatched a seismic vessel to shoot a survey in the inlet. The U.S. Fish & Wildlife Service stepped in to halt the seismic activity, and Richfield turned its attention to the onshore prospect along the Swanson River.

Richfield brought in a rig to test the Swanson River prospect in spring 1957, having leased 71,680 acres in the area. Its 1 Swanson River Unit reached total depth of 11,170 ft, the first deep test in the basin, and found a significant oil accumulation. The discovery flowed oil at rates of more than 900 b/d. It was Richfield's first Alaska well.

Richfield's strike spurred a leasing and drilling boom in Alaska. By yearend 1958, much of the Cook Inlet basin was leased, and interest was spreading to other subarctic Alaskan provinces.

In December 1957, Richfield and Standard of California entered into an agreement for joint exploration and development of leases on the Kenai Peninsula. Under the agreement, Socal committed to spend $30 million during a period of several years and became operator of the venture. Richfield soon opened an office in Anchorage.

Further drilling at Swanson River Unit and in nearby Soldotna Creek Unit helped push production to about 14,000 b/d by 1960. The venture in late 1959 announced plans to build a pipeline from Kenai to a tidewater marine terminal at Nikiski on the eastern side of Cook Inlet. In 1960, Socal loaded its first tanker of crude from Swanson River, a cargo of 100,000 bbl destined for its refinery at Richmond, Calif.

Unocal and Ohio Oil Co. (forerunner of Marathon Oil Co.) in October 1959 drilled the first major gas discovery in the Cook Inlet region. Their 14-6 Kenai Unit proved Kalifornsky Beach gas field, testing at a rate of 15 MMcfd. Union-Ohio subsequently drilled three wells to develop the field with enough capacity to service a 20 year contract with Anchorage Natural Gas Corp. (ANGC).

ANGC ran into problems building a pipeline 85 miles from the Kenai Peninsula across Turnagain Arm to Anchorage, stymied for a year by the notorious tides and abrasive silt of Cook Inlet. ANGC acquired a pipelay barge, service barge, and cement coated pipe to finish the underwater crossing at a cost of more than $3 million. It was the state's first natural gas pipeline system.

Discovery of oil in Cook Inlet waters by Pan American Oil Co. in 1962 triggered a fresh drilling boom leading to the discovery of several giant fields that made Cook Inlet basin a significant producing province. Cook Inlet oil production peaked in the early 1970s at about 200,000 b/d but has declined to about 50,000 b/d.

Developing oil and gas in Cook Inlet proved a major challenge for industry technology in the 1960s. Problems included the inlet's treacherous tides and ice forces in winter. That led to the installation of some special design fixed platforms, such as the conical monopod Unocal installed in Trading Bay field. Unocal started up the oil field in 1967 and operates it today on behalf of itself and Marathon.

Granite Point, just to the northeast, also started up in 1967. Amoco, ARCO, Texaco, and Unocal share operatorship of the oil field, which produces from three platforms.

Unocal 2 years later started up production from the inlet's most significant discovery, giant McArthur River field. To date, the field has produced more than a half billion bbl of oil. Its production peaked at more than 45,000 b/d in 1970-71.

The other offshore Cook Inlet oil field, Middle Ground

TRANS-ALASKA PIPELINE section is lowered into place in the Brooks Mountain Range about 175 miles south of Prudhoe Bay. TAPS has been called the greatest solely private feat of engineering in history.

SEAWATER TREATMENT PLANT at Prudhoe Bay treats 1 million gal/day of water from the Beaufort Sea for a massive waterflood. Prudhoe partners are spending billions of dollars to extend the productive life of North America's biggest oil field. Photo courtesy BP.

Shoal, started up in 1967, and has produced more than 150 million bbl of oil from four platforms.

North Cook Inlet, Offshore Cook Inlet's only significant gas field, was discovered and brought on stream in 1969 by Phillips. Together with onshore Kenai gas field, it feeds an LNG plant at Nikiski, which is the source of the only significant volume of hydrocarbons ever to be exported from Alaska to another nation. Phillips and partner Marathon have a long term contract to supply a group of Japanese utilities with LNG.

In all, companies have discovered 7 oil fields and 13 gas fields in the Cook Inlet area. A number of the gas fields remain shut in for lack of market. Earlier plans, abandoned in the early 1980s, to develop an Alaskan-Indonesian LNG supply network to the U.S. West Coast, could have spurred their development.

The Kenai Peninsula also supports or has supported refineries, topping plants, petrochemical facilities, and a gas processing plant.

Despite its decline, Cook Inlet region still attracts interest. Subsequent lease sales sparked added drilling onshore and offshore. Most recently, Unocal has begun development of gas reserves in Cannery Loop field estimated at 300 bcf. It discovered the field in 1986.

There remains enough significant potential to warrant installation of another platform in Cook Inlet. Marathon and Unocal installed Steelhead platform in McArthur River field in 1986 to develop shallow Grayling gas sands at a cost of about $100 million. Production of about 160 MMcfd was to have begun in 1987, but was postponed when a blowout and fire damaged the platform, the first installed in the inlet in 15 years.

For the most part, however, Cook Inlet operators at the end of the 1980s were mulling choices ranging from abandonment of platforms to launching tertiary recovery projects.

Except for Alaska's Beaufort Shelf/North Slope province and despite the vast state's much vaunted potential and billions of dollars spent in exploration, no other Alaskan basin outside the Cook Inlet has produced commercial hydrocarbons to date.

But persistence in the 1960s paid off on the frigid North Slope, and Alaska's petroleum future changed forever.

Prudhoe and TAPS

Events leading to the development of North America's greatest oil field began inauspiciously enough: a decade of dry holes amid a sprinkling of noncommercial discoveries.

With statehood imminent and a drilling boom under way in the Cook Inlet region, companies began to clamor for access to the North Slope, drawn by minor oil and gas finds by USGS and the Navy.

In 1958, the Bureau of Land Management opened a large tract of land to commercial leasing between the Naval Petroleum Reserve and what was to be set aside as the Arctic National Wildlife Range, later known as the Arctic National Wildlife Refuge. The first federal North Slope lease sale was held in 1958.

In 1963, British Petroleum spudded the first exploratory well by a private company on Alaska's North Slope. It was BP's belief that the province was the geological analog to Iran, where the company had its origin at the turn of the century.

The drilling equipment for that first effort was barged to the slope via the MacKenzie River in Canada's Northwest Territories, a small harbinger of the massive North Slope summer sealifts to come.

BP's rig joined a second that had been used in the petroleum reserve.

Large surface structures in the foothills of the Brooks Range first drew BP's attention, yielding dry holes at five sites and noncommercial gas finds at Gubik and East Umiat. A large seismic anomaly on the Colville River delta yielded only traces of oil. After $30 million and eight failures, BP stacked its North Slope rigs.

In 1964 and 1965, Alaska offered its own leases for sale on the slope in small, selected areas, notably near a desolate stretch of arctic coastline facing Prudhoe Bay. The first three wells on state leases also yielded dry holes.

In March 1968, ARCO and Exxon electrified the world with the news that their Prudhoe Bay State No. 1 had tapped a major new oil field. The following June, ARCO and Exxon

FIRST OFFSHORE ARCTIC PRODUCTION of oil commercially occurred at giant Endicott oil field northeast of Prudhoe Bay. BP operates the field, now producing about 100,000 b/d. Main production island is shown under construction in the Beaufort Sea, site of industry's highest hopes for Offshore Alaska. Photo courtesy BP.

BERING SEA drilling has proven a major disappointment, with no successes resulting from a multibillion dollar effort during 2 years in three basins. Here, the Sedco 708 semisubmersible awaits deployment in Dutch Harbor on the Alaska Peninsula before drilling a well for ARCO Alaska Inc. in the St. George basin. Offshore western Alaska focus now shifts to the Chukchi Sea and North Aleutian basin. Photo courtesy ARCO.

confirmed the strike with the Sag River State No. 1.

In late October of that year, ARCO and Exxon, together with BP, a major leaseholder in the Prudhoe Bay area, announced plans to study the feasibility of building a pipeline transport Prudhoe Bay oil.

On Feb. 10, the three companies announced their decision to build an 800 mile trans-Alaska pipeline from Prudhoe Bay to Valdez in South Central Alaska.

The next month, BP announced its Put River 1 discovery, which later flowed 21,000 b/d on test. That confirmed the true magnitude of the find—at recoverable reserves of more than 10 billion bbl, the 18th largest field in the world.

Within months, Alaska was to get a taste of the riches that would someday comprise more than 90% of its income. That September, a further sale of Prudhoe Bay state leases netted Alaska more than $900 million in high bids. One parcel that was ignored 3 years earlier drew a $72 million high bid.

In April 1970, the newly formalized Trans-Alaska Pipeline System project was shut down when environmental and other organizations filed suits to block TAPS construction. TAPS owner companies, now eight, set up a formal corporate organization to deal with legal and political problems. They formed Alyeska Pipeline Service Co. as agent for the owners to design, build, and operate TAPS.

TAPS was proposed the same year as the Santa Barbara Channel oil spill off California, which gave birth to the modern environmental movement and left the oil industry tagged in some minds as despoilers of ecology. It was shortly after the TAPS proposal was made public that the National Environmental Policy Act was enacted, under which the environmental groups challenged the project.

The major concern voiced by the project's opponents was that the pipeline could prove disastrous to caribou, Dall sheep, and moose. Since then, wildlife populations have flourished in the region during 20 years of industry activity.

While industry grappled with permitting and legal snarls, Prudhoe operators proceeded with development plans, bringing up the first small shipments of equipment via barge in 1969. The 1970 sealift of equipment, however, involved a flotilla of 70 barges carrying more than 175,000 tons of equipment.

On Nov. 16, 1973, about a month after war erupted in the Middle East and the beginning of calls for an Arab oil embargo against the U.S., the Trans-Alaska Pipeline Authorization Act passed in the Senate by one vote and became law with President Nixon's signature. Controversy still plagued the project, nevertheless.

By January 1974, construction began on the mammoth project. It involved building 360 miles of roadway, 800 miles of 48 in. pipeline, eight pump stations, and a marine terminal at Valdez, all carved out of pristine wilderness. At the project's peak in its third year, the main construction force totaled 21,600. About 2,000 contractors and specialty contractors employed more than 28,000 people in work related to the project. It required in all 515 federal permits and 832 state permits, although its right of way occupies less than 14 sq miles in a state of 586,000 sq miles.

A major sealift in 1975, carrying Prudhoe's monster central compression plant, was stymied when early freezeup caught many of the barges in the ice. Thirteen arrived at Prudhoe, but so late they were frozen 1 mile offshore. The operators carved out the Beaufort Sea ice in giant cubes, stacking them to the side, and installed a gravel road in the makeshift alley to the barges. The modules then were moved via crawler to shore without disrupting the schedule.

On May 31, 1977, the pipeline was completed at a cost of $8 billion when the final weld was completed in the Brooks Mountain Range. On June 20, the first oil entered the pipeline at Prudhoe Bay at 9:06 a.m. local time. That first shipment reached the Valdez terminal on July 28 at 11:02 p.m. On Aug, 1, the ARCO Juneau, the first tanker to load Prudhoe crude departed Valdez. That month, Prudhoe flow climbed to 500,000 b/d.

The early years of Prudhoe Bay and TAPS were not without incident. In 1977, an explosion and fire at Pump Station 8 near Fairbanks killed a technician and knocked out the station for 7 months. In early 1978, sabotage caused a pipeline leak north of Fairbanks that was cleaned up.

Prudhoe production built until it reached its decreed limit of 1.5 million b/d in 1980. It was only a few years before Prudhoe operators ARCO and Standard Alaska launched the first of several mammoth projects to boost recovery from Prudhoe Bay.

In August 1984, injection of Beaufort seawater began in a waterflood project expected to add 1 billion bbl to Prudhoe recovery.

The giant seawater treatment plant treats 1 million gal/day seawater for injection.

On Sept. 15, 1986, the 5 billionth bbl of North Slope oil transported through TAPS, including production from Prudhoe and other fields, arrived at Valdez. That surpassed supergiant East Texas field's cumulative production, on

MILNE POINT is the best example of how fragile North Slope economics can be. The 100 million bbl reserve field, just north of Kuparuk River field and the TAPS line, has been under a warm shutin since oil prices collapsed in 1986. Rig shown is drilling one of Milne Point's development wells for operator Conoco Inc. in 1985. Photo courtesy Conoco.

stream since 1930. The energy content of 5 billion bbl of oil would equate to Hoover Dam running at full throttle for 256 years, or enough to take care of Chicago's total energy needs for 35 years.

The start-up of commercial scale tertiary recovery operations at yearend 1986 marked the third generation of reserves explotiation at Prudhoe Bay. ARCO, Sohio, and partners began a fivefold expansion of the Prudhoe NGL/ EOR project, a pilot program that began in 1982. It involves alternating injection of water and NGL enriched gas in a miscible gas flood that is expected to add about 115 million bbl of incremental oil recovery. Installed in December 1986, the 2.7 bcfd NGL plant also yields more than 40,000 b/d of NGL for market in addition to the NGL provided for gas enrichment.

On Mar. 5, 1987, Prudhoe bay field production topped 5 billion bbl, half the field's estimated producible total. Another 12 billion bbl cannot be recovered with existing technology.

On Apr. 19, 1987, the Atigun Pass became the 7,000th tanker to load North Slope crude at Valdez. If lined up, that many tankers would stretch from Houston to slightly beyond Cleveland.

On June 19, 1987, Prudhoe Bay celebrated its tenth anniversary. Efforts by field operators ARCO and Sohio and their partners to wring additional oil out of Prudhoe Bay have pushed the field's natural decline back another 5 years.

Originally expected to begin in 1984, the decline's beginning was still the subject of debate in 1989. ARCO and Sohio disagreed as to whether the field's decline would begin in 1989 or 1990. Both dates are beyond projections made only 2 years earlier, when the expectation was that Prudhoe flow would drop from about 1.55 million b/d to

about 1 million b/d in 1990. Given industry's track record, that day might yet be forestalled again.

Other
North Slope

The enormous potential remaining on the North Slope has spurred operators there to remain stubborn about pushing the economic envelope of discoveries.

Reserves potential totaling almost 15 billon bbl of oil and almost 30 tcf of gas have been identified on the North Slope. An untapped resource potential of another 40 billion bbl and 200 tcf of gas has been identified on the slope.

Kuparuk River field is the best example of how persistence brought a huge resource to bear. Only a year after the discovery at Prudhoe Bay, BP and ARCO predecessor Sinclair Oil Co. brought in the Ugnu 1 discovery between the Colville River and Prudhoe Bay.

Kuparuk River at first was deemed uneconomic to develop, despite an original oil in place estimate of 4-5 billion bbl of oil and its proximity to Prudhoe Bay. Rising crude prices in the 1970s in tandem with continued appraisal drilling in the field led ARCO to decide upon developing a 20 sq mile portion of the field. The subsequent decontrol of oil prices and Kuparuk's exemption from the "windfall profits" tax enabled ARCO to expand field development to 200 sq miles.

Kuparuk River started up Dec. 13, 1981, with flow of 50,000 b/d. Development would come in three phases, involving as many as 800 producing and injection wells and a total outlay of more than $8 billion.

The first U.S. arctic field placed on stream outside Prudhoe Bay, this one time uneconomic field surpassed its expected peak of 260,000 b/d by more than 40,000 b/d in

ARCTIC NATIONAL WILDLIFE REFUGE may hold the best remaining chance for the U.S. to find more giant fields to replace declining production, if environmental opposition can be overcome. Shown is Chevron Corp.'s 1 KIC Jago River—ANWR's sole wildcat—drilled on native lands and probably the tightest hole in history. Photo courtesy Chevron.

1988, making it the second biggest producer in North America.

Similar "marginally economic" reserves on the North Slope recently began to make contributions to Alaska's production. Lisburne reservoir, a 3 billion bbl OOIP fractured limestone carbonate reservoir underlying the main Sadlerochit pay at Prudhoe, yielded flow rates of more than 1,000 b/d in the 1970s. That was handsome by Lower 48 standards but not by the tough economics of arctic Alaska.

In 1984, ARCO and Sohio launched a 180 well initial development of Lisburne that would cost about $2 billion and build to a peak of 100,000 b/d in 1987. However, the collapse of oil prices intervened in 1986, slowing the project to where it produced on average only about 45,000 b/d in 1988.

What could prove to be the biggest North Slope project of all, development of supergiant West Sak reservoir, got a preliminary start in 1984. West Sak, a thin Cretaceous sand overlying the Kuparuk River sand in Kuparuk River field and extending offshore over an area of about 250 sq miles, holds about 15-25 billion bbl of OOIP. Using a hot waterflood, ARCO started output of about 1,000 b/d from 13 wells in a $75 million pilot that later doubled in production. Commercial scale-up of 100,000-200,000 was envisioned in the late 1980s until the price collapse.

The North Slope's third commercial development, Milne Point, came on stream at the beginning of 1986. An extension of Kuparuk River and West Sak sands north of Kuparuk River field, the 100 million bbl field needed years of fine tuned engineering and North Slope operating expertise as well as higher oil prices and lower drilling costs before operator Conoco could convince other leaseholders to pro-

ceed with development. The $800 million project made it only about halfway to its peak of 30,000 b/d when the oil price collapse rendered it uneconomic. It remains under warm shut-in, awaiting the next rise in oil prices.

Beaufort:
the move offshore

Start-up of Endicott oil field in shallow waters of the Beaufort Sea in late 1987, several months ahead of schedule, heralded the first commercial production of oil from arctic waters anywhere.

Sohio and partners invested more than $2 billion to develop finds stretching back into the 1970s in state and federal waters about 15 miles north of Prudhoe Bay. Production climbed to 100,000 b/d in 1988.

Giant structures in the Beaufort Sea have lured explorationists since the Outer Continental Shelf Lands Leasing Act became law and onshore North Slope leases were largely picked over in the late 1970s.

That interest culminated in high bids totaling more than $1 billion, a record at the time, for federal and state leases in a joint sale there on Dec. 11, 1979. Again, environmental opposition intervened, with lawsuits blocking the award of leases through 1980.

The Beaufort Sea campaign got under way in earnest in 1982, with as many as a dozen wells originally planned.

The initial excitement tempered greatly in December 1983, when Sohio and partners drilled the "dry hole that was heard around the world," at Mukluk Island in the central Beaufort Sea. The companies ponied up more than $1.5 billion for leases covering the giant structure in Harrison Bay, and a premature optimism hinted at reserves potential

TAPS line is running at design capacity today, with North Slope production burgeoning in recent years. But Prudhoe Bay will begin to decline sharply soon, as will other North Slope fields, without enough other productive capacity available on the horizon to forestall a big shortfall in TAPS throughput in the 1990s. Photo courtesy Alyeska.

of as much as 5 billion bbl before the well was spudded. Drilling costs alone at Mukluk topped $100 million.

Mukluk not only deflated industry's high hopes for the Beaufort, its occurrence came when oil prices were starting to decline and drilling costs were peaking. The combination led many companies eventually to veer away from the megabuck elephant hunt in frontier areas to replace reserves and start considering acquisitions, EOR, infill drilling, restructuring, staff cuts, and other common aspects of the industry today.

Despite the setback, however, companies nevertheless offered more than $1 billion in leases at Beaufort Sea Sale 87 in 1984. And Exxon was to spud another elephant class probe in the central Beaufort only a year after Mukluk, drilling a $45 million dry hole in search of Mukluk oil that may have migrated to the northwest. Its Antares prospect leases cost the Exxon group more than $125 million.

The litany of disappointments off Alaska was to continue for several years, dashing the hopes of many who thought the state's offshore basins held the best prospects for giant fields in the U.S.

The Bering Sea, long prized for its potential, drew more than 20 wildcats in the first campaigns there in 1984-85, without a single success. The first three lease sales there netted more than $1.3 billion in high bids. Declining oil prices discouraged a return to the costly Navarin, Norton, and St. George basins for looks at lesser plays. However, the recent

end to litigation that freed the leases awarded in the North Aleutian basin sale and the suprisingly strong Chukchi Sea sale in 1988 could spur a revival of Bering Sea drilling as early as 1989.

A $40 million wildcat by ARCO heralding industry's return to the Gulf of Alaska failed to revive that moribund province.

But the more recent exploratory results off Alaska have been encouraging, even if oil prices haven't. Shell and partners remain optimistic that their early 1985 discovery of more than 300 million bbl of oil at Seal Island northwest of Prudhoe Bay will someday prove commercial. Earlier discoveries by Exxon and others onshore and offshore in the Point Thomson-Flaxman Island area east of Prudhoe Bay have confirmed more than 350 million bbl of condensate and 5 tcf of gas.

Beaufort strikes by Unocal at its Hammerhead prospect and Shell at Corona nearby await delineation and a higher oil price. Lesser finds by Shell at Sandpiper Island and BP at Niakuk near Sag Delta are stymied by marginal commerciality or environmental concerns.

ANWR:
the future

Even with all the offshore potential yet to be tapped, Alaska explorationists now are betting that the Coastal Plain of the Arctic National Wildlife Refuge is the best remaining prospect for a giant oil field discovery in North America.

The Coastal Plain of ANWR may hold as much as 9.2 billion bbl of reserves potential, virtually another Prudhoe Bay. It contains the largest single untapped structure in North America and is less than 100 miles from TAPS.

Whether or not industry gains access to this tantalizing prospect will probably be the hottest environmental dispute of the next few years. It is up to Congress to decide whether further exploration and any development occur in the region. Only one wildcat on native lands and some limited seismic and surface mapping have occurred within the Coastal Strip.

In winter 1985, Chevron Corp. spudded the 1 KIC Jago River 14 miles east of the village of Kaktovik. BP is partner in the well with 50%. Chevron wrapped up operations the following winter after spending more than $50 million.

If Mukluk was the most resounding dry hole ever, surely Chevron's wildcat is history's tightest hole. Even the chairman of Chevron, George Keller, disclaimed knowledge of the well results.

The Unocal and Shell finds off ANWR in the Beaufort Sea tend to support industry's suspicion that the region features an interplay of geology extending from the Prudhoe Bay to the west and from the Canadian Beaufort discoveries to the east.

Even though natives in the region support ANWR development, environmental groups are vehement in their opposition.

The Bush administration has leaned cautiously in favor of determining ANWR's petroleum potential. But even if Congress approves, lawsuits will be inevitable.

Should industry get the opportunity, ANWR development probably won't yield oil before the turn of the century. But the activity could improve the economics of existing discoveries onshore and offshore in the region, giving the green light to other developments in the interim.

And there remains the prospect of bringing North Slope natural gas to market, seemingly forever the subject of debate. ANWR action could enhance the prospects for North Slope gas as well.

As gas demand accelerates in the U.S. and the nation's imports continue to rise steadily, approaching the 50% level, the hurdles of low prices and environmental opposition may yet be overcome.

If so, full exploitation of all the North Slope and Beaufort Sea petroleum resources could keep Alaska the bright spot of the U.S. petroleum industry for many years to come. ∎

International Active Rig Count

Region	1981	1982	1983	1984	1985	1986 Land	1986 Off.	1986 Total	1987 Land	1987 Off.	1987 Total	1988 Land	1988 Off.	1988 Total
NORTH AMERICA														
Canada	263	200	201	259	311	174	4	178	179	2	181	194	2	196
United States*	3,969	3,117	2,229	2,429	1,976	847	123	970	816	120	936	784	152	936
Subtotal	**4,232**	**3,317**	**2,430**	**2,688**	**2,287**	**1,021**	**127**	**1,148**	**995**	**122**	**1,117**	**978**	**154**	**1,132**
LATIN AMERICA														
Argentina	70	67	73	82	81	47	0	47	60	1	61	63	0	63
Bolivia	11	11	8	6	5	4	0	4	6	0	6	5	0	5
Brazil	81	94	83	70	76	46	31	77	35	25	60	28	11	39
Chile	7	7	6	6	6	3	3	6	4	3	7	2	3	5
Colombia	22	23	18	16	21	17	0	17	14	0	14	19	0	19
Costa Rica	0	0	1	1	0	0	0	0	1	0	1	0	0	0
Ecuador	4	6	5	4	2	3	0	3	3	0	3	5	1	6
Mexico	218	202	187	196	196	132	31	163	112	31	143	121	34	155
Peru	24	23	16	17	12	9	4	13	11	4	15	6	5	11
Trinidad	14	13	10	9	10	7	4	11	5	4	9	3	4	7
Venezuela	63	69	41	30	33	16	13	29	9	9	18	15	10	25
Other	7	7	3	2	1	1	0	1	1	0	1	0	0	0
Subtotal	**521**	**522**	**451**	**439**	**443**	**285**	**86**	**371**	**261**	**77**	**338**	**267**	**68**	**335**
FAR EAST														
Australia	23	33	25	34	31	9	6	15	13	3	16	14	5	19
Bangladesh	5	5	2	4	5	5	0	5	5	0	5	2	0	2
Brunei	7	10	10	9	7	1	4	5	0	3	3	1	3	4
Burma	30	32	36	33	33	32	1	33	29	0	29	26	0	26
China-offshore	9	10	10	9	14	0	7	7	0	4	4	0	8	8
India	44	50	56	57	62	55	15	70	94	22	116	102	25	127
Indonesia	86	92	88	82	80	42	20	62	25	12	37	33	11	44
Japan	16	22	20	18	16	12	3	15	10	0	10	11	0	11
Malaysia	12	13	10	9	8	0	8	8	0	8	8	0	9	9
New Zealand	3	1	2	2	4	3	1	4	2	1	3	1	1	2
Pakistan	15	17	17	17	18	17	0	17	14	0	14	13	0	13
Papua New Guinea	0	0	1	1	2	1	0	1	2	0	2	3	0	3
Philippines	13	14	8	3	2	1	0	1	2	0	2	3	1	4
Taiwan	11	4	5	6	7	5	2	7	2	2	4	2	2	4
Thailand	7	8	13	11	9	3	1	4	3	3	6	3	3	6
Other	0	0	0	0	0	0	0	0	0	0	0	0	1	1
Subtotal	**281**	**311**	**303**	**295**	**298**	**186**	**68**	**254**	**201**	**58**	**259**	**214**	**69**	**283**
AFRICA														
Algeria	91	81	54	27	35	41	0	41	40	0	40	32	0	32
Angola	5	11	12	7	11	0	7	7	1	9	10	0	10	10
Congo	7	6	4	3	3	0	2	2	0	3	3	0	2	2
Gabon	13	12	7	6	7	1	2	3	0	3	3	3	2	5
Kenya	0	0	0	0	0	0	0	0	1	0	1	1	0	1
Libya	33	31	24	26	30	20	0	20	12	0	12	14	2	16
Nigeria	23	27	17	11	10	7	3	10	6	5	11	7	6	13
South Africa	2	2	2	2	3	0	3	3	0	3	3	0	4	4
Tunisia	11	11	7	6	4	2	1	3	3	1	4	2	1	3
Other	28	27	20	17	15	7	1	8	3	0	3	3	2	5
Subtotal	**213**	**208**	**147**	**105**	**118**	**78**	**19**	**97**	**66**	**24**	**90**	**62**	**29**	**91**
EUROPE														
Austria	9	9	9	7	7	6	0	6	4	0	4	4	0	4
Denmark	2	2	5	3	4	0	3	3	0	3	3	0	2	2
France	21	22	17	20	22	15	0	15	8	0	8	8	0	8
Germany, Fed. Rep.	26	31	25	19	20	19	0	19	12	0	12	9	0	9
Greece	6	5	3	4	1	0	0	0	2	0	2	2	0	2
Holland	17	21	17	18	24	4	11	15	3	7	10	4	10	14

International active rig count—continued

Region	1981	1982	1983	1984	1985	1986 Land	1986 Off.	1986 Total	1987 Land	1987 Off.	1987 Total	1988 Land	1988 Off.	1988 Total
Italy	32	38	26	26	40	24	9	33	20	6	26	17	7	24
Norway	12	13	10	10	13	0	12	12	0	12	12	0	15	15
Spain	11	12	8	9	10	4	3	7	2	0	2	1	1	2
United Kingdom	58	58	42	60	63	7	36	43	4	40	44	5	52	57
Yugoslavia	22	24	21	25	27	27	3	30	28	3	31	28	3	31
Other	2	2	2	2	2	0	1	1	0	1	1	0	1	1
Subtotal	**218**	**237**	**185**	**203**	**233**	**106**	**78**	**184**	**83**	**72**	**155**	**78**	**91**	**169**
MIDDLE EAST														
Abu Dhabi	31	41	35	29	21	8	7	15	3	6	9	2	5	7
Dubai	2	3	7	7	5	1	2	3	0	2	2	0	3	3
Egypt	27	36	35	36	37	17	16	33	14	9	23	11	10	21
Iran	0	3	13	20	20	16	2	18	18	0	18	18	0	18
Iraq	14	11	23	19	28	21	0	21	10	0	10	23	0	23
Jordan	0	0	3	3	2	2	0	2	2	0	2	2	0	2
Kuwait	3	4	5	5	6	6	0	6	6	0	6	6	0	6
North Yemen	0	0	0	0	1	4	0	4	4	0	4	3	0	3
Oman	10	12	11	12	13	12	1	13	10	0	10	9	0	9
Qatar	7	5	3	2	3	1	0	1	1	0	1	2	1	3
Saudi Arabia	29	29	26	16	11	5	1	6	4	1	5	3	1	4
Syria	16	15	15	23	26	27	0	27	22	0	22	24	0	24
Turkey	26	34	26	24	25	27	0	27	26	0	26	20	1	21
Other	5	7	4	1	3	2	0	2	1	0	1	0	0	0
Subtotal	**170**	**200**	**206**	**197**	**201**	**149**	**29**	**178**	**121**	**18**	**139**	**123**	**21**	**144**
Total	**5,635**	**4,795**	**3,722**	**3,927**	**3,580**	**1,825**	**407**	**2,232**	**1,727**	**371**	**2,098**	**1,722**	**432**	**2,154**

* U.S. offshore includes inland water rigs: 23 in 1986, 24 in 1987, 30 in 1988. Source: Baker Hughes Inc.

International seismic crew count

	1983 Land	1983 Marine	1983 Total	1984 Land	1984 Marine	1984 Total	1985 Land	1985 Marine	1985 Total
					(Average crews operating)				
United States	430	49	479	441	47	488	319	45	364
Canada	72	7	79	82	5	87	86	4	90
Mexico	28	1	29	32	0	32	32	0	32
Central & South America	81	5	86	79	9	88	64	8	72
Europe	71	26	97	81	35	116	78	35	113
Middle East	33	1	34	36	2	38	40	2	42
Africa	60	10	70	53	6	59	55	6	61
Far East	95	15	110	85	12	97	92	9	101
Total	**870**	**114**	**984**	**889**	**116**	**1,005**	**766**	**109**	**875**

	1986 Land	1986 Marine	1986 Total	1987 Land	1987 Marine	1987 Total	1988* Land	1988* Marine	1988* Total
					(Average crews operating)				
United States	161	22	183	154	24	178	162	30	192
Canada	53	1	54	58	1	59	78	0	78
Mexico	29	0	29	23	0	23	21	0	21
Central & South America	48	5	53	46	5	51	55	4	59
Europe	65	26	91	52	25	77	49	25	74
Middle East	33	4	37	35	3	38	41	2	43
Africa	44	4	48	37	5	42	35	7	42
Far East	74	9	83	87	9	96	90	11	101
Total	**507**	**71**	**578**	**492**	**72**	**564**	**531**	**79**	**610**

*First half.
Source: Society of Exploration Geophysicists

VENEZUELAN CRUDE is delivered via tanker to 160,000 b/d Champlin Refining Co. (CRC) refinery at Corpus Christi, Tex. Venezuela's state owned oil company, Petroleos de Venezuela SA, early in 1989 completed purchase of Union Pacific Corp.'s 50% interest in CRC for $89 million. CRC is a joint venture of UP and Pdvsa that purchased the refinery from UP in 1987. Pdvsa supplies the refinery all of its crude under a program to secure downstream foreign outlets for 700,000 b/d of Venezuela's crude. (Photo by Union Pacific Corp.)

OPEC reintegration aims downstream

THE LATE 1980s HAVE USHERED IN A FLURRY OF acquisitions of downstream interests in oil importing nations by members of the Organization of Petroleum Exporting Countries

OPEC reintegration marks a fundamental change in the structure of the petroleum industry. Whether it helps to stabilize oil markets, undermine them, or has no appreciable effect, is a critical question for the industry in the years ahead.

The intent is to ensure placement of oil via equity or other joint ventures with foreign refiners. That would serve as a hedge for revenues in a persistently volatile market.

OPEC integration first got under way in the late 1970s, with a push to develop domestic export refineries—notably by Saudi Arabia. Despite many alarms over the potential competition for importing nations' refineries, the OPEC export refineries had little effect on them in comparison with market fundamentals such as demand and crude prices.

The collapse of crude oil prices in 1986 and the market shakiness that followed has revitalized the interest in OPEC reintegration. Faced with persistently resilient competition from non-OPEC producers, OPEC members have been venturing into foreign downstream businesses seeking to diversify their sources of revenue to avoid the kind of economic devastation that a crude price collapse causes.

It has culminated in the biggest such deal of all, the Saudis' acquisition of some Texaco Inc. refining/marketing interests in the U.S. for about $2 billion.

In addition, state oil companies in China, Norway, and Mexico have acquired or are seeking to acquire downstream interests abroad.

Differing views

Industry officials have been mixed in their views of the effects of reintegration on the health of the petroleum industry. Some think reintegration will wholly transform the industry in the 1990s, diminishing OPEC's influence.

The idea is that OPEC divisiveness would increase because the group would split into two opposing sides, those heavily integrated and those with little or no downstream capacity. The heavily integrated ones—which also are those members with some of the longest lived reserves—would have less concern about the absolute level of crude prices and therefore less to lose in a price collapse,

the argument goes.

Others contend that securing outlets for their production would tend to encourage OPEC members to be less inclined to allow volatility in oil prices.

OPEC's foreign downstream partners have enjoyed the economic benefits of the ventures. But others remain concerned that such ventures could spur dumping of cheap crude to hike market share or sustain steady revenues should crude prices fall.

In the long term, reintegration could depress crude prices and some refiner/marketer margins, plus boost oil imports in net consuming nations through the 1990s, some observers contend.

Whether that happens or OPEC reintegration proves to be primarily just a hedge against volatile prices could depend on just how well OPEC and non-OPEC crude exporters are able to share the burden of oil production restraint.

Others worry about the entry into a strategic industry by entities of a foreign government whose political interests might not always be friendly.

Melville A. Conant wrote in Geopolitics of Energy, "There is, for instance, a U.S. argument that its American controlled refining capacity must always be equal to mobilization requirements.

"Since the U.S. seems destined to import more of its oil needs, it could well happen that product imports will represent an even larger share of its foreign supply. The same could be true for Europe and Japan."

Reintegration justified

Reintegration provides crude exporters market insurance when oil prices and demand are soft, said Duke Ligon, Bankers Trust Co. managing director.

Integration, via acquisition or joint venture, can benefit producers and refiners by providing security of supply or market, Ligon said. The venture lets each partner enjoy smoother profits and cash flows, avoiding the historic countercycling of upstream vs. downstream profits by linking crude supplies directly to product market conditions.

Smoother cash flow cuts the refiner's operating risk, providing greater leverage for financing and thereby improving stock values, Ligon said. In addition, consuming nations' crude slates will become more variable as oil imports climb. So linking with a producer lets a refiner ensure crude slate

OPEC's foreign downstream ventures

●SAUDI ARABIA. Acquired 50% of Texaco refining/marketing assets in 23 states and D.C. for almost $2 billion.

●VENEZUELA. Wants to acquire foreign downstream interests for 700,000 b/d of its crude production. Joint ventures cover 50% in Unocal's 147,000 b/d refinery at Lemont, Ill., 100% of Champlin's 160,000 b/d refinery at Corpus Christi, Tex., 50% of Citgo's 290,000 b/d refinery at Lake Charles, La., 150,000 b/d representing 50% of Veba's Ruhr Oel refineries in West Germany, 40,000 b/d representing 50% of Sweden's Nynas Petroleum, 5% of Petromed's 123,000 b/d Castellon, Spain, refinery.

●KUWAIT. Acquired former Gulf assets in Europe in 1983. Still wants to penetrate French and West German markets. Now has more than 4,800 retail outlets under the Q8 brand in Belgium, Denmark, Italy, Luxembourg, Netherlands, Sweden, and the U.K. Holds 24.9% interest in Spain's biggest chemical company—and an 80,000 b/d refiner—Union Explosivos Rio Tinto SA.

Seeking a stake in Spain's Cepsa, which has 320,000 b/d of refining capacity.

●LIBYA. Libya's state oil company will earn a 50% interest in a Coastal Corp. unit's 90,000 b/d Hamburg refinery for a firm crude supply arrangement, among other considerations. Libya ships 75,000 b/d to the refinery at a netback price. A Libyan controlled company in Italy has about 6% of Italy's retail gasoline market and a 105,000 b/d refinery at Cremona.

●UNITED ARAB EMIRATES. Abu Dhabi holds a 15% stake in Cepsa and has sought to boost that to 20%.

●NIGERIA. Made overtures to Sun Co. for interests in its refining/marketing business and for the Bantry Bay refinery in Ireland.

---

stability, thus enhancing refinery planning, efficiency, and inventory control.

Ligon suggested the refiner in such a venture seek to:

• Obtain a secure, long term crude supply with a call option on netback type pricing, tying crude prices to market values of products, when refining margins are low.

• Avoid surrendering equity in the refinery to the producer.

• Limit the venture to the refinery, giving the refiner freedom to operate its marketing business.

• Find a producer with compatible crudes, goals, and operating style.

And the producer in such a venture should:

• Obtain a secure, long term market for its crude, with a put option to sell crude to the highest bidder when prices are high and to place it with the refiner to avoid discounting prices when markets are flooded.

• Make the smallest possible cash investment in the refinery.

• Achieve compatibility.

Both sides can buoy long term mutual interests through an equity participation venture, thus resolving short term price conflicts, Ligon said.

Such deals are limited, however, he said. They don't necessarily boost cash flows—ruling out unprofitable plants, and they require a sizable, flexible refining capability and a strong marketing system.

Charles River Associates' Philip K. Verleger Jr. contends that the economic benefits to oil exporting nations from investment in downstream activities are small but not negligible.

"A country that is fully integrated should experience a per barrel after tax loss of between 75¢ and 80¢ from every dollar drop in the price of crude rather than losing the full $1.

"The same country will also find that fluctuations in income had been smoothed because changes in retail prices tend to lag behind changes in crude prices."

Refiners' views

Refiner/marketers in equity ventures with crude exporters show good results.

Citgo Petroleum Corp. reported profits in first quarter 1988 among the best in company history. Its earnings jumped 68% to $41.9 million from the same period the year before. The company cited excellent margins, more than 85% yields of higher value products, and boosts in runs of low cost Venezuelan and Mexican crudes.

However, Mobil Corp. Chairman Allen E. Murray, said that if OPEC members investing downstream in the U.S. "come in solely as investors, fine. But if the refineries are supplied with crude at discounted prices, then I'm worried because we won't be able to compete."

ARCO Pres. Robert E. Wycoff said, "Since we don't view refining and marketing in most of the world as having attractive stand alone economics, we see this integration as a potential mechanism for OPEC members to cheat on crude oil production and marketing agreements."

Still, the uncertain outlook for refining, especially for smaller companies, could pressure more U.S. downstream companies to look at joint ventures with foreign and domestic oil producers.

Ronald W. Williams, president, Gary-Williams Oil Producer Inc., Englewood, Colo., contends that a lack of financial flexibility is hindering many U.S. independent refiners from achieving the kind of attractive margins that the downstream units of major integrated companies are enjoying.

Unlike integrated companies, independent refiners don't have the upstream cash flow to spend downstream and the equity crude to run through the refinery.

Independent refiners have always had to finance capital improvements out of cash flow. Joint venture programs, he said, represent a method to obtain expansion capital quickly.

Market boon?

Downstream acquisitions by key oil exporting nations are welcome news for the market, said East-West Center's Fereidun Fesharaki.

He contends that the trend "should not be feared as a negative move which will cause unfair competition and depress refining margins."

OPEC's downstream push is part of a second phase in a structural market change that entails reintegration and centralization and less room for independent refiners, he said. The first phase involved OPEC's buildup of domestic downstream capacity.

"Contrary to previous expectations, delays, together with OPEC's careful handling of products, meant that the entry of products did not cause much interruption in the market."

More than half the Middle Eastern petroleum products found markets in Europe, and the U.S. has imported large volumes from Venezuela. Asian markets have taken about 20% of Middle Eastern product exports.

Fesharaki welcomes OPEC member downstream acquisi-

Table 1

How OPEC members have boosted refining capacity

	Refining capacity/(access)* 1,000 b/d			Percent change	Oil production 1,000 b/d			Percent change
	1983	1985	1987	1983-87	1983	1985	1987	1983-87
Algeria	471.2	471.2	471.2	—	685.0	647.1	652.4	− 4.8
Ecuador	94.6	94.6	94.6	—	233.0	275.7	173.9	− 25.4
Gabon	44.0	44.0	21.0	− 52.3	148.3	152.5	157.3	+ 6.1
Indonesia	473.4	837.0	837.0	+ 76.8	1,301.1	1,242.3	1,185.8	− 8.9
Iran	670.0	615.0	615.0	− 8.2	2,466.7	2,241.7	2,309.2	− 6.4
Iraq	215.5	365.5	365.5	+ 69.6	883.3	1,447.5	2,135.0	+ 141.7
Kuwait	594.0	614.0	720.0	+ 21.2	887.5	862.5	1,120.8	+ 26.3
Access	—	132.5	132.5	—	—	—	—	—
Libya	117.0	315.0	342.0	+ 192.3	1,018.3	1,047.5	1.093.8	+ 7.4
Access	105.0	105.0	105.0	—	—	—	—	—
Neutral Zone	—	—	—	—	391.2	352.5	387.1	− 1.0
Nigeria	234.0	234.0	243.0	+ 3.8	1,233.8	1,481.7	1,254.5	+ 1.7
Access†	—	—	(156)	—	—	—	—	—
Qatar	13.0	63.0	63.0	+ 384.6	294.4	298.8	280.3	− 4.8
Saudi Arabia	765.0	1,190.0	1,490.0	+ 94.8	4,850.0	3,283.3	4,025.0	− 17.0
Access	—	—	(886)	—	—	—	—	—
United Arab Emirates	121.5	162.0	162.0	+ 33.3	1,109.2	1,141.7	1,399.6	+ 26.2
Access	—	—	346.0	—	—	—	—	—
Venezuela	1,323.1	1,224.2	1,224.2	− 7.5	1,775.8	1,645.8	1,580.8	− 11.0
Access	294.0	614.0	1,354.9	+ 360.8	—	—	—	—
Subtotal	**5,136.3**	**6,229.5**	**6,648.5**	**+ 29.4**	**—**	**—**	**—**	**—**
Access subtotal	**399.0**	**851.5**	**1,938.4**	**+ 385.8**	**—**	**—**	**—**	**—**
Total	**5,535.3**	**7,081.0**	**8,586.9**	**+ 55.1**	**17,277.6**	**16,120.6**	**17,755.5**	**+ 2.8**

*Generally, access to refinery capacity outside country; refers to total capacity of outside refineries, even if equity interest or crude supply deal represents only a portion of total capacity. †Hypothetical, not included in total. Represents Nigeria National Petroleum Co. overtures for Sun Oil Co. refining assets in U.S. and Bantry Bay refinery in Ireland.

tions because they:
- Eliminate incentives to build new grassroots refineries.
- Teach OPEC members about end-use products markets, thereby making them much more reasonable sellers of crude.
- Rein OPEC member incentives to cut supplies.
- Force OPEC to keep official prices in line with the market—"antiquated" government selling prices will slowly disappear.
- Improve rationalization of the refining industry.

Or market bane?

Although diversification will pay short term dividends for OPEC members, in the long term, said Verleger, "exporting countries will conclude that their success or failure in the joint control over the level of output will be the primary determinant of their economic well being. It is in this area that downstream integration poses the greatest threat."

Verleger thinks that the acquisition of downstream assets by the major crude exporting nations will reduce their ability to coerce cooperation from smaller exporters because the big ones will lack their traditional diversified marketing channels.

If a crude exporter becomes fully integrated, then the traditional multinational buyers of that country's exports will be forced to find other sources of supply, he contends.

Should that fully integrated crude exporter then decide to confront increased output by smaller OPEC and non-OPEC nations by boosting production, it would find that its former buyers have turned to other suppliers and are no longer willing to take oil from them, Verleger said.

"Over time, we expect that smaller exporting countries will realize that the larger exporting nations have lost much of their former flexibility by developing integrated systems.

"The nonintegrated producers will use this loss of flexibility to their advantage by making term arrangements with both integrated and nonintegrated companies that are now owned by producing nations."

In that event, production quotas will then become essentially meaningless to the smaller exporters, Verleger said.

"Since flexibility, particularly to adjust output, was required to make the cartel work and to hold prices at official levels, the move downstream seems sure to undercut efforts to hold oil prices above the marginal cost of production."

OPEC demise omen?

That scenario points to other analysts' contention that the push downstream is an omen of OPEC's coming demise.

John Roberts, Middle East Institute senior adviser, said, "As the trend toward downstream integration intensifies, the principal division among oil producers is likely to move from that between OPEC and non-OPEC to the distinction between countries with integrated oil industries and secure outlets for their crude and products and those that lack such a structure and facilities."

He sees principal divisions being headed by Saudi Arabia and Kuwait in the integrated group and Iran and Iraq in the nonintegrated group.

Without access to products markets to unload surplus crude, nonintegrated producers will become marginal producers. That likely will make crude prices more volatile as more producers compete to cover what should be a relatively narrow gap between world demand and supply available from nonintegrated sources, Roberts wrote in a monograph by the University of Colorado's International Research Center for Energy and Economic Development.

"The effect will be to push prices sharply downward at times of excess output."

Verleger said that OPEC's push downstream intensifies the problem of overproduction.

"The oil exporting nations will undoubtedly be disappointed with their foray into downstream activities. In the long run, their actions will probably depress, not increase, their incomes."

CASUALTY OF GULF CONFLICT between Iran and Iraq was this tanker, Tiburon, of Panamanian registry. An Iraqi air attack killed eight crewmen, injured three others, and inflicted damage resulting in a total loss of the vessel. Smit International of Holland conducted firefighting and salvage operations. Photo courtesy of Smit International.

Increased demand for tankers focuses on larger vessels; extent remains uncertain

NON-COMMUNIST DEMAND FOR TANKERS IS HEADED for an increase, paced by demand for large vessels.

Shipping consultants generally agree on that outlook. But continued volatility of oil prices keeps the amount of increase uncertain.

These forecasts were made in 1988:

• Ocean Shipping Consultants, Weybridge, England, said non-Communist tanker demand, as measured in metric ton-miles, will climb to 7.971 trillion metric ton-miles/year in 2000, up 46% from 1986 demand.

• Drewry Shipping Consultants Ltd., London, said tanker employment by 1992 will rise by 12.7-23.9% from the 1988 level, depending oil on how oil prices behave.

Data compiled by Overseas Shipholding Group Inc., New

York, showed that non-Communist tanker capacity totaled 229 million dwt at the end of 1987, down only 2% from 1986. And shrinkage of capacity, under way since 1978, could end in 1988, the company said.

ULCCs to set pace for demand growth

Ocean Shipping predicted that non-Communist demand will improve for all classes of tankers, but volume of growth will vary by vessel size.

For example, demand for ultralarge crude carriers (ULCCs) with capacities of more than 300,000 dwt will jump 35% to 571 billion metric ton-miles/year in 1990. But demand for small crude and products tankers will rise only about 4% to 1.098 trillion metric ton-miles/year in 1990 (Table 1).

ULCC demand is expected to be 744 billion metric ton-miles/year in 2000, 75% higher than in 1986. The huge tankers will benefit most from the long term trend of increasing haul lengths in crude oil trading, Ocean Shipping said.

Very large crude carrier (VLCC) demand will increase 52% to 2.506 trillion metric ton-miles/year in 2000.

Total world tanker trade will climb to 22.3 million b/d of crude in 2000, up from 20.2 million b/d in 1986, and more than 10 million b/d of products in 2000, up from about 6 million b/d in 1986, the firm predicted.

Ocean Shipping said Middle East crude exports will grow 3% in the late 1980s and then by an average of nearly 3.5%/year during the 1990s.

Crude oil shipments from South America will grow, and exports from Africa will decline as traditional crude exporters continue the trend toward greater downstream activity.

North American crude oil imports will increase to 5.6 million b/d by the mid-1990s from 4.5 million b/d in 1986, while western European imports will fall to 6.4-6.5 million b/d through most of the 1990s from 7 million b/d.

Japan's crude oil imports will decline to 2.35 million b/d in 2000 from 3.35 million b/d in 1986.

Liberalization of Japan's product market will result in a near-doubling of products imports. North America and western Europe, which import about two thirds of the petroleum products in world trade, can expect moderate growth of 1.5-2%/year, Ocean Shipping predicted.

Drewry underscores oil price role

Uncertainty over demand for crude tanker shipping is fueled by oil price volatility and questions about the Organization of Petroleum Exporting Countries' ability to rein oil production, Drewry noted in developing two price forecasts for 1992 (Table 2).

Under one scenario, oil prices remain flat with Saudi light at $14.50/bbl in 1986 dollars through 1992.

Under the other, OPEC trims production and restores balance to the market.

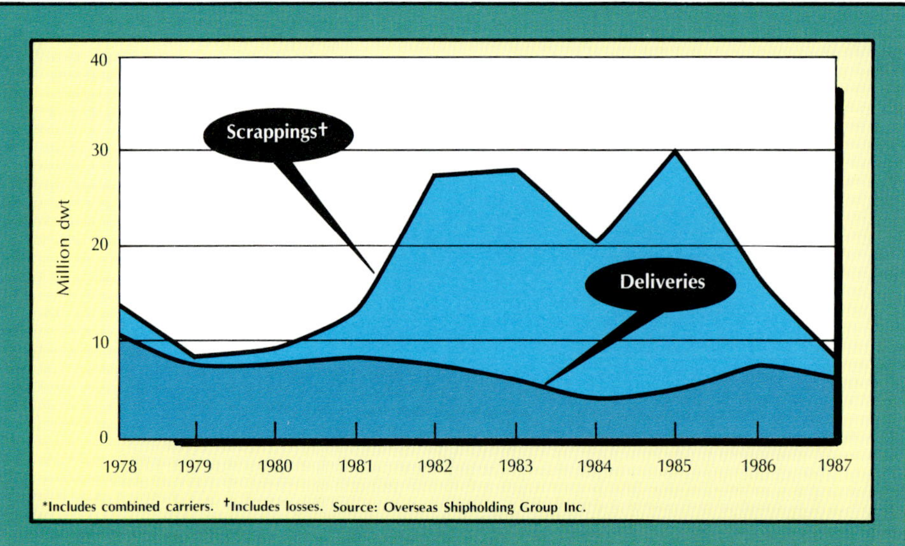

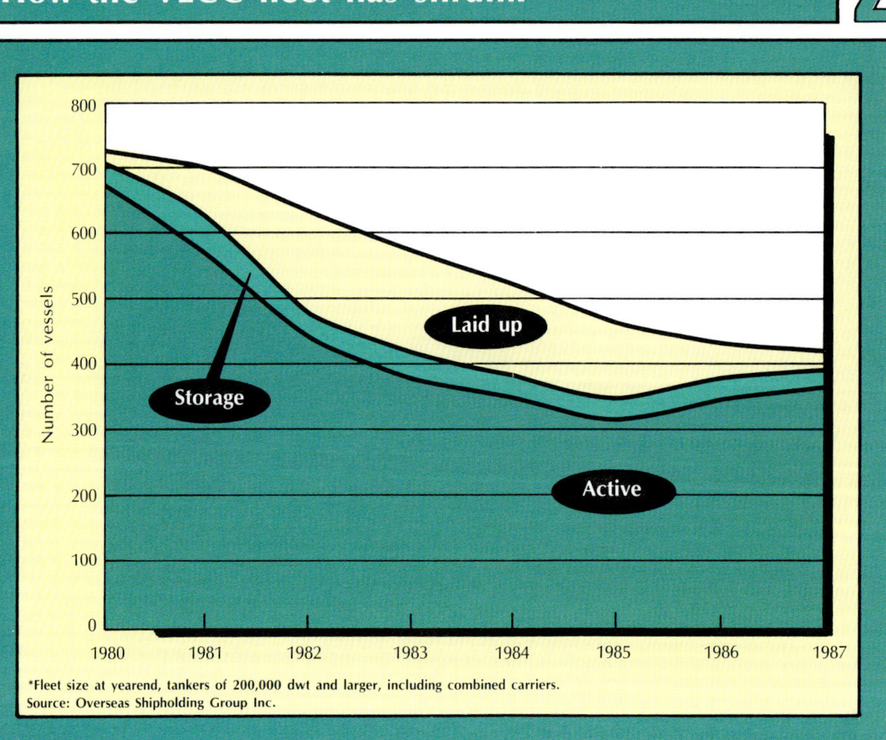

Drewry calculated that would push average crude oil prices in 1986 dollars to $17.72/bbl in 1988, $18.32/bbl in 1989, $18.92/bbl in 1990, $19.52/bbl in 1991, and $20.12/bbl in 1992.

The company's best guess was that oil prices will fluctuate between the low and high cases.

With that assumption, it is unlikely that the surge in seaborne trade and tanker employment seen in 1986 will be repeated.

The healthy 12% recovery in seaborne oil movements in 1986, stemming from a surge in demand sparked by low oil prices, was followed by a rise of only 0.5% in 1987.

The effect of flat oil prices from 1987 to 1992 would be sharp jumps of 12% in non-Communist oil production and 13.5% in oil product consumption, with a consequent 17.6% rise in seaborne crude movements.

The world's main oil shipments by sea, 1987*

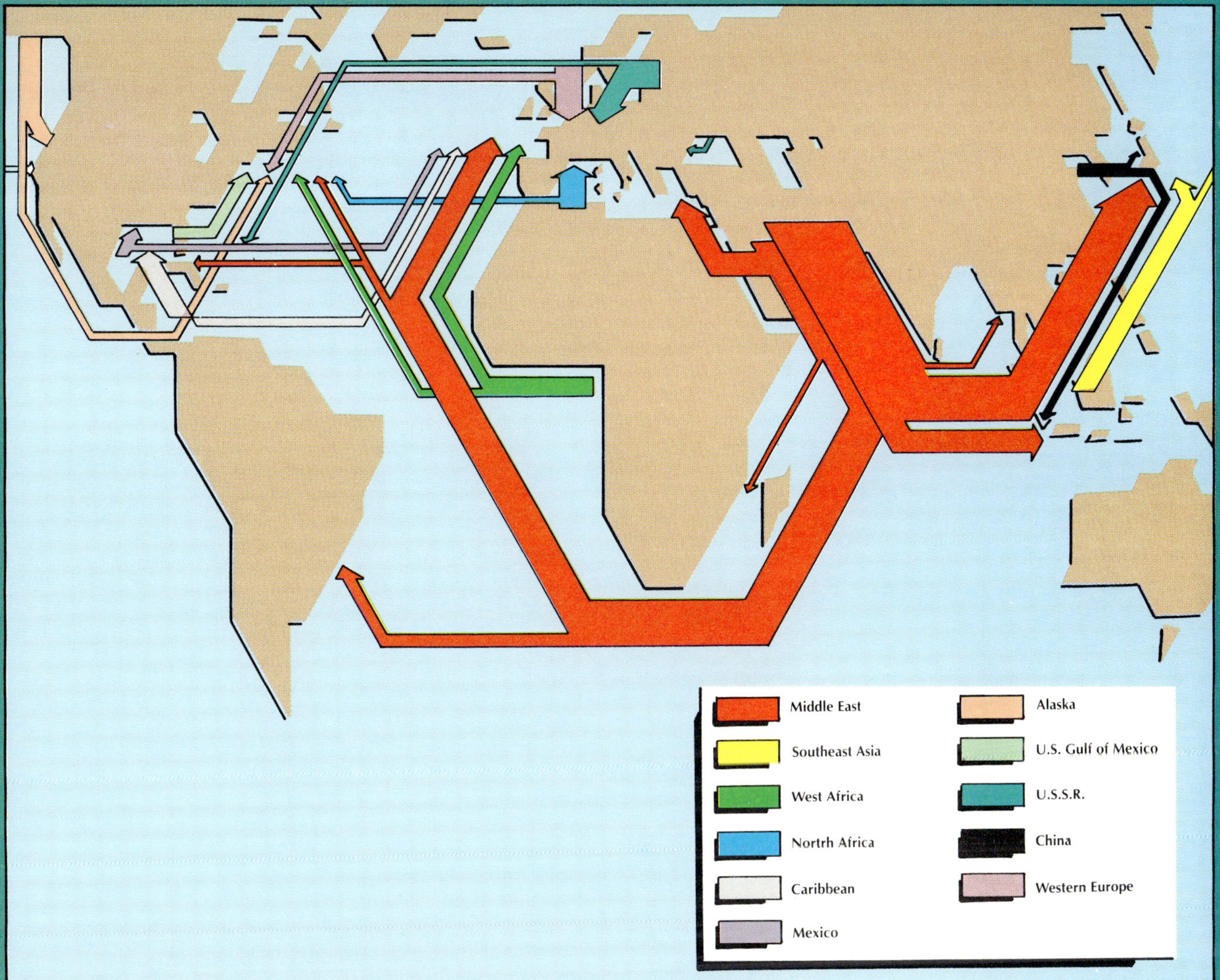

*Width of arrow is proportional to the volume of oil exported. Movements of less than 10 million metric tons/year are not shown. Bunkers and intraregion movements, for example between countries in western Europe, are not included. Source: British Petroleum Co. plc

If the rising price forecast held true, non-Communist oil flow would increase by 3%, oil product consumption by 3.9%, and seaborne crude movements by 7.3%.

Tanker employment was expected to slip slightly in 1988 from 1987, which was flat with 1986 levels.

Meantime, non-Communist refining capacity rose by 1.1% in 1987 from 1986, posting the first increase since 1981. That reflects the gain in crude and products imports owing to lower prices.

During the 1980s, refining capacity has plummeted among oil importing nations and climbed among exporting nations.

Drewry's figures showed the decline in the world crude carrier fleet almost halted in 1987, slipping by only 3.8 million dwt from 234.7 million dwt at yearend 1986.

Further, the amount of idle tonnage in the fleet dropped as well, by 7.9 million dwt to 29 million dwt. VLCCs and ULCCs—vessels of more than 175,000 dwt always used in crude transport—accounted for almost all of that dip in idle tonnage.

Drewry placed the pure crude carrier share of the 1987 fleet at about 82% of total tanker supply.

However, there was little increase in vessel employment in 1987 vs. 1986, although more tonnage was available. This decline in efficiency stemmed largely from war tension and convoy sailings in the Persian Gulf, effectively requiring more tankers to ship the same amount of oil.

Regional pattern of crude shipments

The Middle East will account for a disproportionate share of added seaborne crude exports under either Drewry price scenario, mainly because most of the other oil suppliers will be producing at or near capacity, the firm predicted.

Expansion of the Iraq-Turkey export pipeline and growing crude exports from Syria will mean a sizable jump in eastern Mediterranean seaborne trade, offsetting some of the potential growth in the Middle East seaborne exports.

Caribbean crude exports will rise sharply because some export refineries in the area are likely to be closed or scaled back, making more crude available for export.

Boosts in crude oil production in Southeast Asia will feed into export traffic, supporting steady growth in trade, with some regional exceptions.

North Africa will devote less crude to seaborne trade because refinery capacity has grown sizably there, and

Breakout of tanker demand forecast

Table 1

VESSELS AND TRAFFIC				
Vessel size 1,000 dwt	1986	1990	1995	2000
	—— Billion metric ton-miles/year ——			
<60	1,228	1,280	1,443	1,671
60-100	931	1,098	1,224	1,379
100-200	1,227	1,387	1,521	1,671
200-300	1,650	2,036	2,229	2,506
>300	424	571	644	744
Total	**5,460**	**6,372**	**7,061**	**7,971**

SEABORNE TRADE VOLUME*				
	1986	1990	1995	2000
	—— Million metric tons ——			
Exports				
Middle East	521	586	704	784
Central/S. America	183	222	231	246
Imports				
North America	343	350	418	406
Western Europe	443	421	432	440
Total	**1,263**	**1,343**	**1,489**	**1,584**

*Crude and products.

Source: Ocean Shipping Consultants

Breakout of 1987 shipments

Table 3

	Imports		Exports	
	Crude	Products	Crude	Products
	1,000 b/d			
U.S.	4,625	1,620	150	595
Canada	270	75	625	5
Latin America	585	415	2,080	955
Western Europe	6,104	1,765	550	505
Middle East	80	35	8,640	1,675
North Africa	10	55	1,940	495
West Africa	10	30	1,810	70
East, southern Africa	395	45	—	5
South Asia	425	65	—	55
Southeast Asia	1,545	475	970	385
Japan	3,200	925	—	30
Australasia	125	95	85	55
U.S.S.R., eastern Europe, China*	790	105	1,750	1,160
Destination unknown†	400	285	—	—
Total	**18,600**	**5,990**	**18,600**	**5,990**

*Excludes intraregional trade within the Communist bloc. †Includes changes in the volume of oil in transit, transit losses, minor movements not otherwise shown, unidentified military use, etc.

Source: British Petroleum Co. plc

Drewry's oil market/tanker outlook

Table 2

	1987	1988		1992	
		Case 1	Case 2	Case 1	Case 2
Average crude oil price (Saudi light-1986 $/bbl)	17.12	17.72	14.50	20.12	14.50
	—— Billion metric tons/year ——				
World oil product consumption	1.888	1.936	2.049	1.946	2.142
World refinery throughput	1.901	1.923	1.933	2.044	2.116
Non-Communist oil production					
OPEC	0.928	0.897	0.917	1.049	1.175
Non-OPEC	1.226	1.239	1.250	1.189	1.239
Total	2.154	2.136	2.167	2.238	2.414
OPEC share (%)	43	42	42	47	49
Seaborne crude oil movements	1.156	1.135	1.151	1.240	1.359
Export zones					
Middle East	0.454	0.426	0.436	0.509	0.575
Eastern Mediterranean	0.107	0.132	0.133	0.131	0.143
North Africa	0.094	0.074	0.079	0.068	0.101
West Africa	0.087	0.084	0.085	0.098	0.109
Caribbean	0.132	0.118	0.117	0.146	0.146
Import zones					
North America	0.277	0.271	0.285	0.321	0.388
Europe	0.378	0.375	0.371	0.396	0.421
Pacific	0.163	0.162	0.159	0.167	0.171
	—— Million dwt ——				
Crude oil tanker employment	167.3	154.9	157.9	174.6	195.7
VLCC/ULCC	90.0	79.5	81.0	93.2	106.1

Source: Drewry Shipping Consultants Ltd.

throughput is likely to follow suit.

Europe's small exports of crude will fall further with declining oil production, even more sharply under a low price scenario in which plans to develop marginal fields are abandoned.

Drewry said U.S. crude imports will continue to rise without further significant discoveries or price rises to boost marginally economic reserves. European imports will follow the same track.

Crude imports in Japan, Australia, and South/Southeast Asia will remain flat, mostly because there will be little change in refinery throughputs.

South America will be a mixed bag, as countries with sharply rising production like Brazil slash crude imports and countries with sharply rising demand and flat production hike crude imports.

Stable capacity seen for tanker fleet

Overseas Shipholding said at the beginning of second quarter 1988 the non-Communist tanker fleet size would remain "more or less constant" if 1988 scrappings matched 1987 levels.

At yearend 1987, the newbuilding tanker backlog for the next several years totaled 21 million dwt, compared with 15 million dwt at yearend 1986. Deliveries scheduled for 1988 total about 8 million dwt, about equal to 1987's scrapping total (Fig. 1).

Deliveries in 1987 totaled about 5 million dwt.

The share of total capacity carrying oil was shrinking. Attracted by higher freight rates, many combined carriers—vessels capable of carrying liquid or dry bulk cargoes—shifted to dry bulk.

The combined carrier fleet totaled about 34 million dwt at yearend 1987, about constant with 1986, but the portion active in oil trade fell from about 64% at the beginning of the year to 35% at yearend.

Combined with the decline in tanker capacity, the shift of combined carriers to dry bulk trade reduced the capacity of the oil carrying fleet by 13 million dwt.

Demand for oil carrying tonnage, measured as total tonnage employed, declined in 1987—but less than supply.

Employment of tankers, excluding combined carriers, increased 5 million dwt.

Laid up tanker tonnage fell to 10 million dwt at yearend 1987—the lowest level in 7 years—from 14 million dwt in 1986 and 36 million dwt in 1985.

An additional increase of 5 million dwt in tanker employment resulted from efficiency measures such as less slow

Table 4

Eleven years of world oil trade

	1977	1978	1979	1980	1981	1982	1983	1984	1985	1986	1987	% change 1986-87	1987 % of total
IMPORTS						1,000 b/d							
U.S.	8,710	8,225	8,410	6,735	5,950	5,040	4,990	5,380	5,065	6,045	6,245	+ 3.3	25.4
Western Europe	13,295	13,090	13,080	11,825	10,245	9,350	8,665	8,595	8,365	8,860	7,905	−10.8	32.1
Japan	5,510	5,335	5,605	4,985	4,460	4,155	4,145	4,305	4,045	4,140	4,125	− 0.4	16.8
Rest of world	7,240	7,400	8,255	8,390	8,000	7,020	6,555	6,470	6,645	7,220	6,315	−12.5	25.7
Total	**34,755**	**34,050**	**35,350**	**31,935**	**28,655**	**25,565**	**24,355**	**24,750**	**24,120**	**26,265**	**24,590**	**− 6.4**	**100.0**
EXPORTS													
U.S.	250	365	485	555	605	815	740	720	780	765	745	− 2.6	3.0
Canada	540	445	455	445	455	485	545	655	685	675	630	− 6.7	2.6
Latin America	3,115	3,400	3,645	3,885	4,380	4,135	4,065	4,100	3,565	3,600	3,035	−15.7	12.3
Middle East	20,730	19,730	20,435	17,510	14,605	11,660	10,355	9,845	9,340	10,880	10,315	− 5.2	41.9
North Africa	3,295	3,370	3,425	2,820	2,175	2,145	2,180	2,290	2,415	2,515	2,435	− 3.2	9.9
West Africa	2,370	2,115	2,645	2,475	1,730	1,500	1,425	1,670	1,765	1,980	1,880	− 5.1	7.6
Southeast Asia	1,915	1,790	1,915	1,705	1,660	1,480	1,400	1,600	1,555	1,550	1,355	−12.6	5.5
U.S.S.R., E. Europe, China	1,975	2,200	1,740	2,005	2,195	2,245	2,655	2,865	2,920	3,020	2,910	− 3.6	11.8
Rest of world	565	635	605	535	850	1,100	990	1,005	1,095	1,280	1,285	+ 0.4	5.4
Total	**34,755**	**34,050**	**35,350**	**31,935**	**28,655**	**25,565**	**24,355**	**24,750**	**24,120**	**26,265**	**24,590**	**− 6.4**	**100.0**

Source: British Petroleum Co. plc

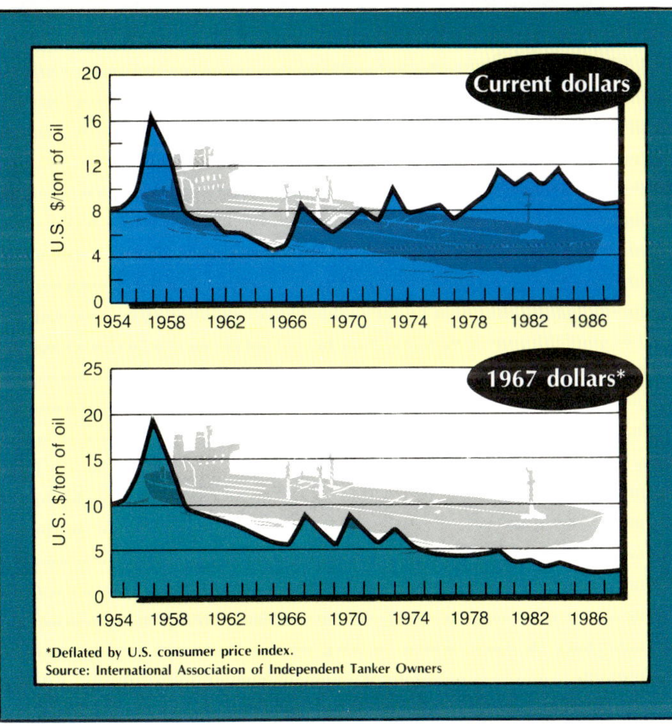

steaming.

But the shift to dry bulk by combined carriers was enough to offset those employment gains for a small demand decline on balance.

VLCCs, ships of 175,000 dwt and larger, made up most of the capacity surplus, which Overseas Shipholding estimated at 26% of the fleet at yearend 1987, compared with 29% a year earlier.

Most of the surplus was in slow steaming, tankers used for storage, and idle time between voyages.

The share of total surplus capacity in layup was shrinking, and more than 90% of the laid-up tonnage involved VLCCs and ULCCs (Fig. 2).

"Because of their age and condition, it is likely that many of these vessels will never resume trading," Overseas Shipholding said. Seaborne trade in crude oil and petroleum products totaled about 1.3 billion dwt in 1987, the same as in 1986.

Start-up of Iraq's 500,000 b/d crude oil pipeline to Ceyhan, Turkey, on the Mediterranean Sea moderated average voyage distances, Overseas Shipholding said. Offsetting that were minor changes in the proportion of oil moving by sea and shifts in geographic distribution of production and consumption.

Biggest exporters, biggest importer

British Petroleum Co. plc figures highlight the dominant role of the Middle East in world shipments of oil by tankers (Fig. 3).

The Middle East also has long dominated oil exports by tanker and pipeline, accounting for 42% of world exports in 1987. The U.S. is easily the world's biggest oil importer, with incoming cargoes heavily weighted toward crude oil shipments.

BP data show that total oil exports slipped 6.4% in 1987 after a big runup in 1986 as a result of plunging oil prices. The runup was due in large part to a netback pricing system used by large exporters.

And although oil prices have bounced around during the years, freight rates in constant dollars have declined since the mid-1970s (Fig. 4).

That's because competition in the tanker trade led to more efficient vessels, particularly through economies of scale, said the International Association of Independent Tanker Owners (Intertanko), Oslo.

An Intertanko report in September 1988 showed that the average current dollar freight rate from the Middle East to western Europe was $8.20/ton in 1954.

The corresponding figure during first half 1988 was $8.75/ton.

"In other words," Intertanko said, "the freight rate is still nearly the same as it was 34 years ago. And this has happened at a time when high inflation has been one of the most serious problems."

In constant 1967 dollars, deflated by the U.S. consumer price index, the Persian Gulf-western Europe freight cost in 1954 was $10.25/ton. That fell to $2.51/ton in first half 1988.

Tanker technology moves forward

TECHNOLOGY IN THE TANKER MOORING BUSINESS IS moving forward. In The Norwegian sector of the North Sea, Den norkse stats oljeselskap a.s. (Statoil), has successfully operated a subsea tanker loading buoy on the Statfjord field.

The unit, known as the Ugland-Kongsberg Offshore Loading Systems (UKOLS) was designed and installed by a group comprising a partnership of Ugland and SBM, Kongsberg Offshore and Ugland Engineering A/S.

Statoil commissioned the novel subsea system to replace the Statfjord A loading buoy. UKOLS comprises a rigid riser from the base of the old Statfjord A buoy to a buoy with a swivel yoke suspended 180 ft below the surface. A 510 ft catenary riser extends to a manifold coupling on the surface.

Only shuttle tankers with dynamic positioning can load from the subsea buoy in all weathers. Seven of the ten Statfjord tankers now have this facility. Statoil says the unit has only been used as a back-up to the two main surface loading buoys, Statfjord B and C.

However the system has been rated as a success as the risk of collision between the shuttle tanker and the surface buoy has been eliminated. Statoil says the unit is used about once a month and has operated successfully in difficult weather conditions.

Statoil originally opted for the subsea system on the grounds of cost. The total bill for UKOLS was NK 40 million ($6.1 million) about 10% of the cost of a replacement for a surface loading buoy.

Statfjord has a remarkable record for tanker loading. Statoil says there has been a 98% availability for the loading systems, better than some pipelines. Since start-up in 1979 production has never been suspended because the tanker loading system was not available. Each of the three concrete platforms has storage in the gravity base which allowed production to continue on the few occasions when weather has prevented loading operations.

A record daily output of 853,000 b/d was achieved in 1987 and Statoil is now regularly producing around 720,000-730,000 b/d. On several occasions about 1.5 million b/d was loaded in a single day with two tankers filling simultaneously. Normally, only one tanker loads at a time.

The UKOLS may also form part of a subsea concrete storage tank successfully tested by Norwegian Contractors for fields with floating production systems. NC has tested models of 500,000 to 1 million bbl capacity units with the UKOLS fitted to the top of the tank.

Across the median line in the British sector of the North Sea, production was cut when the biggest offshore tanker loading operation, through a permanently moored storage tanker, was shut-down when the FSU broke away from its moorings in a storm on Christmas Eve 1988.

The failure of the latch plates that secured the universal joint to the seabed was responsible for the incident. The 210,000 dwt converted tanker Medora, operated by Shell UK Exploration and Production, acted as a gathering station and transshipment point for production from Shell/Esso's Fulmar and Auk fields and from Britoil plc's Clyde field.

The FSU loaded with 480,000 b/d of crude and with the yoke and the 196 ft SALM still attached, drifted close to several producing platforms before it was taken in tow and moved to Stavanger, Norway, for repairs.

Loss of the vessel cut Britain's offshore production by about 220,000 b/d of crude. Gas supplies through a pipeline from Fulmar to St. Fergus were also curtailed.

The SALM was secured to the piled seabed base by four 24x10 in. latch plates. Inspection showed that three of the four plates had fractured allowing the SALM to break free from the seabed structure. In an effort to discover the cause of the fracture, Shell is recovering the base from the seabed.

In the field of new developments, IMODCO has built a full-scale working prototype of a recently developed high pressure fluid swivel. The program was supported by British Petroleum, Mobil North Sea Ltd, Statoil, Sun Oil Co, Petro-Canada, and Petrobras.

The high pressure swivel prototype was designed to handle any fluid product at pressures up to 6250 psi working pressure. It is stackable which allows multiple flow paths to be combined into a fluid swivel stack having a common axis of rotation.

To assure long uninterrupted service, the swivel was designed with a pressure seal isolation system. It isolated the pressure seals from the product by using a massive self-renewing clean liquid barrier between the product and the pressure seals.

The swivel was shipped to BP's Sunbury research center in London where it began dynamic testing for a five-year working life using 95degree C methane at 5500 psi, their highest rated pressure. Tests started in late April 1988 and ended in early 1989.

IMODCO was also awarded a $26 million contract by BHP Petroleum Pty Ltd, Melbourne, Australia, for the development of a mooring system to be installed on the Challis field in the Timor Sea, 150 miles offshore and 400 miles west of Darwin.

Challis is 12 miles south of the Jabiru field where BHP uses a disconnectable turret mooring system for the converted 150,000 dwt ton tanker that acts as a production storage unit. This is designed for rapid disconnect in the cyclone season.

The SBM bow-mounted riser turret system on Jabiru enables the tanker to disconnect in less than two minutes in case of sudden severe storms. Disconnection is achieved automatically in two steps. The first separates the riser flowlines followed by the disengagement of the riser via a large hydraulically operated collet-type connector.

On Challis, BHP has opted for a custom-built 116,000 dwt vessel linked by a rigid yoke to a SALM installed in 348 ft of water. IMODCO will supply the mooring base, riser structure, base universal joint, triaxial swivel, mooring yoke, multi-product distribution unit and hydraulic control swivel.

IMODCO was also awarded a contract for a new CALM buoy by Mobil Producing Nigeria to be located at the Qua Iboe terminal under a seven-year lease agreement. The unit, 22 miles of the Nigerian coast near Eket in 85 ft of water, replaces an existing IMODCO buoy.

The turnkey contract is for the design, fabrication, shipment and installation of the CALM with a maximum throughput of 90,000 bbls/hour. It can handle tankers up to 300,000 dwt in a maximum wave height of 16 ft and wind speed of 25 knots.

SBM has been awarded a second by Japan China Oil Development Corp. for a second jacket soft-yoke system for use in China's Bohai Bay. The first contract, awarded at the end of 1986, called for the design and supply of a jacket soft-yoke system for the mooring of a purpose built floating production, storage and offloading barge for the development of the BZ-28-1 field in Bohai Bay.

The mooring arm was fabricated at the Hudong shipyard in Shanghai together with the 52,000 dwt FPSDO barge. The Bohai Oil Corp. yard at Tanggu was responsible for the jacket and mooring head structure. Installation in 77 ft of water 130 miles west of Tanggu was completed in 1988 by BOC. The system includes a special disconnect facility allowing the barge to return to port in case of heavy ice formation, leaving only the specially-designed jacket and

Text continued at end of following table

Single point mooring around the world

Country	Year	Port/Field	Owner	Designer/Contractor	Max. Vessel Size (Dwt)	Hose number-Size/System	Water Depth (Ft.)
Angola	1968	Cabinda	Gulf	SBM	100,000	2-16″/CALM	77
	1979	Quinfuquena	Petrangol	EMH	150,000	2-16″	NA
	1980	Essungo	Texaco/Petrangol	IMODCO	250,000	2-24″ & 1-16″/CALM	115
	1980	Cabinda	Gulf	SBM	55,000	1-4″/CALM	118
	1982	Takula	Gulf	SBM	300,000	1-12″/CALM	200
	1985	Takula II	Gulf	SBM	319,000	CALM	72
	1984	Palanca	Elf	SBM	265,000	S-Y CALM	148
Argentina	1970	Puerto Rosales	YPF	IMODCO	40,000	1-16″ & 1-12″/CALM	59
	1975	Caleta Olivia	YPF	IMODCO	60,000	1-20″ & 1-12″/CALM	113
	1979	Caleta Cordova	YPF	IMODCO	60,000	1-20″ & 1-12″/CALM	89
	1980	Puerto Rosales	YPF	IMODCO	60,000	1-20″ & 1-12″/CALM	NA
	1987	Hidra	Total	SBM	N/A	CALM	N/A
Australia	1970	Botany Bay	Maritime Services	SBM	120,000	3-12″/CALM	63
	1985	Jabiru	BHP	SBM	160,000	Turret/FPS	394
Bangladesh	1967	Chittagong	Chittagong Port Trust	IMODCO	45,000	1-12″/CALM	45
Brazil	1968	Tramandai	Petrobras	SMB	105,000	2-16″/CALM	71
	1970	Tramandai	Petrobras	SBM	200,000	1-24″/CALM	71
	1975	Garoupa	Perobras	CB & 1	53,000	NA	NA
	1975	Garoupa	Petrobras	CB & 1	53,000	NA	NA
	1976	Sao Francisco	Petrobras	IMODCO	200,000	2-20″/CALM	72
	1978	Garoupa	Petrobras	IMODCO	53,000	2-16″/CALM	410
	1978	Enchova 4	Petrobras	SBM	53,000	1-8″/CALM	410
	1979	Tedut & Tefran	CEC/Petrobras	SBM	200,000	2-20″/CALM	72
	1980	Arembepe	Tibras	SBM	4,000	1-16″/CALM	60
	1901	Pampo	Petrobras	SBM	53,00	1-16″/CALM	460
	1981	RJS 28A	Petrobras	SBM	53,000	1-16″/CALM	395
	1981	Badejo	Petrobras	IMODCO	53,000	2-10″/CALM	338
	1982	Corvina	Petrobras	SBM	53,000	2-8″/CALM	492
	1982	Garoupa	Petrobras	Bluewater	100,000	S-Y CALM	
	1986	Albacora	Petrobras	SBM	55,000	SBS	754
Brunei	1971	Seria	Shell	SBM	150,000	2-16″/CALM	73
	1975	Seria	Shell	SBM	210,000	2-20″/CALM	67
	1981	Seria	Shell	Mitsubishi	313,000	1-20″ & 1-16″/CALM	NA
Cameroon	1976	Kole	Elf Serepca	IMODCO	150,000	1-12″/CALM	80
	1979	Kole	Elf Serepca	Bluewater	250,000	1-24″/CALM	102
	1979	Limboth Point	Sonara	SBM	150,000	2-16″ & 1-16″/CALM	75
	1981	Kole	Elf Serepca	IMODCO	250,000	2-24″ & 1-16″/CALM	85
	1982	Victoria Marine	Total	EMH	280,000	CALM	183
Canada	1969	St. John, N.B.	Irving Oil	SBM	350,000	1-24″ & 1-16″/CALM	149
	1987	Saint John	Canaport	SBM	N/A	CALM	N/A
Chile	1971	Quintero Bay	Enap	SBM	209,000	2-20″/CALM	154
China	1985	Weitzhou	Total	EMH	250,000	Fix tower	123
	1985	Bohai Bay	JCODC	Bluewater	100,000	FPSO/TOWSY	72
	1985	Weitzhou	CSOFL	SBM	250,000	Tank Conv.	250
	1988	Block 16/08 S. China Sea	ACT	SBM	250,000	Turret-disc	380
	1988	BZ-34-2 Bohai	JCODC	SBM	52,000	SY-Jacket	65
	1986	BZ-28-1, Bohai	JCODC	SBM	52,000	SY-Jacket	75
Colombia	1985	Covenas	Occidental	Bluewater	120,000	CALM	80
	1985	Covenas	Occidental	Bluewater	350,000	CALM S-Y	110

Single point mooring around the world—*continued*

Country	Year	Port/Field	Owner	Designer/ Contractor	Max. Vessel Size (Dwt)	Hose number- Size/System	Water Depth (Ft.)
Congo	1972	Djeno	Elf Congo	SBM	250,000	2-20″/CALM	75
Denmark	1971	Dan	Danbor	SBM	70,000	1-12″/CALM	150
	1979	Gorm	Danbor	SBM	70,000	2-12″/CALM	128
	1982	Frederickshaven	DDCS	SOFEC	35,000	2-20″/CALM	50
Dominican Republic	1971	Santo Domingo	Refineria Domincia	SBM	150,000	2-16″/CALM	85
Ecuador	1971	Esmeraldas	Gulf/Texaco	SBM	100,000	2-20″/CALM	124
	1971	Esmeraldas	Gulf/Texaco	SBM	100,000	1-24″ & 1-20″/CALM	124
	1978	Esmeraldas	Texaco	IMODCO	100,000	1-24″ & 1-20″/CALM	118
Equatorial Guinea	1963	Bata	CEPSA	IMODCO	20,000	1-16″/CALM	39
Egypt	1968	Ras-el-Shaqiq	Wepco	SBM	100,000	2-16″/CALM	78
	1974	Suez	Sumed	SBM	120,000	2-20″/CALM	82
	1974	Alexandria	Sumed	SBM	120,000	2-20″ & 1-16″/CALM	85
	1974	Alexandria	Sumed	SBM	120,000	2-20″ & 1-16″/CALM	82
	1974	Alexandria	Sumed	SBM	120,000	2-20″ & 1-16″/CALM	82
	1976	Suez	Sumed	IMODCO	250,000	2-24″/CALM	80
	1976	Alexandria	Sumed	IMODCO	250,000	2-24″ & 1-20″/CALM	108
	1976	Alexandria	Sumed	IMODCO	250,000	2-24″ & 1-20″/CALM	108
	1976	Agami	EGPA	SBM	100,000	2-16″/CALM	98
	1978	El Alamein	Wepco	SBM	100,000	2-16″/CALM	70
	1980	Ras Budran	Suez Oil Co.	SBM	250,000	1-24″ & 1-20″/CALM	115
	1981	Suez	Sumed	SBM	500,000	2-24″ & 1-16″/CALM	118
	1982	Sidi Kerir	Sumed	IMODCO	250,000	2-20″ & 1-20″/CALM	108
	1983	Zeit Bay	SUCO	SBM	155,000	2-20″/CALM	82
	1983	Ras Shukeir	EGPC	SBM	100,000	CBM	
	1983	Wadi Feiran	Petrojet	NA	100,000	NA	
	1984	East Zeit Bay	Esso	IMODCO	85,000	CALM	200
	1984	Geisum	Conoco	IMODCO	100,000	ENG FPSO	
	1985	Sadat area	Petrojet	IMODCO	60,000	CBM	60
Falklands	1984	Port	M.O.D.	IMODCO	45,000	1-10″ CALM	65
France	1968	Bay of Biscay	Elf	CFEM	100,000	NA	NA
	1973	Frontignan	Mobil	IMODCO	270,000	2-20″/CALM	102
Gabon	1965	Gamba	Shell	SBM	100,000	1-16″/CALM	63
	1967	Lucina	Shell	SBM	165,000	1-16″/CALM	112
	1979	Gamba	Shell	SBM	140,000	1-16″ & 1-10″/CALM	63
	1983	Inguessi	AMOCO	Bluewater	—		
	1980	Mayumba	Elf Gabon	Bluewater	70,000	1-4″/CALM	79
Ghana	1978	Bonsu	Agripetco	SOFEC	64,000	1-10″/CALM	79
India	1975	Bombay High	ONGC	SBM	100,000	1-16″ & 1-8″/CALM	240
	1975	Bombay High	ONGC	SBM	100,000	1-16″/CALM	240
	1977	Gulf of Kutch	IOC	SBM	300,000	2-24″/CALM	115
	1982	Bombay High	ONGC	IMODCO	2,500	SBBM	262
	1983	Bombay High	ONGC	IMODCO	150,000	1-16″/CALM	NA
	1984	Gulf of Kutch	IOC	IMODCO	100,000	CALM	101
	1985	Panna	ONGC	IMODCO	115,000	CALM	120
	1987	South Bassein	ONGC	SBM	N/A	2 SVMs	N/A
	1987	D-18	ONGC	IMODCO	115,000	1-16″ CALM	300

Country	Year	Port/Field	Owner	Designer/Contractor	Max. Vessel Size (Dwt)	Hose number-Size/System	Water Depth (Ft.)
Indonesia	1970	Pang Kalan Susu	Pertamina	IMODCO	100,000	2-12"/OALM	NA
	1971	Balikappan	Union Oil	SBM	250,000	2-20" & 1-16"/CALM	102
	1971	Java Sea	ARCO	McDermott	45,000	1-8"	NA
	1971	Java Sea	IIAPCO	IMODCO	55,000	2-12"/CALM	120
	1972	Java Sea	ARCO	IMODCO	133,000	1-12" & 2-16"/CALM	135
	1972	Java Sea	IIAPCO	IMODCO	133,000	2-12" & 2-20"/CALM	130
	1973	Djatibarang	Pertamina	IMODCO	150,000	3-20"/CALM	75
	1973	Ardjuna	ARCO	SBM	150,000	3-16"/CALM	138
	1973	Ardjuna	ARCO	SBM	200,000	2-16"/CALM	126
	1974	Ardjuna	ARCO	SBM	60,000	SBS	140
	1974	Bekapai	Total	SBM	100,000	2-8"/CALM	123
	1975	Poleng	Cities Service	SBM	55,000	2-8"/SBS	180
	1976	Handil	Total	SBM	125,000	2-20"/CALM	105
	1977	Balongan	Pertamina	SBM	150,000	2-20" & 1-16"/CALM	74
	1978	Udang	Conoco	SBM	93,000	2-8"/SBS	302
	1980	Cinta	IIAPCO	IMODCO	133,000	2-12" & 2-20"/CALM	130
	1980	Semarang	Pertamina	SBM	17,000	1-16"/CALM	39
	1981	Krisna	IIAPCO	IMODCO	230,000	2-20", 2-16" & 1-6"/CALM	NA
	1981	Ardjuna	ARCO	SBM	56,000	1-6"/CALM	135
	1981	NA	LETJES	SOFEC	10,000	2-8"	60
	1981	Balikpapan	Pertamina	SBM	150,000	CALM	102
	1983	Lalang	BP	Bluewater	140,000	Yoke Tower	NA
	1983	Balongan	Pertamina	SBM	35,000	1-16"/CALM	46
	1983	Arun	Mobil	Bluewater	100,000	CALM	
	1984	Kakap	Marathon	SBM	140,000	SBS	295
	1985	West Madura	Kodeco	Bluewater	100,000	CALM	330
	1985	Kepiting	Conoco	Comb. Eng.	100,000	FPSO/CBM	330
Iran	1970	Cyrus	Ipac	SBM	140,000	2-16"/CALM	141
	1971	Iman Hasan	SIRIP	IMODCO Ltd	250,000	2-16"/CALM	82
	1983	Iman Hassan	NIOC	SBM	250,000	CALM	160
	1985	Bushire	Smit Tak.	SBM	150,000	CALM	118
	1985	Perslan Gulf	NIOC	IMODCO Ltd	250,000	2-24" CALM 1-16"	85
	1985	Persian Gulf	NIOC	IMODCO Ltd	250,000	2-24" CALM 1-16"	100
	1985	Persian Gulf	NIOC	IMODCO Ltd	250,000	2-24" CALM 1-16"	94
	1985	Persian Gulf	NIOC	IMODCO Ltd	250,000	2-24" CALM 1-16"	Stored as spare
	1985	Persian Gulf	NIOC	IMODCO Ltd	250,000	2-24" CALM 1-16"	Stored as spare
	1985	Persian Gulf	NIOC	IMODCO Ltd	250,000	2-24" CALM 1-16"	Stored as spare
	1985	Persian Gulf	NIOC	IMODCO Ltd	250,000	2-24" CALM 1-16"	Stored as spare
Iraq	1981	Arabian Gulf	Brown & Root	IMODCO Ltd	270,000	2-24"	NA
	1981	Arabian Gulf	Brown & Root	IMODCO Ltd	270,000	2-24"	NA
	1981	Arabian Gulf	Brown & Root	SBM	270,000	2-24"	NA
	1981	Arabian Gulf	Brown & Root	SBM	270,000	2-24"	NA
Italy	1963	Fimicino	PURFINA	IMODCO	50,000	2-12"/CALM	49
	1963	Ravenna	SAROM	IMODCO	50,000	2-12"/CALM	NA
	1964	Fiumicino	PURFINA	Dalmine	100,000	1-16"/RMD	NA
	1970	Porto Torres	Sardoil	SBM	255,000	2-20"/CALM	101
	1972	Ancona	API	Lavore Adriatica Maritime	300,000	NA	NA

Country	Year	Port/Field	Owner	Designer/Contractor	Max. Vessel Size (Dwt)	Hose number-Size/System	Water Depth (Ft.)
	1972	Genoa	Port Authority	CIDONIO	500,000	2-20″ & 2-12″	NA
	1978	Nilde	Agip	SBM	80,000	1-6″/SALS	312
	1981	Rospo Mare	Elf Italiana	EMH	35,000	2-10″	NA
	1981	Genoa	C.A.P.	Technomare	270,000	SALM (Inv).	
	1984	Vega	Montedison	EMH	250,000	SALMRA	394
	1985	Mila	Montedison	Bluewater	70,000	CALM/SYFPS	90
	1985	Nilde II	Agip	IMODCO	138,000	SPT 3-4″ 12-1″ hydraulic	329
	1985	Rospo Mare	Elf	SBM	140,000	FSO/turret	230
Ivory Coast	1980	Bouet	Societe Invoirienne de Raffinage	IMODCO	250,000	3-24″/CALM	164
	1981	Espoir	Phillips	Bluewater	240,000	1-12″/CALM	279
Japan	1961	Niigata	Shell	Niigata	100,000	1-20″/CALM	NA
	1963	Oita	Kyushu Oil	IMODCO	100,000	2-12″/CALM	177
	1965	Yokkaichi	Showa-Yokkaichi	Mitsubishi	100,000	2-16″ & 1-16″/CALM	NA
	1965	Yokkaichi	Showa-Yokkaichi	Mitsubishi	100,000	2-16″ & 1-16″/CALM	NA
	1965	Chiba	Maruzen Oil	IMODCO	100,000	3-12″/CALM	NA
	1967	Koshiba	US Army	IMODCO	100,000	2-12″/CALM	65
	1968	Hakozaki	US Army	IMODCO	100,000	2-16″ & 2-12″/CALM	60
	1968	Kawasaki	Showa-Mitsubishi	Mitsubishi	265,000	2-24″/CALM	NA
	1968	Hakodate	Asia Oil	IMODCO	32,000	1-16″/CALM	49
	1968	Yokkaichi	Daikyo Oil	Mitsubishi	200,000	2-20″/CALM	NA
	1968	Yokkaichi	Daikyo Oil	Mitsubishi	230,000	2-20″/CALM	NA
	1969	Toyama	Japan Sea Oil	IMODCO	150,000	2-16″/CALM	NA
	1969	Yokohama	Ogishima Oil	Mitsubishi	200,000	2-20″/CALM	NA
	1969	Ube	Seibu Oil	Mitsubishi	200,000	2-20″/CALM	NA
	1970	Atsumi	Chubu Electric	Mitsubishi	200,000	2-20″/CALM	NA
	1970	Himeji	Idemitsui Oil	IMODCO	220,000	2-20″/CALM	NA
	1970	Buckner Bay	Toyo Gasoline	IMODCO	100,000	2-16″/CALM	NA
	1970	Tengan	US Army	IMODCO	50,000	2-12″/CALM	70
	1971	Nakagusuki Bay	Esso	Esso, Van Houten	250,000	2-24″/SALM	85
	1973	Ube	Seibu Oil	Mitsubishi	200,000	2-20″/CALM	NA
	1974	Kawasaki	Showa-Mitsubishi	Mitsubishi	300,000	2-20″/CALM	NA
	1975	Yokohama	Oghishima Oil	Mitsubishi	200,000	2-20″/CALM	NA
	1976	Yokkaichi	Showa-Yokkaichi	Mitsubishi	300,000	1-20″ & 1-20″/CALM	NA
	1981	Mutsu-Ogawara	JNOC	SOFEC	350,000	2-24″/SALM	145
	1983	Fuqui	JNOC	SOFEC	240,000	2-20″/SALM	148
S Korea	1987	Inchon	NCS/Kudong Oil Co.	SBM	N/A	CALM	N/A
Kuwait	1967	Ras Al Khafji	Arabian Oil	McDermott	150,000	1-16″ & 1-12″	NA
	1972	Ras Al Khafji	Arabian Oil	McDermott	250,000	1-24″ & 1-10″	NA
	1977	Mina-Al-Ahmadi	KOC	ETPM	532,000	2-24″ & 1-16″/RMD	NA
	1980	Ras al Kahfji	Arabian Oil	Mitsubishi	300,000	CALM	NA
Lebanon	1985	Beruit	Elec du Liban	EMH	40,000	SALM	118
Libya	1962	Brega	Esso	Esso	100,000	Fixed Tower/ Loading Arm	NA
	1965	Ras-es-Sider	Oasis	SBM	100,000	3-16″/CALM	101
	1968	Zueitina	Occidental	SBM	100,000	1-24″/CALM	100
	1968	Zueitina	Occidental	SBM	150,000	2-24″/CALM	105
	1968	Zueitina	Occidental	SBM	150,000	2-24″/CALM	105
	1969	Brega	Esso	Esso Van Houten	300,000	1-24″/SALM	140
	1985	Marsa el Brega	Sirte Oil Co	IMODCO Ltd	300,000	2-24″ CALM	148
	1969	Ras Lanuf	Mobil	SBM	255,000	2-24″/CALM	96
	1969	Ras-es-Sider	Oasis	SBM	255,000	2-24″/CALM	101
	1974	Azzawiya	NOC	DBV/Woodfield	100,000	2-16″	NA
	1974	Azzawiya	NOC	DBV/Woodfield	30,000	1-8″ & 1-10″	NA
	1976	Azzawiya	NOC	SBM	140,000	2-20″/SALM	85
	1976	Azzawiya	NOC	SBM	100,000	2-20″/SALM	98

Country	Year	Port/Field	Owner	Designer/Contractor	Max. Vessel Size (Dwt)	Hose number-Size/System	Water Depth (Ft.)
	1979	Zueitina	Occidental	IMODCO Ltd	275,000	2-24″/CALM	99
	1980	Ras-es-Sider	Oasis	SBM	300,000	2-24″/CALM	102
	1983	Ras-es-Sider	Oasis	IMODCO Ltd	NA	CALM	NA
Malaysia	1959	Miri	Shell	SBM	70,000	2-8″ & 1-6″/CALM	48
	1959	Miri	Shell	SBM	70,000	2-12″ & 1-16″/CALM	48
	1962	Miri	Shell	SBM	70,000	2-8″ & 1-6″/CALM	48
	1962	Port Dickson	Shell	SBM	100,000	2-16″/CALM	90
	1963	Miri	Shell	SBM	70,000	2-12″ & 1-6″/CALM	53
	1974	Labuan	Shell	SBM	313,000	2-20″/CALM	98
	1974	Tembungo	Exxon	SOFEC	94,000	1-10″/SALM	300
	1974	Pulai	Exxon	SBM	190,000	1-12″/SALM	240
	1977	Pulai	Exxon	SBM	166,000	SALS	216
	1978	Bintulu	Shell	SBM	210,000	2-20″/CALM	67
	1981	Trengganu	Petronas	SOFEC	85,000	2-20″/SALM	60
	1981	Trengganu	EPMI	SBM	250,000	1-24″/SALM	92
	1981	Trengganu	EPMI	SBM	250,000	1-24″/SALM	85
	1981	Paka	JGC Corp.	SOFEC	85,000	CALM	65
	1982	Port Dickson	Shell	Mitsubishi	100,000	CALM	
	1984	Lutong	Shell	SBM	125,000	CALM	77
	1984	Lutong II	Shell	SBM	125,000	CALM	77
Mexico	1973	Tuxpan	Pemex	IMODCO	60,000	2-16″/CALM	60
	1974	Tuxpan	Pemex	IMODCO	60,000	1-10″ & 3-16″/CALM	67
	1976	Salina Cruz	Pemex	IMODCO	60,000	3-16″ & 1-12″/CALM	75
	1976	Rosarito Beach	Pemex	IMODCO	60,000	1-20″ & 2-16″/CALM	77
	1978	Rabon Grande	Pemex	IMODCO	150,000	1-16″ & 2-20″/CALM	90
	1980	Cayo Arcas	Pemex	IMODCO	151,000	2-24″ & 1-16″/CALM	144
	1980	Doc Bocas	Pemex	SBM	250,000	2-24″ & 1-16″/CALM	93
	1981	Cayo Arcas	Pemex	SBM	250,000	2-20″ & 1-16″/CALM	151
	1981	Dos Bocas	Pemex	SBM	250,000	2-24″ & 1-16″/CALM	93
	1981	Salina Cruz	Pemex	Mitsubishi	250,000	2-24″ & 1-16″/CALM	NA
	1981	Cayo Arcas	Pemex	EMH	285,000	2-20″/RMD	131
	1981	NA	Pemex	SOFEC	2,000	SBBM (21)	150
	1985	Unknown	Pemex	Mitsubishi	150,000	CALM	93
Morocco	1971	Mohammedia	RAPC	IMODCO	100,000	1-8″ & 1-20″/CALM	72
New Caledonia	1977	Noumea	Le Nickel	SBM	100,000	1-20″/CALM	62
New Zealand	1971	Waipipi Point	Marcona Corp.	IMODCO	75,000	1-12″/CALM	64
	1972	Taharoa	New Zealand Steel	IMODCO	70,000	2-12″/CALM	77
	1976	Taharoa	New Zealand Steel	SBM	150,000	2-12″/CALM	105
Nigeria	1967	Lagos	Nidogas	IMODCO	2,000	1-3″ & 1-4″/CALM	13
	1968	Escravos	Gulf	IMODCO	100,000	2-16″/CALM	70
	1968	Forcados	Shell	SBM	210,000	1-20″/CALM	83
	1968	Forcados	Shell	SBM	210,000	-124″/CALM	83
	1971	Forcados	Shell	SBM	313,000	1-24″/CALM	91
	1971	Bonny	Shell	SBM	313,000	1-24″/CALM	91
	1971	Bonny	Shell	SBM	313,000	1-24″/CALM	91
	1972	Brass River	Agip	IMODCO	200,000	2-20″/CALM	96
	1974	North Apoi	Texaco	SBM	50,000	2-20″ & 1-12″/CALM	87
	1974	North Apoi	Texaco	SBM	250,000	2-20″/CALM	93
	1976	Brass River	Agip	IMODCO	250,000	2-20″/CALM	95
	1977	North Apoi	Texaco	SBM	250,000	2-24″/CALM	94
	1977	Escravos	Gulf	SBM	300,000	2-24″/CALM	102
	1979	Qua Iboe	Mobil	IMODCO	300,000	2-24″/CALM	95
	1984	Andanga	Ashland	SBM	285,000	SY-Jacket	134

Country	Year	Port/Field	Owner	Designer/Contractor	Max. Vessel Size (Dwt)	Hose number-Size/System	Water Depth (Ft.)
	1985	Qua Iboe	Mobil	IMODCO	380,000	2-24″ CALM	- 95
	1985	Bonny	Shell	Bluewater	350,000	CALM	92
	1987	Brass	NAOC	SBM	N/A	CALM	N/A
North Yemen	1986	Salif	YEPCO	SOFEC	N/A	Turrent moored FSO	N/A
	1987	Ras Isa	YEPCO	SBM	N/A	CALM	N/A
Norway	1970	Ekofisk	Phillips	SBM	60,000	1-12″/CALM	232
	1970	Ekofisk	Phillips	SBM	150,000	1-12″/CALM	208
	1979	Statfjord B	Statoil	SBM	150,000	1-20″/ALP	480
	1982	Statfjord C	Statoil	SBM	150,000	1-20″/ALP	478
	1987	Statfjord A	Statoil	UKOLS	150,000	1-20″/subsea	480
	1986	Gullfaks A	Statoil	SBM	150,000	1-20″/ALP	445
	1987	Gullfaks B	Statoil	SBM	150,000	1-20″/ALP	445
Oman	1963	Mina-al-Fahal	P.D. Oman	SBM	100,000	2-8″/CALM	71
	1966	Mina-al-Fahal	P.D. Oman	SBM	165,000	2-16″ & 1-8″/CALM	115
	1966	Mina-al-Fahal	P.D. Oman	SBM	165,000	2-20″ & 2-8″/CALM	124
	1973	Mina-al-Fahal	P.D. Oman	SBM	500,000	2-20″ & 2-8″/CALM	155
	1979	Mina-al-Fahal	P.D. Oman	SBM	350,000	2-20″ & 2-12″/CALM	124
	1984	Kuria Muria	Shell	SBM	250,000	CALM	125
Panama	1981	Chiriqui Grande	Petroterminal de Panama	EMH	150,000	2-20″/CALM	NA
	1981	Chiriqui Grande	Petroterminal de Panama	EMH	150,000	2-20″/CALM	NA
	1983	Chiriqui Grande	Petroterminal de Panama	EMH	150,000	2-20″/CALM	NA
Philippines	1967	Subic Bay	US Navy	IMODCO	108,000	2-16″ & 1-10″/CALM	85
	1977	S. Nido	Cities service	SBM	70,000	SBS	200
	1980	Bataan	Bataan Refining	SOFEC	300,000	2-24″/SALM	94
	1980	Cadlao	Amoco	SBM	127,000	SBS	318
	1982	Nido	Cities Service	Bluewater	55,000	2-10″/CALM	219
Qatar	1965	Halul	QGPC	SBM	100,000	2-16″/CALM	97
	1972	Halul	QGPC	McDermott	300,000	2-24″	NA
	1972	Umm Said	QGPC	IMODCO	300,000	1-20″ & 1-24″/CALM	66
	1976	Doha	QGPC	SBM	65,000	1-6″/CALM	88
	1979	Halul	QGPC	SBM	540,000	2-20″ & 1-10″/CALM	120
	1983	Umm Said	QGPC	IMODCO	500,000	CALM	NA
Saudi Arabia	1970	Zuluf	Aramco	SBM	250,000	2-24″ & 1-16″/CALM	111
	1970	Zuluf	Aramco	SBM	450,000	2-24″/CALM	129
	1974	Ju'aymah	Aramco	McDermott	500,000	2-24″ & 1-12″	NA
	1974	Ju'aymah	Aramco	McDermott	500,000	2-24″ & 1-12″	NA
	1974	Ju'aymah	Aramco	McDermott	500,000	2-24″ & 1-12″	NA
	1975	Ju'aymah	Aramco	SOFEC	750,000	2-24″ & 1-12″/CALM	135
	1975	Ju'aymah	Aramco	SOFEC	750,000	2-24″ & 1-12″/SALM	117
	1981	Ras Al Khafji	Aramco	Mitsubishi	300,000	1-24″ & 1-10″/CALM	NA
	1987	Duba	ODR	SBM	N/A	CBM	N/A
Singapore	1971	Singapore Harbor	Esso	IMODCO	252,000	2-24″/CALM	88
	1974	Pulau Bukom	Shell	McDermott	300,000	2-24″	NA
	1980	Singapore Harbor	PSA	IMODCO	300,000	2-24″/CALM	105
South Africa	1969	Durban	Shell	SBM	210,000	2-20″/CALM	150
	1974	Durban	Shell	SBM	210,000	1-20″/CALM	150
South Korea	1963	Ulsan	KOC	IMODCO	75,000	2-12″, 1-8″ & 1-4″/CALM	65
	1969	Yosu	Honam Oil	IMODCO	100,000	2-16″/CALM	112
	1969	Ulsan	KOC	IMODCO	200,000	2-16″ & 1-12″/CALM	88
	1977	Onsan	KIPCO	IMODCO	250,000	2-24″/CALM	90
	1980	Pusan	KOC	Mitsubishi	325,000	2-24″ & 1-16″	NA

Single point mooring around the world—*continued*

Country	Year	Port/Field	Owner	Designer/Contractor	Max. Vessel Size (Dwt)	Hose number-Size/System	Water Depth (Ft.)
Spain	1966	Huelva	Rio Tinto	SBM	100,000	2-16"/CALM	75
	1971	Amposta	Shell	SBM	33,530	2-10"/CALM	209
	1973	Tarragona	Enpetrol	SBM	325,000	2-24" & 1-12"/CALM	135
	1975	Algeciras	Cepsa	DBV/Woodfield	500,000	2-20" & 1-12"	NA
	1976	Castellon	Shell	SBM	60,000	SALS	383
	1979	Badalona	Campsa	SOFEC	50,000	2-12"/SALM	107
Sri Lanka	1986	Columbo Harbor	Ceylon Pet. Co.	IMODCO	180,000	CALM 1-24"	95
Sudan	1983	Marsa Nimeiri	Chevron	SBM	250,000	CALM (HEX)	160
Sweden	1959	Dalaro	Swedish Navy	IMODCO	3,000	1-4"/CALM	150
Taiwan	1967	Tai-Chung	US Air Force	IMODCO	50,000	1-12"/CALM	76
	1968	Kaohsiung	CPC	IMODCO	100,000	2-16" & 1-10"/CALM	68
	1968	Tai-Chung	US Air Force	IMODCO	50,000	1-12"/CALM	76
	1971	Kaohsiung	CPC	SBM	250,000	2-20"/CALM	100
	1972	Kaohsiung	CPC	SBM	250,000	2-20"/CALM	115
	1979	Chu-Wei	CPC	IMODCO	250,000	2-20"/CALM	121
	1981	Ta-Lin-Po	CPC	SBM	250,000	2-24" & 1-16"/CALM	130
Tanzania	1971	Dar-Es-Salaam	Tanzanian Harbour Authorities	SBM	100,000	2-20"/CALM	78
	1982	Dar Es Salaam	T.H.A.	IMODCO	300,000	CALM	
Thailand	1979	Gulf of Thailand	Union	Bluewater	70,000	1-8"/CALM Yoke	NA
	1981	Erawan	Union	Bluewater	85,000	1-8"/CALM	219
Trinidad	1971	Galeota Point	Amoco	SBM	250,000	2-20"/CALM	95
	1973	Point-A-Pierre	Texaco	IMODCO	260,000	1-12" & 2-24"/CALM	78
	1976	Galeota Point	Amoco	IMODCO	250,000	2-20"/CALM	95
Tunisia	1972	Ashtart	Elf	SBM	170,000	SBS	220
	1974	Ashtart	Elf	SBM	100,000	1-20" CALM	220
	1978	Ashtart	Serept	SBM	137,000	1-20" & 1-10"/SBS	220
	1982	Tazerka	Shell/Tunirex	SBM	210,000	SALS	470
United Arab Emirates	1968	Fateh	Conoco	SBM	150,000	2-16"/CALM	103
	1972	Das Island	BP	IMODCO	300,000	2-24" & 1-16"/CALM	94
	1972	Fateh	Dubai Petroleum	McDermott	300,000	2-24"	NA
	1972	Mubarras Island	ADOC	SBM	200,000	1-24"/CALM	58
	1973	Mubarek	Crescent	SBM	350,000	2-20" & 1-16"/CALM	160
	1974	Abu-Al-Bu-Koosh	Total/ABK	SBM	100,000	2-12"/CALM	97
	1977	Das Island	ADMA	IMODCO	400,000	2-24" & 1-16"/CALM	94
	1977	Arzanah	Amerada Hess	SBM	252,000	1-12"/CALM	107
	1979	Fateh	Dupetco	IMODCO	120,000	2-20"/CALM	130
	1980	Zirku	Zadco	IMODCO	350,000	2-20" & 1-20"/CALM	95
	1980	Zakum Zirku	Zadco	IMODCO	350,000	2-20" & 1-20"/CALM	95
	1981	Sharjah	Amoco	SBM	300,000	1-20"/CALM	65
	1981	Abu-Al-Koosh	Total	EMH	232,799	1-10"	NA
	1983	Salah	Gulf Ras al Khaimah	IMODCO	300,000	CALM	100
United Kingdom	1965	Nore Estuary	BP	BP/Harland Wolff	100,000	NA	NA
	1970	Humber River	Conoco	SBM	210,000	1-24"/CALM	74
	1972	Auk	Shell/Esso	Shell/SBM	42,000	1-10"/ELSBM	280
	1973	Argyll	Hamilton Bros.	SBM	100,000	1-12"/CALM	252
	1975	Brent	Shell/Esso	Shell/HC	90,000	NA	NA
	1975	Anglesey	Shell	SBM	750,000	2-24" & 1-16"/CALM	127

SHUTTLE TANKER NEPTUNE LEO moored in tandem to the floating production facility Jabiru Venture (Photo courtesy BHP).

Single point mooring around the world—*continued*

Country	Year	Port/Field	Owner	Designer/Contractor	Max. Vessel Size (Dwt)	Hose number-Size/System	Water Depth (Ft.)
	1975	Scapa Flow	Occidental	Micoperi	250,000	NA	NA
	1975	Scapa Flow	Occidental	Micoperi	250,000	NA	NA
	1975	Thistle	BNOC	SBM	80,000	2-16″/SALM	535
	1977	Brent	Shell	SBM	80,000	1-16″/CALM	460
	1978	Buchan	BP	IMODCO	107,000	1-12″/CALM	369
	1979	Ninian	Chevron	IMODCO	3,500	SBBM	433
	1980	Ninian	Chevron	IMODCO	3,500	SBBM(4)	433
	1982	Maureen	Phillips	EMH	85,000	1-16″/AC	NA
	1980	Humber River	Conoco	SBM	280,000	1-24″/CALM	74
	1982	Beryl B	Mobil	EMH	80,000	1-16″/ALP	NA
	1984	Cardigan Bay	M.O.D.	Havron	10,000	SPLM	100
	1985	Fulmar	Shell	Bluewater	100,000	ALPRA/HEAD	360
	1985	Beryl A	Mobil	EMH	100,000	ALP	
United States	1967	Louisiana	Kerr-McGee	McDermott	8,000	1-4″	NA
	1978	Worldwide (portable)	US Navy Ltd	IMODCO	70,000	2-10″/CALM	65-200
	1980	Louisiana	Loop	SOFEC	700,000	2-24″/SALM	110
	1980	Louisiana	Loop	SOFEC	700,000	2-24″/SALM	112
	1980	Louisiana	Loop	SOFEC	700,000	2-24″/SALM	114
	1981	Santa Ynez	Exxon	IMODCO	50,000	1-12″, 1-8″, 6-1″ & 1-6″/SALM	494
	1984	Unknown	Brown & Root	IMODCO	550,000	Chain SALM	125
	1984	Unknown	Brown & Root	IMODCO	550,000	Chain SALM	125
Uruguay	1975	Jose Ignacio	Ancap	SBM	150,000	2-24″/CALM	64
	1985	Jose Ignacio	Ancap	SBM	150,000	CALM	64
Venezuela	1968	Moron	CVP	SBM	100,000	2-16″/CALM	63
West Germany	1962	Cuxhaven	West Germany Navy	IMODCO	2,500	1-4″/CALM	26
Western Sahara	1961	El Aaiun	CEPSA	IMODCO	3,000	1-6″/CALM	28
Zaire	1975	Zaire	Gulf	IMODCO	100,000	2-16″/CALM	71
	1976	Zaire	Gulf	IMODCO	79,200	2016″/CALM	79

mooring arm on location. Because of the high pour point of the waxy crude, the piping systems are equipped with insulation and heat-tracing equipment.

The second contract, awarded in February 1988, calls for a similar system for the BZ-34-2 field in 65 ft of water in Bohai Bay. The system is due to be installed during 1989 as part of a field development using two wellhead platforms and a water injection platform. The mooring system will cater for the future addition of gaslift facilities and the tie-in of a third wellhead platform.

World upstream spending to show slight gain this year

What's influencing E&P spending

	1987	1988	1989
		Percent	
MAJORS			
Oil prices	—	—	61
Gas prices	—	—	22
Energy prices	50	19	—
Economics	23	6	17
Reserve replacement	18	31	17
Cash flow	18	—	17
Development commitments	14	31	6
Leases	5	13	6
Prospect availability	—	13	44
INDEPENDENTS			
Oil prices	—	—	47
Gas prices	—	—	44
Energy prices	49	29	—
Cash flow	43	19	25
Economics	23	33	15
Financial position	17	10	7
Gas markets	16	12	5
Prospect availability	6	21	14
Reserve replacement	2	14	6
Development spending	1	4	12
Capital availability	—	4	6

Source: Salomon Bros. Inc.

Breakout of exploration/production spending*

	1989	1988
	(Million U.S. $)	
MAJORS, U.S.†		
Amerada Hess	108	100
Amoco	1,400	1,500
ARCO	1,350	1,305
BP	925	970
Chevron	1,075	1,100
Coastal	50	50
Du Pont	475	400
Elf Aquitaine	83	139
Exxon	1,300	1,400
Kerr-McGee	112	102
Mobil	960	1,010
Murphy	33	26
Occidental	245	245
Pennzoil	240	210
Phillips	430	514
Royal Dutch/Shell	1,950	1,950
Sun E&P	435	490
Tenneco	0	250
Texaco	700	860
Total Minatome	82	74
Union Pacific	280	300
Unocal	400	440
USX	450	465
SELECTED INDEPENDENTS, U.S.†		
Adobe Oil & Gas	49	44
Amax Oil & Gas	30	12
Anadarko Petroleum	185	135
Apache Corp.	97	54
Arkla	55	48
Brooklyn Union	24	17
Burlington Resources	200	200
Cabot	50	47
Columbia Gas System	90	54
Consolidated Natural	244	244
Dekalb Energy	26	22
Diversified Energies	30	40
Enron	98	77
Enserch	95	113
Equitable Resources	58	65
Forest Oil	80	65
Freeport-McMoRan	100	100
Helmerich & Payne	25	18
Invexco	20	20
Kelley Oil	20	15
Louisiana Land	95	134
Maxus Energy	53	100
Mesa Petroleum	40	25
Mitchell Energy	55	49
Nerco Oil & Gas	25	5
Nicor	30	30
Noble Affiliates	50	48
Northern Michigan	50	29
Norcen Energy	51	44
Odeco	65	48
Pacific Enterprises	140	150
Placer Dome	20	18
Questar	31	23
Santa Fe Energy	45	45
Smith Co.	35	28
Sonat	55	49
Southdown	57	57
Transco Exploration	65	65
Union Texas	130	141
Wolverine	90	110
CANADA†		
Alberta Energy	45	45
Amerada Hess	35	35
Amoco	380	345
Anderson Exploration	27	30
Bow Valley	52	51
Canadian Hunter	131	115
Canadian Occidental	44	59
Chevron	265	220
Columbia Gas	35	20
Du Pont	30	30
Gulf Canada	185	286
Imperial	337	383
Mobil	160	160
Murphy Oil	24	26
Norcen Energy	74	91
North Canadian	59	76
Numac Oil	25	29
PanCanadian	220	244
Poco Petroleum	38	76
Sceptre Resources	29	38
Shell Canada	405	295
Suncor	50	54
Texaco Canada	279	248
Total Petroleum NA	30	20
Union Pacific	50	35
Unocal	70	80
OTHER INTERNATIONAL†		
Adobe Oil & Gas	11	13
Amerada Hess	223	225
Amoco	1,050	1,000
Ashland Oil	30	30
ARCO	315	330
Bow Valley	71	84
BP	2,805	2,090
Canadian Occidental	21	18
Chevron	675	600
Du Pont	380	440
Elf Aquitaine	1,420	1,450
Exxon	2,375	2,225
Gulf Canada	42	34
Hamilton Oil	75	64
Kerr-McGee	89	89
Lasmo plc	164	91
Louisiana Land	48	34
Maxus Energy	94	74
Mobil	755	630
Murphy	21	20
Norcen Energy	41	23
Norsk Hydro	540	600
Occidental	450	435
Odeco	32	19
Phillips	297	258
Ranger Oil	42	34
Renaissance Energy	59	114
Royal Dutch/Shell	2,100	2,100
Saga Petroleum	200	200
Statoil	747	648
Sun Co.	108	74
Texaco	450	450
Texas Eastern	100	75
Total SA	500	533
Triton Energy	70	68
Union Texas	100	94
Unocal	280	280
USX	320	310

*Estimates for both years. †Companies spending at least $20 million in either year.

Source: Salomon Bros. Inc.

Stratigraphic charts

TEN STRATIGRAPHIC COLUMNS from around the world appear on these pages.

They represent sedimentary basins in Argentina, the Netherlands, Nevada, New Mexico, Gabon, Papua New Guinea, Florida, Texas, Colorado, and Nova Scotia.

Argentina has 1,337,750 sq km of sedimentary basins, 80% of which has not been fully explored. Despite more than 70 years of exploration and production, only Argentina's Northwest, Cuyo, Neuquen, San Jorge, and Austral basins have production.

The Northwest basin produced 174 million bbl of oil and condensate and 1.6 tcf of gas through 1985. Production comes from Devonian, Carboniferous, and Cretaceous reservoirs, and oil has been generated in Devonian and Cretaceous-Paleocene source rocks.

About one third of the Cuyo basin is not fully explored. Cumulative production of 786 million bbl of oil left 99 million bbl of reserves at the end of 1985. The oil is accumulated in sandstone and conglomeratic fluviatile Triassic deposits and predominantly trapped in north-south trending asymmetric anticlines and homoclines.

Neuquen basin production of 904 million bbl of oil and more than 2 tcf of gas makes it Argentina's second most prolific basin. Major oil and gas accumulations are found in Jurassic and Cretaceous reservoirs and occasionally on top of volcanic reservoirs.

San Jorge is the oldest and most prolific basin in Argentina. It has produced more than 1.7 billion bbl of oil mostly from nonmarine Cretaceous reservoirs. Other minor production comes from marine Tertiary strata. The oil is accumulated in combination traps and was generated in Jurassic and Cretaceous black shales. The oil fields are in the eastern part of the basin, and similar conditions are found

Papua New Guinea

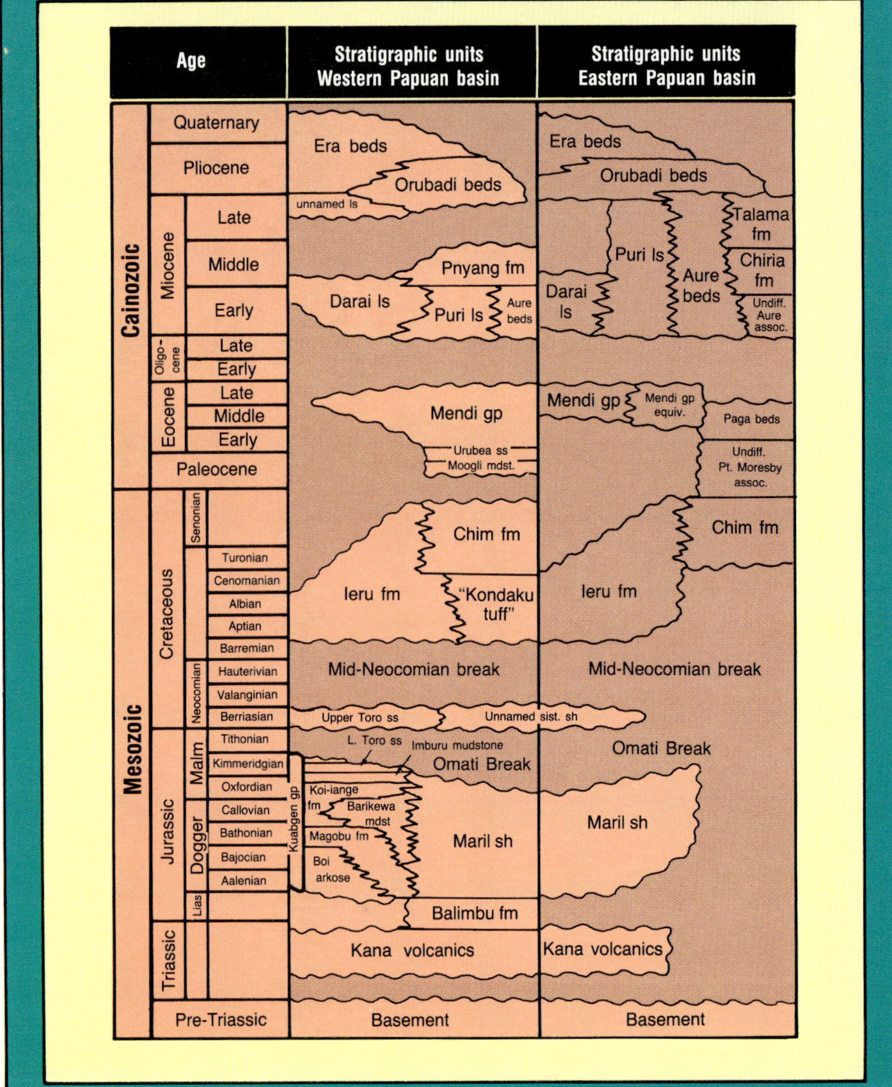

Gabon, Grondin field

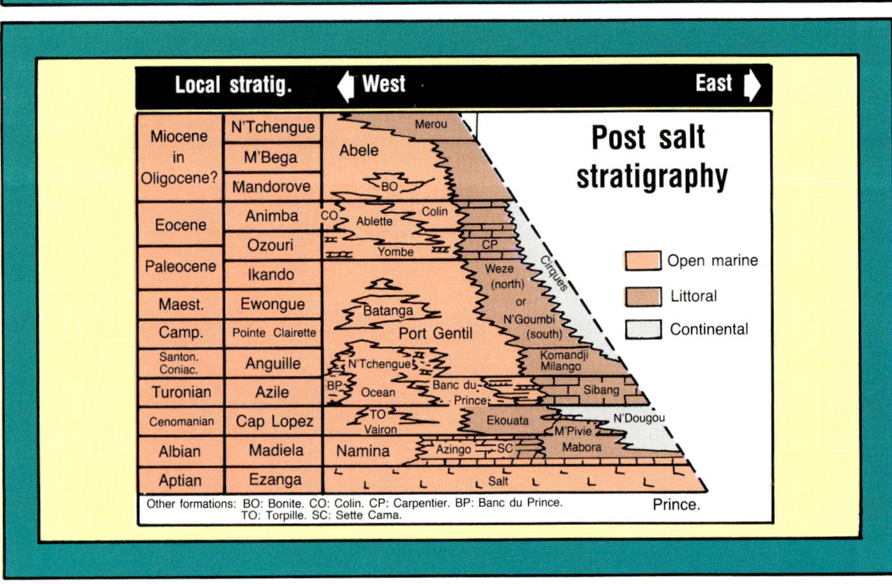

The Netherlands and continental shelf

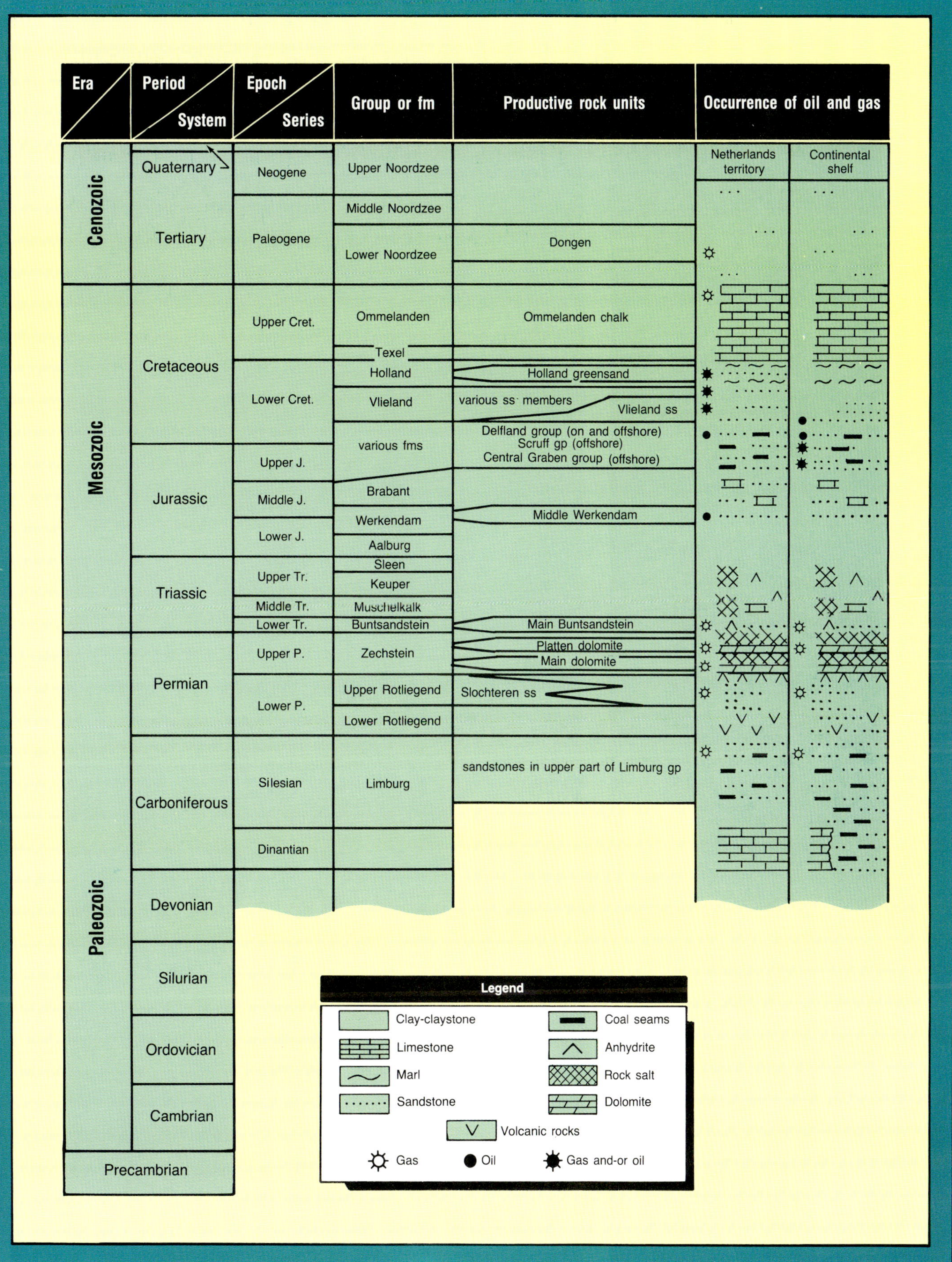

Era	Period / System	Epoch / Series	Group or fm	Productive rock units	Occurrence of oil and gas (Netherlands territory)	Occurrence of oil and gas (Continental shelf)
Cenozoic	Quaternary	Neogene	Upper Noordzee			
Cenozoic	Tertiary	Paleogene	Middle Noordzee			
Cenozoic	Tertiary	Paleogene	Lower Noordzee	Dongen		
Mesozoic	Cretaceous	Upper Cret.	Ommelanden	Ommelanden chalk		
Mesozoic	Cretaceous	Lower Cret.	Texel			
Mesozoic	Cretaceous	Lower Cret.	Holland	Holland greensand		
Mesozoic	Cretaceous	Lower Cret.	Vlieland	various ss members / Vlieland ss		
Mesozoic	Jurassic	Upper J.	various fms	Delfland group (on and offshore) / Scruff gp (offshore) / Central Graben group (offshore)		
Mesozoic	Jurassic	Middle J.	Brabant			
Mesozoic	Jurassic	Middle J.	Werkendam	Middle Werkendam		
Mesozoic	Jurassic	Lower J.	Aalburg			
Mesozoic	Triassic	Upper Tr.	Sleen			
Mesozoic	Triassic	Upper Tr.	Keuper			
Mesozoic	Triassic	Middle Tr.	Muschelkalk			
Mesozoic	Triassic	Lower Tr.	Buntsandstein	Main Buntsandstein		
Paleozoic	Permian	Upper P.	Zechstein	Platten dolomite / Main dolomite		
Paleozoic	Permian	Lower P.	Upper Rotliegend	Slochteren ss		
Paleozoic	Permian	Lower P.	Lower Rotliegend			
Paleozoic	Carboniferous	Silesian	Limburg	sandstones in upper part of Limburg gp		
Paleozoic	Carboniferous	Dinantian				
Paleozoic	Devonian					
Paleozoic	Silurian					
Paleozoic	Ordovician					
Paleozoic	Cambrian					
Paleozoic	Precambrian					

245

offshore.

The Austral basin is about 55% explored. It has produced 168 million bbl of oil and more than 2 tcf of gas mainly from sandstone beds of the marine-continental Cretaceous Springhill formation. Minor production comes from underlying fractured tuffs of the Tobifera formation. Remaining reserves are 69 million bbl of oil.

In the Netherlands North Sea, Rotliegende production requires the combination of Rotliegende aeolian sands being capped by overlying Zechstein salt. Gas is generated in the Coal Measures of the Carboniferous, migrates upward, and is trapped in structures where salt overlies Rotliegende sands.

In the southern Netherlands offshore, the Zechstein changes from salt to dolomite and finally sand. When this occurs, the Rotliegende is not sealed, and gas continues to migrate upwards until trapped in Triassic Bunter sands.

Rotliegende and Bunter production are usually mutually exclusive.

In the northern North Sea, major fields have been found in Mid-Jurassic sands sourced from organically rich Upper Jurassic Kimmeridgian shales. In the Dutch offshore, there is some oil production from Mid-Jurassic Dogger age sands, but the source for oil in the Netherlands is organically rich Lower Jurassic-Lias shales. In southern Norway and Denmark, significant fields produce from Upper Cretaceous chalk sealed by overlying Tertiary shales. The oil source beds are also Jurassic shales.

F/3 field is the only commercial Dogger production off the Netherlands. It is a residual salt anticline, turtle structure. F/3 field has reserves of about 50 million bbl of oil and 500 bcf of gas.

The highest flowing well in the onshore 48 states in 1988, 3 Grant Canyon Unit, is in the Nevada portion of the Great Basin. The well makes 3,000-4,000 b/d of oil.

Grant Canyon field reserve estimation is difficult. Using 240 acres, 400 ft of oil column and pay in Devonian Guilmette dolomite, and 135 bbl/acre-ft, 13 million bbl is a reasonable estimate.

Similar fields are believed to be present not only in Railroad Valley but in other valleys in East Central Nevada.

The first reservoir of importance on the stratigraphic column of Northwest New Mexico's San Juan basin is the Upper Cretaceous Pictured Cliffs sandstone. Pictured Cliffs, the upper gas reservoir in Blanco field, overlies thick Lewis shale, beneath which is the Mesaverde group of sandstones. Me-

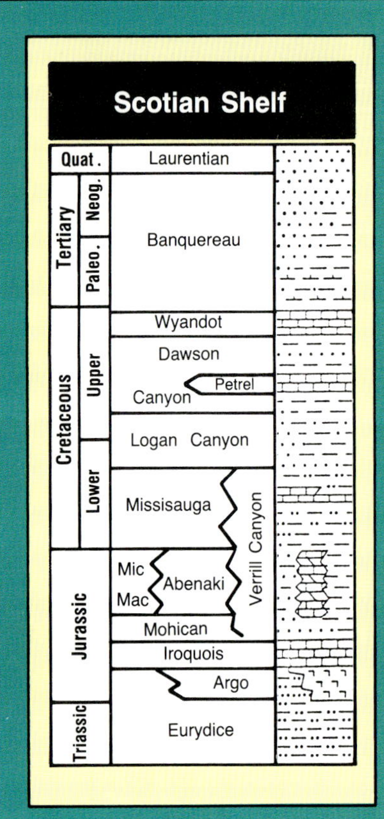

saverde reservoirs are gas productive in Blanco field over an area of about 1 million acres. The main gas producing sandstones are the Cliff House at the top of the group and the Point Lookout at the bottom. These are not blanket sands but are discontinuous shoreline deposits.

Beneath the Mesaverde group is the Mancos shale. Within the Mancos are sand bars, collectively known as Gallup sandstone, from which most of the oil so far produced in Northwest New Mexico has come. The Mancos shale itself contains some oil in fractures. Under the Mancos is the widespread Dakota series of sands, which is gas productive in Basin Dakota gas field over a probable area of 1.4 million acres.

Oil and gas are produced from carbonate reservoirs in the Pennsylvanian Hermosa formation. The principal accumulations are in anticlinal traps high on the western flank of the San Juan basin. Helium has also been found in this reservoir as well as in the deeper Devonian Ouray limestone.

In recent years commercial methane production has been established from Upper Cretaceous Fruitland coal beds over large expanses of San Juan and Rio Arriba counties, and Mancos, Gallup, and Niobrara oil pays have been opened in many fields dotting the huge Blanco Mesaverde gas field.

Offshore from the Ogooue River area in central Gabon, two major episodes of deltaic sedimentation during Late Cretaceous have resulted in the occurrence of a series of oil fields with initial reserves of more than 1 billion bbl.

More than 30 fields have been discovered with reserves of as much as 190 million bbl.

The first episode of deltaic deposition lasted into Early Campanian and deposited the sandstone reservoirs of the N'Tchengue Ocean formation. The second deltaic episode, which deposited the Batanga sandstone, commenced in Late Campanian and continued through to Maastrichtian.

The thick overburden of deltaics brought the source rock, the Banc du Prince formation, to maturity.

The prime reservoirs in the western half of Papua New Guinea's Papuan basin are in Jurassic Toro sandstone. The Magobu, Koi-lange, and Ieru formations locally contain good quality sands that would constitute at least secondary targets over much of the western Fly platform.

In the central basin, Miocene atoll reefs form the reservoir for two discoveries, Pasca and Uramu.

In the Papuan fold belt, fractured Puri limestone is the reservoir in the Kuru, Bwata, and Puri discoveries.

The reservoir potential of Eocene carbonates remains virtually untested, but the Darai limestone contains a number of potential reservoirs.

The 1984 discovery of Bluff Springs field in Escambia County, Fla., brought the number of oil fields in the Florida Panhandle to five.

All fields produce from Jurassic Smackover dolomites and limestones, and Mt. Carmel field also produces from underlying Jurassic Norphlet sandstone.

The fields of Northeast Texas occur mainly in fault traps, reservoir wedgeouts by truncation as illustrated by East Texas field, anticlinal traps, facies change traps, and along piercement salt domes.

Minor oil production has been found in Claiborne. The Upper Cretaceous section has been the most prolific source of oil, mainly because of the presence of the Woodbine Sand at the base.

Woodbine is the reservoir rock at East Texas, Van, Hawkins, Powell, and Mexia fields, all giants.

Other important reservoirs are the Nacatoch, Taylor, Blossom, and Sub-Clarksville.

The upper part of Lower Cretaceous contains Washita, Fredericksburg, and Trinity reservoirs. Buda limestone in the upper Washita produces oil from fractures in Houston and Madison counties.

Paluxy is near the top of the Trinity group; the upper, middle, and lower Glen Rose and the Travis Peak formation

Argentina

Era	Age	Noroeste basins — Santa Barbara	Noroeste basins — Sierras Subandinas y Chaco Salteno	Cuyana basin	Neuquina basin	Golfo San Jorge basin	Austral basin — Santa Cruz Sur	Austral basin — Tierra del Fuego
Cenozoic	Tertiary	Jesus Maria fm Anta fm Rio Seco fm Lumbrera fm ☀ Maiz Gordo fm Mealla fm Olmedo fm	☀ Chaco fm ☀ Candado o Tranquitas fm El Madrejon fm	☀ Papagayos fm		Santa Cruz fm Patagonia fm Sarmiento fm ☼ Rio Chico fm ☀ Salamanca fm	Santa Cruz fm Magallanes gp	Serie arenosa Margosa sup. Glauconite A Margosa medium and inferior Senonian
Mesozoic	Cretaceous	☀ Yacoraite fm Lecho fm Pirgua fm			☼ Neuquen gp ☼ Rayoso fm / ☀ Centenario fm ☀ Huitrin fm ☀ Agrio fm ☀ Mulichinco fm / Chachao fm ☀ Quintuco fm / Loma Montosa fm	☀ Chubut gp ☀ Pozo d 129 fm	Palermo Aike ☀ Springhill fm	Cabeza de Leon fm Arroyo Alfa fm Nueva Argentina fm Pampa Rincon fm ☀ Springhill fm
Mesozoic	Jurassic				☀ Vaca Muerta fm ☼ Q. del Sapo fm ☼ Tordillo fm / Sierras Blancas fm ☼ Auquilco fm / ☀ Punta Rosada fm ☀ Barda Negra fm ☀ Lotena fm ☼ Los Molles fm	Pozo Co. Guadal Lonco Trapial gp Varied fm	☀ Tobifera or efusive complex	Lemaire fm
Mesozoic	Triassic		Cuevo gp { Vitiacua fm … Cangapi fm }	Punta D. L. Bardas fm ☀ Barrancas fm ☀ Rio Blanco fm Cacheuta fm ☀ Potrerillos fm ☀ Las Cabras fm Rio Mendoza fm	Planicie Morada fm			
Paleozoic	Permian			Choiyoi fm	● Choiyoi gp			
Paleozoic	Carboniferous		San Telmo fm ☀ Las Penas fm ☀ Tarija fm Itacuami fm ☀ Tupambi fm		● Granite			
Paleozoic		Arroyo Colorado fm	Iquiri fm — Los Monos fm Tonono fm ☀ Santa Rosa fm — Rio Pescado fm ☼ Michicola fm Porongal fm	Villavicencio fm	Plutonite		Rio Lacteo fm Bahia de La Lancha fm	
Paleozoic	Silurian	Cachipunco fm Zapla fm	Baritu fm — Lipeon fm ☀ Kirusillas fm — Zapla fm					
Paleozoic	Ordovician	Santa Victoria fm	☀ Santa Victoria gp	San Juan fm				
Paleozoic	Cambrian		Meson gp					
	Precambrian		Canani fm					

complete the Trinity and the Lower Cretaceous section.

In the Glen Rose are the highly productive Bacon, Rodessa, James, and Pettet reservoirs. Hosston sand is the principal hydrocarbon source in the Travis Peak formation.

Deep Jurassic rocks include, top to bottom, Cotton Valley, Buckner, and Smackover formations, highly important producers to the northeast in southern Arkansas and northern Louisiana. Of the three, Cotton Valley and Smackover have been the most productive in Northeast Texas.

Multimillion barrel reserves at relatively shallow depths make the Pennsylvanian Morrow point bar sands of Cheyenne County and elsewhere in eastern Colorado economically attractive. Continued success will depend on accurate understanding of the stratigraphy and careful log interpretation. Modern, high quality seismic data can give valuable information away from well control.

The Scotian basin reaches from eastern Georges Bank to the western Grand Banks. The oldest sediments are Triassic and lower Jurassic red clastics, evaporites, and sabkha rift basin deposits formed during the early stages of the breakup of the North American and African continental plates. These are overlain by thick Middle Jurassic to Lower Cretaceous regressive deltaic sequences and shelf carbonates.

Salt mobilization since the Middle Jurassic has resulted in numerous diapirs and pillows as well as an extensive complex of down to basin faults. Such structures have provided the trapping mechanism for significant gas and oil discoveries. Other plays on the shelf include basement structures, the carbonate front, and stratigraphic traps. IPE

Great Basin, Nevada

Age		Formation	Member	General stratigraphy
Tertiary	Pliocene	Muddy Creek fm		
	Miocene	Red sandstone		
		Horse Spring and Thumb fms	Rocks of Lovell Wash	
			Limestone of Bitteridge	
			Clastic unit	
			Rocks of Rainbow Gardens	
Tertiary and Cretaceous		Baseline ss		
Cretaceous		Willow Tank fm		
Jurassic		Aztec ss		
Triassic		Moenave-Kayenta		
		Chinle fm		
		Moenkopi fm		
Permian		Kaibab ls		
		Toroweap fm		
		Red beds		
Pennsylvanian		Callville ls		
		Bird Spring fm		
Mississippian		Monte Cristo ls		
Devonian		Sultan ls		
		Dunderberg sh		
Cambrian		Bonanza King fm		
		Pioche sh		
		Tapeats ss		
Precambrian		Gneiss, schist, granite		

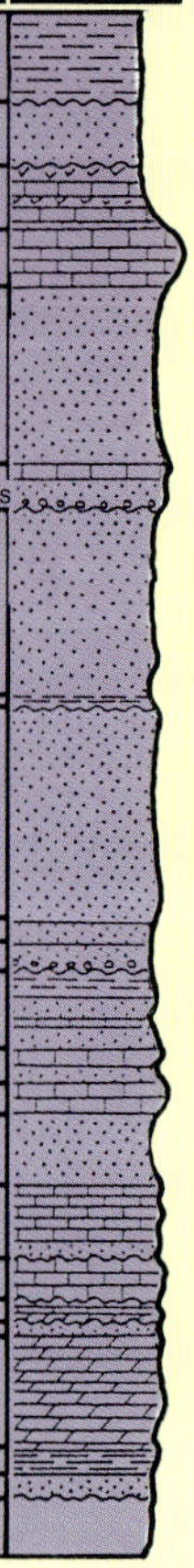

Northwest New Mexico

Geologic time units			Formations		
Cenozoic (to 70 million years)	Quaternary	Recent	Alluvium		
		Pleistocene	Gravel, sand, clay, and volcanic deposits		
	Tertiary	Pliocene	Santa Fe group		Chuska ss
		Miocene			
		Oligocene	May be present?		
		Eocene	Galisteo fm (Eocene and questionably Oligocene)		
			Baca fm (questionably Eocene) San Jose fm		
		Paleocene	Nacimiento fm Animas fm (Upper Cretaceous and Paleocene)		
Mesozoic (70–225 million years)		Cretaceous	Ojo Alamo ss		
			McDermott fm		
			Kirtland sh and Fruitland fm		
			Pictured Cliffs ss		
			Lewis sh		
			Cliff House ss		All of the Mesaverde group
			Menefee fm		
			Point Lookout fm		
			Crevasse Canyon fm		
			Mancos sh with Gallup ss		
			Dakota ss		
		Jurassic	Morrison fm	San Rafael group	Zuni ss
			Bluff ss		
			Summerville fm		
			Todilto fm		
			Entrada ss		
		Triassic	Glen Canyon group		
			Wingate ss		
			Chinle fm		
		Permian	San Andres ls		
			Glorieta ss		
			Yeso fm		Cutler fm
			Abo fm		Rico fm
Paleozoic (225–600 million years)	Carboniferous	Pennsylvanian	Madera fm		Hermosa fm
					Paradox member
			Sandia fm		Molas fm
		Mississippian	Arroyo Penasco fm		Present in subsurface (includes the Leadville ls)
		Devonian	Present in subsurface		Ouray ls (includes the Elbert fm)
		Silurian	Absent		
		Ordovician	Absent		
		Cambrian	Present in subsurface (includes the Ignacio quartzite)		
		Precambrian	Granite, quartzite, pegmatites, etc.		

Eastern Colorado

Geologic time units	Subsurface SE and E. Colorado	Margin of mountain province
Permian	Nippewalla gp / Blaine fm / Sumner gp / Chase gp / Council Grove gp / Admire gp	
Pennsylvanian	Wabaunsee gp / Shawnee gp / Douglas gp / Lansing and Kansas City gps / Marmaton gp / Cherokee gp / Atoka series / Morrow series	Fountain fm
Mississippian	Ste. Genevieve ls / St. Louis ls / Salem and Warsaw lms / Keokuk and Burlington lms / Kinderhook series	Leadville ls
Devonian	Absent in much of region	Chaffee fm
Silurian	Absent	Absent
Ordovician	Viola ls / Simpson gp	Fremont ls / Harding ss / Manitou fm
Cambrian	Arbuckle gp / Bonneterre dolo / Lamotte ss	Peerless fm / Sawatch quartzite

Florida Panhandle

Series	Stages	Groups and formations
Lower Cretaceous	Berriasian	Cotton Valley gp
Upper Jurassic	Tithonian	Cotton Valley gp
Upper Jurassic	Upper Kimmeridgian	Haynesville fm
Upper Jurassic	Lower Kimmeridgian	Buckner member (L. Haynesville)
Upper Jurassic	Oxfordian	Smackover fm
Middle Jurassic	Callovian	Norphlet fm
Middle Jurassic	Bathonian	Louann Salt

Northeast Texas

Age	Rock unit	Lithology	Reservoirs	Section
Tertiary	Wilcox		●	Claiborne Carrizo
Tertiary	Mid-way			
Upper Cretaceous	Nav-arre		✸	Nacatoch
Upper Cretaceous	Taylor		●	
Upper Cretaceous	Austin		✸	Blossom
Upper Cretaceous	E F		●	Eagle Ford sh containing Sub-Clarksville ss
	Woodbine		●	Woodbine
Lower Cretaceous	Washita		●	Buda
Lower Cretaceous	Washita		●	Georgetown
Lower Cretaceous	Washita		✸	Fredericksburg
Lower Cretaceous	Fb		✸	Paluxy
Lower Cretaceous	Trinity			Anhydrite — Glen Rose fm
Lower Cretaceous	Trinity		●	Bacon
Lower Cretaceous	Trinity		✸	Rodessa (Hill and Gloyd reservoirs)
Lower Cretaceous	Trinity		●	James
Lower Cretaceous	Trinity		✸	Pettet
Lower Cretaceous	Trinity		✸	Hosston — Travis Peak fm
Jurassic	Smackover-Buckner-Cotton Valley		✸	
Jurassic	Smackover-Buckner-Cotton Valley		✸	Buckner fm / Smackover fm
U. or Perm.	E.M.			Salt — Eagle Mills fm

Worldwide oil and gas at a glance

COUNTRY	ESTIMATED PROVED RESERVES 1-1-89		OIL PRODUCTION			REFINING Capacity on 1-1-89				
	Oil (1,000 bbl)	Gas (bcf)	Producing wells* 12-31-87	Estimated 1988 (1,000 b/d)	% change from 1987	No. of ref.	Crude	Thermal operations	Catalytic cracking	Catalytic reforming
									b/cd	
ASIA-PACIFIC										
Australia	1,673,089	16,633	926	552.3	+ 0.4	10	644,100	...	190,300	159,600
Bangladesh	500	12,715	10	0.9	...	1	31,200	...	...	1,650
Brunei	1,400,000	11,600	608	138.3	− 0.6	1	10,000	...	...	...
Burma	54,000	9,460	450	15.0	− 21.1	2	26,300	1,700	...	...
China, Taiwan	5,000	885	81	2.6	...	2	570,000	13,500	22,500	56,250
India	6,354,200	22,861	2,670	631.8	+ 3.7	12	1,051,441	110,670	133,953	26,030
Indonesia	8,250,000	83,590	6,065	1,137.5	− 4.1	6	714,200	81,700	12,600	61,500
Japan	54,500	1,410	387	12.3	+ 0.8	41	4,362,750	82,800	596,950	551,650
Korea, South	...	...	...	...	...	6	880,000	49,000	...	62,450
Malaysia	2,922,000	51,700	448	540.0	+ 8.4	4	209,300	...	...	21,980
New Zealand	182,650	5,150	40	28.0	− 3.4	1	88,400	...	...	21,900
Pakistan	170,438	17,722	77	47.0	+ 11.9	3	130,050	...	...	5,450
Papua New Guinea	200,000	3,030	...	...	...	...	...	...	...	...
Philippines	15,800	9	11	9.2	+ 64.3	4	254,000	...	23,100	39,400
Singapore	...	...	...	...	...	5	852,000	137,700	...	58,300
Sri Lanka	...	...	...	...	...	1	50,000	12,500	...	3,750
Thailand	85,200	3,885	245	38.4	+ 20.0	3	191,045	16,920	8,100	26,280
Total Asia-Pacific.......	**21,367,377**	**240,650**	**12,018**	**3,153.3**	**+ 0.9**	**102**	**10,064,786**	**506,490**	**987,503**	**1,096,190**
WESTERN EUROPE										
Austria	100,000	420	1,198	23.2	+ 10.5	1	204,000	17,000	24,000	32,000
Belgium	...	...	...	...	...	4	630,500	62,800	102,000	80,300
Cyprus	...	...	...	...	...	1	17,100	...	...	4,250
Denmark	861,000	4,320	NA	95.5	+ 2.2	3	176,500	70,670	...	31,700
Finland	...	...	...	...	...	2	241,000	46,300	41,800	42,900
France	206,030	1,180	625	68.7	+ 5.7	14	1,875,970	152,850	319,200	247,000
Germany, West	408,000	9,393	3,088	77.9	+ 4.1	15	1,518,200	289,500	179,400	275,100
Greece....................	20,000	140	16	21.8	− 3.1	4	384,500	41,120	50,500	51,300
Ireland....................	...	1,800	...	...	...	1	56,000	...	...	11,000
Italy-Sicily	739,000	10,240	174	92.9	+ 59.3	19	2,450,200	360,400	278,300	284,300
Netherlands...............	205,590	62,507	478	83.7	− 2.3	7	1,380,700	154,360	125,500	166,400
Norway....................	10,435,000	85,500	215	1,069.1	+ 7.9	3	239,400	77,300	...	30,100
Portugal...................	...	...	...	...	...	3	313,300	...	10,100	52,200
Spain	25,955	840	47	30.4	− 4.1	10	1,285,000	173,000	164,000	171,600
Sweden	...	...	...	...	...	5	426,500	67,000	25,000	71,000
Switzerland	...	...	...	...	...	2	132,000	20,000	...	26,000
Turkey	381,000	1,037	571	50.0	− 2.5	5	724,654	4,500	34,128	63,127
United Kingdom	5,175,000	22,740	787	2,376.3	− 3.1	15	1,803,100	152,000	442,000	307,500
Total Western Europe..	**18,556,575**	**200,117**	**7,199**	**3,989.5**	**+ 1.1**	**114**	**13,858,624**	**1,688,800**	**1,795,928**	**1,947,777**
MIDDLE EAST										
Abu Dhabi	92,205,000	183,500	1,001	1,012.6	+ 4.6	2	180,000	...	...	30,055
Bahrain....................	126,000	6,700	274	42.3	+ 0.5	1	243,000	20,000	39,000	18,000
Dubai......................	4,000,000	5,015	150	355.3	− 5.9	...	...	...	...	...
Iran	92,850,000	494,400	361	2,207.5	− 4.4	4	530,000	80,800	...	63,845
Iraq	100,000,000	95,000	868	2,679.2	+ 23.5	8	318,500	...	...	43,500
Israel	1,619	30	10	0.3	...	2	180,000	70,000	20,000	26,000
Jordan.....................	5,000	980	3	0.5	...	1	100,000	...	4,410	8,640
Kuwait.....................	91,920,000	42,500	363	1,254.2	+ 11.8	4	817,000	54,000	42,000	33,000
Lebanon...................	...	...	...	...	...	2	37,000	...	7,250	7,492
Neutral Zone	5,210,000	12,300	492	316.3	− 18.7	...	...	...	...	...
Oman......................	4,071,160	9,320	1,004	596.7	+ 7.1	1	76,932	...	...	15,386
Qatar	3,150,000	156,700	174	349.2	+ 23.8	1	62,000	...	...	12,030
Ras al Khaimah	400,000	2,000	6	10.0	...	...	...	...	...	...
Saudi Arabia	169,970,000	145,848	588	4,708.3	+ 16.9	7	1,375,000	32,000	76,000	180,900
Sharjah....................	1,500,000	11,000	20	65.0	...	...	...	...	...	...
Syria......................	1,730,000	13,130	963	273.3	+ 18.8	2	243,744	36,384	...	18,994
Yemen, North.............	1,000,000	3,700	12	159.9	+ 788.3	1	10,000	...	...	2,500
Yemen, South	3,380,000	...	...	12.5	+ 316.7	1	161,500	...	...	10,800
Total Middle East.......	**571,518,779**	**1,182,123**	**6,289**	**14,043.1**	**+ 11.7**	**37**	**4,334,676**	**293,184**	**188,660**	**471,142**

EDITOR'S NOTE: All reserves figures except those for the U.S.S.R. (and gas for Canada) are reported as proved reserves recoverable with present technology and prices. U.S.S.R. figures are "explored reserves" which include proved, probable, and some possible. Canadian gas figure, under criteria adopted by Canadian Petroleum Association in 1980, includes proved and some probable. *Does not include shut in, injection, or service wells. †Estimates based on capacity as of Jan. 1, 1989, plus known 1988 closures and expansions. §Includes Albania, Bulgaria, Cuba, Czechoslovakia, East Germany, Hungary, Mongolia, North Korea, Poland, Romania, Viet Nam, and Yugolsavia.

	ESTIMATED PROVED RESERVES 1-1-89		OIL PRODUCTION			REFINING Capacity on 1-1-89				
COUNTRY	Oil (1,000 bbl)	Gas (bcf)	Producing wells* 12-31-87	Estimated 1988 (1,000 b/d)	% change from 1987	No. of ref.	Crude	Thermal operations	Catalytic cracking	Catalytic reforming
								b/cd		
AFRICA										
Algeria	8,400,000	104,200	840	666.8	+2.7	4	464,700	...	...	55,600
Angola-Cabinda	2,024,000	1,900	396	449.3	+24.8	1	32,100	...	...	1,900
Benin	100,000	...	6	4.6	−13.2	...	...	...	...	...
Cameroon	400,000	3,880	175	170.0	−2.3	1	42,000	...	...	7,000
Congo	710,000	2,440	289	134.8	+9.6	1	21,000	...	...	2,000
Egypt	4,300,000	11,470	821	851.3	−5.0	8	489,203	16,470	...	30,940
Ethiopia	...	...	...	...	...	1	18,000	...	...	2,270
Gabon	720,000	600	299	175.0	+12.9	1	24,000	7,200	...	1,400
Ghana	1,000	...	0	...	...	1	26,600	...	...	5,850
Ivory Coast	120,000	3,530	17	12.9	−24.1	2	60,000	...	...	16,600
Kenya	...	...	...	...	...	1	90,000	...	...	9,000
Liberia	...	...	...	...	...	1	15,000	...	...	2,000
Libya	22,000,000	25,700	664	1,012.5	−7.1	3	329,400	...	...	13,982
Madagascar	...	...	...	...	...	1	16,350	7,000	...	2,600
Morocco	2,600	100	5	0.4	...	2	154,600	...	5,600	26,800
Mozambique	...	2,290	...	...	...	...	...	...	...	...
Nigeria	16,000,000	85,000	1,522	1,358.3	+9.4	4	414,500	...	84,000	71,000
Senegal	...	...	...	...	...	1	29,800	...	...	2,870
Sierra Leone	...	...	...	...	...	1	10,000	...	...	...
Somalia	...	210	...	...	...	1	10,000	...	...	...
South Africa	...	1,760	...	...	...	4	433,500	63,500	84,400	62,800
Sudan	300,000	3,000	...	...	...	1	21,440	...	...	1,700
Tanzania	...	4,100	...	...	...	1	13,500	...	...	3,000
Tunisia	1,790,000	3,100	152	102.8	−0.9	1	34,000	...	...	3,300
Zaire	96,170	30	110	29.5	−4.8	1	16,700	...	...	3,500
Zambia	...	...	...	...	...	1	20,594	...	...	5,070
Total Africa	**56,963,770**	**253,310**	**5,296**	**4,968.2**	**+2.5**	**44**	**2,786,987**	**94,170**	**174,000**	**331,182**
WESTERN HEMISPHERE										
Argentina	2,268,000	26,700	8,389	449.6	+5.5	11	690,400	125,100	124,800	41,000
Barbados	3,300	7	80	1.1	...	1	3,000	...	...	...
Bolivia	176,200	5,354	300	18.8	−0.5	3	57,500	...	...	14,900
Brazil	2,550,000	3,700	5,909	555.6	−1.8	13	1,407,300	33,800	350,500	24,800
Canada	6,785,514	95,100	37,882	1,604.6	+4.7	27	1,855,800	88,250	387,600	369,600
Chile	287,000	4,200	325	24.8	+1.2	3	146,800	20,000	36,670	10,000
Colombia	2,028,000	3,943	2,711	346.7	−9.9	4	227,400	60,000	91,000	6,000
Costa Rica	...	...	...	...	...	1	16,200	6,500	....	1,200
Dominican Republic	...	...	...	...	...	2	47,000	...	...	7,500
Ecuador	1,350,000	4,020	942	310.1	+77.5	3	123,300	25,200	16,000	2,780
El Salvador	...	...	...	...	...	1	17,000	...	...	3,800
Guatemala	80,830	20	15	3.7	...	1	16,000	...	...	3,000
Honduras	...	...	...	...	...	1	14,000	...	...	1,800
Jamaica	...	...	...	...	...	1	34,200	...	...	3,240
Martinique	...	...	...	...	...	1	12,800	...	...	7,410
Mexico	54,110,000	74,831	3,768	2,527.3	−0.5	9	1,354,000	82,000	267,000	157,800
Netherlands Antilles	...	...	...	...	...	1	320,000	67,000	42,000	15,000
Nicaragua	...	...	...	...	...	1	14,500	...	...	2,800
Panama	...	...	...	...	...	1	100,000	...	...	7,500
Paraguay	...	...	...	...	...	1	7,500	...	...	...
Peru	456,806	623	3,529	141.7	−13.4	6	172,438	...	22,540	1,680
Puerto Rico	...	...	...	...	...	2	123,000	...	12,000	5,800
Suriname	28,700	...	74	2.6	+13.0	...	...	...	...	...
Trinidad & Tobago	528,000	10,500	3,255	149.3	−4.9	2	300,000	18,000	28,000	31,000
Uruguay	...	...	...	...	...	1	33,000	...	4,100	2,500
Venezuela	58,083,900	102,243	28,101	1,658.0	+4.5	6	1,201,100	138,000	193,100	6,000
Virgin Islands	...	...	...	...	...	1	545,000	80,000	...	125,000
United States	26,500,000	187,200	613,430	8,165.9	−2.2	†182	†15,556,700	†1,877,400	†5,336,100	†3,839,400
Total Western Hemisphere	**155,236,250**	**518,441**	**708,710**	**15,959.8**	**+0.2**	**286**	**24,395,938**	**2,621,250**	**6,911,410**	**4,691,510**
Total non-Communist	**823,642,751**	**2,394,641**	**739,512**	**42,113.9**	**+4.2**	**583**	**55,441,011**	**5,203,894**	**10,057,501**	**8,537,801**
COMMUNIST AREAS										
China	23,550,000	31,700	39,500	2,690.3	+0.3	40	2,200,000	NA	NA	NA
U.S.S.R.	58,500,000	1,500,000	128,000	12,476.8	...	39	12,300,000	NA	NA	NA
Other§	1,750,000	29,000	NA	422.0	−1.9	45	3,298,135	NA	NA	NA
Total Communist	**83,800,000**	**1,560,700**	**167,500**	**15,589.1**	**...**	**124**	**17,798,135**	**NA**	**NA**	**NA**
TOTAL WORLD	**907,442,751**	**3,955,341**	**907,012**	**57,703.0**	**+3.0**	**707**	**73,239,146**	**5,203,894**	**10,057,501**	**8,537,801**

Worldwide Production

• Offshore   (e) Estimated   (c) Condensate   (NA) Not available

WORLDWIDE PRODUCTION

	Name of field, discovery date	Depth, ft	No. of wells Producing	No. of wells Total	Production 1987 average B/d	Production Cumulative to Dec. 31, 1987 Bbl	°API gravity
ABU DHABI DEPARTMENT OF PETROLEUM	Asab, 1965	8,000- 8,700	106	248	242,900	1,476,700,000	40.8
	Bab, 1958	8,600- 9,300	100	232	...	680,100,000	40.6
	Bu Hasa, 1962	8,300- 9,200	201	371	335,500	2,785,700,000	39.0
	Sahil, 1967	9,500- 9,800	22	52	7,700	57,000,000	39.7
	• Umm Shaif, 1958	9,800-12,000	142	203	139,400	1,398,600,000	37.0
	• Zakum Lower, 1963	8,000-9,000	131	254	103,700	1,212,700,000	39.0
	• Zakum Upper, 1963	7,000-8,000	142	262	167,300	364,700,000	35.0
	• Umm Addalkh, 1969	7,700-8,500	32	54	11,200	9,300,000	32.5
	• Satah, 1975	8,550-9,500	10	21	4,600	1,700,000	39.8
	• Arzanah, 1973	11,000-11,300	14	34	11,500,000	38,900,000	44.6
	• Abu Al Bukhoosh, 1969	8,900-10,900	42	48	21,300	300,200,000	32.0
	• Mubarraz, 1969	8,300-10,000	28	28	15,300	99,500,000	37.0
*This represents only Abu Dhabi's share of 50% in the field.	• Bunduq,* 1964	9,400-10,900	23	40	11,700	22,800,000	40.0
	Total Abu Dhabi		1,001	1,863	1,083,400	8,458,900,000	
ALGERIA(e) SONATRACH	Zarzaitine, 1957	4,700					42.2
	Edjeleh, 1956	1,500- 2,800					37.2
	Tiguentourine, 1966	2,160- 4,265					44.4
	El Adeb Lareche, 1958	4,100					43.3
	Assakaifaf N. & S., 1958	2,800					42.7
	Tan Emellel N. & S., 1960	2,800- 4,500					42.5
	Edeyen, 1964	5,580					41.5
	Hassi Mazoula S., 1963	4,430					40.2
	Hassi Mazoula B., 1965	4,920					42.0
	Acheb W. & Kreb, 1963 & 1960	7,360					45.6
	Tin Fouye N. +F6, 1962	4,265					39.6
	El Borma, 1967	8,200					42.9
	Keskassa, 1968	8,860					42.9
	Mereksen, 1974	9,420					45.8
	Gara, 1962	9,420					42.9
	Hassi Messaoud N., 1956	10,500					45.9
	Hassi Messaoud S., 1956	10,500					46.1
	Gassi-Touil & Gassi-Touil E., 1961	6,890					50.6-50.8
	Hassi Chergui	8,200					50.6
	Nezla N., 1965	8,530					51.5
	Rhourde El Baguel, 1962	9,850					40.2
	Mesdar E. & W., 1972	11,155					40.3
	El Gassi-El Agreb, 1959	10,170	840	840	649,000	7,537,780,000	47.9
	Rhourde Nouss & E., 1972	8,530					39.7
	Tin Foye Tabankort, 1966	4,600- 6,560					41.1-41.3
	Djoua W., 1967	5,580- 6,240					41.1
	Amassak, 1970	6,565					41.1
	Tamendjelt, 1970	5,585- 6,890					41.1
	Timedratine, 1964	8,155					42.7
	Ohanet N. & S., 1960	7,878					43.8
	Askarene, 1962	8,200					41.7
	Guelta, 1962	8,960					41.7
	Stah, 1971	9,515					45.8
	Ras Toumb, 1976	9,190					35.7
	Haoud Berkaoui-Ben Kahla, 1965	11,000					43.6
	Guellala-Guellala N.E., 1969	11,320					42.2-42.3
	N'Goussa, 1974	11,715					42.2
	Kef El Argoub, 1973	12,570					42.2
	Oued Noumer, 1969	8,860					48.3
	Djorf, 1974	9,025					48.3
	Djebel Onk, 1960	3,940					37.0
	Alrar FJT	...					...
	Hassi R'Mel(c)	...					...
	Gassi Touil(c)	...					...
	Total Algeria		840	840	649,000	7,537,780,000	
ANGOLA FINA ANGOLA	Benfica, 1955	8,300	...	14	...	2,267,118	30.0
	Cacuaco, 1958	7,870	1	12	26	255,153	30.0
	Galinda, 1959	7,220	...	8	...	2,245,401	30.0
	Tobias, 1961	1,970	1	12	55	28,824,501	31.0
	Mulenvos, 1966	5,900- 6,600	7	16	154	5,012,135	25.0
	Quenguela, 1968	6,000	24	58	2,089	37,131,908	31.0
	Legua, 1972	6,300	1	3	105	661,375	31.0

WORLDWIDE PRODUCTION

	Name of field, discovery date	Depth, ft	No. of wells Producing	Total	Production 1987 average B/d	Cumulative to Dec. 31, 1987 Bbl	°API gravity
	Bento, 1972	7,350	1	5	276	4,901,014	26.0
	C. Cobra, 1969	4,000	9	15	1,068	3,012,952	41.0
	Quinguila, 1972	3,400	18	31	8,035	33,117,951	36.0
	N'Zombo, 1973	5,000- 6,300	24	37	11,302	87,556,808	33.0
	Quinfuquena, 1975	6,800- 7,100	14	18	4,659	24,271,920	32.0
	Ganda, 1975	5,000	4	5	3,046	5,842,561	36.0
	Sereia, 1974	6,200	3	4	607	1,073,417	27.0
	Lumueno, 1977	4,600	8	23	2,792	2,676,720	37.0
	Pambo, 1982	4,600	4	4	722	1,681,821	39.0
	Luango, 1977	3,800	3	6	682	1,942,609	38.0
	Pinda, 1966	6,709	...	4	8	2,824	26.0
CHEVRON ●	Malongo North, 1966	1,170- 9,250	58	73	13,411	210,363,571	30.0
●	Malongo West, 1969	1,680- 8,800	22	49	18,485	230,464,506	22.5
●	Malongo South, 1966	1,150- 1,440	29	33	9,108	72,399,649	25.0
●	Kali, 1969	11,400-11,660	...	1	288	4,222,347	30.0
●	Kungulo, 1975	4,000	16	17	10,305	43,099,612	32.2
●	Limba, 1969	2,660- 7,350	10	14	10,768	85,107,068	31.6
●	Kambala, 1971	10,500	4	5	6,389	17,552,178	32.8
●	Livuite, 1979	8,000	4	7	445	2,476,298	31.7
●	Takula, 1971	3,070- 9,090	64	72	121,883	155,574,940	31.5
●	Lifuma, 1984	7,800	...	1	...	184,404	31.4
●	Vuko, 1983	3,700	4	6	8,465	6,839,692	35.0
●	Wamba, 1982	2,700	6	7	24,969	9,113,557	29.0
	Numbi, 1980	3,446	15	15	6,358	2,320,774	34.0
TEXACO ●	Cuntala, 1978	7,200	1	3	87	2,184,064	31.0
●	Essungo, 1975	6,500	10	12	4,781	19,387,494	35.0
	Lombo Este, 1983	7,800	4	6	794	189,714	41.0
	Tubarão, 1984	7,600	4	5	1,859	678,537	41.0
	Sulele Oeste, 1985	7,600	1	1	140	51,130	41.0
ELF ANGOLA ●	Palanca, 1981	8,860	12	23	45,100	39,702,361	40.0
●	Pacassa, 1982	11,500	10	13	40,164	18,206,769	38.0
	Total Angola		**396**	**638**	**359,425**	**1,162,696,943**	

ARGENTINA(e)

	Name of field, discovery date	Depth, ft	Producing	Total	1987 average B/d	Cumulative to Dec. 31, 1987 Bbl	°API gravity
YPF	Chubut, 1907	1,908- 9,168	2,171	2,220			21.1-26.0
	Jujuy, 1969	2,461-16,732	9	10			46.5-47.0
	La Pampa, 1968	3,609- 4,921	180	186			31.0
	Mendoza, 1932	2,178-16,778	1,087	1,113			23.7-32.0
	Neuquen, 1918	1,312-12,498	986	1,049	426,000	4,394,925,000	24.0-41.0
	Rio Negro, 1960	2,953- 7,546	794	853			22.5-41.0
	Salta, 1928	6,551-15,092	41	60			29.0-59.0
	Santa Cruz, 1946	1,468-17,060	2,935	3,014			21.1-30.0
	Tierra del Fuego, 1959	5,249- 7,824	182	196			43.4-46.0
	Formosa, 1984	11,466-11,490	4	5			42.0-42.9
	Total Argentina		**8,389**	**8,705**	**426,000**	**4,394,925,000**	

AUSTRALIA

	Name of field, discovery date	Depth, ft	Producing	Total	1987 average B/d	Cumulative to Dec. 31, 1987 Bbl	°API gravity
BARRACK ENERGY	Mount Horner, 1965	3,800- 5,800	2	7	26	20,240	38.0
BHP PETROLEUM PTY. LTD ●	Jabiru, 1983	4,786	2	2	14,352	7,565,029	42.3
BOND CORP., PETROLEUM DIVISION ●	Harriet, 1983	6,230	10	10	10,760	6,668,000	38.2
CSR OIL & GAS DIV. BRISBANE	Anabranch, 1965	4,188	...	1	4	13,701	44.0
	Duarran, 1964	4,105	...	1	...	14,036	37.0
	Maffra, 1965	4,251	1	2	17	28,086	48.0
	Richmond, 1963	3,643	...	1	2	16,345	43.0
	Snake Creek, 1964	4,990	...	1	4	12,658	67.0
	Sunny Bank, 1962	6,139	...	1	...	706	40.0
	Trinidad, 1965	4,599	...	2	3	63,446	46.0
ESSO-BHP ●	Barracouta, 1965	4,550	2	2	3,690	33,277,000	62.8
●	Marlin, 1966	5,100	1	4	11,410	54,629,500	50.0
●	Kingfish, 1967	7,500	26	37	43,440	934,636,400	46.9
●	Halibut, 1967	7,700	10	17	64,290	689,006,000	43.3
●	Mackerel, 1969	7,700	18	18	80,790	347,077,800	45.6
●	Tuna, 1968	6,500	20	22	6,780	32,786,700	40.5
●	Snapper, 1968	4,350	16	21	4,710	11,906,400	47.0
●	Cobia, 1972	7,700	10	12	49,620	65,075,900	44.6
●	West Kingfish, 1968	7,500	15	18	41,800	46,622,780	46.9
●	Fortescue, 1978	7,700	23	25	91,930	142,895,000	43.0
●	Flounder, 1968	8,200	8	10	16,960	13,271,900	46.7
HARTOGEN GROUP	Kooroopa, 1985	3,300	1	1	9	12,312	42.0
	Talgeberry, 1985	3,900	3	3	356	217,882	46.0
	Washpool, 1985	5,250	1	1	4	715	35.0
	Toobunyah, 1985	3,500	5	5	300	198,683	43.0
	Takyah, 1985	3,800	2	2	11	8,708	43.0
	Ipundu, 1986	2,700	1	1	77	44,129	43.0
	Borah Creek, 1982	4,900	1	3	36	11,584	34.0
	Kincora, 1980	4,700	2	2	33	296,342	34.0
	Sandy Creek, 1982	4,900	1	1	4	8,685	34.0

WORLDWIDE PRODUCTION

	Name of field, discovery date	Depth, ft	No. of wells Producing	Total	1987 average B/d	Cumulative to Dec. 31, 1987 Bbl	°API gravity
	Waratah, 1982	5,350	4	4	71	262,047	34.0
	Tintaburra, 1984	3,500	6	6	449	424,864	43.0
	Monlor, 1987	4,150	1	1	68	5,875	49.0
	Cranstoun, 1987	4,250	1	1	140	4,071	48.0
HOME ENERGY	Blina, 1981	4,300	6	6	438	1,058,202	37.0
	Sundown, 1983	3,600	4	4	65	165,626	39.0
	West Terrace, 1985	3,800	1	1	55	105,686	33.0
	Lloyd, 1987	4,806	1	1	94	14,419	37.0
LASMO	Bodalla South, 1984	4,800- 5,230	6	6	2,425	2,758,481	47.2-48.7
	Kenmore, 1985	4,600- 4,980	9	9	2,639	1,959,951	48.2-45.8
	Black Stump, 1986	5,360	1	1	126	118,373	45.2
	Glenvale, 1985	3,800	1	1	37	17,834	49.3
SANTOS LTD.	Jackson, 1981	4,750	32	32	16,354	18,834,000	40.0-48.0
	Jackson South, 1982	4,800	6	6	1,144	1,884,000	40.8
	Chookoo, 1985	6,000	3	3	306	245,000	45.0
	Wilson, 1983	5,000	5	5	458	526,000	40.0
	Naccowlah South, 1983	6,000	7	7	914	2,069,000	40.0-46.0
	Gunna, 1983	6,000	1	1	37	71,000	40.0
	Tinpilla, 1983	6,000	1	1	38	66,000	40.0
	Naccowlah West, 1983	5,700	8	8	1,195	1,546,000	40.0
	Bogala, 1984	6,900	2	2	159	244,000	40.0
	Yanda, 1984	8,000	1	1	92	41,000	40.0
	Mooliampah, 1985	6,000	1	1	76	64,000	40.0
	Watson South, 1985	5,500	1	1	825	511,000	40.0
	Sigma, 1983	5,700	2	2	132	71,000	40.0
	Tickalara, 1984	5,600	4	4	291	149,000	40.0
	Big Lake, 1984	6,600	2	2	298	280,000	45.0-48.0
	Gidgealpa, 1984	6,150	10	11	1,809	2,004,000	45.0-50.0
	McKinlay, 1985	5,100	1	1	36	26,000	41.0
	Strzelecki, 1978	5,800	16	16	4,518	11,305,000	43.0-49.0
	Dullingari, 1979	5,200	16	17	3,787	6,587,000	54.0
	Narcoonowie, 1983	6,000	2	2	134	118,000	52.0-53.0
	Limestone Creek/Biala, 1984	5,100	9	9	1,454	992,000	41.0
	Wancoocha, 1984	6,000	5	5	818	1,060,000	44.0-54.0
	Muteroo, 1985	6,000	3	3	1,785	1,512,000	45.0
	Spencer, 1986	6,500	4	5	1,038	468,000	47.0
	Jena, 1985	5,200	1	1	83	56,000	41.0
	Alwyn, 1985	5,200	1	1	127	77,000	41.0
	Merrimelia, 1983	...	9	9	1,252	2,507,000	...
	Meranji, 1985	...	3	3	401	317,000	...
	Bookabourdie, 1985	...	2	2	116	48,000	...
	Dirkala, 1986	7,000	1	1	697	474,000	56.0
	Nungaroo, 1985	4,150	1	1	74	39,000	40.0
	Ulandi, 1985	5,200	1	1	63	62,000	40.4
	Watson, 1985	...	2	2	206	64,000	...
	Cooroo/Cooroo North, 1986	...	3	3	1,438	473,000	...
	Challum, 1983	...	1	1	19	13,000	...
	Kerinna, 1984	4,300	1	1	29	15,000	41.0
	Kercummura, 1985	...	...	...	...	1,000	50.8
	Cook, 1985	...	...	...	25	23,000	...
	Toby, 1987	...	1	1	5	1,000	...
	Dingera, 1987	...	1	1	NA	...	...
	Pitchery, 1988	...	1	1	NA	...	...
	Munro, 1988	...	1	1	NA	...	...
	Naccowlah East, 1988	...	1	1	NA	...	...
	Mawson, 1987	...	1	1	173	51,000	...
	Pintari North, 1988	...	1	1	NA	...	...
	Taloola, 1988	...	2	2	NA	...	...
	Sturt, 1988	...	1	1	NA	...	...
	Tantanna, 1988	...	3	3	NA	...	...
	Tirrawarra, 1970	9,500	28	35	4,770	7,685,227	52.0
	Moorari, 1971	7,705- 9,400	5	6	615	1,163,619	46.9
	Fly Lake, 1971	9,300	1	3	94	469,454	53.0
	Brolga, 1983	9,547	1	1	39	45,211	53.0
	Woolkina, 1982	9,700	1	1	26	98,546	51.5
TMOC RESOURCES LTD.	Moonie, 1961	5,800	27	36	658	21,679,619	44.5
	Alton, 1964	6,000	...	9	26	1,890,105	49.5
	Bennett, 1965	5,960	1	2	5	115,704	45.0
	Cabawin, 1961	10,800	1	1	6	59,972	49.0
	Mereenie, 1964	4,000	24	36	2,581	3,209,269	48.9
WAPET GROUP	Barrow Island, 1964	1,200- 6,700	407	516	17,447	224,561,262	37.7
	Dongara, 1969	5,250- 5,577	5	9	111	927,370	35.5
WOODSIDE	North Rankin,[c] 1972	9,800	8	13	14,820	10,374,000	52.3
BRIDGE OIL LTD.[e]	Thomby Creek, 1979	5,180	...	4			50.0
	Yellow Bank Creek, 1982	5,064	2	3			52.0
	Louise, 1986	5,900	1	2			59.0
	Narrows, 1986	5,740	1	1			59.0
PANCONTINENTAL PETROLEUM LTD.[e]	Nockatunga, 1983	5,420	4	5	18,372	7,357,000	49.6
	Winna, 1985	4,526	1	1			46.3

WORLDWIDE PRODUCTION

	Name of field, discovery date	Depth, ft	No. of wells Producing	No. of wells Total	Production 1987 average B/d	Production Cumulative to Dec. 31 1987 Bbl	°API gravity
	Koora, 1985	4,822	1	2			41.7
	Thungo, 1986	4,670	1	1			45.7
	Kihee, 1986	3,200	1	1			44.7
STRATA OIL[e]	Woodada,[c] 1980	7,400	...	...			54.0
WESTERN MINING CORP.	North Herald, 1982	3,937	1	1	...	15,914	44.0
	South Pepper, 1982	4,036	1	1	...	345	44.0
	Total Australia		**926**	**1,148**	**550,000**	**2,737,768,713**	
AUSTRIA MOBIL	Voitsdorf, 1963	7,000	18	36	398	21,568,397	36.0
	Sattledt, 1971	5,700	12	20	441	5,023,043	37.0
	Kemating, 1979	6,100	8	9	940	3,675,281	34.0
	Gaiselberg, 1938	5,200	32	71	774	32,123,128	27.0
	Other		88	117	1,596	26,108,744	32.0
OEMV	Hochleiten, 1977	3,100- 5,000					21.3
	Matzen, 1949	6,100					25.7
	Pirawarth, 1957	1,600- 6,500	1,040	1,040	15,795	571,329,890	28.4
	St. Ulrich, 1938	1,600- 4,300					31.1
	Other						...
	Total Austria		**1,198**	**1,293**	**19,944**	**659,828,483**	
BAHRAIN BANOCO	Awali, 1932	1,850- 4,600	274	374	39,290	784,000,000	33.0
	Total Bahrain		**274**	**374**	**39,290**	**784,000,000**	
BARBADOS BARB. NOC	Woodbourne, 1966	1,665- 6,701	80	135	1,361	4,590,303	27.0
	Total Barbados		**80**	**135**	**1,361**	**4,590,303**	
BENIN[e] PAN OCEAN	Seme, 1968	6,232	6	6	5,300	11,529,000	22.6
	Total Benin		**6**	**6**	**5,300**	**11,529,000**	
BOLIVIA YPFB	Camiri, 1927	3,936	45	54	485	48,286,707	55.0
	Carand, 1960	3,940	33	37	349	54,114,187	60.0
	Colpa, 1961	9,184	15	15	238	18,168,729	65.0
	Monteagudo, 1966	4,592	25	29	1,600	31,531,126	41.0
	La Pena, 1965	8,856	17	23	4,734	24,974,160	43.0
	Palmar, 1964	10,758	6	6	93	1,686,528	65.0
	Rio Grande, 1961	9,512	37	37	2,604	59,966,393	62.0
	Vuelta Grande, 1978	8,220	6	12	1,050	1,791,995	67.0
	Espino, 1979	14,655	1	3	61	964,698	56.2
	San Roque, 1981	7,170	5	6	726	417,724	71.0
	Santa Cruz, 1986	9,350	5	11	1,524	1,099,149	65.0
	H. Suarez R., 1982	7,820	2	5	140	474,547	35.0
	Villamontes, 1987	9,770	2	4	429	156,668	57.0
	Other		79	79	683	30,785,458	...
OCCIDENTAL	Tita, 1976	6,450	10	15	171	4,332,766	67.9
	Porvenir, 1978	9,600	6	9	2,881	10,330,669	60.5
TESORO BOLIVIA	La Vertiente, 1977	15,230	6	7	1,110	4,065,393	64.1
	Escondido, 1980	11,293	...	3	...	...	60.0
	Taiguati, 1981	10,174	...	1	...	...	63.0
	Los Suris, 1981	13,577	...	1	...	...	50.0
	Total Bolivia		**300**	**357**	**18,878**	**293,146,897**	
BRAZIL ALAGOAS	Furado, 1969	4,440- 5,500	50	58	2,315	13,793,400	35.0
	Pilar, 1981	4,100- 5,100	73	78	6,820	9,695,800	40.0
	Tabuleiro de Martins, 1957	3,900- 4,950	27	28	460	3,835,000	27.0
	● Mero, 1974	6,500- 7,400	...	...	...	150,950	35.0
	Other		20	24	350	1,769,400	...
AMAZONAS	Igarapé Cuia, 1985	3,600- 3,700	...	1	...	25,250	41.0
	Rio Urucú, 1986	7,500- 8,300	...	...	...	...	44.0
BAHIA	Agua Grande, 1951	820- 4,265	129	135	5,220	273,575,800	40.0
	Aracas, 1965	3,280- 8,856	170	172	7,080	109,178,300	41.0
	Buracica, 1959	1,476- 1,840	222	225	6,910	135,632,600	35.0
	Faz. Imbe, 1964	2,130- 5,575	65	70	1,825	14,686,100	40.0
	Faz. Boa Esperanca, 1966	7,215- 8,036	19	22	640	1,510,300	39.0
	Faz. Panelas, 1962	2,400- 5,100	13	14	400	6,415,300	37.0
	Malombe, 1966	3,280	18	21	430	6,824,600	41.0
	Mata/Remanso, 1971	4,445- 4,560	77	91	60	19,498,400	35.0
	Miranga, 1965	3,280- 3,936	228	232	9,670	186,914,000	40.0
	Rio Dos Ovos, 1979	2,410- 3,750	13	16	350	1,901,200	35.0
	Santana, 1962	2,400- 4,000	7	7	240	5,533,700	37.0
	Sesmaria, 1963	3,100- 4,600	38	39	900	2,847,600	40.0
	Candeias, 1941	3,280- 7,415	36	36	2,770	77,415,500	30.0
	Cassarongongo, 1959	1,475- 4,755	80	86	1,300	14,484,800	32.0

WORLDWIDE PRODUCTION

Name of field, discovery date	Depth, ft	No. of wells Producing	Total	1987 average B/d	Cumulative to Dec. 31, 1987 Bbl	°API gravity
Dom Joao Terra, 1947	920- 1,350	71	71	900	20,336,800	38.0
Riacho da Barra, 1982	2,910- 3,998	58	58	4,120	10,410,900	39.0
Taquipe, 1958	3,115- 3,935	107	108	4,790	84,255,600	40.0
● Candeias Mar, 1941	3,300- 7,400	...	15	490	6,889,200	30.0
● Dom Joao Mar, 1947	920- 3,000	293	296	4,440	81,049,100	38.0
● Ilheus, 1979	1,500	...	...	...	895,470	41.0
Cexis, 1983	1,670- 5,230	21	23	1,940	2,944,800	30.0
Faz. Balsamo, 1983	2,500- 4,000	141	153	8,250	6,254,300	31.0
Rio Pojuca, 1982	5,600- 6,200	25	25	1,780	3,653,300	35.0
Faz. Alvorada, 1983	2,100- 4,900	81	92	6,110	4,644,300	31.0
Rio do Bú, 1984	2,900- 3,000	42	48	2,570	1,730,300	33.0
Other	...	165	170	610	20,912,200	...
CEARÁ Faz. Belem, 1980	1,150- 2,500	286	295	2,220	4,008,100	13.0
● Curima, 1978	6,888	16	17	6,240	17,046,400	32.0
● Xareu, 1977	5,717- 5,766	27	28	3,600	13,478,200	39.0
● Atum, 1983	5,800- 6,750	9	10	3,590	6,715,500	25.0
● Espada, 1983	5,750- 6,900	6	6	1,770	3,382,700	26.0
ESPIRITO SANTO Faz. Cedro, 1962	5,248	31	33	1,160	9,683,000	28.0
Sao Mateus, 1969	5,750	78	81	1,490	4,090,750	27.0
Lagoa Parda, 1977	5,576	43	44	3,780	20,621,250	30.0
Rio Itaunas, 1977	5,400	41	43	1,080	3,326,300	28.0
● Cacao, 1978	9,184	4	5	1,030	8,493,000	32.0
Lagoa Suruaca, 1982	6,400	15	24	1,530	4,516,300	30.0
Lagoa Piabanha, 1984	6,500	7	11	640	1,564,750	27.0
Rio Preto/Oeste/Sul, 1976	...	38	40	880	1,813,600	...
Other	...	67	72	1,820	1,997,740	...
RIO DE JANEIRO ● Enchova, 1976	6,973- 7,885	20	26	31,010	72,174,800	23.0
● Bonito, 1977	6,950- 8,000	14	17	10,500	28,425,200	27.0
● Garoupa, 1974	11,382	16	20	21,780	62,421,500	30.0
● Namorado, 1975	10,083	34	42	62,510	189,920,000	28.0
● Garoupinha, 1974	12,693	3	4	4,040	10,416,000	30.0
● Pampo, 1977	6,215	27	32	58,640	93,758,300	21.0
● Area RJS-236 Trilha, 1983	8,800-11,000	4	4	5,210	9,094,000	28.0
● Linguado, 1981	9,000-11,000	12	15	31,715	53,547,300	20.0
● Badejo, 1981	8,500-10,500	10	12	7,760	15,694,300	28.0
● Bicudo, 1982	8,000-11,000	10	12	11,510	31,208,100	22.0
● Corvina, 1983	9,000-11,000	7	8	17,920	24,591,400	25.0
● Pirauna, 1982	8,000-10,500	7	9	12,470	20,957,600	27.0
● Sul de Pampo, 1978	9,300- 9,600	...	...	...	...	20.0
● Cherne, 1982	9,900-10,200	19	23	31,660	39,985,100	27.0
● Bagre, 1983	8,700- 9,000	8	10	6,010	4,501,400	25.0
● Parati, 1982	8,900- 9,200	3	5	1,590	4,714,200	25.0
● Viola, 1981	9,200- 9,900	7	8	6,915	8,050,900	26.0
● Marimba, 1983	8,900- 9,200	3	5	12,970	5,866,800	28.0
● Anequim, 1981	8,200- 8,500	3	3	2,540	3,181,400	27.0
● Area RJS-150	...	...	...	...	296,070	26.0
● Moreia, 1984	8,200- 9,500	6	6	4,640	1,716,900	22.0
● Albacora, 1984	8,430	4	15	2,083	760,500	26.0
RIO GRANDE DO NORTE Alto do Rodrigues, 1981	990- 1,120	290	302	4,410	5,845,400	18.0
● Ubarana, 1973	7,216- 8,365	87	89	18,050	47,159,500	40.0
● Agulha, 1975	5,412	4	4	365	6,380,000	33.0
Estreito/Rio Panon, 1983	650- 820	441	449	5,600	8,424,100	16.0
Macau/Terra, 1982	3,000- 3,300	2	12	150	712,250	31.0
Lorena, 1984	1,200- 1,700	18	20	890	593,500	28.0
Upanema, 1985	1,110- 1,800	16	24	1,700	1,574,000	28.0
● Area 1-RNS-36, 1982	...	...	...	...	178,290	32.0
● Macau-Mar, 1986	3,000	5	5	200	372,750	31.0
Faz. Pocinho, 1982	1,450- 2,000	191	202	4,705	8,669,700	35.0
Palmeiras, 1982	1,450- 1,800	...	...	...	...	25.0
Serraria, 1982	3,200- 3,900	36	37	1,625	2,298,500	27.0
Guamaré, 1983	1,400- 1,600	32	36	600	1,435,400	25.0
Canto do Amaro, 1986	2,500- 3,600	114	128	6,015	2,550,700	25.0
Livramento, 1986	2,500- 3,300	31	33	5,100	2,324,100	41.0
Other	...	81	96	200	419,000	...
PARÁ ● Área 1-PAS-11, 1983	...	...	...	...	421,300	41.0
MARANHÃO São João, 1983	4,950- 6,200	4	5	30	72,060	40.0
SERGIPE Atalaia Sul, 1961	1,450	5	7	100	2,679,500	24.0
Carmopolis, 1963	2,460	771	798	23,700	161,371,000	24.0
Riachuelo, 1961	1,475	239	245	3,840	20,501,800	28.0
Siririzinho, 1967	1,640	187	196	5,790	34,408,000	30.0
● Caioba, 1969	6,232- 6,890	12	12	2,490	29,573,700	36.0
● Camurim, 1970	6,232	48	50	5,000	15,359,200	25.0
● Dourado, 1970	3,620- 3,635	5	5	1,010	3,955,000	36.0
● Guaricema, 1968	3,335- 4,430	12	12	4,480	35,025,500	41.0
● Robalo, 1973	3,300- 4,500	4	5	130	818,600	30.0
Brejo Grande, 1967	1,670	12	13	625	1,735,200	26.0
Other	...	88	93	1,420	3,506,080	...
Total Brazil		5,909	6,277	556,268	2,310,047,860	

WORLDWIDE PRODUCTION

	Name of field, discovery date	Depth, ft	No. of wells Producing	No. of wells Total	Production 1987 average B/d	Production Cumulative to Dec. 31, 1987 Bbl	°API gravity
BRUNEI SHELL	● Fairley, 1969	10,740	37	54	13,157	107,200,000	40.0
	● Ampa SW, 1963	7,068- 8,155	98	219	43,694	515,500,000	41.0
	● Magpie, 1975	...	16	23	17,718	66,900,000	31.0
	● Champion, 1970	4,300	106	210	39,846	285,800,000	23.0
	Seria, 1929	...	346	585	20,277	973,300,000	34.0
	Rasau, 1929	...	5	9	4,508	4,100,000	30.0
	Other	...	...	...	...	13,000,000	...
	Total Brunei		**608**	**1,100**	**139,200**	**1,965,800,000**	
BURMA(e) MYANMA	Chauk-Lanywa, 1902	1,500- 4,000					36.0
	Mann, 1970	1,600- 7,000					37.5
	Myanaung, 1964	2,000- 4,000	450	450	19,000	489,685,000	40.5
	Prome, 1965	850- 5,000					36.0
	Yenangyaung, 1902	100- 6,700					36.0
	Total Burma		**450**	**450**	**19,000**	**489,658,000**	
CAMEROON ELF(e)	● Betika, 1972	5,300					42.5
	Ekoundou S., 1975	6,200					24.0
	Kole, 1974	5,500					31.0
	Kombo Center, 1976	5,900					37.0
	Kombo North, 1978	7,200					35.0
	Ekoundou Horst plus Center, 1972	7,200					29.0
	Boa, 1976	5,900					35.0
	Bavo South, 1976	5,600					29.3
	Betika S. Marine, 1979	6,300	166	170	167,619	364,898,935	30.0
	Ekuondou N. Marine, 1977	5,250					36.0
	Mokoko NE plus Abana, 1977	4,600					29.0
	Asoma Marine, 1973	3,600					19.0
	Kita E. Marine, 1978	6,900					40.8
	Mokoko S. Marine, 1977	3,300					17.0
	Inova Marine, 1976	6,200					32.8
TOTAL-CFP	● Moudi, 1979	5,250- 5,900	4	4	4,088	9,632,140	40.0
	● Moudi D., 1980	3,300	5	5	2,293	2,336,836	21.3
	Total Cameroon		**175**	**179**	**174,000**	**376,867,911**	
CANADA ALBERTA	Acheson, 1950	3,390	62	118	12,986	120,066,226	38.0
	Bantry, 1940	2,574- 3,300	150	184	6,647	46,537,461	...
	Bellshill Lake, 1955	2,950	419	432	12,980	58,646,234	...
	Bonnie Glen, 1952	3,959- 6,779	144	170	23,229	496,104,770	34.0-44.0
	Carson Creek N., 1958	8,787- 8,935	41	74	9,967	136,947,486	44.0
	Cessford, 1950	2,800- 3,028	196	272	3,523	29,739,639	...
	Clive, 1951	6,074- 6,208	120	169	5,436	49,182,090	...
	Countess, 1951	2,858- 4,272	116	148	7,258	61,650,607	...
	Fenn-Big Valley, 1950	3,900- 5,400	145	318	18,300	312,219,206	32.0
	Gilby, 1962	4,201- 7,000	85	155	3,116	46,423,510	...
	Golden Spike, 1949	3,570- 5,961	69	84	2,371	180,961,066	36.0
	Harmattan East	...	87	149	4,131	70,926,216	...
	Harmatton Elkton, 1955	8,796-10,757	62	76	4,581	60,444,189	...
	Innisfail, 1956	6,730- 8,440	33	74	5,752	73,761,997	...
	Joarcam, 1949	3,179	204	288	5,382	100,259,782	...
	Joffre, 1953	4,983- 6,779	88	167	3,892	81,882,361	...
	Judy Creek, 1951	8,307- 8,701	190	266	18,915	388,265,226	43.0
	Kaybob, 1957	4,870- 9,577	56	88	5,856	100,629,149	...
	Kaybob S., 1958	5,600-10,042	71	100	9,592	76,445,982	...
	Leduc-Woodbend, 1947	3,062- 5,380	348	737	4,604	375,198,288	40.0
	Lloydminster, 1933	1,690- 1,945	500	836	5,863	47,200,642	...
	Medicine River, 1954	5,435- 7,600	246	347	10,220	67,323,063	...
	Mitsue, 1964	5,908	250	386	25,401	271,263,665	43.0
	Nipisi, 1965	5,648- 5,726	244	340	31,912	263,581,751	41.0
	Pembina, 1953	3,000- 6,133	3,315	4,055	93,541	1,204,156,195	32.0-37.0
	Provost, 1946	2,760- 2,898	949	1,308	17,938	70,388,031	...
	Rainbow, 1965	4,994- 6,160	261	339	48,992	497,656,724	38.0-42.0
	Rainbow South, 1965	6,102- 6,400	50	81	7,029	62,063,923	...
	Red Earth, 1956	4,184- 4,878	174	259	4,942	36,024,951	...
	Redwater, 1948	2,012- 3,200	587	848	15,145	784,068,674	35.0
	Simonette, 1958	10,500	33	46	2,460	37,367,554	...
	Snipe Lake, 1962	8,500	48	95	3,785	52,574,155	...
	Sturgeon Lake S., 1953	4,912- 8,471	92	146	15,992	137,010,616	37.0
	Sundre, 1954	8,700- 9,000	54	69	3,182	35,791,800	...
	Swan Hills, 1957	8,100	589	820	40,545	667,109,678	40.0
	Swan Hills S., 1959	8,400	178	198	18,009	338,479,554	...
	Turner Valley, 1913	3,100- 9,150	101	158	2,655	138,926,474	39.0
	Utikuma Lake, 1963	5,624	104	138	4,461	43,380,838	...
	Virginia Hills, 1957	9,210	126	183	16,958	141,966,553	34.0
	Wainwright, 1925	1,903- 2,200	617	760	9,684	65,562,771	...
	Westerose, 1952	6,818	28	35	15,048	125,517,746	...
	Willisden Green, 1956	5,157	572	797	9,416	114,056,076	...
	Wizard Lake, 1951	4,044- 5,973	38	48	14,372	318,444,912	...

WORLDWIDE PRODUCTION

	Name of field, discovery date	Depth, ft	No. of wells — Producing	Total	1987 average B/d	Cumulative to Dec. 31, 1987 Bbl	°API gravity
	Zama	3,702	136	242	5,914	73,391,284	33.0-37.0
	Other	...	8,770	13,103	358,948	1,501,088,849	...
BRITISH COLUMBIA	Boundary Lake, 1957	3,418- 4,575	287	348	11,171	167,805,481	40.0
	Other	...	391	829	24,749	253,385,577	...
MANITOBA	All fields	...	1,434	1,624	13,479	161,286,112	...
SASKATCHEWAN	Steelman, 1950	4,600	660	721	9,188	251,994,739	...
	Weyburn, 1955	4,600	809	878	17,793	280,897,407	...
	Other	...	12,378	16,144	180,528	1,647,494,821	...
ONTARIO	All fields	...	1,175	1,494	2,293	61,049,607	...
NEW BRUNSWICK	All fields	...	...	...	9	802,951	...
NORTHWEST TERRITORIES	All fields	...	...	...	26,413	54,571,770	...
	Total Canada		37,882	51,744	1,236,553	12,839,976,429	
CHILE ENAP-MAINLAND	Canadon, 1962	6,090	13	63	236	14,468,812	38.4
	Daniel, 1960	5,806	4	119	165	34,434,471	25.4
	Daniel Este, 1961	5,523	5	64	490	28,311,574	25.0
	Dungeness, 1962	5,251	3	19	107	3,915,136	28.6
	Posesion, 1960	5,622	33	75	838	21,176,251	63.6
	Other	...	22	120	2,336	22,192,977	...
TIERRA DEL FUEGO	Calafate, 1955	6,045	13	61	1,008	25,561,355	49.3
	Catalina Sur, 1961	5,779	4	43	25	6,360,443	44.0
	Cullen, 1954	5,730	16	107	382	40,980,042	41.8
	Tres Lagos, 1957	5,724	18	84	366	15,826,176	40.1
	Catalina, 1956	5,756	10	26	1,788	6,204,654	38.7
	Gorrión, 1986	6,025	6	6	947	375,381	39.0
	San Sebastián Norte, 1986	6,870	4	4	227	87,726	37.0
	Other	...	33	255	1,538	57,773,725	...
STRAIT OF MAGELLAN	Ostion, 1977	7,156	15	18	1,468	7,724,254	33.0
	Posesion, 1977	6,727	12	39	1,102	5,169,663	34.5
	Spiteful, 1977	6,298	35	82	5,036	41,014,552	34.5
	Spiteful Norte, 1978	7,150	13	55	799	9,213,168	39.0
	Daniel Este-Dungeness, 1977	6,126	28	36	3,140	3,423,262	35.9
	Pejerrey, 1984	7,229	36	75	7,855	8,274,097	39.0
	Jaiba, 1978	6,650	2	2	871	25,251	46.5
	Total Chile		325	1,353	30,724	352,512,970	
CHINA (e) WESTERN REGION: JUNGGAR BASIN, TARIM BASIN, TURPEN BASIN	Karamay, 1955	...	...	...			...
	Dushanzi, 1941	...	...	...			...
	Other	...	...	...			...
CENTRAL REGION: JIUQUAN BASIN	Yumen, 1938	...	...	...			...
CENTRAL REGION: SICHUAN BASIN	All fields	...	...	...			...
CENTRAL REGION: SHAN-GAN-NING BASIN	Chang Qing, 1978	...	...	...			...
	Yanchang, 1907	...	...	...			...
CENTRAL REGION: QAIDAM BASIN	Qinghai, 1955	...	...	...			...
	Lenghu 3,4,5, 1958	...	...	...			...
	Other	...	...	...			...
EASTERN REGION: SONGLIAO BASIN	Daqing, 1959	...	39,500	NA	2,682,000	12,695,513,000	...
	Fuyn, 1958	...	...	...			...
	Jilin, 1978	...	...	...			...
EASTERN REGION: BOHAI BASIN	Shengli, 1962	...	...	...			...
	Renqiu, 1975	...	...	...			...
	Liaohe, 1964	...	...	...			...
	Dagang, 1964	...	...	...			...
	Zhongyuan, 1981	...	...	...			...
	Itu, 1970	...	...	...			...
EASTERN REGION: JIANGHAN BASIN	Qianjiang, 1966	...	...	...			...
	Other	...	...	...			...
EASTERN REGION: NANXIANG BASIN	Nanyang, 1971	...	...	...			...
EASTERN REGION: SUBEI BASIN	All fields	...	...	...			...
	Total China		39,500	NA	2,682,000	12,695,513,000	
CHINA, TAIWAN CHINESE PETROLEUM	Tiechengshan,(c) 1959	...	25	33	1,060	15,306,500	51.0
	Chingtsohu,(c) 1967	...	5	18	48		50.0
	Yunghoshan,(c) 1971	...	23	33	270	6,301,067	45.0
	Chuhuangkeng,(c) 1960	...	24	26	146		50.0
	● CBK,(c) 1986	...	4	9	1,089	417,981	47.0
	Total China, Taiwan		81	119	2,613	22,025,548	

WORLDWIDE PRODUCTION

| | Name of field, discovery date | Depth, ft | No. of wells | | Production | | °API gravity |
			Producing	Total	1007 average B/d	Cumulative to Dec. 31, 1987 Bbl	
COLOMBIA ECOPETROL[e]	Galán, 1945	3,200	62	71			19.0
	Infantas, 1918	3,200	263	307			25.8
	La Cira, 1925	3,250	612	703			24.0
	Lis-Nutria, 1957	9,500	130	137			31.0
	Llanito, 1960	7,300	60	61			21.0
	Cantagallo, 1941	6,800	11	13			19.7
	Casabe, 1941	3,800	330	416			20.7
	Yarigui, 1954	8,700	58	59			19.5
	Boquete, 1961	8,000	12	20			43.0
	Cicuco, 1956	...	5	22	76,988	1,954,497,703	...
	Petrolea, 1933		6	21			...
	Tibú, 1940	...	189	261			...
	Ortega, 1951		6	6			28.0
	Apiay, 1981	10,500	9	22			23.0
	Orito, 1963	6,600	58	95			40.0
	Other	...	129	215			...
SHELL[e]	Sta. Rosa, 1984	...	1	1			...
ARGOSY	Bourdine, 1974	11,000	2	4	679	3,069,634	26.7
	Máxime, 1976	11,000	1	1	2	111,911	11.0
	Nancy, 1974	11,000	1	1	312	1,880,073	26.0
ESSO COLOMBIANA LTD.	Bonanza, 1963	4,000	17	17	858	12,727,667	30.0
	Provincia, 1962	8,000	98	98	15,059	156,221,504	33.0
	Arauca, 1980	18,000	3	3	2,210	4,473,706	41.0
	Other	...	...	...	...	667,000	...
HOCOL (TENNECO)	Dina K, 1969	6,500	23	30	8,632	26,702,576	22.0
	Dina T, 1962	3,600	47	58	4,965	19,933,643	20.0
	Palogrande, 1971	6,500	24	32	7,802	16,533,954	20.0
	Cebu, 1981	6,500	8	10	2,240	4,348,329	19.0
	Brisas, 1973	4,200	4	5	797	4,494,143	23.5
	Pijao, 1981	7,000	6	7	734	3,483,747	21.0
	Hato Nuevo, 1984	6,900	3	4	545	1,421,601	36.0
	Tello, 1972	7,500	27	33	7,447	38,818,837	22.0
	La Canada, 1966	700	5	7	133	1,051,654	22.0
	San Francisco, 1985	2,800	33	35	13,286	9,844,179	27.0
	Tenay, 1985	11,500	2	2	2,180	1,403,466	36.3
	Loma Larga, 1985	3,500	1	1	44	24,857	22.4
	La Jagua, 1986	5,300	1	1	141	128,625	22.0
	Santa Clara, 1987	2,100	1	1	163	14,350	19.0
ELF	Trinidad, 1974	9,000	3	3	2,678	2,980,624	32.4
	Barquerena, 1982	9,300	2	2	1,879	2,382,765	32.3
	Cano Garza Norte, 1983	7,900	2	2	798	1,366,054	40.2
	Cano Garza, 1979	7,900	3	3	560	929,196	40.2
	Tocaria, 1980	13,500	2	2	829	648,661	32.1
	Cravo Sur, 1982	10,600	2	2	2,056	1,790,917	40.2
	Gloria Norte, 1983	12,100	1	2	632	551,144	17.2
	Cravo Este, 1987	11,376	1	1	10	3,481	42.0
OCCIDENTAL	Payoa, 1962	8,000	24	39	1,067	69,319,875	27.0
	La Salina, 1936	2,500	49	71	2,868	12,438,352	28.0
	Caño Limón, 1983	7,600	36	38	182,875	107,666,248	29.4
	Redondo, 1984	7,600	3	3	9,410	291,708	30.2
TEXACO	Cocorná, 1963	2,100	44	54	2,392	13,171,138	12.0-13.0
	Tisquirama, 1963	8,000	5	6	467	4,668,322	16.0-28.0
	Velasquez, 1946	7,500	119	144	4,130	164,756,697	19.0-24.0
TEXACO-ECOPETROL	Teca, 1981	2,100	139	154	13,768	19,649,607	12.0-13.0
	Nare, 1981	2,100	4	6	152	237,107	11.0-12.0
	Los Angeles, 1983	6,400	2	4	464	350,905	14.0-15.0
	Other	...	3	20	41	4,260,355	11.2-26.1
CHEVRON	Castilla, 1969	6,500	17	18	11,250	20,395,052	13.5
	Chichimene, 1969	6,500	1	1	557	393,556	21.3
LASMO	Santiago, 1985	9,600	2	2	1,200	693,000	29.0
	Total Colombia		**2,711**	**2,436**	**385,300**	**2,690,832,415**	
CONGO AGIP	● Loango, 1972	3,000	52	62	23,960	81,500,000	26.8
ELF CONGO	● Emeraude, 1969	820	93	98	22,000	160,000,000	22.3
	● Likouala, 1972	4,270	28	42	17,800	51,000,000	33.1
	● Yanga, 1979	2,950	48	63	20,700	62,000,000	29.4
	● Sendji, 1973	3,940	49	75	34,900	44,000,000	29.5
	Pointe Indienne, 1959	4,460	3	3	180	7,000,000	36.9
	Mengo, 1979	6,500	3	3	180	600,000	34.0
	Kunsi, 1983	4,000	2	2	130	500,000	35.8
	● Tchibouela, 1983	2,100	11	12	3,200	1,100,000	28.0
	Total Congo		**289**	**360**	**123,000**	**407,700,000**	

WORLDWIDE PRODUCTION

	Name of field, discovery date	Depth, ft	No. of wells Producing	No. of wells Total	Production 1987 average B/d	Production Cumulative to Dec. 31, 1987 Bbl	°API gravity
DENMARK — DUC	● Dan, 1971	6,000- 6,400	...	...	21,236	38,584,436	30.7
	● Gorm, 1971	6,500- 7,000	...	...	25,842	67,044,548	34.7
	● Skjold, 1977	5,000	...	...	20,897	26,503,654	29.6
	● Rolf, 1981	...	...	...	10,923	6,935,733	37.7
	● Tyra,[c] 1986	...	...	...	14,523	11,510,668	56.6
	Total Denmark		...	...	93,421	150,579,039	
DUBAI[e] — DUPETCO	● Fateh, 1966	7,900- 9,000					31.8
	● S.W. Fateh, 1970	7,500- 9,000					30.3
	● Falah, 1972	8,100	141	156	377,600	1,922,775,000	25.5
	● Rashid, 1973	9,400-11,500					38.0
ARCO	Margham, 1981	11,200	9	14			43.5
	Total Dubai		150	170	377,600	1,922,775,000	
ECUADOR — TEXACO-CEPE	Lago Agrio, 1967	9,600- 9,900	20	35	4,605	107,874,969	28.5
	Sacha, 1969	8,700- 9,800	91	101	34,440	337,853,986	28.0
	Shushufindi, 1969	8,800- 9,200	48	61	66,322	494,732,056	30.0
	Aguarico, 1969	9,000- 9,300	4	10	3,286	42,352,887	27.2
	Auca, 1970	8,900-10,100	21	23	8,842	64,868,855	28.0
	Auca Sur, 1980	9,100-10,300	2	2	453	1,138,109	25.6
	Atacapi, 1968	9,200- 9,500	4	5	774	11,567,397	31.0
	Parahuacu, 1968	8,700- 9,600	1	5	503	6,036,575	30.0
	Yuca, 1970	9,500- 9,900	5	7	1,965	10,192,208	25.2
	Yuca Sur, 1979	9,600-10,000	1	1	70	880,500	14.5
	Yulebra, 1980	8,800-10,000	2	2	964	2,193,240	24.3
	Culebra, 1973	8,800-10,000	2	2	418	1,707,251	15.2
	Cononaco, 1972	9,700-10,200	11	11	8,149	20,743,356	33.3
	Dureno, 1969	8,900-10,000	1	1	373	666,428	31.0
	Guanta, 1986	8,900- 9,900	9	9	3,499	1,855,863	29.0
	Rumiyacu, 1982	10,100-10,600	...	1	...	184,939	28.7
CLYDE PETROLEUM	Mariann, 1971	7,700	5	5	456	3,515,214	20.0
	Fanny, 1972	7,700	8	8	2,650	9,319,890	22.5
	Tarapoa, 1973	8,300	2	2	212	919,579	20.0
CEPE-ORIENTE[e]	Shushuqui, 1980	9,200	6	6			28.8
	Shuara, 1980	9,100	8	8			29.6
	Secoya, 1980	9,100	15	15			30.3
	Charapa, 1967	10,000	12	12	36,719	177,633,235	30.0
	Cuyabeno, 1972	8,050	4	4			28.8
	Sansahuari, 1979	8,100	1	1			28.8
	Tetete, 1980	9,300	5	5			28.0
	Bermejo, 1967	5,000	9	9			34.0
CEPE-PENINSULA[e]	Ancon, 1921	4,000	475	475			35.6
	Carpet	...	70	70			...
	Cautivo	...	100	100			...
	Total Ecuador		942	996	174,700	1,296,236,537	
EGYPT — GUPCO/AMOCO	● El Morgan, 1965	5,100- 6,400	89	112	119,412	1,010,500,000	27.0-32.0
	● July, 1973	8,100-10,000	37	44	78,928	451,000,000	31.0-34.6
	● Ramadan, 1974	11,300	23	25	42,507	362,100,000	31.7
	● October, 1978	11,000	23	27	123,102	298,800,000	28.0
	● Shoab Ali, 1978	5,500	15	47	13,865	126,200,000	30.0
	● SG-300, 1976	5,876	1	2	2,326	17,700,000	24.8
	● Sidki (382), 1977	10,500	5	7	7,542	38,200,000	33.9
	● GS (365) 1980	10,140	...	1	...	490,000	32.0
	● GS (277), 1980	8,610	1	1	877	2,200,000	34.0
	● SB (305), 1981	8,100	1	1	585	2,020,000	30.0
	● Badri (315), 1981	6,400	16	35	38,735	42,700,000	27.0
	● Younis (347), 1981	4,300	4	5	6,870	11,980,000	32.0
	● GH (376), 1981	8,600	2	5	3,184	5,800,000	25.0
	● SB (339), 1981	4,600	1	1	626	571,000	29.5
	● Nessim (336), 1981	6,700	2	2	865	8,700,000	26.9
	● Hilal (404), 1982	9,200-10,000	5	6	29,869	36,100,000	32.0
	● Waly (356), 1982	6,900	3	3	976	2,980,000	32.0
	● GS (345/346), 1982	7,500	1	2	131	1,130,000	38.0
	● SB (367), 1984	7,000	1	1	455	1,180,000	38.0
	● GS (381), 1985	8,700	...	1	...	920,000	37.0
	Abu Gharadig, 1972	9,500	18	25	11,951	50,390,000	36.0
	WD-33, 1972	9,800-10,800	1	1	56	2,110,000	32.0
	Razzak and E. Razzak, 1972	7,000	9	17	3,475	50,940,000	36.0
	● GS-327, 1983	5,600	1	1	3,202	1,500,000	20.0
	● SB-294, 1986	11,600	1	1	1,099	692,000	30.5
	NAG-9/9-1, 1985	10,000	2	2	1,318	986,000	35.0
	NEAG-9/15, 1984	10,635-12,128	1	1	194	203,000	35.0
	WD-33/15, 1985	5,800	2	2	1,863	683,000	30.0
AGIP	● Belayim Marine, 1961	9,100	55	65	133,493	557,392,209	29.0
	Belayim Land + Minori, 1954	8,500	96	103	40,613	413,674,357	21.5
	Meleiha, 1972	5,950	21	23	11,263	5,157,963	41.6

WORLDWIDE PRODUCTION

	Name of field, discovery date	Depth, ft	No. of wells Producing	No. of wells Total	Production 1987 average B/d	Production Cumulative to Dec. 31, 1987 Bbl	°API gravity
WEPCO/PHILLIPS PETROLEUM	El Alamein, 1966	8,700	17	19	1,822	66,182,919	33.6
	Umbarka, 1976	10,700	4	5	2,580	5,713,603	43.9
	Yidma, 1971	9,000	3	5	1,185	21,760,274	44.0
OSOCO/TOTAL	● Shukheir Bay, 1978	6,300	1	1	467	1,916,875	34.5
	● S. Ramadan, 1982	10,000	2	2	886	723,666	36.5
GEYSUM	● Geysum, 1980	5,500	5	5	...	...	17.0
SUESSO (ESSO)	● East Zeit, 1982	11,000	6	6	15,730	9,065,353	38.0
KHALDA	Salam, 1983	9,000	19	22			43.0
	Khalda, 1983	7,000	5	6			43.0
	Hayat, 1986	9,000	3	3	20,228	7,405,340	43.0
	Safir, 1986	12,000	1	1			43.0
	Tut, 1986	9,000	1	1			43.0
	Safir North, 1986	11,600	2	2			...
BADR PETROLEUM/SHELL	Badr el din, 1982	10,800	9	9	11,040	15,730,000	38.0
EGPC[e]	Gharib, 1938	2,200- 4,500	81	215			24.3
	Bakr/W. Bakr, 1958/1978	3,200- 1,500	48	83			18.0-19.0
	Umel Yusr, 1968	3,000- 3,300	19	37			21.5
	Amer, 1965	3,500	21	39			21.5
	El Ayoun, 1968	3,300	5	12			22.0
	Kareem, 1958	2,300	6	35			16.6
	Kheir, 1973	1,050	1	4			28.5
	Shukheir, 1966	2,500	5	11			34.0
	Ras El Behar, 1983	4,500	6	11			32.5
	Esh Elmalaha	3,700	2	2			44.0
	Sudr, 1946	2,600	12	20			21.6
	Asl, 1947	3,000	3	6			21.6
	Matarma, 1948	2,000	2	2	162,680	691,653,441	14.6
	El Khaligue, 1980	1,600	...	1			29.4
	Abu Senan T&Y, 1981	6,000	12	20			...
	Harghada, 1913	...	NA	NA			...
SUCO[e]	● Ras Budran, 1978	11,400	16	16			25.0
	● LL 87-2, 1975	11,000	...	1			35.0
	● Ras Fanar, 1978	2,300	6	6			35.0
	● Zeit Bay, 1981	4,200	22	22			33.0
EPEDECO[e]	West Bakr, 1978	1,300- 1,500	20	25			18.0
EGYPTCO[e]	Meleha, 1972	6,000	12	12			40.0
SHUKAIR MARINE CO.[e]	O. Suco	...	1	1			...
EL ALAMEIN[e]	Mors	...	1	1			...
	Total Egypt		**821**	**1,244**	**896,000**	**4,325,151,000**	

FRANCE

	Name of field, discovery date	Depth, ft	No. of wells Producing	No. of wells Total	Production 1987 average B/d	Production Cumulative to Dec. 31, 1987 Bbl	°API gravity
ALSACE	All fields	...	18	11	347	2,929,089	...
	Abandoned	...	...	...	...	24,337,343	...
AQUITAINE BASIN	Lacq Superieur, 1949	2,170	42	27	904	25,563,610	22.3
	Parentis, 1954	7,710	58	45	5,553	196,115,891	33.4
	Lugos, 1956	5,250	21	5	732	9,933,193	20.8
	Cazaux, 1959	7,550	36	35	4,381	67,779,650	34.2-37.0
	Lavergne, 1962	11,480	4	2	389	12,146,393	41.7
	Pecorade, 1974	10,660	10	10	2,470	9,041,862	29.3
	Castéra-Lou, 1976	9,600	8	3	427	2,330,489	21.7
	Vic Bilh, 1979	7,100	30	15	5,775	14,742,496	26.6
	Lagrave, 1984	6,500	3	...	4,152	4,061,327	42.5
	Other	...	7	5	244	10,231,603	...
	Abandoned	...	...	...	...	8,848,691	...
PARIS BASIN	Coulommes, 1958	6,230	31	18	391	13,213,378	32.6
	Châteaurenard, 1958	1,800	25	22	354	4,734,307	27.5
	St. Martin, 1959	4,590	12	4	320	8,351,063	33.0
	Villeperdue, 1959	6,100	93	27	14,594	11,434,528	34.4
	St. Firmin, 1960	1,800	34	28	513	9,799,700	27.5
	Chuelles, 1961	1,800	35	33	453	6,857,012	27.5
	Soudron, 1976	4,600	20	10	446	1,925,558	34.4
	Donnemarie, 1979	8,040	2	5	800	1,828,126	36.2
	Chaunoy, 1983	7,350	43	20	15,767	16,141,341	37.1
	St. Germain, 1984	7,200	7	5	1,103	1,433,088	36.0
	Vert-le-Grand, 1986	6,230	4	...	1,288	494,891	34.8
	Other	...	82	86	3,781	12,605,903	...
	Abandoned	...	...	...	...	13,476,998	...
SOUTHEAST + JURA	Abandoned	...	...	...	...	223,063	...
	Total France		**625**	**416**	**65,184**	**490,580,593**	

GABON

	Name of field, discovery date	Depth, ft	No. of wells Producing	No. of wells Total	Production 1987 average B/d	Production Cumulative to Dec. 31, 1987 Bbl	°API gravity
ELF[e]	● Anguille, 1962	9,200					31.2
	● Anguille NE, 1968	7,500					32.0
	● Anguille NNE, 1971	7,500					32.0
	● Barbier, 1972	6,000					29.5
	● Breme, 1976	5,000					36.5

WORLDWIDE PRODUCTION

	Name of field, discovery date	Depth, ft	No. of wells — Producing	Total	Production — 1987 average B/d	Cumulative to Dec. 31, 1987 Bbl	°API gravity
	Clairette, 1956	4,400					29.0
	● Girelle, 1973	9,400					32.0
	● Gonelle, 1972	5,800					24.2
	● Grondin, 1971	7,500					16.4
	● Mandaros, 1972	4,900					19.8
	Olende, 1976	6,700					33.8
	● P.G.S. Marine, 1975	7,500	214	330	113,323	906,840,000	32.5
	● Port Gentil Ocean, 1964	6,500					33.0
	Tchengue, 1957	4,100					26.8
	● Torpille, 1968	8,200					35.8
	● Baliste, 1974	4,900					28.3
	● M'Bya Sud, 1969	6,700					35.4
	● Anguille SE, 1962	8,200					32.8
	● Ayol Marine, 1979	7,700					22.8
	● Baudroie N., 1980	7,500					34.5
	● Baudroie Marine, 1968	7,500					33.3
	Other	...					...
SHELL GABON	Gamba/Ivinga, 1963	3,000- 3,100	50	84	12,800	254,472,000	31.6
	Lucina-Marine, 1971	5,900	14	22	9,620	48,311,000	38.4
	Lucina-Marine W., 1982	NA	...	...	1,220	1,645,000	37.0
AMOCO	Oguendjo B and C, 1981	5,000- 6,000	15	15	12,907	27,656,000	36.0
TENNECO	● Pelican, 1985	9,300	3	3	4,980	1,818,000	31.0
	● Octopus, 1984	10,300	3	3	1,150	420,000	33.0
	● Obando, 1984	8,650	...	1	...	...	32.0
	Total Gabon		**299**	**458**	**156,000**	**1,241,162,000**	
GHANA(e) AGRI-PETCO	● Saltpond, 1977	8,500	...	3	...	3,938,000	36.0-39.0
	Total Ghana		**...**	**3**	**...**	**3,938,000**	
GREECE(e) N. AEGEAN PET. CORP.	● Prinos, 1974	8,000- 8,800	16	16	22,500	52,325,000	30.6
	Total Greece		**16**	**16**	**22,500**	**52,325,000**	
GUATEMALA	Rubelsanto, 1974	5,300- 7,000	7	7	1,061	5,345,739	26.0-30.0
	West Chinaja, 1977	3,200- 3,800	3	3	924	6,795,954	35.0-43.0
	Caribe, 1981	6,830- 7,280	3	3	1,600	3,185,709	22.0-24.0
	Tortugas	2,300	...	...	...	67,804	35.0
	Tierra Blanca	11,230	...	...	...	175,113	19.5
	San Diego, 1981	17,150	1	1	105	120,953	34.7
	Xan, 1985	...	1	1	...	60,082	17.5
	Total Guatemala		**15**	**15**	**3,690**	**15,752,354**	
INDIA GUJARAT STATE	Ankleshwar, 1960	1,969- 6,564	114	198	23,917	409,840,000	44.0-51.0
	Kalol, 1961	4,595-10,830	114	204	6,254	36,663,000	38.0
	Sanand, 1962	4,267-10,174	21	40	2,408	8,047,000	26.0
	Nawagam, 1963	4,595-10,174	71	110	2,690	23,440,000	31.0
	Kadi, N., 1968	3,610- 7,220	152	201	13,882	42,780,000	21.0
	Kadi, S., 1968	6,892- 9,846	3	16	448	3,210,000	24.0
	Sobhasan, 1968	4,595- 9,846	59	88	6,290	17,236,000	26.0-40.0
	Viraj, 1978	4,595- 4,923	7	24	1,468	2,776,000	20.0
	Gandhar, 1984	9,486-12,145	9	21	2,950	1,106,000	47.0
ASSAM STATE	Rudrasagar, 1960	7,877-13,128	44	71	9,124	28,457,000	25.0
	Lakwa-Lakhmani, 1964	6,564-14,113	176	260	34,292	120,488,000	33.0
	Geleki, 1968	9-518-14,769	74	127	7,628	22,536,000	30.0
NAGALAND	Changpang, 1973	8,861-10,502	7	12	1,721	2,690,000	34.0
TAMIL NADU	Narimanam, 1985	6,236- 7,549	2	3	152	87,000	29.0-45.0
WEST COAST	Bombay High, 1974	3,282- 7,877	205	323	362,462	885,306,000	39.0
	Panna (North Bassein), 1976	4,923- 7,549	5	11	998	325,000	38.0
OTHER FIELDS(e)	Wavel, 1962	4,595- 7,220					
	Kosamba, 1963	1,313- 8,533					
	Kholka, 1966	4,365- 5,577					
	Bakrol, 1966	2,626- 9,846	1,573	1,652	104,054	365,560,000	...
	Kathana, 1965	5,250- 5,906					
	Digboi, 1980	300- 7,500					
	Moran, 1958	11,000					
	Nakhonkatiya, 1953	9,400					
	Total India		**2,670**	**3,413**	**609,000**	**2,002,164,000**	
INDONESIA PERTAMINA UEP I(e)	Besitang, 1977	3,973					51.5
	Gebang, 1977	2,917					54.0
	K. Simpang, 1971	3,740- 5,019	204	226			45.0
	P. Tabuhan, 1937	3,350- 5,893					50.8

WORLDWIDE PRODUCTION

	Name of field, discovery date	Depth, ft	No. of wells Producing	Total	1987 average B/d	Cumulative to Dec. 31, 1987 Bbl	°API gravity
	Rantau, 1929	2,764					45.3
	Serang Jaya, 1926	3,701					47.5
PERTAMINA UEP II[(e)]	Benuang, 1942	6,103					35.1
	Benakat East, 1973	NA					NA
	Tempino, 1931	1,647					23.0
	Gn. Kemala, 1938	5,448					38.0
	Limau, 1928	3,896- 4,626					NA
	Ogan Timur, 1943	3,793					27.42
	Talang Jimar, 1937	4,123	412	1,419			31.0
	T. Miring, 1935	640					NA
	Tanjung Tiga, 1938	3,750					28.0
	Kenali Asam, 1931	2,077					29.0
	Belimbing	5,371					...
	Kuang, 1940	5,249					25.1
PERTAMINA UEP III[(e)]	Cemara, 1976	3,964- 7,104					...
	Jatibarang, 1969	3,063	88	184			29.0
	Tugu Barat, 1979	3,014					...
PERTAMINA UEP IV[(e)]	Bunyu, 1923	2,077					33.2
	Sangatta, 1939	1,842					33.0
	Tanjung, 1938	2,296	161	334	69,668	2,928,974,192	39.9
	Tapian Timur, 1967	4,396					38.8
	Warukin S., 1966	2,067					26.9
PERTAMINA UEP V[(e)]	Klamono, 1936	574					19.0
	Linda, 1977	3,281	47	55			19.0
	Sele, 1951	2,156					35.0
PERTAMINA (FROM STANVAC)[(e)]	Sago, 1940	1,800					34.0
	Lirik, 1939	1,600					34.0
	Molek, 1956	2,600					34.0
	N. Pulai, 1941	1,800					34.0
	S. Pulai, 1941	1,800	354	354			34.0
	Talang Akar, 1922	2,800					35.0
	Benakat, 1932	1,600					35.0
	Jirah, 1930	2,300					35.0
	Rajah, 1940	6,000					38.0
	Abab, 1951	6,000					36.0
PERTAMINA (N.E. KALIMANTAN)[(e)]	Sembakung, 1976	2,600	17	19			36.0
	Bangkudulis, 1980	3,250	...	1			39.5
PERTAMINA (FROM PHILLIPS)[(e)]	Salawati, 1976	5,749	7	13			39.2
LEMIGAS[(e)]	All fields, 1893-1909	2,155- 2,297	71	419			...
ARCO[(e)]	Ardjuna, 1969	2,380- 7,250	266	330			37.0
	Arimbi, 1972	2,943	10	12			33.0
CSR	Bula, 1897	280- 1,300	61	70	1,299	14,632,000	23.0
CAL ASIATIC	Batang, 1975	690	12	20	310	916,000	19.2
	Buaya, 1978	2,035	...	2	...	96,000	28.0
	Lindai, 1971	1,100	16	18	550	3,022,000	30.7
	Menggala S., 1968	3,765	1	1	480	7,669,000	33.5
	Tanjung Medan, 1976	2,770	2	7	380	890,000	37.6
	Benua, 1978	2,400	19	22	4,930	7,357,000	41.4
	Beruk, 1974	1,750	17	23	5,310	22,080,000	38.0
	Bungsu, 1976	2,050	7	8	750	5,303,000	38.4
	Damar, 1974	4,650	5	6	270	2,225,000	33.5
	Dusun, 1979	2,550	6	7	1,310	2,332,000	42.3
	Gatam, 1977	2,385	3	3	...	314,000	33.5
	Kasikan, 1972	660	18	24	470	2,894,000	27.3
	Beruk NE, 1976	1,800	3	5	90	796,000	40.0
	Osam, 1978	900	4	5	150	384,000	31.8
	Paitan, 1978	2,315	1	1	110	305,000	31.6
	Pedada, 1973	950	30	31	5,680	32,503,000	33.1
	Pusaka, 1977	2,250	19	23	2,990	5,101,000	34.6
	Sabak, 1974	2,350	14	15	880	8,805,000	35.8
	Terantam, 1973	900	4	4	80	335,000	31.7
	Zamrud, 1975	3,600	53	54	11,460	28,945,000	40.3
	Langgak, 1976	1,380	17	22	640	2,056,000	31.7
ASAMERA (NORTH SUMATRA)	Geudondong, 1965	3,192	...	...	...	...	53.1
	Julu Rayeu, 1968	2,832	9	65	221	12,284,056	53.6
	Iee Tabeu, 1971	2,721	3	73	47	11,891,664	50.3
	Alur Cimon, 1972	3,098	...	34	...	...	50.8
	Tualang, 1973	2,631	13	75	2,602	29,601,774	49.9
	Peudawa, 1980	3,083	...	30	...	...	48.1
ASAMERA (SOUTH SUMATRA)	Kluang, 1913	2,591	30	88	595	33,674,788	42.6
	Mangunjaya, 1934	2,700	8	110	138	22,980,078	32.0
	Tanjung Laban, 1982	3,590	16	24	5,609	4,567,078	38.1
	Ramba, 1982	3,150	60	65	28,360	35,936,986	37.0
	Rawa, 1985	4,038	...	18	1	57,289	38.7
	Tempino, 1931	4,493	3	17	153	1,259,127	40.9
	Bentayan, 1932	4,446	6	13	825	672,555	22.0
	Panerokan, 1976	4,915	3	8	36	178,846	39.4

WORLDWIDE PRODUCTION

| Name of field, discovery date | Depth, ft | No. of wells | | Production | | °API gravity |
		Producing	Total	1987 average B/d	Cumulative to Dec. 31, 1987 Bbl	
MAXUS ENERGY ● Cinta, 1970	3,500	36	46	13,158	171,187,861	34.0
● Gita, 1972	5,000	6	9	1,036	6,452,248	34.5
● Kitty, 1973	2,700	6	7	1,161	13,615,268	18.0
● Nora, 1973	3,200	3	5	391	8,667,165	29.0
● Rama, 1974	3,200	44	75	7,381	88,607,804	31.0
● Selatan, 1971	4,000	8	19	1,277	20,402,184	19.5
● Zelda, 1971	7,500	27	43	6,303	34,764,756	33.0
● Krisna, 1979	4,500	28	41	7,413	57,130,762	37.0
● Yvonne, 1980	4,500	9	11	4,016	11,317,588	36.0
● Farida, 1982	8,000	10	22	3,118	10,581,595	32.0
● Titi, 1982	8,000	...	4	...	1,654,547	33.0
● Sundari, 1982	4,400	12	15	3,371	9,700,365	25.0
● Duma, 1983	2,300	3	6	158	657,997	16.0
● Karmila, 1983	4,800	7	9	10,585	25,813,993	36.0
Wanda, 1984	4,800	2	5	110	1,549,429	36.0
PETROMER TREND Canderawasih, 1976	3,500	11	17	1,268	11,697,058	27.8
Jaya, 1973	3,050	14	19	1,785	35,366,509	42.0
Kasim, 1972	3,350	36	49	4,790	45,774,419	38.0
Walio, 1975	2,750	224	268	17,803	146,754,005	34.2
Other	...	11	22	1,883	11,882,221	...
CALTEX Aman, 1974	4,700	10	10	2,343		38.6
Antara, 1978	1,425	5	5	700		35.9
Balam SE, 1972	4,500	13	13	671		29.0
Balam South, 1969	1,600	48	48	8,446		33.4
Bangko, 1970	1,950	75	85	26,341		33.7
Batang CPI, 1975	690	6	7	286		19.0
Bekasap, 1955	2,950	53	60	20,392		33.7
Bekasap S., 1968	3,900	8	11	736		34.1
Benar, 1973	2,450	25	26	4,429		33.2
Cebakan, 1974	4,700	4	4	619		30.2
Duri, 1941	770	1,073	1,370	80,105		21.9
Garuk, 1980	4,600	1	2	4		30.6
Hitam, 1975	6,690	4	5	1,343		38.8
Intan, 1977	3,350	6	6	864		32.9
Jorang, 1972	5,500	16	18	9,123		38.0
Kerang, 1977	3,900	11	13	1,044		38.7
Kopar, 1974	5,300	13	18	2,582		37.2
Kotabatak, 1952	5,500	67	113	18,384		28.8
Kulin, 1970	2,050	55	61	4,843		19.8
Libo SE, 1973	5,900	24	27	10,328		41.6
Libo, 1968	6,000	9	10	1,711		34.6
Lincak, 1981	2,570	4	4	728	6,021,389,337	35.2
Menggala N., 1968	3,850	14	17	4,136		32.5
Menggala S. CPI, 1968	3,765	7	8	1,852		33.6
Minas, 1944	2,600	465	604	254,990		34.7
Mindal, 1971	3,400	2	3	74		36.5
Mutiara, 1976	3,500	2	2	36		34.1
Nella, 1977	5,000	3	3	47		34.5
Obor, 1978	7,370	1	1	149		39.9
Pager, 1974	4,300	23	25	6,544		37.5
Pelita, 1977	6,600	5	5	621		39.0
Pematang, 1959	3,750	37	43	8,683		31.9
Pematang Bow, 1969	5,300	5	9	851		33.0
Pemburu, 1981	2,300	8	9	3,709		35.3
Perkebunan, 1977	2,300	...	1	...		29.4
Petani, 1964	4,750	33	34	14,478		32.8
Petapahan, 1971	4,500	23	28	5,736		28.1
Pinang, 1971	3,950	19	20	2,355		29.8
Pinggir, 1972	3,300	5	6	373		34.5
Pinggir S., 1973	3,300	2	2	56		38.6
Pudu, 1972	5,950	8	12	1,614		35.3
Puncak, 1979	2,400	14	14	3,769		35.9
Pungut, 1951	3,400	27	27	4,850		37.9
Rangau, 1968	6,100	3	10	104		42.0
Rantaubais, 1972	1,070	14	16	543		18.3
Rokiri, 1979	6,100	...	2	...		41.3
Seruni, 1972	3,000	22	23	4,821		35.3
Sikladi, 1975	4,340	18	18	2,593		39.8
Singa, 1977	3,200	1	1	55		31.0
Sintong, 1971	3,400	19	20	2,005		32.3
Sintong SE, 1973	4,500	...	1	...		36.1
Suram, 1971	1,750	4	6	252		27.7
Tandun, 1969	3,000	9	9	740		33.2
Telinga, 1975	4,205	7	8	925		33.3
Topaz, 1977	5,200	7	7	1,992		33.2
Topi, 1979	3,630	1	1	63		36.7
Tunas, 1983	4,819	2	2	370		35.0
Ubi, 1976	4,350	20	21	1,584		35.1
Waduk, 1982	6,100	10	11	324		37.6
STANVAC (CENTRAL SUMATRA) Binio, 1972	1,600	20	21	454	11,677,943	34.0
Pekan, 1976	3,700	22	22	795	7,573,274	35.0

WORLDWIDE PRODUCTION

	Name of field, discovery date	Depth, ft	No. of wells — Producing	No. of wells — Total	Production 1987 average B/d	Production Cumulative to Dec. 31, 1987 Bbl	°API gravity
	Merbau, 1979	2,700	26	26	1,299	6,334,383	36.0
	Kerumutan, 1980	3,000	22	23	1,163	4,711,496	35.0
	N. Merbau, 1980	2,600	8	9	272	2,750,090	36.0
	Panduk, 1981	2,500	6	6	228	1,401,271	46.0
	Kayuara, 1982	3,700	9	10	559	1,878,810	35.0
	Gemuruh, 1983	2,800	17	17	1,505	1,928,259	34.0
	E. Kayuara, 1983	3,500	45	49	5,698	5,238,225	35.0
	Mutiara, 1985	2,700	2	2	79	97,163	34.0
	Other	4,600	...	...	...	9,175	...
STANVAC (SOUTH SUMATRA)	Ibul, 1970	5,700	12	14	760	21,667,050	36.0
	Rambutan, 1972	4,500	5	7	29	3,210,751	42.0
	• Tabuan (PSC)	4,000	10	10	1,220	1,178,193	34.0
	Jene, 1985	6,500	8	10	9,564	4,085,391	35.0
	Pian, 1986	6,000	1	1	217	26,512	36.0
	• Kerang (PSC), 1984	2,900	2	3	744	271,616	35.0
	Lagan, 1986	2,400	4	4	290	111,815	56.8
	Other	4,500	3	3	196	868,406	...
ROY M. HUFFINGTON INC.	Badak, 1972	5,190-10,600	36	107	5,362	37,953,108	32.0
	Nilam, 1974	7,500-11,500	36	95	5,640	18,470,239	32.0
	Pamaguan, 1974	900- 5,100	14	27	1,962	5,898,117	30.0
	Mutiara, 1984	900- 5,800	12	19	2,599	2,961,789	30.0
	Semberah, 1974	2,900-10,100	...	16	...	7,054	35.0
TOTAL CFP	• Tambora, 1980	8,000-13,000	15	34	4,109	5,667,498	33.0
	• Handil, 1974	2,900- 9,000	219	272	115,033	617,340,703	33.0
	• Bekapai, 1972	4,600- 5,600	42	45	15,845	152,065,282	40.0
UNOCAL	• Attaka, 1970	7,500	75	88	47,116	433,197,460	39.0
	• Melahin, 1972	7,500	3	6	888	6,217,272	23.1
	• Kerindingan, 1972	7,000	...	8	20	5,038,615	23.8
	• Sepinggan, 1973	11,150	16	18	5,855	41,580,242	34.0
	• Yakin, 1976	5,100	21	28	8,917	18,290,674	18.8
	• Rajah, 1981	5,500	3	3	1,672	3,596,380	33.8
CONOCO GROUP	• Udang, 1974	5,595	22	28	8,646	53,319,790	40.0
	• Kepiting, 1982	6,000	2	2	2,570	1,094,050	37.0
MOBIL OIL	Arun, 1971	10,050	40	57	110,819	293,625,359	55.0
TESORU INDONESIA PETROLEUM	Samboja, 1909	360- 4,380	28	126	487	64,387,468	21.0
	Sanga Sanga, 1897	275- 6,000	107	240	2,941	266,631,554	32.0
	Tarakan, 1906	200- 6,540	130	1,165	1,730	210,151,159	20.0
MARATHON	• Kakap KH, 1980	7,000	11	12	16,150	9,343,000	45.0
HUDBAY (MALACCA STRAIT)	• Lalang, 1980	3,100- 3,600	13	16	10,870	30,550,000	39.6
	• Mengkapan, 1981	3,850- 4,450	10	10	22,300	8,440,000	42.0
	• Melibur, 1984	900- 1,300	12	13	8,680	3,210,000	35.6
	• Kurau, 1986	4,500- 5,000	7	7	...	...	47.4
TARAKAN PETROLEUM	Tarakan (Mamburungan field), 1984	5,227- 5,291	2	4	...	...	31.1
	Total Indonesia		**6,065**	**10,677**	**1,186,000**	**12,330,062,000**	

IRAN(e)

	Name of field, discovery date	Depth, ft	Producing	Total	1987 average B/d	Cumulative to Dec. 31, 1987 Bbl	°API gravity
NIOC	Agha Jari, 1936	7,440	58	89			34.0
	Ahwaz Asmari, 1958	8,150	73	85			32.0
	Ahwaz Bangestan, 1958	12,000	...	7			24.3
	Bibi Hakimeh, 1961	4,570	19	33			29.9
	Binak, 1959	10,440	...	4			29.9
	Chillingar, 1974	3,350	...	1			39.3
	Chesmeh Khush, 1967	10,700	...	4			27.7
	Dehluran, 1972	11,900	...	6			31.0
	Gachsaran, 1937	3,400	46	62			31.0
	Haft Kel, 1927	1,980	2	11			37.7
	Karanj, 1963	5,360	7	7			33.9
	Kharg, 1961	10,900	...	...			31.6
	Kupal, 1965	10,500	...	7			32.8
	Lab-e Safid, 1968	4,350	...	7			35.3
	Mansuri, 1963	7,100	4	8			28.4
	Marun, 1963	9,400	55	112			32.6
	Masjid-e Suleiman, 1908	1,600	7	24			39.3
	Naft Safid, 1938	5,000	5	19	2,309,000	35,852,300,000	36.8
	Paris, 1964	4,670	22	27			38.8
	Par-e Siah, 1964	4,250	...	1			37.7
	Pazanan, 1961	7,250	44	71			37.2
	Rag-e Safid, 1964	7,230	19	29			28.6
	Ramin, 1966	12,200	...	...			32.9
	Ramshir, 1962	8,750	...	3			27.6
	Sulabedar, 1971	3,400	...	1			30.4
	Naft-e Shah, 1923	2,000- 3,000	...	...			43.0
	• Bahregansar, 1960	9,000	...	...			30.0
	• Hendijan, 1968	11,000	...	...			23.0
	• Nowruz, 1966	8,200	...	...			21.0
	• Rakhsh, 1969	7,650	...	...			35.8
	• Rostam, 1966	7,000	...	...			34.4
	• Ardeshir, 1969	2,800	...	...			27.0

WORLDWIDE PRODUCTION

	Name of field, discovery date	Depth, ft	No. of wells Producing	No. of wells Total	Production 1987 average B/d	Production Cumulative to Dec. 31, 1987 Bbl	°API gravity
●	Cyrus, 1962	7,000	...	...			18.3
●	Darius, 1961	11,000	...	...			34.0
●	Fereidoon, 1966	7,000	...	...			28.5
●	Sassan, 1965	7,500- 8,100	...	...			34.0
	Total Iran		**361**	**618**	**2,309,000**	**35,852,300,000**	

IRAQ

Name of field, discovery date	Depth, ft	Producing	Total	1987 average B/d	Cumulative to Dec. 31, 1987 Bbl	°API gravity
Ain Zalah, 1939	5,200- 6,500	10	28			31.0
Bai Hassan, 1953	4,800- 5,400	62	77			34.0
Butmah, 1953	3,900	6	17			31.0
Hamrin, 1973	2,000	...	14			...
Jambur, 1954	5,500-12,500	32	50			38.0
Kirkuk, 1927	2,800- 4,200	65	238			36.0
Naft Khaneh, 1909	3,500	6	11			42.5
Rumaila North, 1958	7,300-10,300	291	433	2,360,000	19,630,000,000	33.0
Rumaila South, 1953	7,300-10,200	126	201			34.0
Zubair, 1949	7,400-10,200	53	108			35.0
Abu Ghurab, 1971	9,800	15	16			24.0
Buzurgan, 1970	12,000	19	22			24.0
Jabal Fauqui, 1974	10,000	20	23			25.5
Luhais, 1961	8,200	16	20			33.0
Qaiyanah, 1929	650	42	50			16.0
East Baghad, 1975	6,000- 7,000	72	79			17.0-23.0
Suffayah, 1978	3,900	28	39			25.0
Nahr Umar, 1949	8,200	5	13			42.0
Total Iraq		**868**	**1,439**	**2,360,000**	**19,630,000,000**	

ISRAEL

	Name of field, discovery date	Depth, ft	Producing	Total	1987 average B/d	Cumulative to Dec. 31, 1987 Bbl	°API gravity
LAPIDOTH	Heletz-Brur, 1955	5,200	3	18	71	12,275,749	29.2
	Kochav, 1962	5,500	5	13	179	4,176,980	27.9
OEIL	Ashdod, 1976	8,400	2	2	37	142,300	32.5
EMOG	Gurim, 1984	3,280	...	...	...	6,950	17.0
	Total Israel		**10**	**34**	**287**	**16,601,979**	

ITALY

	Name of field, discovery date	Depth, ft	Producing	Total	1987 average B/d	Cumulative to Dec. 31, 1987 Bbl	°API gravity
AGIP	Cavone, 1979	10,663-18,068	13	15	2,248	6,595,652	24.0
	Malossa/S. Bartolomeo, 1973	17,881-21,231	3	11	1,272	25,405,322	55.0
	Pisticci-S. Cataldo, 1960	6,820- 6,880	7	11	1,376	7,444,887	11.4
	Gela, 1956	11,280-11,350	45	68	8,704	99,768,192	10.0
	Perla, 1982	7,850	4	4	732	1,558,244	13.3
	Cela Mare, 1959	10,300	6	17	1,386	14,019,078	10.0
	Ragusa, 1954	5,760	41	41	4,790	126,201,977	19.5
	Nilde, 1977	4,545	2	3	11,802	18,935,805	38.9
	S. Maria, 1980	7,590	3	3	181	258,168	15.9
	Emilio S., 1985	9,215	1	1	...	337,622	13.0
	Narciso, 1985	2,650	1	1	...	989,278	21.1
	Other	...	25	63	4,001	19,949,277	2.8-55.9
ELF ●	Santa Maria a Mare, 1974	7,900	9	10	1,695	17,240,000	23.0
●	Rospo Mare, 1975	4,600	9	9	17,880	10,644,00	11.0
●	Sarago Mare, 1979	5,200	5	5	4,830	7,917,000	8.0
	Total Italy		**174**	**262**	**60,897**	**357,264,502**	

IVORY COAST

	Name of field, discovery date	Depth, ft	Producing	Total	1987 average B/d	Cumulative to Dec. 31, 1987 Bbl	°API gravity
ESSO COTE D'IVOIRE	● Belier, 1975	6,500	11	18	4,184	16,750,000	35.0
PHILLIPS-AGIP-SEDCO-PETRIOCI	● Espoir, 1980	6,500-11,000	6	8	11,990	28,085,648	32.3
	Total Ivory Coast		**17**	**26**	**16,174**	**45,835,648**	

JAPAN(e)

	Name of field, discovery date	Depth, ft	Producing	Total	1987 average B/d	Cumulative to Dec. 31, 1987 Bbl	°API gravity
JAPEX	● Aga-Oki, 1972	7,590	7	7			24.8
	Higashi-Niigata, 1965	4,500- 9,200	18	28			53.9
	Sarukawa, 1958	2,000- 3,200	62	134			32.3
	Yoshii, 1968	8,000	9	25			61.8
	Yurihara, 1976	1,310- 7,610	9	14			43.5
	Katakai, 1960	3,310-16,300	5	16			50.0
	Other	...	56	77			...
IDEMITSU OIL DEVELOPMENT CO. LTD.	● Agaoki-Kita, 1982	4,950	10	10	12,200	139,623,000	33.0
TEIKOKU OIL	Yabase, 1933	1,100- 5,800	61	73			33.0
	Minami-Aga, 1964	7,300- 9,600	10	26			38.5
	Minami-Nagaoka, 1984	13,200-15,100	5	11			55.7
	Hagashi-Kashiwazaki, 1970	6,000- 8,900	6	13			62.0
	Kubiki, 1959	1,100- 6,400	94	125			29.7
	Other	...	35	47			...
	Total Japan		**387**	**606**	**12,200**	**139,623,000**	

JORDAN

	Name of field, discovery date	Depth, ft	Producing	Total	1987 average B/d	Cumulative to Dec. 31, 1987 Bbl	°API gravity
NRA	Hamzah, 1984	9,515	3	3	500	295,437	30.0
	Total Jordan		**3**	**3**	**500**	**295,437**	

WORLDWIDE PRODUCTION

		Name of field, discovery date	Depth, ft	No. of wells Producing	No. of wells Total	1987 average B/d	Cumulative to Dec. 31, 1987 Bbl	°API gravity
KUWAIT(e)	KOC	Burgan, 1938	4,800	210	393			30.8
		Ahmadi, 1952	4,800	11	84			31.7
		Magwa, 1951	4,800	71	113			33.0
		Raudhatain, 1955	8,600	41	53			34.4
		Bahra, 1956	8,500	...	2	1,122,000	22,649,998,000	...
		Sabriya, 1957	8,300	9	44			36.3
		Minagish, 1959	10,000	1	21			34.4
		Umm Gudair, 1962	9,000	20	33			26.9
		Total Kuwait		**363**	**743**	**1,122,000**	**22,649,998,000**	
LIBYA	AGIP	Bu Attifel, 1968	14,300	28	37	136,758	770,690,617	40.8
		Rimal Katib, 1965	14,000	3	7	6,412	5,252,690	35.0
	ELF	El Meheiriga	10,500	1	2	829	22,599,371	46.0
	LNOC(e)	Arshad, 1966	10,100					42.0
		Jebel, 1962	8,100					38.0
		Majid, 1968	9,000- 9,800					...
		Don Mansour Unit	5,600					...
		Nasser, 1959	5,500- 7,600					38.0
		Lehib-Dor Marada Unit, 1965	9,500					50.0
		Raguba, 1961	5,400					43.0
		Beda, 1959	4,000					36.0
		Bualawn, 1972	6,000					40.0
		Dor, 1967	6,400					33.2
		Kotla, 1963	5,500					34.0
		Sahabi B & D, 1961	9,000-10,000					39.0
		Sarir, 1961	9,000					37.2
		Amal, 1959	9,900					37.5
		Ora, 1962	7,250					40.2
		Ghani, 1978	6,000					40.0
		Bahi, 1968	3,500- 3,750					43.4
		Belhedan, 1962	7,000					33.6
		Dahra, 1959	3,100- 3,700					37.0-41.0
		Defa, 1969	5,400- 5,900					35.6
		Gialo, 1961	2,200- 6,300					35.7
		Harsh, 1978	7,400	491	491	946,001	16,329,782,322	34.6
		Khalifa, 1978	5,400- 7,000					34.6
		Masrab, 1978	10,000					39.6
		Samah, 1961	9,000					33.4
		Waha, 1960	6,000					36.0
		Zaggut, 1962	6,000					36.4
		Balat, 1983	5,100- 5,400					32.2
		Ali, 1975	3,900	1	2			50.0
		Almas, 1975	4,000	5	8			49.5
		Intisar A, 1967	9,750	8	19			45.0
		Intisar B, 1974	9,600	2	2			50.0
		Intisar C, 1967	9,650	1	6			37.5
		Intisar D, 1967	9,400	19	27			39.0
		Intisar E, 1968	5,300- 7,100	8	11			34.7
		Nafoora-Augila, 1966	2,200-10,000	13	55			35.5
		Aswad, 1978	6,250	8	8			46.0
		Sabah, 1977	5,500	36	38			42.0
		Zella, 1977	6,750	27	31			48.1
		Other	7,900- 8,900	1	2			40.4
		Fidaa, 1979	7,000	9	9			41.7
		Hakim, 1978	6,700	3	3			48.5
		Total Libya		**664**	**758**	**1,090,000**	**16,128,325,000**	
MALAYSIA	ESSO MALAYSIA	● Bekok, 1976	5,500	23	30			47.0
		● Pulai, 1973	4,300	23	32			42.3
		● Tapis, 1975	6,700	70	106			45.0
		● Kepong, 1979	6,000	3	10			47.0
		● Tiong, 1978	5,800	28	42			47.0
		● Tinggi, 1980	4,400	26	31	268,000	759,820,000	48.0
		● Irong Barat, 1979	3,300	10	12			35.0
		● Semangkok, 1980	3,700	5	5			42.5
		● Guntong, 1978	...	9	14			...
		● Palas, 1979	...	11	11			...
		● Tabu, 1978	...	9	13			...
	PETRONAS CARIGALI(e)	● Tembungo, 1971	6,000	8	16	10,620	31,276,000	40.0
		● Fairley Baram, 1973	8,655	...	2	...	5,845,000	39.5
	SARAWAK SHELL	● D18, 1981	4,000	5	5	4,160	1,618,000	35.0
		● West Lutong, 1966	5,600	12	27	10,110	121,690,000	39.5
		● Baram, 1963	9,500	12	36	7,640	135,989,000	40.5
		● Bakau, 1971	11,000	3	3	130	7,047,000	39.0
		● Temana, 1972	3,100	18	31	9,070	36,211,000	35.0
		● Bayan, 1976	4,200	6	7	13,750	21,719,000	35.0
	SARAWAK SHELL/PETRONAS CAG.	● Baronia, 1967	7,900	19	40	15,260	148,770,000	42.0
		● Betty, 1967	7,900	13	17	14,320	62,327,000	38.0

WORLDWIDE PRODUCTION

	Name of field, discovery date	Depth, ft	No. of wells		Production		°API gravity
			Producing	Total	1987 average B/d	Cumulative to Dec. 31, 1987 Bbl	
	• Tukau, 1966	5,000	26	48	22,600	91,749,000	29.2
	• Bokor, 1972	2,000	18	19	24,140	34,011,000	35.0
	• Siwa, 1973	4,100	4	4	1,360	1,496,000	25.0
SABAH SHELL	• Barton, 1971	2,400	4	4	3,920	9,331,000	37.0
	• South Furious, 1974	3,300	7	15	3,010	13,499,000	30.0
	• St. Joseph, 1975	2,200	11	14	16,360	20,271,000	32.0
	• Erb West, 1977	6,785	15	19	6,630	21,620,000	30.0
	• Samarang, 1972	5,700	49	70	53,370	235,380,000	37.0
	• Ketam, 1977	4,800	1	6	550	901,000	33.0
	Total Malaysia		**448**	**689**	**485,000**	**1,760,571,000**	

MEXICO(e)

	Name of field, discovery date	Depth, ft	Producing	Total	1987 average B/d	Cumulative to Dec. 31, 1987 Bbl	°API gravity
NORTH ZONE, NE	Misión, 1945	5,440	3	3			...
	Monterrey, 1950	6,510	5	6			...
	Tigrillo, 1971	3,342	...	11			...
NORTH ZONE, N.	• Arenque, 1970	11,362	17	17			26.0
	Constituciones, 1956	6,350	176	176			17.0
	Ebano-Pánuco, 1901	1,450	324	324			12.0
	Tamaulipas	4,200	152	152			19.0
NORTH ZONE, S.	Chicontepec, 1973	4,264	220	220			34.1
	• Isla de Lobos, 1963	6,875	9	9			40.0
	• Marsopa, 1974	10,198	7	7			39.0
	Naranjos-C. Azul, 1909	1,800	206	206			22.0
	Tres Hermanos, 1959	6,960	34	34			21.0
	Other	...	...	...			...
CENTRAL ZONE, POZA RICA	• Atún, 1966	9,040	7	7			37.0
	• Bagre, 1973	10,919	8	8			33.0
	Escolín, 1942	7,216	95	95			26.3
	Hallazgo, 1955	10,170	36	36			25.0
	Jiliapa, 1958	7,390	29	29			24.0
	Mecatepec, 1941	7,544	32	32			29.8
	Miquetla, 1959	6,480	57	57			35.0
	Poza Rica, 1930	7,090	132	132			35.0
	Pres. Alemán, 1949	8,036	86	86			26.0
	Remolino, 1962	10,745	36	36			29.5
	San Andrés, 1956	10,410	123	123			29.0
	Other	...	315	316			...
CENTRAL ZONE, CUENCA DEL PAPALOAPAN	Matapionche, 1974	11,129	29	45			35.6
	Other	...	22	31			...
SOUTHERN ZONE, NANCHITAL	Ixhuatlán Ote., 1965	1,960	20	34			22.6
	Moloacán, 1962	500	170	207			22.3
	Santa Rosa, 1952	1,312	6	9			25.4
CENTRAL ZONE, N. FAJA DE ORO	Acuatempa, 1955	4,085	6	6			21.0
	Muro, 1965	3,966	11	19			17.0
	Ordónez, 1952	5,220	8	8			21.0
	Ocotepec, 1953	3,737	13	13			20.0
	Santa Agueda, 1953	4,780	3	3			16.0
	Other	...	57	57			...
SOUTHERN ZONE, EL PLAN	Agata, 1956	3,830	9	15			...
	Bacal, 1976	3,500	24	40			35.0
	Concepción, 1974	1,600	3	15			31.0
	Cuichapa, 1935	2,200	63	123			30.0
	El Plan, 1931	1,700	67	86			30.0
	Lacamango, 1973	1,700	18	27			26.3
	Los Soldados, 1953	4,492	18	22			32.0
	Other	...	26	38			...
SOUTHERN ZONE, AGUA DULCE	Blasillo, 1967	7,216	31	58	2,541,000	14,676,444,000	40.0
	Cinco Presidentes, 1960	6,862	73	134			35.0
	Burro, 1931	2,200	23	38			26.0
	Venta, 1954	4,730	20	83			41.0
	Ogarrio, 1957	5,790	57	194			38.0
	Otates, 1965	7,469	12	18			39.0
	Rodador, 1971	11,398	14	14			26.0
	Sánchez Magallanes, 1957	4,240	73	345			27.0
	San Ramón, 1967	9,517	29	65			30.0
	Tonalá, 1928	1,770	31	61			28.0
	Other	...	3	15			...
SOUTHERN ZONE, COMALCALCO	Agave, 1977	13,450	7	16			41.7
	Arroyo Zanapa, 1978	14,599	1	2			38.9
	Arteza, 1977	11,800	3	5			26.4
	Ayapa, 1972	8,200	3	5			7.2
	Cacho López, 1977	14,250	1	1			29.0
	Cáctus, 1972	14,100	14	30			31.5
	Carrizo, 1962	3,500	18	35			23.3
	Castarrical, 1967	10,080	38	59			29.3
	Comoapa, 1979	14,432	4	6			44.5
	Copano, 1977	11,890	9	10			43.9
	Cunduacán, 1974	13,775	12	42			28.9

WORLDWIDE PRODUCTION

	Name of field, discovery date	Depth, ft	No. of wells — Producing	No. of wells — Total	Production 1987 average B/d	Production Cumulative to Dec. 31, 1987 Bbl	°API gravity
	El Golpe, 1963	3,500	93	135			25.7
	Giraldas, 1977	15,225	21	22			37.0
	Iride, 1974	13,775	12	16			28.5
	Muspac, 1982	9,676	1	13			57.0
	Mecoacán, 1957	2,200	17	30			8.6
	Mora, 1981	...	61	63			38.1
	Mundo Nuevo, 1977	11,800	4	5			46.0
	Níspero, 1974	14,100	4	18			34.4
	Paredón, 1978	15,690	12	13			39.8
	Oxiacaque	11,150	5	21			29.1
	Platanal, 1978	15,900	...	...			30.2
	Río Nuevo, 1975	14,950	5	5			34.8
	Samaria (Cret), 1973	14,209	26	49			31.0
	Santuario, 1966	9,617	28	37			37.0
	Sitio Grande, 1972	13,766	19	22			35.0
	Sunuapa, 1978	12,887	2	7			32.8
	Tapijualpa, 1982	15,613	1	12			39.8
	Tintal, 1968	5,904	4	6			22.0
	Topén, 1978	11,172	2	3			25.4
	Tupilco, 1959	9,685	67	101			27.0
	Bellota, 1982	17,056	9	12			41.0
	Cardenas, 1979	17,548	26	45			38.0
	Chiapas, 1979	12,136	5	7			42.0
	Fenix, 1979	17,876	4	7			40.0
	Iris, 1979	14,432	8	8			39.2
	Jujo, 1980	17,548	18	19			32.0
	Caparroso, 1982	18,368	1	1			41.2
	Tecominoacan, 1983	19,519	1	5			37.5
	Jolote, 1983	18,119	1	1			39.4
	Artesa Terc., 1984	...	2	3			28.0
	Carmito, 1980	10,050	1	1			45.6
	Edén, 1983	17,590	1	2			38.5
	Samaria Terc., 1965	4,920	16	23			13.9
	Other	...	16	37			...
GULF OF CAMPECHE	● Cantarell, 1976	8,528	63	75			21.3
	● Abkatun, 1978	11,800	25	35			30.0
	● Ku, 1979	10,000	8	15			22.0
	● Pol, 1979	12,600	8	8			35.0
	● Chuc, 1982	13,100	1	3			35.0
	● Ixtoc, 1984	...	1	2			35.0
	Total México		3,768	5,120	2,541,000	14,676,444,000	
MOROCCO	Essaouira	...	4	4	370	16,157,000	...
	Rharb, 1951	4,590	1	1	28		...
	Total Morocco		5	5	398	16,157,000	
NETHER-LANDS	UNOCAL ● Helm, 1979	4,400	7	8	2,390	6,999,012	17.9
	● Helder, 1980	4,800	12	13	8,536	16,807,755	21.6
	● Hoorn, 1981	5,000	7	7	6,258	17,385,582	25.8
	NAM DeLier, 1955	5,200	22	33	480	16,775,000	34.8
	Rotterdam, 1984	NA	2	2	680	848,000	32.0
	Berkel, 1954	NA	17	17	2,330	4,750,000	20.0
	Ysselm, 1956	5,200	59	86	4,500	88,943,000	22.0
	Pynacker, 1955	NA	4	5	500	5,983,000	22.0
	Wassenaar, 1956	4,200	28	41	1,740	46,535,000	18.7
	Zoetermeer, 1957	3,200	30	30	510	9,786,000	16.5
	Werkendam, 1958	NA	...	...	20	207,000	33.0
	Schoonebeek, 1944	2,500	262	481	9,580	232,097,000	35.6
	CONOCO Kotter, 1980	6,500	9	11	19,667	24,911,460	33.0
	Logger, 1982	6,500	7	8	12,390	8,746,740	33.5
	AMOCO ● Rijn, 1982	6,600	12	25	12,900	10,980,550	35.0
	Total Netherlands		478	767	82,481	492,205,099	
NEUTRAL ZONE(e)	AOC ● Khafji, 1961	4,400-10,300	111	130			28.2
	● Hout, 1969	5,300-10,500	47	67			33.0
	CHEVRON & KOC Wafra, 1953	1,100- 7,000	312	392	389,000	4,304,878,000	18.9-23.5
	S. Fuwaris, 1963	6,300	6	9			25.0
	S. Umm Gudair, 1966	8,900	16	18			24.3
	Total Neutral Zone		492	616	389,000	4,304,878,000	
NEW ZEALAND	SHELL-BP-TODD ● Kapuni, 1959	11,700	8	9			54.0
	● Maui, 1969	11,000	12	14			58.0
	PCNZ McKee, 1980	7,848- 8,251	17	19			38.5
	Kaimiro, 1982	11,900	1	1	29,000	78,539,000	46.4
	Tariki, 1986	...	1	1			...
	Ahuroa, 1987	...	1	1			...
	Total New Zealand		40	45	29,000	78,539,000	

Name of field, discovery date	Depth, ft	No. of wells — Producing	Total	Production 1987 average B/d	Cumulative to Dec. 31, 1987 Bbl	°API gravity
NIGERIA						
AGIP						
Akri, 1967	9,600-10,600	2	12	848	59,375,372	43.8
Akri West, 1972	9,900-10,200	...	2	...	482,258	31.4
Ashaka, 1968	9,700-12,000	1	1	588	2,343,257	42.2
Beniku, 1974	12,000	...	2	...	3,309,326	29.5
Ebegoro, 1976	11,500-12,000	6	9	8,303	59,359,488	40.5
Ebocha, 1965	8,000-10,900	4	16	5,508	107,790,257	40.4
Idu, 1973	7,750-10,700	...	10	...	23,862,581	28.0
Kwale, 1967	10,000-11,500	6	10	4,381	4,694,547	42.4
M'Bede, 1966	7,300- 9,400	8	20	15,735	157,971,247	44.9
Obama, 1973	11,600-14,300	6	8	16,240	92,445,019	42.7
Obiafu, 1967	9,300-12,200	14	20	11,075	79,977,676	45.0
Obrikom, 1976Ii	8,000-10,000	6	11	5,527	53,807,211	46.7
Odugri, 1972	12,500	2	3	333	9,556,180	42.9
Ogbogene, 1972	10,000	...	2	...	3,302,763	37.0
Ogbogene W., 1982	13,000	...	1	...	353,136	56.2
Okpai, 1968	10,000-12,000	8	9	3,129	17,281,882	47.3
Omoku W., 1975	11,500	1	1	83	4,746,492	38.3
Oshi, 1972	9,600-11,300	9	10	8,733	62,932,362	40.6
Tebidaba, 1975	10,300-13,300	7	10	17,661	112,234,712	40.1
Umuoru, 1975	14,100	2	4	2,049	4,446,722	43.4
Agwe, 1975	10,800	...	2	12	2,002,300	49.4
Clough Creek, 1977	10,000-12,400	9	9	7,801	19,738,108	44.4
● Beniboye N., 1978	9,000-10,000	9	9	8,437	8,313,466	35.5
ASHLAND						
Izombe-Ossu, 1974	9,500	17	17	12,641	4,613,854	41.0
● Akam, 1980	6,700	8	8	8,315	3,035,036	35.2
● Adanga, 1980	10,611-17,586	8	8	13,262	10,645,430	35.0
ELF						
Aghigo, 1972	...	18	19	3,992	22,770,998	24.8
Okpoko, 1967	...	20	22	7,617	29,653,693	24.8
Obodo-Jatumi, 1966	...	21	25	17,910	63,216,952	24.8
Upamini, 1965	...	13	14	3,842	15,580,567	24.8
Obagi, 1964	...	81	92	43,140	350,235,015	24.6
Erema, 1972	...	4	5	3,448	9,122,120	24.6
GULF NIGERIA[(e)]						
Abiteye, 1970	5,750- 9,400	12	19			39.7
● Delta, 1965	5,600- 9,500	22	27			37.3
● Delta South, 1965	7,100-10,179	18	27			38.4
● Isan, 1970	5,900- 9,000	4	11			40.4
Makaraba	7,100-12,005	12	26			27.7
● Malu, 1969	4,800- 6,300	12	18			40.4
● Mefa, 1965	8,570-12,030	4	5			38.1
● Meji, 1965	5,200-10,900	19	22			31.9
● Meren, 1965	5,000- 7,500	28	51			31.9
● Okan, 1964	5,500- 9,245	36	67			38.1
● Parabe/Eko, 1968	4,500- 8,200	11	23			40.4
Utonana, 1971	7,400- 9,165	2	5			20.4
● W. Isan, 1971	7,825-10,229	4	9			40.4
Yorla South, 1974	11,389-12,635	...	2			41.0
Jisike, 1975	6,300- 7,600	3	4			41.1
Robertkiri, 1964	11,484-13,190	5	12			40.2
● Tapa, 1978	8,150-10,842	10	10			39.5
MOBIL[(e)]						
● Adua, 1967	6,970	4	5			30.3
● Asabo, 1966	5,600	8	14			32.4
● Ekpe, 1966	8,200	5	17			30.3
● Ekpe-WW, 1977	6,810	6	6			30.3
● Eku, 1966	5,420	1	6			30.8
● Enang, 1968	6,600	18	23			35.1
● Etim, 1968	6,200	7	9			32.9
● Idoho, 1966	9,020	3	4			30.8
● Inim, 1966	5,850	8	10			37.8
● Mfem, 1967	5,200	3	3			36.1
● Ubit, 1968	5,400	30	45			36.1
● Unam, 1968	5,180	6	9			33.5
● Utue, 1966	5,700	4	7			36.8
● Isobo, 1968	7,345	3	4			30.8
● Iyak	...	6	6			38.7
● Asabo 'D'	...	1	2			35.0
● Ata	...	...	...			...
● Oso	...	...	...			...
● Usari	...	...	...			...
PAN OCEAN (NIGERIA)[(e)]						
Ogharefe, 1973	9,900	8	11			47.4
SHELL[(e)]						
Ugh-Ogini, 1964	5,860	7	17			19.0
Ugh-Uzere-East, 1960	8,500	11	13			23.0
Ugh-Uzere-West, 1964	8,500	10	11			23.0
Ugh-Olomoro, 1963	7,000-10,000	28	32			22.0
Ugh-Oweh, 1964	12,300	11	11			28.0
Ugh-Kokori, 1960	8,000- 9,800	22	27			42.0
Ugh-Afiesere, 1966	8,000- 9,000	31	33			22.0
Ugh-Eriemu, 1961	12,500	15	16			26.0
Ugh-Ughelli-East, 1959	11,800	9	11			35.0
Ugh-Ughelli-West, 1963	7,400-10,200	8	10			21.0
Ugh-Utorogu, 1964	9,000	13	19			25.0

WORLDWIDE PRODUCTION

Name of field, discovery date	Depth, ft	No. of wells Producing	Total	Production 1987 average B/d	Cumulative to Dec. 31, 1987 Bbl	°API gravity
Ugh-Oroni, 1964	12,000	7	7			22.0
Ugh-Warri-River, 1961	12,264	2	3			31.0
Ugh-Evwreni, 1968	10,900	7	11			38.0
Ugh-Isoko, 1960	...	1	2			19.0
Forc-Odidi, 1967	10,980	24	29			37.0
Forc-Opukushi, 1963	7,823	10	14			29.0
Forc-Jones Creek, 1967	7,000- 9,000	31	34			29.0
Forc-Opuama, 1973	...	3	5			41.0
Forc-Egwa, 1967	9,350	13	22			37.0
Forc-Forcados/Yokri, 1968	10,859	63	83			22.0
Forc-Batan, 1968	11,000	8	9			27.0
Forc-Escravos Beach, 1969	8,176	9	13	953,413	10,172,279,000	36.0
Forc-Otumara, 1969	8,176	25	28			25.0
Forc-Saghara, 1970	8,590	3	6			31.0
Forc-Ajuju, 1970	13,324	1	2			27.0
Forc-Benisede, 1973	...	15	19			23.0
Node-Sapele, 1970	12,788	13	19			42.0
Node-Oben, 1972	12,036	18	23			40.0
Forc-Opukushi N., 1977	...	1	2			30.0
Abura	...	1	3			44.8
Rapele	...	1	2			44.4
Amukpe	...	2	2			38.9
Osioka	...	2	2			...
Phl-Bomu, 1958	6,500- 7,500	14	39			36.0
Phl-Imo River, 1959	5,800-10,000	32	47			32.0
Phl-Nkali, 1963	12,000	4	10			42.0
Phl-Elelenwa, 1959	11,000	6	11			39.0
Phl-Umuechem, 1969	5,800-10,700	8	19			36.0
Phl-Apara, 1960	9,000	4	8			36.0
Phl-Bodo-West, 1959	9,700	9	11			29.0
Phl-Afam, 1956	6,000	8	15			45.0
Phl-Yorla, 1970	11,917	9	13			44.0
Phs-Bonny, 1959	12,254	9	15			36.0
Phs-Cawthorne Channel, 1963	11,000	21	30			38.0
Phs-Ekulama, 1958	10,483	22	29			32.0
Phs-Soku, 1958	11,500	15	23			31.0
Phs-Orubiri, 1971	...	2	6			38.0
Egb-Oguta, 1965	10,300	9	21			46.0
Eade-Utapate South, 1974		6	11			44.0
Eade-Opobo South, 1974	...	4	6			44.0
Ibigwe	...	2	2			...
Korokoro	...	6	8			34.3
Otamini	...	3	5			21.8
Ugada	...	1	1			...
Ahia	...	8	12			38.2
Akpor	...	2	3			27.5
Alakiri	...	12	21			38.2
Assa	...	2	2			20.7
Egbema West	...	7	16			42.2
Ebubu	...	2	8			22.2
Egbema	...	6	7			35.5
Cesw-Nun River, 1960	...	4	6			39.0
Cesw-Etelebou, 1971	12,000	9	9			33.0
Cesw-Kolo Creek, 1960	12,000	15	25			42.0
Cesw-Diebu Creek, 1966		10	12			43.0
Phl-Isimiri, 1964	5,900-11,000	4	6			28.8
Phl-Onne, 1965	10,384	3	3			29.0
Phl-Obigbo-North, 1963	6,500-10,000	26	37			22.0
Phl-Ajokpori, 1967	...	2	2			29.0
Phl-Agbada, 1960	8,000-12,000	37	48			32.1
Phl-Akuba, 1967	...	1	1			29.0
Cesw-Adibawa, 1967	11,950	11	23			26.0
Cesw-Adibawa NE, 1973	...	3	3			26.0
Cesw-Ubie, 1961	14,380	4	5			28.0
Phl-Obeakpu, 1975	...	2	4			45.0
Cosw-Nembe Creek, 1973	...	38	38			35.0
Phs-Akaso, 1979	...	1	1			30.0
Bugama Creek	...	3	6			43.0
Kalaekule	...	7	7			39.6
Krakama	...	6	10			23.8
Odeama Creek	...	5	5			33.3
Tai	...	2	2			37.4
Obele	...	5	5			...
Rumuekpe	...	2	3			...
Enwhe	...	2	2			...
Akaso	...	1	1			...
TEXACO-CHEVRON ● Pennington, 1965	5,000-10,400	2	9	1,980	24,377,203	38.2
● Middleton, 1972	5,000- 7,000	3	7	2,710	14,948,284	36.6
● North Apoi, 1973	4,000- 8,100	23	26	31,150	150,632,247	35.8
● Funiwa, 1978	5,000- 7,000	20	22	22,130	35,788,219	38.2
● Sengana, 1967	...	1	2	7	666,120	46.7
Total Nigeria		**1,522**	**2,179**	**1,242,000**	**11,797,891,200**	

WORLDWIDE PRODUCTION

			No. of wells		Production 1987 average B/d	Production Cumulative to Dec. 31, 1987 Bbl	°API gravity
	Name of field, discovery date	Depth, ft	Producing	Total			
NORWAY PHILLIPS	● Ekofisk, 1968	10,000	35	38	82,076	910,326,654	41.1
	● West Ekofisk, 1970	10,300	12	12	3,620	81,829,312	43.9
	● Cod, 1968	9,500	6	6	923	19,353,892	55.0
	● Tor, 1970	10,000	8	13	9,624	114,531,350	38.0
	● Albuskjell, 1972	10,400	15	15	2,362	49,738,965	45.0
	● Edda, 1972	10,480-10,600	7	7	803	22,859,793	41.5
	● Eldfisk, 1970	9,350-10,000	32	32	58,106	223,241,326	37.6
MOBIL	● Statfjord, 1974	7,700	55	86	*602,657	*1,022,483,814	39.0
AMOCO	● Valhall, 1972	8,200	20	20	59,841	79,826,000	35.0
CONOCO	● Murchison, 1975	9,900	20	28	14,863	55,950,200	38.0
	● Ula, 1976	11,000	5	5	78,978	34,136,970	39.0
	Total Norway		215	262	913,853	2,614,278,276	
OMAN PDO	Lekhwair, 1968	4,352	40	57	17,598	57,132,000	37.8
	Natih, 1963	4,788	35	65	23,021	341,981,000	31.0
	Shibkah, 1976	2,591	2	2	296	761,366	31.0
	Fahud, 1964	2,499	138	215	66,586	751,259,000	32.7
	Yibal, 1962	8,465	164	170	145,028	613,558,000	40.2
	Al Huwaisah, 1969	5,320	37	39	16,460	165,710,000	37.0
	Burhaan, 1984	4,011	3	3	610	1,327,000	36.7
	Ghaba North, 1972	2,227	18	24	11,177	49,722,000	35.8
	Saih Nihaydah, 1972	8,700	19	26	29,620	146,651,000	39.0
	Saih Rawl, 1973	9,826	15	24	9,592	87,444,000	41.1
	Qarn Alam, 1972	1,354	4	7	1,050	18,983,000	14.8
	Habur, 1975	5,800	1	2	107	4,315,000	14.8
	Mahjour, 1979	11,700	1	4	579	2,057,000	37.2
	Al Ghubar, 1974	1,916	6	7	2,220	3,699,000	23.5
	Barik, 1974	8,856	3	4	2,000	1,447,000	42.1
	Anzauz, 1984	7,098	3	3	5,321	5,026,000	47.8
	Hasirah, 1984	8,708	5	6	4,850	2,024,000	48.8
	Zauliyah, 1981	8,177	10	15	6,944	10,819,000	45.2
	Wafra, 1983	5,510	2	6	2,025	1,352,000	40.6
	Suwaihat, 1982	6,452	8	11	8,473	7,976,000	36.0
	Bahja, 1984	5,697	4	7	5,120	4,428,000	44.5
	Fayyadh, 1983	3,608	1	1	491	308,000	27.0
	Sayyala, 1981	3,690	11	11	27,645	21,046,000	49.7
	Zareef, 1985	3,264	6	7	6,699	2,963,000	46.9
	Rima, 1979	3,247	34	34	61,623	105,722,000	32.8
	Rasha, 1982	4,310	4	5	1,120	1,025,000	32.8
	Nimr, 1980	3,296	78	78	24,393	10,881,000	21.3
	Karim West, 1982	3,546	15	15	12,215	10,108,000	23.5
	Amal/Amal South, 1973	4,638	29	31	9,888	7,485,000	22.5
	Birba, 1978	8,528	...	2	...	7,280,000	34.0
	Kaukab, 1983	8,282	...	1	761	925,000	35.8
	Marmul, 1977	5,087	174	181	42,558	110,364,000	22.0
	Rahab, 1977	3,237	19	25	3,390	5,372,000	22.0
	Qaharir East, 1978	4,936	1	3	109	667,000	30.6
	Qaharir, 1978	4,536	22	28	7,186	21,048,000	30.6
	Qata, 1979	4,707	9	14	2,007	3,623,000	26.2
	Runib, 1978	3,215	10	11	2,441	925,000	20.3
	Jalmud, 1979	4,236	3	3	710	264,000	21.0
	Jalmud North, 1980	4,265	2	2	341	138,000	21.0
	Jawdah, 1984	3,570	3	3	2,487	1,001,000	22.0
	Mawhoob, 1985	6,562	1	1	131	48,000	25.7
	Ishan, 1984	5,578	2	2	309	113,000	27.5
	Jameel, 1984	5,906	2	2	258	94,000	30.2
	Al Burj, 1987	6,890	1	1	6	2,000	25.7
	Mukhaizna, 1975	3,478	...	1	31	13,000	15.0
	Dhiab, 1985	5,414	6	6	730	270,000	26.5
	Thuleilat, 1986	3,773	5	5	586	214,000	24.0
	Warad, 1985	4,922	3	3	879	331,000	21.4
	Simsim, 1985	5,250	6	7	2,671	975,000	31.5
	Zahra, 1986	6,562	1	1	398	145,000	32.1
ELF GROUP	Sahmah, 1978	12,400	6	6	5,174	22,739,617	45.1
	Ramlat	...	1	1	106	93,569	...
OCCIDENTAL	Safah, 1983	6,500	31	32	5,762	4,180,000	37.8
	Total Oman		1,004	1,220	581,782	2,618,034,552	
PAKISTAN PAKISTAN OILFIELD LTD.	Balkassar, 1946	9,398	2	4	411	31,530,084	26.0
	Joya Mair, 1944	6,900	2	3	373	5,763,597	15.0
	Dhulian, 1936	8,899	14	41	29	41,314,779	48.0
	Khaur, 1915	9,306	1	3	9	4,167,399	33.0
	Meyal, 1968	12,513	8	10	3,228	29,596,654	41.3
PAKISTAN PETROLEUM LTD.	Adhi,[(c)*] 1978	9,118- 9,190	...	4	...	1,416,737	56.1
OGDC	Toot, 1968	14,765	14	15	1,500	9,428,741	38.0
	Fimkassar, 1982	12,337	1	1	17	31,809	39.0
	Tando Alam, 1984	8,203	9	14	3,434	4,586,054	41.4

*Norwegian production share.

WORLDWIDE PRODUCTION

	Name of field, discovery date	Depth, ft	No. of wells — Producing	Total	Production 1987 average B/d	Cumulative to Dec. 31, 1987 Bbl	°API gravity
	Ghotana, 1986	7,874	2	2	243	116,202	42.1
	Chak-Naurang, 1986	9,187	2	4	150	54,808	14.3
	Thora, 1987	7,546	2	3	784	286,311	...
UNION TEXAS	Khaskeli, 1981	3,600	5	9	1,525	6,791,500	39.0
	Laghari, 1983	2,500	5	7	7,591	9,966,525	42.0
	Dabhi, 1984	5,500	3	4	1,520	1,406,679	39.0
	Mazari, 1985	3,520	4	4	5,309	2,230,581	43.0
	Other	...	...	12	...	...	31.0-46.0
OCCIDENTAL	Dhurnal, 1984	13,100	3	6	15,585	15,719,082	38.0
	Total Pakistan		**77**	**146**	**41,708**	**164,407,542**	
PERU PETROMAR S.A.	● Peña Negra, 1960	3,000- 8,500	168	284	12,296	104,179,142	38.0
	● Lobitos, 1960	3,500- 7,500	115	249	8,322	50,006,041	38.0
	● Litoral, 1955	1,500- 6,000	92	138	2,642	36,409,940	36.0-40.0
	● Providencia, 1967	3,000- 7,000	36	80	1,757	20,834,644	38.0-40.0
	Other	...	4	11	125	197,346	...
OCCIDENTAL	● Bartra, 1979	8,000- 8,930	10	16	5,240	17,049,168	11.6
	● Capahuari Sur, 1973	11,800-12,900	17	26	10,337	128,708,116	33.2
	● Dorissa, 1978	9,800-10,600	7	12	8,782	34,591,373	30.8
	● Forestal, 1973	9,000- 9,900	8	10	6,663	32,660,815	18.1
	● Huayuri Sur, 1977	10,000-10,800	9	12	4,397	17,406,645	27.1
	● San Jacinto, 1978	7,200- 9,760	12	22	7,027	26,908,128	13.1-23.5
	● Shiviyacu, 1973	9,100-10,100	15	22	23,012	58,561,369	20.4
	● Jíbaro, 1974	10,110-10,155	5	6	3,222	6,338,698	10.7
	● Jibarito, 1981	9,000- 9,650	4	5	6,311	9,651,813	10.5
	Other	...	3	11	1,029	2,210,471	...
OCCIDENTAL/BRIDAS	Talara area	1,343- 6,080	770	1,219	7,226	44,628,171	33.2
PETROPERU	Talara-Lima area, 1869	2,000- 9,000	2,152	5,975	26,347	945,336,088	34.1
	Corrientes, 1971	10,000-12,600	32	36	22,823	86,601,829	29.8
	Pavayacu, 1972	9,600-11,300	10	14	3,301	15,315,532	30.2
	N. Esperanza, 1980	10,200-10,700	2	3	1,149	4,549,394	44.0
	Yanayacu, 1974	11,400-13,600	7	8	1,295	3,771,578	19.1
	Capirona, 1972	9,800-12,000	5	5	723	3,274,242	25.3
	Valencia, 1975	10,500-11,200	...	2	93	534,339	44.0
	Selva Central	...	46	53	1,460	25,296,888	...
	Total Peru		**3,529**	**8,219**	**165,579**	**1,675,021,770**	
PHILIPPINES ALCORN PHILIPPINES	● South Nido, 1977	6,576- 6,885	3	5	684	15,200,730	27.1
	● Matinloc, 1978	6,656- 6,750	5	5	2,684	8,853,496	43.8
ALCORN PALAWAN	● Cadlao, 1977	5,734- 5,881	2	2	2,085	9,203,358	45.6
	South Tara, 1987	4,335- 4,436	1	1	52	19,078	40.4
	Total Philippines		**11**	**13**	**5,505**	**33,276,662**	
QATAR(e) QGPC	● Bul Hanine, 1970	6,000	14	15			35.0
	Dukhan, 1940	6,500	116	161			41.1
	● Idd El Shargi, 1960	4,500- 8,250	25	31	282,000	4,044,985,000	35.0
	● Maydan-Mahzam, 1963	7,000	19	24			38.0
	Total Qatar		**174**	**231**	**282,000**	**4,044,985,000**	
RAS al KHAIMAH(e) CHEVRON	● Saleh, 1983	14,500	6	6	10,000	12,831,000	50.0
	Total Ras al Khaimah		**6**	**6**	**10,000**	**12,831,000**	
SAUDI ARABIA(e) ARAMCO	Abqaiq, 1940	6,690					37.0
	Abu-Hadriya, 1940	9,340					35.0
	Abu Jiffan	...					...
	● Abu-Safah, 1963	6,600					30.0
	Bakr	...					...
	● Berri, 1964	8,300					32.0-34.0
	Dammam, 1938	4,800					35.0
	Dhib	...					...
	Dibdibah	...					...
	Duhul	...					...
	El Haba	...					...
	Fadhili, 1949	8,050					39.0
	Farhah	...					...
	Faridah	...					...
	Ghawar, 1948	6,920					34.0
	Habari	...					...
	● Hamur	...					...
	Harmaliyah, 1971	8,430					35.0
	● Harqus	...					...
	● Hasbah	...					...
	Jaham	...					...

*Field was shut in. Normal maintenance operations continue.

WORLDWIDE PRODUCTION

	Name of field, discovery date	Depth, ft	No. of wells — Producing	Total	Production — 1987 average B/d	Cumulative to Dec. 31, 1987 Bbl	°API gravity
	Jaladi	...					...
•	Jana	...					...
	Jauf	...	588	588	4,027,000	54,147,214,000	...
	Jawb	...					...
	Juraybi'at	...					...
•	Jurayd	...					...
•	Karan	...					...
	Khurais, 1957	5,200					33.0
	Khursaniyah, 1956	6,560					31.0
•	Kurayn	...					...
•	Lawhah	...					...
	Lughfah	...					...
•	Maharah	...					...
•	Manifa, 1957	7,950					...
•	Marjan, 1967	6,800					33.0
	Mazalij	...					...
	Qatif, 1945	7,100					33.0-34.0
	Qirdi	...					...
•	Ribyan	...					...
	Rimthan	...					...
•	Safaniya, 1951	5,100					27.0
	Sahba	...					...
	Samin	...					...
	Sharar	...					...
	Suban	...					...
	Tinat	...					...
	Wari'ah	...					...
	Watban	...					...
•	Zuluf, 1965	5,800					32.0
	Total Saudi Arabia		**588**	**588**	**4,027,000**	**54,147,214,000**	

SHARJAH

	Name of field, discovery date	Depth, ft	Producing	Total	1987 average B/d	Cumulative to Dec. 31, 1987 Bbl	°API gravity
CRESCENT[e] •	Mubarek,[c] 1972	...	4	4	23,460	85,659,800	37.0
AMOCO	Saaja and Moveyid,[c] 1980	10,200-12,300	16	16	41,540	91,958,000	50.0
	Total Sharjah		**20**	**20**	**65,000**	**178,617,800**	

SPAIN

	Name of field, discovery date	Depth, ft	Producing	Total	1987 average B/d	Cumulative to Dec. 31, 1987 Bbl	°API gravity
SHELL ESPANA •	Amposta Marino Norte, 1970	10,000	2	3	1,050	55,379,101	37.0
•	Tarraco (Castellon B5/8), 1973	9,000	...	3	...	14,547,234	35.0
CHEVRON	Ayoluengo, 1964	4,600- 5,200	35	51	1,328	14,709,350	33.0
REPSOL •	Casablanca, 1975	8,900	7	9	29,368	79,407,563	32.6
•	Dorada, 1975	6,700	...	4	...	16,565,228	21.5
•	Gaviota,[c] 1980	8,300	3	3	1,242	598,443	52.7
•	Salmonete, 1984	12,600	...	1	...	1,529,323	44.0
	Angula, 1984	12,600	...	1	...	46,636	44.0
	Total Spain		**47**	**75**	**32,988**	**182,782,878**	

SURINAME

	Name of field, discovery date	Depth, ft	Producing	Total	1987 average B/d	Cumulative to Dec. 31, 1987 Bbl	°API gravity
STAATSOLIE	Tambaredjo, 1981	1,010	74	74	2,176	2,244,700	15.9
	Total Suriname		**74**	**74**	**2,176**	**2,244,700**	

SYRIA

	Name of field, discovery date	Depth, ft	Producing	Total	1987 average B/d	Cumulative to Dec. 31, 1987 Bbl	°API gravity
AL FURAT OIL CO./SHELL	Thayyem, 1983	NA	...	...	61,450	34,629,000	36.5
	Al Kharrata, 1986	NA	...	...	380	139,000	37.5
SPC[e]	Alian, 1974	5,900	25	37			19.0
	Gbebeh, 1976	2,300	54	75			18.0
	Jebisseh, 1968	2,000- 3,300	132	219			18.0
	Karachok, 1956	6,500	103	119	168,170	999,732,000	20.0
	Rumilan, 1962	6,500	59	66			24.0
	Soudie, 1959	6,500	532	579			24.0
	Tichrine, 1976	2,600	58	73			18.0
	Total Syria		**963**	**1,168**	**230,000**	**1,034,500,000**	

THAILAND

	Name of field, discovery date	Depth, ft	Producing	Total	1987 average B/d	Cumulative to Dec. 31, 1987 Bbl	°API gravity
UNOCAL •	Erawan,[c] 1973	7,000	88	119	6,356	14,446,200	54.6
•	Baanpot,[c] 1980	7,000	20	21	438	1,180,400	54.6
•	Platong,[c] 1976	8,500	29	43	3,647	4,084,700	59.1
•	Satun,[c] 1980	8,500	56	58	4,728	4,020,500	55.1
THAI SHELL	Sirikit, 1981	5,258- 5,976	50	61	16,180	27,865,000	41.0
	Sirikit West, 1983	5,500	2	2	500	474,000	41.0
	Pru Krathiam, 1984	...	...	2	30	11,000	19.0
	Total Thailand		**245**	**306**	**31,879**	**52,081,800**	

TRINIDAD

	Name of field, discovery date	Depth, ft	Producing	Total	1987 average B/d	Cumulative to Dec. 31, 1987 Bbl	°API gravity
AMOCO •	Teak, 1971	15,191	57	97	33,861	227,983,000	29.3
•	Samaan, 1971	11,780	34	54	12,901	174,129,000	36.8
•	Poui, 1974	11,650	33	57	20,454	150,192,000	34.0
•	Cassia, 1973	12,700	7	8	6,081	10,531,000	45.0
	Mora, 1986	8,500	3	5	677	414,000	...

WORLDWIDE PRODUCTION

	Name of field, discovery date	Depth, ft	No. of wells — Producing	No. of wells — Total	Production 1987 average B/d	Production Cumulative to Dec. 31, 1987 Bbl	°API gravity
TNA	● Soldado, 1955	11,000	354	655	37,648	462,680,000	...
	● Fortin Offshore (FOS), 1954	...	12	35	287	7,001,000	...
TRINTOPEC	Palo Seco/Erin/McKenzie, 1926	12,718	519	1,576	8,598	117,001,000	...
	Fyzabad/Apex/Quarry, 1920	11,000	323	1,039	4,130	170,550,000	...
	Coora/Quarry, 1936	14,000	186	733	2,133	91,180,000	...
	Galeota, 1963	6,304	53	105	2,930	15,218,000	...
	Central Los Bajos, 1973	...	136	218	2,557	8,488,000	...
	Other	...	199	914	2,407	60,884,000	...
TRINTOC	Area IV & Guapo, 1913	10,626	83	192	1,326	38,709,000	...
	Parrylands, 1913	10,626	124	508	1,667	40,127,000	...
	Point Fortin C & W, 1907	10,626	125	551	2,318	40,707,000	...
	Point Fortin E., 1929	11,000	...	168	...	26,353,000	...
	Penal, 1936	11,067	70	289	975	62,042,000	...
	Catshill, 1950	9,693	32	134	337	23,027,000	...
	Brighton, 1908	7,500	62	619	678	72,675,000	...
	Barrackpore, 1911	11,100	67	369	2,283	30,817,000	...
	Guayaguayare, 1902	10,750	107	699	1,729	86,477,000	...
	Forest Reserve, 1913	11,000	294	2,042	4,976	259,156,000	...
	Oropouche, 1944	9,100	32	128	171	6,639,000	...
	Palo Seco, 1929	12,700	87	933	2,207	93,122,000	...
	Trinity, 1956	9,700	18	95	233	15,131,000	...
	Other	...	139	813	969	80,637,000	...
PCOL	All fields	...	99	494	645	20,568,000	...
	Total Trinidad		**3,255**	**13,530**	**155,178**	**2,392,438,000**	

TUNISIA

	Name of field, discovery date	Depth, ft	Producing	Total	1987 average B/d	Cumulative to Dec. 31, 1987 Bbl	°API gravity
ELF-ETAP	● Ashtart	10,500	14	19	19,880	169,685,984	29.8
ELF-SEREPT	Douleb/Semmama	3,937	13	15	779	17,700,000	40.0
	Tamesmida	3,937	3	3	223	2,018,000	36.9
CFPT	Sidi El Itayem, 1971	7,500- 7,800	21	23	3,054	21,411,533	44.0
SHELL	● Tazerka, 1979	4,111- 4,232	5	6	4,720	13,323,000	30.0
ETAP-AGIP-FINA (SODEPS)	Makrouga, 1980	6,230	3	4	3,328	4,096,740	45.0
	Larich, 1979	7,550	2	2	102	243,372	40.7
	Debech, 1980	7,550	1	1	64	233,099	42.6
AGIP-TUNISIA (SITEP)	El Borma, 1964	8,250- 8,900	87	113	70,580	496,300,000	42.5
	Chouech es Saida, 1971	12,600	1	1	372	2,400,000	42.9
TENNECO	El Hajeb, 1982	6,890	2	2	980	1,010,000	30.5
	Total Tunisia		**152**	**189**	**104,082**	**728,421,728**	

TURKEY

	Name of field, discovery date	Depth, ft	Producing	Total	1987 average B/d	Cumulative to Dec. 31, 1987 Bbl	°API gravity
SHELL	Barbes, 1971	4,600	17	19	1,808	15,700,000	29.7
	Baysu, 1985	3,250	3	3	578	428,000	33.3
	Bektas, 1985	3,250	1	1	246	264,000	33.7
	Beykan, 1965	2,700	37	41	3,403	58,790,000	33.2
	Cobantepe, 1975	4,000	1	1	8	176,000	32.8
	Katin, 1971	4,800	3	3	107	3,688,000	29.5
	Kayakoy, 1961	3,500	14	20	1,241	19,313,000	38.2
	Kayakoy-West, 1965	2,900	18	18	3,014	28,700,000	34.7
	Kervan, 1982	3,800	1	1	25	22,000	30.3
	Kurkan, 1962	2,125	29	31	2,467	48,465,000	31.4
	Kurkan-South, 1968	2,900	5	5	489	6,550,000	34.5
	Malatepe, 1970	2,350	14	14	941	6,522,000	32.7
	Piyanko, 1968	4,200	1	1	79	1,485,000	35.5
	Sahaban, 1966	2,900	7	10	655	12,910,000	34.5
	Sahaban South East, 1983	2,900	2	2	177	316,000	34.5
	Sincan, 1980	2,125	9	9	980	2,830,000	31.4
	Yatir-East, 1974	1,908	9	10	592	5,057,000	30.9
	Yenikoy-East, 1974	3,700	2	2	860	3,925,000	31.3
	Yesildere, 1985	3,250	2	2	366	240,000	34.7
MOBIL	Selmo, 1964	5,800	21	21	4,451	65,323,748	34.2
	Bulgurdağ, 1961	4,500	2	3	41	2,467,145	37.9
TPAO	Raman, 1940	4,500	82	87	4,818	48,683,893	18.7
	Bati Raman, 1961	4,200	74	85	3,396	34,037,592	13.3
	Garzan, 1956	4,700	21	21	1,460	33,115,229	26.2
	Germik, 1968	6,500	8	10	407	3,293,783	18.8
	Mağrip, 1961	5,700	10	11	681	14,591,351	18.4
	Silivanka, 1962	8,200	10	10	836	7,633,478	23.0
	Beycayir, 1976	7,700	1	1	26	186,155	26.1
	Celikli, 1964	10,500	11	12	2,408	2,955,167	35.2
	Sezgin, 1970	5,600	...	...	...	87,559	17.0
	Oyuktas, 1972	7,600	...	...	...	592,005	31.0
	Devecatak, 1973	4,800	1	1	23	222,722	37.0
	K. Osmancik, 1971	3,800	2	2	204	2,004,177	37.6
	Adiyaman, 1971	5,800	10	11	702	8,025,831	27.6
	G. Adiyaman, 1977	5,000	...	...	...	115,836	20.4
	Kuzey Adiyaman, 1977	9,500	...	...	...	59,325	32.0
	Bölükyayla, 1977	10,300	...	...	...	167,663	35.0

WORLDWIDE PRODUCTION

	Name of field, discovery date	Depth, ft	No. of wells — Producing	Total	1987 average B/d	Cumulative to Dec. 31, 1987 Bbl	°API gravity
	Saricak, 1973	5,250	3	3	195	2,550,062	31.5
	G. Saricak, 1973	5,250	11	12	659	3,819,626	31.5
	Yeniköy, 1973	6,400	16	16	1,370	8,437,870	31.4
	G. Kayaköy, 1976	8,600	5	5	4,675	2,032,804	30.4
	Sivritepe, 1977	8,200	2	2	94	509,731	33.4
	Camurlu, 1976	5,600	18	18	477	863,514	12.2
	Ikiztepe, 1976	4,900	...	...	...	88,051	11.3
	G. Sahaban, 1978	5,450	13	13	1,429	2,525,466	33.2
	G. Dincer, 1981	5,350	10	11	727	4,281,313	16.7
	B. Selmo, 1981	5,500	1	1	198	470,682	34.0
	Mehmetdere, 1982	6,700	...	...	...	149,057	31.0
	Kartaltepe, 1982	6,550	4	4	342	751,850	32.0
	Cemberlitas, 1983	6,500	18	21	4,195	6,001,201	30.5
	Alcik, 1983	6,500	1	1	24	69,399	32.0
	Cukurtas, 1985	10,000	1	1	26	200,804	36.0
	B. Kozluca, 1985	5,000	18	21	1,667	787,648	12.0
	Akpinar, 1984	10,500	2	4	141	99,062	31.6
	B. Firat, 1986	8,150	5	7	1,315	633,800	34.8
	Kurtalan, 1961	5,750	1	1	22	236,217	31.9
	S. Sinan, 1987	3,900	1	1	12	4,250	13.0
ALADDIN MIDDLE EAST LTD.[(e)]	Kahta, 1958	3,500	10	23	130	4,041,000	11.0
	Yasince, 1974	6,400	2	4	10	4,650	31.0
	Total Turkey		**571**	**638**	**55,197**	**477,501,716**	

UNITED KINGDOM

	Name of field, discovery date	Depth, ft	No. of wells — Producing	Total	1987 average B/d	Cumulative to Dec. 31, 1987 Bbl	°API gravity
AMOCO	● Montrose, 1969	8,200	16	20	11,345	74,712,198	40.5
	● NW Hutton, 1975	12,000	20	29	32,370	81,694,979	36.0
BP	Wytch Farm Bridport Sand, 1974	2,910	11	16	6,000	13,576,000	37.0
	Wytch Farm Sherwood Sandstone, 1977	5,115					38.0
	East Midlands, 1939	3,500	71	129	1,280	24,941,994	36.0
	Kimmeridge, 1960	1,800	...	...	...		40.0
	Welton, 1983	5,000	14	19	2,310	1,601,240	36.0
	Nettleham, 1983	8,850	1	2	346	334,601	35.0
	● Forties and Echo, 1970	7,000	78	103	327,000	1,802,600,000	37.0
	● Buchan, 1974	8,700	9	9	21,000	59,157,572	33.0
	● Magnus, 1974	9,500	11	20	127,000	194,576,010	39.0
BRITOIL	Thistle, 1973	9,200	33	60	56,050	316,328,503	38.5
	Beatrice (A + B), 1976	6,750	25	44	33,332	89,225,571	38.4
	Clyde, 1978	12,000	10	30	32,568	11,920,031	38.1
	Deveron, 1972	13,500	3	*...	5,836	8,159,530	38.0
CHEVRON	● Ninian, 1974	10,000	69	102	153,352	759,533,710	35.0
CONOCO	● Statfjord,† 1974	9,500	57	88	114,623	193,754,230	39.0
	● Murchison,† 1975	9,900	20	28	52,090	182,883,940	38.0
	● Hutton, 1973	10,000	16	26	61,033	79,950,045	33.0
	● Dunlin, 1973	9,100	23	34	51,743	254,272,195	36.0
HAMILTON BROS. OIL & GAS LTD.	● Argyll, 1971	9,000- 9,400	7	7	9,050	63,400,000	38.0
	● Duncan, 1980	9,600-10,000	3	5	4,840	16,500,000	36.0
	● Innes, 1983	12,300-12,600	2	2	2,740	4,300,000	45.0
MARATHON	● South Brae Crude Oil, 1975	12,700-13,437	19	27	97,535	141,917,809	35.0
MOBIL	● Beryl, 1972	9,900-11,500	43	56	91,830	355,234,746	37.4
	Ness, 1986	9,900-10,112	2	2	15,576	2,237,306	37.5
OCCIDENTAL	● Claymore, 1974	8,100	25	27	79,233	310,655,000	30.1
	● Piper, 1973	8,100	25	27	165,396	810,261,000	37.0
	● Scapa, 1981	8,100	3	3	20,600	12,308,000	32.0
SHELL	● Auk, 1971	NA	10	11	11,080	76,644,000	38.0
	● Brent, 1971	NA	59	66	338,900	1,060,198,000	37.8
	● Cormorant S. & C., 1972	NA	8	13	39,890	93,560,000	36.0
	● Cormorant North, 1974	NA	14	15	87,270	162,854,000	34.9
	● Dunlin A, 1973	NA	29	31	51,530	250,708,000	36.0
	● Fulmar, 1975	NA	9	14	152,240	263,468,000	40.0
PHILLIPS	● Maureen, 1973	8,000- 8,700	3	12	68,649	113,760,238	36.0
TEXACO NORTH SEA	● Tartan, 1974	10,217-12,614	8	15	35,110	60,211,713	37.0-42.0
	● Highlander, 1976	8,759- 9,372	4	5	28,060	27,699,604	34.0-37.0
	● Petronella, 1975	7,269- 8,037	1	1	10,372	4,036,700	40.1
UNOCAL	● Heather, 1973	10,800	26	37	18,512	74,501,902	34.3
	Total United Kingdom		**787**	**1,135**	**2,417,781**	**8,052,678,367**	

*Included in Thistle total.
†U.K. share of production.

VENEZUELA

	Name of field, discovery date	Depth, ft	No. of wells — Producing	Total	1987 average B/d	Cumulative to Dec. 31, 1987 Bbl	°API gravity
PDVSA/ZULIA	Boscan, 1946	...	500	520	27,000	730,129,000	10.4
	Bachaquero, 1930	...	4,103	4,223	252,560	5,105,880,000	23.2
	Cruces-Manueles, 1916	3,000- 8,000	266	284	2,300	173,117,000	39.2

WORLDWIDE PRODUCTION

	Name of field, discovery date	Depth, ft	No. of wells Producing	No. of wells Total	1987 average B/d	Cumulative to Dec. 31, 1987 Bbl	°API gravity
	Cabimas, 1917	2,200	832	832	60,000	468,973,000	22.8
	Centro, 1959	9,600-11,000	259	279	119,000	838,640,000	36.2
	La Concepcion, 1953	3,148- 8,000	218	238	4,330	136,255,000	34.5
	La Paz, 1925	4,268- 8,000	199	199	19,000	855,108,000	30.6
	Lago, 1959	11,450	117	116	15,000	1,700,474,000	32.0
	Lagunillas, 1926	3,000	5,393	5,581	142,380	3,902,898,000	25.6
	Lama, 1957	8,320	296	310	136,000	2,445,142,000	32.5
	Lamar, 1958	13,000	220	241	65,000	189,063,000	35.5
	Mara, 1945	5,240	147	148	4,400	406,198,000	24.0
	Mene Grande, 1914	4,130	779	779	5,500	637,648,000	18.4
	Motatan, 1952	...	36	36	10,600	71,846,000	20.2
	Sibucara, 1948	13,450	10	10	402	43,406,000	30.8
	Tia Juana, 1928	3,000	3,520	3,551	221,000	10,360,221,000	18.2
	Urdaneta, 1970	10,000	40	285	23,000	122,508,000	26.7
	West-Tarra, 1942	4,250- 5,500	17	44	2,000	61,690,000	39.2
	Ceuta	...	194	203	54,000	446,771,000	30.6
	Other	...	43	124	41,300	107,205,000	...
PDVSA/ANZOATEGUI	Boca, 1971	9,500	107	112	800	79,748,000	32.2
	Caico Seco, 1946	6,500- 7,300	189	192	2,800	8,826,000	33.1
	Chimire, 1948	7,000- 7,200	267	292	4,500	257,059,000	34.6
	Dacion, 1957	6,700	159	171	5,370	223,932,000	21.8
	Elias, 1954	5,000- 6,470	249	265	6,670	424,932,000	37.5
	El Roble, 1939	3,500-11,500	70	70	2,700	38,250,000	50.0
	Guara, 1946	5,000-10,000	625	648	12,000	424,932,000	25.7
	Inca, 1948	7,250	71	72	1,000	1,545,000	28.2
	Leona, 1938	2,200-12,800	423	429	7,300	50,828,00	24.7
	La Ceiba, 1946	9,450	45	45	2,460	47,582,000	42.4
	Mata, 1954	8,970-10,516	485	547	8,800	18,310,000	32.5
	Merey, 1937	5,400- 5,700	775	790	19,700	34,938,000	12.4
	Nipa, 1945	6,000- 8,500	394	412	13,000	163,421,000	28.0
	Oscurote-Norte, 1952	9,513	83	88	700	109,739,000	19.2
	Oficina, 1937	5,900	976	1,047	10,750	381,194,000	22.6
	Soto, 1950	9,500	85	87	1,400	95,643,000	37.8
	Sta. Ana, 1936	8,500	134	143	4,400	119,683,000	42.6
	San Joaquin, 1939	6,550	137	137	5,400	108,031,000	48.8
	Sta. Rosa, 1941	8,500	242	271	13,900	390,012,000	50.4
	Yopales, 1937	4,600	390	406	11,400	127,328,000	23.6
	Zanjas, 1958	13,270	18	21	2,600	40,252,000	41.0
	Zapatos, 1955	11,500	74	83	3,600	256,253,000	35.9
	Zorro, 1953	11,100	30	36	2,000	74,630,000	35.9
	Zulus, 1957	12,710	28	29	1,250	23,286,000	42.1
	Zumo, 1954	9,200	38	41	1,300	77,486,000	26.3
	La Ceibita, 1963	9,878	56	66	2,400	64,889,000	41.8
	Adjuntas, 1957	5,000	28	717	...	...	13.9
	Arecuna	...	141	142	2,200	7,662,000	11.2
	Bare	...	279	286	1,450	15,159,000	10.3
	El Toco	...	52	53	2,000	30,111,000	52.2
	Guario	...	82	83	1,700	43,045,000	43.8
	Isla	...	26	27	2,700	57,905,000	43.5
	Other	...	70	432	14,000	392,013,000	...
PDVSA/MONAGAS	Acema, 1960	...	74	75	15,000	22,325,000	29.0
	Aguasay, 1955	8,100-13,400	17	43	4,800	14,382,000	31.3
	Jobo, 1956	3,600- 4,000	452	465	30,300	269,850,000	11.0
	Morichal, 1958	3,312	237	253	3,600	168,847,000	8.7
	Orocual, 1953	2,954	79	81	3,000	29,043,000	25.1
	Oritupano, 1950	7,657	269	279	3,400	39,681,000	19.1
	Pirital, 1958	450- 1,100	42	81	330	27,372,000	30.2
	Quiriquire, 1928	7,000- 7,200	661	671	1,150	760,239,000	23.5
	Sta. Barbara, 1941	5,020- 6,500	254	268	1,000	160,292,000	24.2
	Tacat, 1953	1,820- 3,670	151	157	150	44,691,000	18.2
	Temblador, 1936	3,500- 4,500	128	135	1,700	109,106,000	19.5
	Aguasay N.	...	60	64	6,200	5,300,000	45.4
	Acema-Casma	...	75	77	18,200	27,407,000	24.1
	Cerro Negro	...	323	324	14,500	11,541,000	8.6
	Mata Acema	...	26	26	11,000	287,667,000	24.0
	Onado	...	29	30	4,500	17,472,000	27.7
	Oritupano N.	...	38	40	5,200	23,586,000	28.4
	Pilon	...	42	147	12,000	150,505,000	12.3
	Furrial, 1985	16,000	5	5	20,000	6,660,000	28.0
	Musipan, 1985	16,500	1	1	7,000	655,000	32.0
	Other	...	20	200	5,000	120,768,000	....
PDVSA/BARINAS	Hato, 1965	9,550	13	18	2,200	43,357,000	24.4
	Maporal, 1957	10,950	15	17	2,900	15,004,000	26.7
	Paez, 1963	...	59	70	8,000	52,325,000	19.6
	Silvan, 1949	10,860	15	23	3,600	48,539,000	29.0
	Silvestre, 1948	8,860	29	58	4,300	136,253,000	23.6
	Sinco, 1953	8,500- 9,100	56	93	10,300	264,843,000	23.7
	Other	...	...	...	...	6,300,000	...

WORLDWIDE PRODUCTION

	Name of field, discovery date	Depth, ft	No. of wells Producing	Total	Production 1987 average B/d	Cumulative to Dec. 31, 1987 Bbl	°API gravity
PDVSA/GUARICO	Budare, 1958	4,523	55	58	2,840	66,442,000	29.6
	Las Mercedes, 1942	4,500	377	445	82	91,318,000	22.8
	Ruiz, 1949	4,500	64	90	50	33,185,000	29.9
	Barzo	4,500	12	12	10	4,005,000	29.0
	Bella Vista	4,500	91	112	40	12,347,000	21.6
	Jobal, 1960	4,500	15	16	10	2,126,000	36.3
	Saban	4,500	128	163	50	22,044,000	41.1
	Guafita, 1984	8,500	14	14	56,000	12,838,000	29.0
PDVSA/AMACURO	Tucupita, 1945	...	6	96	54	67,074,000	16.6
PDVSA/FALCON	All fields, 1920	500-12,000	152	165	...	107,133,000	25.0-52.0
	Total Venezuela		**28,101**	**31,115**	**1,635,488**	**36,274,363,000**	

WEST GERMANY

	Name of field, discovery date	Depth, ft	No. of wells Producing	Total	Production 1987 average B/d	Cumulative to Dec. 31, 1987 Bbl	°API gravity
WINTERSHALL A.G.[e]	Emlichheim, 1944	3,116	74	130			24.5
	Ruehlertwist, 1949	2,919	72	134			25.0
	Aldorf, 1952	4,100	48	106			33.0
	Bockstedt, 1954	4,264	34	48			27.4
	Dueste, 1954	3,936	33	71			33.0
	Pattensen, 1954	2,165	5	7			29.3
	Landau, 1955	3,936	73	78			36.9
	Sohnedeneck, 1956	5,182	19	29			27.5
	Ruelzheim, 1984	7,544	2	2			37.7
	Rheinzabern, 1959	5,576	1	1			37.9
	Moenchsrot, 1958	5,248	15	23	21,253	497,243,766	38.1
	Kirchdorf, 1982	5,737	1	1			38.5
	Oberschnarzach, 1978	4,920	4	4			37.7
	Arlesried, 1964	4,920	11	20			38.5
	Maximiliansau, 1959	5,363	1	1			41.0
	Hauerz, 1985	8,639	5	5			40.4
	Meckelfeld, 1954	6,396	21	28			27.5
	Dickel, 1953	3,936	37	60			28.3
	Nienhagen, 1925	3,729	12	12			30.0
	Wathlingen, 1939	2,624	8	9			30.0
	Eicklingen, 1937	2,460	12	13			30.0
	Schwabmunchen, 1976	4,428	1	1			25.7
	Aitingen, 1977	6,232	4	4			31.1
OTHER OPERATORS[e]	Scheerhorn, 1949	3,550- 3,750					30.0
	Reitbrook, 1937	1,450- 2,100					29.0
	Oelheim-Sued, 1968	7,650	932	1,630			33.0
	Sinstorf, 1960	6,750- 7,150					...
	Vorhop, 1952	2,950- 6,900					36.0
DEUTSCHE TEXACO AG	Boostedt, 1952	5,900	25	41	662	18,178,547	32.0
	Ploen-Ost, 1958	8,850	31	62	1,160	50,716,673	33.0
	Preetz, 1962	8,200	19	30	345	11,527,580	26.0
	● Schwedeneck-See, 1978	4,900	15	15	7,749	7,617,042	28.0
	Hankensbuettel-Sued, 1954	4,900	43	82	2,376	81,818,000	27.0-35.0
	Hardesse, 1957	8,200	31	69	549	8,667,754	40.0
	Oerrel-Sued, 1954	4,600	24	46	1,081	11,404,076	26.0
	Hohne, 1951	5,250	29	55	834	40,639,319	33.0
	Lieferde, 1956	2,600	22	42	348	20,261,379	33.0
	Wesendorf, 1943	4,650	18	36	304	17,394,007	25.0-40.0
	Mittelplate, 1987	7,500- 9,350	4	4	836	305,271	19.3
	Other, 1935	2,300- 8,200	27	69	518	30,513,974	19.0-37.0
MOBIL ERDGAS-ERDÖL GMBH (MEEG)	Hofolding, 1980	11,100	2	3	468	581,148	28.0-30.0
	Bavaria	4,660- 6,190	2	31	1	7,526,952	20.0-25.0
	Hannover	980- 6,630	46	70	859	28,243,280	25.0-31.0
	Viogtei, 1953	980- 1,900	77	106	633	25,923,571	32.0-34.0
	Süd-Oldenburg, 1950	2,630- 6,800	76	153	1,187	39,518,517	31.0-39.0
	Weser Ems	2,000- 2,450	16	25	260	5,984,259	24.0
BEB	Annaveen, 1963	4,428	3	4	17	490,240	33.8
	Barenburg, 1953	2,706	45	73	2,240	38,862,348	31.1
	Barver, 1963	3,640	3	5	39	484,081	32.8
	Dodenteich, 1960	4,641	1	2	15	350,491	31.1
	Eich, 1983	5,995	5	8	725	780,891	38.9
	Eilte-West, 1956	6,051	13	16	94	3,526,095	32.8
	Eldingen, 1949	5,028	38	45	275	20,659,451	35.5
	Elsfleth, 1956	5,175	3	13	44	2,843,467	25.7
	Eystrup 3-6, 1959	1,075	4	4	149	1,109,904	17.4
	Eystrup 2, 1960	3,526	1	1	14	440,007	17.4
	Eystrup-Verden, 1977	3,526	1	4	7	172,143	17.1
	Georgsdorf, 1943	2,653	212	369	5,600	100,618,330	23.8
	Gross-Lessen, 1969	3,280	7	11	834	21,223,931	29.8
	Hankensbüttel, Mitte, 1954	5,182	6	8	158	4,280,434	33.0
	Hankensbüttel-Nord, 1955	5,182	4	4	327	6,242,108	33.0
	Hankensbüttel, Ost, 1954	5,182	1	1	55	1,254,636	33.0
	Hebelermeer, 1955	3,899	10	15	130	4,492,425	27.3

WORLDWIDE PRODUCTION

	Name of field, discovery date	Depth, ft	No. of wells Producing	Total	Production 1987 average B/d	Cumulative to Dec. 31, 1987 Bbl	°API gravity
	Hillerse-Nord, 1958	2,683	1	1	7	764,693	31.5
	Hillerse-West, 1957	6,077	1	1	3	28,790	43.1
	Hohenassel, 1943	1,705	13	28	43	3,511,514	30.4
	Hohnebostel, 1969	8,341	3	6	70	1,181,756	41.0
	Königsgarten, 1984	6,235	0	3	108	63,854	38.1
	Lingen-Dal. (incl. Wd), 1942	3,129	89	148	216	16,882,629	31.1
	Lüben, 1955	4,100	8	11	749	11,998,436	33.0
	Lüben-West, 1958	4,690	5	6	147	1,463,259	36.9
	Meerdorf, 1954	6,077	16	27	99	1,593,134	43.1
	Meppen-Schwefingen, 1959	4,257	22	24	937	12,134,969	29.8
	Mölme-Feldbergen, 1934	1,410	2	2	11	508,466	31.1
	Mölme-Wachtel, 1934	3,886	2	18	2	758,642	36.9
	Nienhagen, 1909	3,079	8	30	107	26,835,081	29.6
	Nienhagen-Dachtmissen, 1959	3,427	6	8	90	998,833	27.8
	Pfullendorf-Ostrach, 1962	3,558	15	24	129	2,498,576	37.7
	Pötrau, 1959	5,248	…	2	…	30,187	14.3
	Rühlermoor-Malm, 1962	4,519	4	10	121	6,408,949	31.7
	Rühlermoor-Valendis, 1949	2,381	283	453	9,069	132,109,053	24.6
	Rühme, 1954	2,361	39	48	740	10,468,482	33.4
	Steimbke-Alt + Nord, 1935	1,246	29	41	102	7,698,424	20.6
	Steimbke-Lichtenmoor, 1943	2,722	17	34	124	3,885,967	17.4
	Steimbke-Ost, 1959	4,428	4	11	51	3,798,042	30.5
	Stockstadt, 1952	5,402	9	22	119	7,721,109	39.6
	Suderbruch, 1949	5,175	49	73	527	23,336,008	31.1
	Sulingen, 1973	3,345	9	9	553	5,036,459	30.0
	Varel, 1957	5,084	6	15	135	5,712,723	25.5
	Varloh, 1984	4,129	6	6	226	118,388	31.5
	Volkensen, 1960	7,540	4	5	75	743,953	22.9
	Wald, 1966	3,116	…	4	…	51,025	28.2
	Wehrbleck, 1957	3,342	56	67	722	16,289,366	32.0
	Wietingsmoor, 1954	2,820	24	32	540	12,968,076	30.5
DST	Bramberge, 1958	2,600	58	73	6,075	101,649,164	28.0
	Fronhofen, 1964	3,800	9	15	238	2,360,472	37.0
	Illimensee, 1967	3,800	2	6	39	983,981	37.0
	Memmingen-M, 1958	5,000	…	7	…	446,484	37.0
	Markdorf, 1960	7,700	…	1	…	13,912	37.0
	Total West Germany		**3,088**	**5,210**	**74,800**	**1,535,219,000**	

YEMEN, NORTH

	Name of field, discovery date	Depth, ft	Producing	Total	1987 average B/d	Cumulative to Dec. 31, 1987 Bbl	°API gravity
HUNT OIL	Alif, 1984	5,700	12	76	18,800	9,514,000	40.0
	Azal, 1987	6,850	…	9	…	…	40.0
	Total North Yemen		**12**	**85**	**18,800**	**9,514,000**	

ZAIRE

	Name of field, discovery date	Depth, ft	Producing	Total	1987 average B/d	Cumulative to Dec. 31, 1987 Bbl	°API gravity
CHEVRON	● GCO, 1970	7,000- 9,000	6	8	244	6,228,697	29.0
	● Mibale, 1973	5,000- 6,000	7	8	13,824	81,266,684	30.0
	● Mwambe, 1979	6,150	2	2	174	638,862	30.4
	● Motoba, 1976	11,098	4	4	2,848	1,693,960	44.3
	● Lukami, 1983	12,400	3	5	3,336	6,170,507	35.0
	● Libwa, 1981	5,500	…	2	…	…	33.0
ZAIREP (FINA)	East Mibale, 1978	6,000	…	1	…	397,637	34.0
	Kinkasi, 1972	3,700	40	42	3,134	6,287,706	31.0
	Liawenda, 1972	3,900	27	32	2,056	4,764,380	31.0
	Nsiamfumu, 1977	8,300	…	1	…	5,347	34.0
	Kifuku, 1983	3,400	1	1	25	37,688	31.0
	Makelekese, 1983	3,600	8	8	491	737,007	31.0
	Muanda, 1972	3,600	6	6	547	536,119	31.0
	Tshiende, 1978	6,000	6	7	4,693	3,930,966	34.0
	Total Zaire		**110**	**127**	**31,372**	**112,695,560**	

USSR

Field, year discovered	No. wells	1988 prod. (million bbl)	Cum. prod. 1-1-89 (million bbl)	Pay, ft.
Arlan, 1955	4,000	50	2,500	Carb., 4,429
Baku Archipelago, 1963	250	20	490	Pliocene, 8,773
Fedorovskoye, 1971	2,500	245	2,645	L. Cret., 5,900
Kholmogorskoye, 1973	1,100	80	630	L. Cretaceous
Mamontovo, 1965	2,000	195	2,425	Cretaceous, 6,300
Neftianye Kamni, 1949	1,110	15	1,130	Pliocene, 5,406
Pravdinsk, 1964	800	45	965	L. Cret., 9,580
Romashkino, 1948	8,000	310	14,310	Devonian, 5,791
Samotlor, 1965	7,500	780	14,805	L. Cret., 7,316
Sovetskoye, 1962	500	38	958	Cret., 11,130
28th of April, 1979	80	44	154	Pliocene, 11,000
Usa, 1963	1,250	70	1,000	Paleozoic, 6,510
Ust-Balyk, 1961	1,400	65	1,725	L. Cret., 8,845
Uzen, 1961	2,200	70	1,695	Jurassic, 2,667

U.S. fields with reserves exceeding 100 million bbl

State Field	Disc. date	1988 prod.	Cum. prod. 1-1-89	Est. rem. reserves	Est. No. wells
		1,000 bbl			
ALABAMA					
Citronelle, 1955		1,827	147,861	7,000	435
ALASKA					
Endicott, 1978		36,098	43,098	324,902	37
Granite Point, 1965		2,787	109,535	16,213	29
Kuparuk River, 1969		112,055	503,397	994,945	328
Lisburne, 1967		14,600	35,600	175,689	49
McArthur River, 1965		7,040	529,040	34,960	76
Middle Ground Shoal, 1962		2,737	155,889	5,263	42
Prudhoe Bay, 1967		576,335*	6,053,018†	3,533,665	691
Swanson River, 1957		2,154	209,490	8,846	29

*Includes about 30.66 million bbl of condensate. †Includes about 138.66 million bbl of condensate.

State Field	Disc. date	1988 prod.	Cum. prod. 1-1-89	Est. rem. reserves	Est. No. wells
ARKANSAS					
Smackover, 1922		2,664	556,907	9,097	2,100
CALIFORNIA					
San Joaquin Valley					
Belridge South, 1911		60,583	677,083	435,286	6,000
Buena Vista, 1909		1,574	647,274	38,271	895
Coalinga, 1890		10,212	753,845	161,188	2,174
Coalinga Nose, 1938		1,285	498,685	17,370	78
Coles Levee North, 1938		442	160,432	2,748	85
Cuyama South, 1949		469	218,191	6,938	105
Cymric, 1909		8,479	199,308	40,000	1,013
Edison, 1928		1,470	134,391	26,670	674
Elk Hills, 1911		39,144	893,374	579,776	1,099
Fruitvale, 1928		577	115,553	13,709	274
Greeley, 1936		237	112,647	1,763	27
Kern Front, 1912		1,500	173,066	55,230	950
Kern River, 1899		46,899	1,204,479	743,000	6,709
Kettleman North Dome, 1928		172	456,648	1,298	44
Lost Hills, 1910		5,627	178,293	61,603	1,634
McKittrick, 1896		2,551	266,822	90,919	931
Midway-Sunset, 1894		57,497	1,879,347	373,863	9,180
Mount Poso, 1926		6,620	263,250	79,280	411
Rio Bravo, 1937		151	116,051	1,319	15
Yowlumne, 1974		6,570	81,368	27,015	65
Coastal Area					
Carpinteria, 1966		2,690	88,101	29,289	114
Cat Canyon E. & W., 1908		2,736	288,236	46,930	512
Dos Cuadras, 1969		4,371	212,804	54,140	140
Elwood, 1928		317	105,705	2,283	7
Hondo, 1969		9,952	89,621	112,223	20
Orcutt, 1901		906	165,874	10,238	136
Point Pedernales, 1982†		6,515	11,715	331,544	10
Rincon, 1927		1,118	148,759	14,832	240
San Ardo, 1947		4,641	408,351	122,903	600
Santa Maria Valley, 1934		1,751	198,131	40,408	172
South Mountain, 1916		727	145,984	12,035	352
Ventura, 1919		7,072	894,742	97,026	570
Los Angeles Basin					
Beta, 1976		6,013	40,423	173,825	60
Beverly Hills, 1900		2,034	122,616	42,356	116
Brea Olinda, 1880		2,143	383,894	54,787	720
Coyote East, 1909		584	108,663	13,212	103
Coyote West, 1909		808	249,969	7,481	113
Dominguez, 1923		607	269,266	7,448	113
Huntington Beach, 1920		5,816	1,066,368	72,044	990
Inglewood, 1924		2,790	345,453	54,149	366
Long Beach, 1921		2,466	909,757	17,216	410
Montebello, 1917		526	192,357	10,267	155
Richfield, 1919		1,494	189,165	27,412	203
Santa Fe Springs, 1919		987	613,077	9,036	151
Seal Beach, 1924		876	203,118	14,118	168
Torrance, 1922		1,693	212,083	35,847	366
Wilmington, 1932		29,921	2,292,229	495,633	2,050
COLORADO					
Rangely, 1933		12,492	739,418	35,000	488

State Field	Disc. date	1988 prod.	Cum. prod. 1-1-89	Est. rem. reserves	Est. No. wells
		1,000 bbl			
FLORIDA					
Jay, 1970		4,676	360,612	55,944	121
ILLINOIS					
Clay City, 1938		2,448	393,860	6,000	2,600
Lawrence, 1906		2,919	394,521	5,500	2,700
Louden, 1938		1,345	388,237	3,555	1,340
Main, 1906		2,066	233,273	5,000	3,356
New Harmony, 1939		1,072	153,545	4,000	1,140
Salem, 1938		2,167	386,983	4,500	1,325
KANSAS					
Bemis-Shutts, 1928		1,169	244,247	4,405	973
Chase-Silica, 1930		1,018	301,003	4,499	1,103
El Dorado, 1915		839	296,734	2,616	822
Hall-Gurney, 1931		1,032	145,677	4,051	1,130
Trapp, 1929		1,200	225,888	4,643	1,000
LOUISIANA					
Offshore					
Bay Marchand Blk. 2, 1949		5,547	596,972	53,758	120
Eugene Island Blk. 330, 1930		7,859	269,080	55,920	169
Grande Isle Blk. 16, 1948		1,659	263,729	85,645	44
Grande Isle Blk. 43, 1956		4,312	272,256	85,688	126
Mississippi Canyon Blk. 194, 1980		4,929	116,868	76,311	44
Main Pass Blk. 41, 1957		2,985	237,854	23,384	112
Main Pass Blk. 306, 1969		1,776	201,869	78,335	94
South Pass Blk. 27, 1954		1,659	125,317	73,198	118
South Pass Blk. 61, 1968		8,140	152,151	45,000	156
South Pass Blk. 62, 1965		3,331	106,364	81,594	73
South Pass Blk. 65, 1969		4,032	100,875	89,252	61
Ship Shoal Blk. 204, 1968		1,591	66,070	38,930	40
Ship Shoal Blk. 207, 1967		1,023	87,818	38,000	25
Ship Shoal Blk. 208, 1962		4,117	160,169	65,274	67
South Timbalier Blk. 21, 1939		1,324	216,239	47,133	45
South Timbalier Blk. 135, 1956		1,390	139,337	25,663	37
West Delta Blk. 30, 1949		6,754	446,083	47,375	153
West Delta Blk. 73, 1962		4,469	188,700	86,291	73
Onshore South					
Bay de Chene, 1941		390	96,382	17,952	24
Bay St. Elaine, 1928		321	164,808	25,292	18
Bayou Sole, 1941		769	161,369	3,417	20
Black Bay West, 1953		1,971	144,799	10,432	94
Caillou Island, 1930		2,308	602,231	74,020	136
Cote Blanche Bay West, 1940		741	181,594	46,374	86
Delta Farms, 1944		293	115,351	7,019	13
Garden Island Bay, 1934		1,406	221,261	31,954	144
Golden Meadow, 1938		894	135,439	4,290	176
Grand Bay, 1938		454	170,514	3,760	41
Hackberry, East, 1927		777	109,087	7,603	60
Hackberry, West, 1928		1,771	141,895	6,492	107
Iowa, 1931		134	99,303	697	25
Jennings, 1901		315	116,409	700	181
Lafitte, 1935		1,665	255,304	9,550	111
Lake Barre, 1929		941	204,023	20,061	31
Lake Pelto, 1929		477	117,000	17,415	23
Lake Washington, 1931		2,353	242,390	16,376	93
Leeville, 1931		447	141,646	7,221	37
Paradis, 1939		725	126,650	8,600	31

State Field	Disc. date	1988 prod.	Cum. prod. 1-1-89	Est. rem reserves	Est. No. wells
Quarantine Bay, 1937 ...		761	172,672	1,654	71
Venice, 1937		858	181,878	7,590	59
Vinton, 1910		287	161,001	800	96
Weeks Island, 1945		816	225,939	21,356	33
West Bay, 1940		1,349	228,246	16,355	79
North					
Caddo-Pine Island 1905		3,310	360,899	12,888	10,688
Delhi, 1944		673	211,707	34,651	58
Haynesville, 1921		761	168,237	2,373	166
Homer, 1919		434	98,375	1,908	198
Rodessa, 1935		331	106,027	1,506	67

MISSISSIPPI

State Field	1988 prod.	Cum. prod. 1-1-89	Est. rem reserves	Est. No. wells
Baxterville, 1944	2,609	239,164	10,391	316
Heidelberg, 1944	2,829	174,809	12,171	316
Tinsley, 1939	834	29,677	3,197	173

MONTANA

State Field	1988 prod.	Cum. prod. 1-1-89	Est. rem reserves	Est. No. wells
Bell Creek, 1967	958	128,836	22,984	91
Cut Bank, 1926	994	162,459	37,169	575
Pine, 1951	1,302	105,145	5,687	96

NEW MEXICO

State Field	1988 prod.	Cum. prod. 1-1-89	Est. rem reserves	Est. No. wells
Denton, 1949	652	138,693	3,000	179
Empire-Abo, 1957	1,424	219,983	50,017	405
Eunice-Monument, 1929	2,500	125,414	10,894	879
Hobbs, 1928	8,480	297,482	20,000	613
Maljamar, 1926	1,869	144,961	5,070	851
Vacuum, 1929	12,359	432,242	40,000	1,556

NORTH DAKOTA

State Field	1988 prod.	Cum. prod. 1-1-89	Est. rem reserves	Est. No. wells
Beaver Lodge, 1951	1,704	111,574	16,684	132
Billings Nose, 1978	2,560	61,669	51,744	153
Little Knife, 1977	3,392	51,437	57,924	131
Mondak, 1976	439	13,165	82,887	96

OKLAHOMA

State Field	1988 prod.	Cum. prod. 1-1-89	Est. rem reserves	Est. No. wells
Burbank, 1920	1,135	536,746	8,377	1,105
Eola-Robberson, 1920	746	133,124	8,210	603
Fitts, 1934	2,963	199,894	12,000	589
Glenn Pool, 1905	1,170	327,865	5,000	714
Golden Trend, 1945	4,333	474,100	28,000	1,396
Healdton, 1913	1,938	334,645	9,563	1,000
Hewitt, 1919	3,152	268,371	13,615	923
Oklahoma City, 1928	802	816,170	5,966	174
Postle, 1958	1,216	106,393	14,921	226
Sho-Vel-Tum, 1905	18,398	1,167,379	60,000	7,616
Sooner Trend, 1945	4,147	295,544	20,000	4,746

TEXAS

District 2

State Field	1988 prod.	Cum. prod. 1-1-89	Est. rem reserves	Est. No. wells
Greta, 1933	748	147,570	12,530	100
Lake Pasture, 1953	2,184	87,784	12,644	143
Tom O'Connor, 1934	10,380	747,848	55,000	646
West Ranch, 1938	2,640	380,034	8,504	307
District 3				
Anahuac, 1935	1,068	284,888	15,112	106
Conroe, 1931	3,864	727,215	33,728	278
Giddings, 1971	8,868	278,570	148,032	2,281
Hastings, 1934	3,084	697,237	72,764	206
Magnet Withers, 1936	1,884	110,960	5,000	150
Oyster Bayou, 1941	864	160,204	18,096	39
Thompson, 1931	3,972	472,540	27,360	262
Tomball, 1933	400	121,056	9,868	86
Webster, 1937	5,304	573,192	20,000	243
District 4				
Agua Dulce-Stratton, 1928	360	146,623	24,751	94
Borregos, 1945	192	114,021	20,186	40
Kelsey, 1938	200	114,859	36,247	60
Plymouth, 1925	400	122,823	3,300	60
Seeligson, 1925	156	271,483	55,544	42
TCB, 1944	456	112,700	52,358	30
White Point E, 1938	60	104,034	6,340	21
District 5				
Alabama Ferry, 1983	3,600	14,600	86,400	253
Van, 1928	3,096	521,960	15,000	366
District 6				
East Texas, 1930	40,597	5,008,747	988,758	9,363
Fairway, 1960	2,424	192,691	17,777	100
Hawkins, 1940	8,244	821,644	42,372	442
Neches, 1953	1,416	103,980	6,036	163
Quitman, 1948	1,680	121,346	8,440	208
District 8				
Andector, 1946	1,500	181,888	6,500	28
Block 31, 1945	3,556	220,068	12,000	325
Cowden, N., 1930	13,596	488,092	40,000	1,210
Cowden, S, Foster, Johnson, 1932	9,696	500,796	40,000	1,593
Dollarhide, 1945	2,568	193,780	11,448	202
Dune, 1938	2,928	183,000	18,704	771
Fullerton, 1942	7,452	348,200	22,000	870
Goldsmith, 1934	7,248	755,516	33,752	2,036
Howard Glasscock, 1925	5,844	404,508	26,000	2,208
Iatan, E., 1926	3,372	142,956	12,000	1,304
Jordan, 1937	480	128,976	2,520	138
Keystone, 1930	1,872	313,080	9,374	812
McElroy, 1926	7,884	465,549	56,800	1,600
Means, 1934	7,020	228,598	20,000	713
Midland Farms, 1944	4,608	241,980	18,616	415
Sand Hills, 1931	2,760	248,207	21,600	1,312
TXL, 1944	2,004	263,516	6,500	600
Waddell, 1927	708	100,712	3,792	188
Ward Estes, N., 1929	3,408	364,177	76,436	1,562
Westbrook, 1923	2,000	88,000	16,000	718
Yates, 1926	33,540	1,171,820	782,685	1,146
District 8-A				
Anton-Irish, 1944	3,564	175,910	24,190	239
Cogdell Area, 1949	1,656	292,218	41,204	103
Diamond M., 1948	1,908	239,418	16,063	474
Kelly-Snyder, 1948	11,592	1,234,962	115,038	806
Levelland, 1938	17,328	484,144	50,000	3,012
Prentice, 1951	6,216	161,036	20,000	437
Salt Creek, 1950	10,404	247,340	12,552	173
Seminole, 1936	17,004	524,768	36,000	624
Slaughter, 1936	20,364	1,029,800	50,000	3,001
Spraberry Trend, 1951	22,212	653,388	50,000	7,321
Wasson, 1936	28,656	1,711,898	60,000	2,152
Welch, 1942	3,324	144,228	14,000	651
District 10				
Panhandle, 1921	7,812	1,423,286	41,240	11,643

UTAH

State Field	1988 prod.	Cum. prod. 1-1-89	Est. rem reserves	Est. No. wells
Altamont, 1955	3,047	89,493	231,216	227
Aneth, 1956	5,340	354,004	30,000	461
East Anschutz Ranch, 1979	12,756	80,365	727,799	28
Red Wash, 1951	1,075	77,511	13,377	146

WYOMING

State Field	1988 prod.	Cum. prod. 1-1-89	Est. rem reserves	Est. No. wells
Brady, 1960	2,438	54,637	49,251	22
Byron, 1918	1,119	120,688	7,656	67
Elk Basin, 1915	2,968	446,698	25,750	170
Frannie, 1928	1,122	116,020	5,000	63
Garland, 1906	2,625	154,885	6,000	225
Grass Creek, 1914	2,414	185,120	9,000	288
Hamilton Dome, 1918	2,853	268,109	6,000	239
Hartzog Draw, 1976	6,647	66,722	283,739	157
Hilite, 1969	541	75,416	55,885	93
Lance Creek, 1918	163	107,805	400	24
Little Buffalo Basin, 1914	2,666	118,653	9,389	154
Lost Soldier, 1916	2,308	192,073	6,000	71
Oregon Basin, 1912	8,669	388,180	30,000	500
Painter Reservoir, 1979	1,739	31,674	80,674	31
Salt Creek, 1906	5,210	629,689	25,000	1,217
Wertz, 1920	3,500	99,695	15,000	65
Whitney Canyon, 1980	1,652	10,379	105,485	29

Offshore industry not looking for sustained recovery

DURING THE FIRST HALF OF 1988, THE INTERNAtional offshore community eagerly predicted the beginning of long-awaited revival in the exploration and production industry. While most segements of the industry saw improvement last year, the level of activity and corresponding prices did not rise to the levels envisioned twelve months earlier.

With the memory of that unrealized prediction fresh on their minds, offshore oilmen have substantially tempered their outlook for 1989. With the exception of the North Sea, operators and service company executives say they are not expecting a major increase in activity and profits this year. Many believe most segments of the industry will see improvement - albeit slight.

At mid-year 1988, the worldwide offshore rig utilization, which is the most telling barometer of activity, had increased dramatically over the like period the previous year, and helped continue to fuel the optimism. Despite slight seasonal declines, the rig count held fairly steady during the year, but never reached the levels anticipated when 1988 began.

In January 1988, the international offshore mobile rig utilization rate stood at 71.6 percent, which was a substantial increase of the 54.3 percent usage rate for the beginning of 1987. By July, it remained in the 72 percent range, which was 10 percentage points ahead of the 1987 mid-year figure. At this time, contractors felt the worldwide utilization rate could hit the mid 80's before 1988 drew to a close. That never happened.

At the end of the year, the international utilization rate stood at 71.2 percent, but operators and contractors warned it would probably begin to drop steadily during the winter of 1989. The sharpest decline at year's end was in the highly promising Gulf of Mexico, which had a total utilization of only 58.8 percent, which compared to the 60 percent average as 1987 closed.

Throughout 1988, the North Sea, particularly the Southern Gas Basin of the UK, saw the best jackup market in five years. At the same time, the Gulf of Mexico was described by one contractor as being "wimpy."

Jackups were very close to the 100 percent utilization rate at the end of 1988, while dayrates were in the $30,000 range. By contrast, day rates in the Gulf of Mexico for 300 ft. cantilever jackups were ranging between $13,000 to $14,000 with smaller units going for around $10,000 to $12,000 a day.

"Jackups are really going over there, but I don't know if it's a reflection of increased activity in the North Sea, or the fact that the gulf if simply flat," a spokesman for Zapata Offshore said late last year. "Nobody felt the gulf would be this wimpy."

"The jackup activity and rates in the North Sea are better than they've been in five years," added Bob Williams, a vice president for Rowan Companies. "It is very strong, and getting stronger. Less than a year and a half ago, we were drilling wells there for $9,000 a day, and now we're getting $30,000 a day or more."

As in years past, 1988 was highlighted by intense speculation on the price of oil and gas and its impact on expensive offshore projects. Most industry leaders said last year that a price level of $21-$25/bbl was needed to stimulate drilling activity in the capital intensive offshore sector. When it became clear that price range was not in the cards for the forseeable future, engineers returned to the drawing boards to develop cost-effective ways of producing from the offshore environment. Armed with their revised cost scenarios, operators said there were profits to be made at around $15/bbl oil, even in the deepwater.

Deepwater draws interest

"Deepwater production is profitable at $15/bbl, providing companies break the geometric increase of cost with water depth," said Dennis E. Gregg, general manager of international production for Conoco. "In fact, we need to do much better than that. We can't afford for costs to increase even linearly with depth. "

"Clearly, subsea and floating production looks like the technological choice, particularly in the Gulf of Mexico," added Robert Hauptfuhrer, chairman and chief executive officer of Sun Exploration and Production. "By the turn of the century, we will have moved into 3,000 ft. water depth for production in the Gulf of Mexico."

Conoco still maintains, however, that tension leg platforms and subsea systems are the two best-developed methods of extracting oil from waters deeper than 1,000 ft. Brazilian state company Petrobras has advanced the subsea production concept in the Campos Basin and Norsk Hydro's Togi project off Norway will be completely subsea-based.

The greatest advantage of the TLP over a subsea system, Conoco contends, is the ability to place the wellheads on the platform deck.

Rig workovers on subsea wells are very expensive. The industry estimates the cost to mobilize a rig over a subsea well in 2,000 ft of water in the Gulf of Mexico, to run the riser and BOP's, remove them at the end of the job and to demobilize the rig is about $500,000."

"Subsea completions can be a technologically feasible means of development, because it's been proven in Brazil and the North Sea," said Clyde Hewett, project manager for Ocean System Engineering's (OSE) Offshore Production Systems Division. "The thrust of our (subsea) study revolves around cost-effective means of maintenance and workover, once the system has been installed. There are capital advantages to subsea completions."

OSE and Amoco last year commissioned the first phase of the study, which broadly identified the problems inherent to subsea completion maintenance and workover. Omega Marine Engineering and Oil Industry Engineering Inc. joined with OSE to complete the more specific second phase of the study. That phases addressed such issues as motion compensation systems, vessel environmental and design criteria, project management, remotely operated vehicle tasks, and workover and maintenance systems for semisubmersibles, along with supply and diving support vessels.

Those interested in subsea completions kept a close watch on Placid Oil Company, which last year shattered the water depth record for a subsea completion. In late August, oil began flowing from the company's Green Canyon Block 31 Well No. 4, which is producing from 2,243 ft. of water in the Gulf of Mexico. The start-up was delayed after two 10-mile-long flowlines collided with outcroppings in February during tow-out to the site.

Significant deepwater wells

In the highly touted deepwater frontier of the Gulf of Mexico, waters south of New Orleans in the Mississippi Canyon area were home to some of the industry's most significant drilling programs last year.

Shell sank a pair of record-breakers on Block 657 in 7,520 ft and 7,638 ft. of water. Those were followed by a well on Block 786 in 7,467 ft. of waters.

Following on Shell's heels was Texaco, which hit paydirt on Mississippi Canyon Block 285 where two wells have been drilled in 3,085 ft of waters, along with a well on southerly adjacent block 329 covered by 3,290 ft. waters.

No less than 26 different Mississippi Canyon tracts have been drilled by mobile rigs this year, with as many as 10 rigs working the region at any one time.

Additionally, a total of seven separate combinations of tracts have been unitized encompassing 24 blocks, including Exxon's Block 280 Lena guyed tower development, which takes up four blocks, and Shell's Block 194 Cognac platform that produces from four tracts.

Just past the northeastern border of Mississippi Canyon, Amoco late last year announced a significant strike on Viosca Knoll Block 956 in 3,560 ft. of water. Amoco described the strike as "what could be one of the larger deepwater finds to date in the gulf." Amoco said the reservoir extends across several undrilled Viosca Knoll tracts with water depths ranging from 2,500 ft. to 4,000 ft.

BP Exploration, formerly Standard Oil, will concentrate this year on the Gulf of Mexico's so-called Flexure Trend in waters from 600 to 2,000 ft.

With Brazil, the Gulf of Mexico, and West Africa leading the way, deepwater exploration is expected to continue to increase this year. Contractors, meanwhile, are cautiously optmistic that it will increase sufficiently enough to bring with it an increase in rates.

In both the North Sea and gulf, the market for deepwater units, such as semisubmersibles, was showing signs of solidfying. Contractors, however, said activity still had not reached the level where substantial increases in day rates could be enjoyed.

Although the overall activity in the gulf was disappointing to contractors hoping for a banner year, there were 11 wells drilled in 1,000 ft. of water in August, 1988, which was more than in any month prior. By the end of the year, the increased activity had not yet translated into increased day rates, but contractors say they are guardedly optimistic charters will bring higher prices this year.

Minimal platforms accepted

This is not to say, however, that operators and contractors are turning their backs on the reserve potential in the shallower waters. In fact, a trend taking on a new head of steam last year was the design and installation of production systems for mature structures closer to shore.

Minimal platforms, which typically use the well drive pipe/caisson as part of the deck support structure or as an installation vehicle, are gaining popularity in relatively calm-weather regions, particularly the Gulf of Mexico. The units are applicable in between 30 to 300 ft. of water.

Growing interest in these minimal structures was spawned by the wave of cost-consciousness, which continues to sweep the industry. Increased offshore activity among tight-budget independents has also provided impetus to this growing market.

Another motivating factor that developed last year centers around concern among major operators over retaining valuable personnel. Over the past few years, the number of offshore projects, both large and small, fell in parallel with oil prices. Minimal platforms offer operators a way to keep a continuous string of projects underway by making marginal properties more economically acceptible. Thus, valuable personnel resources are kept intact.

In many cases, to take advantage of the lower costs afforded by minimal platforms, operators must accept reduced structural versatility. Should the topsides load need to be expanded beyond the facilities originally installed - to increase pumping capacity, for example - lack of structural versatility will severely restrict the operator's options.

There also is the continual risk of unforseen installation costs and accidental damage. Minimal platforms feature none of the structural redundancies found in more conventional four-pile platforms. If certain designs undergo substantial vessel contact damage to a pile or a brace, there is a strong chance serious problems will develop resulting in costly repair bills.

This threat was highlighted in 1988 when Hurricane Gilbert, the largest such storm ever recorded in North America, had its eye on the prime producing area of the Gulf of Mexico. Minimal platforms are engineered against 100-year storm conditions, which include winds of 138 miles per hour. At its height, Hurricane Gilbert packed sustained winds of 175 to 180 mph. Few standard Gulf of Mexico platforms could have survived such conditions, especially minimal structures. Fortunately, the major producing regions in the gulf was spared the impact of the storm.

One of the larger and newer designs on the market is a chopped-tripod developed by Houston-based Mustang Engineering. The first chopped-tripod was installed in the fall of 1988 for Mark Producing in the Gulf of Mexico on East Cameron Block 300. The unit was placed in 183 ft. of water.

Another is the so-called Quadpod design by Thronson Engineering, which was developed to operator in any water depth accessible by a jackup. This unit can support five wells in water depths of up to 250 ft., but the company also has designed a phased structural development technique to address the initial requirements of a satellite platform and provide expansion flexibility as platform requirements change.

Moss I is a conventionally-piled two-leg jacket that is clamped and welded to the well drive pipe to form a tripod. It is applicable in water depths to 125 ft., and can support up to six wells with a single 40 ft. by 60 ft. deck and helipad.

The larger Moss II unit is suitable for waters up to 200 ft. and can support a maximum of three wells internal to the drill pipe. The design consists of a caisson connected to a pair of diagonal braces above the water line and running to a piling point on the seafloor. As of December 1988, 22 Moss structures have been installed.

Last year, CBS Engineering unveiled its Moss III, which features a larger jacket-type substructure than the Moss II, and can support up to four wells in 200 ft. waters. Installation time from mobilization to demobilization is estimated at eight days, depending on the water depth involved. Fabrication time is estimated at six to 10 weeks.

Another Houston-based design firm, Petro-Marine Engineering has developed and overseen the installation of six of its Guardian platforms in the Gulf of Mexico. The Guardian is based on the time-proven "breasting dolphin" structure and is not considered patentable.

One of the oldest minimal platform designs is the Seahorse developed by Atlantia Engineering, also of Houston. This is a four-pile unit that forms a pyramid support for the surrounded drive pipe or caisson. Various designs will support one to four wells in waters up to 300 ft. deep. Two decks can handle facilities adequate for 10 to 15 MMcfd of gas for a two-slot platform and 18 to 20 MMcfd of gas for a four-slot platform.

At year's end, the first of five T-Horses was under construction and more than likely will be installed in 128 ft. of water offshore Ecuador. The other four, which are being built on speculation, were to have been delivered this February.

Southeast Asia is expected to play host to several minimal structures in the near future. Waters off Indonesia, Malaysia, and Thailand and in the South China Sea in particular are highly suitable for these type of structures.

Though some operators continue to steer away from low-tonnage or minimal platforms as a matter of policy, popularity and water depth applications of these structures are expected to grow. Reliability over time will be the key convincing factor.

As the designs improve, the tonnage required for a given water depth will decrease. This will reduce development costs and improve the economics of marginal development in moderate water depths. That will provide more business thoroughout the offshore development industry - from operators to engineering companies and from fabrication yards to installation contractors.

Contractors to merge?

One of the most critical situations in the marine industry is the presence of un-utilized mobile rigs and owners trying to hold out for better times. The situation is worse for onshore contractors, but marine drillers have larger investments tied up and less flexibility concerning the disposal of units.

Mobile rig contractors will have to consolidate, merge, or agree to stack rigs in 1989 in order to prevent situations that could result in unsafe conditions and major accidents.

Rigs with deepwater drilling capacity or floor automation equipment that will permit faster and safer drilling were the only units at the end of 1988 with large utilization rates and the corresponding more attractive day rates. Unfortunately, few rig contractors have such units and any profitability derived from these rigs cannot be spread over sub-profit rig contracts.

One industry analyst went so far last year as to say the offshore contractor segement may not survive.

At the end of 1988, there were 100 companies working offshore worldwide with a total fleet of 750 rigs. Nearly 40 percent of those rigs are tied up in complex debt restructuring.

With an average age of 10.5 years, the worldwide fleet is growing older and older. Most feel a rig is ending its life span at 15 years.

In late 1988, rig activity off Northwest Europe was at its highest level in three years. By the end of that year, operators were to have drilled an estimated 210 wells, which is slightly below the record for this decade, which was set in 1984 when 220 wildcat and appraisal wells were drilled.

During the first nine months of 1988, a total of 178 exploration and appraisal wells were spudded in the seven-nation region, according to figures supplied by London analyst James Capel & Co. Capel analyst Greg McNab said 58 of those were spudded in the third period, which was overshadowed only by the 60 wells spudded in the like period of 1984.

Transportation rates static

The 1989 forecast appears somewhat brighter for the marine transportation industry, but most do not envision much of a jump in charter rates, if any.

Since the worldwide transportation fleet has been steadily reduced over the past few years, most say any continuing improvement, perhaps for the rest of this decade, will be supply led and not demand led.

Smit-Lloyd B.V. of the Netherlands ended 1988 with an "acceptable overall utilization rate," primarily because of a combination of increased activity in the North Sea, Australia and the Far East, and its fleet reduction. However, the company's chief executive said this increased usage did not correspond into higher prices.

Once more, boat operators say they will see no sustained improvement until oil prices do the same.

Ironically, as 1988 drew to a close, the overall Gulf of Mexico utilization rate for vessels equipped with liquid mud was averaging 94.2%. At the same time, the plateau of $2,500-$2,600/day for 180-ft supply boats had disappeared, with some boat operators reportedly accepting contracts in the $2,100/day level.

Underwater business deteriorates

By the end of last year, the business conditions within the subsea contracting industry had deteriorated to an alarming rate, forcing many contractors into a major re-assessment of strategy and position.

"The current dilemma faced by all contractors," says Ric

Wharton, managing director of Wharton Williams, "is to identify the mechanism, or events, that are required to return the market to stability. Shareholder confidence in the underwater industry has largely evaporated and one must question whether it will be possible for any company to sustain an across-the-board underwater capability to react to emergencies and major requirements in the future."

According to most subsea contractors, the biggest challenge facing the underwater industry is to enhance the capabilities and sophistication of remotely operated vehicles. This is particularly relevant in a time when divers are becoming a rare species in most of the major operating theaters.

In the Gulf of Mexico, for instance, the number of divers available for work last year had dwindled to the point that many jobs went begging. Diving company executives say the turmultous situation in the industry had forced many qualified oilfield divers to seek more stable positions elsewhere.

A survey by one Louisiana-based diving company showed 1,750 divers available for assignments in the Gulf of Mexico in 1982. By early 1988, that figure had been slashed to 410. Furthermore, in 1981, 55 diving companies were in operation in the gulf and by the end of last year only 10 remained.

"Many of the qualified divers have gone to the (U.S.) East Coast for inshore work," said Johnny Johnson, a vice president for Oceaneering International. "We can handle business to a certain level and after that the resources are limited."

Construction
jump seen

The most promising geographic area for offshore fabricators is the North Sea, and Norway in particular.

Although low oil prices may force abandonment of some projects in the North Sea, fabricators say development of the gigantic Sleipner and Troll fields in Norway will proceed on schedule because of contractual demands.

Refurbishment is especially gaining acceptance in the Gulf of Mexico. One independent last year said his company has installed 10 such structures and intends to do more.

This trend also has spawned a commodity market for used and refurbished production platforms in the Gulf of Mexico. Companies that own and operate small to medium-sized lifting equipment are carving out a niche in the used platform market by buying platforms directly from operators and refurbishing them for speculative resale, typically to independents.

The oil price instability facilitates this trend by reducing profit margins for existing marginal fields. Thus, some of relatively new jackets (five to 10 years old) are being removed. These make ideal candidates for refurbishment.

Hall-Houston Oil Co., for instance, has installed a number of such platforms in the gulf in water depths up to 200 ft. At year-end 1988, the company was still considering a refurbished jacket for 300 ft. of water.

In the gulf, there are more than 3,500 producing structures with 90 percent located in waters 200 ft. deep or less. Two-thirds of these are less than 20 years old.

Geographically, the strongest activity in 1989 will be centered in the North Sea, followed by the Gulf of Mexico, Australia, Brazil, and Southeast Asia.

No rebound
seen in Gulf

While most Gulf of Mexico watchers expect a third consecutive year of increased drilling activity, few are prepared to say a major rebound is in the making.

The number of wells drilled in the U.S. Gulf began to move up slightly in 1987 and last year, and most observers say at least a 10% increase is expected for this year. That's welcome news, but it's still below the levels forecast at the beginning of last year.

According to the January issue of **Offshore Magazine,** operators indicate at least 1,067 wells will be drilled in the Gulf of Mexico this year, as opposed to the 971 expected to be completed when all of 1988 tallies are recorded. If the 1989 scenario proves accurate, it will be the first time since 1985 that 1,000-plus wells were drilled in the Gulf of Mexico.

One of the most interesting trends unfolding in the Gulf of Mexico last year was the increase in wildcat drilling. Up to October of last year, 62% of the wells drilled in the Gulf of Mexico were exploratory in nature, which reverses the historical trend in the region, which over the past five years saw only 18% to 24% of the wells drilled being exploratory. This year, operators' intentions indicate 61% of the wells completed will be exploratory and 39% will be development.

In the coming years, exploration and development drilling will probably balance near the 50% level, and later drop back to historical levels. This reflects the declining quality of undrilled acreage on the continental shelf. While deepwater drilling may provide better acreage for explorationists in the future, widespread exploration there will depend both on improved oil prices and sizeable discoveries.

As 1988 drew to a close, operators were looking at a total of 2,643 undrilled leases in the OCS - all of which, have exploration dates on or before 1998.

A total number of leases still in effect at the beginning of 1988, was 4,225 and during 1988, 1,003 of these were set to expire. By Oct. 31, 1988, 501 of the 1,003 leases had been drilled. An additional 292 drilled and undrilled leases were relinquished, and 210 had expired undrilled.

Of the 1,003 leases set to expire last year, 197 were being held as producible on OCS Order No. 4 and 150 were already in production. Of the 4,225 leases set to expire, including the 1,003 expiring in 1988, 337 were on OCS Order No. 4 producible hold and 244 were already in production.

Operators indicate 20% of the leases expiring in 1989, 1990 and 1991 will probably not be drilled. It is estimated that an average of 555 exploration wells will be drilled each year through 1991 on these leases.

Reserve base
substantial

Meanwhile, despite reports to the contrary, substantial reserves remain in the Gulf of Mexico.

"We estimate the ultimate resource recovery from the gulf at about 68 billion bbl of oil and gas equivalent with only about one-third of that already produced," said Dr. William Fisher, director of the Bureau of Economic Geology at the University of Texas.

Of the remaining 45 billion bbl of hydrocarbons yet to be produced in the gulf, two-thirds is yet to be discovered, according to Fisher. About 20% of that is proven and nearly 15% will come from reserve growth in existing fields. Natural gas comprises most of the remaining potential, Fisher asserts.

"By any measure, the gulf looks solid in terms of remaining potential," Fisher said late last year, disputing some claims that the gulf, being a maturing basin, does not hold tremendous potential.

The potential Fisher spoke of is drawing the attention of international operators as more continue to stream in the Gulf of Mexico.

Last year alone, Brazil's Petrobras and Total Minatome both indicated they are looking to increase their activities in the U.S. Gulf. These two operators alone committed more than $65 million on separate exploration and production projects. Both indicated they were primarily interested in deepwater prospects.

"The big fields are out there, or else we would not be moving into the gulf," said Petrobras America President Jose Barbosa. "It is very promising. We'd like to bring some floating production systems into the deepwater gulf."

Total, likewise, said from a foreign operator's perspective, the deepwater Gulf of Mexico is extremely attractive.

"We want to get out where the big boys are - deepwater," Total Minatome President Jean-Pierre Donnet said earlier last year. "We intend to continue buying up acreage leases in the gulf."

Donnet indicated his company would like to obtain 10 to 12 leases.

U.S.-based companies were no less hesitant to increase their holdings in the gulf. At year's end, Exxon had finalized a deal with the financially troubled Hunt family of Dallas, which own Placid and Penrod, to acquire a 60% interest in 58 Gulf of Mexico blocks.

Exxon said it intended to begin drilling on the new holdings at the beginning of 1989, many of which are located in waters as deep as 3,800 ft. The majority of the blocks acquired from the Hunts lie in Green Canyon, Mississippi Canyon and Garden Banks.

Not included in the package was the Hunt's interest in Green Canyon blocks 29 and 31, which is home to the Placid floating production system - the first in the gulf.

The Hunts spent more than $80 million to acquire the tracts during OCS lease offerings between 1984-86. With its latest acquisition, Exxon holds about 170 gulf blocks in waters deeper than 1,000 ft.

Conversely, others were more interested last year in liquidating many of the tracts acquired in past OCS offerings.

Bullwinkle
on schedule

Along with the aforementioned Placid Green Canyon floating production system, which shattered the world's water depth record for production last year, many eyes were also focused on Shell's mammoth Bullwinkle project.

Development drilling began on Nov. 5, 1988 with production from the 1,615-ft structure expected to begin the first quarter of this year. The platform is located in 1,353 ft of water on Green Canyon Block 65. Shell has not made reserve or production figures available.

The total cost of the projected is $500 million. At the end of 1988, Shell officials said if the decision was to be made at that time, the company would continue to go with the fixed jacket scheme as opposed to the floating and subsea production systems gaining popularity in the Gulf of Mexico.

Besides Placid, another floating production scenario raising eyebrows was the decision by Standard and Kerr-McGee to launch a deepwater production project in the Gulf of Mexico, which brought to six the number of deepwater developments on the boards, or underway.

In October 1989, the Standard-Kerr McGee consortium said it will begin design engineering on the project, which would combine four blocks. A Standard spokesman said Viosca Knoll Blocks No. 989 and 990, and Mississippi Canyon Blocks 27 and 28, which are located in 1,300 ft of water, are slated for development under the project.

Besides the Placid and Bullwinkle projects, other deepwater developments underway at the end of 1988 were Exxon's Lena guyed tower and Conoco's Jolliet tension leg well platform.

At the end of the year, Standard and Kerr McGee were still debating the type of production system to utilize for their deepwater project. The companies' engineers have examined conventional, subsea, floating production and compliant tower.

The consortium paid over $22 million to obtain the four blocks involved. Mississippi Canyon Block 28 was purchased in last year's OCS Central Gulf of Mexico offering for a bid of $13.5 million.

Conoco expects to have its Jolliet platform, which is located in 1,760 ft of water on Green Canyon Block 184, on stream by September 1989. Far East Livingston Shipyard is scheduled to deliver the platform into the gulf by mid-year.

Closer to shore, operators indicate they will continue their push to examine the Jurassic Norphlet play that has already yielded one of the largest gas fields in the nation. Mobil's Mary Ann field, located off Mobil Bay, Ala., hit a production level of more than 45 MMcfd last summer when a fourth well came on stream.

As the fourth well was beginning to flow gas, Mobil plunked down $25.178 million to acquire an adjoing block during an Alabama State Lease Sale in mid-July. In fact, Mobil and a consortium comprised of Shell Offshore and Amoco Production were the only companies participating in the sale, which offered 106 tracts.

Activity hot
off Australia

Although not known for sizeable discoveries, the Australian/New Zealand region has been the site of some of the world's hottest offshore plays. Exploration activity was intense in 1988, and if operator commitments are any indication, 1989 should be the most active year on record.

Australia's Resources Minister Peter Cook said at the conclusion of last year that operators indicate they plan to sink a record 33 wells this year in Australian waters, primarily in the rejuveneated Timor Sea. The trend began in 1988 and during the third quarter, 10 offshore wells were completed which represented the largest number of completions for a single quarter in more than two and one-half years.

The Timor Sea/Bonaparte Gulf area saw the most activity during last year and the most successes. Discoveries made in the region last year included the Jabiru-7, Challis-6, Skua-4, Oliver-1, Cassini-1 and the Montara-1. At year's end a consortium led by BHP Petroleum was drilling the Allaru-1, Kalyptea-1, Challis-7, and the Skua-5.

Brian Loton, BHP's chief executive, told **Ocean Oil Weekly Report** last year that the Timor Sea will probably be the next major offshore development area in Australia. The Timor is rapidly replacing the once-prolific Bass Strait as Australia's major exploration and producing sector.

BHP and its partners will be the most active offshore Australia this year, with plans to drill at least 23 wells. Most of those will be spudded in the Timor Sea.

Already a significant producing area, the Timor Sea is rivalling the onshore Cooper-Eromanga basin, and this year is expected to surpass it. Whether the Timor Sea will replace the massive investments that continue to pour into Bass Strait will depend solely on the discoveries made this year, oilmen say.

While commercial discoveries in the Timor Sea thus far have averaged only 30 to 50 million bbl, BHP said it believes much larger fields are possible.

"I think we will continue to find fields typically of the size we have found thus far," said BHP's Exploration Manager John Froning. "But, there is the potential for a field significantly larger. It is still a very new area of exploration, and this is what we have to remember. BHP believes no other Australia area offers the same potential as the Timor Sea."

Bass Strait
projects

Australia's major producing area, Bass Strait, has seen its reserves dwindling over the past few years. With that in mind, the government last year approved several development schemes designed to improved the output from the region, but not by much.

A total of 3.8 billion bbl of oil and condensate has been discovered in the Bass Strait, and operators say that only 1.4 billion bbl remain.

In October, Esso and BHP released plans to develop the small Perch, Dolphin, Seahorse, Tarwhine and Whiting fields. The fields hold combined reserves of only 51 million bbl of oil.

Esso officials said the technology to be employed for Perch and Dolphin will serve to maximize profits from the small fields. Rather than using a full-scale, or a mini-platform, the consortium will use unmanned single-well monotowers. Production would be controlled onshore.

Each 600-ton monotower will be held in position by 1,400 tons of iron ore ballast, which Esso claimed would make the Perch and Dolphin installations the world's first gravity-based monotowers.

Meanwhile, the Whiting mini-platform will carry a minimum of production equipment with most of its facilities located on the nearby Snapper platform. The Maersk Giant gorilla-class jackup was contracted for last year to drill five wells planned for the structure at a total cost lower than drilling from the mini-platform, Esso claimed.

Ironically, while operators were making plans to increase Bass Strait production, at the same time, they shut-in much of the output already flowing into the market. The cutback was in dispute to the government tax regime which they claim to be too rigid.

As threatened earlier, Esso and BHP by November had shut in 65,000 b/d, or 20% of the total output from the nation's major producing sector. The two operators had initially threatened to close in as much as 80,000 b/d until the government formulates a more relaxed tax structure.

John Schubert, chairman and managing director of Esso Australia, said last year that "hundreds of million bbl of oil" could potentially be recovered in the Bass Strait if operators had a more relaxed tax regime under which to work.

"We have identified a large number of projects - in the order to 20 that are possibilities - and there is a great deal of oil within those projects," he said.

Despite the tax regime and small discoveries to date, Australian waters are attracting the interest of several foreign operators. This was reflected in last December's awards of new exploration licenses.

When bids were reviewed, the government noted Petrofina, Total and British Petroleum had returned to Australia after several years' absence. Newcomers into Australian waters included Trafalgar House of the UK and Canada's Coachwood of Calgary.

Renewed interest was also shown by American-based operators Texaco, Chevron, Occidental, Conoco, Arco, and a number of Japanese explorers and trading houses.

Brazil
pushing deeper

This South American nation is as synonymous to deepwater exploration and production as Houston is to humidity and traffic jams. Last year, state company Petrobras did much to build on that reputation.

As Petrobras continues to push the technological limits of deepwater production, especially in the Campos Basin, a number of new development schemes are being introduced. When 1988 drew to a close, the operator had completed designs for three floating production systems, and as examining an additional 11 different designs for future projects.

The three systems completed and planned for installation in Campos Basin would pump an additional 310,000 b/d of oil onto the mainland by 1993. Brazil's total production would then be raised to one million b/d.

According to the Petrobras research facility, CENPES, the first system to be installed will be a converted semisubmersible, which is expected to draw 60,000 b/d from the Marimba field. The system will be on location in 1990 in water depths of 1,312 to 1,968 ft.

That same year, the converted flotel Fortuna Ugland will be installed in 2,132 ft. of water in the Marlim field. That system will have a production capacity of 50,000 b/d.

Also on the Marlim field, Petrobras is looking at a concept for two GVA class semisubmersibles designed specifically for production. These would be built inside Brazil and placed in waters ranging from 2,624 to 3,280 ft. deep. Petrobras said this concept could be on stream by 1992 and would have the capacity to produce 100,000 b/d.

Long-range projects being studied by CENPES could push Brazil's oil production to 1.5 million b/d by 1997. Petrobras at year's end said it was looking at 11 designs, 10 of which were submitted by foreign companies.

Marimba
sets record

A major milestone was reached offshore Brazil last year when the Marimba field established a world's record for subsea completion and production.

The record was set upon completion of well 3-RJS-376D in 1,614 ft. of water. Production of 5,000 b/d of oil and 65 MMcfd of gas is being transferred to the Petrobras XV, which integrates the floating production system of the Pirauna and Miramba fields through a bundle of production lines, gas-lift injection and electric-hydraulic control.

Rigs pour
into West Africa

Drilling activity off West Africa had reached such strong levels late last year that operators were uncertain as to whether any qualified rigs would be available to handle the demand.

The level of activity in 1989 promises more of the same.

Sources cited the tight supply of qualified rigs, semisubmersibles in particular, as the reason Elf delayed the spudding of a well off Angola. The well was to have been drilled beginning last August.

Elf late last year also received bids from contractors for a three-well drilling program off Cameroon, which was to have started in late November.

Amoco said it would require a floater for a drilling program off the Congo, which was to have begun in early 1989. In November, 1988, Arco was sending out tenders for a one-well program scheduled for January of this year off Ghana.

In December, 1988, Elf said that while it would reduce its exploration activities, primarily off Angola, it would continue an aggressive development drilling program. The French operator at the same time said it would drill at least two wildcats off the Congo beginning in May or June, and no less than four development wells.

In Angola, Conoco was seeking a floater at year's end for two wells to be spudded after February 1989. Chevron, likewise, was scheduled to begin drilling a well off Cabinda in February.

China records
largest flow

Prospects offshore China was heightened tremendously in mid-1988 when the largest flow rate to date was recorded in the South China Sea. However, by the end of the year, the outlook dimmed to the point that the government said it would significantly reduce its exploration spending in 1989.

Elsewhere in Southeast Asia, India appears to be increasing its activities this year, while Indonesia and Malaysia are expected to either maintain 1988 levels of activity, or reduce them slightly.

Meanwhile, despite the new discovery off China by a Texaco-led group, at least one expert on the area urged operators to proceed with caution. He warned a number of geological problems could dampen the commercial outlook for the region.

ACT Operators in May said its discovery came from a wildcat in Contract Area 16/08. The Huizhou 26-1-1 well tested oil from seven separate zones at rates ranging from 741 to 5,925 b/d. The tests yielded a cumulative production rate of more than 26,000 b/d of oil.

According to Texaco, the discovery is located 170 miles southeast of Guangzhou in the Pearl River Basin and about 15 miles southwest of the Huizhou 21-1 oilfield.

At the end of the year, Texaco was still evaluating the test results and other data achieved from the Huizhou 26-1-1. The consortium, comprised of Texaco, Agip and Chevron, said additional drilling will begin through this year to determine the potential of the prospect.

While one analyst described the prospect as an "incredibly good well," an expert on China's offshore sector warned that any development in the area would be extremely costly.

"This well tested from seven zones and multiple zone completions are very complex," said Dr. Kim Woodard, president of Washington D.C.-based China Energy Ventures. "This is a history of discontinuity in the structures offshore China. There is a lot of fracturing in the fields there."

Traps questionable

He said the biggest question mark hanging over China is the type of trap structure found offshore. He said most of the traps in China are stratigraphic, and if this structure continues offshore, it could pose development problems.

"The problem you have with wildcats there is the discovery well looks good and then the step-outs tend to drop down. This is a nice hit, but I don't feel we're ready to go out and say we have another Persian Gulf," he said.

Many China watchers believe the Bohai Sea may someday replace the South China Sea as the center of the nation's offshore activity.

Much of that projection centered around Bohai Oil Corp's SZ36-1 discovery in Liaodong Bay, about 31 miles offshore Jinxi. Water depth in the region averages 98 ft. The Chinese have since indicated 300 to 400 wells may be drilled over the next decade to evaluate and develop the structure.

At the end of last year, eight wells had been drilled there, which led some Chinese bureaucrats to say the field could hold at least 879 million bbl of oil.

This year, Bohai Oil said it would like to begin drilling several wells on its Jz20-2 gas/condensate field in the northernmost fringes of Liaodong Bay in the Bohai. The exact number of wells was not released.

China estimated reserves in that structure to be 706 bcf of gas, 22 million bbl of condensate and 73 million bbl of oil. That field could be on stream by 1991.

Elsewhere, Indian state company ONGC outlined plans last year to substantially increase its offshore production, especially in the much explored Bombay High region.

The company said late last year that it has devised a plan to increase total Bombay High oil output from 146 million bbl to 197.91 million bbl during the next eight plan period (1990-1995). ONGC sources said the company plans to concentrate on newly discovered fields in the Bombay offshore region. To that end, ONGC invited bids in December for a turnkey 3D seismic program in the Krishna Godavari basin. The survey is to begin in April 1989 and continue for one year. It will be carried out between Kakinada and Surasanyanam covering between 2,000 to 3,000 line km.

Last December, ONGC recorded a new gas discovery off the East Coast near Kakinada. It marked the 10th well drilled in the Ravva structure in the Krishna-Godavari basin.

ONGC's Bombay region early this year was scheduled to secure two drillships to work from mid-September until May 1990, and then resume work from mid-September 1990 until May 1991.

Hibernia finally underway

All the attention on Canada's east coast will be directed this year to the mammoth Hibernia project, which finally got underway last year - nine years after its discovery.

The development program will involve a concrete gravity-base structure, two articulated loading platforms, and three ice-strenghtended shuttle tankers. The earliest possible start-up date for production is 1993, according to operator Mobil. An agreement to finally proceed with the project was signed late last summer by the Hibernia consortium and the Canadian provincial and federal governments.

The field is located 315 km east of St. John's in 80 meters of water. Total development cost is about $7 billion.

Betting oil prices will increase and stabilize in six years, the Canadian government, in an unprecedented move, agreed to contribute $830 million in grants and up to $1.3 billion in loan guarantees to get the long-delayed platform in the water. Most of the contracts for the giant structure are to have been awarded by mid-1989.

In agreeing to help fund the project, the Canadian government is predicting healthy oil prices by the mid-1990's when first oil is expected to flow from the estimated 500-million-bbl field. A price of $20/bbl is required for the field to be profitable.

Offshore production systems planned, installed in '88

Operator Platform name	Fabrication yard/ Location	Type or Function (No. legs/main piles/skirt piles)	Water Depth/No. Well Slots	Install. Date
U.S. GULF OF MEXICO (45)				
Amoco				
	Jacket (4/-/-)	NA	6/	Design
	South Timbalier 221		155	NA
A	Jacket (4/-/-)	NA	12/	Planned
	South Timbalier 245		NA	NA
Anadarko Petroleum				
	Jacket (4/-/-)	NA	NA	Pending
	Vermillion 78'A'		NA	NA
Arco				
	Well Protector (1)	Apex	1	Construction
	Main Pass 128		95	NA
	Jacket (4/-/-)	Raymond-	6/	Construction
	Main Pass 129	McDermott-Zachry	74	NA
Chevron				
	Jacket (4/-/-)	NA	0/	Bidding
	Main Pass 69 M		14Feb. 89	
	Jacket (4/-/-)	NA	0/	Design
	Ship Shoal 108 D		3589	
	Jacket (satellite) (4/-/-)	NA	6/	Design
	Ship Shoal 100 DA		2289	
	Jacket (caisson) (-/-/-)	NA	3/	Design
	West Cameron 555		185 89	
	Jacket (caisson) (-/-/-)	NA	3/	Design
	East Cameron 298		180 89	
AA	Jacket (satellite) (4/-/-)	NA	6/	Design
	Main Pass 299		210	NA
A	Jacket (satellite) (4/-/-)	NA	4/	Design
	Matagorda Island 784		183	NA
	Jacket (3/3/1)	NA	1/	Fabrication

Offshore production systems planned, installed in '88 — continued

Operator / Platform name	Fabrication yard / Location	Type or Function (No. legs/main piles/skirt piles)	Water Depth/No. Well Slots	Install. Date
	South Timbalier 35 #6		50	NA
	Jacket (caisson)	NA	3/	Fabrication
	South Marsh Island 48		100	Nov. 88
Conoco				
	Tension Leg Well	NA	24/	Design
	Platform, Green		1,850	89
	Green Canyon 184			
	Jacket (4/-/-)	NA	NA	Design
	Canyon 52		615	89
A	Jacket (16/-/-)	NA	24/	Planned
	Green Canyon 184		NA	NA
Corpus Christi Oil & Gas				
	Jacket (-/-/-)	NA	NA	Planned
	Main Pass 98		NA	NA
Exxon				
Alabaster	Jacket (-/-/-)	NA	NA	Design
	Miss. Canyon 397		NA	NA
	Well Protector (4/-/-)	McDermott	6/	Fabrication
	NA	Morgan City, LA	51	NA
Forest Oil				
EI 325	Jacket (-/-/4)	NA	12/	Design
	NA		25589	
EI 326	Jacket (-/-/4)	NA	6/	Design
	NA		25589	
EI 366	Jacket (-/-/4)	NA	12/	Design
	NA		34589	
SS 277	Jacket (-/-/4)	NA	12/	Design
	NA		21489	
Hall-Houston				
A	Jacket (4/-/-)	NA	8/	Construction
	Main Pass 165		NA	NA
Kerr-McGee				
	Well Protector (4/-/-)	Apex Offshore	3	In progress
Maxus				
	Jacket (4/ /)	NA	0	Planned
	Main Pass 31		15	2nd Qtr 88
Mobil (Moepsi)				
	Well protector (-/3/-)	NA	2/	Design
	West Cameron 533-B		175Apr. 89	
	Jacket (-/4/-)	NA	3/	Design
	Vermilion 271-A		165Jun. 89	
	Jacket (-/4/-)	Gulf Island	6/	Fabrication
	Vermilion 326-A		219Nov. 88	
	Jacket (-/4/-)	NA	NA	Design
	West Cameron 421-A		98Feb. 89	
	Jacket (-/4/-)	McDermott	9/	Fabrication
	S. Marsh Island 205-A		442Nov. 89	
	Vermilion 326-A		217	Dec. 88
	Jacket (8/-/-)	McDermott	8/	Installed
	Mobile Bay 95-E		23	Pending
Pelto Oil Co.				
	Jacket (4/-/-)	NA	4/	Under study
	Main Pass 95		60	NA
Samedan				
	Jacket (4/-/-)	NA	6/	Fabrication
	Eugene Island 248		160	NA
3 Units	Jacket (4/-/-)	NA	9/	Fabrication
	Brazos A-52, A-53		NA	NA
Shell				
C SS 274	Jacket (8/0/-)	Avondale	24/NA	
	NA		218NA	
A MP 252	Jacket (4/4/-)	McDermott	24/NA	
	NA		272NA	
JA PN 969	Jacket (4/0/-)	McDermott	9/NA	
	NA		140NA	
1 MB 113	Well Protector (4/0/-)	NA	3/NA	
	NA		23NA	
2 MB 132	Well Protector (4/0/-)	NA	3/NA	
	NA		39NA	
Tenneco Oil				
HI 128 #1	Jacket (caisson) (1/-/-)	NA	1/	Bid phase
	NA		50NA	
WC 168A	Jacket (4/-/-)	NA	1/	Bid phase
	NA		46NA	
Walter Oil & Gas				
	Jacket (-/2/-)	McDermott	1/	Fabrication
	Vermillion 95 A	27		4th Qtr. 88
U.S WEST COAST (13)				
Arco				
Coal Point	Jacket (8/-/-)	NA	NA	Design
	Santa Barbara, CA		218	NA
	Jacket (12/-/-)	NA	NA	Planned
	Complex Heron		220	NA
	Santa Barbara chnl			
	Jacket (12/-/-)	NA	0/	Planned
	Complex Heron		220	NA
	Santa Barbara Chnl			
	Two Units (8/-/-)	NA	NA	Planned
	Complex Hawk		220	NA
	Santa Barbara Chnl			
Chevron				
Hermosa	Jacket (8/-/-)	Hitachi	48	In progress
	Point Arguello field	Ariake	NA	NA
Rocky Point	Jacket (-/-/-)	NA	NA	Under study
Unit	Off Point Arguello		300	NA
Hacienda	Jacket (-/-/-)	NA	48	Planned
	Point Arguello field		308	Summer 89
Exxon				
	Jacket (8/-/-)	NA	60/	Design
	Point Pedernales		281	NA
	OCS-P-0438, 0440			
	Jacket (8/-/-)	NA	60/	Design
	Santa Ynez Unit		1,200	NA
	OCS-P-0190			
	Jacket (8/-/-)	NA	60/	Design
	Santa Ynez Unit		1,075	NA
Independence	Jacket (8/-/-)	Wakamatu	NA	Suspended
	Santa Maria basin	Kita-Kyushu	NA	NA
		Japan		
Shell				
Hercules	Jacket (8/-/-)	NA	NA	Awaiting permit
	Molino field		237	NA
	Santa Barbara			
	Channel			
NORTH SEA (40)				
Amoco				
HOD	Jacket (4/-/-)	NA	8/	Under study
	NA		23090	
Arco				
	Jacket (4/4/0)		9/Design	NA
	West Cameron 248		80NA	
Britoil				
Amethyst	Jacket (quarters)			
	(8/-/-)	NA	0	Under study
	UK 47/14		90	NA
Ettrick	Floating Production	NA	NA	Under study
	UK 20/2		360	NA
British Petroleum				
Gyda	Jacket (6/-/-)	Not awarded	30/	Planned
	Norway		216	2nd Qtr 89
Bruce	Jacket (-/-/-)	NA	NA	Under study
	UK		394	NA
Miller	Jacket (8/-/-)	NA	NA	Under study
	UK		354	NA
Elf Petroland				
L7-H	Jacket (4/4/0)	ETPM/Lummus	6/	Fabrication
	NA		100Jul. 89	
HGB Ltd.				
DP6	Jacket (4/12/-)	Contract	14/	Design
	UK Blk 110/2a	not awarded	85	Spring 89
DP8	Jacket (4/12/-)	Contract	14/	Design
	UK Blk 110/2a	not awarded	85	Spring 89
Hamilton Bros.				
	Jacket (-/-/-)	NA	NA	Under study
	UK Bruce field		NA	90

Operator / Platform name	Fabrication yard/ Location	Type or Function (No. legs/main piles/skirt piles)	Water Depth/No. Well Slots	Install. Date
Hydrocarbons GB Ltd.				
DP6	Jacket (4/-/-)	NA	14/	Fabrication
Morecambe Bay		85Spring 89		
DP8	Jacket (4/-/-)	NA	14/	Fabrication
Morecambe Bay		85Spring 89		
Maersk (A. P. Møller)				
	Wellhead (-/-/-)	NA	NA	Planned
Danish Roar field			NA	89
	Jacket (processing) (-/-/-)	NA	NA	Planned
Danish Roar field			NA	89
Mobil Producing Netherlands				
	Jacket (-/-/-)	NA	4/	Under study
The Netherlands, P/12-SW		9091		
	Jacket (-/-/-)	NA	4/	Under study
The Netherlands P/12-C		9091		
NAM				
L	Jacket (satellite) (-/-/-)	NA	NA	Planned
Dutch Ameland field			26	89
FB	Jacket (satellite) (-/-/-)	NA	4	Planned
Dutch K/8			102	89
FA	Jacket (-/-/-)	NA	8	Planned
Dutch L/2			118	90
Norsk Hydro				
Oseberg B	Jacket (-/-/-)	NA	NA	Planned
Norway 30/6			NA	NA
Norske Shell				
Four Units	Jacket (Qtrs) (-/-/-)	NA	NA	Under study
Troll field		NA	900	92
Norway 31/2, 3, 5, 6				
	Jacket (piled steel) (-/-/-)Condeep			
(concrete) or T-300				
Norwegian Contractors				
Gullfaks	NA (-/-/-)	Statoil	52/NA	
Vats, Norway		716		May 89
Sleipner	NA (-/-/-)	Statoil	NA	Construction
Stavanger, Norway		346Jul. 89		
Troll	NA (-/-/-)	Norske Shell	NA	Under study
NA		NA	NA	
Draugen	NA (-/-/-)	Norske Shell	NA	Under study
NA		NA	NA	
Occidental Int. Oil				
	NA (-/-/-)	NA	NA	Under study
Piper field		474NA		
	NA (-/-/-)	NA	NA	Under study
Chanter field		470NA		
Wintershall Noordzee				
K10-V	Unknown (-/-/-)	NA	NA	Under study
NA		1021991-92		
Petroland				
L7N	Jacket (satellite) (-/-/-)	Heerema Haven	NA	Construction
Netherlands, bedrijf B.V.		NA		Autumn 88
Phillips				
	Jacket (-/-/-)	NA	NA	Under study
Toni-Thelma field, UK				
16/17		Tiffany	430NA	
Shell Expro (Shell/Esso)				
Tern	Jacket (-/-/-)	RGC	30	Design
UK Tern field		Methil	550	89
Kittiwake	Jacket (-/-/-)	NA	16/	Under study
UK Kittiwake field			279	91
Draugen	Jacket (Qtrs) (-/-/-)	Not awarded	12/	Under study
Draugen field (Nor)		Concrete	853	91
Statoil				
Gullfaks C	Jacket (Qtrs) (-/-/-)	Norwegian Cont.	52/	Construction
Norway 34/10,		Stavanger	716	May 89
Veslefrikk	Jacket (4/-/-)		20/	Engineering
Veslefrikk field			566	Spring 89
Sleipner A	Jacket (integrated) (-/-/-)NA		NA	Planned
Sleipner field			346	Summer 92
Zeepipe Riser	Riser on Pipeline	NA	NA	Planned
NA			NA	92-93
CANADA (1)				
Mobil				
Hibernia	Jacket (-/-/-)	Unknown, Argentia	NA	Design
Off East Coast		and Adam's	NA	91

SOUTHEAST ASIA/OCEANIA (46)

Operator / Platform name	Fabrication yard/ Location	Type or Function (No. legs/main piles/skirt piles)	Water Depth/No. Well Slots	Install. Date
BHP Petroleum				
Challis	Jacket (floater) (-/-/-)	NA	NA	Under study
Timor Sea			360	NA
Bohai Bay Oil				
	Jacket (-/-/-)	NA	0	Planned
Bohai Bay, China			81	NA
Bond				
	Jacket (satellite) (-/-/-)	NA	NA	Planned
Heera field, Australia		NA	NA	
Brunei Shell Petroleum				
FADP-7	Jacket (4/4/-)	NA	4/	Design
Fairley field		200May 90		
CPWJ-AX	Jacket (4/4/-)	NA	4/	Planned
Champion field		130Jun. 91		
CPWJ-AC	Jacket (4/4/-)	NA	4/	Planned
Champion field		110Jan. 92		
CPDP-AU	Jacket (6/6/-)	NA	12/	Planned
Champion field		105Oct. 91		
CPDP-AS	Jacket (6/6/-)	NA	12/	Planned
Champion field		85Oct. 93		
Carigali/Petronas				
Dulang A	Jacket (4/-/-)	NA	24/	Design
South China Sea, West		Malaysi 249	NA	
Dulang B	Jacket (8/-/-)	NA	24/	Design
South China Sea			249	NA
Dulang C	Jacket (4/-/-)	NA	24/	Design
South China Sea			249	NA
Dulang D	Jacket (4/-/-)	NA	28/	Design
South China Sea			249	NA
Bayan DP-B	Jacket (-/-/-)	Sime Sembawang	15/	Fabrication
South China Sea			104Dec. 88	
Bokor-A	Jacket (gas compression) (-/-/-)	A.Y Engineering	-/	Fabrication
South China Sea			220May 89	
D-35	NA (-/-/-)	NA	NA	Design
South China Sea		NA	NA	
St. Joseph JT-BJacket (-/-/-)		Sime Sembawang	9/	Fabrication
South China Sea			113Dec. 88	
St. Joseph JT-BJacket (-/-/-)		NA	9/	Design
South China Sea			92NA	
St. Joseph JT-CJacket (-/-/-)		NA	7/	Design
South China Sea			128Design	
Erb West DP-B Jacket (-/-/-)		Bina	15/	Fabrication
South China Sea			210Dec. 88	
St. Joseph G-A Jacket (gas injection)		NA	NA	Design
NA		NA	NA	
Seligi-G	Jacket (-/-/-)	NA	NA	Design
South China Sea		NA	NA	
Chengbai Oil				
BZ-32	Jacket (4/-/-)	Nippon Steel	24/	Planned
Bohai Gulf, China		Japan	67	NA
Esso Australia				
Seahorse,	Jacket (-/-/-)	NA	NA	Under study
Whiptail,	Gippsland basin		130-175	NA
Whiting,	Victoria,			
Dolphin,				
Perch,				
Tarwhine				
Hudbay				
	Two Wellhead (-/-/-)	NA	NA	Planned
Mengkapan field, Indonesia			NA	NA
Japan National Oil				
Iwafune-oki		Jacket (8/-/-)	NA	20/Planned
Sea of Japan			115NA	
	Jacket (gravity caisson) (-/-/-)	NA	-/	Under study
Chaivo field, U.S.S.R.			66-98NA	
	Jacket (gravity caisson) (-/-/-)	NA	-/	Under study
Odopt field, U.S.S.R.			98NA	
BZ-28-1	Jacket (floating) (-/-/-)	China	-/	Construction
Bohai Gulf, China			6588	

Offshore production systems planned, installed in '88 — continued

Operator Platform name	Fabrication yard/ Location	Type or Function (No. legs/main piles/skirt piles)	Water Depth/No. Well Slots	Install. Date
BZ34-2/4F Construction	Jacket (floating) (-/-/-)		China -/	
	Bohai Gulf, China 65		89	
Lufeng 13-1	Jacket (floating) (-/-/) NA		-/	Under study
	S. China Sea 460		NA	
Japex Offshore				
JCODC (Japan China Oil Development Co.)				
BZ28-1	Jacket (satellite) (-/-/-) NA		NA	Engineering
	Bohai Gulf off China		NANA	
BZ28-1	Jacket (floating) (-/-/-) NA		NA	Planned
	Bohai Gulf off China		6588	
BZ34-2/4	Jacket (floating) (-/-/-) NA		NA	Under study
	Bohai Gulf off China		65	88
Marathon Petroleum				
Talisman	Jacket (floating) (-/-/-) NA		NA	Under study
	NW shelf, W. Aust. NA		NA	NA
Nippon Steel				
P6A	Jacket (4/4/0)		12/	Under construction
	Amara yard, Thailand	240Apr. 89		
P6B	Jacket (4/4/0)		12/	Under construction
	Amara yard, Thailand	240Apr. 89		
P7A	Jacket (4/4/0)	12/		Under construction
	Amara yard, Thailand	240Oct. 89		
P7B	Jacket (4/4/0)		12/	Under construction
	Amara yard, Thailand	240Oct. 89		
Phillips				
	Jacket (8/8/0) NA		24	Design
	Xijiang 24/3, South China Sea		320	1990
POC				
	Jacket (-/-/-) NA		8	Planned
	3DA structure, Gulf		NA	NA
SSB/Pctronas				
TEJT-T	Jacket Promet (-/-/-)	NA		Fabrication
	S. China Sea	Deck: Sabah yard	90	Nov. 88
Sakhalin Oil Development				
	Jacket (4/-/-) NA		40-60/	Planning
	Odoptu field	100		Middle 90s
	NA (-/-/-) NA		40-56/	Planning
	Chaivo field	80		Middle 90s
Shell/BP/Todd				
	Jacket (4/20/-) NA		10/	Design
	Maui field, New Zealand	39692		
Total Chine				
	Jacket (-/-/-) Nippon Steel		NA	Fabrication
	Weizhou field, Japan	NA		NA
Unocal Thailand				
PWF	Wellhead (4/-/-) Wakamatu		12/	Construction
	Gulf of Siam Kita-Kyushu		213	NA
PWG	Wellhead (4/-/-) Wakamatu		12/	Construction
	Gulf of Siam Kita-Kyushu		213	NA
PWH	Wellhead (4/-/-) Wakamatu		12/	Construction
	Gulf of Siam Kita-Kyushu		213	NA
Vietsovpetro				
	Jacket (-/-/-) Baku, USSR		16	Fabrication
	White Tiger field		NA	NA
	Jacket (serv./supply) (-/-/-) NA		NA	Planned
	White Tiger field		NA	NA
Western Australia Petroleum				
Saladin	Jacket (-/-/-) NA		NA	Under study
	Western Australia NA		NA	NA
Western Mining Corp.				
North Herald	Jacket (-/1/-) McDermott		56/	Under study
	Perth, Australia	NA	NA	
Woodside Offshore				
Goodwyn	NA NA		24	Under study
	NW shelf Australia	420		91

LATIN AMERICA/CARIBBEAN (33)

Operator Platform name	Fabrication yard/ Location	Type or Function (No. legs/main piles/skirt piles)	Water Depth/No. Well Slots	Install. Date
ENAP				
SK-4	Jacket (4/-/-/)	Laredo	9/	Planned
	PTA. Arenas		112Feb. 89	
AN-1	Jacket (4/-/-/)	Laredo	18/	Planned
	PTA. Arenas 197	Laredo	197	Nov. 88
CAN-1	Jacket (4/-/-/)	Laredo	12/	Planned
	PTA. Arenas 27			Sept. 89
CAN-2	Jacket (4/-/-/)	Laredo	9/	Planned
	PTA. Arenas		27Oct. 89	
Enap-Magallanes				
Skua 3	Jacket (4/4/-)	Bahia Laredo	9	Planned
	NA		120 NA	
Skua 4	Jacket (4/4/-)	Bahia Laredo	9	Planned
	NA		114 NA	
Anguila 1	Jacket (4/4/-)	Bahia Laredo	12	Planned
	NA	NA	200	NA
Petrobras				
Pargo-1A	Jacket (utilities) (4/4/8) NA		21/	Construction
Brazil		331Dec. 88		
Pargo-1B	Jacket (Utilities) (4/4/8) NA		NA	Construction
Brazil		331Dec. 88		
Carapeba-1	Jacket (4/4/8)	NA	15/	Construction
Brazil		283Dec. 88		
Carapeba-2	Jacket (4/4/8)	NA	21/	Construction
Brazil		286Dec. 88		
Vermelho-1	Jacket (4/4/8)	NA	21/	Construction
Brazil		262Mar. 89		
Vermelho-2	Jacket (4/4/8)	NA	21/	Construction
Brazil		262Jan. 89		
Vermelho-3	Jacket (4/4/8)	NA	21/	Construction
Brazil		267Mar. 89		
Atum-3	Jacket (4/4/8)	NA	12/	Construction
Brazil		137Jan. 89		
Carapeba-3	Jackup (-/-/-)	NA	18/	Under study
Brazil		300Jun. 91		
Enchova-Oeste	Jacket (8/-/32)	NA	27/	Under study
Brazil		387May 92		
Enchova	Jacket (-/-/-)	NA	18/	Under study
Brazil		394Jan. 92		
Pescada-1	Jacket (Utilities) (4/4/8) NA		NA/	Under study
Brazil		70Jan. 92		
Pescada-2,3, 4 and 5	Jacket (4/4/8) Brazil	NA	24/ 70Jan. 92	Under study
Marlim	Semi (-/-/-)	NA	NA	10/Under study
Brazil		1,970Jan. 91		
Marimba	Semi (-/-/-)	NA	10/	Under study
Brazil		1,500Jan. 91		
Merluza	Jacket (4/4/8)	NA	6/	Construction
Brazil		557Jan. 92		
Tubarao	Semi	NA/	NA	Planned
Brazil		560Jan. 92		
Albacora	Semi	NA	NA	18/Planned
Brazil		1,000Jun. 92		
Camorim-10	Jacket (4/4/4)	NA	6/	Construction
Brazil		68Apr. 89		
Enchova-Oeste	Jacket (8/8/12)	NA	27/	Under study
Brazil		387Jun. 90		
Camorim-11	Jacket (4/4/4)		6/	Under study
Brazil		85May 89		
Camorim-12	Jacket (4/4/4)		6/	Planned
Brazil		85NA		
Pescada-1	Jacket (8/8/-)	NA	NA	Planned NA
Brazil		70NA		
Pescada 2,3,4 5 and 6	Jacket (4/4/-) Brazil	NA	4/ 50-80NA	Planned
Petrobras XVIII	Semi GVA 4500 (-/-/-) NA		NA/	Under study
Brazil		5,250Feb. 92		
Petrobras XIX	Semi GVA 4500 (-/-/-) NA		NA	Under study
Brazil		5,250Jan. 92		
TLP conceptual	Jacket (tension-leg) (-/-/-)	NA	19/	Planned
NA project	Brazil		3,280NA	
Merluza	Jacket (4/-/8)	NA	7/	Under construction
Brazil	360		Jul. 89	

Offshore production systems planned, installed in '88 — continued

Operator / Platform name	Fabrication yard/ Location	Type or Function (No. legs/main piles/skirt piles)	Water Depth/No. Well Slots	Install. Date
Carapeba-3	Jacket (4/-/4)	NA	18/	Planned
Brazil	285NA			
Shell				
	Jacket (-/-/-)	NA	10-11/	Planned
	Strait of Magellan		230	NA
Total				
	Jacket (-/-/-)	NA	9/	Planned
	Hidra field Argentina		90	NA
Total Australia				
Hidra North	Jacket (4/-/-)	NA	12/	Call for bid
	Tierra del Fuego	NA	75	End 88
Hidra Centre	Jacket (4/-/-)	NA	12/	Call for bids
	Argentina	NA	105	End 88
MIDDLE EAST (16)				
Total CFP (Amapetco)				
Shukheir Ma-rine	Jacket (4/-/-)	NA	3/	Planned
	Egypt		100Mid 89	
Hidra Center	Jacket (water injec.) (4/-/-)	NA	12/	Construction
	Buenos Aires		110Feb. 89	
Hidra North	Jacket (water injec.) (4/-/-)	NA	12/	Construction
	Buenos Aires		85Feb. 89	
ARA	Jacket (6/-/-)	NA	12/	Under study
	NA		50Mid 90	
North Aegean Petroleum				
	Jacket (-/-/4)	NA	12/	Under study
	NA		12090	
	Jacket (-/-/4)	NA	NA	Under study
	NA	NA	NA	
Mavanga A	Jacket (4/-/-)	NA	12/	Design
	Angola		110Aug. 90	
Mavanga B	Well Protector (4/-/-)	NA	6/	Design
	Angola		110Aug 90	
Wepco				
P3	Jacket (4/-/-)	NA	2/	Design
	Abu Qir Bay		59	Sept. 89
NAF-I	Jacket (-/-/-)	NA	NA	Planned
	Abu Qir Bay		115Nov. 90	
Udeco				
	NA	NPCC	NA	NA
	Satah Bridge	Sadiyat	NA	NA
	NA (4)	NPCC	NA	NA
	Satah	Sadiyat	NA	NA
PMP	Jacket (6/-/-)	NPCC	NA	NA
	Satah	Sadiyat	NA	NA
2FP	Jacket (4/-/-)	NPCC	NA	NA
	Satah	Sadiyat	NA	NA
2FP	TPWER	NPCC	NA	NA
	Satah	Sadiyat	NA	NA
WHP1	Jacket (4/-/-)	NPCC	6/	NA
	Satah	Sadiyat	45	NA
WHP2	Jacket (4/-/-)	NPCC	9/	NA
	Satah	Sadiyat	36	NA
WHP3	Jacket (4/-/-)	NPCC	6/	NA
	Satah	Sadiyat	48	NA
WHP4	Jacket (water injec.) (4/-/-)	NPCC	6/	NA
	Satah	Sadiyat	31	NA
WHP5	Jacket (water injec.) (4/-/-)	NPCC	6/	NA
	Satah	Sadiyat	39	NA
WHP6	Jacket (4/-/-)	NPCC	9/	NA
	Satah	Sadiyat	14	NA
WHP7	Jacket (water injec.) (4/-/-)	NPCC	6/	NA
	Satah	Sadiyat	44	NA
WHP8	Jacket (4/-/-)	NPCC	9/	NA
	Satah	Sadiyat	46	NA
WHP9	Jacket (gas prod./injec.) (4/-/-)	NPCC	6/NA	
	Satah	Sadiyat	45	NA
PMP	Jacket (6/-/-)	NPCC	NA	NA
	Satah	Sadiyat	44	NA
WEST AFRICA (22)				
AGIP-Recherches				
DP-6	Jacket (4/-/-)	NA	12	Under definition
	Loango field	NA	NA	Mid 89
Chevron Overseas Pet.				
Elf Congo				
TAP	Jacket (4/-/-)	NA	NA	Planned
	Tchibouela		280	Autumn 88
Elf Angola				
BUF F1	Jacket (4/-/-)	Ambrise	16/	Construction
	Bufalo field		280	Spring 88
IPS F1	Jacket (4/-/-)	NA	16/	Planned
	Impala Sud Est		195	89
PAC F3	Jacket (4/-/-)	NA	8/	Under study
	Pacassa field		280	89
Chevron Overseas Pet.				
Numbi A	Jackets (4/4/-)	Petromar	9/	Construction
	Numbi field, Angola		135Dec. 88	
Numbi E	Jacket (4/4/-)	Petromar	9/	Pending
	Numbi field, Angola		13089	
Numbi F & G	Jacket (4/-/-)	NA	9/	Design
	Numbi field, Angola		140Dec. 88	
ALP-AUX	Jacket (gas sweet.) (4/4/-)	NA	-/	Under study
	NA		18590	
GS-N	Jacket (8/-/-)	Petromar	-/	Construction
	Numbi field, Angola		140Jul. 90	
	Well Jackets (4)	NA	9	Awaiting design
	Numbi field, Angola		140	Dec. 89
Alp	Jacket (8/8/-)	McDermott	-/	Construction
	Takula field, Angola		185	Dec. 88
Takula Alp	Jacket (gas comp.) (8/-/-)	Petromar	-	/Construction
	Takula field, Angola		185Apr. 89	
GS-P	Jacket (8/-/-)	Sharjah, UAE	-/	Construction
	Wamba field, Angola		210Apr. 90	
Wamba B	Jacket (4/-/-)	Petromar	9	Construction
	Wamba field, Angola		210	Jan. 89
Takula Wip	Jacket (8/8/-)	Petromar	-/	Construction
	Takula field, Angola		185Jun. 90	
Takula Tap	Jacket (8/-/-)	Petromar	-/	Design
	Takula field, Angola		185Mar. 90	
	Jacket (module) (8/8/-)	NA	-/	Design
	Takula field, Angola	NA	90	
IMP F1	Jacket (4/-/-)	NA	16/	Being designed
	Impala field, Angola		216	Jun. 89
Pac F3	Jacket (4/-/-)	NA	18/	Being designed
	Pacassa field, Angola		286	Mar. 89
Motoba 5 Negotiating	Jacket (3/3/0)	NA	3/	
	Motoba field, Zaire		3589	
Motoba 6	Jacket (3/3/0)	NA	3/	Negotiating
	Motoba field, Zaire		3589	
Motoba 7	Jacket (3/3/0)	NA	3/	Negotiating
	Motoba field, Zaire		3589	
Motoba Flow Sta.	Jacket (gathering sta.) (4/4/0)		0/	Planned
	Motoba field, Zaire		35NA	
Libwa Flow Sta.	Jacket (gathering sta.) (4/4/0)		0/	Planned
	Libwa field, Zaire		20NA	
Texaco				
Mavanga A	Jacket (4/-/-)	NA	12/	Design
	Angola		124Oct. 90	
Mavanga B	Jacket (4/-/-)	NA	6/	Design
	Angola		118Oct. 90	
Chuchupa B	Jacket (4/-/-)	NA	-/	Design
NA			115Jun. 90	
	Jacket (4/-/-)	NA	12/	Design
NA			115Jun. 90	

Source: Annual production system survey by Offshore, Incorporating The Oilman.

Gas production (billions cu ft)

	1988
WESTERN HEMISPHERE	
Argentina	757.4
Barbados	0.7
Bolivia	106.5
Brazil	193.7
Canada	3,990.2
Chile	56.3
Colombia	173.5
Ecuador	3.6
Guatemala	0.3
Mexico	1,308.5
Peru	46.1
Trinidad	156.3
United States	17,504.0
Venezuela	722.9
Total	**25,020.0**
EUROPE	
Austria	44.4
Denmark	84.6
France	113.2
Germany, Fed. Rep.	551.7
Greece	3.0
Italy	703.7
Netherlands	2,412.8
Norway	1,002.9
Spain	33.4
Turkey	3.5
United Kingdom	1,640.5
Total	**6,593.7**
MIDDLE EAST	
Bahrain	184.9
Iran	620.0
Iraq	110.4
Israel	1.5
Kuwait	174.6
Oman	73.2
Qatar	204.2
Saudi Arabia	899.3
Syria	22.2
United Arab Emirates	600.0
Total	**2,890.3**
ASIA-PACIFIC	
Afghanistan	101.4
Australia	544.4
Bangladesh	152.9
Brunei	306.1
Burma	32.0
India	312.3
Indonesia	1,223.7
Japan	71.5
Malaysia	464.8
New Zealand	189.0
Pakistan	441.7
Taiwan, China	37.9
Thailand	194.8
Total	**4,072.5**
AFRICA	
Algeria	1,542.4
Angola	7.8
Egypt	197.3
Gabon	2.4
Libya	198.0
Morocco	2.3
Nigeria	126.0
Tunisia	13.1
Total	**2,089.3**
COMMUNIST	
China	715.3
Romania	1,318.8
USSR	27,192.3
Other	1,087.3
Total	**30,313.7**
WORLD TOTAL	**70,979.5**

World well completions—1987*

	Total wells	Total wildcats	Total producers	Total wildcat producers
United States	35,917	7,003	23,457	1,583
Canada	7,078	2,369	5,293	1,318
Australia	282	167	—	—
Africa, Central South				
Angola	66	24	—	10
Burundi	2	2	—	0
Cameroon	6	0	—	0
Congo	35	10	—	4
Gabon	31	16	—	9
Ghana	11	11	—	0
Malagasy Rep.	2	2	—	1
Kenya	2	2	—	0
Namibia	1	1	—	0
Nigeria	72	44	—	23
Somali Rep.	1	1	—	0
South Africa	17	17	—	2
Tanzania	2	2	—	0
Zaire	4	1	—	0
Zambia	2	2	—	0
Europe				
Austria	34	24	21	11
Denmark	1	1	12	0
W. Germany	38	24	7	7
France	93	38	60	14
Greece	3	3	0	0
Ireland	4	4	1	1
Italy	181	96	99	34
Netherlands	33	11	24	2
North Sea				
U.K.	240	111	145	42
Danish	7	6	3	2
Dutch	34	24	22	12
Norwegian	69	33	51	21
Spain	13	12	3	2
Sweden	8	6	4	2
Switzerland	1	1	0	0
U.K.	33	31	9	8
Yugoslavia	128	3	—	2
Far East				
Bangladesh	2	2	—	—
Brunei	20	12	—	—
Burma	103	3	—	—
India	316	37	—	—
Indonesia	586	75	—	—
Japan	13	12	—	—
Malaysia	80	13	—	—
Pakistan	39	9	—	—
People's Rep. of China	5,721	21	—	—
Philippines	3	2	—	—
Rep. of China	11	9	—	—
Thailand	51	23	—	—
Israel	4	2	2	0
Bahrain	24	0	24	0
Iraq	89	10	—	—
Kuwait	23	2	—	—
Oman	—	36	—	21
Syria	—	10	—	8
Turkey	—	31	—	2
New Zealand	23	—	7	—
North Africa				
Algeria	45	25	—	1
Egypt	130	55	—	21
Libya	41	25	20	10
Morocco	7	7	—	1
Sudan	1	1	—	0
Tunisia	25	16	16	3
Latin America				
Argentina	416	45	—	10
Bolivia	25	10	16	2
Brazil	950	176	704	57
Chile	104	35	59	9
Colombia	162	69	108	27
Costa Rica	3	3	0	0
Ecuador	30	9	23	3
Guatemala	1	0	1	0
Mexico	103	27	68	8
Paraguay	2	2	0	0
Peru	152	13	138	8
Suriname	35	9	23	3
Trinidad-Tobago	142	8	79	4
Uruguay	1	1	1	1
Venezuela	177	21	—	—

*AAPG data

COALBED WELL in southwestern Colorado in the San Juan mountains is equipped with a pumping unit that removes water from coal cleats, reducing reservoir pressure and allowing desorbed gas to flow to the wellbore. Photo courtesy of Amoco Production Co.

Tax credit aids Upper Cretaceous Fruitland coalbed gas drilling in San Juan basin

OPERATORS IN THE SAN JUAN BASIN ARE WELL along on a program to drill perhaps more than one thousand 1,500-4,000 ft wells by the end of 1990 to tap gas in Upper Cretaceous Fruitland coalbeds.

A federal tax credit is the main impetus for the drilling boom.

Records indicate that a few hundred wells have been drilled to tap Fruitland coal gas, most since 1984.

Fruitland contains the largest coal resource, and other coals in the basin are located in the Dakota sandstone and Mesaverde Group.

The operators face several hurdles, including poor roads in a remote locale at elevations of 6,000-7,000 ft, brutal winters, archaeological surveys on federal lands, and operating restrictions to safeguard wildlife.

Most of the drilling is on or adjacent to leases held by production from other formations.

Wells started by the end of 1990 qualify through 2001 for a complicated tax credit that for some operators could represent a significant percentage of 1988 gas prices.

The credit is not available to the Southern Ute Indian Tribe, which owns leased and unleased oil and gas rights to about 1,200 miles in the northern San Juan basin in La Plata, Archuleta, and Montezuma counties, Colo.

The Southern Utes own 139,000 acres of leased or producing oil and gas properties and 192,000 unleased acres.

The New Mexico Oil Conservation Division is expected to issue special rules and operating procedures soon governing coalbed methane production in San Juan and Rio Arriba counties.

Hearings in July 1988 treated pool creation, spacing, locations, density, horizontal wellbores, deviated drilling procedures, venting, flaring, and testing.

**Early blowouts
identified coalbeds**

Practically every early well blew out while drilling the Fruitland coals, but no one saw long term prospect in coalbed methane.

Wells blew so hard and produced so much water it was felt the reservoirs were watered out and that gas production could not be long lived, say Thomas A. Dugan and Barbara L. Williams, Dugan Production Corp., Farmington, N.M.

Amoco's 1977 discovery of Cedar Hills field in New Mexico, the first designated coalbed gas field in the western U.S., initiated industry thought, if not action, regarding dewatering the Fruitland coalbeds for degasification of the coal, says Clarence Harr, a Grand Junction, Colo., consulting geologist to the Southern Utes.

Pre-1964 Colorado Oil and Gas Conservation Commission Fruitland-Pictured Cliffs production records indicate high early flow rates with rapid declines and short producing lives attributable to depletion of free, previously desorbed gas and subsequent water encroachment, Harr said.

Full development of coalbed methane on the reservation may require about 1,400 wells, including about 200 existing wells.

Given the presence of economic and regulatory conditions subject to change, it could take 10-20 years to achieve full production from such a large reservoir.

Sizing up an old
coalbed gas well

One of the oldest continuously producing coal seam wells in the U.S. is the Northwest Pipeline Corp. 6 San Juan 32-7 Unit, San Juan County, N.M.

Cumulative production through November 1987 is 1.364 bcf of gas from an open hole section that includes about 40 ft of Fruitland coal since 1957, say Brent Hale and Carolyn Firth, Northwest Pipeline. A 7 ft clean sand at 3,203-10 ft may also be contributing gas.

Production has remained constant at 150-180 Mcfd of gas.

The well unloads a large slug of water following each shut-in and then produces gas with little additional water.

Gas composition has averaged 94.24% methane, 4.97% CO_2, 0.36% ethane, and small amounts of nitrogen and heavier hydrocarbon gases. BTU content varied from 955-991 BTU/scf and averaged 969 BTU/scf the past 10 years.

"The gas composition is distinctly different from the average San Juan basin gas well producing from a sandstone reservoir, which is 87.87% methane, 5.85% ethane, and 1.5% CO_2," Hale and Firth said.

Attention to ownership
legalities is wise

Increased attention is being paid to the legal questions of ownership of and access to coalbed gas on federal, Indian, state, and fee coal deposits.

Denver Attorneys J. Hovey Kemp and Kurt M. Peterson said answers are emerging, but the issues are far from being settled, and coalbed gas developers must consider certain legal risks.

Any judicial examination of coalbed gas ownership turns on resolving two issues: Is coalbed gas a gas or an intrinsic part of the coal, and what were the intents of the parties to the particular grants, reservations, or conveyances of the coal rights on the one hand and the gas rights on the other.

Amoco's pioneering
Cedar Hill development

Amoco drilled 1 Cahn in May 1977 to open Cedar Hill field, which has produced 8.6 bcf of coalbed gas through 1987.

The field contains 16 producing wells and three pressure monitor wells, say David Decker, Resource Enterprises Inc., Grand Junction, Colo., and other authors in a paper produced from public record data.

The coal reservoir is water saturated and overpressured with a gradient of 0.58 psi/ft. Over time, wells have exhibited the classic coalbed methane production profile: increasing gas production with decreasing water production.

Applied geochemical techniques have been effective at distinguishing bicarbonate-rich coal-derived water from chloride rich water produced from sandstone reservoirs. A low C_2+ hydrocarbon composition with the presence of CO_2 identifies gas originating from coal.

The field is significant because:

• Commercial levels of coalbed gas have been established and a long production history is available.

• The field serves as a case history to examine favorable geologic and reservoir characteristics that may be helpful in identifying other commercial coalbed gas fields.

• Many varied completion and stimulation techniques were employed documenting an evolution of drilling, completion, and production practices.

Cedar Hill is the only multiple well coal degasification field in the San Juan basin with a relatively long production history.

Amoco, Union Texas, and Meridian have developed the field, in which the basal Fruitland coal seam contains an

estimated 313 million tons of coal and 89 bcf of coalbed methane in place.

The volume of Fruitland gas is small relative to other parts of the basin, but commercial production has been achieved due to unusually high permeability which should allow the existing wells to recover most of the gas in place.

No supporting data relate production levels to completion practices or coal thickness. Deliverability is probably tied to structural influence that relieves in situ stress and maintains higher cleat permeability in the coals.

As production at 1 Cahn continued, a constant decrease in reservoir pressure observed in two monitor wells indicated a drainage radius at least 2,200 ft away from the wellbore. In effect, the single producing well significantly reduced reservoir pressure over a large area.

Production increases
as pressure declines

By yearend 1985, three producing wells had reduced pressure in the Cedar Hill reservoir to 900 psi from 1,580 psi, yet combined gas production from the three wells continued to increase, most likely the result of strongly nonlinear desorption behavior.

The field's first five wells were cased above the top of the basal coal, then deepened to expose the coal without penetrating underlying Pictured Cliffs sandstone.

It is suspected that problems with borehole stability and production of coal fines in several of the open hole completions led to installation of gravel packs and prepacked screened liners.

After several of these gravel packs were removed, production increased, suggesting that gravel packs may plug with fines.

Meridian, at more recent open hole completions, drilled the coal zone with air and completed open hole with installation of pre drilled liners to provide wellbore stability.

The other Cedar Hill wells were cased, perforated, and fracture stimulated.

Relatively high permeabilities of 36-75 md for coal at depths of 2,800 ft indicate Cedar Hill field is in an area of enhanced permeability, possibly due to low effective net stress.

Fairly low porosities of 1-2% are consistent with values found elsewhere in the basin. Lower porosities generally result in shorter dewatering periods.

A consequence of shutting in coalbed gas wells, especially those completed open hole, may be increased water production, hole instability, and reduced gas production as evidenced by the production history of two Cedar Hill field wells.

Marketing coalbed gas
requires diligence

Successful marketing of San Juan basin gas requires aggressive pursuit of the most attractive markets.

Four main systems take San Juan basin gas. They are El Paso Natural Gas Co., Northwest Pipeline Corp., Sunterra Gas Gathering Co., and Western Gas Supply Co.

El Paso's are the most attractive markets, says David M. Posner, Ladd Gas Marketing Inc., Denver.

"In many instances, access to markets for new drilling will be accomplished most effectively by construction of producer owned gathering lines or small gathering systems built by independent gatherers that connect into the El Paso system."

Construction of independent gathering systems will mean timely connection and avoid payment of second gathering fees such as WestGas and Northwest for delivery to El Paso.

Using an independent gas marketer is often essential because many California utilities require minimum bids of 10,000 Mcfd.

Amoco is trying to line up power generation and cogeneration companies, which would be year round, baseload customers. The level loads would supplement the more conventional seasonal markets.

Water disposal a big operation, expense

Only one company, Meridian Oil Inc., has laid extensive water lines. Seven of its wells produced 407 MMcf of gas during November 1987 and 73,188 bbl of water.

Piped to a central tank battery, the water is used in drilling or disposed of into Jurassic Entrada and Morrison sandstones.

Amoco pipes and trucks water to injection wells.

Horizontal drilling is being attempted, but wellbore caving occurs in coal zones even at vertical wells. One company plans to counter caving by displacing mud with a gas while maintaining circulation pressure.

No single water disposal method is appropriate for all situations. Successful disposal systems must be designed on the basis of producstion or operational concerns, economics, environmental factors, and regulatory framework, said Gerald Zimpfer, Yates & Auberle Ltd., Denver, and others.

Groundwater varying from fresh to highly mineralized occurs in the interbedded coal and sandstone layers in Fruitland.

Water has less than 1,000 mg/l total dissolved solids on the western and northwestern parts of the Southern Ute reservation. Concentrations can reach 10,000-15,000 mg/l near the reservation's southern boundary.

At 10,000 mg/l, this is a sodium bicarbonate fresher than most oilfield brines.

Water production is 14-1,000 b/d/well and averages about 70 b/d/well on the reservation.

Rundown of the basin's water injection wells

Northern New Mexico has about 23 operating underground injection wells, most private and accepting water produced with oil.

Meridian Oil Inc. operates one well each in San Juan and Rio Arriba counties for Fruitland water disposal. Cumulative injection is 800,000-1 million bbl of water.

Five injection wells are operating in the Colorado San Juan, and several more are proposed or being tested.

Amoco, which purchased a substantial acreage position from William Perlman, Houston, in 1985, has drilled three injection wells and has proposed several more.

The company's Salvador and McCaw injection systems dispose of the majority of the coalbed brines produced in Colorado. Plastic or fiberglass pipe serves as gathering lines.

Injection exceeded 1 million bbl by Jan. 1, 1988, into the Salvador well 3 miles south of Ignacio. Maximum injection rate is 4,800 b/d. Fourteen wells are connected.

The McCaw well near the La Plata County airport dis-

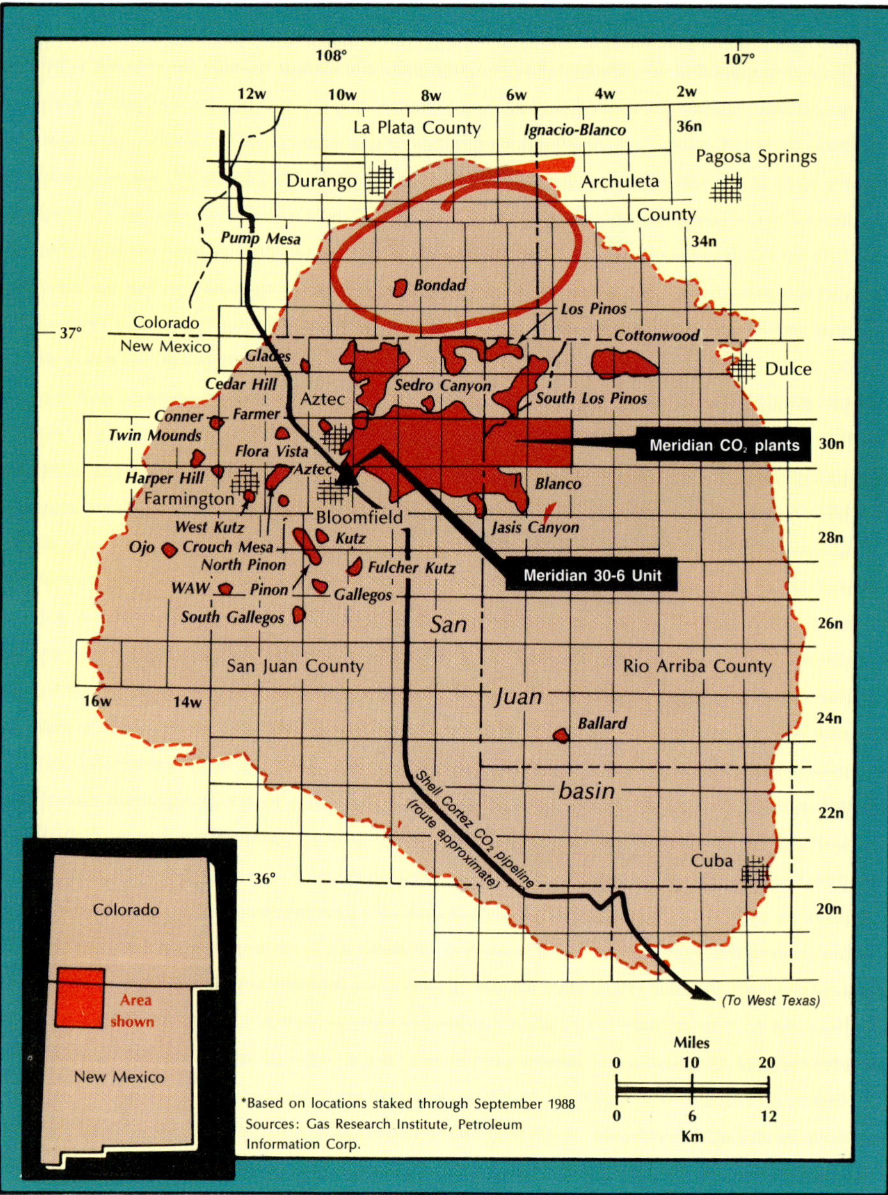

posed of 1.5 million bbl and is connected to 27 wells. Capacity is 9,600 b/d.

Independent Jay Magness Inc. operates 1-12 Echols-Ute in 12-34n-7w, which simultaneously produces Fruitland coal gas and injects water into the deeper Point Lookout. Cumulative injection is 47,200 bbl of water and production is 67.8 MMcf of gas.

Pipeline collection of water is more capital intensive initially and requires more sophisticated control and personnel but in the long term less expensive than trucking on a unit cost basis.

Future research and experience may provide cost effective alternatives, including water processing equipment allowing surface discharge of produced formation waters.

Operators plan multiwell coalbed drilling programs

Several larger operators have defined their drilling programs through yearend 1989, and many more are conducting smaller programs.

Amoco, most active coalbed methane operator in the basin, planned to drill 100-140 wells in 1988 and 100 or more in 1989.

Most of Amoco's drilling will take place in La Plata County, Colo., but other drilling will take place in New Mexico.

The company had 25% of the basin's coal gas acreage under lease before its October 1988 purchase of Tenneco Oil Co.'s Rocky Mountain division for $900 million. More than 90% of the division's assets are in the New Mexico San Juan.

Amoco plans to lay 420 miles of gathering and secondary lines this year and next to connect its wells to gas transmission systems.

Amoco through mid-July 1988 had drilled or acquired 210 wells in southern Colorado. It operated eight wells in Cedar Hill field. Of the 210 wells, 90 were connected to sales. The rest were awaiting pipeline hook-up.

Excluding the Tenneco purchase, the company through mid-1988 had spent nearly $100 million on coal gas drilling and development since 1978, two thirds of it in the latest 2 years.

Amoco expects economic well lives of at least 50 years.

Amoco said coal wells produce as much as 1,000 b/d of water initially and relatively little gas.

Meridian Oil conducts
major coalbed development

Burlington Northern Inc.'s Meridian Oil Inc. affiliate has initiated a large coalbed methane program in the basin.

The program involves a large number of new wells and work overs, gas and produced water gathering systems, and a CO_2 stripping plant.

Meridian by mid-1988 had received state permits for about 150 locations Fruitland coal gas wells or reworks in San Juan County, N.M., and about 66 in Rio Arriba County.

Meridian was constructing a CO_2 stripping plant near Bloomfield, N.M., perhaps for injection into the Cortez pipeline to West Texas.

Meridian booked 800 bcf of San Juan basin coalbed methane reserves in January 1989 in addition to the 46 bcf it added from a 20 well Fruitland pilot program in 1986.

The sharply expanded program includes properties in which San Juan Basin Royalty Trust, Fort Worth, owns an interest. The trust said the tax credit, which it estimates at about 80¢/BTU for 1988, may be allocated to individual unitholders, depending on their tax positions.

Other coalbed
methane programs

Other operators are conducting smaller Fruitland programs.

ARCO Oil & Gas Co. could drill as many as 135 wells and spend about $100 million by the end of 1989. In fall 1988 it was drilling the last six wells of a 38 well, $30 million program.

ARCO was to complete its 1988 drilling program in November. Drilling was to resume in April 1989, but fracturing and well testing usually continue during winter.

Northwest Pipeline had a four well program in late 1988 and planned 40 or more wells in 1989.

Burlington Northern's Southland Royalty Co. unit caused 12 gross coal seam gas wells to be drilled on trust properties through Sept. 15, 1988, at a combined cost of $6.5 million.

Production start-up is expected within 6-9 months for spot market sales, with gas transported by El Paso Natural Gas Co.

Southland indicated it may drill 20 gross wells on trust properties by yearend 1989 at a cost of about $400,000/well. Additional costs will be incurred to equip the wells, dispose of water, extract CO_2, and construct pipeline connections.

Whether the 20 wells will be drilled depends on numerous factors, including results of other coal seam wells, gas demand, gas prices, drilling costs, and other factors, the trust said.

Other Fruitland operators include Blackwood & Nichols Co. Ltd., Durango; Union Texas Petroleum Corp., Houston, McKenzie Methane/Carbon River Energy Partnership, Seattle; Flag-Redfern Oil Co., Midland; Palo Petroleum Co., Dallas; Mesa Limited Partnership, Amarillo; Dugan Production Corp., Jerome P. McHugh, and Nassau Resources Co., Farmington, N.M.; Caulkins Oil Co., Bloomfield; Northwest Pipeline Corp., Salt Lake City; and National Cooperative Refinery Association, Bowen-Edwards Associates, and Vince Allen & Associates, Denver.

Kerr-McGee Corp. acquired Flag-Redfern in late 1988.

Early coalbed
methane completions

Mesaverde, Menefee, and Fruitland contain sandstone, shale, and coal beds.

Sandstones were the main target of early completions, but over the years the coals have become significant contributors to San Juan basin gas production.

Coal probably produced initially through open hole completions, and cumulative coalbed methane production figures aren't known. Producers began researching and designing large scale coal gas projects more than 13 years ago.

In early completions 7 in. casing was set to the top of the Cliff House formation, and the well was shot with solid nitroglycerin. Coalbed gas pressure collapsed the casing at some wells.

Stanolind Oil and Gas Co., now Amoco, probably was the first to try perforated completions. It drilled wells, some of which also produced from Pictured Cliffs, in Ignacio Blanco field, 32n-33n, 6w-7w, La Plata County, Colo.

At about 11¢/Mcf for gas, wells that began producing too much water were plugged.

First Fruitland coal drilling in New Mexico opened Gallegos, Aztec, La Jara, and West Kutz fields in 1952.

Phillips Petroleum Co. 3 San Juan Unit 32-7, discovery well of North Los Pinos field in San Juan County, N.M., in 1954, has produced 430 MMcf of gas. Six wells in the field produced a combined 1.74 bcf of gas through Jan. 1., 1987.

Phillips' South Los Pinos field has produced more than 10 bcf of gas from 39 wells, including 1.3 bcf of gas from the discovery well.

A few fields were discovered in the early 1950s, but exploration showed producing areas to be spotty. Development drilling continued during 1955-65.

Southwestern basin
wells behave differently

Coalbed gas wells in the southwestern San Juan basin have behaved differently from northern San Juan basin wells in several ways.

Performance at the shallow southwestern wells has been unpredictable. Production fell dramatically at some for no apparent reason, Dugan and Williams said.

Plugging of production equipment by coal fines has plagued northern wells and rarely affected southwestern wells. Water so evident in deeper northern coal wells seems absent from shallower southwestern coals.

The shallow wells have not produced extremely large quantities of gas like the northern wells, but they have been easy to maintain and comparatively inexpensive to operate, Dugan and Williams said. In the 1960s and 1970s small operators opened South Gallegos, Harper Hill, WAW, and Ojo fields, starting with sandstone production. Dugan Production discovered Ojo Fruitland-Pictured Cliffs field in 1971, drilling slim holes and completing through 2⅞ in. pipe. A two stage compressor is required to produce the low pressure field, which has produced more than 1 bcf of gas.

Completion, operation
of Fruitland coal wells

The best Fruitland coalbed gas producing wells have

Coal seam methane tax credit

Year of Production	Inflation adjustment factor	GNP deflator	Oil price, $/bbl	Tax credit $/MMBTU	Oil price required for phaseout		
					Partial	Total	
1979	1.0000	1.0000	—	—	23.50	29.50	
1980	1.0896	1.0896	33.03	0.0000	25.61	32.14	Actual
1981	1.1900	1.0921	32.19	0.2513	27.97	35.10	
1982	1.2676	1.0652	28.50	0.6556	29.79	37.39	
1983	1.3197	1.0411	26.19	0.6826	31.01	38.93	
1984	1.3673	1.0361	25.88	0.7072	32.13	40.34	
1985	1.4211	1.0393	24.08	0.7351	33.40	41.92	
1986	1.4555	1.0242	12.66	0.7528	34.20	42.94	
1987	1.4949	1.0271	15.41	0.7732	35.13	44.10	
1988	1.57	1.041	—	0.81	36.89	46.31	Projected
1989	1.66	1.056	—	0.86	39.01	48.97	
1990	1.75	1.057	—	0.91	41.12	51.62	
1991	1.82	1.040	—	0.94	42.77	53.69	
1992	1.89	1.040	—	0.98	44.48	55.84	
1993	1.97	1.040	—	1.02	46.26	58.07	
2000	2.59	1.040	—	1.34	60.90	76.43	

Source: Northwest Fuel Development Inc.

been open hole completions, said William F. Clark, petroleum engineer, Blackwood and Nichols Co. Ltd., Durango, and Tom Hemler, geologist, Carbon River Energy, Seattle.

The operator usually cements 7 in. casing above the Fruitland and drills out using a fluid.

When 100-300 ft of uncased wellbore is open to several coal zones, a liner is usually used. Slotted liners and gravel pack completions have been tried with mixed results. Both have plugging problems.

Most Fruitland wells require artificial lift. Sucker rod pumps are the most common.

Determining which pool a well's hydrocarbons are coming from can be a problem in some areas because of the intertonguing of the Fruitland formation and the Pictured Cliffs sandstone, Clark and Hemler said.

Horizontal drainholes have been successfully placed into Rocky Mountain coal seams using short and medium radius directional drilling techniques, says Terry L. Logan, manager of operations, Resource Enterprises Inc., Grand Junction, Colo.

The cost to place a horizontal drainhole using either technique is about $17,600/day.

Success of horizontal drainhole drilling projects required a team approach with fully integrated expertise in well design, field supervision, drilling, cementing, and well site geology, Logan said.

Coalbed reservoirs produce a distinctive bicarbonate type connate water and have higher reservoir pressures than adjacent sandstones, say Dudley D. Rice and three other USGS authors.

Also, coal generated gas is usually contained in the coal beds, neither expelled nor migrated into adjacent sandstone reservoirs.

These factors and differences in composition of the gas indicate that coal beds are a closed reservoir system created by the gas, water, and associated pressures being trapped in the micropore structure, Rice said.

Federal tax credit boosts Fruitland drilling

A bill passed in the closing days of the 1988 congress extended to the end of 1990 from the end of 1989 the drilling time to take advantage of the unconventional energy tax credit.

The credit was part of the Crude Oil Windfall Profits Tax of 1980. WPT was repealed in summer 1988, but the tax credit continues because it was in a part of the Internal Revenue Code not repealed.

The credit should be 78¢/MMBTU in 1987 and rise to 91¢ in 1990 and $1.34 in 2000, according to one estimate.

Tax credit phaseout, possible if crude oil prices rise too much, has not been a factor since 1981 and is not expected to play a role for at least several years, if ever. For instance, the price of oil would have to exceed $35.40/bbl in 1987 before the phaseout factor lowered the credit from 78¢.

The credit is applicable only for production from new wells on new property as defined by the Tax Reform Act of 1986.

The credit may not apply if the gas is sold at incentive prices allowed under Natural Gas Policy Act Sec. 107. This has been an academic consideration for several years

Several IRS letter rulings conclude that producers can switch back and forth between accepting incentive prices and the tax credit but not double dip.

The wells cannot be drilled into reservoirs where commercial gas was produced from coal seams prior to 1980. Production from valid wells is eligible for the tax credit until Jan. 1, 2001.

The credit has to be taken in the year the gas is produced and, once taken, is not subject to recapture like investment tax credits.

If the owner of coalbed methane production does not have a current tax liability against which the credit can be taken, there are ways of reorganizing projects where the credit could be taken, says Peet Soot, Northwest Fuel Development Inc., Portland, Ore.

The credit's benefits are apportioned among owners of the production according to their respective interests in gross sales.

Fruitland, San Juan basin geology notes

Diameter of the San Juan basin is about 100 miles, and surface area is about 7,500 sq miles, says James E. Fassett, U.S. Geological Survey, Lakewood, Colo.

In the conterminous U.S., the San Juan basin gas field is second in size only to Hugoton field of Oklahoma, Kansas, and Texas.

Cumulative basin production through 1982 was 190 million bbl of oil, 14 tcf of nonassociated gas, 425 bcf of associated gas, and 60 million bbl of condensate.

Only about 90 bcf of gas has come from the Fruitland, and about half that came from Ignacio Blanco field.

Two authors suggest that most of the basin's gas is coalbed methane that migrated into sandstones. Various authors estimate 26-50 tcf of coalbed methane in Fruitland in the basin. No recovery factor has been applied.

It would take a hypothetical 1,500 wells producing an average of 200 Mcfd/well for 30 years to recover 3.3 tcf of gas.

That would result in average reserves of 2.1 bcf/well.

Various operators estimate well lives of 30-50 years, depending on pipeline takes.

Fruitland coal beds range from thin stringers to beds more than 40 ft thick. Coal zones may contain several coal beds that total as much as 80 ft of net thickness.

Analysis of Fruitland coal, gas properties

Due to minor transgressive episodes during deposition, the Pictured Cliffs locally intertongues with the Fruitland, said Bruce Kelso and Donald Wicks, ICF-Lewin Energy Division, Fairfax, Va.

Heating values of the coal generally range from 9,000 BTU to more than 13,000 BTU/ton.

Coal ash content of 8-30% seems to increase eastward across the basin. Moisture content is 2-5%. Sulfur content is 0.5-2.5% and averages 1%.

Coal bearing Fruitland underlies about 700 sq miles of the Southern Ute reservation. The resource on the reservation is estimated to total 16.446 billion short tons of bituminous coal in beds at least 2 ft thick. Most of the coal is at least 2,000 ft deep, says Dorothy Sandberg, USGS, Lakewood.

Thick coal beds generally are in the lower half of the Fruitland, but thinner coal beds are common throughout it. Thickest beds are generally in the lowest part, in places lying directly on top of underlying Pictured Cliffs sandstone, Sandberg says.

The geologic setting that most favors coalbed methane extraction is where the coal seams have enhanced permeability and the reservoir is overpressured, saids Kelso and Wicks.

Additional site specific geological and geophysical research and analysis of well drilling and test data are required to establish accurate boundaries for these favorable areas for Fruitland coals in the basin, they said.

Economic coalbed methane production depends on four important coal seam characteristics, says Chester R. McKee, In-Situ Inc., Laramie, Wyo.

They are gas pressure, gas content, coal seam thickness, and permeability.

Operators have published permeability on only a few wells.

Hydraulic fracturing in Fruitland coals

Most hydraulic fracturing research and application has been targeted towards elastic and carbonate reservoirs, and practically all fracturing techniques, fluids, additives, and propping agents have been developed to stimulate sandstones and carbonates.

Compared with knowledge of sandstone reservoirs, little is known about completing, fracturing, and producing coal seam gas reservoirs, say John Ely and two other authors with S.A. Holditch & Associates, College Station, Tex.

The purpose of a hydraulic frac job in a coal seam must be to connect the cleat system, where the natural permeability of the reservoir exists, to the well bore.

It is critical to ascertain formation properties such as permeability, gas in place, and elastic properties before attempting a stimulation.

Frac design for coal seams will vary greatly based on in situ stress of the coal and surrounding zones, seam thickness, and presence of multiple layers.

Smaller size proppants, such as 20-40 or 16-30 mesh, can provide adequate conductivity for most coal reservoirs.

Substantial productivity increases have been obtained through the use of very small, 100 mesh proppant, although significant sand production has been observed from many of the wells stimulated in this manner, Ely said.

High proppant concentrations of 5-6 lb/gal are commonly pumped in San Juan basin fracs.

Inability to achieve good coalbed isolation can cause a completion attempt to fail. It is often very difficult to treat just the coal, and the adjacent Pictured Cliffs sandstones accept frac fluid as readily as the coal.

Field experience has indicated it is desirable to initiate fractures within the coal seam.

Most operators prefer cased hole, perforated completions due to frequency of large workover costs to clean coal fines from wells completed open hole or with slotted liners or gravel packs, Ely said.

Planning for high Fruitland gas recoveries

Gas storage is dominantly by adsorption in coal rather than by gas compression in the voids as in sandstone and limestone, say Arfon Jones and Gregory Bell, Terra Tek Inc., Salt Lake City, and Richard Schraufnagel, Gas Research Institute, Chicago.

A large fraction of the adsorbed gas is released at low pressures, whereas in conventional gas wells the gas produced is proportional to the pressure drop.

To recover a large fraction of gas in place, it is important to plan for gas production to low pressures.

Coal gas reservoirs initially tend to be highly water saturated, and water pumping is normally required to produce gas. With the large fraction of the pore volume residing in the micropores and mesopores, the irreducible water saturations are found to be greater than 70%, resulting in a low relative permeability to gas throughout the producing life of the reservoir.

For this reason, hydraulic fracture stimulations are effective for increasing gas production, and high fracture conductivities are required for production to low reservoir pressures. Drainholes drilled in the coal seam should be equally effective.

Coal frac jobs are usually performed at high pressures, possibly because weak coal fails at the wellbore and near the fracture tip.

This failed coal can later clog downhole pumps and surface equipment.

Logging of Fruitland coalbed gas wells

Coal beds do not lend themselves readily to conventional well log analysis, says Michael J. Mullen, senior sales engineer for Halliburton Co.'s Welex affiliate.

Coals are relatively easy to identify and correlate on well logs.

The bulk density measurement from the spectral density log readily flags coal beds because surrounding shales have a density of 2.65 g/cm^3 and coal has an average density of 1.5 g/cm^3.

Mullen recommends running spectral density log, dual-spaced neutron II log, and the dual induction log with gamma ray and spontaneous potential logs.

The logs can be run stacked on one trip if enough rat hole is provided.

Different logging procedures using high resolution processing with the spectral density log are recommended to improve the vertical resolution of the density measurement and thereby improve the accuracy of bed thickness determination and coal quality analysis in new wells.

To predict coalbed productivity, the relation between productivity potential and sustained coalbed gas production has been used. Gas content in coal can be calculated from wire line log measurements to approximate direct measurements from coal chips or cores. [IPE]

EOR projects down, but CO$_2$ ups production

ALTHOUGH U.S. ENHANCED OIL PRODUCTION WAS still rising in early 1988, enhanced oil recovery (EOR) has obviously suffered from the crude oil price collapse of 1986.

This is evident in the Journal's worldwide, biennial EOR survey when compared with the previous one conducted in late 1985 and early 1986, largely before it was evident a new oil era had arrived. However, the 1988 survey also shows that a noteworthy amount of investment is still targeted for EOR projects in the U.S. and Canada.

The mass of data presented in the tables in this section boils down to reveal that the number of U.S. EOR projects has decreased markedly, but production was at the 637,000 b/d level the first of this year (Table 1). This is a 5.4% gain over the level for the same time in 1986. The production trend is presented in Fig. 1.

Looking at the U.S. on the basis of total EOR projects of all kinds, the survey reveals that the number declined from 512 in 1986 to 366. This is a loss of 146 or a decline of some 28%. Table 2 breaks down the active projects by type in the four main EOR categories. In some cases operators combined a number of projects and shortened the list or shut down pilot operations for reasons that may have had nothing to do with the price of crude oil. However, in contrast to the trend in projects going generally up since 1971, the year the Journal conducted its first EOR survey, the direction since 1986 has been a thunderous down.

Nevertheless, investment data available for 76 projects planned for 1988 and 1989 in the U.S. and in foreign fields show that $960 million is expected to be spent on EOR projects during the 2 year period. The projects marked with an asterisk in Tables A and B are the ones where investment figures were available and are the source of this total.

Scope, guide

The intent of the Journal's biennial survey is to report, among other things, the additional production that would not be available from a reservoir unless it was stimulated by added energy by means of, for example, steam or carbon dioxide injection. The accompanying diagram, Fig. 2, represents the various recovery mechanisms. The intent of the Journal survey is in general only to cover the bottom tier of mechanisms, i.e., thermal, gas, chemical, and others.

But it must be emphasized that this diagram represents a traditional production sequence. There are, for example, reservoirs that skip the primary stage shown in the diagram and begin production with the assistance of injected natural gas or water. The intended scope of this survey and some of the complexities involved in defining enhanced oil recovery are discussed in the accompanying box.

As an aid to locating different types of information in the massive tables, which are lettered instead of numbered, a guide is also presented in this section listing the content of each table and the abbreviations used.

Outside the U.S.

Venezuela and Canada both have lively EOR programs. Canada had 75 active projects as of the first of this year. They are reported in Table C.

Reported enhanced production in Canada is 148,000 bo/d

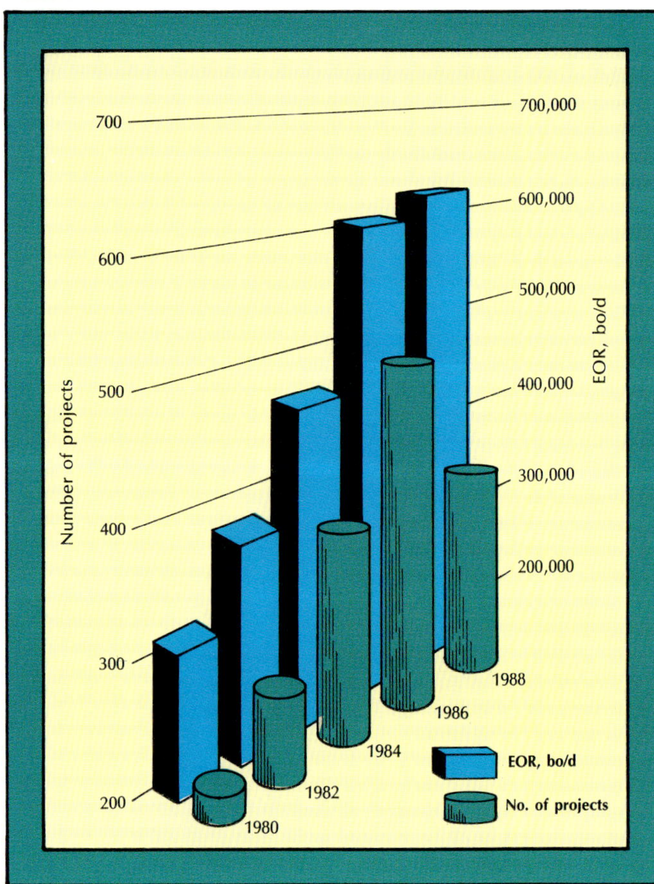

EOR production climb moderated

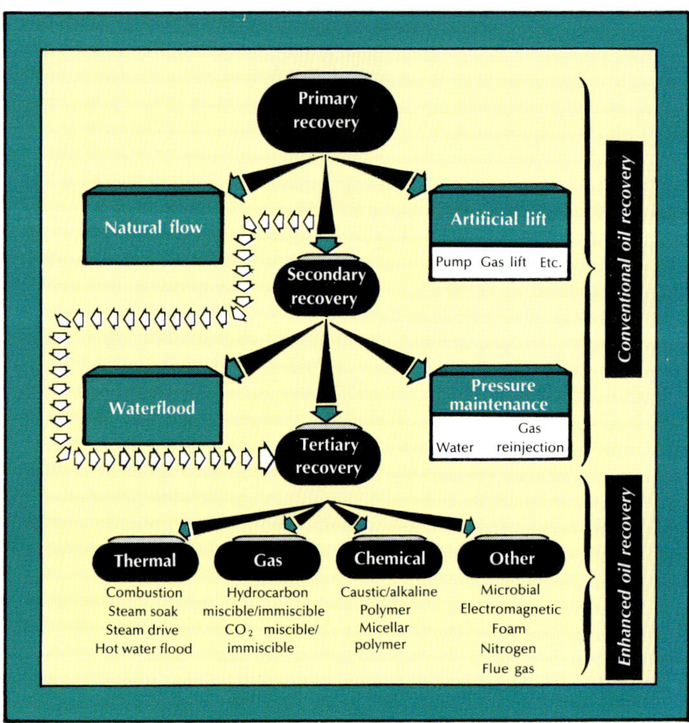

Recovery mechanisms

Abbreviations

Formation type

S: Sandstone
LS: Limestone
Dolo.: Dolomite
Congl.: Conglomerate
Tripol.: Tripolite
US: Unconsolidated sand

Project maturity

JS: Just started
HF: Half finished
NC: Nearing completion
C: Completed
PP: Postponed
Term.: Terminated

Previous production

Prim.: Primary
WF: Waterflood

Project evaluation

TETT: To early to tell
Prom.: Promising
Succ.: Successful
Disc.: Discouraging
None: Not evaluated

Project scope

P: Pilot project
FW: Field wide
LW: Lease wide
RW: Reservoir wide
Exp. L: Expansion likely
Exp. UL: Expansion unlikely

with 42 hydrocarbon miscible projects accounting for 115,000 bo/d of that.

Venezuela has fewer and less diverse projects but is the hemisphere's second largest EOR producer with 216,000 bo/d. Details on these and other foreign projects are listed in Table D.

Unfortunately, up-to-date data covering one of the world's major steam flood projects could not be obtained. That is P.T. Caltex Pacific Indonesia's steam project in the Duri field of Sumatra. It has lifted production there by about 65,000 bo/d. Original plans called for the project to cover the entire field and to lift production from 40,000 bo/d to some 300,000 bo/d by 1985. The pace of the originally planned program is apparently falling short of that but the project could still become one of the world's largest.

Another major project was announced for Brazil last year. Petroleo Brasileiro SA says it plans to inject CO_2 in Aracas field in Bahia and boost recovery by 5.3 million bbl of oil in 14 years. This project is not yet listed in planned foreign projects.

U. S. EOR
production

As Table 1 shows, enhanced production from thermal projects declined, while chemical and gas projects logged increases.

Defining EOR

By some definitions, determining whether a particular production operation is enhanced oil recovery or not should be relatively straightforward. It is not.

First, it is generally agreed that " enhance" means to make something that already exists or is happening better. Therefore, reservoirs that have never produced oil and that do not flow by natural pressure or that must be heated to be pumped are not having their production "enhanced" by stimulation. Therefore, they should not be designated as EOR projects. Canada has over 20 such projects. The viscosity of the reservoir oil in most of these projects ranges from 25,000 to 800,000 cp at temperatures in the range of 55-70° F. Most are designated experimental heavy oil projects. They are not listed in the Journal's survey.

A problem of semantics arises, however, because many nonflowing, heavy-oil reservoirs are being stimulated by combustion or steam injection. These are definitely recognized as EOR processes. That means an EOR process can be used on a project that should not bear the designation of an EOR project.

Earlier, another definitional trail was blazed with useage of the terms primary, secondary, and tertiary recovery. That approach is shown in the accompanying Fig. 2 and has some merit if one keeps in mind that it has some exceptions and it represents the traditional concept of EOR. With world crude heading toward its price spike in the early 1980s, the practicality of starting primary commercial production with a tertiary process was in general an idea that got better as the price of crude increased.

The National Petroleum Council in its 1984 study "Enhanced Oil Recovery" defined EOR for the purposes of the study as:

"The incremental ultimate oil that can be economically produced from a petroleum reservoir over that which can be economically recovered by conventional primary and secondary methods. Primary methods rely on the natural reservoir energy to drive the oil through reservoir rock to the production well. Over time, this natural energy drive dissipates, and energy must be added to the reservoir to produce significant amounts of additional oil."

That is a lucid and sensible definition that has utility. But later in discussing the scope of the study the NPC says certain EOR processes have been referred to as tertiary processes.

A following statement then puts the neat preceding definition of EOR in jeopardy:

"However, the term 'enhanced oil recovery,' as used in this report, is considered to have a broader meaning than 'tertiary oil recovery,' in that potential from reservoirs that are not suitable for the application of conventional and primary and secondary techniques is included."

That statement then bumps up against the accepted meaning of "enhanced," and requires some qualification.

The thrust of the Journal surveys has been, and still is, to cover projects fitting the first part of the NPC definition. We are confident that essentially all of our respondents understood our intent and that the survey faithfully reflects this.

The processes listed at the bottom of the diagram of recovery mechanisms, Fig. 2, under thermal, gas, and chemical, are the ones Journal respondents were asked specifically to identify. They had the option of also listing a number of of minor processes under the "other" category and designating them by type. Respondents listed the five different types shown.

Finally, it should be noted that the word "project" is used in this report to mean that the operation is moving toward a goal of recovering a certain amount of oil.

Of total EOR U.S. production in 1988, thermal output accounted for 72.9% compared to 79.3% in 1986; chemical output was 3.5% compared to 2.8% in 1986; and gas production was 23.6 vs. 17.9%

The most barrels of EOR production lost was in the steam category and most of that was in venerable California fields. Table H shows steam and other type projects terminated or completed since the 1986 survey. California steam projects were obviously pummeled by low crude prices.

One small California producer with a 160-acre project started in 1965 was employing 34 producers and 20 injector wells to make 140 b/d of 13° API oil.

He shut down the project in 1986 because he needed $14/bbl to stay in operation but was only getting $9/bbl.

A production manager working for a larger company in Kern County, Calif., reported earlier this year that he had turned off the steam, his reservoir had cooled down, and he was "just pumping the dregs" trying to keep the office open a bit longer. This operation was getting $12 to $13/bbl for crude of 13 to 16° API gravity oil. However, this manager reported that in December 1987 his lease costs were running $21/bbl. That is how much he would need per barrel to break even.

But steam EOR is also characterized by huge projects. The largest of all steam projects is Shell Oil's Belridge project in California making 101,000 bo/d. Second largest is

Table A

U.S. planned EOR projects by type

Operator	Project type	Field	State	Size, acres	Pay zone	Depth, ft	Gravity, °API	Target date
Mobil*	Alkaline	Luling	Tex.	200	Edwards Limestone	2,100	28.0	6/89
Nat. Coop. Refinery Assn.*	Alkaline/polymer	W. Moorcroft	Wyo.	640	Muddy 55	3,600	32.0	5/88
Mobil	CO_2	Postle	Okla.	23,000	Morrow	6,200	40.0	1991
Mobil*	CO_2	Shiloh	Tex.	3,900	Upper Pettit	7,100	38.0	12/89
Amoco*	CO_2 miscible	Lost Soldier	Wyo.	1,400	Tensleep	5,000	35.0	1988
Amoco*	CO_2 miscible	Anton Irish	Tex.	8,545	Clearfork	5,900	28.0	1989
Amoco	CO_2 miscible	Slaughter	Tex.		San Andres	4,950	27.0	
Chevron*	CO_2 miscible	North Ward Estes	Tex.	3,840	Permian Yates	2,900	37.0	4/89
Conoco	CO_2 miscible	E. Huntley	Tex.	1,308	San Andres	3,180	36.0	1989
Conoco	CO_2 miscible	W. Sussex	Wyo.	1,740	Shannon	3,040	38.0	1991
Conoco	CO_2 miscible	North El Mar	N.M.	2,000	Delaware	4,600	41.0	1989
Conoco	CO_2 miscible	McCallum	Colo.	813	Pierre 'B'	800	28.0	Late 1988
Conoco	CO_2 miscible	Maljamar	N.M.	3,840	Grayburg/San Andres	3,665	38.0	1/88
Mobil*	CO_2 miscible	Slaughter	Tex.	2,495	San Andres	4,900	32.0	7/89
Texaco*	CO_2 miscible	Wasson	Tex.	13,655	San Andres	5,100	33.0	6/89
Texaco*	CO_2 miscible	Paradis	La.	44	9500-Foot	9,800	38.0	3/88
Texaco*	CO_2 miscible	Paradis	La.	160	16 Sand	9,900	38.0	7/88
ARCO Alaska*	HC immiscible	Kuparuk River	Alaska	7,680	Kuparuk A & C	6,000	24.0	3/88
ARCO Alaska*	HC miscible	Prudhoe Bay	Alaska	4,000	Sadlerochit	8,800	27.0	9/88
ARCO Alaska*	HC miscible	Prudhoe Bay	Alaska	5,000	Sadlerochit	8,800	27.0	9/89
ARCO Alaska*	HC miscible	Kuparuk	Alaska	5,120	Kuparuk	6,000	23.0	6/88
Kerr-McGee*	HC miscible	North Buck Draw	Wyo.	4,800	Dakota	12,500	45.0	10/88
Louisiana Land & Expl.*	HC miscible	Buck Draw	Wyo.	4,800	Dakota	12,500	46.0	1/89
Texaco	Hot water	San Ardo	Calif.	130	Aurignac	2,200	13.0	7/89
ARCO	Polymer	Dundee Healdton Unit	Okla.	660	Penn	1,000	31.0	1988
Citronelle Unit*	Polymer	Citronelle	Ala.	160	Rodessa	10,600	42-45.0	Open
Enron Oil & Gas*	Polymer	Big Piney	Wyo.	2,789	Mesaverde	3,500	44.0	7/87
Mitchell Energy*	Polymer	Alba	Tex.	512	Sub-Clarksville	4,010	15.5	7/87
Sun E&P*	Polymer*	Dora Roberts	Tex.	150-700	San Andres	2,100	29.0	3/88
Sun E&P*	Polymer*	Stephens County Regular	Tex.	1,000	Caddo	3,200	39.0	4/88
Texaco*	Polymer	Inglewood	Calif.	20	VMR	2,800	18.0	7/89
Unocal*	Polymer	Hueneme	Calif.	135	Hueneme	5,400	14.0	1/89
Chevron*	Steam	Midway Sunset	Calif.	22	Sub Lakeview	1,000	11.0	1989
Chevron*	Steam	Coalinga	Calif.	150	Temblor	950	13.0	1989
Chevron	Steam	Coalinga Sec. 31A	Calif.	100	Temblor	900	14.0	1/87
Chevron*	Steam	Whittier	Calif.	10	1st Zone	1,000	15.0	1988
Chevron*	Steam	Homer	La.	40	Nacotoch	1,250	37.0	4/89
Chevron*	Steam	Coalinga	Calif.	100	Temblor	900	14.0	1988
DCR Petroleum*	Steam	Camp Hill	Tex.	35	Carrizo	400	18.4	7/88
DCR Petroleum*	Steam	O'Connell	N.M.	1.2	O'Connell	350	17.0	5/88
DCR Petroleum*	Steam	T-4	N.M.	1.2	O'Connell	700	17.0	5/88
Lawrence-Allison*	Steam	Teapot Dome NPR-3	Wyo.	60	Shannon	300	33.0	10/88
M.H. Whittier*	Steam	Coalinga	Calif.	25	Temblor	1,500	14.0	1/89
MacPherson Oil*	Steam	Mt. Poso	Calif.	243	Bedder	2,200	16.0	1989
Mobil	Steam	Midway Sunset	Calif.	20	Monarch	950	13.0	1988
Mobil*	Steam	North Midway Sunset	Calif.	300	Potter	1,200	12.0	1/89
Mobil*	Steam	Talco	Tex.	5	Paluxy	4,300	19.0	6/88
Mobil	Steam	West Newport	Calif.	12	A Zone	600-1,450	13.0	1/88
Mobil*	Steam	Bayou Bleu	La.	22	Sch. No. 1 RD	1,500	18.0	12/89
Mobil*	Steam	Bayou Bleu	La.	140	Sch. No. 1 RA	1,700	18.0	2/88
Oxy USA	Steam	South Kern Front	Calif.	80	Etchegoin chanac	1,500	14.0	1/90
Oxy USA	Steam	North Kern Front	Calif.	342	Etchegoin	1,300-1,500	14.0	1/89
Tenneco*	Steam	Midway-Sunset	Calif.	15	Metson	1,400	12.0	1988
Tenneco*	Steam	Placerita	Calif.	55	L. Kraft	1,800	12.0	1988
Tenneco*	Steam	Kern River	Calif.	24	Kern River	600	13.5	1988
Texaco*	Steam	San Ardo	Calif.		Lombard	1,900	11.0	
Texaco*	Steam	Kern River	Calif.	111	Kern River Series	1,000	13.0	9/88
Texaco*	Steam	Cat Canyon	Calif.	800	S1B	2,500	10.0	4/88
Texaco*	Steam	Midway	Calif.	12	Potter	1,000	13.0	1/89
Texaco*	Steam	Poso Creek	Calif.	55	Basal Etchegoin	2,400	12.3	2/88
Texaco*	Steam	McKittrick	Calif.	12	Potter	1,100	13.0	1/89
Texaco*	Steam	Cat Canyon	Calif.	460	S2-S8	2,600	10.0	4/88
Texaco*	Steam	Sour Lake	Tex.	5	Miocene	1,400	18.0	1/88
Texaco*	Steam	Midway	Calif.	5	Potter	950	12.0	7/88
Union Pacific Resources*	Steam	Wilmington	Calif.	95	Tar	2,500	14.0	1/89
Unocal*	Steam	Cymric (Walport)	Calif.	80	Tulare	1,200	13.0	1989
Unocal	Steam	Cymric	Calif.	150	Antelope Shale	2,400	13.0	1989
Unocal*	Steam	Midway Sunset	Calif.	16	Hoyt	1,750	13.0	6/88
Unocal*	Steam	S. Tapo Canyon	Calif.	15	4th Sespe	2,100	16.0	3/88
Unocal	Steam	Brea	Calif.	150	Pliocene	2,500	16.0	1990

*Project for which investment data were available.

Texaco's Kern River project with 87,600 bo/d. These two projects account for 41% of the steam enhanced production in the U.S.

Steam EOR is also a key element in what is perhaps a unique EOR project. A $125 million, 49.5×10^6 w coal-fired cogeneration plant is under construction on a 5-acre site about 12 miles north of Bakersfield, Calif. The coal will come by train from Utah. The steam produced will be used for generating electricity and for injection in leases that Macpherson Oil Co. operates in the nearby Mt. Poso field. Macpherson is also co-owner of the cogeneration facility.

The company says that primary recovery from the Mt. Poso field, west area, is now 30% of the original oil in place and that the thermal operations will increase total expected recovery to in excess of 60%.

Polymer projects

Although there was a major drop in polymer projects during the period 1986-1988, there was a healthy gain in output (Table 1 and Table G).

The reasons are that of the some 80 polymer projects shut down since 1986 over half of them had not previously reported any enhanced production.

Prior to 1986, with the higher price of crude there was an incentive to launch projects to escape the windfall profit tax with enhanced production. Secondly, there are now simply more larger projects. There are seven capable of over 1,000 b/d in this compilation vs. only two in 1986. The largest is Marathon Oil's Oregon basin, Wyo., Setpa project rated at 3,000 b/d of enhanced production.

Carbon dioxide

The large jump in CO_2 miscible EOR of nearly 36,000 bo/d, representing a 125% increase over 1986, has been in the offing since three large trunk lines began trans-

---

Guide to EOR project tabulations

Table A: U.S. planned EOR projects by type

Table B: Planned projects outside U.S. by country

Table C: Producing Canadian EOR projects

Table D: Producing EOR projects outside U.S. and Canada

Table E: Producing thermal-EOR projects in U.S.

Table F: Producing CO_2, gas-EOR projects in U.S.

Table G: Producing chemical-EOR projects in U.S.

Table H: Completed, terminated U.S. projects

Table B

Planned projects outside U.S. by country

Operator	Project type	Field	Country	Size, acres	Pay zone	Depth, ft	Gravity, °API	Target date
Vikor	CO2 miscible	Joffre	Canada/Alta.	640	Viking - Phase II	5,000	42.0	1988
Shell Canada*	CO2 miscible	Harmatton East	Canada/Alta.	2,210	Turner Valley	8,600	38.6	Mid 1988
Canada Northwest Energy*	Electromagnetic heating	Frog Lake	Canada/Alta.	120	Sparky	1,500	12.0	9/88
Amoco Canada*	HC miscible	Pembina	Canada/Alta.	1,400	Cardium	5,100	38.0	Mid 1989
Amoco Canada*	HC miscible	Nipisi	Canada/Alta.	8,640	Gilwood	5,500	41.0	4/88
Canada Northwest Energy*	HC miscible	Morinville	Canada/Alta.	160	D3	5,200	28.0	4/89
Canterra Energy*	HC miscible	Bigoray K	Canada/Alta.	200	Nisku	7,500	35.0	Mid 1988
Canterra Energy*	HC miscible	Rainbow South S	Canada/Alta.	100	Keg River	6,000	42.0	Mid 1988
Chevron Canada	HC miscible	Bigoray D	Canada/Alta.	400	Nisku D	11,000	38.0	1/89
Chevron Canada	HC miscible	Brazeau River I	Canada/Alta.	250	Nisku I	10,000	45.0	3/88
Chevron Canada*	HC miscible	Kaybob	Canada/Alta.	4,100	Beaverhill Lake A	10,000	36.0	7/88
Chevron Canada	HC miscible	Mitsue	Canada/Alta.	2,132	Gilwood-Stage 1 Expan.	5,000	41.0	7/88
Chevron Canada	HC miscible	Mitsue	Canada/Alta.	5,152	Gilwood-Stage 3	5,000	41.0	7/89
Gulf Canada*	HC miscible	Fenn-Big Valley	Canada/Alta.	2,246	Nisku	5,249	32.8	1989
Gulf Canada*	HC miscible	Fenn-Big Valley	Canada/Alta.	3,000	Nisku	5,249	32.8	1991
Gulf Canada*	HC miscible	Fenn-Big Valley	Canada/Alta.	1,300	Nisku	5,249	32.8	1992
Home Oil*	HC miscible	Swan Hills	Canada/Alta.	8,240	Beaverhill Lake	8,700	39.0	5/88
Mobil Oil Canada*	HC miscible	Carson Creek North	Canada/Alta.	5,760	Beaverhill Lake	8,700	43.7	Late 1989
Shell Canada*	HC miscible	Virginia Hills	Canada/Alta.	6,080	Beaverhill Lake	9,200	38.0	Late 1989
Canada Northwest Energy*	Steam	Atlee Buffalo	Canada/Alta.	40	Glauconite	2,900	14.5	1/89
Mobil Oil Canada	Steam	Cold Lake	Canada/Alta.	160	Lower Grand Rapids	1,200	12.0	3/88
Norcen*	Steam	Cold Lake	Canada/Alta.	50	Clearwater	1,450	11.0	12/88
Suncor*	Steam	Burnt Lake	Canada/Alta.	660	Clearwater	1,600	11.0	
Canada Northwest Energy*	Electromagnetic heating	Tangleflags	Canada/Sask.	80	Sparky	1,700	13.8	4/88
Canada Northwest Energy*	Electromagnetic heating	Northminster	Canada/Sask.	160	Sparky	1,750	14.9	4/88
Mobil Oil Canada	Steam	Celtic	Canada/Sask.	160	Gen. Petroleum/ Waseca	1,650	12.0	1988
Phillips Petroleum	Steam	Coleville	Canada/Sask.		Bakken	2,600	14.0	1989
NKFV	C2 rich gas inj.	Algyo	Hungary	400	Tisza-2	5,700	47.0	1988
OKGT	CO2 gas cap	Nagylengyel	Hungary	3,000	I-IV Rudistid	6,000	27.0	1988
NKFV	Steam	Demjen-West	Hungary	15	Rupelien 3/B	1,000	30.0	1988
Trintoc*	Steam	Point Fortin	Trinidad	30	Cruse Sands	1,600-1,800	16-18.0	1988
Trintoc*	Steam	Forest Reserve	Trinidad	52	Forest Sands	1,000-1,200	16-18.0	1989
Corpoven	Steam	El Pao	Venezuela	15	U-1,2,3	1,800	8.5	6/88
Mobil Oil AG*	Polymer	Voigtei	West Germany	870	Cornbrash	1,300-1,600	35.0	1989
BEB	Steam	Ruhlermoor SD 3	West Germany/ Emsland	370	Valanginian	1,670-2,165	25.0	1988
Wintershall AG*	Hot water	Emlichheim	West Germany/ Saxony	430	Valenginian	2,600	24.5	1988

*Project for which investment data were available.

porting Colorado and New Mexico carbon dioxide into the Permian basin in 1984. Table F includes details on existing CO_2.

The three major pipelines moving CO_2 into the Permian basin are the 30-in. Cortez line operated by Shell Pipe Line Co., the 24-in. Sheep Mountain line operated by ARCO, and the 20-in. Bravo Dome line operated by Amoco Pipeline Co. The latter brings in the gas from New Mexico. All three terminate in the Denver City, Tex., area. Further pipeline construction within the basin has spurred carbon dioxide EOR projects in the area.

Enron has built the Central Basin line stretching from the Denver City area in Yoakum County some 150 miles south to the Yates field in Pecos County. It ties in to the 16-in. Sacroc CO_2 line which could be tied in and make a major source of CO_2 available to an even larger part of West Texas. Chevron's Sacroc line originally took carbon dioxide from a Pecos County gas processing plant to Kelly-Snyder and Diamond M fields in Scurry County, Tex.

Big Three Industries Inc.'s West Texas Carbon Dioxide Pipeline is also taking carbon dioxide south from the Denver City area. It terminates at another CO_2 line laid to transport carbon dioxide to Conoco's Ford Geraldine field in Reeves County. The route of this pipeline opened another part of the Permian basin to CO_2 EOR.

This will be a major factor there for the next 20 to 30 years, and with so much infrastructure in place it is likely production will rise even as the price of crude languishes. With a little help from oil markets, this region could "go critical" in CO_2 EOR activity and production.

Carbon dioxide, which can in the wrong location still spoil the economics of a gas field, is becoming a valuable commodity, certainly in West Texas. Other oil regions, like the Canadian province of Saskatchewan, would also welcome an abundent and affordable source of CO_2.

Amoco Corp., for example, now gives carbon dioxide reserves and production figures in its annual report. Production last year was 222 MMcfd compared to 33 MMcfd in 1984. Amoco is also tapping other sources of carbon dioxide in West Texas. It has a plant in the Wasson field in the Permian basin separating carbon dioxide for injection from natural gas. It has two separation plants (one nearing start-up early last month) in the nearby Slaughter field even

Table 1

U.S. EOR production

	1980	1982	b/d 1984	1986	1988	Change from 1986, %
Thermal						
Steam	243,477	288,396	358,115	468,692	455,484	−2.8
Combustion in situ	12,133	10,228	6,445	10,272	6,525	−36.5
Hot water	—	—	—	705	2,896	+310.8
Total thermal	255,610	298,624	364,560	479,669	464,905	−3.1
Chemical						
Micellar-polymer	930	902	2,832	1,403	1,509	+7.6
Polymer	924	2,927	10,232	15,313	20,992	+37.1
Caustic/alkaline	550	580	334	185	0	−100.0
Total chemical	2,404	4,409	13,398	16,901	22,501	+33.1
Gases						
Hydrocarbon miscible/immiscible	—	—	14,439	33,767	25,935	−23.2
CO_2 miscible	21,532	21,953	31,300	28,440	64,192	+125.7
CO_2 immiscible	—	—	702	1,349	420	−68.9
Nitrogen	—	—	7,170	18,510	19,050	+2.9
Flue gas (miscible and immiscible)	—	—	29,400	26,150	40,450	+54.7
Total gases	74,807	71,915	83,011	108,216	150,047	+38.7
Other						
Carbonated waterflood	—	—	—	—	—	—
Grand total	332,821	374,948	460,969	604,786	637,453	+5.4

Producing Canadian EOR projects*

Operator	Field	Province	Start date	Area, acres	Number wells — Prod.	Number wells — Inj.	Pay zone	Formation type	Porosity, %	Permeability, md
Alkaline										
Dome Petroleum	David	Alta.	1986	245	22	7	Lloydminster	S	29.0	1,480
CO_2 miscible										
Shell Canada	Midale	Sask.	10/85	5	3	4	Midale	LS/Dolo.	16.0	3
Vikor	Joffre Viking	Alta.	12/82	320	5	5	Viking	S	13.0	500
Vikor	Joffre Viking	Alta.	8/85	640	6	3	Viking	S	13.0	500
Caustic										
Dome Petroleum	Cessford	Alta.	7/84	680	19	7	Basal Colorado	S	24.0	350
Caustic-polymer										
Pancanadian	Horsefly Lake	Alta.	12/84	34	5	4	Lower Mannville	S	18.0	260
Combustion										
Husky	Tangleflags	Sask.	1985	240	19	3	GP	US	33.0	4,000
Mobil Oil Canada	Battrum	Sask.	10/66	4,920	87	13	Battrum/Roseray	S	26.0	1,265
Mobil Oil Canada	Battrum	Sask.	8/67	2,400	47	5	Battrum/Roseray	S	25.0	930
Mobil Oil Canada	Kitscoty	Alta.	1977	1,040	84	7	Sparky	S	34.0	2,000
Mobil Oil Canada	Fosterton Northwest	Sask.	1/70	200	6	4	Roseray	S	29.0	958
Mobil Oil Canada	Battrum	Sask.	11/65	680	24	3	Roseray	S	27.0	930
Pancanadian	Countess Upper Mannville B Pool	Alta.	1/83	40	4	2	Glauconite S.S.	S	25.0	800
Murphy	Evehill	Sask.	9/81	180	16	9	Cummings Waseca	S	34.0	6,000
Electromagnetic heating										
Canada Northwest Energy	Wildmere	Alta.	1/86	120	3		Lloydminster	S	30.0	800
Foam flood										
Signalta Resources	Pembina	Alta.	10/86	2,240	13	1	Ostracod	S	12.0	70

Active U.S. EOR projects

Table 2

	1971	1974	1976	1978	1980	1982	1984	1986	1988	Change from 1986, %
Thermal										
Steam	53	64	85	99	133	118	133	181	133	− 26.5
Combustion in situ	38	19	21	16	17	21	18	17	9	− 47.1
Hot water	—	—	—	—	—	—	—	3	10	+233.3
Total thermal	91	83	106	115	150	139	151	201	152	− 24.4
Chemical										
Micellar-polymer	5	7	13	22	14	20	21	20	9	− 55.0
Polymer	14	9	14	21	22	55	106	178	111	− 37.6
Caustic/alkaline	—	2	1	3	6	10	11	8	4	− 50.0
Total chemical	19	18	28	46	42	85	138	206	124	− 39.8
Gases										
Hydrocarbon miscible-immiscible	21	12	15	15	9	12	16	26	22	− 15.4
CO_2 miscible	1	6	9	14	17	28	40	38	49	+ 28.9
CO_2 immiscible	—	—	—	—	—	—	18	28	8	− 71.4
Nitrogen	—	—	—	—	—	—	7	9	9	0
Flue gas (miscible and immiscible)					8	10	3	3	2	− 33.3
Total gases	—	—	—	—	34	50	84	104	90	− 13.5
Other										
Carbonated waterflood								1	0	−100.0
Grand total					226	274	373	512	366	− 28.5

though the Bravo line cuts right through this field.

The Permian basin is now the world leader in CO_2 EOR. Even further infrastructure for EOR production is being added. Chevron U.S.A. kicked off the world's first CO_2 EOR in 1972 at the Sacroc unit, but earlier this year announced a three stage project for the North Ward Estes field of Ward and Winkler Counties, Tex. The new project could ultimately recover 65 million bbl of oil and 32 bcf of gas with a miscible carbon dioxide EOR project.

The $75 million first stage, which will involve 6 sq miles in the field, will require a 26-mile, 12-in. pipeline to the field from the existing Central Basin CO_2 pipeline.

Texas, of course, doesn't have a monopoly on CO_2 EOR projects. Exxon Co. U.S.A.'s 112-mile, 20-in. CO_2 line from Green River to Bairoil, Wyo., opened the door for Amoco to begin injection into the Wertz field Tensleep formation in 1986. The Lost Soldier field in that area is targeted for injection this year.

Outlook for the future

In 1986, the Journal's EOR survey was virtually complete before the full import of the crude oil price plunge was known. It is therefore no wonder that of the 108 projects listed for completion during 1986 and 1987, few survived. Since the beginning of 1986 to early 1988, only 31 new projects were initiated. Planned polymer projects took the biggest hit. Almost half of those delayed or canceled were polymer projects.

Of the 38 planned, only one has been started. Six still remain as future projects in the current survey. One additional polymer project not in the 1986 survey was started in 1987. Planned immiscible CO_2 projects also took their lumps. Some 27 were cancelled by Texaco offshore Louisiana. The largest cancellation in this category in terms of acreage was Amoco's 16,340-acre Levelland, Tex., project. Of the 43 CO_2 projects listed as planned in 1986, only 9 have survived (5 were started up and four are included in the 15 still listed as planned).

Steam projects, however, bucked the trend. There are now 38 planned compared with 25 two years ago. Of the 38 now listed, 15 of them were also listed 2 years ago which raises a question about the viability of some of them.

Table C

Depth, ft	Gravity, °API	Reservoir oil cp	°F.	Previous prod.	Residual saturation % Start	End	Project maturity	Total prod., b/d	Enhanced prod., b/d	Project eval.	Profit	Project scope
2,450	22.0	34	75	Prim./WF	60.0		JS	1,360		TETT		FW
4,600	28.0	3.00	150	Prim./WF	35.0		NC					P
4,900	42.0	1.14	133	WF	38.0	25.0	HF			Prom.		P (Exp. L)
4,900	42.0	1.14	133	WF	38.0	27.0	JS			Prom.		
3,000	24.0	24	84	Prim./WF	62.0		HF	340	100	Prom.	Yes	P (Exp. L)
3,200	23.0	13.40	94	WF	51.0	40.0	NC					
1,700	14.5	4,000	74	Prim.	77.0		JS	550	550	Succ.	Yes	P (Exp. UL)
2,900	18.0	70.30	110	Prim.	66.0		HF	3,585	3,585	Succ.	Yes	FW
2,900	18.0	70.00	110	Prim.	62.0		HF	1,460	1,460	Succ.	Yes	FW
1,800	12.2	15,000	70	Prim.	85.0	60-68	HF	650	325	Disc.		P
3,100	24.0	16.00	125	WF	62.0	40.0	HF	138	138	Prom.	No	P (Exp. UL)
2,900	18.0	70.00	110	Prim.	70.0		HF	1,400	1,400	Succ.	Yes	FW
3,550	28.0	6.00	100	WF	45.0	25.0	NC					
2,450	14.3	2,750	70	Prim.	100.0	80.0	HF	350	350	Prom.	Yes	P (Exp. UL)
2,100	12.0	8,000	75	Prim.	85.0		JS	80	60	Prom.	Yes	P (Exp. L)
5,659	42.0	0.35		Gas injection	78.0	31.0	HF	168	168	Prom.	Yes	FW

Producing Canadian EOR projects* — continued

Operator	Field	Province	Start date	Area, acres	Number wells Prod.	Number wells Inj.	Pay zone	Formation type	Porosity, %	Permeability, md
Hydrocarbon miscible										
Amoco Canada	Ante Creek Unit 1	Alta.	1968	6,080	13	9	Beaverhill Lake	LS	6.0	9
Amoco Canada	Nipisi	Alta.	2/84	16,640	105	26	Gilwood	S	18.0	200
Amoco Canada	South Swan Hills	Alta.	1973	28,480	166	49	Beaverhill Lake	LS	8.0	49
Amoco Canada	Bigoray Nisku B	Alta.	2/80	251	3	1	Nisku D-2	LS	5.0	1,130
Amoco Canada	Rainbow S.	Alta.	8/72	490	5	2	Keg River	Dolo.	6.0	40
Canterra Energy	Rainbow KRD Pool	Alta.	3/76	85	1	1	Keg River	LS	10.0	200
Canterra Energy	Rainbow H Pool	Alta.	6/73	165	3	1	Keg River	LS	9.4	200
Canterra Energy	Rainbow Lake	Alta.	6/84	2,500	28	9	Keg River	LS	8.0	300
Canterra Energy	Rainbow E Pool	Alta.	6/72	173	5	1	Keg River	LS	12.0	300
Canterra Energy	Rainbow G Pool	Alta.	10/72	163	2	1	Keg River	LS	8.0	300
Canterra Energy	Rainbow O Pool	Alta.	2/70	703	9	1	Keg River	LS	6.0	150
Canterra Energy	Rainbow KRA Pool	Alta.	12/68	633	17	2	Keg River	LS	10.1	100
Canterra Energy	Rainbow EEE Pool	Alta.	4/70	150	2	1	Keg River	LS	16.8	500
Chevron Canada Resources	Mitsue, Stage 1 & 2	Alta.	5/85	17,280	71	27	Gilwood	S	13.4	282
Chevron Canada Resources	Pembina Nisku F	Alta.	12/87	200	3	1	Nisku F	LS	12.7	660
Chevron Canada Resources	Acheson D-3A	Alta.	7/87	610	15	1	Leduc D3	LS	12.0	5,000
Chevron Canada Resources	Bigoray Nisku F	Alta.	3/87	150	2	1	Nisku F	LS	10.0	250
Chevron Canada Resources	W. Pembina Nisku A	Alta.	8/81	200	2	1	Nisku A	LS	10.0	750
Chevron Canada Resources	W. Pembina Nisku D	Alta.	5/81	320	2	1	Nisku D	LS	10.0	300
Chevron Canada Resources	Pembina Nisku A	Alta.	2/84	400	3	1	Nisku A	LS	8.0	250
Chevron Canada Resources	W. Pembina Nisku C	Alta.	4/84	130	2	1	Nisku C	LS	11.0	1,100
Chevron Canada Resources	Pembina Nisku D	Alta.	11/85	400	5	1	Nisku	LS	12.0	750
Dome Petroleum	Kaybob South	Alta.	7/84	8,000	40	9	Triassic	Dolo.	11.5	92
Dome Petroleum	Willesden Green	Alta.	1977	26,000	110	46	Cardium	S	11.0	5
Gulf Canada	Fenn-Big Valley	Alta.	4/83	1,268	26	7	Nisku	Dolo.	8.0	400
Gulf Canada	Goose River	Alta.	10/86	2,847	14	5	Beaverhill Lake	LS	8.0	240
Home Oil	Swan Hills	Alta.	10/85	11,520	86	23	Beaverhill Lake	LS	8.5	54
Mobil Oil Canada	Rainbow	Alta.	7/83	110	4	1	Keg River	Dolo./LS	9.8	1,000-4,000
Mobil Oil Canada	Rainbow	Alta.	9/72	360	14	2	Keg River	Dolo./LS	8.6	7,000
Petro-Canada	Brazeau River	Alta.	9/81	640	3	1	Nisku	LS	7.0	50
Petro-Canada	Caroline	Alta.	9/84	12,620	27	12	Cardium	S	10.0	12
Petro-Canada	Brazeau River	Alta.	9/80	640	2	1	Nisku	LS	10.0	396
Petro-Canada	Brazeau River	Alta.	11/80	640	2	1	Nisku	LS	10.0	35
Shell Canada	Simonette	Alta.	11/86	150	4	1	Leduc D-3	Dolo.	7.0	10-50
Texaco Canada Resources	Wizard Lake	Alta.	1969	2,725	49	14	Leduc D-3A	Dolo.	9.6	1,375
Texaco Canada Resources	Pembina 'K' Pool	Alta.	1980	126	2	1	Nisku	Dolo.	12.7	2,020
Texaco Canada Resources	Pembina 'Q' Pool	Alta.	1985	301	2	1	Nisku	Dolo.	9.8	1,970
Texaco Canada Resources	Pembina 'M' Pool	Alta.	1983	192	2	1	Nisku	Dolo.	9.0	540
Texaco Canada Resources	Pembina 'L' Pool	Alta.	1985	625	4	2	Nisku	Dolo.	10.5	1,060
Texaco Canada Resources	Pembina 'O' Pool	Alta.	1983	346	1	1	Nisku	Dolo.	11.8	3,100
Texaco Canada Resources	Pembina 'G' Pool	Alta.	1981	328	3	1	Nisku	Dolo.	8.0	900
Texaco Canada Resources	Pembina 'P' Pool	Alta.	1983	420	3	1	Nisku	Dolo.	10.3	2,400
Natural gas immiscible										
Home Oil	Silverdale	Sask.	12/85	160	9	3	Sparky	S	29.0	2,000
Polymer										
Chevron Canada Resources	Taber Mannville D Unit 1	Alta.	6/87	480	7	2	Mannville D	S	22.3	1,140
Encor Energy	Rapdan Unit	Sask.	1/86	440	12	5	Upper Shaunavon	S	17.0	85
Steam										
Amoco Canada	Elk Point	Alta.	1987	33	9	4	Lower Cummings	S	30.0	1,500
Canada Northwest Energy	Atlee Buffalo	Alta.	9/86	640	10		Glauconite	S	30.0	2,000
Home Oil	Kitscoty	Alta.	1981	160	9	9	Sparky B	S	27.6	905
Husky	Westhazel	Sask.	1985	0	0	0	GP	US	33.0	5,000
Husky	Aberfeldy	Sask.	1979	63	24	5	Sparky	US	30.0	1,700
Koch Exploration Canada	Fort Kent	Alta.	12/85	160	13		Upper Grand Rapids	S	35.0	
Mobil Oil Canada	Celtic	Sask.	4/85	50	5		General Petro.	S	35.0	2,000
Mobil Oil Canada	Cold Lake	Alta.	1982		3		Lower Grand Rapids	S	35.0	1,000
Mobil Oil Canada	Celtic	Sask.	10/80	125	25		Waseca	S	35.0	1,000
Norcen	Provost	Alta.	1985	20	9	9	McLaren	S	27-30.0	4,000
Phillips Petroleum	Coleville	Sask.	1984	35	2	2	Bakken	S	25.0	400
Sceptre Resources	Tangleflags North	Sask.	12/87	17	1	3	Lloydminster Sand	US	32.5	
Texaco Canada Resources	Lone Rock	Sask.	11/84	45	13	3	Sparky	S	34.4	2,800
Westmin Resources	Lindbergh	Alta.	1985	40	8	1	Basal Mannville	S	30.0	3,000
Westmin Resources	Lindbergh	Alta.	1985	160	16		Basal Mannville	S	30.0	3,000

Producing EOR projects outside U.S. and Canada

Country	State/area	Operator	Project type	Field	Start date	Area, acres	Number wells Prod.	Number wells Inj.	Pay zone	Formation type	Porosity, %	Permeability, md
Austria		OMV	Caustic	Matzen/Scho	7/80	33	5	1	Torton 9th	S	23.5	300
Austria		OMV	Steam	Maustrenk	7/80	123	19	3	Schlier		24.0	147
Colombia	Middle Magdalena Basin	Texas Petroleum	Steam	Teca	2/84	3,000	200		Oligocene A & B	S	28.0	1,200
Colombia	Middle Magdalena Basin	Texas Petroleum	Steam	Cocorna	3/87	15	8	3	Oligocene B	S	28.0	700
Congo		Elf Congo	Steam	Emeraude-Offshore*	1985	9	8	2	Senonien	S	25.0	80

Depth, ft	Gravity, °API	Reservoir oil cp	°F.	Previous prod.	Residual saturation % Start	End	Project maturity	Total prod., b/d	Enhanced prod., b/d	Project eval.	Profit	Project scope
11,000	40.0	0.14	233	Prim.	90.0	60.0	HF	1,230	1,230	Succ.	Yes	FW
5,500	41.0	0.84	131	WF	30.0	5.0	JS	17,000	3,000	Succ.	Yes	FW
8,500	38.0	0.40	225	Prim./WF	35.0	13.0	NC	17,925	17,925	Succ.	Yes	FW
7,500	34.0	1.40	169	Prim.	0.7	0.3	HF	800	800	Succ.	Yes	FW
6,200	40.0	0.30	183	Prim.			NC	1,900	1,900	Succ.	Yes	FW
6,310	40.0	0.426	195	Prim.	90.0	27.0	NC	180	180	Succ.	Yes	FW
6,210	39.0	0.596	187	Prim.	92.0	28.0	HF	1,500	1,500	Succ.	Yes	FW
6,000	39.0	0.83	180	WF	52.0	23.0	JS	8,000	3,000	Prom.	Yes	FW
5,932	39.0	0.439	175	Prim.	91.0	27.0	HF	3,200	3,200	Succ.	Yes	FW
6,151	39.0	0.485	185	Prim.	92.0	28.0	HF	750	750	Succ.	Yes	FW
6,053	42.0	0.28	180	Prim.	87.0	26.0	NC	950	950	Succ.	Yes	FW
6,380	43.0	0.29	195	Prim.	90.0	30.0	HF	5,030	5,030	Succ.	Yes	FW
6,085	37.0	0.542	180	Prim.	95.0	29.0	NC	200	200	Succ.	Yes	FW
5,000	41.0	0.65	147	WF	30.0	19.0	JS	10,250	4,350	Succ.	Yes	FW
8,000	40.0	0.50	208	WF	75.0	43.3	JS	285	0	TETT	TETT	P (Exp. L)
5,000	39.0	0.85	139	Prim.	90.0	29.0	JS	2,400	1,400	TETT		FW
10,500	37.0	0.80	175	WF	90.0	5.0	JS	4,000	1,100	Succ.	Yes	FW
10,000	42.0	0.30	228	Prim.	90.0	5.0	HF	3,700	2,000	Succ.	Yes	FW
10,000	42.0	0.20	218	Prim.	90.0	5.0	HF	3,000	1,500	Succ.	Yes	FW
10,000	41.0	0.30	210	WF	80.0	5.0	JS	3,650	900	Succ.	Yes	FW
10,300	42.0	0.30	218	WF	75.0	9.0	HF	6,000	2,500	Prom.	Yes	FW
9,000	41.0	0.70	176	WF	90.0	5.0	JS	6,400	1,700	Succ.	Yes	FW
6,981	40.0	0.419	190	WF	46.0	31.0	HF	4,500	950	Prom.	Yes	FW
6,400	38.0	0.55	136	Prim.	87.0	73.0	NC	2,000	700	Disc.	No	FW
5,249	32.8	1.34	136	Prim.	33.0		JS	623	623	Succ.	Yes	FW
9,200	41.0	0.4	234	WF			JS	3,800	0	TETT		FW
8,700	39.0	0.4		WF	30.0	5.0	JS	13,000	3,800	Succ.	Yes	FW
5,500	39.0	0.46	190	WF	0.4	0.1	HF	283	283	Prom.	Yes	FW
5,500	39.0	0.47	184	Gas/WF	0.8	0.1	HF	2,500	2,500	Succ.	Yes	FW
10,091	42.0	0.38	216	Prim.	90.0	15.0	HF	3,100	3,100	Succ.	Yes	FW
8,200	46.0	0.25	173	Prim.	85.0	45.0	HF	1,900	1,550	Succ.	Yes	FW
10,215	44.0	0.40	218	Prim.	90.0	10.0	HF	7,400	7,400	Succ.	Yes	FW
10,544	46.0	0.20	222	Prim.	90.0	15.0	HF	2,500	2,500	Succ.	Yes	FW
11,600	47.0	0.13		WF			JS	170	0	TETT		
6,500	38.0	0.88	167	Prim.	93.0	3.8	HF	28,000	28,000	Succ.	Yes	FW
9,469	43.6	0.37	198	Prim	82.0	5.0	JS	2,010	860	Prom.	Yes	FW
9,421	41.3	0.42	196	Prim.	91.0	5.0	JS	2,700	1,410	Prom.	Yes	FW
9,336	41.1	0.14	198	Prim.	93.0	5.0	JS	2,460	1,150	Prom.	Yes	FW
9,415	40.9	0.42	199	WF	88.0	5.0	JS	3,770	1,470	Prom.	Yes	FW
9,332	43.4	0.32	190	Prim	84.0	5.0	JS	1,250	540	Prom.	Yes	FW
9,541	43.2	0.33	204	Prim.	80.0	5.0	JS	2,500	1,100	Prom.	Yes	FW
9,531	45.4	0.36	200	Prim.	87.0	5.0	JS	4,110	1,920	Prom.	Yes	FW
560	16.0	5,000		Prim.	82.0	74.6	HF	75	75	Prom.	Yes	P (Exp. UL)
3,200	18.0	55	121	WF	92.0	7.0	JS	17,600	16,400	Succ.	Yes	FW
4,500	23.0	10	132	WF	55.0	43.0	JS	470	350	Prom.	Yes	P (Exp. L)
2,000	12.0	30,000	73	Prim.			JS	400		TETT		P
2,900	14.5	800	80	Prim.	80.0	1.0	JS	1,000	1,000	Succ.	Yes	P (Exp. L)
2,066	13.0	3,000		Prim.	80.0	52.0	HF	725	575	Succ.	Yes	P (Exp. L)
1,575	13.6	18,000	70	Prim.	82.0		JS			TETT	TETT	P (Exp. UL)
1,700	17.0	800	70	Prim./WF	71.0	43.0	HF	1,000	1,000	Prom.	No	P (Exp. UL)
1,000	11.0	21,300		None	12.0		HF	0	0	Prom.		P
1,576	13.0	7,000	67	Prim.	69.0	46.0	JS	290	290	Prom.	Yes	P (Exp. L)
1,200	10-12.0	20000-100000	60	None	65.0	50.0	NC	100	100	Prom.		
1,500	13.0	5,380	61	Prim.	70.0	47.0	HF	630	630	Prom.	Yes	P (Exp. L)
2,430	12.0	2,000	80	Prim.	82.0	45.0	HF	600	600	Prom.	No	P (Exp. L)
2,600	14.0	1,000		WF			NC	60	22	Prom.	No	P
1,476	12-13.0	13,000	66	Prim.	80.0	15.0	JS			TETT		P
1,900	16.0	800	70	Prim.	82.0		HF			Prom.	TETT	P (Exp. L)
1,900	12.0	1,200		Prim.	70.0	55.0	JS	500		TETT	TETT	P (Exp. UL)
1,900	12.0	1,200		Prim.	70.0	55.0	JS	500	400	TETT	Yes	P (Exp. UL)

Table D

Depth, ft	Gravity, °API	Reservoir oil cp	°F.	Previous prod.	Residual saturation % Start	End	Project status†	Total prod., b/d	Enhanced prod., b/d	Project eval.	Profit	Project scope
4,330	21.0	12.7		Prim./WF	62.0	43.0	NC	78	21	Succ.	Yes	P (Exp. L)
3,018	26.0	7.81		Prim./WF	51.0	46.0	NC	84	54	TETT	No	FW
1,600	12.8	2,965		Prim.	60.0		HF	17,000	8,500	Succ.	Yes	FW
2,100	12.8	2,965		Steam soak	57.0	19.0	JS	300	50	Prom.		P (Exp. L)
600	22.0	100	85	Prim.			HF	1,600	1,600	Prom.		P

Producing EOR projects outside U.S. and Canada — continued

Country	State/area	Operator	Project type	Field	Start date	Area, acres	Number wells Prod.	Number wells Inj.	Pay zone	Formation type	Porosity, %	Permeability, md
England		BP	Micellar-polymer	Bothamsall	7/83	10	3	1	Sub-Alton	S	12.0	14
France		Elf Aquitaine	Micellar-polymer	Chateaurenard	1984	71	9	4	Neocomian	S	30.0	1,300
France		Elf Aquitaine	Polymer	Chateaurenard	1985	12	4	1	Neocomian	S	30.0	1,000
France		Elf Aquitaine	Steam	Lacq Superieur	1977	30	8	1	Senonien	LS	15.0	1
Hungary		NKFV	CO$_2$	Nagylengyel	9/80	200	5	1	Triassic	LS		1,000
Hungary		NKFV	CO$_2$	Budafa	1981	1,013	77	60	Zala Kerettye	S	22.0	110
Hungary		NKFV	Combustion	Demjen East	1986	13.6	6	3	Oligocene c,da,df,e	S	19.0	100
Indonesia		Total	Micellar-polymer	Handil	1982	4	1	3		S	30.0	1,000
Trinidad		Trintoc	CO$_2$ push-pull	Forest Reserve	5/84		17		Forest Sands	S	30.0	150-300
Trinidad		Trintoc	Steam	Forest Reserve	8/76	58	4	1	Forest Sand	S	31.0	205
Trinidad		Trintoc	Steam	Point Fortin Central	2/86	32	15		Cruse E	S	31.0	95
Trinidad	Point Fortin	Trintoc	Steam	Parrylands	7/82	90	46		Forest Sands	S	30.0	500
Trinidad		Trintoc	Steam	Forest Reserve	7/65	135	50	14	Forest Zone 5.1	S	33.0	340
Venezuela	Anzoategui	Corpoven S.A.	Gas	Oveja	5/64	1,686	12	2	L-1L	S	30.4	2,642
Venezuela	Anzoategui	Corpoven S.A.	Gas	Oveja	8/61	5,467	31	3	I.2L,3	S	37.0	2,500
Venezuela	Anzoategui	Corpoven S.A.	Gas	Oveja	1/67	1,325	9	1	L-1L	S	31.0	2,940
Venezuela	Anzoategui	Corpoven S.A.	Gas	Oveja	1/64	2,422	18	2	J-3	S	28.0	3,000
Venezuela	Monagas	Corpoven S.A.	Gas	Oritupano	2/79	1,052	8	1	A-9	S	28.0	1,300
Venezuela	Monagas	Corpoven S.A.	Gas	Oritupano	9/76	1,054	6	1	A-7	S	30.0	300
Venezuela	Anzoategui	Corpoven S.A.	Steam	F.P.O.			7	7	U-1 (OS-721)	S	27.9	2,000
Venezuela	Anzoategui	Corpoven S.A.	Steam	Mezones	4/83	57,314	14	14	U-1,2,3	S	34.0	1,200
Venezuela	Anzoategui	Corpoven S.A.	Steam	F.P.O.			13	13	U-23 (MFB-53)	S	31.9	1,400
Venezuela	Anzoategui	Corpoven S.A.	Steam	F.P.O.			4	4	S-3,4 (MBA-8)	S	30.7	5,800
Venezuela	Anzoategui	Corpoven S.A.	Steam	F.P.O.			58	58	U-3,4 (MFB-24)	S	28.6	5,000
Venezuela	Anzoategui	Corpoven S.A.	Steam	Arecuna	12/83	500	15	15	T	S	30.0	2,500
Venezuela	Campo Cerro Negro	Lagoven S.A.	Steam	B.E.P.-Cerro Negro	1984	49,090	144		Ofic. Morichal-Memb.	S	32.0	7,000
Venezuela	Campo Jobo	Lagoven S.A.	Steam	Jobo Picv	3/81	90	22	6	Morichal-1	S	32.0	10,000
Venezuela	Campo Jobo	Lagoven S.A.	Steam	Jobo	12/69	4,329	18		Ofic. Jobo Member	S	32.0	2,500
Venezuela	Campo Jobo	Lagoven S.A.	Steam	Jobo - P.E.T.C.	8/85	267	17		Oficina	S	30.0	8,000
Venezuela	Campo Jobo	Lagoven S.A.	Steam	Jobo	12/69	27,522	99		Ofic. Mor-Member	S	31.0	5,000
Venezuela	Campo Pilon	Lagoven S.A.	Steam	West Pilon	12/69	1,065	29		Oficina-1	S	29.1	5,000
Venezuela	Zulia	Lagoven S.A.	Steam	Bachaquero	12/80	343	2		Bachaquero Superior	S	23.0	1,500
Venezuela	Zulia	Lagoven S.A.	Steam	Lagunillas	2/71	9,343			Bachaquero	S	34.0	4,000
Venezuela	Zulia	Maraven S.A.	Steam	East Tia Juana	3/69	1,773	145		L.L.	S	38.1	
Venezuela	Zulia	Maraven S.A.	Steam	Lagunillas	4/65	420	59		U.L.H.	S	35.0	
Venezuela	Zulia	Maraven S.A.	Steam	East Tia Juana	4/59	339	33		L.L.	S	38.1	
Venezuela	Zulia	Maraven S.A.	Steam	East Tia Juana	8/69	415	24		L.L.	S	38.1	
Venezuela	Zulia	Maraven S.A.	Steam	Lagunillas	11/79	3,101	267		L.L.	S	33.7	
Venezuela	Zulia	Maraven S.A.	Steam	East Tia Juana	9/64	415	36		L.L.	S	38.1	3,000
Venezuela	Zulia	Maraven S.A.	Steam	Main Tia Juana	10/63	873	82		L.L.	S	38.1	
Venezuela	Zulia	Maraven S.A.	Steam	East Tia Juana	8/68	2,629	224		L.L.	S	38.1	
Venezuela	Zulia	Maraven S.A.	Steam	East Tia Juana	8/69	1,642	131		L.L.	S	38.1	
Venezuela	Zulia	Maraven S.A.	Steam	Lagunillas	4/70	2,565	220		U.L.H.	S	35.0	
Venezuela	Zulia	Maraven S.A.	Steam	Main Tia Juana	7/67	1,271	114		L.L.	S	38.1	675
Venezuela	Zulia	Maraven S.A.	Steam	Main Tia Juana	7/67	1,500	134		L.L.	S	38.1	750
Venezuela	Zulia	Maraven S.A.	Steam	Main Tia Juana	6/68	141	15	7	L.L.	S	33.0	
Venezuela	Zulia	Maraven S.A.	Steam	Main Tia Juana	10/67	2,221	197		L.L.	S	38.1	750
Venezuela	Zulia	Maraven S.A.	Steam	Lagunillas	1/70	3,025	147		L.L.	S	33.7	
Venezuela	Zulia	Maraven S.A.	Steam	East Tia Juana	1/74	1,830	135	21	L.L.	S	38.1	780
Venezuela	Zulia	Maraven S.A.	Steam	Bachaquero	10/84	7,795	539		Post-Eoceno	S	37.3	
Venezuela	Zulia	Maraven S.A.	Steam	East Tia Juana	2/61	36	7		L.L.	S	38.0	
Venezuela	Zulia	Maraven S.A.	Steam	Lagunillas	7/67	618	54		U.L.H.	S	35.0	
Venezuela	Zulia	Maraven S.A.	Steam	Main Tia Juana	6/66	141	15		L.L.	S	38.1	
Venezuela	Zulia	Maraven S.A.	Steam	Lagunillas	11/80	3,565	299		U.L.H.	S	35.0	
Venezuela	Zulia	Maraven S.A.	Steam	Lagunillas	8/70	2,114	175		U.L.H.	S	35.0	
Venezuela	Zulia	Maraven S.A.	Steam	East Tia Juana	5/68	2,667	232		L.L.	S	38.1	
Venezuela	Zulia	Maraven S.A.	Steam	East Tia Juana	12/68	1,956	168		L.L.	S	38.1	
Venezuela	Zulia	Maraven S.A.	Steam	Lagunillas	1/70	76	7		L.L.	S	33.7	
West Germany	Lower Saxony	Wintershall AG	Hot water flood	Emlichheim 15	1/67	115	15	2	Valenginian	S	30.0	6,000
West Germany	Lower Saxony	Wintershall AG	Hot water flood	Emlichheim 14	10/67	64	8	1	Valenginian	S	30.0	6,000
West Germany	Lower Saxony	Wintershall AG	Hot water flood	Emlichheim 17	10/74	81.5	10	2	Valenginian	S	30.0	6,000
West Germany		Wintershall AG	Hot water flood	Ruehlertwist	12/86	53	10	2	Valenginian	S	30.0	400
West Germany	Lower Saxony	Wintershall AG	Hot water flood	Emlichheim 10	11/81	41.5	6	1	Valenginian	S	30.0	6,000
West Germany	Lower Saxony	Wintershall AG	Hot water flood	Emlichheim 11	9/73	70	8	1	Valenginian	S	30.0	6,000
West Germany	Lower Saxony	Wintershall AG	Hot water flood	Emlichheim 07	10/67	81.5	10	1	Valenginian	S	30.0	6,000
West Germany		Texaco AG	Polymer	Hankensbuettel Westblock 4	4/77	165	7	2	Upper Dogger-Beta	S	28.0	2,000-4,000
West Germany		Texaco AG	Polymer	Hankensbuettel Westblock 2	5/79		6	3	Upper Dogger-Beta	S	28.0	2,000-4,000
West Germany	Northwest	Preussag AG	Polymer	Edesse-Nord	11/85	4.2	3	2	Wealden	S	24.0	1,200
West Germany	Lower Saxony	Wintershall AG	Polymer	Bockstedt	9/84	57	5	1	Valenginian	S	23.0	4,000
West Germany		Texaco AG	Polymer	Oerrel Westblock 4	1980	410	13	3	Upper Dogger-Beta	S	28.0	2,000-4,000
West Germany	Emsland	BEB	Steam	Ruhlermoor	8/86	300	30	4	Valanginian	S	27-31.0	600-1000
West Germany	Lower Saxony	BEB	Steam	Georgsdorf	1/75	165	19	2	Valanginian	S	25.0	460-1150
West Germany	Lower Saxony	BEB	Steam	Georgsdorf	12/79	144	23	3	Valanginian	S	25.0	1000-1300
West Germany	Lower Saxony	BEB	Steam	Ruhlermoor	11/80	230	24	4	Valanginian	S	28-30.0	300-1000
West Germany	Lower Saxony	Wintershall AG	Steam	Ruehlertwist	12/78	237	24	2	Valenginian	S	28.0	5,000
West Germany	Lower Saxony	Wintershall AG	Steam	Emlichheim 06	5/81	121	25	4	Valenginian	S	30.0	6,000
West Germany		Texaco AG	Surfactant	Hankensbuettel Westblock 4	1982	83	7	3	Lower Dogger-Beta	S	28.0	300-1,000

Depth, ft	Reservoir oil				Previous prod.	Residual saturation %		Project status†	Total prod., b/d	Enhanced prod., b/d	Project eval.	Profit	Project scope
	Gravity, °API	cp		°F.		Start	End						
3,140	41.0	3		110	WF	57.0	50.0	NC	12		Disc.	No	
1,950	27.0	40		86	Prim.	48.0	23.0	NC	118	32	Succ.	No	P (Exp. UL)
1,950	27.0	40			Prim.	55.0	30.0	HF	144	108	Succ.	Yes	P (Exp. UL)
1,950-2,100	22.0	20		150	Prim.	45.0		HF	1,000	400	Succ.	Yes	FW
7,400	26.0	92		248	Prim.			NC	200	200	Succ.		
3,000	42.0	0.8		147	Prim.			NC	700	700	Succ.		
820	39.0	7			Prim.			JS	92	83	TETT	No	P (Exp. L)
4,500	33.0	0.4		115	WF	35.0			400	400			
	20-25.0	66-90			Prim.	60.0		JS	150	150	Prom.		P (Exp. L)
3,000	19.0	32		120	Prim.	57.0	45.0	HF	100	100	Prom.		
1,200-1,600	16-18.0	150			Prim.	58.0	18.0	JS	300	300	Prom.		P (Exp. L)
1,000-1,200	11.5	5,000				75.0	20.0	JS	650	650	Prom.		FW
1,000-1,300	15.7	148			Prim.	70.0	20.0	HF	1,650	1,650	Succ.	Yes	FW
3,350	18.0	12.5		152	WF	90.0	65.0	NC	2,600	2,600	Succ.	Yes	FW
3,150	15.0	38		150	Prim.	90.0	67.0	HF	500	400	Prom.	Yes	FW
3,430	21.0	5.5		154		90.0	63.0	NC	2,200	2,200	Succ.	Yes	FW
3,300	20.0	52		149		90.0	50.0	HF	3,000	3,000	Succ.	Yes	FW
5,200	21.0	4.9		146	Prim.	84.0	69.0	HF	1,200	1,200	Prom.	Yes	FW
5,150	20.5	4.9		146	Prim.	90.0	72.0	HF			Succ.	Yes	FW
3,400	9.8	232			Prim.	78.8		HF	1,358	980	Prom.	Yes	P (Exp. L)
4,300	11.5	800		130	Prim.	70.0	63.0	HF	4,900	2,500	Prom.	Yes	P (Exp. UL)
2,650	9.3	376			Prim.	88.0		HF	11,047	6,700	Prom.	Yes	P (Exp. L)
2,850	10.0	370			Prim.			JS	600	450	TETT	Yes	P (Exp. L)
	9.2	351			Prim.	83.0		HF	46,400	28,000	Succ.	Yes	P
3,150	10.0	1,617		140	Prim.	80.0	72.0	HF	1,390	800	Prom.	Yes	FW
2,800	8.5	5,500			Prim.	80.0		JS	33,120	8,050	Succ.	Yes	
3,600	8.5	1,800			Prim.	82.0		NC	4,000	4,000	Succ.	Yes	P (Exp. UL)
	12.6	80			Prim.	80.0		JS	2,500	180	Succ.	Yes	FW
4,025	8.5	1,850			Prim.	75.0		JS	3,500	3,500	Succ.	Yes	P (Exp. L)
3,600	8.5	1,850			Prim.	85.0		JS	27,100	1,430	Succ.	Yes	FW
3,200	9.5	1,850			Prim.	82.2		JS	4,350	450	Succ.	Yes	FW
3,400	14.0	185			Prim.	85.0		JS	10		TETT	No	P (Exp. L)
2,690	12.0	600			Prim.	84.0		HF	12,420	6,428	Succ.	Yes	FW
1,700	11.9	2,000		111	Prim.	92.7	74.5	HF	1,100	850	Prom.	Yes	
2,100	11.4	3,500		117	Prim.	96.9	58.7	NC	3,900	1,370	Succ.	Yes	
1,500	12.2	1,000		108	Prim.	92.6	72.3	HF	1,800	550	Prom.	Yes	
850	10.5	10,000		95	Prim	99.9	84.6	HF	1,600	1,600	Prom.	Yes	
2,615	15.0	580		126	Prim.	83.3	68.8	HF	12,980	6,000	Prom.	Yes	
1,250	12.0	3,000		104	Prim.	95.6	78.6	HF	1,400	950	Prom.	Yes	
1,750	13.1	750		113		86.0	74.0	NC	1,900	592	Prom.	Yes	
1,000	11.6	4,000		104	Prim.	94.6	79.6	HF	6,300	5,000	Prom.	Yes	
1,250	10.2	12,000		102	Prim.	99.9	89.9	HF	5,800	5,800	Prom.	Yes	
1,750	11.8	2500-11500		115		97.3	89.4	HF	20,000	10,680	Prom.	Yes	
1,220	13.1	6,000		123	Prim.	94.3	80.8	HF	3,500	2,200	Prom.	Yes	
1,400	11.8	1,000		110	Prim.	91.1	76.0	HF	5,400	3,300	Prom.	Yes	
1,250	13.1	1,000		113	Prim.	88.0	68.0		3,100	2,000	Prom.	Yes	
1,746	13.1	4,100		110	Prim.	95.1	80.9	HF	12,600	10,000	Prom.	Yes	
2,500	15.2	500-4500		122	Prim.	86.0	74.3	HF	6,190	2,460	Prom.	Yes	
1,590	12.0	1,000		111		81.0	56.0	HF	20,000	18,000	Succ.	Yes	
2,000	13.0	300-800		117	Prim.	85.0		JS	93,000	41,600	Prom.	Yes	
1,000	9.5	12,000		105	Prim.	100.0	76.9	NC	130	130	Prom.	Yes	
2,100	11.4	3,500		117	Prim.	9.0	7.0	NC	4,100	760	Succ.	Yes	
1,250	13.1	1,300		104	Prim.	92.0	78.0	NC	300	60	Prom.	Yes	
2,398	11.0	2000-3700		125	Prim.	90.6	77.9	HF	38,000	12,870	Prom.	Yes	
1,725	11.4	4000-9000		118	Prim.	97.8		HF	10,600	5,200	Prom.	Yes	
1,200	12.0	500		104		93.6	82.5	HF	6,400	5,100	Prom.	Yes	
1,250	11.7	7,500		106	Prim.	97.5	82.5	HF	7,500	6,200	Prom.	Yes	
2,600	15.2	250		125	Prim.	73.7	63.5	NC	270	220	Succ.	Yes	
2,300	24.5	175		95	WF	72.0	54.0	NC	155	155	Succ.	Yes	
2,550	24.5	175		95	WF	73.0	58.0	NC	135	135	Succ.	Yes	
2,300	24.5	175		95	WF	66.0	55.0	HF	180	150	Succ.	Yes	
2,460	25.0	175			WF	76.0	68.0	JS	65	47	TETT	TETT	P (Exp. L)
2,460	24.5	175		95	WF	70.0	68.0	NC	40	30	Disc.	No	
2,550	24.5	175		95	WF	72.0	68.0	NC	250	230	Succ.	Yes	
2,400	24.5	175		95	WF	70.0	56.0	HF	90	90	Succ.	Yes	
4,900	27.0	13		136	WF	69.0	43.0	NC	1,000	700	Succ.	Yes	
4,900	28.0	13		136	WF	72.0	42.0	NC	1,700	1,300	Succ.	Yes	
1,200	36.0	8		73	WF	53.0	37.0	NC	20	12	Succ.	No	P (Exp. L)
4,100	34.0	11		130	WF	39.0	33.0	JS	150	40	Prom.	TETT	
4,900	25.0	17		136	WF	60.0	45.0	HF	540	400	Prom.	Yes	
1,700-2,100	25.0	120			Prim.	78.0		JS	2,100	1,500	TETT	TETT	FW
2,130-2,790	27.0	120		95	Prim.			HF	1,160	1,000	Succ.	Yes	
2,130-2,790	27.0	120		95	Prim.			HF	1,260	900	Succ.	Yes	
2,070-2,460	25.0	120		96	Prim.			HF	1,400	1,100	Succ.	Yes	
2,650	25.0	175		100	Edge water drive	51.0	42.0	HF	1,370	1,300	Succ.	Yes	
2,400	24.5	175		95	Hot water flood	62.0	45.0	NC	1,670	1,660	Succ.	Yes	
4,900	27.0	11		136	WF	58.0	40.0	HF	100	30	TETT	TETT	

Producing thermal EOR in U.S.

Operator	Field	State	County	Start date	Area, acres	Prod.	Inj.	Pay zone	Formation type	Porosity, %	Permeability, md
Steam											
ARCO	Midway-Sunset	Calif.	Kern	12/84	1.5	3	1	Fairfield Tar Sand	S	32.0	3,030
ARCO	Midway-Sunset	Calif.	Kern	1/72	7.8	22	2	Potter	S	32.0	1,675
ARCO	Midway-Sunset	Calif.	Kern	2/72	3.1	49	5	Potter	S	32.0	2,500
ARCO	Midway-Sunset	Calif.	Kern	8/72	3.2	11	1	Reef Ridge	S	32.0	1,000
Amoco	Winkleman Dome	Wyo.	Fremont	1964	160	21	15	Nugget	S	22.8	481
Angus Petroleum	Kern River	Calif.	Kern	9/83	85	35	35	Kern River	S	31.0	2,000
Celeron Oil & Gas	South Belridge	Calif.	Kern		90	41	30	Tulare	S	38.0	4,000
Celeron Oil & Gas	Midway-Sunset	Calif.	Kern		15	10	10	Monarch Sand	S	30.0	2,500
Chevron	Cymric	Calif.	Kern	5/66	26	18	0	Tar	S	32.0	1,200
Chevron	Cymric 36W	Calif.	Kern	5/75	110	98	47	Tulare/Amnicola	S	33.0	1,800
Chevron	Cymric 62Z	Calif.	Kern	2/86	30	31	19	Tulare	S	33.0	2,900
Chevron	Cymric Sec. 31X	Calif.	Kern	4/79	69	38	17	Amnicola	S	35.0	2,700
Chevron	Edison 27-RT	Calif.	Kern	7/77	30	65	29	Kern River	S	30.0	
Chevron	Fruitvale	Calif.	Kern	5/77	860	142		Chanac	S	29.0	920
Chevron	Kern Bluff	Calif.	Kern	7/66	214	25		Santa Margarita	S	29.0	2,000
Chevron	Kern River KCL 39	Calif.	Kern	10/75	118	67	42	Kern River	S	33.0	
Chevron	Kern River MC1	Calif.	Kern	4/76	132	120	76	Kern River	S	34.0	2,800
Chevron	Kern River MCII	Calif.	Kern	4/71	80	129	30	Kern River	S	34.0	2,800
Chevron	Kern River Sec. 3	Calif.	Kern	9/68	424	345	155	Kern River	S	34.0	3,389
Chevron	Kern River Sec. 4	Calif.	Kern	6/78	156	200	64	Kern River	S	34.0	
Chevron	Kern River-ANO	Calif.	Kern	5/74	160	177	82	Kern River	S	32.0	
Chevron	Lost Hills Sec. 18	Calif.	Kern	8/85	14	8	0	Tulare	S	38.0	2,000
Chevron	Lost Hills Sec. 18	Calif.	Kern	7/83	38	39	11	Tulare	S	38.0	2,100
Chevron	Lost Hills Sec. 30	Calif.	Kern	1/80	190	37	0	Etchegoin A	S	40.0	800
Chevron	Lost Hills Sec. 32	Calif.	Kern	3/85	16	11	4	Etchegoin A	S	40.0	2,500
Chevron	Lost Hills Sec. 32	Calif.	Kern	2/79	150	29		Etchegoin A	S	36.0	
Chevron	McKittrick 6Z	Calif.		4/85	30	29	18	Tulare/Amnicola	S	36.0	2,900
Chevron	McKittrick 9Z	Calif.	Kern	8/78	80	27	11	Amnicola	S	38.0	2,700
Chevron	Midway-Sunset Sec. 15A	Calif.	Kern	5/78	23	36	5	Potter	S	36.0	3,900
Chevron	Midway-Sunset Sec. 25A	Calif.	Kern	6/87	35	3	2	Tulare	S	35.0	1,000
Chevron	Midway-Sunset Sec. 26C	Calif.	Kern	11/75	84	243	81	Monarch	S	32.0	1,100
Chevron	Midway-Sunset Sec. 2F	Calif.	Kern	10/83		44	10				
Chevron	Midway-Sunset. Sec. IF	Calif.	Kern	2/78	19	23	8	Upper 10-10	S	32.0	2,200
Chevron	N. E. McKittrick 16-Z	Calif.	Kern	9/75	87	42	27	Amnicola	S	35.0	2,700
Chevron	Tejon-Grapevine	Calif.	Kern	12/78	40	8		Basal Chanac		30.0	2,160
Chevron	West Coalinga 13-D	Calif.	Fresno	5/73	550	163	93	Temblor	S	35.0	1,000
Chevron	West Coalinga 25-D	Calif.	Kern	5/80	36	131	95	Chanac	S	29.0	1,000
Chevron	Yorba Linda	Calif.	Kern	7/62	80	76	0	Shallow Tar	S	27.0	1,000
Chevron	Kingdom Abo	Tex.	Terry	5/85	2,240	59	48	Abo Reef	Dolo.	4.5	8
DCR Petroleum	Camp Hill	Tex.	Anderson	7/86	7.25	11	4	Carrizo	S	37.0	2,500
Exxon	Edison	Calif.	Kern	1965	1,100			Kern River	S	28.0	1,100
Lawrence-Allison	Teapot Dome NPR-3	Wyo.	Natrona	10/85	20	22	4	Shannon	S	18.0	63
M. H. Whitter	Midway-Sunset	Calif.	Kern	1965	200	340		Potter	US	35.0	2,500
Marathon Oil	Garland	Wyo.	Big Horn	5/86	35	6	1	Madison	LS	15.5	10.4
Mobil	Cymric McKittrick Fee	Calif.	Kern	7/65	180	40	0	Welport Amnicola	S	35.0	3,000
Mobil	Edison	Calif.	Kern	12/81	100	11	0	Kern River	S	30.0	1,000
Mobil	Kern Front	Calif.	Kern	5/74	480	43	0	Etchegoin	S	37.0	1,000
Mobil	Midway-Sunset	Calif.	Kern	10/70	400	104	0	Monarch	S	34.0	4,000
Mobil	North Midway-Sunset	Calif.	Kern	11/67	410	300		Potter	S	35.0	1,000
Mobil	San Ardo	Calif.	Monterey	6/68	230	80	15	Aurignac	S	34.5	3,000
Mobil	San Ardo	Calif.	Monterey	3/80	63	31	9	Lombardi	S	32.5	3,000
Mobil	South Belridge	Calif.	Kern	1965	1,600	42		Tulare	S	35.0	3,000
Mobil	South Belridge	Calif.	Kern	1969	900	370	135	Tulare	S	35.0	3,000
Mobil	Saratoga	Tex.	Hardin	6/87	54	10	5	Miocene	S	35.0	3000
Oxy USA	Kern Front	Calif.	Kern	11/81	342	120	8	Etchegoin Chanac	S	33.0	3,250
Phillips	Smackover	Ark.	Ouachita	10/71	985	91	5	Nacatoch	S	35.0	2,000
Santa Fe Energy	Coalinga	Calif.	Fresno	3/79	280	250	60	Lower Temblor	S	30.0	100-3,000
Santa Fe Energy	Kern River	Calif.	Kern	1/80	180	143	68	Kern River	S	30.0	100-3,000
Santa Fe Energy	Midway	Calif.	Kern	1964	1,000	1,000		Potter Cyc.	S	30.0	2,500
Santa Fe Energy	Midway	Calif.	Kern	1970	400	450		Spellacy	S	30.0	2,250
Santa Fe Energy	Midway	Calif.	Kern	6/82	800	200		Tulare	S	33.0	1,300
Santa Fe Energy	Midway	Calif.	Kern	1/85	100	160	25	Potter S.D.	S	30.0	2,500
Shell	Arroyo Grande	Calif.		1/87	45	16	25	M6	S	31.0	
Shell	Belridge	Calif.	Kern	1961	2,880	4,350	984	Tulare	S	36.0	2,400
Shell	Brea Olinda	Calif.	Orange	6/85	19	20	2	Middle A Sand	S	20.0	200
Shell	Cat Canyon	Calif.	Santa Barbara	3/85	1,000	50	0	Basal Sisquoc	S	32.0	750
Shell	Coalinga	Calif.	Fresno	1981	300	150	12	Temblor Zone II	S	30.0	
Shell	Coalinga	Calif.	Kern	1980	300	199	39	Temblor Zone I	S		30
Shell	Coalinga	Calif.	Fresno	1984	300	60	1	Temblor Zone II	S	27.0	
Shell	Coalinga	Calif.		11/87	223	57	17	Etchegoin	S	34.0	
Shell	Cymric	Calif.	Kern	10/85	185	27	24	Tulare	S	36.0	
Shell	Kern River	Calif.	Kern	1970	600	550	220	Kern River Series	S	30.0	
Shell	McKittrick	Calif.	Kern	11/85	226	40	26	Tulare	S	36.0	
Shell	Midway-Sunset	Calif.	Kern	1971	375	637	50	Potter	S	30.0	
Shell	Midway-Sunset	Calif.	Kern	1980	85	146	13	Monarch	S	25.0	
Shell	Midway-Sunset	Calif.		1981	95	60	0	Upper Spellacy	S	28.0	
Shell	Midway-Sunset	Calif.	Kern	1983	143	182	24	Sub-Hoyt	S	31.0	
Shell	Midway-Sunset	Calif.	Kern	6/86	110			San Joaquin	S	33.0	
Shell	Mount Poso	Calif.	Kern	1971	2,800	253	46	Vedder	S	33.0	
Shell	Poso Creek	Calif.	Kern	6/80	360	70	0	Etchegoin	S	33.0	
Shell	Yorba Linda	Calif.	Orange	1971	310	300	44	Upper Conglomerate	S	22.0	6,000
Shell	White Castle	La.	Iberville	8/76	64	42	13	V type	S	35.0	3,000
Shell	White Castle	La.	Iberville	8/76	31	42	13	U	S	35.0	3,000
Shell Western E&P	Cymric	Calif.	Kern	12/86	185	27	24	Tulare	S	36.0	2000
Shell Western E&P	McKittrick	Calif.	Kern	6/87	226	40	26	Tulare	S	36.0	2000
Sun	Cymric	Calif.	Kern	1981	135	57	0	Tulare	S	35.0	1,500
Sun	Midway-Sunset	Calif.	Kern	3/64	106	205	14	Potter/Tulare	S	35.0	3,000
Sun	Midway-Sunset	Calif.	Kern	8/65	40	249		Potter	S	35.0	3,000
Sun	Midway-Sunset	Calif.	Kern	1/84		95	6	Potter/Marvic/Tulare	S		
Sun	Midway-Sunset	Calif.	Kern	7/84		174	0	Potter	S		
Sun	Midway-Sunset Sec. 15	Calif.	Kern	5/64	80	106	0	Potter	S	35.0	3,000

Table E

Depth, ft	Gravity, °API	cp	°F.	Previous prod.	Residual saturation % Start	End	Project maturity†	Total prod., b/d	Enhanced prod., b/d	Project eval.	Profit	Project scope
1,050	14.0	1,500	95	Prim.			HF	20	15	Succ.	Yes	P (Exp. L)
1,200	12.0	2,000	100	Steam soak			HF	475	315	Succ.	Yes	P (Exp. L)
1,400	14.0	1,500	100	WF			NC	610	275	Succ.	Yes	FW
1,000	13.0	1,500	100	Steam soak			HF	110	83	Succ.	Yes	P (Exp. L)
1,225	14.0	1,000	81	Prim.	71.0		NC	400	400	Succ.		FW
1,200	12.0	5,000		Prim.	55.0		JS	300	300	Succ.	Yes	FW
1,250	13.0	450			63.0	49.0		2,000	2,000	Succ.	Yes	FW
1,500	12.0	11,000	95					600	600	Succ.	Yes	FW
700	11.0	6,500	81	Prim.	99.0		NC	332	273	Succ.	Yes	FW
1,200	13.0	4,000	100		52.0		NC	4,261	4,261	Succ.	Yes	FW
1,200	12.0	5,500	100	Cyclic	51.0	15.0	JS	1,850	1,650	Succ.	Yes	FW
1,300	12.0	3,000	100	Prim.	50.0		NC	1,033	1,033	Succ.	Yes	FW
1,000	14.0	2,000	90	Prim.	50.0		NC	1,868	1,868	Succ.	Yes	FW
3,300	19.0	78.0	131	Prim.	71.0		HF	1,168	920	Succ.	Yes	FW
950	15.0	1,000	84	Prim.	74.0		NC	58	58	Succ.	Yes	
1,400	14.0	2,000	90		65.0		HF	1,043	1,043	Succ.	Yes	FW
960	14.0	2,000	90		59.0		HF	3,235	3,155	Succ.	Yes	FW
960	14.0	2,000	90	Prim.	59.0		HF	2,121	2,121	Succ.	Yes	FW
775	14.0	2,000	90	Prim.	55.0		HF	9,718	9,475	Succ.	Yes	FW
850	14.0	2,000	90		60.0		HF	4,251	3,980	Succ.	Yes	FW
1,000	14.0	2,000	90	Prim.	55.0		HF	5,107	5,107	Succ.	Yes	FW
350	12.0	4,600	90	Cyclic	55.0	20.0	JS	19	9	Disc.	No	P (Exp. UL)
300	14.0	400	75	Cyclic	86.0		JS	708	708	Succ.	Yes	P (Exp. L)
550	18.0	300	90	Prim.			HF	534	478	Prom.		
850	14.0	1,400	100	Cyclic	50.0	20.0	JS	170	125	Succ.	Yes	FW
900-1,150	18.0	200	90	Prim.	72.0		HF	516	75	Succ.	Yes	
1,300	13.0	2,700	100	Cyclic	48.0	15.0		1,912		Prom.		
1,300	13.0	4,000	100	Cyclic	47.0		HF	580	580	Succ.	Yes	FW
1,400	14.0	900	120	Cyclic	61.0	20.0	NC	1,416	965	Succ.	Yes	P
1,000	11.0	10,000	90	Cyclic	53.0		JS	75	0	TETT		P
1,300	·14.0	1,500	150	Cyclic	58.0	20.0	HF	8,195 2,797	7,638	Succ.	Yes	FW
1,600	14.0	1,500	105	Cyclic	55.0	15.0	NC	498	418	Succ.	Yes	FW
1,300	12.0	4,000	100	Cyclic	57.0		HF	634	634	Succ.	Yes	FW
2,600	16.0	167	115	Prim.				41		Disc.	No	
1,200	13.0	1,000	90	Prim.	46.0	15.0	NC	2,741	2,550	Succ.	Yes	FW
2,500	12.0	2,300	100	Prim.	50.0	12.0	HF	5,911	5,500	Succ.	Yes	FW
250-1,650	13.0	6,400	97	Prim.	95.0		NC	829	656	Succ.	Yes	FW
7,500	28.5	3.40	120	WF	58.0	40.0	JS	3,180	0	TETT		FW
400	18.4	1,200	75	Prim.	50.0		HF	150	150	Succ.	Yes	P (Exp. L)
1,100	12.0	8,000	95	Prim.	45.0	20.0	HF	525	525	Succ.	Yes	
250	33.0	10.0		Prim.	50.0	15.0	HF	350	300	Succ.	Yes	P (Exp. L)
1,000-1,500	13.0	2,200	100	Prim.	60.0	30.0	HF	5,900	5,900	Succ.	Yes	FW
4,250	22.0	29.0	140	Prim.			HF	419	81	TETT	No	P (Exp. L)
500	13.0	1,500	95	Prim.	87.5	57.7	NC	375	375	Succ.	Yes	FW
1,100	17.0	800	100	Prim.	87.0	33.0	JS	30	20	Succ.	Yes	FW
1,800	14.0	800	100	Prim.	86.0	67.0	NC	200	200	Succ.	Yes	FW
950	13.0	800	85	Prim.	64.0	53.0	HF	700	700	Succ.	Yes	FW
1,000	12.0	2,000	100	Prim.	65.0	30.0	HF	6,000	6,000	Succ.	Yes	FW
2,300	12.0	300	130	Steam soak	55.0	27.0	HF	1,900	1,900	Succ.	Yes	FW
2,100	11.0	2,000	125	Steam soak	55.0	27.0	HF	2,500	2,500	Succ.	Yes	P(Exp. L)
1,000	14.0	1,600	95	Prim.	64.0	60.0	NC	580	450	Succ.	Yes	FW
1,000	14.0	1,600	95	Prim.	60.0	35.0	HF	24,030	22,900	Succ.	Yes	FW
500	15.0	500	74	Prim.	50.0	20.0	JS	380	380	Prom.	Yes	FW
1,300-1,500	14.0	1,525	95	Prim.	50.0	15.0	HF	1,000	950	Succ.	Yes	FW
1,920	20.0	75.0	110	Prim.	71.0	50.0	NC	800	130	Succ.	Yes	FW
850-1,800	14.5	1,900	75	Prim.	35.0	15.0	JS	3,200	2,800	Succ.	Yes	FW
700	13.0	4,000	85	Prim.	45.0	15.0	NC	3,200	2,600	Succ.	Yes	
1,300	12.0	4,000	100	Prim.	60.0	33.0		17,000	15,000	Succ.	Yes	FW
900	11.5	6,500	90	Prim.	60.0	33.0		4,300	3,500	Succ.	Yes	FW
1,200	11.0	5,000	100	Prim.	50.0	43.0		1,800	1,400	Succ.	Yes	FW
1,300	12.0	4,000	100	Prim./cyclic	60.0	20.0		3,200	1,400	Succ.	Yes	FW
500-1,000	14.0	3,000		Prim.			JS	250	200	Prom.		FW
400-1,400	13.0	1,900	95	Prim.	75.0	20.0	HF	101,000	101,000	Succ.	Yes	FW
2,600	14.0	500	120	Steam soak	49.0		JS	320	280	Succ.	Yes	FW
4,400	10.8	2,896	130	Prim.	55.0		JS	1,500	500	TETT	Yes	FW
1,600	18.0	150	100	Steam soak			HF	2,500	2,200	Succ.	Yes	FW
1,000	13.0	6,000	110	Prim.			HF	3,400	3,200	Succ.	Yes	FW
1,800	18.0	150	100	Prim.			JS	900	720	Succ.	Yes	P (Exp. L)
950	10.6	10,000		Prim.			JS	380	380	Prom.		FW
1,400	13.0	2,000	100	Prim.			JS	1,700	1,400	Prom.	Yes	
300-1,500	13.0	6,000	85	Prim.			HF	12,000	11,436	Succ.	Yes	FW
500	12.0	10,000	100	None			JS	15	0	TETT	TETT	
1,200	11.0	8,000	95	Prim.			HF	13,000	13,000	Succ.	Yes	FW
1,300	13.0	5,000	95	Steam soak			HF	7,000	6,800	Succ.	Yes	FW
900	13.0	2,235	90	Prim.			JS	1,500	1,200	Succ.	Yes	FW
1,080	13.0	3,000	100	Prim.			JS	6,200	5,800	Succ.	Yes	FW
1,200	12.0	2,900	95	Prim.			JS	800	750	Prom.	Yes	FW
1,800	16.0	277	110	Prim.			NC	16,600	16,050	Succ.	Yes	FW
2,500	12.5	600	110	Prim.			JS	800	500	Succ.	Yes	FW
650	12.0	6,400	85	Steam soak	66.0		NC	2,800	2,800	Succ.	Yes	FW
1,400	16.0	312	92	Prim.			HF	2,300	2,300	Succ.	Yes	
1,100	16.0	145	88	Prim.			HF	250	250	Succ.	Yes	
1,000	13.0	1,800	105	Prim.			JS	1,700	1,400	Prom.		FW
600	11.0	15,300	100	Prim.			JS	15	0	TETT		FW
600-1,500	12.0		80	Prim.			HF	1,239	1,239	Succ.	Yes	FW
1,200	12.0			Prim.			HF	7,795	7,795	Succ.	Yes	FW
1,700	12.0			Prim.			HF	5,312	5,312	Succ.	Yes	FW
1,100	11.0			Prim.			JS	1,014	1,014	Succ.	Yes	FW
0				Prim.			JS	5,028	5,028	Succ.	Yes	FW
1,200	12.0			Prim.			HF	2,277	2,277	Succ.	Yes	FW

Producing thermal EOR in U.S. — continued

Operator	Field	State	County	Start date	Area, acres	Number wells Prod.	Number wells Inj.	Pay zone	Formation type	Porosity, %	Permeability, md
Sun	Midway-Sunset	Calif.	Kern	7/84		36	0	Potter	S		
Sun	Midway-Sunset	Calif.	Kern	7/84		60	2	Potter	S		
Sun	Midway-Sunset	Calif.	Kern	1/84		127	0	Potter	S		
Tenneco	Kern River	Calif.	Kern	1972	50	41	20	Kern River Series	S	31.0	2,000
Tenneco	Kern River	Calif.	Kern	1970	80	112	33	Kern River Series	S	31.0	2,000
Tenneco	Kern River	Calif.	Kern	9/72	38	32	16	Kern River Series	S	31.0	2,500
Tenneco	Kern River	Calif.	Kern	1970	95	72	36	Kern River Series	S	33.0	4,000
Tenneco	Midway-Sunset	Calif.	Kern	1981	40	70	10	Monarch	S	30.0	3,000
Tenneco	Midway-Sunset	Calif.	Kern	4/85	30	26	5	Potter	S	34.0	3,700
Tenneco	Midway-Sunset	Calif.	Kern	1969	50	60	15	Metson	S	30.0	3,000
Tenneco	Midway-Sunset	Calif.	Kern	1984	15	12	0	Sub Lakeview	S	30.0	4,500
Tenneco	Midway-Sunset	Calif.	Kern	1983	34	13	2	L. Monarch	S	30.0	2,000
Tenneco	Midway-Sunset	Calif.	Kern	4/84	25	24	2	Potter	S	34.0	4,000
Tenneco	Midway-Sunset	Calif.	Kern	1985	22	17	5	U. Monarch	S	29.0	2,000
Tenneco	Midway-Sunset	Calif.	Kern	3/73	24	27	2	Potter	S	34.0	4,100
Tenneco	Placerita	Calif.	Los Angeles	1986	55	50	11	L. Kraft	S	28.0	3,000
Texaco	Cat Canyon	Calif.	Santa Barbara	1964	1,700	30		S1B	S	31.0	3,000
Texaco	Cat Canyon	Calif.	Santa Barbara	1977	390	80		S1A-S6	S	31.0	1,400
Texaco	Kern River	Calif.	Kern	8/62	5,070	4,457	1,469	Kern River Series	S	31.0	4,000
Texaco	Lost Hills	Calif.	Kern	1977	49	23	16	Tulare	S	38.0	2,000
Texaco	Lost Hills	Calif.	Kern	1977	35	46	11	Etchegoin	S	40.0	2,000
Texaco	McKittrick	Calif.	Kern	1984	18	16	9	Tulare	S	35.0	2,800
Texaco	Midway-North Midway	Calif.	Kern	11/81	40	61	16	Potter	S	30.0	3,400
Texaco	Midway-Reward	Calif.	Kern	2/77	10	35	9	Potter	S	33.0	3,000
Texaco	Midway-Santiago Creek	Calif.	Kern	12/83	10	10	4	Tulare	S	34.0	2,500
Texaco	Midway-Sec 35	Calif.	Kern	1984	7.5	16	5	Potter	S	35.0	2,000
Texaco	Midway-Sec 35	Calif.	Kern	1977	10	13	6	Potter	S	35.0	2,000
Texaco	Midway-Security	Calif.	Kern	11/77	30	43	20	Potter	S	33.0	5,000
Texaco	San Ardo	Calif.	Monterey	7/87	51	24	7	Lombardi	S	32.0	5500
Texaco	San Ardo	Calif.	Monterey	1965	1,108	231	13	Aurignac	S	39.0	2,200
Texaco	Sour Lake	Tex.	Hardin	3/85	8	12	4	Miocene OB-2	S	33.3	1,000
Texaco	Sour Lake	Tex.	Hardin	11/81	19.8	24	12	Miocene OB-1	S	35.5	1,436
Texaco	Sour Lake	Tex.	Hardin	6/84	7	8	4	Miocene OB-4	S	35.0	900
Union Pacific Resources	Wilmington	Calif.	Los Angeles	7/82	20	9	4	Tar	S	30.1	2780
Unocal	Cymric	Calif.	Kern	1964	300	179	0	Tulare	S	35.0	1,000
Unocal	Gato Ridge	Calif.	Santa Barbara	1970	130	16	0	Sisquoc Sib & S2	S	32.0	2,500
Unocal	Guadalupe	Calif.	San Luis	1955	2,970	114	0	Sisquoc	S	35.0	1,550
Unocal	McKittrick	Calif.	Kern	8/83	17	21	8	Tulare	S	36.0	1000-2000
Unocal	McKittrick	Calif.	Kern	6/77	131	37	15	Amnicola	S	37.0	1000-2000
Unocal	Midway-Sunset	Calif.	Kern	1983	80.5	61	0	Tulare	US	31.0	1,146
Unocal	Midway-Sunset	Calif.	Kern	6/77	39	48	18	Potter	S	24.0	1,500
Unocal	Midway-Sunset	Calif.	Kern	1/85	5.2	9	4	Potter	US	34.5	3,700
Unocal	Midway-Sunset	Calif.	Kern	1974	80.5	46	17	Potter	US	34.5	3,700
Unocal	Midway-Sunset	Calif.	Kern	5/69	161	184	0	Potter	S	24.0	1,500
Unocal	North Belridge	Calif.	Kern	11/65	160	12	0	Tulare	S	38.0	2,500
Unocal	South Belridge	Calif.	Kern	3/85	167	12	0	Tulare	S	38.0	4,000
Unocal	Tensleep	Wyo.	Natrona	5/81	18	5	2	Tensleep	S	18.0	500

Hot water

Operator	Field	State	County	Start date	Area, acres	Prod.	Inj.	Pay zone	Formation type	Porosity, %	Permeability, md
ARCO	Midway-Sunset	Calif.	Kern	8/85	6	6	2	Tulare	S	33.0	2,500
Mobil	South Belridge	Calif.	Kern	11/82	246	58	22	Tulare	S	35.0	3,000
Tenneco	Kern River	Calif.	Bakersfield	1986	50	45	10	Kern River Series	S	31.0	3,000
Tenneco	Kern River	Calif.	Bakersfield	1986	28	42	11	Kern River Series	S	31.0	2,500
Texaco	Lost Hills	Calif.	Kern	1/86	38	31	13	Etchegoin	S	40.0	2000
Texaco	Lost Hills	Calif.	Kern	1/86	11	13	4	Tulare	S	38.0	2000
Texaco	McKittrick	Calif.	Kern	1981	12	18	9	Potter	S	35.0	4,000
Texaco	Midway-Sec 35	Calif.	Kern	3/87	10	16	6	Potter	S	35.0	2000
Texaco	San Ardo	Calif.	Monterey	9/86	95	41	9	Aurignac	S	33.0	3000
Unocal	McKittrick	Calif.	Kern	6/87	70	25	14	Amnicola	S	37.0	1,500

Combustion Projects

Operator	Field	State	County	Start date	Area, acres	Prod.	Inj.	Pay zone	Formation type	Porosity, %	Permeability, md
Bayou State	Bellevue	La.	Bossier	1970	180	82	15	Nacatoch	S	32.0	650
Chevron	W. Heidelberg Cotton Valley Unit	Miss.	Jasper	12/71	362	9	3	Cotton Valley 4 & 5	S	14.0	85
Greenwich Oil	Forest Hill	Tex.	Wood	9/76	1,900	100	21	Harris Sand	S	28.0	950
Mobil	Midway-Sunset	Calif.	Kern	1/60	150	32	6	Moco	S	36.0	1,500
Mobil	Lost Hills	Calif.	Kern	4/61	164	33	10	Tulare	S	42.6	1,790
Mobil	West Newport	Calif.	Orange	1958	300	139	36	Miocene	S	37.0	500-1,000
Oxy USA	Bellevue	La.	Bossier	1971	180	186	48	Nacatoch	S	33.9	700
Santa Fe Energy	Midway-Sunset	Calif.	Kern	1982	24	40	10	Potter	S	32.5	200-4,000
Texaco	Bellevue	La.	Bossier	9/63	385	183	28	Nacatoch	S	33.0	960

Producing CO$_2$, gas EOR in U.S.

Operator	Field	State	County	Start date	Area, acres	Number wells Prod.	Number wells Inj.	Pay zone	Formation type	Porosity, %	Permeability, md
CO$_2$ miscible											
ARCO	North Coles Levee	Calif.	Kern	6/81	70	7	1	Stevens	S	19.5	9
ARCO	Sho-Vel-Tum	Okla.	Stephens	9/82	1,100	69	41	Sims	S	16.0	70
ARCO	Sable	Tex.	Yoakum	3/84	825	32	21	San Andres	Dolo.	8.4	1.5
ARCO	Wasson-Willard	Tex.	Yoakum	1/86	8,000	286	225	San Andres	Dolo.	9.0	1.3
Amoco	Levelland	Tex.	Hockley	8/78	15	1	4	San Andres	Dolo.	11.8	3.8
Amoco	Levelland	Tex.	Hockley	3/73	13.2	2	6	San Andres	Dolo.	11.5	3.5

Depth, ft	Gravity, °API	Reservoir oil cp	°F.	Previous prod.	Residual saturation % Start	End	Project maturity†	Total prod., b/d	Enhanced prod., b/d	Project eval.	Profit	Project scope
0				Prim.			JS	346	346	Succ.	Yes	FW
0				Prim.			JS	700	700	Succ.	Yes	FW
1,100	11.0			Prim.			JS	2,736	2,736	Succ.	Yes	FW
800	13.0	8,000	90	Steam soak	60.0	15.0	NC	1,600	1,500	Succ.	Yes	FW
600	13.0	5,000	78	Prim.	63.0	15.0	HF	1,100	1,100	Succ.	Yes	FW
900	13.0	7,000	84	Prim.	70.0	15.0	HF	650	650	Succ.	Yes	FW
1,200	13.0	8,000	90	Steam soak	55.0	20.0	NC	1,200	1,100	Succ.	Yes	FW
1,200	13.0	30.0	200	Steam soak	45.0	20.0	HF	1,300	700	Succ.	Yes	FW
1,500	11.5	10,000	95	Steam soak	62.0	15.0	JS	900	700	Succ.	Yes	FW
1,100	11.4	1,200	130	Prim.	75.0	57.0	NC	800	250	TETT	TETT	FW
1,300	13.0	4,000	90	Prim.	60.0	20.0	JS	120	100	Succ.	Yes	FW
1,500	12.0	5,000	86	Cyclic steam	60.0	30.0	JS	300	200	Succ.	Yes	FW
1,600	11.5	10,000	95	Steam soak	67.0	15.0	HF	700	500	Succ.	Yes	FW
1,300	12.5	5,000	86	Steam soak	60.0	30.0	JS	600	500	Prom.	Yes	FW
1,600	11.5	7,000	95	Steam soak	65.0	15.0	NC	900	850	Succ.	Yes	FW
1,800	12.0	7,500	100	Prim.	62.0	15.0	JS	420	420	TETT	TETT	FW
2,300	8.0	100,000	100	None			NC	200	200	Succ.	Yes	
2,500	7.0	50-100000	110	None			NC	2,800	2,800	Succ.	Yes	FW
1,000	13.0	4,060	90	Prim.	70.0	30.0	HF	87,600	87,600	Succ.	Yes	FW
200	13.0	20.0	220	Steam stimulation	70.0	30.0	NC	530	520	Succ.	Yes	FW
400	13.0	10.0	250	Steam stimulation	63.0	30.0	NC	950	900	Succ.	Yes	FW
1,000	13.0	35.0	220	Steam stimulation	60.0	30.0	HF	80	70	Disc.	No	P (Exp.UL)
1,000	13.0	1,500	90	Steam stimulation	60.0	20.0	HF	2,480	1,370	Succ.	Yes	FW
1,300	13.0	1,500	90	Steam stimulation	60.0	20.0	HF	1,125	900	Succ.	Yes	FW
1,300	12.0	1,500	90	Steam stimulation	60.0	20.0	HF	300	300	Prom.	Yes	P (Exp. L)
1,600	12.0	1,500	90	Steam soak	55.0		HF	650	450	Succ.	Yes	FW
1,600	12.0	6,500	90	Steam soak	55.0	20.0	NC	330	180	Succ.	Yes	P (Exp. L)
950	12.0	1,500	90	Steam stimulation	60.0	20.0	NC	2,040	800	Succ.	Yes	FW
1,900	11.0	1,200	135	Cyclic steam	58.0	15.0	JS	1,057	1,057	TETT	No	P (Exp. L)
2,200	13.0	4,500	100	Cyclic	58.0	24.0	NC	6,876	6,876	Succ.	Yes	FW
2,000	16.2	100	95	Prim.	50.0	20.0	NC	140	140	Succ.	Yes	FW
1,000	14.7	665	90	Prim.	53.0	17.0	NC	300	300	Succ.	Yes	P (Exp. L)
1,600	16.2	107	100	Prim.	50.0	20.0	NC	200	200	Succ.	Yes	FW
2,563	14.0	283	123	WF	55.0	20.0	HF	400	400	Succ.	Yes	P (Exp. L)
1,100	13.0			Prim.	65.0	45.0	NC	5,230	4,340	Succ.	Yes	FW
1,900	10.5	1,450	130	Prim.	65.0	49.0	HF	60	5	Disc.	No	FW
3,000	9.0	580	85	Prim.	80.0	60.0	HF	1,200	500	Succ.	Yes	FW
850	11.4	1,000	140	Steam soak	60.0	27.0	NC	650	545	Succ.	Yes	FW
1,100	11.3	500	170	Steam soak	60.0	36.0	NC	610	425	Succ.	Yes	FW
700	11.4	10,000	100	Prim.	54.0	37.8	HF	730	425	Succ.	Yes	FW
1,000	11.3	100	220	Steam soak	55.0	34.0	NC	2,200	1,960	Succ.	Yes	FW
1,100	11.4	10,000	110	Steam drive	52.0	28.0	NC	719	109	Succ.	Yes	P (Exp. L)
1,100	11.4	10,000	110	Steam soak	57.2	32.3	HF	2,780	2,550	Succ.	Yes	FW
1,000	11.3	500	170	Prim.	57.0	35.0	NC	3,600	2,600	Succ.	Yes	FW
800	13.6	450	130	Prim.	63.0	49.0	HF	108	0	Disc.	No	FW
1,250	13.2	450	130	Prim.	66.0	52.0	JS	103	0	Disc.	No	FW
2,500	14.0	580	90	Prim.	67.0	30.0	HF	370	270	Prom.	No	P
1,000	11.0	24,000	100	Steam flood			HF	54	11	Succ.	Yes	P (Exp. UL)
1,000	14.0	1,600	95	Steam flood	40.0	30.0	JS	2,250	2,080	TETT	Yes	P
900	13.5	8,000	90	Steam drive	30.0	20.0	JS	250	150	Succ.	Yes	FW
600	13.5	5,000	78	Steam drive	35.0	20.0	JS	200	50	TETT	Yes	FW
400	13.0	10.0	250	Steam flood			JS	110	100	TETT	Yes	P (Exp. L)
200	13.0	20.0	250	Steam flood			JS	170	165	Succ.	Yes	P (Exp. L)
1,100	13.0	40.0	210	Steam flood	50.0	30.0	HF	90	20	Disc.	No	P (Exp. UL)
1,600	12.0	38.0	200	Steam flood			JS	400	220	TETT	Yes	P (Exp. L)
2,200	13.0	4,000	100	Steam drive	38.0	31.0	JS	570	100	TETT	Yes	P (Exp. L)
1,100	12.0	30.0		Steam drive	41.0	31.0	JS	290	0	TETT		FW
400	19.0	660		Prim.	94.0	49.0	HF	400	400	Succ.	Yes	FW
11,300	18-27.0	6.00	221	Prim.	80.0		NC	565	565	Succ.	Yes	FW
5,000	10.0	1,006	185	Prim.	63.0	32.0	JS	1,050	1,050	Prom.	No	FW
2,700	14.0	110	125		74.0	37.0	HF	790	790	Succ.	Yes	FW
300	15.0	410	95	Prim.	63.0	22.0	NC	520	520	Succ.	Yes	FW
1,600	13.0	750	100	Prim.	86.0	37.0	HF	1,100	1,100	Succ.	Yes	FW
300-400	19.0	676	75	Prim.	72.0	32.0	NC	300	300	Succ.	Yes	FW
1,700	11.5	2,770	110	Prim./drive/cyclic	70.0	<10.0	JS	1,000	800	Prom.	TETT	
350	19.0	675	130	Prim.	60.0	28.5	NC	1,000	1,000	Succ.	Yes	FW

Table F

Depth, ft	Gravity, °API	Reservoir oil cp	°F.	Previous prod.	Residual saturation % Start	End	Project status†	Total prod., b/d	Enhanced prod., b/d	Project eval.	Profit	Project scope
9,000	36.0	0.45	235	Prim.	34.0	25.8	HF	80		Disc.	No	P (Exp. UL)
6,200	25.0	3.30	115	WF	59.0	42.0	HF	2,200	1,100	Succ.	Yes	FW
5,200	32.0	1.46	107	WF			JS	685	100	Prom.	Yes	FW
5,100	32.0	2.01	110	WF			JS	4,393	0	TETT		FW
4,900	30.0	2.30	105	WF	43.0		HF	5	5	Succ.		P (Exp. L)
4,900	30.0	2.30	105	WF	74.0		HF	38	38	Succ.		P (Exp. L)

Producing CO₂, gas EOR in U.S, — continued

Operator	Field	State	County	Start date	Area, acres	Number wells Prod.	Inj.	Pay zone	Formation type	Porosity, %	Permeability, md
Amoco	North Cowden Unit	Tex.	Ector	6/85	12	2	6	Grayburg	Dolo./LS	10.0	5.4
Amoco	Slaughter	Tex.	Hockley	1984	6,412	229	71	San Andres	Dolo./LS	10.0	9
Amoco	Slaughter	Tex.	Hockley	12/84	1,600	72	50	San Andres	Dolo./LS	10.0	4
Amoco	Slaughter	Tex.	Hockley	12/84	5,700	215	127	San Andres	Dolo./LS	13.0	7
Amoco	Wasson	Tex.	Yoakum	12/84	7,800	321	282	San Andres	Dolo./LS	9.0	5
Amoco	Wertz	Wyo.	Carbon, Sweetwater	10/86	1,400			Tensleep	S	10.0	20
Chevron	Rangely Weber Sand	Colo.	Rio Blanco	10/86	15,000	234	70	Weber SS	S	12.0	5-50
Chevron	Quarantine Bay	La.	Plaquemines	10/81	57	2	1	4 Sand Reservoir	S	30.0	100-1,000
Chevron	Timbalier Bay	La.	Lafourche	4/84	37	2	1	S-2B	S	30.0	9,244
Chevron	Kurten	Tex.	Brazos	8/81	672	5	4	Woodbine	S	12.0	0.4
Chevron	Pittsburg	Tex.	Camp	6/85	43	4	1	Pittsburg	Limey Sand	11.0	2.4
Chevron	SACROC Unit	Tex.	Scurry	1/72	49,900	732	387	Canyon Reef	LS	3.9	19.4
Conoco	Ford Geraldine Unit	Tex.	Reeves & Culberson	2/81	3,850	86	60	Delaware	S	23.0	64
Enron Oil & Gas	Twofreds	Tex.	Loving, Ward & Reeves	1/74	4,392	42	33	Delaware	S	19.5	32
Exxon	Cordona Lake	Tex.	Crane	12/85	3,516	33	13	Devonian	Dolo.	22.0	3.5
Exxon	Means (San Andres)	Tex.	Andrews	11/83	6,700	368	214	San Andres	LS	9.0	20
Exxon	Slaughter	Tex.	Hockley	5/85	569	20	11	San Andres	Dolo.	12.5	6.3
Exxon	Wasson (Cornell Unit)	Tex.	Yoakum	7/85	1,923	62	38	San Andres	Dolo.	8.6	1.9
George R. Brown Prtshp.	Rose City South	Tex.	Orange	1/83	900	8	2	Hackberry	S	37.0	4,500
George R. Brown Prtshp.	Rose City North	Tex.	Orange	4/81	800	8	5	Hackberry	S	37.0	4,500
Mitchell Energy	Alvord South	Tex.	Wise	1980	2,291	17	3	Caddo	Congl.	12.8	55
Mobil	GMK South	Tex.	Gaines	1982	1,143	26	27	San Andres	LS	9.8	3.1
Mobil	McElmo Creek Unit	Utah	San Juan	2/85	13,440	170	100	Ismay Desert Creek	LS	14.0	5
Mobil	Wasson	Tex.	Yoakum	10/85	640	29	14	San Andres	Dolo.	13.0	6.2
Oxy USA	Welch	Tex.	Dawson	2/82	2,675	140	119	San Andres	LS	9.3	9
Oxy USA	Northeast Purdy	Okla.	Garvin	9/82	8,320	106	102	Springer	S	13.0	44
Pennzoil	Richardson	W.V.	Calhoun, Roane	8/86	78	8	4	Berea	S	14.0	40
Pennzoil	Tinsley	Miss.	Yazoo	11/81	1,338	15	15	Perry	S	26.4	49
Phillips	E. Vacuum	N.M.	Lea	2/81	4,900	192	100	San Andres	Dolo.	11.7	11
Santa Fe Energy	Raymond	Mont.	Sheridan	8/83	685	2	1	Nisku	LS	8.2	13
Shell Offshore	West Mallalieu	Miss.	Lincoln	11/86	5,760	9	9	Lower Tuscaloosa	S	25.0	20
Shell Offshore	Olive	Miss.	Pike	10/87	1,280	16	6	Lower Tuscaloosa	S	26.0	50
Shell Western E&P	Crossett (N. Cross)	Tex.	Crane & Upton	4/72	1,500	23	11	Devonian	Tripol. chert	22.0	5
Shell Western E&P	Little Creek	Miss.	Lincoln & Pike	12/85	6,200	55	20	Lower Tuscaloosa	S	23.0	33
Shell Western E&P	Wasson	Tex.	Yoakum	4/83	20,000	1,195	805	San Andres	Dolo.	12.0	8
Shell Western E&P	Wasson	Tex.	Gaines	1/86	3,500	110	43	Clearfork	Dolo.	6.0	3
Stanberry Oil	Hansford Marmaton	Tex.	Hansford	6/80	2,010	9	10	Marmaton	S	16.5	52
Stanberry Oil	Farnsworth, North	Tex.	Ochiltree	6/80	1,431	9	5	Marmaton B	LS	11.5	140
Texaco	Paradis	La.	St. Charles	2/82	347	7	4	Lower 9000-Foot	S	26.0	770
Texaco	Paradis	La.	St. Charles	2/82	320	4	3	No. 8	S	27.0	795
Texaco	Bay St. Elaine	La.	Terrebonne	1/81	9	2	1	8000-Foot	S	32.9	1,480
Union Texas Petroleum	Wellman	Tex.	Terry	7/83	1,400	32	7	Wolfcamp	LS	9.2	>100
Unocal	Dollarhide	Tex.	Andrews	5/85	6,183	83	66	Devonian	Dolo./Tripol.	13.5	9

CO₂ immiscible

Operator	Field	State	County	Start date	Area, acres	Number wells Prod.	Inj.	Pay zone	Formation type	Porosity, %	Permeability, md
Chevron	Pittsburg	Tex.	Camp	11/83	40	1	0	Sub-Clarksville	S	23.0	460
Chevron	Heidelberg	Miss.	Jasper	12/83	40	1	0	Eutaw	S	25.0	74
Chevron	Timbalier Bay	La.	Lafourche	12/85	422	10		4900 Foot Sand	S	32.0	500-2,500
Exxon	Pewitt Ranch	Tex.	Titus	6/83		1		Paluxy	S	24.0	1000-1500
Marathon Oil	Yates	Tex.	Pecos & Crockett	11/85	14,300	958	27	Grayburg/San Andres	Dolo.	17.0	175
Shell Western E&P	Weeks Island	La.	New Iberia	1978	8	2	1	S RES B	S	26.0	1,800
Union Pacific Resources	Wilmington	Calif.	Los Angeles	3/81	41	3	4	Tar	S	24.0	465
Union Pacific Resources	Wilmington	Calif.	Los Angeles	2/84	156	10	12	Tar	S	24.0	465

Hydrocarbon miscible

Operator	Field	State	County	Start date	Area, acres	Number wells Prod.	Inj.	Pay zone	Formation type	Porosity, %	Permeability, md
ARCO	Prudhoe Bay	Alaska	Prudhoe Bay	12/82	3,650	42	11	Sadlerochit	S	23.0	300
ARCO	Chandeleur Sound Block 25	La.	St. Bernard	1/83	342	11	3	4800 ft. Sand	S	21.3	1,980
ARCO	Chandeleur Sound Block 25	La.	St. Bernard	1/83	941	48	17	BB	S	33.0	1,680
ARCO	South Pass Block 61	OCS		8/83	26	1	1	LL RG	S	33.0	10
ARCO	South Pass Block 61	OCS		8/83	55	1	1	I RK2	S	30.0	25
ARCO	South Pass Block 61	OCS		5/81	54	4	2	MM RBB	S	30.0	300
ARCO	South Pass Block 61	OCS		5/81	112	5	2	UM RBB	S	30.0	300
ARCO	South Pass Block 61	OCS		5/81	320	6	3	UM RAAO, 2, 3	S	30.0	180
ARCO	South Pass Block 61	OCS		8/83	39	2	1	UJ RL	S	33.0	50
ARCO	South Pass Block 61	OCS		8/85	54	2	1	LL RBB1, 2	S	30.0	175
ARCO	South Pass Block 61	OCS		5/81	179	5	3	MM RAAO, 2, 3	S	30.0	180
ARCO	South Pass Block 61	OCS		2/84	37	2	1	LL ART	S	30.0	50
ARCO	South Pass Block 61	OCS		10/84	68	3	3	LK2 RU	S	29.0	100
ARCO Alaska/Standard Alaska	Prudhoe Bay	Alaska		2/87	13,500	157	43	Sadlerochit	S	22.0	500
Amoco	Levelland	Tex.	Hockley	5/71	1,200	40	28	San Andres	Dolo.	10.2	2.1
Exxon	South Pass Block 89	Tex.	Gulf of Mexico	12/83		12	9	X and Y Series	S	31.0	10-1,500
Hunt	Fairway	Tex.	Anderson/Henderson	3/66	22,618	118	64	James	LS	12.6	11
Phillips	Bridger Lake	Utah	Summit	4/70	3,800	7	1	Dakota	S	12.8	79
Sun E&P	Fordoche W-12	La.	Pt. Coupee	5/80	3,400	5	5	Wilcox	S	19.0	4.6
Sun E&P	Fordoche W-8	La.	Pt. Coupee	5/80	3,300	8	3	Wilcox	S	20.0	8.6

Depth, ft	Gravity, °API	Reservoir oil cp	°F.	Previous prod.	Residual saturation % Start	End	Project status	Total prod., b/d	Enhanced prod., b/d	Project eval.	Profit	Project scope
4,300	34.0	1.67	94	WF			JS	20	20	Succ.		P
4,900	31.0	1.40	105	WF			JS	4,400	1,200	Prom.	Yes	FW
4,950	31.0	1.40	105	WF			JS	2,715	400	Succ.	Yes	FW
4,950	31.0	1.40	105	WF			JS	8,311	1,500	Succ.	Yes	FW
5,100	32.0	1.30	110	WF			JS	15,438	3,500	Succ.	Yes	FW
6,000	35.0	1.116	163	WF			JS	10,000	6,600	Prom.	Yes	FW
5,500-6,500	35.0	1.70	160	Prim./WF	38.0	29.0	JS	32,650	4,500	Prom.		FW
8,120	32.0	0.99	183	Prim.	55.0	15.0	HF			Prom.	No	
7,400	39.0	0.39	180	Prim.	29.0		HF	0	0	TETT	No	P (Exp. L)
8,300	38.0	0.40	230	Prim.	40.0		JS	182	156	Prom.		P
8,000	41.0		205	WF	59.0		NC	130	36	Prom.	Yes	P (Exp. L)
6,700	41.0	0.35	130	Prim./WF	63.3	46.8	NC	32,416	17,634	Succ.	Yes	FW
2,680	40.0	1.40	83	Prim./WF			JS	1,150	1,150	Succ.	No	FW
4,900	36.0	1.50	105	WF			NC	700	700	Succ.	Yes	FW
3,200	40.0	0.50	101	WF			JS			TETT	TETT	
4,300	29.0	6.00	97	WF			JS			TETT	TETT	
4,900	32.0	1.30	110	WF			JS			TETT	TETT	
4,500	33.0	1.00	106	WF			JS			TETT	TETT	
8,200	37.0	2.00	180	WF	50.0	35.0	HF	811	600	Prom.		FW
8,200	37.0	2.00	180	WF	50.0	35.0	HF	160	160	Prom.		FW
5,700	44.0	0.39	154	WF	60.0	52.0	HF	490	215	Prom.	No	FW
5,400	30.0	2.57	101	WF	55.0	28.0	JS	2,000	200	Prom.	Yes	LW
5,600	41.0	0.50	125	Prim./WF	50.0		JS	5,300	900	Prom.	Yes	FW
5,100	33.0	0.97	110	WF	54.4	39.2	JS	1,654	150		Yes	FW
4,890	34.0	2.15	96	WF	30.0	18.0	HF	1,940	155	TETT	TETT	FW
9,400	38.0	1.20	148	WF	46.0	40.0	NC	3,500	1,800	Succ.	TETT	FW
2,300	46.0	3.00	80	WF	48.0		JS	80	40	Prom.	No	P
4,800	39.0	1.50	175	WF	65.0	38.0	HF	300	300	Prom.	No	FW
4,500	38.0	1.00	101	Prim.	70.0	50.0	JS	9,500		Succ.	Yes	FW
7,900	40.0	0.40	178	Prim.			JS	90	50	Prom.	Yes	
10,365	38.0	0.50		Prim.	15.0	1.0	JS	400	400	TETT	TETT	FW
10,500	39.0	0.34		WF	17.0	2.0	JS			TETT	TETT	FW
5,300	44.0	0.36	106	Residue gas inj.	34.0	22.0	HF	1,925	1,925	Succ.	Yes	FW
10,640	38.0	0.40	248	WF	21.0	2.0	JS	3,500	3,500	Prom.	TETT	
5,200	33.0	1.30	105	WF	40.0	27.0	JS	43,600	11,000	Succ.	Yes	FW
6,700	33.0	1.20	105	WF	60.0		JS	6,200	0	Prom.		FW
6,500	44.0	1.56	135	Prim.	43.0		HF	553	553	Succ.	Yes	FW
6,400	44.0	1.61	135	WF	57.0		HF	470	470	Succ.	Yes	FW
10,400	37.0	0.50	205	Prim.	62.0	48.0	HF	400	400	Prom.	No	PW
8,600	39.0	0.40	192	Prim.	30.0	20.0	NC	125	125	Prom.	No	RW
7,400	36.0	0.67	170	Prim.	20.0	5.0	NC	10	10	Disc.	No	RW
9,800	43.5	0.54	151	WF	35.0	10.0	JS	3,600	400	TETT	TETT	FW
8,000	40.0	0.44	122	Prim./WF	35.0	22.0	JS	3,000	2,200	TETT	TETT	FW
3,800	14.0	2,200	120	Prim.	65.0		NC	15	5	Prom.	No	P
5,060	20.0	15.0	150	Prim.			NC	25		Disc.	No	P (Exp. UL)
4,878	26.0	1.70	138	Prim.	60.0		HF	124	95	Prom.	Yes	FW
4,500	19.0	30.0	160	Prim.			JS					
1,100-1,700	30.0	5.50	82	Gas inj.			JS	72,316		TETT	TETT	FW
12,760	32.0	0.50	225	WF	22.0	2.0	NC	20	20	Succ.	No	P (Exp. L)
2,500	14.0	283	123	WF	51.0	30.0	HF	50	50	Succ.	No	P
2,500	14.0	283	123	WF	51.0	30.0	JS	250	250	Succ.	No	FW
8,800	27.0	0.90	200	Prim.	65.0	25.0	JS	50,000	5,000	Prom.		P (Exp. L)
4,800	27.0	3.50	120	Prim.	45.0	30.0	HF	863	786	Succ.	Yes	FW
5,450	27.0	3.45	132	Prim.	45.0	30.0	HF	3,430	2,428	Succ.	Yes	FW
6,800	43.0	0.40	170	Prim.	73.0	31.0	HF	50	5	Succ.	Yes	FW
4,900	39.0	0.70	150	Prim.	65.0	25.0	NC	57	6	Succ.	Yes	FW
6,400	37.0	0.70	165	WF	78.0	27.0	HF	2,150	1,040	Succ.	Yes	FW
6,300	37.0	0.70	165	WF	78.0	27.0	HF	1,200	1,200	Succ.	Yes	FW
9,000	38.0	0.30	190	WF	78.0	25.0	HF	2,867	2,867	Succ.	Yes	FW
5,500	40.0	0.40	150	Prim.	77.0	26.0	NC	237	24	Succ.	Yes	FW
5,750	29.0	1.70	158	Prim.	66.0	28.0	HF	256	148	Succ.	Yes	FW
9,100	38.0	0.30	190	WF	76.0	26.0	HF	2,431	2,431	Succ.	Yes	FW
8,500	33.0	0.86	190	Prim.	70.0	30.0	HF	162	41	Succ.	Yes	FW
7,900	36.0	0.50	180	Prim.	67.0	27.0	HF	745	209	Succ.	Yes	FW
8,800	27.0	0.90	200	WF	50.0	25.0	JS	300,000	0	TETT		FW
4,900	30.0	2.30	105	Prim.	90.0		HF	3,000	3,000	Succ.		P (Exp. UL)
11,270	38-40.0	0.4-0.6	165	Prim.			JS			TETT	TETT	
9,900	48.0	0.15	260	Prim.	73.0	36.0	NC	6,600	6,600	Succ.	Yes	FW
15,600	40.0	0.36	225	Prim.	70.0	61.0	NC	600	150	Succ.	Yes	FW
13,650	45.0	0.13	274				HF	320		Succ.	Yes	FW
13,200	44.0	0.13	267				HF	697		Succ.	Yes	FW

Producing CO$_2$, gas EOR in U.S, — continued

Operator	Field	State	County	Start date	Area, acres	Number wells — Prod.	Inj.	Pay zone	Formation type	Porosity, %	Permeability, md
True Oil	Red Wing Creek	N. Dak.	McKenzie	1/82	640	8	1	Mission Canyon	LS	10.0	0.1
Hydrocarbon immiscible											
ARCO Alaska	Kuparuk River	Alaska		2/86	15,360	49	47	Kuparuk "A" Sand	S	22.0	50
Nitrogen											
Chevron	East Painter	Wyo.	Uinta	11/83	1,500	17	7	Nugget	S	11.0	3
Chevron	Stonebluff	Okla.	Wagoner	3/82	540	18	15	Dutcher	S	16.0	200
Chevron	Painter	Wyo.	Uinta	6/80	1,360	39	19	Nugget	S	11.9	4
Exxon	Blackjack Creek	Fla.	Santa Rosa	2/81	4,600	8	8	Smackover	LS	16.0	112
Exxon	Jay-Little Escambia Creek	Fla./Ala.	Santa Rosa/Escambia	1/81	14,415	60	33	Smackover	LS	14.0	35
Exxon	Fanny Church	Ala.	Escambia	1985	600	3	1	Smackover	LS	12.0	4.5
Phillips	Binger	Okla.	Caddo	1977	12,960	62	30	Marchand	S	7.5	0.2
Phillips	Andector Ellenburger	Tex.	Ector	11/81	1,931	22	3	Ellenburger	Dolo.	3.8	2,000
Unocal	Chunchula Fieldwide Unit	Ala.	Chunchula	4/82	23,279	37	8	Smackover Carbonate	Dolo.	12.4	10
Flue gas											
ARCO	Block 31	Tex.	Crane	6/49	7,840	191	74	Devonian	LS	12.0	5.4
Exxon	Hawkins	Tex.	Wood	4/77	10,590	491	41	Woodbine	S	28.0	3,400

Producing chemical EOR projects in U.S.

Operator	Field	State	County	Start date	Area, acres	Number wells — Prod.	Inj.	Pay zone	Formation type	Porosity, %	Permeability, md
Polymer											
ARCO	So. Eunice/Langlie Mattix	N.M.	Lea	8/82	2,240	39	27	Seven Rivers Queen	S	11.0	1.3
ARCO	Healdton Area III	Okla.	Carter	10/83	1,500	193	44	Penn	S	23.0	115
ARCO	Walnut Bend Unit I	Tex.	Cooke	11/83	2,487	22	18	Winger	S	17.0	103
ARCO	C-H	Wyo.	Campbell	8/84	40	3	1	Minnelusa	S	20.6	80
ARCO	Hamilton Dome	Wyo.	Hot Spring	10/84	127	14	5	Tensleep	S	14.0	58
ARCO	West Nelson Rozet	Wyo.	Campbell	3/84	440	2	2	Minnelusa	S	17.0	170
ARCO	West Rozet	Wyo.	Campbell	9/84	120	3	2	Minnelusa	S	17.0	170
Amoco	Sleepy Hollow	Neb.	Red Willow	2/85	1,600	41	10	Reagan Sand	S	24.0	2,900
Anderson Oil	North Glo	Wyo.	Campbell	10/87	292	4	2	Minnelusa	S	15.9	150
Cenex	Sage Spring Creek-Unit A	Wyo.	Natrona	8/78	1,456	8	5	Dakota	S	13.0	50
Chevron	W. Bay	La.	Plaquemines	10/81	42	4	1	3A Sand	S	34.0	4,600
Chevron	W. Bay	La.	Plaquemines	10/81	64	3	2	3A Sand	S	30.0	40
Chevron	N. Stanley	Okla.	Osage	2/76	810	18	11	Burbank	S	18.0	300
Chevron	S. Stanley	Okla.	Osage	6/83	1,280	10	16	Burbank	S	18.0	300
Chevron	C-Bar	Tex.	Crane	6/83	2,600	56	32	San Andres	Dolo.	10.0	6
Chevron	Dune	Tex.	Crane	3/82	410	26	14	San Andres	Dolo.	14.0	28
Chevron	Goldsmith 5600	Tex.	Ector	9/83	3,840	139	51	Clearfork	Dolo.	15.0	28
Chevron	McElroy	Tex.	Crane	10/82	3,200	242	56	Grayburg	Dolo.	13.0	6
Chevron	N. Ward Estes	Tex.	Ward	12/81	17,986	785	666	Yates	S	18.0	10
Conoco	Vacuum-State 35 Lease	N.M.	Lea	3/83	240	8	9	Grayburg/San Andres	S/LS	10.9	13.7
Enron Oil & Gas	Isenhour	Wyo.	Sublette	7/80	173	4	2	Almy	S	15-20.0	1-25
Enron Oil & Gas	Long Island-Star Corral	Wyo.	Sublette	8/69	173	5	2	Almy	S	15.5	1-50
Enron Oil & Gas	McDonald Draw 1TB	Wyo.	Sublette	1/72	435	1	1	Almy	S	16.7	2-59
Enron Oil & Gas	McDonald Draw 4TB	Wyo.	Sublette	7/72	420	13	10	Almy	S	18.0	1-200
Enron Oil & Gas	Ruben	Wyo.	Sublette	12/70	800	12	6	Almy	S	14.1	1-56
Enron Oil & Gas	Saddle Ridge Unit	Wyo.	Sublette	12/74	376	9	8	Mesaverde	S	22.0	10-95
Exxon	Loudon	Ill.	Fayette	1985		760	554	Chester	S	19.0	30-500
Exxon	Omaha	Ill.	Gallatin	1984		15	4	Chester	S	18.0	20-387
Exxon	Roland	Ill.	White	1985		16	14	Chester	S	15.7	16-626
Exxon	Roland	Ill.	White/Gallatin	1985		12	8	Chester	S	19.5	113-279
Exxon	Trapp	Kan.	Russell	1984		15	9	Lansing/Kansas City	LS	18.4	0.1-300
Exxon	W. Yellow Creek	Miss.	Wayne	9/76	4,813	51	24	Eutaw	S	24.0	112
Exxon	Camrick	Okla.	Beaver/Texas	1984		43	27	Upper Morrow	S	15.0	1-250
Exxon	Hewitt	Okla.	Carter	3/81	886	85	52	Hewitt	S	21.0	175
Exxon	Conroe	Tex.	Montgomery	1985				2B	S		50-5,500
Exxon	Webster	Tex.	Harris	1985		7	3	Frio 1A	S	28.0	206
Fina Oil and Chemical	Garza	Tex.	Garza	10/82	120	10	5	San Andres	LS	19.8	4.1
Fina Oil and Chemical	Westbrook	Tex.	Mitchell	12/80	4,740	153	38	Clearfork	LS	7.4	6.3
Fina Oil and Chemical	South Cowden	Tex.	Ector	6/84	666	20	7	Grayburg	Dolo.	10.1	4.1
Gallagher Drilling	Stewart East Minnelusa Unit	Wyo.	Campbell	7/82	200	4	1	Minnelusa	S	15.6	250
Graham Resources	Carthage	Tex.		12/82		16	9	Morrow	S	19.0	200
Hunt	East Texas	Tex.	Rusk	8/82	508	78	14	Woodbine	S	22.0	300
Marathon Oil	Main Consolidated	Ill.	Crawford	7/81	130	8	12	Robinson	S	18.0	115
Marathon Oil	Yates	Tex.	Pecos	12/83	6,900	313	270	Queen Grayburg/San Andres	Dolo.	15.0	62
Marathon Oil	Byron	Wyo.	Big Horn	12/82	1,500	51	39	Embar/Tensleep	LS/S	13.9	4.5/78
Marathon Oil	Grass Creek	Wyo.	Hot Springs	10/85	740	40	18	Phosphoria/Tensleep	LS/S	21.6/13.1	20/112
Marathon Oil	Kinney Coastal	Wyo.	Park/Big Horn	11/84	1,500	36	23	Tensleep	S	11.2	50
Marathon Oil	Oregon Basin-North Embar Tensleep	Wyo.	Park	11/83	3,000	152	79	Embar/Tensleep	LS/S	20.2/14.7	68/193
Marathon Oil	Oregon Basin-SETPA	Wyo.	Park	1/85	6,108	180	33	Embar/Tensleep	LS/S	15.9/19.5	39/140
Mitchell Energy	Alba Northeast Unit	Tex.	Wood	7/80	410	10	2	Sub-Clarksville	S	21.4	471
Mitchell Energy	Alba SEFB Unit	Tex.	Wood	2/72	731	13	4	Sub-Clarksville	S	23.3	525
Mitchell Energy	Jacksboro S.	Tex.	Jack	5/82	270	7	8	Strawn	S	19.9	86.0
Mitchell Energy	Lucy N.	Tex.	Borden	10/83	550	8	4	Pennsylvanian	LS	9.7	30

Depth, ft	Gravity, °API	Reservoir oil cp	°F.	Previous prod.	Residual saturation % Start	End	Project status†	Total prod., b/d	Enhanced prod., b/d	Project eval.	Profit	Project scope
9,000	40.0	0.25	241	Prim.	60.0	20.0	JS	1,000		Succ.	Yes	FW
6,000	24.0	2.20	158	WF	57.0	54.0	JS	48,000		Prom.	Yes	P (Exp. L)
12,000	46.0	0.20	185	HC gas cycling	98.5	52.0	HF	10,000	1,000	TETT		FW
1,200	28.0	5.00	85	Prim.	67.0		HF	80	50	TETT		FW
11,500	46.0	0.20	174	HC gas cycling	47.0		HF	3,800	2,550	TETT		FW
15,700	48.0	0.30	286	WF	59.0		JS	3,600	1,500	Prom.	Yes	
15,400	51.0	0.20	285	WF	59.0	46.0	JS	20,800	7,600	Prom.	Yes	
15,565	51.0	0.15	288	Prim.	76.0		JS			TETT	TETT	
10,000	38.0	0.30	190	Prim.	98.0	75.0	HF	2,300	2,300	Succ.	Yes	FW
8,835	44.0	0.60	132	Prim.	43.8	37.0	HF	4,420		Succ.	Yes	FW
18,500	54.0	0.18	325	Prim.	80.0	45.0	JS	9,200	4,050	Succ.	Yes	FW
8,600	46.0	0.30	130	Prim.			NC	8,000	6,400	Succ.	Yes	FW
4,530	24.0	3.70	168	Prim.			HF		15,000	Succ.	Yes	

Depth, ft	Gravity, °API	Reservoir oil cp	°F.	Previous prod.	Residual saturation % Start	End	Project maturity	Total prod., b/d	Enhanced prod., b/d	Project eval.	Profit	Project scope
3,700	35.0	2.60	90	WF	63.0	60.0	HF	420		Prom.	Yes	
1,000	31.0	13.0	87	WF	37.0	35.5	HF	850		Succ.	Yes	FW
5,500	32.0	2.80	143	WF	39.0	26.0	NC	410		Prom.		FW
7,500	20.7	26.0	135	WF			NC	94		Prom.		P (Exp. UL)
2,750	20.0	39.0	135	WF			HF	900		Prom.	Yes	FW
8,700	21.0	12.0	197	WF			HF	200		Prom.		P
8,700	22.0	24.0	170	WF			HF	150		Prom.		P
3,500	31.0	20.0	95	WF			JS	2,400	1,200	Prom.	Yes	FW
7,900	20.2	29.7		Prim.	74.4	33.3	JS			TETT		FW
7,400	36.0	1.40	160	WF	45.0	21.0	NC	400	200	Succ.	Yes	FW
7,100	30.0	2.40	165	Prim.	65.0		HF				No	
7,100	41.0	2.30	120	WF	36.0	32.0	HF			TETT	TETT	
2,900	36.0	2.36	107	WF	51.0		NC	280	80	Succ.		FW
2,900	32.0	2.50	165	Prim.	54.0		HF	90	15	NE		FW
3,500	36.0	5.00	107	WF	39.0		NC	240		Succ.	Yes	FW
3,350	32.0	3.50	95	WF	64.0		NC	250		Succ.	Yes	FW
5,600	32.0	3.50	100	WF	49.0		NC	1,310	390	Succ.	Yes	FW
2,800	32.0	2.70	88	WF	58.0		HF	5,000		Succ.	Yes	FW
2,600	33.0	1.50	95	WF	42.0		NC	5,700		Succ.	Yes	FW
4,500	37.0	1.30	101	Prim./WF			HF	791		Succ.	Yes	FW
3,600	41.5	1.20	100	Prim.	38.1	27.0	HF	175	11	Succ.	Yes	FW
3,550	44.0	1.20	100	Prim.	43.6	23.9	NC	50	2	Succ.	Yes	FW
3,050	42.0	1.40	90	WF	40.0	36.7	NC	30	1	Succ.	Yes	FW
3,200	41.8	1.45	95	WF	43.8	29.1	NC	180	9	Succ.	Yes	FW
3,300	43.0	0.95	95	Prim.	42.0	25.4	NC	220	11	Succ.	Yes	FW
1,800	44.0	2.33	85	Prim.	37.5	27.3	NC	200	12	Succ.	Yes	FW
1,800	36.0	6.00	75	WF			JS			TETT	TETT	
1,950	28.0	17.0	78	WF			JS			TETT	TETT	
3,100	36.0	10.0	87	WF			JS			TETT	TETT	
3,000	32.0	10.0	83	WF			JS			TETT	TETT	
3,215	38.0	1.40	97	WF			JS			TETT	TETT	
4,950	20.0	20.0	150	WF	84.0	75.0	NC			Succ.	Yes	
7,530	39.0	1.50	142	WF			JS			TETT	TETT	
2,400	35.0	8.70	96	WF	43.0		HF			Prom.	Yes	
4,900			170	Prim.			JS			TETT	TETT	
5,900	29.0	1.30	165	WF			JS			TETT	TETT	
2,900	36.0	2.50	90	WF	40.0		NC	680		NE	Yes	
3,000	26.0	9.10	90	WF	50.0		NC	1,700		NE	Yes	
4,500	35.0	1.90		WF	57.0	53.0	HF	1,250		Prom.	Yes	FW
8,000	20.0	20.0	124	Prim.	65.0	37.7	HF	170	120	Succ.	Yes	FW
4,500	39.0			Prim./WF	50.0	40.0	NC	900	550	Succ.	Yes	FW
3,500	37.0	1.90	142	Prim.	57.0	33.0	NC	530	355	Succ.	Yes	LW
1,100	30.0	18.0	70	Prim.	51.0		NC	20	20	Disc.	No	P (Exp. UL)
1,500	30.0	5.50	82	Prim.			HF	15,567		Succ.		
5,600	23.0	17.0	121	Prim./WF			NC	3,200	413	Succ.	Yes	FW
4300/4500	24.0/24.0	15/15	105/110	Prim./WF			HF	2,050	1,000	Succ.	Yes	FW
4,240	22.1	20.0	140	Prim./WF			JS	1,538	365	Prom.	Yes	FW
3370/3520	22.5/22.1	8.2/10.0	108/110	Prim./WF	45.0		HF	12,500	2,600	Succ.	Yes	FW
3600/3840	21.0/21.0	15.7/15.7	105/180	Prim./WF	54.0	39.0	NC	10,400	3,000	Succ.	Yes	FW
4,100	15.1	75.0	150	WF	62.6	58.7	NC	54	54	Disc.		FW
4,200	15.5	75.0	150	Prim.	69.5	60.7	NC	63	63	Succ.	Yes	FW
1,975	40.0	1.73	100	Prim.	50.0	22.0	NC	209	209	Succ.	Yes	FW
7,640	40.0	0.44	140	Prim.	52.6	35.7	HF	185	30	TETT	TETT	FW

Producing chemical EOR projects in U.S. — continued

Operator	Field	State	County	Start date	Area, acres	— Number wells — Prod.	Inj.	Pay zone	Formation type	Porosity, %	Permeability, md
Mobil	Vacuum	N.M.	Lea	5/84	320	10	4	San Andres	Dolo.	10.4	3-100
Mobil	Cement	Okla.	Caddo	11/81	1,510	96	23	Fortuna/Noble-Olson	S	19.2	18
Mobil	Cement	Okla.	Caddo	9/81	1,120	71	23	Fortuna/Noble-Olson	S	36.1	74
Mobil	Cushing	Okla.	Creek	5/84	120	4	2	Bartlesville	S	19.0	290
Mobil	Fitts	Okla.	Pontotoc	7/82	1,520	148	87	Cromwell 60	S/LS	10-25.0	6-60
Mobil	Golden Trend	Okla.	Garvin/Grady	10/85	800	14	4	Springer	S	12.5	58
Mobil	Healdton	Okla.	Jefferson/Carter	7/81	1,045	151	41	Healdton	S	24.2	386
Mobil	Postle, Hough, Morrow 'A' Unit	Okla.	Texas	7/82	3,360	24	15	Morrow	S	15.8	45
Mobil	Postle, Hovey, Morrow Unit	Okla.	Texas	9/82	7,881	34	17	Morrow	S	16.8	45
Mobil	Postle, Postle Upper Morrow Unit	Okla.	Texas	11/82	3,944	29	19	Morrow	S	15.4	45
Mobil	Postle, West Hough, Morrow Unit	Okla.	Texas	4/83	7,601	39	24	Morrow	S	15.7	45
Mobil	Sho-Vel-Tum	Okla.	Carter	7/82	780	35	29	Deese	S	18.8	95.3
Mobil	Sho-Vel-Tum	Okla.	Carter	10/82	450	21	14	Deese	S	17.0	45-860
Mobil	Sho-Vel-Tum	Okla.	Stephens	11/82	1,980	100	43	Deese	S	17.4	44.8
Mobil	Sho-Vel-Tum	Okla.	Carter	8/82	451	30	6	Deese	S	20.0	192
Mobil	Sho-Vel-Tum	Okla.	Stephens	8/82	350	20	11	Deese	S	16.0	4.5-123
Mobil	Sho-Vel-Tum	Okla.	Carter	7/82	2,750	136	56	Tatums Stray Pennington	S	22.3	787
Mobil	Sho-Vel-Tum	Okla.	Stephens	8/82	1,360	71	52	Deese Springer	S	18.1	66
Mobil	Sho-Vel-Tum	Okla.	Carter	7/82	1,700	109	53	Deese	S	17.2	100
Mobil	Salt Creek	Tex.	Kent	12/84	12,100	176	85	Canyon Reef	LS	12.0	20
Natl. Coop. Refinery Assn.	Stewart Ranch	Wyo.	Campbell	2/72	1,200	14	14	Minnelusa	S	16.5	92
Oxy USA	Harmony Hill	Kan.	Rooks	10/84	130	5	3	Lansing/Kansas City	LS	12.5	
Oxy USA	Hoof	Kan.	Graham	6/82	960	18	8	Lansing/Kansas City	LS	12.0	74-113
Phillips	South Cowden	Tex.	Ector	9/83	2,050	54	23	Grayburg	Dolo.	11.9	3.7
Phillips	Vacuum	N.M.	Lea	8/83	320	14	8	San Andres	Dolo.	11.5	17
Phillips	North Burbank	Okla.	Osage	9/80	1,440	64	36	Burbank	S	15.5	50
Santa Fe Energy	Candy Draw	Wyo.	Campbell	5/87	482	9	1	Minnelusa	S	16.0	
Shell	Big Mineral Creek-Barnes Sand Unit	Tex.	Grayson	12/84	1,065	14	18	Barnes Sand	S	18.5	59
Shell	Big Mineral Creek-S Sand Unit	Tex.	Grayson	5/85	1,600	22	14	S Sand	S	13.0	28
Sun	Fitts	Okla.	Pontotoc	1982	3,030	88	102	Viola	LS	13.6	1-36
Sun	Hitts Lake	Tex.	Smith	10/80		22	8	Paluxy	S	18.9	300
Sun	Stephens County Regular	Tex.	Stephens	2/80	3,900	67	62	Caddo	LS	14.2	11
Sun	Stephens County Regular	Tex.	Stephens	2/80	700	11	12	Caddo	LS	14.2	11
Sun E&P	Stephens County Regular	Tex.	Stephens	3/86	3,026	63	75	Caddo	LS	14.0	10
Tesoro Petroleum	Hospah Sand Unit	N.M.	McKinley	1/83	1,061	31	16	Upper Hospah	S	24.7	467
Tesoro Petroleum	S. Hospah Lower Sand Oil Pool	N.M.	McKinley	10/83	594.6	50	13	Lower Hospah	S	27.3	1,282
Tesoro Petroleum	Elaine (San Miguel)	Tex.	Dimmit	12/81	1,260	11	13	San Miguel	S	21.7	6
Texaco	Vacuum Grayburg-San Andres	N.M.	Lea	7/82	1,486	37	29	Grayburg/San Andres	Dolo.	11.0	18
Texaco	Carthage N.E.	Okla.	Texas	4/83	1,024	7	4	Morrow 'A'	S	18.6	92
Texaco	Hewitt	Okla.	Carter	4/84	220	8	5	Hoxbar	S	21.0	84
Texaco	Hewitt	Okla.	Carter	8/82	300	6	5	Hoxbar	S	19.0	270
Texaco	Hewitt	Okla.	Carter	11/85	100	6	3	Lone Grove	S	18.0	50
Texaco	Naval Reserve	Okla.	Osage	2/82	3,360	90	28	Bartlesville	S	15.8	12
Texaco	Sho-Vel-Tum	Okla.	Carter	2/83	225	14	5	Deese	S	17.0	59
Texaco	Sho-Vel-Tum	Okla.	Carter	1/83	1,390	73	34	Hoxbar Deese	S	20.3	200
Texaco	Sho-Vel-Tum	Okla.	Carter	5/85	570	32	22	Deese	S	17.0	70
Texaco	Sho-Vel-Tum	Okla.	Stephens	8/81	190	7	4	Deese	S	17.4	216
Texaco	Black Diamond	Tex.	Jim Hogg	4/83	200	4	3	Pettus II	S	25.0	212
Texaco	Slaughter	Tex.	Cochran	12/81	7,406	275	90	San Andres	Dolo.	10.7	3
Texaco	Slaughter	Tex.	Cochran	2/84	1,887	33	15	San Andres	Dolo.	11.2	6.02
Toco Corp.	Clareton	Wyo.	Weston	10/85	200	6	1	Newcastle	S	15-18.0	1-15
Toco Corp.	Mush Creek	Wyo.	Weston	1969	900	8	6	Newcastle	S	15-18.0	1-15
Unocal	Dos Cuadras	Calif.	Santa Barbara	6/83	320	37	10	EP/FP	S	27-31.0	260-680
Unocal	Healdton	Okla.	Carter	12/82	2,010	162	105	Healdton Sands	S	26.2	346
Unocal	Osage-Hominy	Okla.	Osage	4/83	2,880	97	6	Miss. Chat	Chert. LS	30.0	27
Unocal	Farnsworth	Tex.	Ochiltree	9/83	13,005	35	40	Morrow	S	14.5	21.8
Unocal	Howard Glasscock	Tex.	Howard	1/83	169	12	11	Queen	S	22.0	37
Unocal	Southwest Van Unit	Tex.	Van Zandt	9/84	529	24	29	Lewisville	S	26.0	100

Micellar-polymer

Operator	Field	State	County	Start date	Area, acres	— Number wells — Prod.	Inj.	Pay zone	Formation type	Porosity, %	Permeability, md
Chevron	Berry Hill (Kiefer)	Okla.	Creek	9/82	82	12	21	Glenn	S	20.0	150
Gary	S. Johnson	Ill.	Clark	6/81	49	36	22	Partlow	S	18.3	302
Marathon Oil	Lawrence Robins 102 B Maraflood	Ill.	Lawrence	8/82	25	12	5	Bridgeport	S	21.0	398
Marathon Oil	Lawrence-Robins 202-B Project	Ill.	Lawrence	10/85	1.25	4	1	Bridgeport Main	S	20.6	352
Marathon Oil	Main Consolidated M-1 Project	Ill.	Crawford	2/77	407	56	60	Robinson	S	19.0	102
Marathon Oil	Main Consolidated M-2 Project	Ill.	Crawford	7/80	331	72	53	Robinson	S	19.0	150
Oxy USA	Madison	Kan.	Greenwood	8/82	40	15	16	Bartlesville	S	18.9	42
Texaco	Salem Consolidated	Ill.	Marion	8/81	60	20	12	Benoist	S	15.4	154
Texaco	Sho-Vel-Tum	Okla.	Stephens	8/83	8.75	3	3	Sims		17.0	83

Alkaline

Operator	Field	State	County	Start date	Area, acres	— Number wells — Prod.	Inj.	Pay zone	Formation type	Porosity, %	Permeability, md
Conoco	San Miguelito	Calif.	Ventura	11/82	316	16	11	Third Grubb	S	15.1	36
Unocal	South Van Unit	Tex.	Van Zandt	9/80	800	6	12	Lewisville	S	26.0	100
Unocal	Carroll Unit	Tex.	Van Zandt	11/81	212	3	7	Lewisville	S	26.0	100
Shell Offshore	White Castle	La.	Iberville	9/87	1	2	1	Q RA SU	S	31.0	1,000

Depth, ft	Gravity, °API	Reservoir oil cp	°F.	Previous prod.	Residual saturation % Start	End	Project status†	Total prod., b/d	Enhanced prod., b/d	Project eval.	Profit	Project scope
4,700	37.0	1.02	101	Prim./WF	38.0	30.0	HF	300		TETT	Yes	P (Exp. UL)
2,000-3,200	35.0	5.00	95	WF	52.0		HF	350	60	Succ.	Yes	FW
2,400-3,325	35.0	6.00	95	WF	57.0		HF	615	70	Succ.	Yes	FW
2,700	39.0	4.00	94	WF	31.0		JS	125	35	Succ.	Yes	FW
2,500-4,000	40.0	4.00	115	WF	57.0		HF	2,775	740	Succ.	Yes	FW
9,400	38.0	3.00	160	WF	49.0		JS	250	30	Succ.	Yes	P (Exp. L)
1,000	32.0	10.0	84.2	WF	46.2		HF	708	100	Succ.	Yes	FW
6,200	41.0	1.03	147	WF	39.0	37.0	NC	624	180	Succ.	Yes	FW
6,200	41.0	1.03	146	WF	45.0	44.0	NC	500	20	Succ.	Yes	FW
6,100	40.0	1.03	147	WF	40.0	38.0	NC	260	100	Succ.	Yes	FW
6,500	40.0	1.03	145	WF	43.0	41.0	NC	740	120	Succ.	Yes	FW
3,500	30.0	15.7	112	WF	47.9		HF	413	30	Succ.	Yes	FW
3,500	28.0	20.0	80	WF	49.0		HF	153	20	Succ.	Yes	FW
4,900	26.0	13.6	120	WF	47.0		HF	580	10	Disc.	No	FW
3,900	34.0	10.5	89	WF	53.0		HF	570	313	Succ.	Yes	FW
4,000	31.0	5.60	105	WF	36.0		HF	40	5	Disc.	No	FW
2,750	25.0	68.0	83	WF	48.0		HF	1,847	380	Succ.	Yes	FW
4,400	30.0	16.0	117	WF	55.0		HF	1,052	170	Succ.	Yes	FW
3,350	30-38.0	3-5-10	102	WF	52.4		HF	1,330	100	Succ.	Yes	FW
6,300	39.2	0.85	129	WF	60.0	58.0	NC	27,699	2,800	Succ.	Yes	FW
8,100	20.0	25.0	136	WF	68.0	49.0	NC	1,013	1,013	Succ.	Yes	FW
3,130	38.6	3.70	105	Prim.	49.0	34.0	JS	65	65	Succ.	Yes	FW
3,700	41.0	2.30	120	WF	36.0		JS	62	62	TETT	NE	FW
4,750	35.0	3.40	96	WF	67.0	59.0	NC	1,000		Disc.	No	FW
4,500	37.0	1.20	101	Prim.	74.9	58.2	NC	1,400		Succ.	Yes	FW
2,900	39.0	3.00	120	WF	53.0	47.0	NC	575	355	Succ.	Yes	FW
7,300	26.0			Prim.			JS	1,300	1,300	Prom.		
5,000	37.0	1.50	120	WF	50.0	47.0	JS	300	0	Disc.	No	P (Exp. UL)
5,200	37.9	1.67	124	WF	44.0	37.0	JS	570	0	Disc.	No	FW
	39.0	3.20	119	WF			HF	2,469		Succ.	Yes	FW
7,250	26.0	2.71	210	WF			HF	1,135		Succ.	Yes	FW
3,300	39.0	1.80	131	WF			HF	2,366		Succ.	Yes	FW
3,300	39.0	1.80	131	WF			HF	174		Succ.	Yes	FW
3,200	39.0	3.00	130	WF			HF	1.430		Prom.		FW
1,600	30.0	13.0	95	WF	69.0	68.2	JS	163	23	Succ.	Yes	FW
1,600	24.0	35.0	95	WF	82.5	65.9	JS	483	25	Succ.	Yes	FW
4,450	36.0	3.00	140		80.2	67.0	HF	117	82	Succ.	Yes	FW
4,720	38.0	1.47	105	WF	45.0	44.0	NC	3,650	0	Succ.	Yes	FW
4,400	40.0	2.00	113	WF	45.0	44.5	C	209	60	Succ.	Yes	FW
2,400	33.0	10.0	94	WF	39.0	37.0	C	102	46	Succ.	Yes	FW
2,350	35.0	6.00	98	WF	40.0	39.8	C	41	10	Succ.	Yes	FW
4,000	30.0	4.30	115	WF	39.0	37.0	C	135	31	Succ.	Yes	FW
2,650	38.0	3.50	90	WF	27.2	26.7	C	463	15	Succ.	Yes	FW
4,400	20-34.0	4.30	102	WF	38.0	37.2	C	94	8	Disc.	Yes	FW
3,100	25.0	20.0	95	WF	49.0	47.8	C	1,520	200	Succ.	Yes	FW
3,500	34.0	10.0	110	Prim.	57.0	55.0	NC	1,059	320	Succ.	Yes	FW
5,500	22.0	60.0	124	Prim.	64.0	56.0	C	88	29	Succ.	Yes	FW
4,550	48.0	0.40	164	Prim.	68.0	41.0	C	202	178	Succ.	Yes	FW
5,000	31.0	1.47	107	WF	49.4	45.6	C	6,150	900	Succ.	Yes	FW
5,000	31.0	1.47	110	WF	53.0	50.0	NC	291	0	Disc.	Yes	FW
6,000	40.0		160	Prim.			JS	5	5	TETT	TETT	P (Exp. UL)
4,400	40.0		115	Prim.			NC	90	90	Succ.	Yes	FW
1,280-1,650	25-26.0	15-24	95-103	WF	54-59	44-47	JS	3,100	0	TETT	TETT	FW
800	33.0	10.0	80	WF	56.0	47.0	HF	1,270	80	Succ.	Yes	FW
2,880	39.0	3.00	100	WF	51.0	48.0	HF	1,090	55	Succ.	Yes	FW
7,400	38.0	1.00	141	WF	43.0	40.0	HF	190	10	Disc.	Yes	FW
1,600	32.0	10.0	90	WF	50.7	43.4	JS	76	32	Succ.	Yes	FW
2,600	35.0	2.30	130	WF			JS	762	5	Prom.	Yes	FW
1,300	39.0	4.40	100	WF	50.0		HF	1,000	990	Succ.		FW
475	30.0	20.0	65	WF	46.0	28.0	HF	60	40	TETT	No	
950	29.0	18.0	70	Prim./WF	45.0		HF	250	250	TETT	TETT	P (Exp. UL)
900	29.0	18.0	70	WF	45.0		HF	20	20	TETT	TETT	P (Exp. UL)
1,000	36.0	6.00	70	Prim./WF	40.0		NC	100	100	Prom.	No	FW
1,000	35.0	7.00	70	Prim./WF	40.0		JS	50	50	TETT	TETT	FW
1,900	38.0	2.90	97	Prim.	34.0		NC			NE	No	P (Exp. UL)
1,750	36.0	36.0	72	WF	30.5	15.3	C	110	110	Prom.	No	P (Exp. UL)
7,000	28.0	12.0	135	WF	50.0	45.0	C	30	30	Disc.	No	P (Exp. UL)
8,011	31.0	0.70	205	Prim./WF	70.0	55.0	JS	950			TETT	FW
2,600	35.0	2.30	130	WF			NC	88	0	Disc.	Yes	FW
2,600	33.0	2.30	130	WF			NC	9	0	Disc.	Yes	FW
5,600-5,800	25.0	3.00		Aquifer influx	22.0	16.0	HF			TETT		P

Completed, terminated U.S. projects*

Operator	Field	State	County	Start date	Area, acres	Number wells Prod.	Number wells Inj.	Pay zone	Formation type	Porosity, %	Permeability, md
Alkaline											
Chevron	Quarantine Bay	La.	Plaquemines	11/81	200	5	3	3 (RB)SU	S	27.0	50-2,000
Chevron	Whittier	Calif.	Los Angeles	2/83	373	25		2nd & 3rd Zones	S	28.0	350
Mobil	Torrance	Calif.	Los Angeles	6/81	604	45	21	Main	S	31.0	220
Biopolymer											
Apache	Dry Creek	Neb.	Hitchock	7/85	500	9	5	Lansing/Kansas City	LS	13.0	13
CO_2 immiscible											
Phillips	Lick Creek	Ark.	Bradley-Union	1/76	1,640	39	13	Meakin	S	30.3	1,200
Texaco	Bayou Sale	La.	St. Mary	5/84	564	5	2	St. Mary	S	31.0	500
Texaco	West Delta Block 109	La.	Offshore	6/85	75	1	1	13100-Foot	S	29.7	68
Texaco	West Delta Block 109	La.	Offshore	6/85	74	1	1	10800-Foot	S	30.0	1,900
Texaco	Magnet Withers Pierce Estates B&C	Tex.	Wharton	7/83	500	4	0		S	23.0	1,700
Texaco	Pickett Ridge	Tex.	Wharton	5/83	726	4	0	Pickett Ridge	S	30.0	1,200
Texaco	Thompson	Tex.	Fort Bend	1/83	100	2	0	Frio	S	27.0	100-1,000
Texaco	Paradis	La.	St. Charles	3/84	110	2	2	Main Pay	S	28.0	1,033
Texaco	Manvel	Tex.	Brazoria	11/83	128	1	0	Oligocene	S	30.0	1,000
Texaco	Cote Blanche Bay West	La.	St. Mary	3/84	55	3	3	14 Sand	S	29.0	322
Texaco	West Delta Block 109	La.	Offshore	6/85	48	1	1	10200-Foot	S	27.0	205
Texaco	South Marsh Island Block 6	La.	Offshore	6/85	82	2	1	Rob E-S	S	29.6	323
Texaco	Lake Barre	La.	Terrebonne	3/84	1,164	12	4	Upper M	S	25.0	139
Texaco	Manvel	Tex.	Brazoria	10/82	43	1	0	Oakville	S	30.0	400
Texaco	Lafitte	La.	Jefferson	8/84	271	5	2	8900-Foot	S	27.0	250
Texaco	Pierce Ranch	Tex.	Wharton	1/83	480	8	0	Pierce Ranch	S	31.8	534
Texaco	West Columbia	Tex.	Brazoria	6/83	33	3	0	PTSD	S	30.0	560
Texaco	Magnet Withers BH&S State Tracts J.E. Broussard A.C. Thompson	Tex.	Wharton	10/83	1,224	1	0	Magnet Withers	S	23.0	1,700
Texaco	Withers North	Tex.	Wharton	3/83	454	10	0	Withers N.	S	25.0	1,050
Texaco	West Delta Block 109	La.	Offshore	6/85	78	1	1	12500-Foot	S	27.0	1,032
CO_2 miscible											
ARCO	Garber	Okla.	Garfield	10/81	80	9	4	Crews	S	19.0	12
Chevron	University Waddell	Tex.	Crane	5/83	920	49	13	Devonian	Dolo.	12.0	14
Citronelle Unit	Citronelle	Ala.	Mobile	5/81	13,640	42	10	Rodessa	S	15.0	13
Grace Petroleum	South Bishop Ranch	Wyo.	Campbell	1/82	640	5	2	Minnelusa	S	16.0	50
Grace Petroleum	South Bishop Ranch	Wyo.	Campbell	1/82	1,280	5	4	Minnelusa	S	15.0	150
Carbonated waterflood											
Amoco	San Andres	Calif.	Fullerton	10/85	1,600	40	32	San Andres	Dolo./LS	13.0	1.4
Combustion											
Bayou State	Caddo Pine Island	La.	Caddo	1979	20	13	3	Nacatoch	S	30.0	500
Lawrence-Allison	Teapot Dome NPR-3	Wyo.	Natrona	9/80	10	9	8	Shannon	S	18.0	63
Marathon Oil	Main Consolidated	Ill.	Crawford	10/61	517	51	5	Robinson	S	19.7	320
Mobil	South Belridge	Calif.	Kern	1964	245	24	0	Tulare	S	35.0	3,000
Texaco	Camp Hill	Tex.	Anderson	9/85	17.4	29	6	Carrizo	S	38.0	4,000
Unocal	Brea-Olinda	Calif.	Orange	3/72	17.1	12	2	1st Miocene	S	29.0	300
Hydrocarbon miscible											
Amoco	Slaughter	Tex.	Hockley	4/72	12.2	2	6	San Andres	Dolo.	10.5	4.3
Chevron	Swanson River	Alaska		11/83	2,660	33	8	Hemlock	Congl.	21.0	143
Exxon	Means	Tex.	Andrews	11/83	8,700	438	184	San Andres	Dolo.	9.0	20
Mitchell Energy	Alvord (3,000 ft Strawn)	Tex.	Wise	7/80	217	6	3	Bryson	S	20.0	46
Micellar-polymer											
Amoco	Torchlight	Wyo.	Big Horn	1981	6.4	1	4	Tensleep	S	14.2	61
Gary	Bell Creek Unit 'A'	Mont.	Powder River	2/81	179	2	0	Muddy	S	24.9	1,218
Marathon Oil	Main Consolidated 219-R Project	Ill.	Crawford	10/75	113	2	0	Robinson	S	20.8	105
Marathon Oil	Main Consolidated 119-R Project	Ill.	Crawford	9/68	40	27	18	Robinson	S	20.0	200
Mitchell Energy	Wizard Wells	Tex.		1981	700	11		Caddo	S	14.0	50
Mobley O.B.	Lewisville	Ark.	Lafayette	7/80	458	8	3	Smackover	LS	21.8	24
Pennzoil	Bradford	Pa.	McKean	7/82	220	132	57	Bradford 3rd	S	21.0	188
Phillips	N. Burbank	Okla.	Osage	1975	90	16	9	Burbank	S	15.5	50
Texaco	Main Consolidated	Ill.	Crawford	12/81	10.88	9	0	Robinson		21.8	100
Unocal	East Coalinga Extension	Calif.	Fresno	12/80	90	18	4	Gatchell	S	14.2	150
Nitrogen											
Texaco	Vealmoor East	Tex.	Borden/Howard	12/81	3,353	34	8	Canyon	LS	9.5	38
Polymer											
ARCO	Camrick District	Okla.	Beaver	9/84	2,980	12	10	Upper Morrow	S	15.0	14
ARCO	Horseshoe Gallup	N.M.	San Juan	11/83	450	10	8	Gallup	S	15.0	14
ARCO	Roehrs	Wyo.	Campbell	9/84	324	1	1	Minnelusa	S	17.0	92
ARCO	East Texas (Kinney Strickland Leases)	Tex.	Gregg & Rusk	3/85	380	59	14		S	25.0	2,000
Conoco	South Elk Basin	Wyo.	Park	11/84	540	6	1	Embar/Tensleep	S	15.1	90
Conoco	Frannie	Wyo.	Park	9/85	2,020	64	50	Phosphoria/Tensleep	S	13.6	130
Grace Petroleum	High Five	Mont.	Roseland	7/81	800	9	2	Stensvad	S	18.0	200
Lawrence-Allison	Teapot Dome NPR-3	Wyo.	Natrona	10/82				Shannon	S	18.0	63
Mobil	Sho-Vel-Tum	Okla.	Stephens/Carter	12/82	1,285	32	8	Springer	S	15.4	17.9
Oxy USA	Beverly Hills/West Pico	Calif.	Los Angeles	12/84				Repetto/Main	S	26.6-28.0	77-193
Phillips	Lick Creek	Ark.	Bradley-Union	9/84	1,640	39	23	Meakin	S	30.3	1,200

Depth, ft	Gravity, °API	Reservoir oil		Previous prod.	Residual saturation %		Project status†	Total prod b/d	Enhanced prod., b/d	Project eval.	Profit	Project scope
		cp	°F.		Start	End						
8,200	33.0	5.00	182	Prim.	51.0					Disc.	No	
2,300	20.0	30.0	120	WF	27.0	22.0		175	175	Disc.	No	
3,500	21.0	17	165	WF	40.7	39.6	Term. 1986	2,450		Disc.	No	
4,100	31.0	9	120	Prim.			Term.			None	No	P
2,500	17.0	160	118	Prim.	55.0	46.0	Term.	450	400	Succ.	Yes	FW
10,000	34.0	0.4	194	Prim.	50.0	45.0	Term.	1,500	0	Succ.	Yes	RW
11,325	37.0	0.3	210	Prim.	50.0	45.0	Term.	1,600	0	Disc.	Yes	RW
10,500	36.5	0.28	200	Prim.	50.0	45.0	Term.	400	0	Succ.	Yes	RW
5,500	26.0	2.3	154	Gas inj.	35.0	34.0	Term.	193	0	Disc.	Yes	FW
4,600	25.0	2.5	138	Prim.	29.0	28.0	Term.	93	0	Disc.	Yes	FW
5,100	25.2	2.7	120	Prim.	42-65		Term.	66	0	Disc.	Yes	FW
10,200	38.0	0.5	195	WF	50.0	45.0	Term.	350	0	Succ.	Yes	RW
5,000	26.7	7.2	149	Prim.	45.0	20.0	Term.	598	0	Disc.	Yes	FW
8,000	32.0	1.3	184	WF	50.0	45.0	Term.	100	0	Succ.	Yes	RW
10,000	37.3	0.26	194	Prim.	50.0	45.0	Term.	675	0	Disc.	Yes	RW
11,200	34.0	0.3	208	Prim.	50.0	45.0	Term.	250	0	Succ.	Yes	RW
13,000	33.0	0.4	236	WF	50.0	45.0	Term.	760	0	Succ.	Yes	RW
4,000	25.0	4.4	149	Prim.	42-65		Term.	122	0	Disc.	Yes	FW
8,900	34.0	0.7	185	Prim.	50.0	45.0	Term.	370	0	Succ.	Yes	RW
4,900	24.4	4.57	155	Prim.	48.0	18.0	Term.	404	0	Disc.	Yes	FW
2,600	30.0	8	116	Prim.	36-48	10-20	Term.	198	0	Disc.	Yes	FW
5,500	26.0	2.3	154	Gas inj.	35.0	31.0	Term.	37	0	Disc.	Yes	FW
5,250	25.7	2.45	145	Prim.	35.0	32.0	Term.	298	0	Disc.	Yes	FW
12,000	34.1	0.25	218	Prim.	50.0	45.0	Term.	2,300	0	Disc.	Yes	RW
1,900	44.0	1	100	WF	30.0	16.0	Term.	35		Succ.	Yes	
8,500	43.0	0.45	140	WF	38.0		Term.	1,750	0	Disc.	No	FW
10,600	42-45.0	0.46		Prim./WF	63.0		PP 2/83	3,519	357	TETT		P
9,200	35.0	1.14	220	WF								
9,400	34.0		180	WF								
4,500	32.0		100					1,300		TETT	TETT	FW
1,050	21.0	620	80	Prim.	96.0	48.0	Term. 6/86	100	84	Prom.	No	
270-425	33.0	10	65	Prim.	50.0	0.0	Term. 4/86	35	15	Succ.		P (Exp. UL)
900	29.0	35	65	Prim./WF	55.0	40.0	Term. 11/86	80	80	Prom.	Yes	
1,000	14.0	1,600	95	Prim.	64.0	35.0	Term. 3/86	210	140	Succ.	Yes	FW
500	18.0	1,200	75	Prim.	59.0	25.0	Term.	29	0	Disc.	No	P (Exp. UL)
2,200	16.0	90	109	Prim.			Term.	150	50	Disc.	Yes	P (Exp. UL)
5,000	28.0	1.9	105	WF	58.0		Term.	0	0	Succ.		P (Exp. UL)
10,300	37.0	1	180	HC immisc.	33.0	30.0	Term. 1986	5,383		NE		FW
4,300	29.0	6	97	WF						TETT	TETT	
3,000	41.0	0.95	107	WF	37.0	32.0	Term.	4	4	Disc.	No	
3,100	34.4	41.2	100	WF	36.0			5	2	TETT	No	P (Exp. UL)
4,500	32.0	7	110	WF	33.1	21.5	Term. 1/87	0	0	Succ.	Yes	
1,000	35.0	7	70	Prim./WF	40.0	27.1	Term 12/82	0	0	Disc.	No	P (Exp. UL)
1,000	35.0	7	70	Prim./WF	40.0	25.0	Term 12/76	0	0	Prom.	No	
4,900	42.0	1.2	130	Prim.	82.0	28.0		50	50	Succ.	Yes	
6,800	43.0	0.323	191	WF	87.0			421	8	NE	TETT	
1,900	45.0	5	68	Prim.	40.0	22.0	Term.	60	200	TETT		
2,900	39.0	3	120	WF	30.0	18.0	Term.	19		Prom.	Yes	P (Exp. UL)
1,040	31.0	16	85	WF	40.0	33.6	Term.	31	0	Disc.	No	P (Exp. UL)
7,500	31.4	2.2	200	Prim.	43.0	10.0	Term.	700	0	Disc.	No	P (Exp. UL)
7,350	43.0	1.2	155	WF	66.0	42.0	Term.	2,521	0	Disc.	Yes	FW
7,100	39.0	1.5	149	WF				290		TETT	TETT	
1,200	42.0	1.7	87	WF			Term.	500		Disc.	No	
7,465	25.0	9	177	WF				14	5	Succ.	Yes	
3,650	39.0	1	140	WF			Term.	1,197	350	Succ.	Yes	
7,100	29.0	6	140	Prim./WF			Term.	83	0	Disc.	No	FW
3,300	26.3	15	95	Prim./WF			Term.	3,593		Disc.	No	FW
5,600	40.0		140	Prim.				480		TETT	Yes	
550	33.0	10	65	Prim.	50.0	50.0	Term. 4/86	20	5	Disc.		P (Exp. UL)
5,200	39.0	2.4	110	WF	50.0		Term.	70		Prom.	Yes	
6,800	26-32.0	1	200	WF	37.5		Term. 12/86				TETT	FW
2,500	17.0	160	118	CO2	55.0	46.0	Term.	450	100	Succ.	Yes	FW

Completed, terminated U.S. projects* — continued

Operator	Field	State	County	Start date	Area, acres	Number wells — Prod.	Number wells — Inj.	Pay zone	Formation type	Porosity, %	Permeability, md
Phillips	South Burbank	Okla.	Osage	7/82	720	34	23	Burbank	S	19.5	150
Phillips	Bates Unit	Kan.	Sumner	1/84	640	14	2	Mississippi	LS	15.5	19.7
Phillips	North Burbank	Okla.	Osage	3/83	1,440	58	23	Burbank	S	15.5	50
Phillips	Golden Trend	Okla.	Grady	10/83	720	4	4	Springer	S	9.7	58
Samedan	S. Robertson	Tex.	Gaines	1/81	1,600	56	11	Glorieta/Clfk.	Dolo.	7.9	0.5
Samedan	Kuehne Ranch Unit	Wyo.	Campbell	3/81	645	5	2	Minnelusa	S	14.4	52.8
Texaco	Cogdell	Tex.	Kent & Scurry	6/83	15,378	99	126	Canyon Reef Limestone	LS	9.6	5
Texaco	Loudon	Ill.	Fayette	12/84	800	18	13	Weiler	S	21.0	142
Texaco	Levelland	Tex.	Hockley	4/83	3,595	108	37	San Andres	Dolo.	10.0	0.6
Texaco	Robertson North	Tex.	Gaines	9/83	1,400	27	16	Glorieta	Dolo.	7.7	1.54
Texaco	Balko South	Okla.	Beaver City	8/82	1,320	7	3	Kansas City	LS	21.0	0-1,070
Texaco	Cowden North	Tex.	Ector/Andrews	12/84	2,440	52	61	Grayburg/San Andres	S/LS/Dolo.	10.1	3.8
Texaco	Mabee	Tex.	Andrews/Martin	12/81	13,500	213	230	San Andres	Dolo.	10.5	1.5
Texaco	Jordan	Tex.	Ector & Crane	11/84	1,513	62	30	San Andres	Dolo.	10.5	6
Texaco	Apache Pool	Okla.	Caddo	9/83	1,060	16	8	Bromide	S	13.6	350
Texaco	McElroy	Tex.	Crane/Upton	10/81	7,550	504	333	Grayburg/San Andres	Dolo.	11.0	5
Texaco	Glenn Pool	Okla.	Creek	1/84	480	16	8	Bartlesville	S	19.5	400
Texaco	Hewitt	Okla.	Carter	3/84	240	9	3	Hoxbar	S	21.0	100
Texaco	Penwell	Tex.	Ector	11/84	4,470	106	36	San Andres	Dolo.	7.8	2.2
Texaco	Ventura	Calif.	Ventura	3/83	800	95	56	C-Block	S	20.0	130
Texaco	Spraberry Deep	Tex.	Dawson	7/83	2,448	15	20	Upper Spraberry Sand	S	17.9	10.6
Texaco	Sho-Vel-Tum	Okla.	Stephens	4/85	3,740	121	20	Sims	S	19.1	190
Texaco	Sho-Vel-Tum	Okla.	Stephens	6/84	150	16	12	Permian	S	25.3	128
Texaco	Harris	Tex.	Gaines	10/83	3,800	83	80	Glorieta	Dolo.	8.6	3
Texaco	Hewitt	Okla.	Carter	7/84	1,200	29	11	Lone Grove	S	18.0	50
Texaco	Flowers	Tex.	Stonewall	8/83	5,876	112	67	Canyon	S	14.7	2
Texaco	Blue Buttes	N. Dak.	McKenzie	12/85	12,770	30	19	Madison	LS	9.6	22
Texaco	Salem	Ill.	Marion	7/82	8,800	330	229	Miss./Devonian	S	10.5-17.9	5-2,200
Texaco	Aneth Unit	Utah	San Juan	9/85	17,000	160	167	Paradox	LS	10.6	18.3
Texaco	Cushing	Okla.	Creek	12/83	480	14	10	Bartlesville	S	17.0	100
Texaco	Dale	Ill.	Hamilton	11/85	4,100	35	4	Aux Vases	S	14.5	68
Texaco	Vacuum Grayburg-San Andres	N.M.	Lea	4/83	3,046	77	81	Grayburg/San Andres	Dolo.	11.6	13
Texaco	Tonti	Ill.	Marion	10/84	285	10	11	Renoist Auxvases McClusky	LS	47.3	358
Texaco	Atlantic	Okla.	Osage	5/85	1,920	35	7	Mississippi Chat/Burgess	S	25.0	25
Texaco	Langlie Mattix	N.M.	Lea	2/83	9,360	124	110	Queen/Seven Rivers	S	11.4	7.2
Texaco	Spraberry Deep	Tex.	Dawson	7/83	2,368	12	11	Spraberry Lwr. Sand	S	16.3	5.7
Texaco	Sumatra	Mont.	Rosebud	12/84	4,778	18	21	Tyler	S	15.3	48
Unocal	Dollarhide (Clearfork)	Tex.	Andrews	10/85	2,800	57	33	Clearfork	Dolo.	11.6	8.5
Unocal	South Cowden	Tex.	Ector	1/86	1,600	38		Grayburg	Dolo.	8.5	3.1
Unocal	Smyer	Tex.	Hockley	11/84	4,410	72	38	Clearfork	Dolo.	8.3	1-20
Unocal	Orcutt	Calif.	Santa Barbara	6/84	1,680	78	51	Pt. Sal	S	22.5	78
Unocal	North Riley	Tex.	Gaines	2/84	6,720	105	40	Clearfork	Dolo.	7.7	12
Steam											
Angus Petroleum	Blackwell's Corner	Calif.	Kern	1/84	340	25		Agua	S	37.0	475
Chevron	Lost Hills Sec. 32	Calif.	Kern	4/77	120	30	4	Etchegoin Main	S	40.0	62
Chevron	Kern Front	Calif.	Kern	2/80	36	18	9	Chanac	S	29.0	1,000
Chevron	Bradley Canyon	Calif.	Santa Barbara	4/82	40	16	8	Sisquoc	S	30.0	800
Chevron	McKittrick 21Z	Calif.	Kern	9/77	44	19	0	Tulare	S	34.0	1,700
DCR Petroleum	O'Connell Ranch Unit	N.M.	Guadalupe	8/81	1.2	5		O'Connell	S	20.5	250
DCR Petroleum	T-4 Ranch Unit	N.M.	Guadalupe	10/82	1.2	3	1	O'Connell	S	20.5	250
Drilling & Production	Midway-Sunset	Calif.	Kern	1965	160	34	20	Monarch	S	22.0	300
Elf Aquitaine	Poso McVan	Calif.	Kern	1980	120	32	19	Etchegoin	S	30-40.0	5,000
Elf Aquitaine	Poso Creek	Calif.	Kern	1980	50	35	19	Etchegoin	S	35.0	1,000
Mobil	Kern Front	Calif.	Kern	10/82	21	10	4	Etchegoin	S	37.0	1,000
Santa Fe Energy	Brea Olinda	Calif.	Orange	10/85	3	4	1	Anaheim	S	27.0	1,000
Shell	Coalinga	Calif.	Fresno	8/84	240	85	12	Temblor Zone I	S		
Shell	Cat Canyon (East)	Calif.	Santa Barbara	8/85	35	7		Brooks Sand	S	30.0	1,400
Shell	Arroyo Grande	Calif.	San Luis Obispo	9/84	208			M2	S	31.0	300
Shell	Coalinga	Calif.	Fresno	1982	57	21	10	Temblor Zone II	S		
Tenneco	Poso Creek	Calif.	Kern	1968	154	14		Etchegoin	S	32.0	2,500
Texaco	South Belridge	Calif.	Kern	1985	20	25	16	Tulare	S	37.0	2,700
Texaco	Poso Creek	Calif.	Kern	3/83	69	26	15	Basal Etchegoin	S	31.0	1,945
Texaco	Cat Canyon	Calif.	Santa Barbara	4/82	80	20	12	S1B	S	35.0	4,500
Texaco	Sour Lake	Tex.	Hardin	4/82	1.5	5	2	Miocene OB-1	S	32.0	1,500
Texaco	Cat Canyon (East)	Calif.	Santa Barbara	11/77	30	18	3	S16B Thomas Sand	S	31.5	404-1,749
Texaco	Sour Lake	Tex.	Hardin	4/80	2,645	1		Miocene	S	25-36.0	1,600
Texaco	Cat Canyon	Calif.	Santa Barbara	1965	300	12		S8-516	S	30.0	2,000
Unocal	Santa Maria Valley	Calif.	Santa Barbara	3/81	260	18	0	Pt. Sal	S	27.0	100
Unocal	Guadalupe	Calif.	San Luis	3/81	644	82	12	Sisquoc	S	35.0	1,550
Unocal	McKittrick	Calif.	Kern	12/82	70	25	14	Amnicola	S	37.0	1,500
Unocal	Belridge	Calif.	Kern	10/64	50	6	0	Tulare	S	37.0	2,500
Unocal	North Belridge	Calif.	Kern	8/64	35	6	0	Tulare	S	37.0	2,500
Unocal	Midway-Sunset	Calif.	Kern	1969	80	2		Tulare	S	26.0	300
Unocal	Jesus Maria	Calif.	Santa Barbara	8/82	20	2	0	Monterey	Fract Chert-Dolo	7.5	2
Unocal	McKittrick	Calif.	Kern	8/71	70	27	0	Tulare	S	37.0	1000-2000

* Data on acreage, wells and production are the most recent reported. They do not necessarily reflect the project's status at termination. † PP = Postponed

Depth, ft	Gravity °API	Reservoir oil cp	°F.	Previous prod.	Residual saturation % Start	End	Project status†	Total prod., b/d	Enhanced prod., b/d	Project eval.	Profit	Project scope
2,750	39.0	3	120	WF	41.0	37.0	Term.	175		Disc.	No	FW
3,700	42.0	0.6	117	WF	50.0	48.0	Term.	200		TETT	TETT	FW
2,900	39.0	3	120	WF	48.0	44.0	Term.	80		Prom.	Yes	FW
9,800	38.0	2.9	155	WF	54.0	53.0	Term.	60		TETT	TETT	FW
5,800	34.0	1	107	WF	64.0	NA		2,080		Succ.	Yes	FW
7,900	26.0	8	134	Prim./WF				150	10	Succ.	Yes	FW
6,800	41.7	0.62	128	WF	28.0	26.0	C	5,022	0	Succ.	Yes	FW
1,550	37.0	5	88	WF	21.0	20.0	Term.	63	0	Disc.	No	FW
4,720	30.5	1.47	107	WF	45.0	44.0	C	2,017	0	Succ.	Yes	FW
5,809	30.8	0.67		WF	45.0	44.0	C	1,261	0	Succ.	Yes	FW
6,100	40.0	1.8	125	WF	40.0	38.3	C	65	0	Disc.	Yes	FW
4,450	34.0	1.63	94	WF	47.0	45.0	Term.	499	0	Disc.	Yes	P (Exp. UL)
4,700	32.0	2.38	106	WF	43.0	38.0	C	5,007	0	Disc.	Yes	FW
3,600	34.0	2.8	95	WF	54.0	53.0	C	677	0	Disc.	Yes	FW
3,900	39.0	3.2	96	WF	36.0	25.0	C	578	0	Disc.	Yes	FW
3,000	32.0	2.6	95	WF	59.0	33.0	C	6,624	0	Succ.	Yes	FW
1,500	37.0	6	90	WF	60.0	58.0	C	90	0	Disc.	No	FW
2,600	35.0	8	92	WF	45.0	44.0	C	66	0	Disc.	No	FW
3,600	33.0	1.8	95	WF	54.0	53.0	C	1,250	0	Disc.	Yes	FW
5,500	30.5	1.3	160	WF			Term. 2/86	6,300		TETT	TETT	
6,500	40.6	0.92	103	WF	37.0	34.0	C	231	0	Succ.	Yes	FW
4,000-7,000	24-29.0	5	105-135	WF	50.0	47.1	Term.	3,179	0	Disc.	No	P (Exp. UL)
1,100	28.0	50	90	WF	48.9	47.8	C	227	0	Disc.	Yes	FW
5,818	30.8	3.1		WF	45.0	44.0	C	5,912	0	Succ.	Yes	FW
4,500	35.0	4	115	WF	54.0	52.0	Term.	155	0	Disc.	No	P (Exp. UL)
4,200	42.0	1.9	130	WF	40.0	35.0	Term.	680	0	Disc.	Yes	P (Exp. UL)
9,400	42.0	0.3	240	WF	41.0	40.0	Term.	1,427	0	Disc.	No	P (Exp. UL)
1,700-3,600	40.0	3-4.4	82-104	WF	27.7	27.5	Term.	3,115	300	Succ.	Yes	FW
5,300	47.0	0.55	134	WF	52.0	46.0	Term.	5,823	0	Disc.	Yes	P (Exp. UL)
2,670	41.0	2.5	100	WF	57.0	54.0	C	103	0	Disc.	Yes	FW
2,500	36.0	4	100	WF	34.0	31.0	Term.	35	0	Disc.	No	P (Exp. UL)
4,720	38.0	1.47	105	WF	45.0	44.0	C	6,073	0	Succ.	Yes	FW
1,900-2,200	39.5	4	83	WF	30.9	28.5	Term.	78	0	Disc.	No	FW
2,500	39.0	4	95	WF	34.0	32.0	C	155	0	Disc.	No	FW
3,650	35.0	2.8	94	WF	50.7	46.2	C	1,088	0	Succ.	Yes	FW
7,100	39.3	0.885	130	WF	42.0	34.0	C	316	0	Succ.	Yes	FW
4,900	33.4	4.8	159	WF	29.8	28.7	Term.	463	0	Disc.	Yes	FW
6,500	37.0	0.56	110	WF	55.5	46.9	Term.	1,800	0	Disc.	No	P (Exp. UL)
4,500	34.0	3.5	103	Prim./WF	43.6	42.4	Cancelled	720		TETT	TETT	
5,900	27.0	5	112	WF	49.3	43.5	Term.	1,100	0	Disc.	No	FW
3,750	23.5	8	165	WF	46.0	40.0	Term.	1,640	0	Disc.	No	FW
6,300	32.0	2.55	104	WF	58.0	48.0	Term.	4,300	0	Disc.	No	FW
1,300	13.0	4,000	100	Prim.	40.0	15.0	Term.		600	TETT	Yes	
1,150	20.0	50	100	Prim.	72.0			138	96			
2,500	12.0	2,300	100	Prim.	50.0	12.0	Term. 1986	195	129	Disc.	No	FW
3,000	11.0	17,600	120	Prim.	90.0	40.0	Term.	70	34	Disc.	No	P
1,000	12.0	7,700	97	Prim.	47.0	12.0	PP 9/87	45	45	Disc.	No	FW
350	17.0	6,000	70	None	55.0	27.5				TETT	No	
709	17.0	6,000	65	None	50.0	25.0				TETT	No	
1,400	13.0	13.3	90	Prim.			Term.	140	30	Prom.	Yes	
1,300	16.0	310	95	Prim.				1,000	1,000	Succ.	Yes	
1,100	13.0	8,000	105	Prim.	20.0			550	520	No		
1,800	14.0	800	100	Steam soak	67.0	25.0	Term.	200	150	Disc.	No	P (Exp. UL)
900	13.0	4,000	100	Prim.	55.0	30.0	Term.	100	70	Prom.	Yes	
1,200	13.0	2,000	95	Prim.				1,800	1,630	Succ.	Yes	
3,300	6-10.0	7,000	135	None	68.0		Term.			TETT	No	
200-500	15.0	4,000	100	Prim.	63.0	55.0	Term.	700	350	TETT	Yes	
1,800	12.0	700	100	Prim.				700	500	Succ.	Yes	
2,400	13.0	4,000	100	Prim.	65.0	55.0	Term.	170	134	Prom.	Yes	
950	14.0	2,340	100	Steam stimulation	58.0	25.0	Term.	125		TETT	TETT	
2,000	13.0	1,800	122	Steam stimulation	50.0	20.0	Term. 8/86	1,250	1,250	Prom.	Yes	
2,300	8.0	100,000	110	None			Term. 6/86	689	689	Succ.	Yes	
500	13.0	1,680	90	Prim.	55.0	22.0	Term.	15	15	Succ.	Yes	P (Exp. UL)
2,610	10.4	13-25000	110	Cyclic steam stim.			Term. 4/86	378	378	Prom.	Yes	
400-2,700	11-24.0			Prim.	53.0	17.0	Term.	25	0	Prom.	Yes	P (Exp. UL)
3,000	10.0	14,000	130	Prim.			Term.	103	103	Succ.	Yes	
1,900	12.0	75	110	Prim.	40.0	31.0	Term.	140	0	Disc.	No	P (Exp. UL)
3,000	9.0	580	85	Cyclic	65.0	61.0	Term.	1,900	200	Disc.	No	P (Exp. UL)
1,100	11.3	300	180	Steam soak	60.0	30.0	Term.	290	0	Succ.	Yes	FW
740	13.7	450	130	Prim.	62.0	41.0	Term.	25	0	Succ.	Yes	FW
770	13.6	450	130	Prim.	59.0	40.0	Term.	25	0	Succ.	Yes	FW
1,900	14.0	700	106	Prim.	50.0	35.0	Term.	5	0	Disc.	No	P (Exp. UL)
3,500	9.0	20,000	110	Prim.	60.0	45.0	Term.	0	0	Disc.	No	P (Exp. UL)
1,200	13.0	700	150	Steam soak	55.0	35.0	Term.	160	0	Succ.	Yes	FW

PLATFORM A on Alabama Tract 76 in Mobile Bay's Mary Ann field is flowing large volumes of natural gas for Mobil Exploration & Producing U.S. Inc. from a Jurassic Norphlet deep gas trend. In the background, Global Marine Inc.'s Glomar High Island IV jack up works over Mobil's 77-1 on Alabama Tract 77 Platform B. The trend, discovered by Mobil in 1979, could be the largest accumulation of gas in the U.S. outside Hugoton field in southwestern Kansas, northeastern Oklahoma, and the Texas Panhandle.

Natural gas is major world energy source

NATURAL GAS HAS BECOME A MAJOR WORLD ENERGY source during the last 20 years.

World reserves of gas are estimated to be more than 3,571.43 tcf—equivalent to about 650 billion bbl of oil or about three-quarters of total world oil reserves, according to Royal Dutch/Shell Group data.

The Royal Dutch/Shell data show world gas production increased to about 63.04 tcf in 1987, compared with 59.64 tcf in 1986 (Table 1). The 6% increase is equivalent to about 32 million bbl of oil equivalent.

World gas consumption increased to about 63.04 tcf in 1987, compared with 59.46 tcf in 1986 (Fig. 1, Table 2). Gas today accounts for 20% of world primary energy demand.

About 8.4 tcf—equivalent to 13% of total consumption—was traded internationally in 1987, an increase of about 7% from 1986.

Most of the increase, said Royal Dutch/Shell, came from trade in pipeline gas that accounted for about three-quarters of total international trade (Fig. 2). Higher exports from the U.S.S.R. to East and West Europe and from Canada to the U.S. were the main contributors.

World liquefied natural gas (LNG) trade increased 6% in 1987 from 1986. Japan's LNG imports account for three quarters of the total trade in LNG. Indonesia is the largest LNG exporter.

Resource base

The geographical distribution of gas reserves is different to oil.

About three-quarters of the world's oil reserves are held by the Organization of Petroleum Exporting Countries, but the equivalent figure for gas in only about one-third.

Almost 100 countries have proven reserves sufficient for commercial development. However, about 80% of the reserves are in just 10 countries.

Nearly 40% of the world's gas reserves are in the

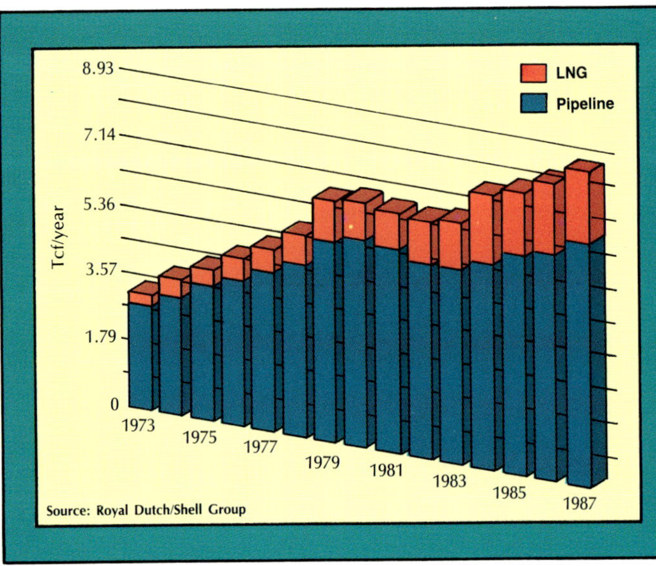

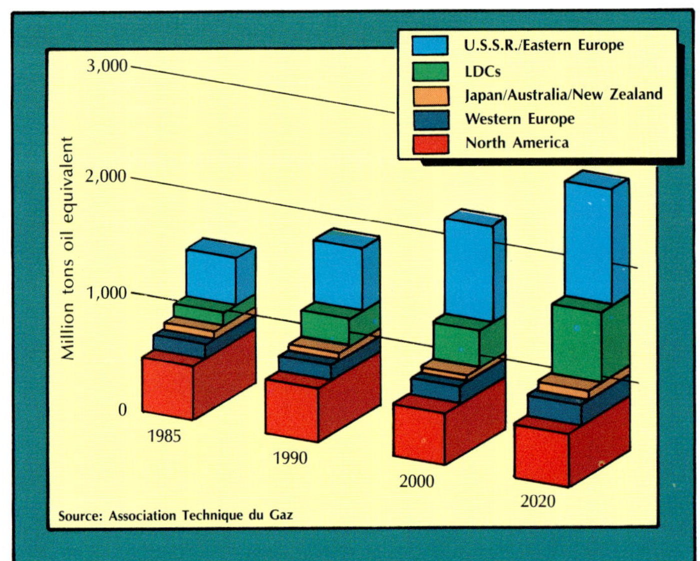

U.S.S.R., about half of which are in supergiant Urengoy, Yamburg, and Bovanenkovskoye fields in West Siberia (Table 3).

A further 30% of the world's gas reserves are in the Middle East, of which about 40% are in Iran.

At the other end of the resource scale, two of the main consuming areas—the U.S. and West Europe—taken together account for only 11% of total world gas reserves.

Royal Dutch/Shell said that about three-quarters of the world's gas reserves are non-associated gas.

In addition to the world's reserves of gas, several countries—particularly the U.S.—contain large quantities of supplies in such non-conventional categories as tight gas sands, coal seam methane, shale gas, ultra deep gas, gas hydrates, and abiogenic gas.

Most non-conventional gas cannot be produced economically today. Estimates of non-conventional gas resources vary widely, but most believe non-conventional resources are significantly greater than current estimates of conventional gas reserves.

Production outlook

Marketed production of gas in the world reached a record level in 1987, continuing the upswing that started earlier this decade following a period of flat production growth.

World gas production now amounts to nearly 60% of oil production, compared to only 45% in 1980, according to a study by Cedigaz and Francais du Petrole.

Production increases affected most geographic zones in the world, including Organization for Economic Cooperation and Development (OECD) countries where the preceding years had been somewhat disappointing.

The U.S., after a period of decline, increased its production, thus recovering part of the market that gas had lost in 1986 when oil prices collapsed. It is questionable, however, whether the U.S. will be able to repeat in 1988.

Western Europe production also rose. The increase, after a mediocre year in 1986, benefited most of the producing countries, but especially the North Sea producers.

In the Middle East and Africa, rational gas production projects and the development of gas markets are being aggressively pursued.

And the Soviet Union again increased output. Siberia now accounts for two-thirds of Soviet production.

The growth differentials between areas have profoundly changed the structure of world production. The share of the industrialized Western countries in the world total has not stopped falling since the 1960s. The trend has benefited other producing countries, especially the planned economies.

Based on Royal Dutch/Shell's estimate of 1987's gas production level, the world's reserves/production ratio is 58 years.

However, reserves/production ratios vary significantly from country to country. In the U.S., for example, the ratio is only 12 years, while for the Middle East it is about 270 years.

Substantial progress has been made in recent years to reduce the amount of gas vented or flared, although flaring associated gas is sometimes inevitable when no demand exists for the gas produced and reinjection is not practicable.

The Cedigaz and Francais du Petrole study shows offshore production represented about 20% of total world production in 1987. The U.S. maintained first place in offshore production, although North Sea production is expanding rapidly.

Developing gas reserves, particularly offshore, can be technologically demanding, costly, and time consuming.

An example is Troll gas field in the Norwegian sector of the North Sea, which contains about 46.43 tcf of reserves representing about 20% of West Europe's gas reserves.

The Troll platform, capable of producing and treating 2.7

RIG 38, owned by Murco Drilling Corp., drills a Lower Cretaceous Travis Peak infill well for Enserch Exploration Partners Ltd. in Opelika field, Henderson County, Tex. U.S. gas markets are improving to the point that such infill drilling projects are once again economic. Photo courtesy Enserch Exploration Inc.

bcfd, will stand in about 1,000 ft of water—more than 300 ft deeper than an existing North Sea platform. The project will cost about $5 billion, including pipelines.

Considerable efforts are being made to reduce the costs of developing and operating offshore fields through platform designs, lower drilling costs, and reducing manning levels. Significant cost reductions have already been achieved allowing development of projects that previously were uneconomic.

Gas markets

About three-quarters of world gas consumption is in three areas. Royal Dutch/Shell data show the U.S.S.R. accounts for 36% of the total, the U.S. 26%, and West Europe 14%.

An important note is that for the world as a whole, nearly 90% of all gas is consumed in the country of production.

U.S.S.R.'s industry

In the U.S.S.R., the gas industry grew rapidly beginning in the 1970s. The country became the world's largest producer and consumer of gas in 1984.

Factors behind the growth include substantial increases in primary energy demand, availability of enormous indigenous gas reserves, a desire to minimize internal oil demand, and a need for hard currency earnings from exports to West

Europe.

About two-thirds of the U.S.S.R.'s hard currency earnings were derived from energy exports in 1987, with gas contributing 15% of the total. Development of reserves for internal consumption and export continues to receive a high priority.

Within the U.S.S.R., gas now meets more than 35% of the country's primary energy requirements. Emphasis has been placed on increasing the use of gas in industry and for power generation. Residential use accounts for about 15% of total consumption.

U.S.
infrastructure

In the U.S., meanwhile, the first practical use of gas began in 1821. But it was not until the 1930s that gas was consumed in large quantities.

The U.S. has had a well established infrastructure covering large areas of the country for decades. The main producing regions are in Louisiana, Oklahoma, Texas, and Gulf of Mexico, whereas the main consuming regions, apart from the Gulf Coast states, are in the Midcontinent and Northeast regions and California.

Since 1951, the U.S. has imported part of its requirements by pipeline from Canada. Since 1969, it has exported relatively small quantities of LNG from Alaska to Japan.

The U.S. was the world's largest gas consuming country until 1984. Its gas industry, which has been subject to strict government regulation since the 1930s, has seen extensive regulatory changes affecting price, transportation, and marketing since the late 1970s.

West Europe
interlinked

Within West Europe, there are a number of distinct, though interlinked, national gas industries that differ considerably in size and structure.

Gas consumption in 1987 accounted for about 15% of West Europe's total primary energy demand.

About 90% of gas consumption in West Europe in 1987 was in five countries—the U.K., Germany, the Netherlands, Italy, and France.

Sweden, Denmark, Spain, and Turkey are in the early phases of developing gas industries, with Greece, Norway, and Portugal planning to introduce gas in coming years.

Nearly half of Europe's gas is consumed in the distribution sector.

A further 40% is used in industry.

However, there are wide variations in consumption patterns between different European countries. For example, almost 20% of gas in the Netherlands and Italy is used for power generation while virtually no gas in the U.K.—Europe's largest gas market—is used for power generation.

West Europe produced about 70% of its total 1987 gas requirement.

The balance of demand was met by imports from the U.S.S.R. and Algeria.

The main West Europe producers were the Netherlands, the U.K., and Norway.

Natural gas industry undergoes sweeping regulatory

The U.S. natural gas industry has undergone a series of sweeping regulatory changes since the late 1970s.

It had been a rigid, tightly structured business, largely insulated from competitive pressures because of regulatory protection and steady market growth.

But conditions changed and the industry was forced into throwing off excess weight of operations not part of its core business while relearning the principles of that core: marketing and transportation.

Federal and state regulators have been struggling to keep up with the dynamic changes.

History of regulation. The U.S. gas industry first became regulated more than 50 years ago when the U.S. Congress passed the Natural Gas Act in 1938. The law was meant to provide consumers inexpensive and efficient access to gas supplies.

The law gave birth to a three pronged industry of producers, pipelines, and distributors. Producers found and produced the gas, pipelines bought gas from producers and resold it to distributors, distributors then resold gas to endusers.

The major sections of the 1938 law were those dealing with price regulation and pipeline entry to and exit from interstate markets. Sixteen years after adoption of the law, the U.S. Supreme Court extended the scope of the regulations to include the wellhead price charged by producers.

Gas prices were kept artificially low in relation to competitive fuels. Problems began when the cost advantage of gas increased in 1974 as the delivered price of heavy fuel oil more than doubled following the Arab oil embargo. Demand for gas skyrocketed due to its low regulated price.

At the same time, however, the low prices reduced producer incentive to find more gas supplies. This, and other factors, led to a supply shortage and severe pipeline curtailments to consumers during the mid-1970s.

The curtailments eventually prompted Congress to pass the Natural Gas Policy Act of 1978, which allowed wellhead prices to increase to provide drilling incentives to producers. Producers responded by stepping up exploration/production efforts, contributing to a U.S. drilling boom that peaked at more than 4,500 active rigs/week at the end of 1981.

Pipelines, believing demand was growing while supply was shrinking, continued to add as much gas as possible to their contracted reserve base. Because most wellhead prices were regulated, pipelines had to use contract provisions other than wellhead price to compete against other pipelines to purchase reserves from producers.

One such contract provision was take or pay, in which the pipeline promised to take a certain percentage of a well's deliverability or pay for that not taken. Take or pay provisions have been in wellhead purchase contracts since at least 1938, but pipelines in the late 1970s and early 1980s were agreeing to unusually high take or pay terms, sometimes promising to take 100% of a well's deliverability capacity in exchange for the contract.

A bubble developes. A marketing problem that most in the industry didn't foresee was a tremendous drop in U.S. gas demand as gas prices rose.

By 1983, gas consumption had fallen to 15.4 tcf/year, compared with 18.2 tcf/year in 1980. The pipeline industry was caught with a 4 tcf/year deliverability surplus, which brought on problems that were magnified by the wellhead purchasing practices.

Pipelines, because they couldn't find markets for gas they had promised to buy, were faced with multibillion dollar take or pay obligations to producers. These costs were soon flowing through to consumers.

But in 1983 the Federal Energy Regulatory Commission issued Order 380, which said customers are required to pay fixed costs, not variable costs associated with minimum bill contract provisions between pipelines and distributors.

The minimum bill provision was similar to the take or pay provision between pipelines and producers. Distributors agreed to take a certain percentage of the gas for which it had contracted or pay for the gas not taken.

FERC's Order 380 excused distributors from the gas purchase cost of the minimum bill.

Japan depends on LNG

Japan is the world's sixth largest gas consumer despite having only very limited indigenous supply. Domestic gas production of about 7 bcf/year supplies only 5% of demand.

Japan alone consumes 75% of the world's supply of LNG. Of the 21 receiving terminals in the world, Japan has ten in service and will add two more to receive Australian LNG during 1988-90. It imports almost all of its requirements as LNG from five countries.

About three-quarters of Japan's total consumption was used for power generation in 1987, equivalent to 22% of total energy consumption in this sector. The role of gas in electricity generation reflects government policies to diversify energy supplies by type and source.

Japan's remaining supplies are distributed by gas utilities to residential, commercial, and industrial consumers.

Other countries

Several other OECD countries have established gas industries based on indigenous reserves, notably Canada, Australia, and New Zealand.

Canadian approval of a free trade agreement with the U.S. will add a dimension to burgeoning energy trade between the two countries.

It will assure producers and consumers on both sides of the border that, except in emergencies, neither country will impose new trade restrictions.

The assurance may have no immediately discernible effect on two way energy trade—at more than $10 billion last year, the world's largest—because growth already is so strong. But it will provide an important planning tool.

Canada exported a net 992 bcf of gas to the U.S. during 1987, 34% of marketed production and 5.77% of U.S. gas demand.

Canadian gas exports into the U.S. won't grow much higher, however, until new pipeline capacity is built. Canadian government data show export capacity was virtually filled during the 1987-88 contract year.

Elsewhere, a number of non-OECD countries in South America, East Europe, Asia, the Middle East, and Africa also have established or are developing gas industries to meet internal energy demands and/or for export.

Saudi Arabia, for instance, is now collecting and using a large volume of associated gas that previously was flared. The first phase of Saudi Arabia's master gas plan was completed in 1983 and the second phase in 1986, raising the total capacity to 1.8 tcf of gas.

Meanwhile, other countries, such as Abu Dhabi, Algeria, Brunei, Indonesia, and Malaysia, have become major exporters of LNG.

The future

The world supply/demand balance varies considerably from country to country. Local circumstances play a major part in future development.

changes

When distributors stopped buying all the gas for which they had contracted because their markets had dwindled, pipelines had no way to pass through the take or pay costs that were growing at the other end of the system.

Spot market growth. The pipeline industry responded in a variety of ways. Many unilaterally abrogated wellhead purchase contracts with producers, a practice for which many are paying dearly today in U.S. courts.

Some, however, developed new marketing programs. One such program was launched by Transco Energy Co. in 1983.

Transco, with FERC approval, began what it called the industrial sales program, which was designed to allow a pipeline with high gas costs to regain lost markets by using its relatively low transportation costs to move spot gas directly to consumers.

Through Transco's program, producers—who needed to move gas in order to boost cash flow—agreed to sell gas on a short term net back basis to consumers who otherwise would switch to lower cost fuel oil. In return for the transportation services rendered, producers would grant Transco some take or pay relief.

The program worked. More pipelines began to reduce their gas merchant function in order to emphasize their transportation services.

The evolution, however, was hit with a District of Columbia Circuit Court of Appeals order on Oct. 31, 1985, that said pipelines had to offer the same transportation service to all customers, not just those that could switch to an alternate fuel. FERC's response was Order 436, of which the requisite nondiscriminatory feature is a key part.

Order 436. The regulatory theory of Order 436 was to force the producing and consuming segments of the industry to compete against the pipelines' merchant function, and to force utilization of the transportation function, in order to lower gas costs to consumers.

Order 436 reregulated the entire industry and, since its adoption, the industry has had to respond. Subsequent orders, such as Orders 490 and 500, have been adopted by FERC to modify certain Order 436 provisions. It appears that industry this year is completing the restructuring process mandated by the regulatory changes.

The new market is characterized by a large spot market for gas in which distributors and consumers purchase gas directly from producers and brokers/marketers under short term, market sensitive contracts.

Stiff competition has developed between interstate pipeline companies to sell or transport gas. There has been short term volatility in spot gas prices resulting from oil price movements and gas-to-gas competition.

Booz, Allen & Hamilton believes the U.S. gas industry has changed permanently. The new industry will correct many of the inefficiencies that existed previously. The value added structure will be redistributed. There will be new rules for success.

The spot market, according to Booz, Allen & Hamilton, peaked either last year or this year and will diminish in the years ahead. But the spot market is not going away as long as there is excess capacity, ability to exchange gas, and the infrastructure interested in preserving spot markets.

The transportation patterns for gas are likely to shift as the market becomes more rational between the sources of supply and loccations of demand.

New types of contracts will be the norm. The industry will move away from the long term contracts of the past and today's short term contracts. It is likely that various mechanisms will be developed to reduce risk, including intermediate term contracts, with price escalators related to a market basket of gas prices.

Booz, Allen & Hamilton said pipelines will need to increase flexibility to maximize takes and minimize seasonal and other fluctuations, which may argue for continued consolidation in the pipeline business, with careful attention to finding areas with lower competitive intensity.

Furthermore, as differences in the weighted average cost of gas among pipelines disappear, the inherent and structural efficiency of the pipelines will become much more critical.

Royal Dutch/Shell said the main factors influencing gas developments include availability and location of reserves, whether supplies are indigenous or have to be imported, price competitiveness of gas in its market sectors, governmental energy policies, climatic conditions, environmental concerns, technological advances, and overall levels of energy demand and prices.

U.S. markets

In the U.S., there is no consensus on future demand levels.

The Energy Modeling Forum believes gas prices will begin to reverse their decline by 1990, reflecting the elimination of excess deliverability. After 1990, gas prices will rise continuously, reflecting higher oil prices and smaller field size and higher exploration/development costs of future reserve additions.

By 2000, inflation adjusted gas prices will reach their peaks of the early 1980s, even with lower oil prices.

Expected industrial gas demand by 2010 is 4-8 tcf/year, depending on future oil price levels. Much of the variation stems from fuel switching and different assumptions about the penetration of new industrial gas fired technologies.

The range in future demand by electric utilities also is large—from 2.9 tcf to 5.7 tcf by 2010—depending upon assumptions about environmental regulations, fuel prices, and load growth.

In the short term, said Royal Dutch/Shell, many large producers expect sales to continue to fall due to a tightening of supply, continued fierce gas-to-gas and interfuel competition, and increased utilization efficiency.

However, others, including some large producers, take a more optimistic view of future demand. They foresee a steady increase.

Whatever the outcome, substantial supplies will be needed from reserves yet to be discovered and developed in the Lower 48 to supplement declining production from known reserves.

Prospects for alternative supplies from, for example, LNG projects, Alaska, or non-conventional gas sources will depend not only on how supply/demand evolves but also on whether marketable prices can be achieved that would make such sources economic.

The potential for significant increases in imports from Canada of gas surplus to local requirements also will depend on economic considerations and the construction of additional transportation capacity.

West Europe diversity

In West Europe, the situation is complicated by the diversity of the gas markets in various European countries.

Environmental concerns have already resulted in tighter legislation in many European countries, a move which is likely to benefit gas in commercial and industrial applications.

It is anticipated that the trend towards higher penetration in the public distribution sector will continue in many countries. In some countries there may also be an increase in gas use for electricity generation stimulated by concerns about nuclear safety and the introduction of combined cycle and cogeneration power plants.

World natural gas net production* — Table 1

	1973	1979	1985 (Tcf/year)	1986	1987
U.S.	21.04	19.04	16.00	15.04	15.32
Canada	2.60	2.75	2.79	2.57	2.75
Mexico	0.43	0.89	0.93	0.93	1.00
North America	**24.07**	**22.68**	**19.71**	**18.54**	**19.07**
South America	**0.89**	**1.14**	**1.64**	**1.71**	**1.75**
Netherlands	2.29	3.00	2.46	2.25	2.29
U.K.	1.04	1.39	1.46	1.57	1.69
Norway	0.00	0.78	0.96	0.96	1.04
Others	1.46	1.43	1.36	1.36	1.50
Western Europe	**4.79**	**6.61**	**6.25**	**6.14**	**6.50**
Indonesia	0.04	0.50	1.11	1.14	1.21
Others	0.61	1.11	2.14	2.32	2.50
Asia/Far East/Australasia	**0.64**	**1.61**	**3.25**	**3.46**	**3.71**
Algeria	0.14	0.68	1.11	1.04	1.29
Others	1.25	1.75	1.96	2.18	2.50
Africa/Middle East	**1.39**	**2.43**	**3.07**	**3.22**	**3.79**
U.S.S.R.	7.39	14.14	22.32	24.14	25.64
Romania	1.00	1.04	1.25	1.25	1.25
Others	0.75	1.14	1.19	1.21	1.21
U.S.S.R./East Europe/China	**9.14**	**16.32**	**24.75**	**26.61**	**28.11**
World	**40.89**	**50.89**	**58.75**	**59.64**	**63.04**

*Excludes reinjected/flared gas and shrinkage due to extraction of NGLs. Totals may not add due to rounding.

Source: Royal Dutch/Shell Gourp

On the supply side, currently contracted supplies will begin to decline during the 1990s. A committed supply/demand deficit will develop that can be covered to some extent by the offtake flexibility available in existing contracts.

Some existing contracts may be renewed or extended, although the availability of reserves to maintain current indigenous production levels in some countries and to extend existing export contracts for Dutch gas could be a constraint. Other exporters—the U.S.S.R., Norway, and Algeria—all have the potential to increase gas exports from large reserves bases, subject to commercial, political, economic, strategic, and other considerations.

The U.S.S.R. has sufficient reserves to increase its exports, but current supplies to some European countries are already at, or close to, the limits set by these countries for non-OECD gas imports. Similar constraints would apply to Algerian gas in certain countries.

Supplies from the Norwegian sector of the North Sea, additional to those already contracted from Troll and Sleipner fields and after allowing for option volumes, are potentially available longer term.

Nigeria has substantial gas reserves. Progress is being made to develop an LNG project aimed at markets of the Atlantic basin for start up in the mid-1990s.

Scandinavia growth

Development of new gas markets in Scandinavia and southern Italy, Portugal, Greece, and Turkey is underway or firmly planned.

Scandinavia's markets are being actively pursued.

Renewed interest in selling gas to Scandinavian countries has been sparked by Sweden's decision to phase out its extensive nuclear power program by 2010.

North Sea gas from Norway and Siberian gas from the Soviet Union are candidates to meet Swedish demand that could rise to 500 MMcfd by 2010 from the current 62 MMcfd if the country maintains its nuclear phaseout policy.

Gas currently plays a minor role in Sweden's total energy supply. A small network in southern Sweden is served by a pipeline from Denmark. First supplies were available in 1985 under a contract with Dansk Olie og Naturgas, Denmark's

World natural gas consumption — Table 2

	1973	1979	1985 (Tcf/year)	1986	1987
U.S.	21.50	19.93	17.07	15.64	16.21
Canada	1.60	1.75	1.93	1.71	1.71
Mexico	0.46	0.89	0.93	0.93	0.93
North America	**23.57**	**22.57**	**19.93**	**18.29**	**18.86**
South America	**0.89**	**1.14**	**1.64**	**1.71**	**1.86**
Netherlands	1.21	1.39	1.29	1.36	1.39
U.K.	1.04	1.68	1.93	1.96	2.04
Germany	1.14	1.89	1.75	1.64	1.89
Italy	0.57	0.96	1.14	1.21	1.39
France	0.57	0.89	1.04	1.00	1.07
Others	0.50	0.75	0.86	0.82	0.93
Western Europe	**5.04**	**7.57**	**8.00**	**8.00**	**8.71**
Asia/Far East/Australasia	**0.57**	**1.68**	**3.32**	**3.64**	**3.86**
Africa/Middle East	**0.96**	**1.50**	**2.25**	**2.36**	**2.61**
U.S.S.R.	7.50	12.86	20.11	21.29	22.50
Others	1.89	2.89	3.64	4.07	4.64
U.S.S.R./East Europe/China	**9.39**	**15.75**	**23.75**	**25.36**	**27.14**
World*	**40.36**	**50.36**	**58.93**	**59.46**	**63.04**

*Totals may not add due to rounding.

Source: Royal Dutch/Shell Gourp

World natural gas reserves — Table 3

	Tcf/year
North America	288.5
Europe	196.9
Pacific	60.6
Total OECD	**546.0**
Latin America	
OPEC	63.0
Non-OPEC	132.3
Total Latin America	**195.3**
Africa	
OPEC	174.7
Non-OPEC	26.4
Total Africa	**201.1**
Middle East	
OPEC	904.8
Non-OPEC	19.2
Total Middle East	**924.0**
Asia Pacific	
OPEC	49.4
Non-OPEC	128.7
CPEs	30.0
Total Asia Pacific	**208.1**
U.S.S.R./Eastern Europe	**1,565.8**
Total	**3,640.3**

Source: Jensen Associates Inc. estimates based on Oil & Gas Journal, Cedigaz, and selected national reports.

state gas company, to supply 222 bcf of gas.

The contract has been extended, and Sweden is importing 62 MMcfd although the system could handle imports of about 195 MMcfd.

By the early 1990s Sweden's gas consumption could rise to 96-145 MMcfd. And if all its nuclear power stations were replaced by gas burning capacity, without energy conservation measures or development of other fuels, the gas load would be about 450 MMcfd.

Sweden does not want a single source for new supplies and is negotiating with Norway and the Soviet Union for 115 MMcfd from each country to meet increased demand prior to the nuclear phaseout. Soviet gas would be delivered through Finland, where a distribution network is under construction by Neste Oy.

Neste Oy, Finland's state energy company, is buying about 135 MMcfd from the Soviet Union and is under contract to step up the flow to 230 MMcfd by the mid-1990s.

Norway is anxious to pursue export opportunities offered by the Swedish market. But its marketing efforts cannot move into top gear without more information from Sweden and formulation of a gas policy by the Norwegian government.

The main objective of the policy would be to provide guidelines for integrating potentially high cost gas fired power generations into the low cost national power grid.

Soviet, Japanese work

The U.S.S.R. is working to boost production to more than 30 tcf/year by 1990 and exceed 35 tcf/year in the 1990s.

Because most of its gas reserves lie in remote regions of western Siberia, long pipelines are required—more than 30,000 miles in the current 5 year plan that ends in 1990.

Gas production in other areas of the country has stabilized, so up to the end of the current century western Siberian gas will be the only source of the increase of gas supply for the national economy, to export, and compensate for the decrease of gas production in the mature regions.

Among the main problems facing the Soviet gas industry is commercial operation of its three supergiant gas fields. The main increase of gas production in the near term will be attained in Yamburg field, where the yield in 1990 will be brought to 4 bcf/year.

In Japan, demand is expected to grow, albeit at a somewhat slower rate than during the 1970s and early 1980s.

Demand for LNG for electricity generation will be limited by the modest growth expected in electricity demand and by competition from nuclear energy and coal.

New LNG supplies from the Australian North West Shelf project is scheduled to begin this year and, assuming that existing contracts that expire during the 1990s are extended, a supply gap for additional LNG is anticipated from the mid-1990s.

The gap is expected to be filled mainly by the expansion of certain existing projects, which have uncommitted reserves, although the introduction of one or two new suppliers remains a possibility.

Technology to help

Technological developments will influence gas demand and improve the cost effectiveness of bringing uncommitted gas reserves to market.

In some cases, technological developments will improve the efficiency and/or competitiveness of gas compared to other fuels. Examples include the use of gas to fuel combined cycle gas turbines for electricity generation that offer significant improvements in thermal efficiency.

Gas utilization developments are expected to result in increased use of gas for powering air conditioning equipment, while new burners designed to produce very low emissions for industrial and domestic applications are becoming available. Compressed natural gas is already used in several countries as an automotive fuel. Others are investigating its potential.

Conversion of gas to liquid hydrocarbons is an important and commercially attractive opportunity for used gas that might otherwise be difficult to develop. Proven technology is now available to produce a range of transportation fuels. In New Zealand, locally produced gas has been converted into gasoline for some years.

There also are a number of plants using gas to produce methanol and various petrochemicals. IPE

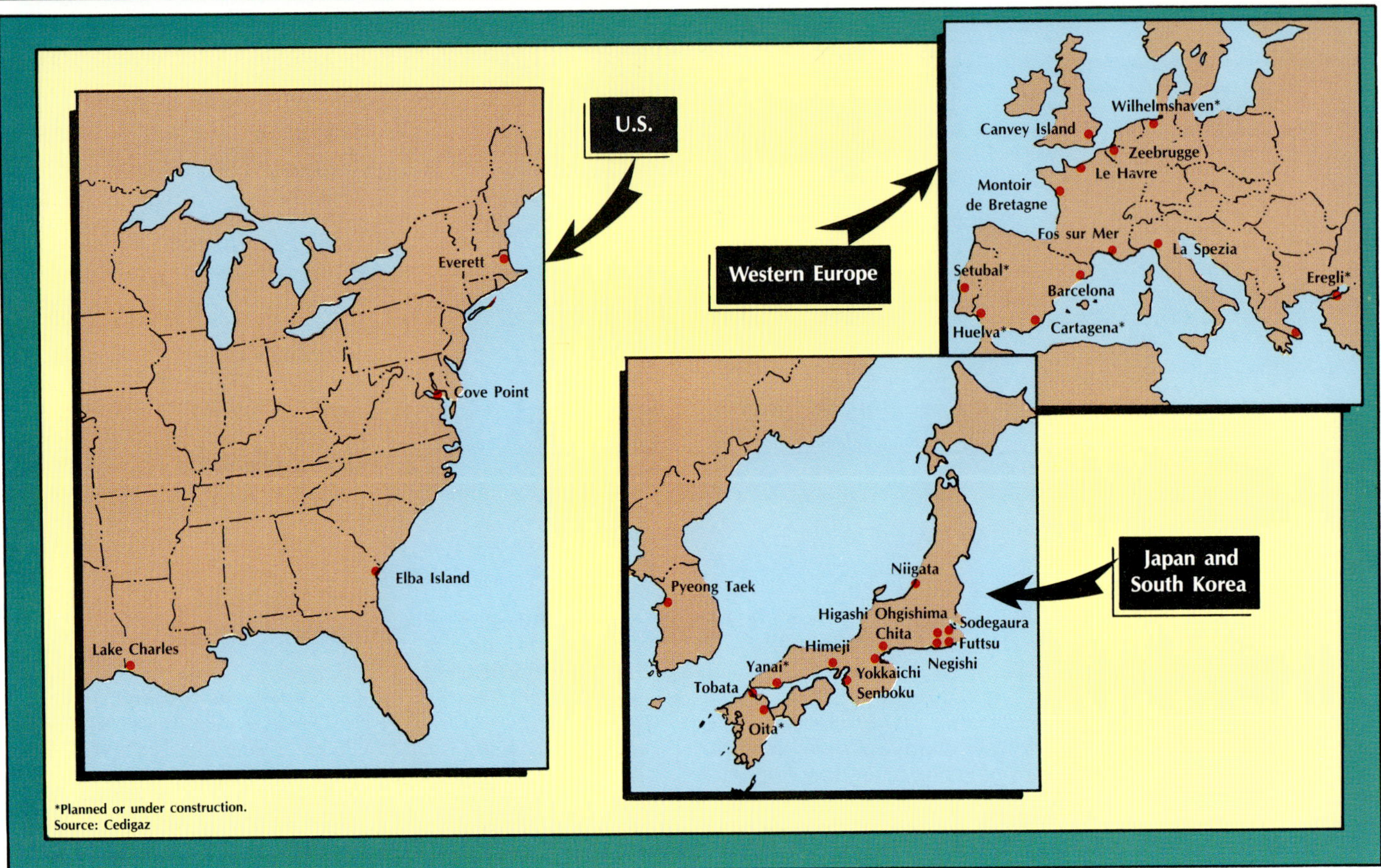

Use of liquefied natural gas expected to grow

WORLD TRADE OF LIQUEFIED NATURAL GAS, AL-ready on the rise, is headed for more growth.

Largely responsible for the increase is burgeoning demand for almost any kind of fuel in energy-poor Japan. That market will receive more LNG in 1989 from the only grassroots gas liquefaction and export project under construction in the world: Australia's Northwest Shelf operation. Several projects are proposed in other countries.

Figures compiled by the American Gas Association (AGA) from British Petroleum Co. plc data show a 28% increase in world LNG shipments during 1983-87 (Table 1).

Although the volume rose to 1.973 quadrillion BTU in 1987, LNG shipments amounted to only about 3% of world gas production that year. Japan, at 1.444 quadrillion BTU of LNG or 98% of its gas consumption in 1987, easily retained its role as the No. 1 LNG customer among countries equipped to receive LNG cargoes.

The share of world gas shipments moved as LNG could increase by 50-60% in the next decade, Pierre Gaussens, senior vice-president of Gaz de France and a director of GdF unit Gaz Transport, predicted in May 1988.

Total gas demand will grow only moderately during the period.

LNG's growth after 2000 could be even more dramatic, Gaussens said. What's more, LNG shipments will grow at a much faster rate than total international gas trade, especially after 2000.

Including replacement of some LNG tankers in service in 1988, that means 30-40 new LNG vessels could be needed by the end of the century at a cost of $6-8 billion.

The result will be further growth for an industry that observed its 30th birthday early in 1989. In January 1959, the Methane Pioneer carried the world's first shipment of LNG from Lake Charles, La., to Canvey Island, U.K.

An overview
of LNG trade

AGA's October 1988 report on world LNG trade underscored the geographic diversity of exporting regions: Africa, Middle East, Asia, and North America.

The most productive region of the world for LNG exports is Southeast Asia. There, Indonesia and Malaysia, the two biggest exporters, shipped 544 trillion BTU or 35% of the world total in 1983. Their exports in 1987 rose to 1.055 quadrillion BTU, a 94% increase, to 53% of the world total.

Indonesia's exports of LNG to Japan, which began in 1977, rank second only to crude oil sales in the importance of Indonesian exports.

Similarly, Malaysia's growth to 292 trillion BTU in 1987 from 61 trillion BTU in 1983 was a dramatic quadrupling of its exports that started at only 4% of the world's LNG trade and by 1987 had risen to 15%.

Brunei has maintained a relatively constant share of the world LNG market at 13-17%. Its sales slipped very slightly to 256 trillion BTU in 1987.

Those three countries of Southeast Asia produced about two thirds of the gas entering world LNG markets.

The second largest LNG producing region is North Africa,

where Algeria and Libya accounted for about one fourth of world LNG shipments in 1987. African production peaked in 1983 before market forces reduced shipments.

Algeria sold most of its LNG to western Europe in 1986 and 1987. Before 1986, Algeria also sold large volumes of LNG to the U.S., but this trade ended in 1985.

Libya sells LNG mainly to Spain.

Japan bought 73% of the world's LNG in 1987. Its LNG imports grew at a hefty 11%/year during 1983-87.

"This growth, albeit at a lower rate, is likely to continue through the end of this century," AGA said.

Japan's Ministry of International Trade and Industry, which places a high priority on "optimum energy mix," said, "Even more use of LNG is expected in the future because of its steady supply and cleanliness."

The growth rate estimated for Japan's LNG import volumes is 3.1%/year through 1993, slipping to 2.5%/year to 2000.

LNG helps provide diversity of supply for western Europe. It has either operating or planned pipeline or LNG access to four of the world's largest gas fields: Groningen in the Netherlands, Troll off Norway, Hassi R'Mel in Algeria, and Urengoi/Yamburg in the U.S.S.R.

Western Europe has reduced its LNG purchases sharply.

The region's gas consumption has been increasing steadily, but other gas supply sources have developed. In 1987, for example, pipeline shipments of Soviet gas to western Europe accounted for about 1.2 quadrillion BTU in the key French, Italian, and West German markets. And North Sea fields increased gas supplies for the U.K. and Norway.

As a result, African LNG shipments fell because they were forced to compete with gas from areas with very large productive capacities.

France held the No. 2 position among world LNG importers in 1987 with 320 trillion BTU, or 16% of world trade.

Gas consumption in France grew at a compound rate of 6.1%/year in 1970-86, but LNG imports remained at about one third of total gas supply. French LNG imports (cif) are more costly than imports by pipeline, but the price spread has been shrinking. Instead of the 47% premium for LNG reported for 1981, differentials were only 12% in 1986 and 15% in 1987. That illustrates the economic pressure on Algerian LNG.

Belgium and Spain are the other sizable LNG importers of western Europe. Most of Spain's LNG comes from Algeria, but in 1987 about one third came from Libya.

"Even though these import levels are relatively static," AGA said, "the pressure of price competition with other gas supplies is important. Soviet and North Sea gas are available by displacement and could replace Algerian and/or Libyan LNG if price differentials become extreme."

U.S. purchases of LNG reached a record 283 trillion BTU in 1979, but the trade declined sharply, and no terminals received regular shipments in 1986 or 1987. That mainly was due to price competition with U.S. and Canadian natural gas.

More LNG,
more tankers

LNG supplies for prime markets in Japan and western Europe are virtually asured until the end of the 1900s, making it unlikely that significant new LNG routes will be opened, Gaz de France's Gaussens said. But beyond 2000, growth will pick up, and ships that will be needed to handle increased LNG trade will have to be built during the next decade.

There also will be some replacement market as many carriers built when LNG trade began reach retirement age. Gaussens cited planned replacement of the two ships on the Alaska-Japan route by a Phillips Petroleum Co.-Marathon Oil Co. joint venture as an example. Those vessels began service in 1969.

Gaussens predicted that half of the 30-40 new vessels ordered will be of the membrane type developed and licensed by Gaz Transport. The design features integrated, double membrane LNG tanks.

To date, 32 carriers of the Gaz Transport design have been ordered, ranging in capacity from 25,000 to 130,000 cu m, and 25 are still operating. They are among the world fleet of 65 LNG tankers, of which 15 were in layup in 1988.

Gaz Transport designed ships made up about 40% of the world's operating LNG tanker fleet in 1988. Cost to build a Gaz Transport design membrane tank ship was about $200 million that year.

Most remaining new orders in the next decade will be for the spherical tank type carrier, a design of Norway's Moss Rosenberg, Gaussens said.

By far the largest current order is for seven ships to move LNG from Australia's Northwest Shelf project to Japan.

Other projects are on the horizon, said Gaussens, that will require new vessels. Korea, for example, planned a second regasification terminal and expected demand for gas to increase.

In addition to possible increases in established ventures that could require more ships, LNG projects were under study for several large gas reserves, including those in Qatar, Nigeria, Norway, and Alaska. Most will have to await further demand growth.

Drewry agrees
demand will rise

Drewry Shipping Consultants Ltd. agreed that seaborne shipments of LNG—LPG, too, for that matter—which staged a recovery in profits and cargo volumes in 1986 and 1987, are headed for even better times.

The London firm predicted in September 1988 that LNG shipments of 87 million cu m in 1987 will rise to 116 million cu m by 1992. LNG benefited from changes in gas price formulas and contract terms since the oil price collapse in 1986.

Beyond 1992, LNG trade will continue to expand as new proposals such as Nigeria's Bonny project start up.

One operator, a unit of Royal Dutch/Shell Group, had purchased or taken options to purchase seven laid-up LNG tankers, Drewry pointed out.

The fleet of such vessels available for new projects has shrunk sharply. The 1988 LNG tanker fleet won't be able to cope with added demand for shipping services, particularly if exports to the U.S. So orders for new LNG carriers cannot be ruled out.

The outlook for LPG shipping is much the same, Drewry said. With the exception of 1983, seaborne LPG shipments increased every year since 1980, rising from 16.3 million metric tons in that year to 27 million tons in 1987.

Despite increased demand for shipping, there have been no LPG tankers built since 1984. The number of companies operating LPG tankers has declined.

That gave a lift to freight rates. Time charter rates for a 50,000 cu m tanker averaged $490,000/month in 1987, up from $350,000/month in 1986.

The biggest project:
Northwest Shelf

Work was progressing on schedule in 1988 for the $9.8 billion Northwest Shelf LNG export project in Western Australia toward a planned October 1989 start of LNG shipments to Japan.

It was billed as Australia's biggest resource project and the world's largest process project under construction.

A group led by Woodside Petroleum Ltd. let contract for design of a platform to develop Goodwyn gas field on the Northwest Shelf off Dampier as part of the project's gas supply.

In addition, the first LNG carrier in the seven ship fleet assigned to the project was launched at a Nagasaki shipyard. Construction was more than 50% complete at a liquefaction plant near Karratha on the Burrup Peninsula.

The Northwest Shelf project will supply Japanese utilities LNG for 20 years. It will go on stream with gas from North Rankin field and tap Goodwyn field later.

The Woodside group (Table 2) let two contracts valued at a combined $54.2 million for design of topsides and substructure for Goodwyn Platform A.

The preliminary design work was intended to ascertain costs and technical details for an early 1989 project approval by partners.

To be installed in 430 ft of water about 14 miles southwest of North Rankin Platform A, Goodwyn A will produce about 900 MMcfd of gas and undisclosed volumes of condensate. North Rankin A produces about 18,000 b/d of condensate.

The topsides contract, valued at about $41.2 million, went to a joint venture of Davy McKee Pacific Pty. Ltd. and McDermott Industries.

The $13 million substructure preliminary design contract went to a joint venture of Australian consulting engineers Hardcastle & Richards and Earl & Wright, San Francisco.

Goodwyn development plans call for drilling 26 wells from the 30 slot platform. It will be set in second quarter 1992.

The platform will have a four leg substructure with a spread base compared with the conventional eight leg North Rankin A platform. That design will involve less materials cost and quicker construction.

Production will come from two identical trains. The platform will include a gas recycling project to boost condensate production in the early years of the project. Gas and condensate will move from Goodwyn A through a 30 in., two phase pipeline to North Rankin A, which supplies gas for domestic use to the State Energy Commission of Western Australia.

Goodwyn is tentatively set to start up in October 1993 from two wells, with another six on production in 1994.

Australian LNG Ship Operating Co., jointly owned by Shell Australia and BHP Petroleum, floated the Northwest Sanderling LNG tanker out of its building dock to complete its fitting for mid-August 1989 delivery.

Built by Mitsubishi Heavy Industries, the 125,000 cu m capacity vessel was among four under construction in 1988. Construction was proceeding on the second Northwest Shelf carrier at Nagasaki. Work was well under way on the third carrier and had just begun on the fourth in the Mitsui Engineering & Shipbuilding yard at Chiba, Japan.

In 1989, Kawasaki Heavy Industries is to start construction of the fifth carrier in Sakaide, Japan. Contracts for the remaining two tankers had not been let by mid-1988.

The Northwest Shelf LNG carriers will be 892 ft long and 155 ft wide with a draft of almost 36 ft. At almost 131 ft diameter, storage spheres on the Northwest Sanderling are the biggest installed on an LNG carrier.

On the Burrup Peninsula, the Woodside group in 1988 completed the LNG complex's first commissioning job with start-up of the first of five 20,000 kw gas turbine power generators. Hydrotesting of all four LNG storage tanks was complete.

The heavy lift program in the main processing train area was almost complete. The big exhaust stacks for the refrigerant compressors were lifted into place and the compressor enclosures clad. Pipewelding to complete the two trains was under way.

The fractionation plant, which will provide LNG refrigerant, was more than 75% complete. Also completed were the plant's seaward approach and in the tank field the emergency flare structure and its two risers.

Work was continuing on hooking up computer process control consoles in the central control room with field auxiliary rooms and plant instrumentation.

The biggest market: Japanese utilities

Japan will need to import increasing volumes of LNG, possibly from new sources, through fiscal 2000. Demand will grow in the electrical power generation market, although LNG's role as a base load fuel is declining, said Lisa Totto and Fereidun Fesharaki of East-West Center, Honolulu.

Their study of Japanese requirements was issued in August 1988 and updated the following November.

Japanese base load power generation comes increasingly from nuclear power, hydropower, and large coal fired plants. LNG, said East-West Center researchers, is becoming an "intermediate load fuel."

Japan uses oil and LPG mainly for peak generation.

Plans for new generating capacity will keep LNG's share of the power market at 20-21% but increase demand in absolute terms.

Capacity fired by oil and LNG will shrink.

Japan consumed 28.8 million tons of LNG in fiscal 1987, which ended Mar. 31. Power generation accounted for 76% of the total, town gas about 21%, and industrial fuel the rest.

Totto and Fesharaki in their August study projected Japanese LNG demand in fiscal 2000 at 36.9-43.2 million tons. In November, Fesharaki said it is almost certain that 5,000-10,000 megawatts of nuclear power capacity will be postponed in the 1990s. This delay, coupled with a faster than expected growth in electricity demand, could easily result in up to 15 million tons of new LNG demand by Japan late in the 1990s.

The original projections assumed LNG-fired capacity of 43 million kw and utilization rates of 45-55%. They assumed LNG demand for power generation of 25.37-37.98 million tons and for town gas and industrial fuel of 8.44 million tons.

LNG-fired generation capacity in fiscal 1986 totaled 29.23 million kw.

Under the Ministry of International Trade and Industry's "flexible fuel system," Japan keeps utilization rates low for power generation capacity fired by imported fuels. The system enables power plants to quickly change fuels in the event of a shortfall of one of the fuels.

Oil-fired capacity typically operates at 30% of capacity or less, LNG-fired capacity at 45-50%. Nuclear capacity, by contrast, usually operates at 74% of capacity, coal-fired capacity at 64%.

If nuclear and coal-fired capacity additions don't come on stream as quickly as planned, the call on spare LNG and oil/LPG capacities will increase accordingly.

For example, if nuclear capacity stopped growing at the 41 million kw level projected for 1995, LNG imports might be 5-10 million tons greater than projected by 2000, Totto and Fesharaki said.

Without that demand kick, Japan still will need to find new sources of imported LNG or boost imports from current sources.

In fiscal 2000, Japan will import 26.8 million tons of LNG under contracts in effect or imminent in 1988. The total includes 5.8 million tons/year from Australia's Northwest Shelf project. It excludes 5.1 million tons/year from Brunei and 2.1 million tons/year from Abu Dhabi under contracts expiring before 2000.

Those projections leave a deficit—to be filled by imports under new or renewed contracts—of 10-16 million tons/year. The deficit will shrink to 3-9 million tons/year if Japan is assumed to renew its contracts with Brunei and Abu Dhabi at 1988 volumes.

All current or planned contracts will expire before 2010.

Suppliers such as Australia, Abu Dhabi, and Indonesia can meet Japan's projected LNG needs.

Current contracts, Totto and Fesharaki said, "are expected to be renewed to last a total of 30-40 years each...The facilities which would have been fully depreciated in the first contract phase can operate much longer."

Table 1

Breakout of world LNG shipments

	1983	1984	1985	1986	1987
			Trillion BTU		
EXPORTERS					
Abu Dhabi	90	105	108	108	104
Algeria	555	461	443	432	482
Brunei	266	259	252	252	256
Indonesia	483	698	738	738	763
Libya	32	40	43	33	29
Malaysia	61	173	209	245	292
U.S.	50	50	50	50	47
Total	**1,537**	**1,786**	**1,843**	**1,858**	**1,973**
Percent of world					
gas production	**2.8**	**2.9**	**3.0**	**2.9**	**2.9**

	1983	1984	1985	1986	1987	1983	1987
						Percent of natural gas consumption	
			Trillion BTU				
IMPORTERS							
Belgium	58	54	84	93	97	17	28
France	324	324	277	277	320	36	32
Italy	0	14	14	0	0	0	0
Spain	86	72	86	90	90	86	87
West Germany	0	0	0	4	4	0	*
Japan	950	1,286	1,357	1,390	1,444	93	98
Korea	0	0	0	4	18	0	*
U.S.	119	36	25	0	0	*	0
Total	**1,537**	**1,786**	**1,843**	**1,858**	**1,973**		

*Less than 0.5%.

Source: American Gas Association from British Petroleum Co. plc data

But new supplies from Qatar, Canada, and Alaska may seek roles in the Japanese market.

"Japan may insist on a lower price in the future, essentially decoupling natural gas prices from the expected high oil price at the turn of the century."

A report compiled by British Petroleum from International Energy Agency data shows that the cif cost of Japan's LNG imports peaked at $5.80 (U.S.)/MMBTU in 1981, then slipped to $3.10/MMBTU in first quarter 1987, the latest period for which figures were available.

Among selected fuels, only crude oil cif in IEA countries hit a higher peak: $6.60/MMBTU, also in 1981.

Japan's Natural Resources & Energy Agency predicted in December 1987 that LNG costs will rise sharply, compared with coal and nuclear power, by 2000. As a result, nuclear plants will remain the cheapest means of electrical power generation.

The agency assumed that crude oil and equivalent LNG prices will jump to $30-40/bbl in 2000 from $18.50 late in 1987, while coal and nuclear fuel costs will rise 1-3%/year.

When NREA issued its report, nuclear power generating costs worked out to 9 yen (7¢ U.S.)/kw-hr. Corresponding costs were 10-11 yen (8-9¢)/kw-hr for power generated with coal as fuel, 11-12 yen (9-9.5¢) with fuel oil or LNG, and 13 yen (10¢) with hydroelectric power.

Norwegian frontiers eyed
for LNG supply

Saga Petroleum AS and other operators were assessing the outlook for a pioneering LNG project as 1989 drew near.

Gas supply source would be the Haltenbanken region of the Norwegian Sea off mid-Norway, a frontier region in which industry has drilled several discoveries. There is no production there.

An earlier study, submitted in March 1987 by a combine of Total-CFP and state owned Statoil to Norway's Petroleum and Energy Ministry, found that LNG exports were a viable option for development of gas found in the Tromsoflaket region, another frontier, off northern Norway.

Saga's Midgard gas field, its 1981 discovery, would be the pillar in any gas development on Haltenbanken, the company said.

The region's first two fields earmarked for development, Heidrun and Draugen, will produce oil and associated gas. Midgard could produce 70-500 bcf/year of gas.

Saga pegged Midgard recoverable reserves at 4.2 tcf out of an estimated 5.5 tcf in place, making the field by far the largest gas reservoir found so far on Haltenbanken. Because of excellent reservoir properties, Saga expected excellent, flexible gas deliverabilities. Saga also expects to recover 110 million bbl of condensate.

Saga was searching for a gas buyer as 1989 approached and said the field can be on stream about 4 years after partners and Norwegian authorities approve the project. Expected start-up of production is 1994-95.

Gas would move by pipeline over iceberg scoured seabottom to shore. It would be used domestically and exported via LNG tankers and other pipelines.

Saga's partners in Midgard field are Agip SpA, ARCO Norway Inc., Deminex (Norge) AS, Norsk Hydro AS, Norske Shell AS, and Den norske stats oljeselskap AS.

Draugen and Heidrun associated gas may be reinjected for several years while a total gas solution for the mid-Norway offshore region is established.

Saga estimated the cost of a 280 bcf/year gas trunk line from the offshore fields to a mid-Norway onshore terminal at $415 million. The terminal would cost $340-850 million, depending on fractionation capacity.

Base gas volumes will depend on markets being served. Probable volumes are 150-200 bcf/year, but Haltenbanken fields are easily capable of producing more than 350 bcf/year.

From the mid-Norway terminal, dry gas would be transported to market by one of three methods: to eastern Norway and Sweden via pipeline, to continental Europe and the U.K. through a connecting pipeline to the existing North Sea gas pipeline grid, or to Europe and the U.S. as LNG.

A link to the North Sea grid would require a pipeline of about 250 miles from the terminal and cost $500-600 million, depending on capacity and distance.

With such a connection, Haltenbanken gas could be shipped to Emden, West Germany, Zeebrugge, Belgium, and St. Fergus, Scotland.

Depending on production volume, investment in a gas liquefaction plant would be $690 million to $1.3 billion.

Saga's studies showed reasonable distances to several LNG receiving terminals, with those at Zeebrugge, Belgium, Montoir de Bretagne, France, and Cove Point in the U.S. considered the most suitable alternatives.

Total-Statoil's study of the Tromsoflaket region found that development could begin in the 1990s, aiming for LNG sales of 5 billion cu m/year from an onshore liquefaction plant to markets in the U.S. and western Europe. Additional gas could provide fuel for power generation.

The study envisioned a floating production system tied to subsea completions beginning in Snohvit field. Gas could move ashore through a 140 km, 28 in. pipeline to a two train liquefaction plant.

Project costs would amount to about $3 billion in 1987 U.S. dollars: $1.3 billion for field development and pipeline construction, $900,000 to $1 billion for the liquefaction plant, and at least $750,000 for five LNG tankers.

Extension approved
for U.S. LNG shipments

In North America's only LNG export project, Phillips 66

Natural Gas Co. and Marathon in summer 1988 obtained U.S. government approval to extend for another 15 years their LNG sales from Alaska's Cook Inlet to Japan.

The Economic Regulatory Administration approved continuing the shipments, which were the first LNG shipments to Asia.

Among other 1988 U.S. action on LNG:

• Shell Oil Co., wholly owned by Royal Dutch/Shell Group, made its first move into LNG operations by agreeing to buy 50% of the stock in a Columbia Gas System Inc. subsidiary that owns an LNG import terminal at Cove Point, Md. Another Royal Dutch/Shell unit is a partner in a proposal for LNG exports from Nigeria.

• Florida was earmarked as a key market for Algerian LNG.

• Cabot Corp. unit Distrigas Corp. and Algeria's Sonatrach agreed to resume long term shipments of LNG to the U.S. under a new contract.

The deal called for Distrigas to purchase as many as 17 cargoes/year, totaling about 51 bcf, of LNG from Sonatrach for 15 years.

Phillips and Marathon in 1988 signed a new agreement with Tokyo Electric Power Co. and Tokyo Gas Co. to continue the LNG sales through 2004.

They signed the first contract, expiring in 1989, with the Tokyo utilities in 1967. Shipments from Cook Inlet to Tokyo began in 1969.

Under the new contract, two new tankers will deliver 1 million tons/year of LNG each in 17 round trips, each 6,600 miles long. The gas is produced by Phillips and Marathon from fields in the Upper Cook Inlet.

The new vessels were scheduled to replace two tankers presently in service.

Phillips operates the LNG plant, and Marathon operates the tankers. Sales and facilities' interests are Phillips 70%, Marathon 30%.

The new contract left the main price formula unchanged. The delivered price of LNG price is determined each month on the basis of the weighted average selling price for the preceding month of the top 20 crude oils, ranked by descending volumes, imported into Japan.

Under the formula, if the weighted average selling price were $17/bbl, the delivered price would be about $2.92/ MMBTU, excluding the adjustment factor.

Under a 1988 agreement in principle that led to signing of the new contract, the parties will limit monthly adjustments to a range of ±30¢/MMBTU to facilitate market responsive pricing.

The agreement in principle called for the contract volume to rise Apr. 1, 1989, from the earlier 50.57 trillion BTU/year to 52 trillion BTU/year the first year.

The volume is likely to rise to 57.5 trillion BTU/year in 1993 or 1994, when the applicants place new, larger LNG tankers in operation.

The buyers may take as little as 10% less or demand as much as 6% more than the contract volume in any year.

Delivery of 50.57 trillion BTU/year requires production of 61.2 bcf of gas. The difference is consumed in LNG manufacture and boiloff during transport to Japan. Net efficiency of the process averages 82.5%.

Gas consumption in LNG manufacture and transportation will rise to 73 bcf/year under the 106% take case from 61.2 bcf/year in 1987-88.

President Reagan early in 1988 signed an executive order finding that export of Alaskan North Slope gas would be in the national interest. That removed one obstacle to Yukon Pacific Corp.'s proposed $11 billion Trans-Alaska Gas System to export North Slope gas as LNG to the Far East.

Northwest Alaskan Pipeline Co., lead sponsor for the competing, shelved Alaska Natural Gas Transportation System, criticized the finding, terming export of Alaskan gas a threat to U.S. energy security and a repudiation of U.S.-Canadian agreements on Angts.

The Shell-Columbia agreement called for transfer of half the stock in Columbia LNG Corp., owner of the Cove Point receiving terminal on Chesapeake Bay. The transfer, in a transaction that could cost Shell about $100 million, was scheduled to occur in the ensuing 2-3 years if "certain significant conditions" are met.

The agreement also provided for a partnership of Shell and Columbia units to import LNG and sell the regasified fuel in the U.S. Cove Point terminal has a design sendout capacity of 1 bcfd of gas.

Columbia estimated that LNG imports under the partnership could begin as early as the winter of 1991-92.

In the Florida LNG deal, Panhandle Eastern Corp. subsidiary Pan National Gas Sales Inc. late in 1988 signed a contract to sell Citrus Corp.'s Citrus Trading Corp. large volumes of Algerian LNG for use in power plants in the state.

Start of deliveries depends on Economic Regulatory Administration and Federal Energy Regulatory Commission approvals. Pan National planned to ask for expedited approval of applications filed in June 1987.

Citrus, owned equally by Enron Corp. and Sonat Inc., planned to use the LNG to satisfy about one fourth of a long term contract it signed with Florida Power & Light Co. That contract called for delivery of about 1.9 tcf of gas during 15 years.

Sonatrading Amsterdam BV, a subsidiary of Algeria's state owned Sonatrach, will supply the LNG.

Citrus will sell FP&L 64-80 MMcfd of regasified LNG. Panhandle's Trunkline LNG Co. will receive, store, and regasify the LNG at Trunkline's idle Lake Charles, La., LNG terminal.

Citrus' Florida Gas Transmission Co. will transport the LNG to FP&L. Citrus will have a one time option to add 40 MMcfd of firm volume for the first 5 years.

The companies said a confidential formula provides market sensitive prices competitive with other fuels.

Reactivation of the Lake Charles LNG terminal will enhance opportunities for added volumes of LNG required to meet market demand for alternate fuel as the U.S. gas surplus begins to fade, Trunkline said.

Lake Charles terminal, built in 1978-82, handled 47 tankerloads of Algerian LNG from September 1982 through December 1983, when Trunkline suspended purchases

Table 2

Who's who in the Northwest Shelf LNG project

EQUAL PARTNERS
Woodside Petroleum Ltd.*
BHP Petroleum Pty. Ltd.
BP Developments Australia Ltd.
California Asiatic Oil Co.
Japan Australia LNG (MIMI) Pty. Ltd.†
Shell Development (Australia) Pty. Ltd.

LNG BUYERS
Tokyo Electric Power Co. Inc.
Chubu Electric Power Co. Inc.
Kansai Electric Power Co. Inc.
Chugoku Electric Power Co. Inc.
Kyushu Electric Power Co. Inc.
Tokyo Gas Co. Ltd.
Osaka Gas Co. Ltd.
Toho Gas Co. Ltd.

**MAIN CONTRACTOR
FOR LNG PLANT**
KJR, a combine of Kellogg Overseas Corp., JGC Corp., and Raymond Engineers Australia Pty. Ltd.

*Operator. †Jointly owned by Mitsubishi Corp. and Mitsui & Co. Ltd.

because of poor markets.

The plant can dock and unload a 125,000 cu m capacity LNG tanker in 12 hr. It has three 600,000 bbl cryogenic tanks with combined storage capacity of 6 bcf of gas.

A 48 mile, 30 in. pipeline connects the LNG terminal to Trunkline's main transmission system.

Citrus predicted that LNG will play a key role during the 1990s in development of long term, base load gas markets represented by the increasing need for electrical power generation.

LNG shipments from Algeria were to resume in fall 1988 to Distrigas of Massachusetts Corp. (Domac) terminal at Everett, Mass.

Distrigas agreed to pay Sonatrach $71 million to resolve a dispute over their earlier LNG purchase contract.

Distrigas first was to pay Sonatrach $60 million to void a pending arbitration between them. The rest was to be paid after LNG trade resumes.

A letter of intent to settle all differences, signed in 1987, led to the first shipment of Algerian LNG to the U.S. in more than 2 years. That cargo was used for utility peak shaving needs.

The new contract eliminated the take or pay obligation in the earlier contract and provided for LNG at flexible, market indexed prices with Distrigas and Sonatrach sharing revenues from LNG sales.

The 1988 document was the third amendment to the original contract, signed in April 1976.

The latest pact created a new price structure under which Sonatrach will receive the highest of three market indexed prices.

The first index price is a minimum for LNG fob Algeria, beginning at $1.475/MMBTU the first year of deliveries and rising to $1.73/MMBTU Sept. 15, 1991, and thereafter. The second is a reference price based on the price of competing fuels. The third is the contract price at which Distrigas sells the gas to its customers.

Other LNG action
around the world

Here is a sample of other LNG activity around the world in 1988:

• Algeria's Sonatrach signed a second contract to supply LNG to countries in the eastern Mediterranean region. Sonatrach will supply Greece a total of 423 bcf of gas during 20 years beginning in 1991. Earlier, Turkey signed a contract for 70 bcf/year of Algerian LNG starting in 1992.

• Indonesia Petroleum Ltd., Tokyo, and five Japanese trading houses formed a firm to finance 60% of the 40 billion yen ($318.5 million) needed for a project to export Indonesian LNG to Taiwan. The money was to be used to build a liquefaction plant at Bontang, East Kalimantan. Exports of 1.5 million metric tons/year could begin in 1990.

Indonesia's Pertamina also signed a 5 year to ship at least 1.45 million tons/year of LNG to Japan's Osaka Gas Co.

Indonesian exports amounted to about 19 million tons/year at mid-1988. PT Arun LNG, a joint venture of Pertamina and Mobil Oil Corp., expected to increase LNG shipments to Japan and South Korea by 5.6% in 1988. Scheduled were 186 cargoes to the two countries, compared with 176 in 1987. Each cargo was about 60,000 tons.

An Arun spokesman called his company the world's most efficient LNG producer at a cost of 14¢/MMBTU, compared with a world average of 23¢/MMBTU.

• Five European gas utilities agreed to negotiate an LNG purchase contract with Nigerian National Petroleum Corp. NNPC, in partnership with Shell Gas BV, Elf Nigeria, and Agip Nigeria, planned a 4 million metric ton/year LNG plant at Bonny to start up in 1995.

The five companies—BEB, Ruhrgas AG, and Thyssengas of West Germany, Enagas of Spain, and SNAM of Italy—were among nine groups that met with Nigeria's minister of

petroleum resources, Rilwanu Lukman, in Britain. Other companies involved in the meeting were British Gas plc, Distrigaz of Belgium, Gaz de France, and representatives of Portugal.

In October 1988 NNPC reported it had found markets in western Europe for 68% of its planned LNG production

Target date for start of construction at Bonny is January 1993. Shell, technical leader for the $2 billion project, began preliminary preparation of the Bonny site and detailed plant design.

The plant will use nonassociated gas from Shell-NNPC's Soku and Bomu fields, which will provide half the supply. The rest will come from Agip-Phillips-NNPC's Ioshi and Idu fields and Elf-NNPC's Ubeta, Obagi, and Ibewa fields.

• A group whose major shareholder is state owned Qatar General Petroleum Co. began development drilling in supergiant North gas field, formerly designated Northwest Dome, off Qatar in the Persian Gulf. Estimates of reserves range from 150 tcf to 380 tcf.

The initial project calls for LNG exports of 6 million tons/year for 20 years or more starting in the mid-1990s. Target markets are in Japan, India, and western Europe.

• East-West Center's Fesharaki predicted that South Korea's demand for LNG as power plant fuel will rise to 3-6 million tons/year by the turn of the century from 2 million tons/year in 1988.

Fesharaki cited rising demand for electricity at a steeper rate than had been forecast. If the environmentalist movement in Taiwan is duplicated in Korea, two nuclear power plants will be postponed, he said. Meantime, siting of coal mines is becoming much more difficult because of environmental opposition.

South Korea in 1986 received its first shipment of Indonesian LNG at its Pyeong Taek terminal, south of Seoul. France's Technigaz built the terminal.

• Fesharaki also predicted that Taiwan's demand for LNG imports will rise to 2.5 million tons/year by 1995 and to 4 million tons/year by 2000.

A contract with Indonesia called for imports of 1.5 million tons/year to begin in 1989-90. Chinese Petroleum Corp. had a reception terminal under construction at Hsingta on Taiwan's southwest coast.

LNG for Taiwan will come from a fifth train at Indonesia's Bontang liquefaction plant, which will have a production capacity of 1.95 million tons/year. A contract for its construction went to Japan's Chiyoda Chemical Engineering & Construction Co. and Mitsubishi Corp.

Meantime, construction of two nuclear power plants in Taiwan was postponed indefinitely because of the "very combative mood" of environmentalists, Fesharaki said. Coal siting was very difficult. Electricity demand was rising at a rate of 10%/year.

• Thai LNG International Co. Ltd. decided to delay work on its proposed LNG project in Thailand because of low oil prices. Following a determination that the project would require oil prices exceeding $25/bbl, the Thai-Japanese joint venture extended the prefeasibility study phase for 2 years from Feb. 14. The current study recommended that the plant be able to produce 3 million metric tons/year of LNG, handling 600 MMcfd of gas. Next stage will be a detailed feasibility study.

• Turkey was negotiating with Algeria and Libya for imports of LNG to reduce dependence on natural gas from the Soviet Union.

It wanted to import a total of 56.5 bcf/year from the North African countries.

Five groups of contractors prequalified to bid on a $150 million contract to build an LNG import terminal at Eregli, Turkey.

Turkish imports of Soviet gas are to reach 211 bcf/year. A pipeline from Bulgaria into western Turkey was to be extended to Ankara.   [IPE]

Prices in the industry

Crude oil prices

%/bbl

	U.S. refiner acquisition cost			U.S. landed cost of imports			Selected countries					Average export prices		
Year	Domestic	Imported	Compo-site	Total	Canada	Mexico	United Kingdom	Nigeria	Saudi Arabia	Vene-zuela	World-wide	OPEC	Non-OPEC
1976	8.84	13.48	10.89	13.34	13.57	NA	NA	13.80	13.04	11.80	NA	NA	NA
1977	9.55	14.53	11.96	14.31	14.21	13.75	NA	15.25	13.61	13.13	NA	NA	NA
1978	10.61	14.57	12.46	14.38	14.50	13.54	NA	14.86	13.92	12.83	NA	NA	NA
1979	14.27	21.67	17.72	21.65	20.43	20.86	22.16	22.96	19.15	18.18	NA	NA	NA
1980	24.29	33.89	28.07	33.95	30.47	31.80	35.88	37.05	30.02	25.86	NA	NA	NA
1981	34.93	37.05	35.24	36.52	32.16	33.78	37.24	39.70	34.19	29.87	34.81	34.42	36.16
1982	31.22	33.55	31.87	33.18	26.92	26.64	34.28	36.17	35.00	24.82	33.33	33.71	32.3
1983	28.87	29.30	28.99	28.93	25.63	25.78	30.87	30.84	29.76	22.94	29.66	29.89	29.13
1984	28.53	28.88	28.63	28.46	26.59	26.87	29.60	30.50	29.50	25.15	28.60	28.61	28.6
1985	26.66	26.99	25.75	26.66	25.71	25.63	28.35	28.96	24.72	24.43	27.60	28.00	26.98
1986	14.82	14.00	14.55	13.49	13.43	12.17	14.63	15.29	12.84	11.52	14.98	15.45	14.21
1987	17.76	18.13	17.90	17.65	17.04	16.69	18.78	19.32	16.81	15.76	17.41	17.31	17.57

From: Oil & Gas Journal Energy Database
Source: U.S. Department of Energy

Petroleum product prices

(¢/gallon)

	Motor Gasoline-Pump Price				Heating Oil Resi-dential	Refiner Resale (Wholesale) Product Prices							
Year	Leaded Regular	Unleaded Regular	Unleaded Premium	Average All-Types		Motor Gasoline	Aviation Gasoline	Kerosene Jet Fuel	Kerosene	No. 2 Fuel Oil	No. 2 Diesel Oil	Propane	Resid
1978	62.60	67.00	NA	65.20	49.00	43.40	53.70	38.60	37.23	35.66	36.50	23.70	23.00
1979	85.70	90.30	NA	88.20	70.40	63.70	72.10	66.00	56.60	54.47	57.40	29.10	33.63
1980	119.10	124.50	NA	122.10	97.40	94.10	112.80	86.80	80.26	78.21	80.10	41.50	44.43
1981	131.10	137.80	147.00	135.30	119.40	106.40	125.00	101.20	101.03	97.20	97.20	46.60	61.17
1982	122.20	129.60	141.50	128.10	116.00	97.30	122.80	95.30	97.18	91.95	91.40	42.70	57.80
1983	115.70	124.10	138.30	122.50	107.80	88.20	117.80	85.40	85.12	80.05	80.80	48.40	57.30
1984	112.90	121.20	136.60	119.80	109.10	83.20	116.50	83.00	84.75	79.62	80.30	45.00	59.14
1985	111.50	120.20	134.00	119.60	105.30	83.50	113.00	79.40	81.69	76.66	77.20	39.80	56.41
1986	85.70	92.70	108.50	93.10	83.60	53.10	91.20	49.50	49.92	44.91	45.20	29.00	36.23
1987	89.70	94.80	109.30	95.70	80.30	58.90	85.90	53.80	56.75	52.25	53.40	25.20	45.36

From: Oil & Gas Journal Energy Database
Source: U.S. Department of Energy

Comparative energy prices

	Natural gas delivered to consumers ($/MCF)					Cost of fossil fuels delivered to utilities (Cents/Million BTU)				Electricity retail prices (Cents/Kilowatt Hour)			
Year	Resi-dential	Com-mercial	Indust-rial	Electric Utilities	Average	Coal	Heavy oil	Natural gas	All fuels	Resi-dential	Com-mercial	Industrial	Total
1973	1.29	0.94	0.50	0.38	0.68	40.5	78.5	33.8	47.6	2.54	2.41	1.25	1.96
1974	1.46	1.07	0.67	0.51	0.84	70.9	189.0	48.2	91.4	3.10	3.04	1.69	2.49
1975	1.71	1.35	0.96	0.77	1.12	81.4	200.5	75.2	104.4	3.51	3.45	2.07	2.92
1976	1.98	1.64	1.24	0.98	1.37	84.8	195.2	103.4	111.9	3.73	3.69	2.21	3.09
1977	2.35	2.04	1.50	1.32	1.66	94.7	219.8	129.1	129.7	4.05	4.09	2.50	3.42
1978	2.56	2.23	1.70	1.48	1.85	111.6	212.5	142.2	141.1	4.31	4.36	2.79	3.69
1979	2.98	2.73	1.99	1.81	2.21	122.4	298.8	174.9	163.9	4.64	4.68	3.05	3.99
1980	3.68	3.39	2.56	2.27	2.80	135.1	426.7	219.9	192.8	5.36	5.48	3.69	4.73
1981	4.29	4.00	3.14	2.89	3.39	153.2	533.4	280.5	225.6	6.20	6.29	4.29	5.46
1982	5.17	4.82	3.87	3.48	4.15	164.70	483.0	337.6	224.9	6.86	6.86	4.95	6.13
1983	6.06	5.59	4.18	3.58	4.82	165.60	457.0	347.4	220.6	7.18	7.01	4.97	6.29
1984	6.12	5.55	4.22	3.70	4.85	166.40	481.0	358.3	219.2	7.54	7.33	5.04	6.52
1985	6.12	5.50	3.95	3.55	4.72	164.8	424.4	343.1	209.6	7.79	7.47	5.16	6.71
1986	5.83	5.08	3.23	2.43	4.13	157.9	240.1	234.4	175.0	7.41	7.13	4.90	6.42
1987	5.54	4.78	2.94	2.32	4.05	150.6	297.6	223.5	170.7	7.41	7.00	4.72	6.32

From Oil & Gas Journal Energy Database
Source: U.S. Department of Energy

Price history, crude oil, natural gas and motor gasoline

| | Actual current prices | | | | | | Inflation adjusted real prices | | | | |
| | Crude oil | | Motor Gasoline | Natural gas | | | Crude oil | | Motor Gasoline | Natural Gas | |
Year	U.S. average wellhead $/bbl	West Texas Inter-mediate posted $/bbl	leaded pump price	U.S. average wellhead $/Mcf	Consumer average $/Mcf	Producer Price index 1982-100	U.S. average wellhead $/BBL	West Texas Inter-mediate posted $/BBL	Leaded pump price $/Gal	U.S. average wellhead $/Mcf	Consumer average $/Mcf
1925	1.68	1.79	0.222	0.094	0.223	17.8	9.438	10.056	1.247	0.528	1.253
1926	1.88	1.90	0.234	0.095	0.229	17.2	10.930	11.047	1.359	0.552	1.311
1927	1.30	1.28	0.211	0.088	0.220	16.5	7.879	7.758	1.278	0.533	1.333
1928	1.17	1.36	0.209	0.089	0.232	16.7	7.006	8.144	1.254	0.533	1.389
1929	1.27	1.45	0.214	0.082	0.216	16.4	7.744	8.841	1.306	0.500	1.317
1930	1.19	0.95	0.200	0.076	0.214	14.9	7.987	6.376	1.339	0.510	1.436
1931	0.65	0.77	0.170	0.070	0.233	12.6	5.159	6.111	1.348	0.556	1.849
1932	0.87	0.69	0.179	0.064	0.247	11.2	7.768	6.161	1.601	0.571	2.205
1933	0.67	1.00	0.178	0.062	0.237	11.4	5.877	8.772	1.563	0.544	2.079
1934	1.00	1.00	0.189	0.060	0.223	12.9	7.752	7.752	1.461	0.465	1.729
1935	0.97	1.00	0.188	0.058	0.224	13.8	7.029	7.246	1.365	0.420	1.623
1936	1.09	1.10	0.195	0.055	0.220	13.9	7.842	7.914	1.399	0.396	1.583
1937	1.18	1.22	0.200	0.051	0.220	14.9	7.919	8.188	1.342	0.342	1.477
1938	1.13	1.02	0.195	0.049	0.218	13.5	8.370	7.556	1.445	0.363	1.615
1939	1.02	1.02	0.188	0.049	0.216	13.3	7.669	7.699	1.410	0.368	1.624
1940	1.02	1.02	0.184	0.045	0.217	13.5	7.556	7.556	1.364	0.333	1.607
1941	1.14	1.17	0.192	0.049	0.221	15.1	7.550	7.748	1.274	0.325	1.464
1942	1.19	1.17	0.204	0.051	0.227	17.0	7.000	6.882	1.202	0.300	1.335
1943	1.20	1.17	0.205	0.052	0.223	17.8	6.742	6.573	1.153	0.292	1.253
1944	1.21	1.17	0.206	0.051	0.215	17.9	6.760	6.536	1.150	0.285	1.201
1945	1.22	1.17	0.205	0.049	0.214	18.2	6.703	6.429	1.126	0.269	1.176
1946	1.41	1.36	0.208	0.053	0.220	20.8	6.779	6.524	0.999	0.255	1.058
1947	1.93	1.84	0.231	0.060	0.232	25.6	7.539	7.188	0.903	0.234	0.906
1948	2.60	2.57	0.259	0.065	0.241	27.7	9.386	9.278	0.934	0.235	0.870
1949	2.54	2.57	0.268	0.063	0.254	26.3	9.658	9.772	1.019	0.240	0.966
1950	2.51	2.57	0.268	0.065	0.266	27.3	9.194	9.414	0.980	0.238	0.974
1951	2.53	2.57	0.272	0.073	0.298	30.4	8.322	8.454	0.893	0.240	0.980
1952	2.53	2.57	0.276	0.078	0.332	29.6	8.547	8.682	0.931	0.264	1.122
1953	2.68	2.72	0.287	0.092	0.355	29.2	9.178	9.290	0.983	0.315	1.216
1954	2.78	2.82	0.290	0.101	0.381	29.3	9.488	9.625	0.991	0.345	1.300
1955	2.77	2.82	0.291	0.104	0.400	29.3	9.454	9.625	0.992	0.355	1.365
1956	2.79	2.82	0.299	0.108	0.415	30.3	9.208	9.307	0.988	0.356	1.370
1957	3.09	3.04	0.310	0.113	0.431	31.2	9.904	9.753	0.992	0.362	1.381
1958	3.01	3.06	0.304	0.119	0.462	31.6	9.525	9.677	0.961	0.377	1.462
1959	2.90	2.98	0.305	0.129	0.477	31.7	9.148	9.385	0.962	0.407	1.505
1960	2.88	2.97	0.311	0.140	0.500	31.7	9.085	9.369	0.982	0.442	1.577
1961	2.89	2.97	0.308	0.151	0.510	31.6	9.146	9.399	0.973	0.478	1.614
1962	2.90	2.97	0.306	0.155	0.514	31.7	9.148	9.369	0.967	0.489	1.621
1963	2.89	2.97	0.304	0.158	0.512	31.6	9.146	9.399	0.963	0.500	1.620
1964	2.88	2.95	0.304	0.154	0.519	31.6	9.114	9.320	0.962	0.487	1.642
1965	2.86	2.92	0.312	0.156	0.522	32.3	8.854	9.040	0.966	0.483	1.616
1966	2.88	2.94	0.321	0.157	0.523	33.3	8.649	8.817	0.964	0.471	1.571
1967	2.92	3.03	0.332	0.160	0.520	33.4	8.743	9.060	0.994	0.479	1.557
1968	2.94	3.07	0.337	0.164	0.504	34.2	8.596	8.977	0.985	0.480	1.474
1969	3.09	3.30	0.348	0.167	0.515	35.6	8.680	9.256	0.978	0.469	1.447
1970	3.18	3.35	0.357	0.171	0.550	36.9	8.618	9.079	0.967	0.463	1.491
1971	3.39	3.56	0.364	0.182	0.590	38.0	8.921	9.368	0.958	0.479	1.553
1972	3.39	3.56	0.361	0.186	0.630	39.8	8.518	8.945	0.907	0.467	1.583
1973	3.89	3.87	0.388	0.216	0.680	45.0	8.644	8.604	0.862	0.480	1.511
1974	6.74	10.37	0.532	0.304	0.840	53.5	12.598	19.383	0.994	0.568	1.570
1975	7.56	11.16	0.567	0.445	1.120	58.4	12.945	19.110	0.971	0.762	1.918
1976	8.14	12.65	0.590	0.580	1.370	61.1	13.322	20.704	0.966	0.949	2.242
1977	8.57	14.30	0.622	0.790	1.660	64.9	13.205	22.034	0.958	1.217	2.558
1978	8.96	14.85	0.626	0.905	1.850	69.9	12.818	21.245	0.896	1.295	2.647
1979	12.51	22.40	0.857	1.178	2.210	78.7	15.896	28.463	1.089	1.497	2.808
1980	21.59	37.37	1.191	1.590	2.800	89.7	24.069	41.661	1.328	1.773	3.122
1981	31.77	36.67	1.311	1.980	3.390	98.0	32.418	37.418	1.338	2.020	3.459
1982	28.52	32.75	1.222	2.460	4.150	100.0	28.520	32.750	1.222	2.460	4.150
1983	26.19	30.25	1.157	2.590	4.820	101.2	25.879	29.891	1.143	2.559	4.763
1984	25.88	29.83	1.129	2.660	4.850	103.6	24.981	28.793	1.090	2.568	4.681
1985	24.09	28.08	1.115	2.510	4.720	103.1	23.366	27.236	1.081	2.435	4.578
1986	12.51	16.44	0.857	1.940	4.130	100.1	12.498	16.421	0.856	1.938	4.046
1987	15.40	18.21	0.897	1.670	4.050	102.8	14.981	17.714	0.873	1.625	3.940

From: Oil and Gas Journal Energy Database
Source: U.S. Department of Energy

Guide to world's major exported crude oils

THIS QUICK REFERENCE GUIDE TO MANY OF THE world's exported crude oils highlights some of the important basic information on each crude oil. Exerpted from the Oil & Gas Journal's exclusive series, "Guide to export crudes for the '80s," published in 1983, and from some crude assays published more recently, the information in the table concentrates on the crude name, the country in which it is produced, the terminal from which it is shipped, and four important properties of the crude.

The table covers the nine major producing areas of the world and lists the crude oils alphabetically by their crude names. The U.S. crudes produced in Alaska are listed at the end of the table because they stand the best chance of any U.S. crudes to be exported, but they are not currently exported.

The physical properties listed are for the raw crude oil only. Properties of the distilled fractions can be obtained from the complete assays published previously.

The pour point and viscosities of the crude oils listed have been converted to degrees F. for the pour point, and to centistokes at a specific temperature in degrees F. for the viscosity. In the original assay series, various units for the two properties are given.

The information on each crude is from the most recent assays received by the Oil & Gas Journal. Because some of the information may not be latest information on a particular crude oil, the table is intended to be used as a guide only.

Under the terminal heading, there are several abbrevia-

tions used to depict the type of mooring facilities for crude oil tankers loading from them. Their definitions are: CBM, conventional buoy mooring; SPM, single point mooring; SBM, single buoy mooring; CALM, caternary anchor leg mooring; SALM, single anchor leg mooring; and P.L. for pipeline.

World Export Crude Oil Properties Quick Reference

Crude	Country	Terminal	Gravity, °API	Sulfur, wt%	Pour Pt. °F.	Visc., cSt	@ F.
MIDDLE EAST							
Aboozar (Ardeshir)	Iran	Kharg Island	26.9	2.48	−30	36.58	68
Abu Al Bu Khoosh	Abu Dhabi	Storage tanker offshore	31.6	2	−24.5	6.7	100
Arabian Heavy	Saudi Arabia	Ras Tanura	27.9	2.85	−20	37	70
Arabian Light	Saudi Arabia	Ras Tanura, Yanbu & Sidon, Lebanon	33.4	1.79	−30	10.2	70
Arabian Light (Berri)	Saudi Arabia	Ras Tanura	37.8	1.19	−25	5.8	70
Arabian Medium (Khursaniyah)	Saudi Arabia	Ras Tanura	30.8	2.4	5	16.2	70
Arabian Medium (Zuluf/Marjan)	Saudi Arabia	Juayma	31.1	2.48	−20	19.4	70
Bahrgansar/Nowruz (SIRIP Blend)	Iran	Ras Bahrgan SBM	27.1	2.45	−33	20.4	100
Basrah Heavy	Iraq	Mina al Bakr sea terminal in Persian Gulf	24.7	3.5	−22	86.3	50
Basrah Light	Iraq	Marine pipeline to Khor al Amaya piled structure in Gulf	33.7	1.95	5	15	50
Basrah Medium	Iraq	Mina al Bakr sea terminal in Persian Gulf (Khor al Khafja)	31.1	2.58	−22	41	105.8
Burgan (Wafra)	Divided Zone	Kharg Island	23.3	3.37	−5	28.2	122
Dorrood (Darius)	Iran	Kharg Island	33.6	2.35	−4.5	5.65	100
Dubai (Fateh)	Abu Dubai	SBMs in offshore Persian Gulf	31.1	2	15.8	12.82	68
Dukhan (Qatar Land)	Qatar	Umm Said terminal CBM & SBM	41.7	1.28	−20	3.8	80
Eocene (Wafra)	Divided Zone	Mina Saud, Divided Zone CBM, also Mina Abdulla, Kuwait	18.6	4.55	−20	15.22	122
Foroozan (Fereidoon)	Iran	Kharg Island	31.3	2.5	−35	15.33	68
Gulf of Suez Mix	Egypt	Ras Shukheir, 3 conventional sea berths	31.9	1.52	35	8.74	100

World Export Crude Oil Properties Quick Reference — continued

Crude	Country	Terminal	Gravity, °API	Sulfur, wt%	Pour Pt. °F.	Visc., cSt	@ F.
Hout	Divided Zone	Sea berth off Ras al Kahfji	32.8	1.91	− 13	9.88	86
Iranian Heavy	Iran	Kharg Island	31	1.65	− 5	16.84	70
Iranian Light	Iran	Kharg Island	33.8	1.35	− 20	10.58	70
Khafji	Divided Zone	Sea berth off Ras al Khafji	28.5	2.85	− 31	21.3	86
Kirkuk Blend	Iraq	Mina al Bakr sea ter−minal in Persian Gulf also Ceyhan, Turkey	35.1	1.97	− 7.6	12.8	50
Kuwait Export	Kuwait	Mina al Ahmadi	31.4	2.52	5	8.36	100
Margham Light	Abu Dubai	Port of Jebel Ali	50.3	0.04	17.6	1.49	68
Mubarek	Sharjah	Floating storage and loading facility in offshore area	37	0.62	10	3.2	100
Murban	Abu Dhabi	Port of Jebel Dhanna 4 CBM berths	40.5	0.78	− 11.2	2.7	104
Oman Export	Oman	Mina al Fahal	36.3	0.79	− 15	10.66	60
Qatar Marine	Qatar	Halul Island	35.3	1.57	10	6.81	60
Rostam	Iran	Lavan Island Terminal	35.9	1.55	− 8.5	3.08	122
Salmon (Sassan)	Iran	Levan Island	33.9	1.91	− 5	6.37	77
Sharjah Condensate	Sharjah	Hamriyah terminal to SBM in Persian Gulf	49.7	0.1	− 25	1.44	68
Sirri	Iran	Offshore terminal in Persian Gulf	30.9	2.3	15.8	20.3	50
Soroosh (Cyrus)	Iran	Offshore terminal in Persian Gulf	18.1	3.3	10	1381.4	68
Souedie	Syria	Tartus/CBMs	24.9	3.82	− 22	88.3	50
Umm Shaif (Abu Dhabi Marine)	Abu Dhabi	Das Island	37.4	1.51	− 22	—	—
Zakum (Lower Zakum/ Abu Dhabi Marine)	Abu Dhabi	Das Island	40.6	1.05	− 5.8	4.28	68

NORTH SEA

Crude	Country	Terminal	Gravity, °API	Sulfur, wt%	Pour Pt. °F.	Visc., cSt	@ F.
Argyll	U.K.	SBM	—	0.18	42.8	4.79	104
Auk	U.K.	SBM	37.2	0.45	122	4.38	104
Beatrice	U.K.	Nigg Bay terminal	38.7	0.05	55	6.85	104
Beryl	U.K.	SBM	36.5	0.42	—		
Brae	U.K.	Forties system	33.6	0.73	21.2	13.6	50
Brent	U.K.	Brent system to Sullom Voe and SPAR	38.2	0.26	26.6	3.72	104
Brent Blend	U.K.	Sullom Voe Terminal Shetlands	38	0.38	5	5.67	68
Buchan	U.K.	CALM	33.7	0.84	42.8	20.37	68
Cormorant, North	U.K.	Brent system to Sullom Voe	34.9	0.71	53.6	5.76	104
Cormorant, South (A)	U.K.	Brent system to Sullom Voe	35.7	0.56	21.2	5	104
Dan	Denmark	Used in Gorm Blend	30.4	0.34	− 45	8.9	104
Dunlin	U.K.	Brent system to Sullom Voe	34.9	0.39	42.8	4.92	104
Ekofisk	Norway	Seal Sands, Teesside, U.K.	43.4	0.14	10	2.11	70
Flotta	U.K.	Flotta terminal, Orkney Islands	35.7	1.14	21	30.8	40
Forties	U.K.	Hound Point terminal Firth of Forth	36.6	0.3	26.6	9.6	50
Fulmar	U.K.	SBM	39.3	0.26	10.4	2.55	104
Gorm	Denmark	SBM	33.9	0.23	− 35	5.44	104
Gullfaks	Norway,		28.6	0.44	− 50	10.14	100
Magnus	U.K.	Ninian system to Sullom Voe	39.3	0.28	26.6	4.56	69.8
Maureen	U.K.		35.8	0.55	45	14.88	70
Montrose	U.K.	SBM	40.1	0.23	15	4.96	68
Murchison	U.K. & Norway	Brent system to Sullom Voe	38	0.27	45	3.6	100
Ninian Blend	U.K.	Ninian system to Sullom Voe	35.6	0.43	35	5.61	104
Statfjord	Norway & U.K.	A & B SPMs	38.4	0.27	40	7.33	59
Tartan	U.K.	Flotta system	41.7	0.56	16	11.6	40
Thistle	U.K.	Brent system to Sullom Voe	37	0.31	53.6	5.87	68

WEST AFRICA

Crude	Country	Terminal	Gravity, °API	Sulfur, wt%	Pour Pt. °F.	Visc., cSt	@ F.
Bonny Light	Nigeria	Bonny shore terminal & Bonny marine SBM	36.7	0.12	—	3.47	100.4

Crude	Country	Terminal	Gravity, °API	Sulfur, wt%	Pour Pt. °F.	Visc., cSt	@ F.
Bonny Medium	Nigeria	Bonny shore terminal & Bonny marine CALM	25.2	0.23	−16.6	12.1	100.4
Brass River	Nigeria	CALM off Brass	40.9	0.09	35	2.68	70
Cabinda	Angola	CALM off Malongo terminal	31.7	0.17	65	18.67	100
Djeno Blend	Congo (Brazzaville)	CALM off Djeno tank farm	26.9	0.33	37.4	220	50
Escravos	Nigeria	CALM	36.2	0.14	10	2.43	100
Espoir	Ivory Coast	CALM with floating storage	31.4	0.32	5	12.22	70
Forcados Blend	Nigeria	CALMs	29.7	0.29	−4	10.6	50
Gamba	Gabon	Buoy offshore from Gamba tank farm	31.8	0.11	73.4	36.7	100.4
Kole Marine Blend	Cameroon	CALM with floating storage	34.9	0.3	21.2	11.1	50
Lucina Marine	Gabon	CALM with floating storage	39.5	0.05	59	16	68
Mandji	Gabon	Cap Lopez terminal, loading berth Prince's Bay	30.5	1.1	37.4	44,50	—
Pennington	Nigeria	CALM with floating storage	36.6	0.07	43	3.06	104
Qua Iboe	Nigeria	CALM	35.8	0.12	45	6.76	59
Salt Pond	Ghana	SALM with permanent storage	37.4	0.1	—	—	—
Soyo Blend	Angola	Quinfequena terminal	33.7	0.23	64.4	10.08	100.4
Zaire	Zaire	CALM with moored stroage	31.7	0.13	80	19.79	100

NORTH AFRICA

Crude	Country	Terminal	Gravity, °API	Sulfur, wt%	Pour Pt. °F.	Visc., cSt	@ F.
Amna (high pour)	Libya	Ras Lanuf/CBMs & SPM	36.1	0.15	75	13.7	100.4
Ashtart	Tunisia	Floating storage in Gulf of Gabes	29	1	48.2	24	68
Brega	Libya	Marsa el Bregal/ CBM, SPM, SBM	40.4	0.21	30	5.58	70
Bu Attifel	Libya	Zueitina/CBM & SBM	43.6	0.03	90	—	—
El Borma	Tunisia	See Zarzaitine	—	—	—	—	—
Es Sider, Libya	Es Sider	1 SBM/3 CBMs	36.7	0.37	45	5.26	100
Saharan Blend	Algeria	Arzew (3 CBMs), Bejaia (4 jetties), Skikda	45.5	0.53	−20	2.38	70
Sarir	Libya	Marsa al Hariga (Tobruk) CBM	38.3	0.18	25	4.73	122
Sirtica	Libya	Ras Lanuf/CBMs & SBM	43.3	0.43	26.6	—	—
Zarzaitine	Algeria	La Skhirra, Tunisia	43	0.07	10.4	6.9	50
Zueitina	Libya	Zueitina/CBMs & SBM	41.3	0.28	45	42.79	60

ASIA

Crude	Country	Terminal	Gravity, °API	Sulfur, wt%	Pour Pt. °F.	Visc., cSt	@ F.
Ardjuna	Indonesia	SBM off North Java	35.2	0.11	75	3.74	100
Arimbi	Indonesia	Balongan sea terminal SBM off North Java	31.8	0.2	100	12.54	100
Attaka	Indonesia	Santan marine terminal SBM off Balikpapan, Kalimantan SBM	42.3	0.09	−10	1.7	70
Badak	Indonesia	Santan terminal off Balikpapan, Kalimantan/ SBM	41.3	0.08	−15	5.72	100
Bekapai	Indonesia	Senipah terminal off Balikpapan, Kalimantan/ SBM	42.3	0.09	−10	1.7	70
Bekok	Malaysia	Shore terminal north of Chukai, Kuala Trengganu	49.1	0.02	26	1.74	100
Bintulu	Malaysia	Bintulu SBM	26.5	0.13	23	4.54	132.8
Bombay High	India	Floating storage in Arabian Sea	39.4	0.17	86	3.28	104
Bunyu	Indonesia	Bunyu terminal, Kalimantan CBM	31.7	0.09	60	2.67	100
Champion Export	Brunei	Seria SBM	23.9	0.12	−32.8	21.1	50
Cinta	Indonesia	SBM and floating storage, Java Sea	33.4	0.08	105	64.34	100
Daqing	China	See Taching	—	—	—	—	—
Duri (Sumatran Heavy)	Sumatra	Port of Dumai	21.1	0.2	54	131.9	122
Handil	Kalimantan	SBM	32.8	0.08	85	4.2	100.4
Jatibarang	Java	SBM	29	0.07	110	83.8	122
Labuan Blend	Sabah	SBM, South China Sea	33.2	0.09	52	4.41	70

Crude	Country	Terminal	Gravity, °API	Sulfur, wt%	Pour Pt. °F.	Visc., cSt	@ F.
Minas (Sumatran Light)	Sumatra	Port of Dumai	34.5	0.08	97	12.4	122
Miri Light	Sarawak	SBMs Miri terminal	36.3	0.08	34	3.65	70
Pulai	Malaysia	Shore terminal Chukai, Kuala Trengganu	42.5	0.02	23	2.12	70
Salawati	West Irian	Salawati terminal	38	0.49	−15	4.2	100
Sanga Sanga	Kalimantan	—	25.7	0.17	5	32.13	100
Sepinggan — Yakin	East Kalimantan	Lawi — Lawi terminal/ SBM off Makasar Straits	31.7	0.11	5	3.16	100
Seria Light	Brunei	Lutong terminal	36.2	0.07	35	2.2	80
Shengli	China	Qingdao terminal	24.2	1	70	99.28	122
Soviet Export Blend	U.S.S.R.	Ventspils and Novorossysk	32.5	1.38	−9.4	11.5	68
Taching (Daqing)	China	Luda terminal	33	0.08	95	—	—
Tapis	Malaysia	Shore terminal Chukai, Kuala Trengganu	44.3	0.02	39	2.69	70
Tapis Blend	Shore terminal Chukai, Kuala Trengganu	Made up of Tapis, Pulai, Bekok, Tiong, Tingge, and Kepong					
Tembungo	Malaysia	Production platform	37.4	0.04	25	2	100
Udang	Indonesia	Storage SBM north of Annambas Island	38	0.05	—	8.71	130
Walio	Indonesia	Kasim terminal CBM	34.1	—	—	5	100

LATIN AMERICA

Crude	Country	Terminal	Gravity, °API	Sulfur, wt%	Pour Pt. °F.	Visc., cSt	@ F.
BCF 24	Venezuela	La Salina, Lake Maracaibo	23.5	1.85	−60	41.17	100
Bachaquero	Venezuela	La Salina	16.8	2.4	−10	294	100
Boscan	Venezuela	Bajo Grande, Lake Maracaibo	10.1	5.5	50	19428	100
Cano Limon	Colombia	Covenas terminal to two CALM	30.8	0.47	60	12.54	100
Ceuta	Venezuela	Puerto Miranda, Lake Maracaibo	31.8	1.2	−35	9.68	100
Galeota Mix	Trinidad Tobago	Point Galeota SBM	32.8	0.27	−5	4.32	100
Isthmus	Mexico	Pajaritos and Salina Cruz	32.8	1.51	−15	10.88	70
Lagomedio	Venezuela	Punta Palmas/ La Salina	31.5	1.17	−15	7.64	100
La Rosa Medium	Venezuela	La Salina	25.3	1.73	−50	24	122
Leona	Venezuela	Puerto La Cruz	24.1	1.5	−45	65.63	122
Loreto	Peru	Puerto Bayovar	34	0.29	34	6.8	80
Maya	Mexico	Pajaritos and Salina Cruz	22	3.32	0	221.4	70
Merey	Venezuela	Puerto La Cruz	17.4	2.2	−10	260.5	100
Oriente	Ecuador	Puerto Balao	29.2	1.01	25	13	104
Tia Juana Light	Venezuela	La Salina	32.1	1.1	−45	9	100
Tia Juana Pesado	Venezuela	Puerto Miranda	12.1	2.7	30	643.1	122

AUSTRALIA

Crude	Country	Terminal	Gravity, °API	Sulfur, wt%	Pour Pt. °F.	Visc., cSt	@ F.
Gippsland (Bass Strait)	Victoria	Jetties at Long Island and Crib Point	45.4	0.1	59	2.93	69.8
Barrow Island	W. Australia	SBM off Barrow Island	36.8	0.05	−22	3.4	50
Jabiru	North. Territory	Floating production	42.3	0.04	—	—	
Northwest Shelf Cond.	W. Australia	Dolphin berth in Withnell Bay	53.1	0.01	−32.8	0.99	68
Jackson Blend	Queensland	Loaded across Ampol Refinery at Lytton	41.9	0.03	—	—	—

NORTH AMERICA

Crude	Country	Terminal	Gravity, °API	Sulfur, wt%	Pour Pt. °F.	Visc., cSt	@ F.
Bow River Heavy	Alberta	Bow River Pipeline	26.7	21	−58	18.3	104
Cold Lake	Alberta	Alberta Oil Sands Pipeline	—	2.91	−76	70.7	59
Federated L&M	Alberta	Federated Pipeline	39.7	2.01	14	3.3	104
Gulf Alberrta L&M	Alberta	Gulf Alberta P.L.	35.1	0.98	−17.5	4.86	104
Lloydminster	Alberta	Husky Oil P.L.	20.7	3.15	−25.6	101	104
Rainbow L&M	Alberta	Rainbow P.L.	40.7	0.5	36.5	3.77	104
Rangeland — South L&M	Alberta	Rangeland P.L.	39.5	0.75	−40	2.67	104
Wainwright — Kinsella	Alberta	Husky Oil P.L.	23.1	1.58	−38.2	32.5	104

SELECTED U.S. CRUDES (NOT EXPORTED)

Crude	Country	Terminal	Gravity, °API	Sulfur, wt%	Pour Pt. °F.	Visc., cSt	@ F.
Alaska North Slope	Alaska	Trans Alaska P.L.	26.4	1.06	0	42.42	60
Kuparuk	Alaska	Trans Alaska P.L.	23	1.76	−55	79.98	60
West Sak	Alaska	—	22.4	1.82	−50	95.92	60

Worldwide construction picture brightens; U.S. companies continue to struggle

PLANS FOR GAS-PIPELINE CONSTRUCTION IN THE Non-Communist world, especially the lucrative markets of Canada and the U.S., have surged this year.

Anticipated strong natural-gas demand in Northeast U.S. and Florida residential and industrial markets and in the thermal-EOR markets of California are behind the growing number of gas-pipeline construction proposals being offered.

Earlier this year, results of an annual survey by Oil & Gas Journal reported that pipeline-operating companies had plans for slightly more than 17,500 miles of new gas-transmission pipe in the U.S. and Canada this year and beyond, compared with the 10,621 miles a year earlier.

This growth is providing a healthy measure of balance to the flat markets for crude-oil or petroleum-product pipeline construction.

For 1989 for all types of petroleum and natural-gas pipeline, operators in the non-Communist world planned 19.7% more miles this year than last.

Cost of this year's work is an estimated $6.2 billion, compared with last year's estimate of $6.66 billion. For 1989 and beyond, total land and offshore construction costs will exceed $23 billion, compared with $19.9 billion last year.

The mileage differences in the two categories of planned 1989 construction and planned 1989-and-beyond construction reflect how pipeline operators change construction plans during the course of a year as business climates for different construction markets change.

Projects for 1989 mileage include only projects in progress at the first of the year, or scheduled to begin during the year, and likely to be completed by yearend. Projects for 1989 and beyond include projects in progress or set to begin this year but likely to be completed in 1989 or later.

Excluded from the projects are those only under study or those planned but not set to begin construction until 1990 or later. In addition, projects are adjusted to eliminate mileage duplications caused by competing projects where only one or two are likely to be built.

Gas-pipeline growth

Of the 34,725 miles of total pipeline construction planned this year and beyond, almost 82% are gas pipeline.

The U.S. accounts for most of the new gas lines—9,284 miles in 1989 and beyond; Canada, 8,233 miles planned.

More large-diameter pipe is planned in Canada, however, than in the U.S. Almost 3,000 miles of pipe larger than 30 in. in diameter are planned in Canada, compared with 2,326 miles in the U.S.

Most of the additional gas pipe planned in Canada is to

THIS LINE is to carry crude production from the Endicott field on Alaska's North Slope to the Trans-Alaska Pipeline (photo courtesy Standard Alaska Production Co.).

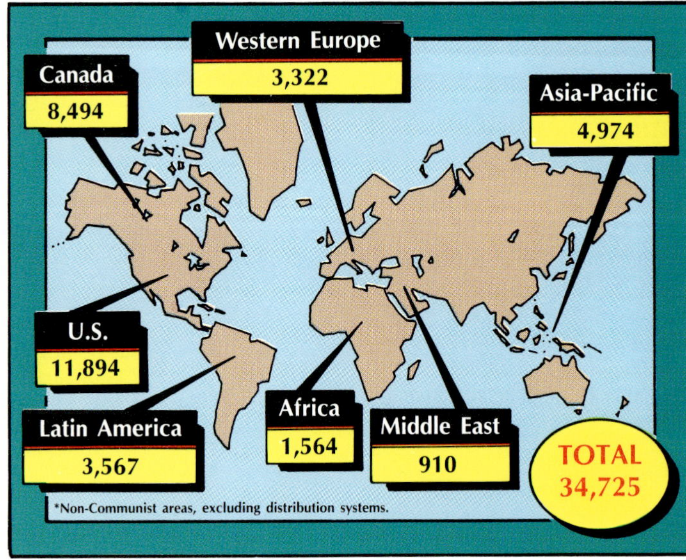

Non-Communist pipeline construction in 1989 and beyond*

Area	4-10 in.	12-20 in.	22-30 in.	30 + in.	Total
			Miles		
GAS PIPELINES					
U.S.	746	2,559	3,653	2,326	9,284
Canada	1,655	1,939	1,660	2,979	8,233
Latin America	806	314	629	666	2,415
Asia-Pacific	690	953	914	1,775	4,332
West Europe	1,119	640	438	910	3,107
Middle East	0	0	0	0	0
Afrida	48	8	350	584	990
Total gas	5,064	6,413	7,644	9,240	28,361
CRUDE PIPELINES					
U.S.	240	61	267	525	1,093
Canada	136	12	0	0	148
Latin America	9	70	67	165	311
Asia-Pacific	20	23	0	409	452
West Europe	13	78	73	0	164
Middle East	0	0	56	820	876
Africa	0	0	26	239	265
Total crude	418	244	489	2,158	3,309
PRODUCT PIPELINES					
U.S.	549	968	0	0	1,517
Canada	78	35	00	0	113
Latin America	110	731	0	0	841
Asia-Pacific	40	0	150	0	190
West Europe	36	15	0	0	51
Middle East	0	0	0	34	34
Africa	0	309	0	0	309
Total product	813	2,058	150	34	3,055
NON-COMMUNIST WORLD TOTALS					
Gas	5,064	6,413	7,644	9,240	28,361
Crude	418	244	489	2,158	3,309
Product	813	2,058	150	34	3,055
Total	6,295	8,715	8,283	11,432	34,725

*Projects planned and under way in 1989 and scheduled for completion beyond 1989.

increase transmission capacity to U.S. markets because current import levels are at capacity of the transmission system (OGJ, Dec. 5, 1988, p. 14).

The Free Trade Agreement between Canada and the U.S. will prompt additional Canadian gas imports into the U.S. As a result, pipeline operators are scrambling to gain permits for proposed projects (OGJ, Feb. 6, p. 21; Jan. 16, p. 20).

Major proposed Canadian gas-pipeline projects which have been on back burners for some time are being looked at again.

The Polar Gas Project, which would involve 1,318 miles of 36-in. pipe, is one of several proposals to bring gas from the Mackenzie Delta and Beaufort Sea area to Alberta. And there is the 330-mile, 10-in. Vancouver Island pipeline project.

U.S., Canada leaders

That U.S. gas-pipeline construction was heating up was evident by mid 1988. For the year ending June 30, 1988, more than 5,000 miles of interstate gas pipeline had been proposed to the U.S. Federal Energy Regulatory Commission (FERC).

This represented a jump of more than 4,200 miles (323%) over mileage (1,304) for projects in the previous 12-month period. Of the 129 land projects proposed in that period ending June 30, 94 were for sites in states ranging northeast from West Virginia and Pennsylvania.

Moreover, during the FERC's "open season" on pipeline proposals to transport gas from the Mobile Bay (Ala.) area

for interconnection with major gas-transmission systems or new markets in Florida, 15 projects were proposed. Four others were already on file.

Most of the planned gas-pipeline construction in the U.S. involves pipe larger than 10 in. in diameter.

Three gas-line projects emerged from the FERC's open season for applications to serve the U.S. Northeast (OGJ, Jan. 30, p. 31). Two of the three projects would mainly serve as U.S. legs of Canadian pipeline expansions.

In the competion to provide gas supply to California's thermal-EOR market, the FERC earlier this year issued final certificates authorizing construction of several proposals. The line that is the most successful in lining up shippers will likely be built first.

The Gulf of Mexico gas-pipeline competition involves transmission of supply from the Jurassic Norphlet deep gas trend off Alabama, Mississippi, and Florida. One major line has already been built.

In Western Europe meanwhile, almost 4,500 miles of new gas line are planned, mainly involving small and large-diameter pipe.

The largest West Europe project involves 515 miles of 38-in. pipe from Statoil's (den Norske Stats Oljeselskap AS) Zeepipe-Sleipner project in the North Sea to a terminal at the Belgian port of Zeebrugge. Project completion is expected by late 1994.

In Asia, Australia's Queensland government is planning 330 miles of 12.2-in. line to connect Surat basin and Denison Trough gas fields.

Bangladesh Oil, Gas & Minerals Corp. is designing a 117-mile, 20-in. line from Kailashtila to Ashanganj.

India's National Gas & Pipelines Corp. has laid 1,094 miles of 18-36 in. line from Bombay High to six fertilizer plants.

Sui Gas Transmission CL is designing a 182-mile, 20-in. loop of the Pakistan gas line from Sui to Karachi.

Flat crude, product plans

Planned crude-oil pipeline construction mileage for 1989 and beyond is 36% less than what was reported a year ago.

The biggest drop occurs in the U.S. and Latin America. Part of the reason for the decline is completion of the first leg of the All American Pipeline in the U.S. and a major line linking a new supply region in Colombia.

Most of the non-Communist world's planned crude-line construction in 1989 and beyond is in the U.S. and the Middle East, mainly Iraq. The U.S. planned mileage accounts for 33% of the total planned in the non-Communist world.

In the U.S., the largest project involves completion of All America Pipeline which currently ends near McCamey in far West Texas. Planned are 525 miles of 30-in. pipe to extend the line to Webster, Tex., just south of Houston. Completion is scheduled in 1990.

Another major U.S. project is the Angeles Pipeline, which involves 130 miles of 30-in. pipe from Emidio, Calif., to Los Angeles-basin refineries. Completion is scheduled in mid-1992.

Middle East lines, meanwhile, include a 602-mile system of 48-56 in. pipe from Iraq to an export terminal near Yanbu, Saudi Arabia, and 62.5 miles of 42-in. pipe in Iran from Gurreh to Taheri.

The planned Middle East projects account for most of the non-Communist world's large-diameter crude-pipeline construction.

In Africa, Sonatrach Inc. is laying 239 miles of 34-in. loops on a crude line from Hooud el Hanra to Skikda.

In Mexico, Petroleos Mexicanos is laying 164.7 miles of 48-in. line from near Nuevo Teapa, Ver., to Salina Cruz, Oax.

Planned product-line construction in 1989 and beyond is

flat compared to estimates made a year earlier.

The U.S. and Latin America account for most of the product lines planned in the non-Communist world.

The biggest product line in the U.S. is a 310-mile, 10-12-in. pipeline that Koch Industries Inc. is building from Rosemount, Minn., to Milwaukee.

Chevron Pipeline Co. is laying 123 miles of 10-in., 52 miles of 12-in., and 20 miles of 8-in. pipe from Sour Lake, Tex., to Fort Worth in addition to its Mesquite Products Pipeline along the same route.

In Latin America, meanwhile, Petroleo Brasileiro SA is laying 180 miles of 16-in. pipe in Brazil from Guararema to Volta Redonda, and Corpoven SA is laying a 373-mile system in Venezuela from Puerto la Cruz refineries to Maturin, San Tome, Ciudad Bolivar, Puerto Ordaz, and Puerto Ayacucho.

In Africa, Nigerian National Petroleum Corp. is designing a 309-mile system of 16 and 12-in. pipe from Port Harcourt, Enugu, and Auchi to Benin City.

Construction costs

More good news for the prospective hike in gas-line activity is that construction costs over the past few years have moderated their rates of increase. For the 12-month period ending June 30, 1988, U.S. FERC data show construction costs in $/mile actually declined over the same period a year earlier.

Estimated costs for all land projects proposed to the Commission averaged $657,388/mile, down barely 0.07% from a year earlier. Proposed offshore projects averaged $755,345/mile, off by less than 1% from the same period a year earlier.

The following cost projections for planned mileage for 1989 and for 1989 and beyond assume that 90% of all construction will be land and 10% offshore. The exceptions are for pipelines 32 in. in diameter or larger, which are assumed to be solely land projects.

Here is a breakout of cost projections:

● Total land construction for 1989 will reach $5.6 billion—$1.9 billion for 4-10 in. pipelines, $2.2 billion for 12-20 in. pipelines, $645 million for 22-30 in. pipelines, and $778 million for 32-in. pipelines and larger.

● Total construction offshore for 1989 will reach $614 million—$252.3 million for 4-10 in. pipelines, $279.5 million for 12-20 in. pipelines, and $82.3 million for 22-30 in. pipelines.

● Total land construction for 1989 and beyond will reach $21.3 billion—$3.7 billion for 4-10 in. pipelines, $5.2 billion for 12-20 in. pipelines, $4.9 billion for 22-30 in. pipelines, and $7.5 billion for pipelines greater than 32 in. in diameter.

● Total construction offshore for 1989 and beyond will reach $1.7 billion—$476 million for 4-10 in. pipelines, $659 million for 12-20 in. pipelines, and $625.4 million for 22-30 in. pipelines.

Company doldrums

The possibility of increased gas-pipeline construction activity--with its implicit promise of increased demand--comes as good news to battered U.S. gas-pipeline operating companies.

Latest available annual-report data for all regulated U.S. interstate petroleum and natural-gas pipelines show that revenues and incomes of gas-pipeline companies took another beating in 1987 as both major and nonmajor natural-gas companies continued to struggle with the industry restructuring wrought by FERC Order 436.

Major gas-pipeline companies are those whose combined gas sold for resale and gas transported or stored for a fee exceeded 50 bcf at 14.73 psi (60° F.) in each of the 3 previous calendar years.

Nonmajors are companies not classified as majors and having had total gas sales of volume transactions exceeding 200 MMcf at 14.73 psi (60° F.) in each of the 3 previous calendar years.

The intent of Order 436 when issued in 1985, and of subsequent orders, was to bring greater competition to the interstate transportation of natural gas. The effect, combined with lingering low real (as opposed to projected) demand for gas, has been to squeeze operating-companies' revenues and incomes.

Revenues and incomes for natural-gas pipeline companies in 1987 continued their steep declines of recent years.

Major gas-pipeline companies' operating revenues for 1987 fell $6 billion (-20.3%) with incomes off $232.5 million.

All natural-gas pipeline companies' revenues for 1987 were off even more, by $8.6 billion (21.3%); all companies' incomes were down $324.2 million (-19.6).

For all natural-gas pipeline companies, net income as a portion of gas-plant investment continued its slide of recent years, falling to 2.7% in 1987 from 3.4% in 1986, 4.5% in 1985, and 8.7% in 1984.

For major gas pipelines, net income as a portion of gas-plant investment mirrored the trend, falling to 2.6% from 3.2% in 1986, 4.3% in 1985, and 7.6% in 1984.

For 1987, all gas companies reported an industry gas-plant investment totaling more than $49.9 billion compared with $49.4 billion for 1986 and $48.3 billion for 1985.

Majors' gas-plant investment in 1987 fell to slightly less than $42 billion from $42.2 billion for 1986.

Similarly, operating revenues for liquids-pipeline companies dropped by $229.8 million (-3.15%). Their incomes, however, showed a significant increase, up $423.7 million (20.6%).

For interstate common-carrier liquids-pipeline companies, net income as a percentage of investment in carrier property was 11.6% in 1987 compared with 9.2% in 1986, 11.3% in 1985, and 13.1% in 1984.

Liquids pipelines' investment in carrier property for 1987 fell by almost $22.4 million (-4.96%) to slightly less than $22.3 billion compared with $22.4 billion in 1986, $21.6 billion in 1985, and $19.4 billion in 1984.

Another measure of the profitability of oil and natural-gas pipeline companies in recent years is reflected in the percentage net income represents of operating income. Trends for liquids-pipeline companies and for natural-gas pipeline companies have been heading in opposite directions for the past 10 years.

For liquids-pipeline companies in 1987, this figure rose to

Liquid-pipeline investment

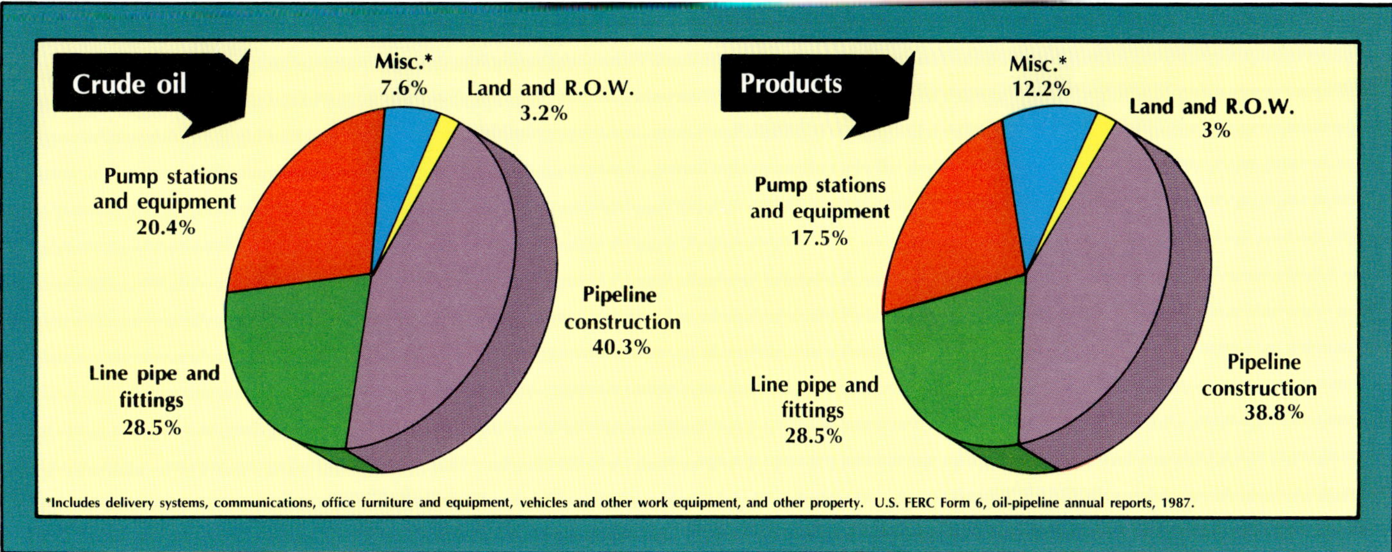

35% from 28% in 1986 and 21.9% in 1978. The figure for 1986 represents the only year the percentage has declined (from 32.6% for 1985).

By contrast for all natural-gas pipeline companies, income as a percentage of operating revenues has been steadily falling in the past 10 years.

At the end of 1987, it stood at slightly more than 4%, fractionally higher than for 1986 but continuing the downward trend since 1978 when it was 8.3%.

U.S. interstate network

Mileage operated by the 264 liquids and natural-gas pipeline companies reporting to the FERC for 1987 also fell. Total mileage was reported at 447,950, off 3,945 miles (0.9%).

Crude-oil and products pipeline companies operated 167,865 miles of gathering and trunk line in 1987, down 2,119 miles (-1.3%). This mileage handled crude oil, crude-oil products, and NGL's.

The decline is entirely made up of drops in product-line mileage.

Mileage for all natural-gas pipeline companies increased fractionally; mileage for major gas-pipeline companies fell, also fractionally.

For any year, however, comparisons of U.S. petroleum and natural-gas pipeline mileages must be done with care for two reasons:

• The number of companies required to file reports with the FERC varies each year.

• Secondly, the FERC's system for classifying interstate natural-gas pipeline companies changed beginning with the 1984 reporting year, as explained previously.

One effect of this change has been that companies classified as nonmajor are exempt from filing certain data, chief among them being mileage and gas-sales or transportation figures. Many companies nonetheless file such data voluntarily. Figures for these nonmajor companies should be excluded from a calculation of total U.S. interstate pipeline utilization for any given year because of their unreliability year to year.

Therefore, combining 1987 mileage data for all regulated liquids-pipeline companies (167,865) with mileage reported by the major natural-gas pipeline companies (42 companies reporting 252,447 miles) yields a total figure of 420,312. This represents less than 1% decrease of 3,343 miles over comparable figures reported for 1986 (170,014 + 253,641

= 423,655). This compares with a similarly small change for 1986 over 1985, a drop of only 156 miles for all liquids and major natural-gas pipelines.

At the end of 1987, reports from liquids-pipeline companies show a total of 35,055 miles of gathering lines, 54,886 miles of crude-oil trunk lines, and 79,924 miles of products pipelines. For liquids-pipeline companies, the criterion determining which company must file an annual report (Form 6) is whether the FERC classifies the company as an interstate common-carrier pipeline.

These reports for 1986 show that gathering lines increased by 846 miles (2.5%) and crude lines, by 733 miles (1.4%) over totals reported for 1986. Products lines decreased by 3,728 miles (-4.6%).

These figures reverse patterns evident in figures reported for 1986 compared with those for 1985 and is in line with the erratic pattern of liquids-pipeline utilization for the previous 7 years. Natural-gas transmission lines were down over 1986 by 3,084 miles (-1.5%) for all companies and by 3,697 miles (-2%) for major companies. Field lines for all companies showed a fractional increase of 47 miles but fell among major companies, by 792 miles.

For all companies as well as majors, storage lines increased in 1987 over 1986: up 3,552 miles (61.7%) for all companies and 3,295 miles (65%) for majors.

Deliveries rise

Throughput for liquids-pipeline companies in 1987 showed a very slight gain over 1986.

Natural-gas pipelines sales fell again in 1987, by 1.9 tcf (-23.6%) for majors and by almost 2.2 tcf (-22.8%) for all companies. For 1987, all regulated U.S. interstate gas pipelines carried 13.5 tcf of gas for others; majors carried nearly 12.9 tcf of that.

Volumes of gas transported for others by all gas-pipeline companies comprised almost 65% of the total volume of gas (20.8 tcf) that moved through U.S. interstate systems in 1987. For major companies (whose total volumes were just shy of 19.2 tcf), the share was a bit larger: slightly more than 67%. Crude-oil and product deliveries in 1987 again exceeded 11 billion bbl, an increase over 1986 of 112 million bbl, or only about 1%.

Crude-oil deliveries actually fell in 1987 by 8.8 million bbl or less than 1%. Product deliveries, reflecting increased U.S. refinery utilization, rose by 200.8 million bbl (4.3%).

Trunkline traffic for U.S. crude-oil and product pipelines

Investment in liquids pipelines

	Company and investment, $					Total, $	%
	A	B	C	D	E		
CRUDE PIPELINES INVESTMENT BY FIVE COMPANIES							
Land	1,787,499	814,497	292,759	979,827	1,189,920	5,064,502	0.30
Right of way	12,169,423	1,004,094	316,392	16,516,482	15,068,015	45,074,406	2.71
Line pipe	134,882,674	25,700,801	7,797,471	96,144,263	153,364,612	417,889,821	25.15
Line pipe fittings	9,490,351	922,085	1,387,778	17,401,123	27,527,324	56,728,661	3.41
Pipeline construction	189,654,709	27,801,794	10,931,879	219,879,456	222,164,700	670,432,538	40.34
Buildings	21,184,793	4,951,147	2,457,791	13,566,759	9,918,148	52,078,638	3.13
Boilers	—	—	—	—	—	—	
Pumping equipment	24,302,627	5,524,747	2,414,882	17,169,861	48,657,207	98,069,324	5.90
Machine tools and machinery	—	—	—	33,140	—	33,140	—
Other station equipment	89,380,146	21,330,115	4,593,316	45,152,887	29,575,598	190,032,062	11.44
Oil tanks	14,494,825	6,038,119	5,678,530	16,653,577	26,446,336	69,311,387	4.17
Delivery facilities	—	—	5,327,265	1,494,974	6,819,985	13,642,224	0.82
Communications systems	710,337	3,225,009	68,631	10,506,049	3,382,127	17,892,153	1.08
Office furniture and equipment	1,689,985	221,313	215,058	1,727,360	891,704	4,745,420	0.29
Vehicles and other work equip.	7,519,510	786,939	515,392	5,852,776	243,243	14,917,860	0.90
Other property	—	1,203,646	—	—	4,659,326	5,862,972	0.35
Total	**$507,266,879**	**$99,524,306**	**$41,997,144**	**$463,078,534**	**$549,908,245**	**$1,661,775,108**	***100.00**
PRODUCT PIPELINES INVESTMENT BY FIVE COMPANIES							
Land	4,095,422	1,763,961	532,035	936,527	3,241,861	10,569,806	0.37
Right of way	28,739,531	10,829,964	20,111,183	8,776,412	6,750,443	75,207,533	2.60
Line pipe	340,532,482	67,590,647	108,516,930	87,679,163	97,788,462	702,107,684	24.29
Line pipe fittings	72,088,338	12,874,224	15,611,443	1,704,032	20,932,923	123,210,960	4.26
Pipeline construction	566,607,912	109,220,477	206,225,286	102,019,195	137,781,102	1,121,853,972	38.81
Buildings	23,930,483	4,026,100	4,997,792	10,566,018	17,617,962	61,138,355	2.11
Boilers	—	—	—	—	—	—	—
Pumping equipment	45,782,626	6,160,023	31,921,749	28,220,704	24,012,357	136,097,459	4.71
Machine tools and machinery	—	—	—	—	363,318	363,318	0.01
Other station equipment	119,871,056	21,184,953	39,746,989	23,304,697	105,543,108	309,650,803	10.71
Oil tanks	48,805,418	17,220,724	9,035,479	17,465,781	51,130,650	143,658,052	4.97
Delivery facilities	—	—	9,594,441	10,811,906	45,787,494	66,193,841	2.29
Communication systems	1,269,854	369,627	2,239,628	8,742,968	—	12,622,077	0.44
Office furniture and equipment	10,248,198	2,374,190	2,468,937	618,384	773,120	16,482,829	0.57
Vehicles and other work equip.	5,498,272	1,809,909	7,911,262	2,235,838	2,041,276	19,496,548	0.67
Other property	72,979,021	—	19,089,144	—	—	92,068,165	3.18
Total	**$1,340,448,613**	**$255,424,799**	**$478,002,298**	**$303,081,625**	**$513,764,067**	**$2,890,721,402**	***100.00**

Source: U.S. FERC Form 6, Annual Report of Oil Pipeline Companies, Dec. 31, 1987.

* Actual total may be inexact due to rounding.

was flat in 1987, advancing by only 33 billion bbl-miles (0.96%) over that for 1986. Crude-oil trunkline traffic increased by a modest 18.7 billion bbl-miles (2%). Product traffic fell by 5 billion bbl-miles (0.3%).

Construction costs

The costs of line pipe and fittings and of laying pipelines make up roughly 70% of the cost of a pipeline system. These elements of pipeline construction costs have risen very little for the past few years, after rapidly increasing before 1982. The average cost per mile for any given diameter may fluctuate from one year to the next as projects' costs are affected by geographic location, terrain, population density, or other factors. Cost-per-mile figures may reveal more about cost trends than aggregate costs.

For proposed U.S. gas-pipeline projects for the 12-month period ending June 30, 1988, the average land cost per mile was $657,388/mile, compared with $707,253/mile for the 1986-87 period, an insignificant decrease of less than 1%.

For offshore projects, the 1987-88 figure was $755,345/mile, compared with $825,395/mile for 1986-87, a decline of 8.5%. Combined land and offshore projects show an average cost of $664,509/mile for this year's coverage; for the 1986-87 period, the figure was $707,634/mile.

Analyses of the four major categories of pipeline construction costs--material, labor, right-of-way and damages, and miscellaneous--can also indicate trends within each group.

"Miscellaneous" generally includes engineering, supervision, contingencies, allowances for funds used during construction, administration and overheads, and FERC filing fees.

For the 132 projects surveyed for the 1987-88 period, cost-per-mile data for the four categories are as follows:

● Material--Land, $217,079/mile; offshore, $377,663/mile; total, $228,753/mile.

● Labor--Land, $281,637/mile; offshore, $209,634/mile; total, $276,402/mile.

● R.O.W. and damages--Land, $40,213/mile; offshore, $12/mile; total, $37,290/mile.

● Miscellaneous--Land, $118,459/mile; offshore, $168,036/mile; total, $122,064/mile.

Average cost per mile for the projects providing these data shows no clear trend related to either length or geographic area.

In general, however, the cost per mile within a given diameter indicates that the longer the pipeline, the lower the cost per mile for construction. Nonetheless, road, highway, and river crossings and marshy or rocky terrain each strongly affects pipeline construction costs.

Compressor-station costs make up another major cost element of natural-gas pipelines. FERC applications for these projects cover the same 12-month period as for pipelines. Costs for new land compressor stations range from a low of $126/hp for a 750-hp station in Texas to a high of $4,347/hp for a 1,000-hp station in New York. Cost-per-

Natural-gas pipeline transmission expenses*

| | Expenses, $ | | Cost/mile, $ | | Cost/MMcf sold, $ | |
	1985-majors†	1986-majors§	1985-majors†	1986-majors§	1985-majors†	1986-majors§
Operation expenses						
Supervision and engineering	141,111,223	159,699,942	534.15	629.59	12.46	20.43
System control and load dispatching	28,003,796	28,679,686	106.00	113.07	2.47	3.67
Communication system expenses	25,198,795	28,265,422	95.39	111.43	2.22	3.62
Compressor station labor and expenses	241,167,595	239,180,275	912.90	942.93	21.29	30.60
Gas for compressor station fuel	748,735,039	532,572,604	2,834.21	2,099.59	66.09	68.14
Other fuel and power for compressor stations	62,883,011	46,066,937	238.03	181.61	5.55	5.89
Mains	194,353,906	182,900,300	735.69	721.06	17.16	23.40
Measuring and regulating station expenses	70,192,395	62,517,614	265.70	246.47	6.20	8.00
Tranmission and compression of gas by others	1,472,161,326	1,267,266,607	5,572.61	4,996.01	129.95	162.13
Other transmission expenses	47,009,813	36,206,154	177.95	142.74	4.15	4.63
Rents	17,089,546	17,142,978	64.69	67.68	1.51	2.18
Total operation expenses	**3,047,906,445**	**2,597,098,519**	**11,537.32**	**10,238.66**	**269.05**	**332.27**
Maintenance expenses						
Supervision and engineering	38,445,814	35,545,963	145.57	140.14	3.39	4.55
Structures and improvements	26,126,983	30,638,479	98.90	120.79	2.31	3.92
Mains	90,064,519	93,717,959	340.92	369.47	7.95	11.99
Compressor station equipment	174,091,227	197,872,245	658.99	780.08	15.37	25.32
Measuring and regulating station equipment	13,635,384	12,539,338	51.61	49.43	1.20	1.60
Communication equipment	13,164,377	13,913,234	49.83	54.85	1.16	1.78
Other equipment	4,216,143	3,138,461	15.96	12.37	0.37	0.40
Total maintenance expenses	**359,754,497**	**387,365,679**	**1,361.78**	**1,527.13**	**31.76**	**49.56**
Total transmission expenses	**$3,407,660,942**	**$2,984,446,198**	**$12,899.10**	**$11,765.79**	**$300.80**	**$381.83**
Total miles of transmission pipeline	**264,178**	**253,656**				
Total natural gas sold, MMcf	**11,328,495**	**7,816,259**				

Source: Statistics of Interstate Natural Gas Pipeline Companies—1985 & 1986. U.S. Department of Energy. *U.S. interstate pipelines for calendar years 1985 & 1986. †45 of 137 companies filing Form 2's or 2A's with the U.S. Federal Energy Regulatory Commission; majors are companies whose combined gas sold for resale and gas transported or stored for a fee exceed 50 bcf (at 14.7 psi; 60°F.) in each of the 3 previous calendar years. §44 of 135 companies filing Form 2's and 2A's with the U.S. FERC for 1986, majors defined in previous note and in U.S. FERC Accounting and Reporting Requirements for Natural Gas Companies, para. 20-011, effective Feb. 2, 1985, beginning with the 1984 reporting year.

horsepower figures show no particular correlation with compressor-station size or location.

Operating costs

Once a natural-gas pipeline is laid and compressor stations constructed, operating costs become the next consideration.

As an aid in estimating this element, transmission expenses for interstate gas pipelines for 1985 and 1986, the latest period for which data are available, are given below.

The data are based on 264,178 miles of pipeline operated and 11,328,495 MMcf sold for 1985 majors and on 253,656 miles of pipeline operated and 7,816,259 MMcf of gas sold for 1986 majors. These represent the most recently available data.

● The highest component of transmission-operating expenses is transmission and compression of gas by others: for majors in 1985, this component was $5,573/mile or $130/MMcf; for majors in 1986, this component was $4,996/mile or $162/MMcf.

● The second highest cost component of transmission-operating expense is gas for compressor-station fuel: for majors in 1985, $2,834/mile or $66/MMcf; for majors in 1986, $2,100/mile or $68/MMcf.

● The highest cost component of maintenance expense covers compressor-station equipment: for majors in 1985, $659/mile or $15/MMcf; for majors in 1986, $780/mile or $25/MMcf.

● The second highest cost component for maintenance is the maintenance of pipelines: for 1985 majors, $341/mile or $8/MMcf; for 1986 majors, $369/mile or $12/MMcf.

● The highest component of transmission expenses for natural-gas pipelines is the operation expense: for 1985 majors, $11,537/mile or $269/MMcf; for 1986 majors, $10,239/mile or $332/MMcf.

● Unit-maintenance expenses for 1985 majors were $1,362/mile or $32/MMcf; for 1986 majors, $1,527/mile or $59/MMcf.

● During 1985, major natural-gas pipeline companies spent almost $3,048 million for operation expenses and nearly $360 million on maintenance. Total expenditures by 1986 majors for operation and maintenance expenses reached more than $3,407 million. Total operation expenses for majors in 1986 fell to slightly more than $2,597 million, a drop of almost $451 million (off a sharp 14.8%).

Total maintenance expenses in 1986, on the other hand, rose to more than $387 million, an increase of almost $28 million (7.7%). Total transmission expenses nonetheless showed a decline between 1985 and 1986. In the latter year, they reached only slightly more than $2,984 million, a drop from 1985 of a little more than $423 million (down a noticeable 12.4%).

● On a unit basis, the total 1985 transmission expense for majors was $12,899/mile or $300/mmcf; for 1986, $11,766/mile or $382/MMcf of gas sold.

PETROCHEMICAL COMPLEX (left) operated by BP at Grangemouth, U.K., is one of many in the world operating at full capacity as the petrochemical industry enjoys record margins and profits due to strong demand, low feedstock costs, and several years of rationalization (photo courtesy BP Chemicals). New products research is critical if the petrochemical industry is to continue enjoying strong financial performance into the 1990s. In the photo at right, a researcher checks the effects of a chemical on a fabric using an electron microscope at Du Pont Co.'s Jackson Laboratory at the company's huge Chambers Works in New Jersey, where more than 1,200 products have been invented or developed (photo courtesy Du Pont Co.).

Petrochemical sector bright spot of industry

THE PETROCHEMICAL SECTOR HAS BEEN THE bright spot of the petroleum industry in the late 1980s.

Petrochemicals have posted dazzling earnings since 1986, bolstering the profits picture for companies otherwise buffeted by sliding oil and gas prices.

Soaring demand brought about by strong economic growth, together with lower feedstock costs created by depressed oil and gas prices, have boosted petrochemical producers' margins to record highs.

Petrochemical companies and units that set record earnings in 1987 saw even that blistering performance outstripped in 1988. The pace of earnings growth seemed ready to cool off only slightly in 1989, forming a foundation for healthy growth and profitability in the 1990s.

At the same time, a flurry of expansion projects announced in 1987-88 threatens to undermine that profitability in the 1990s by recreating the surplus capacity that crippled margins in the late 1970s and early 1980s.

Many companies have responded to that threat by emphasizing debottlenecking and other existing capacity stretching measures instead of launching a grassroots expansion.

However, the continuing pressure of soaring demand, especially for U.S. products overseas, in tandem with plant utilization running at essentially 100%, has forced companies to launch the first grassroots expansions this decade.

Also looming on the horizon for petrochemical producers is the prospect of a possible recession. That would rein demand for consumer goods that has largely driven the explosive growth of commodity and intermediate petrochemicals.

With an eye to a possible shakeout among basic and intermediate petrochemicals should a recession occur, more companies are focusing on niche marketing and developing markets for specialty petrochemicals through new applications of existing products or developing new products.

That sort of diversification strategy is likely to help ease some of the sting of another boom and bust cycle in petrochemicals.

Meantime, petrochemical companies face new pitfalls as well as market opportunities in dealing with environmental concerns.

The controversy over waste disposal, especially in the U.S., is spurring companies to turn more to recycling of products and developing products that are biodegradable, largely to head off legislated mandates against use of their products.

Other environmental concerns may create new market

opportunities for petrochemical companies. The phaseout of fully halogenated chlorofluorocarbons, chief suspect in the destruction of the planet's ozone layer, has chemical companies scrambling to find replacements.

Finally, as a renowed sense of urgency over the degradation of the environment overtakes the world, petrochemical companies face their biggest challenge in the 1990s.

Handling, cleanup, and disposal of toxic materials and related controls to protect citizens and employees from exposure might well prove to be the fastest growing area of capital outlays for petrochemical companies in the 1990s. Even the fat margins of 1988 might disappear under the price tag of a Superfund site cleanup.

Making of a boom

The current boom in petrochemicals followed a depressed period in the late 1970s and early 1980s. That era was marked by sagging demand, soaring feedstock costs, and a surplus of capacity that slashed margins.

A turnaround in the world economy in the mid-1980s revived demand for petrochemicals. The surge in demand was especially strong for U.S. petrochemicals because of a decline in the value of the dollar.

At the same time, industry completed its rationalization of excess capacity in the U.S. and Europe, and plunging oil and gas prices trimmed the cost of feedstocks.

Together with a host of cost containment programs at most plants, those factors enabled petrochemical producers to enjoy the strongest margins ever. Consistently strong margins and record sales volumes have thus pumped up petrochemical profits to record levels in the late 1980s.

An Oil & Gas Journal Survey of 14 U.S. and Canadian companies with a primary focus on oil and gas operations showed an 84% jump in petrochemical profits from 1986 to 1987. A different group of companies surveyed by the Journal in 1988 showed an almost 90% jump in profitability from first half 1987 to first half 1988.

More sophisticated models of petrochemical profits showed comparable rises. Chem Systems. Inc. developed a profitability index based on a weighted average of cash cost margins for petrochemicals, ranging from olefins and aromatics through derivatives and including about 80% of all U.S. petrochemicals. With 1982 serving as an index of 100, U.S. petrochemical profitability by 1988 had virtually doubled in 2 years.

A major factor in improved margins has been the shutdown of about 25% of the world's petrochemical capacity in the 1980s. That has driven up utilization rates for most basic petrochemicals well in the 90s percentage range from an average 70% during the 1970s.

For example, capacity of polyvinyl chloride in western Europe exceeded demand by more than 1 million metric tons/year in 1981, according to Norsk Hydro AS. Resulting capacity closures, mergers, and acquisitions subsequently have trimmed the number of European PVC producers to 16 in 1987 from 28 in 1978.

Supply of some petrochemicals has become so tight that even the temporary loss of a major source of capacity can send shock waves through the industry, as was the case in May 1988 when an explosion and fire at Shell Oil Co.'s Norco, La., refining/petrochemical complex knocked out 2 billion lb/year of ethylene capacity, briefly causing shortages of ethylene and derivatives on the Gulf Coast.

The ethylene supply tightness led Exxon Chemical Co. and Chevron Chemical Co. to form a joint venture to import ethylene into the Houston area during 1988-89.

Outlook for supply/demand

If the economy doesn't suffer more than a mild downturn in 1989-90, then it is likely that strong demand forecasts for petrochemicals will continue.

A study by Chem Systems for the Society of Plastics Industry Inc. showed that demand for plastics resins in the U.S. will soar to 76 billion lb/year by 2000 from 48 billion lb/year in 1985. Much of that growth will come in packaging, building/construction, and convenience/leisure products, mainly in replacement of traditional materials. Greatest volume growth among polymers is likely to occur with polyethylene, PVC, polypropylene, and polystyrene, the study showed.

Industry forecasters see demand for ethylene climbing an average 3%/year in the next few years. Data Resources Inc. pegs demand growth for polypropylene at as much as 13-14%/year and for polystyrene at 5-6%/year. Chemical Market Associates estimates demand will grow for some of the key olefins at an average of about 10% from 1987 to 1990.

Data Resources predicts total demand for ethylene derivatives will rise to about 45 billion lb/year by 1995 from about 35 billion lb/year in 1987. DRI breaks out the demand growth, in ethylene equivalents, for the following derivatives in 1986-90: high density polyethylene 6.9%, low density polyethylene 4.6%, styrene 4.2%, ethylene dichloride 2.2%, vinyl chloride monomer 4%, PVC 2.9%, and other categories down by 10.2%. For 1991-95, DRI sees ethylene derivative growth for HDPE 4.7%, LDPE 3.2%, styrene 2.7%, EDC 3.5%, VCM 3.6%, PVC 2.3%, and others up 3.5%.

Phillips thinks that industry would have to build one world scale ethylene plant each year to handle projected demand growth. For HDPE, industry would have to add two such plants each year to handle expected demand growth.

Phillips remains cautious about adding capacity, although it was the first to announce a major grassroots project. Phillips 66 Co. will almost double ethylene capacity at its Sweeny, Tex., plant to about 4 billion lb/year by 1990. Approval of a 1.5 billion lb/year plant to be added at the Sweeny complex represents the first world scale ethylene plant to be built in the U.S. in the 1980s.

Phillips justified the new plant on the strength of firm sales commitments from outside customers to determine the baseload.

Late in 1988, the Houston Economic Development Council estimated that 79 new petrochemical projects would get under way in the area by 1992 at a combined cost of $3.1 billion. That covers projects that have been announced and are on construction schedules.

In addition, there are 70 projects costing an additional $2.5 billion that were expected to be announced early in 1989.

The recent and expected announcements represent a 61% boost from the previous year's estimates.

The Houston area petrochemical boom also reflects the increasingly global nature of the petrochemical business. Of the projects planned, companies from the Asia-Pacific region account for 9%, and European companies 21%.

In Europe, the growth is expected to be healthy as well. The European Council of Chemical Manufacturers Federations predicted chemical production would slip slightly to 3% in 1989 from 1988's 5.5% increase. It expected demand for chemicals to remain relatively firm in light of a slowing rate of growth in Europe's economy in 1989.

There are disputes within the industry as to the true level of accelerating demand and potential for new capacity.

G.L. McPike, Vice-pres. for polyethylene at Exxon Chemical International, contended in late 1986 that no new polyethylene plants would be needed in the industrial world until the 1990s.

He maintained that industry's forecasts of demand for polyethylene were too optimistic, namely because it has become a mature commodity with an annual growth rate roughly matching that of the world economy.

Another factor is capacity creep, which McPike estimated

Petrochemicals flow chart

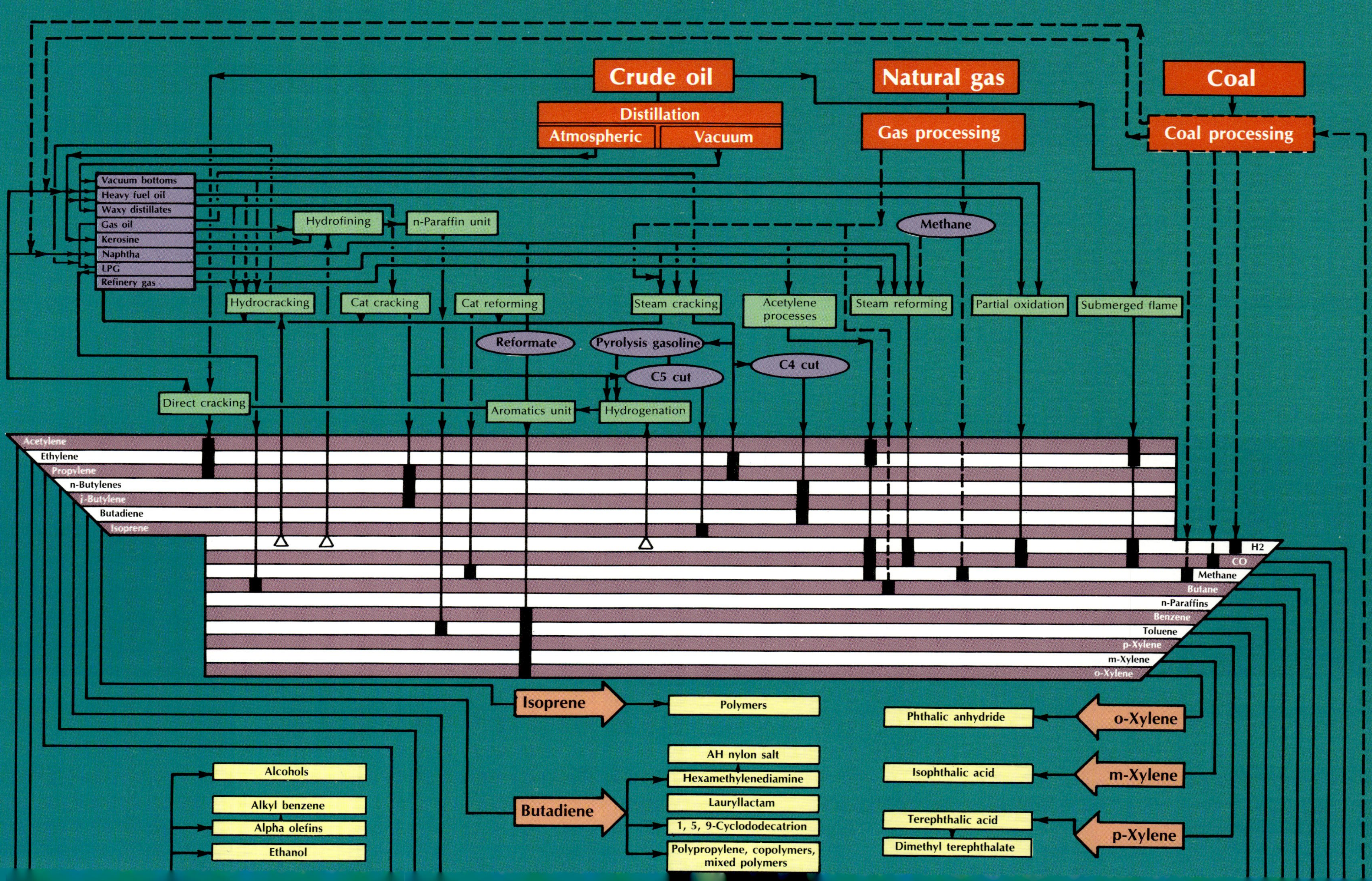

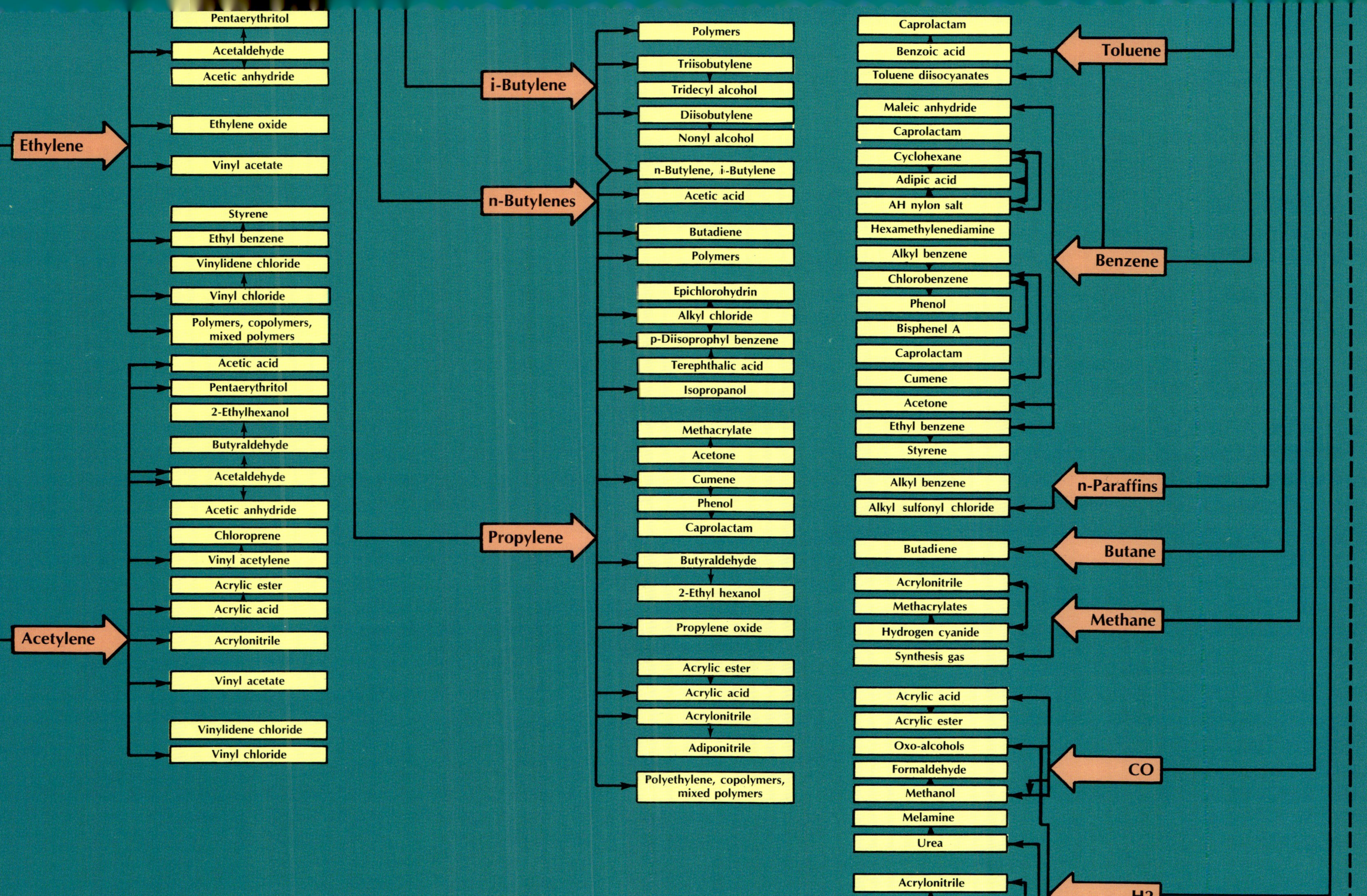

Ethylene
Pentaerythritol
Acetaldehyde
Acetic anhydride
Ethylene oxide
Vinyl acetate
Styrene
Ethyl benzene
Vinylidene chloride
Vinyl chloride
Polymers, copolymers, mixed polymers

Acetylene
Acetic acid
Pentaerythritol
2-Ethylhexanol
Butyraldehyde
Acetaldehyde
Acetic anhydride
Chloroprene
Vinyl acetylene
Acrylic ester
Acrylic acid
Acrylonitrile
Vinyl acetate
Vinylidene chloride
Vinyl chloride

i-Butylene
Polymers
Triisobutylene
Tridecyl alcohol
Diisobutylene
Nonyl alcohol

n-Butylenes
n-Butylene, i-Butylene
Acetic acid
Butadiene
Polymers

Propylene
Epichlorohydrin
Alkyl chloride
p-Diisoprophyl benzene
Terephthalic acid
Isopropanol
Methacrylate
Acetone
Cumene
Phenol
Caprolactam
Butyraldehyde
2-Ethyl hexanol
Propylene oxide
Acrylic ester
Acrylic acid
Acrylonitrile
Adiponitrile
Polyethylene, copolymers, mixed polymers

Toluene
Caprolactam
Benzoic acid
Toluene diisocyanates

Benzene
Maleic anhydride
Caprolactam
Cyclohexane
Adipic acid
AH nylon salt
Hexamethylenediamine
Alkyl benzene
Chlorobenzene
Phenol
Bisphenel A
Caprolactam
Cumene
Acetone
Ethyl benzene
Styrene

n-Paraffins
Alkyl benzene
Alkyl sulfonyl chloride

Butane
Butadiene

Methane
Acrylonitrile
Methacrylates
Hydrogen cyanide
Synthesis gas

CO
Acrylic acid
Acrylic ester
Oxo-alcohols
Formaldehyde
Methanol
Melamine
Urea

H2
Acrylonitrile
Methacrylate
Hydrogen cyanide
Ammonia

adds the equivalent of a world scale polyethylene plant each year through many incremental productivity improvements. For example, a joint venture of Exxon and the state owned Saudi Basic Petrochemicals Corp. (Sabic), Jubail Petrochemical Co., produced 33% more than its nameplate capacity of 260,000 metric tons/year in October 1986.

U.S. ethylene capacity could jump by as much as 29% to more than 47.2 billion lb/year, according to a survey by Bonner & Moore Market Consultants, Houston. Polyethylene is likely to follow suit, with capacity rising by about 28% during 1988-93.

The major concern for Booner & Moore is that the ethylene building boom could lead to surplus particularly if a recession triggers a decline in demand. That in turn could force ethylene utilization rates to the low 80s percentage range from the effectively 100% utilization rate in 1988-89 that sent ethylene prices and petrochemical margins soaring.

With that kind of surplus capacity, the petrochemical industry could face another cyclical downturn like that of the late 1970s and early 1980s. Analysts in early 1989 already were forecasting a decline in market conditions for the early 1990s.

New markets

Even if the boom for commodity and intermediate petrochemicals does begin to wane in the years ahead, there remains significant potential for explosive growth through technology advances that offer new product adaptations.

Developing new petrochemical products that compete more effectively with other materials or creating whole new markets through product introductions will be the focus of growing outlays for research and development during the 1990s.

Companies with an eye on the boom and bust cycles of the 1970s and 1980s are funneling millions of dollars into R&D facilities and market studies for new products and new applications for existing products.

Hence the industry will see a good deal more focus on niche market development. There is a compelling reason, according to Amoco Research Center's director of fine acids research Alan G. Bernis: "Fine acids are low volume, specialty products. Because they have unique performance characteristics, they command a higher price than do commodity chemicals."

Du Pont said that record growth of plastics next decade will be paced by specialty polymers, thermoplastics, and other high performance materials. Du Pont pegs demand growth rates of 10-12%/year for specialty polymers and the other categories at a rate double that of the world's gross national product.

Dow U.S.A. and Exxon Chemical Co. also are bullish on high performance plastics. The two expect world demand for styrenic thermoplastic elastomers to double to 1 billion lb/year by 2000. Dow and Exxon in 1988 formed a 50-50 joint venture to supply and market styrenic TPEs. The venture is constructing its first plant at Plaquemine, La., which is to go on stream in 1990 and produce 70 million lb/year of styrenic TPEs.

The Society of the Plastics Industry's Composites Institute estimated growth for composites at about 6% in 1988, outstripping expectations of lower growth in the wake of the Wall Street stock market crash and reaching deliveries of almost 2.7 billion lb.

Composites are expected to make especially big inroads in aerospace, garnering as much as a 60% share of new aircraft in the 1990s. Stronger per given weight than aluminum or titanium, composites offer weight savings of 10-60%, thereby increasing range, payload, speed, and maneuverability. They also are less prone to fatigue cracks, corrosion, and detection by radar.

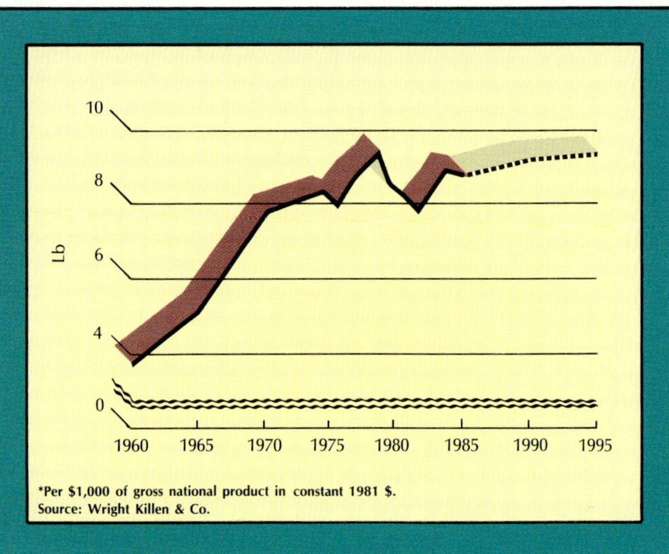

Composites also will build market share in industrial, computer, telecommunications, and chemical processing applications. That kind of potential is why Phillips Petroleum Co. in 1988 opened an advanced composites center for polyarylene sulfide thermoplastics R&D.

Getting an early jump on the tomorrow's chemical markets can require a wholesale revamp of management structure as well as pouring money into R&D and market studies. Amoco Chemical Co. last year restructured into two major business groups, a chemical/specialty products group and a polymer products group. Each group will have its own R&D, planning, business development, marketing, sales, and manufacturing.

Amoco Chemical Pres. Robert D. Cadieux said that his company has "a commitment that at least a fourth of our business a decade from now will be in products we don't make today.

"One of the ways we can meet that high growth goal is through an aggressive, responsive, management structure. We're acting today to make certain we're still an industry leader in the 1990s."

Although some of these new specialty markets will feature growth rates equal to 3-4 times the growth of U.S. GNP, and their margins may be much higher than those for commodity chemicals, the new product lines won't have much of an stimulating effect on demand for the building block petrochemicals that precede them, according to the American Institute of Chemical Engineers.

"For example, one of the larger performance product fields, engineering resins, will probably grow 10-15% annually over the next 10 years. Yet the total tonnage of these products will amount to only 5-10% of annual ethylene volumes, and the actual upstream consumption of basic petrochemicals resulting from growth of these products will have only minor impact on overall demand."

Environmental challenges

Petrochemical companies see perhaps their biggest challenges ahead in the unpredictable arena of environmental issues. There are pitfalls and market opportunities for companies coping with a renewed environmental fervor in the 1990s.

Companies that produce chlorofluorocarbons must contend with a new worldwide consensus that their product is contributing to two major environmental hazards: the degradation of the ozone layer and global warming due to the

U.S ethylene supply/demand

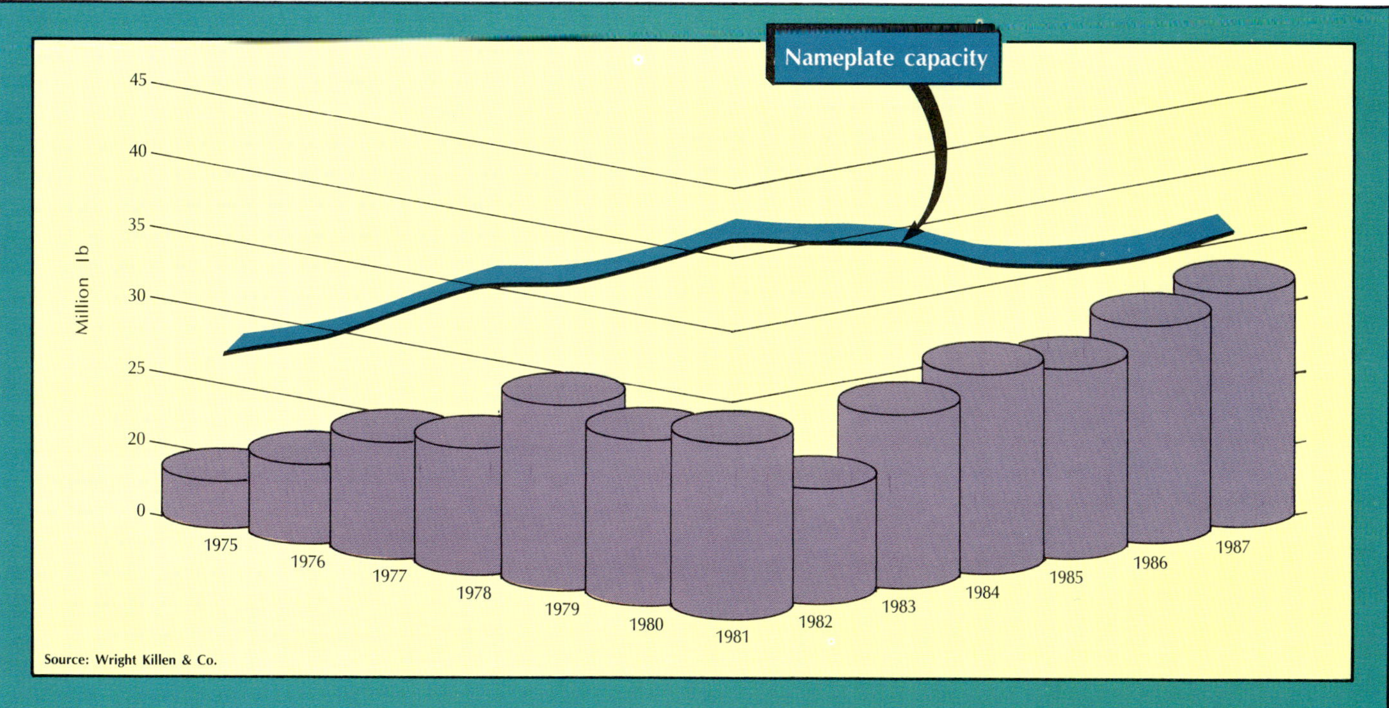

greenhouse effect. They already are scrambling to develop harmless alternatives to avoid loss of market share to competing products.

Under the Montreal Protocol of 1987, the major industrial nations agreed to limit production and emission of CFCs.

ICI will build two $50 million plants to produce a substitute for CFCs, AFC 134a. The first, to be built in Runcorn, U.K., is scheduled for completion in 1991. The second is to be built in the U.S. in 1992.

Du Pont Co is spending more than $25 million for the world's first commercial scale plant to produce HFC 134a. It is intended to replace CFD 12, the working fluid in commercial and home refrigeration and automobile air conditioning, applications which account for about 40% of the U.S. market for CFCs and about 30% of the world market.

Du Pont expects CFCs will be phased out of several major market segments in the early to mid-1990s but complete phaseout could take until 2000.

Burgeoning public concern about dwindling landfill space and nonbiodegradable plastics refuse disposal triggered efforts by competing products to promote themselves as more environmentally benign than plastics. Some municipalities have been approved measures to ban the use of foam and plastic containers. Ethanol producers have pushed a biodegradable plastic made from corn ethanol.

Petrochemical companies have been quick to respond to the challenge. Mobil Chemical Co. established a toll free line for consumers concerned about plastic bags and the environment, notably degradability, recycling, incineration, and landfills.

"The standard plastic grocery bag is degradable in sunlight as well as recyclable, is inert in landfills and will not contaminate groundwater, can be incinerated safety without toxics, and is approved by the Food and Drug Administration as sanitary," Mobil said.

Mobil said it also will supply bags made with degradable additives as a premium product if there is sufficient demand.

Mobil and Genpak Corp. in January 1989 started up in Leominster, Mass., the first plant in the U.S. to recycle used polystyrene foam products. Capacity is about 3 million lb/year, expandable to 6 million lb/year.

Because of legislative efforts to ban polystyrene foam products, Amoco has launched a series of foam recycling projects, although all polystyrene foam food service products used in the U.S. represent less than 0.25% of all municipal solid waste, the company said.

That kind of attitude reflects a growing push by petrochemical companies to improve their public image on environmental issues.

Such attention will be critical in the 1990s as many industry observers expect environmental regulation and enforcement to toughen in the years ahead.

"The most imposing threat to the viability of chemical companies and perhaps to our very existence as an industry is the deep concern of the public over the safety of our products and operations," said Union Carbide Chairman Robert D. Kennedy.

Kennedy stumped for a chemical industry initiative on health, safety, and environmental affairs that would entail developing a set of guiding principles, a set of management practice codes, a public advisory panel to critique codes and help identify public concerns, a self assessment process requiring companies to chart their practices in implementing the codes, and executive leadership programs to give member companies the chance to raise new questions and share their experience and ideas.

Amoco's Cadieux said the chemical industry must assume an aggressive role in setting environmental priorities, starting with those issues that present the greatest risk.

Cadieux said the industry should encourage and support the shift from remedial regulation to preventive regulation, such as its current efforts at voluntary waste minimization programs.

"Taking positive action in addressing these challenges and seeking solutions represent a true commitment, and it will be viewed as such," he said.

Without that commitment, the public is likely to turn to plastics substitutes in the 1990s, and all the emphasis on utilization levels, the economy's direction, and feedstock costs won't mean a thing to petrochemical companies, because the boom in consumer demand that has revived the fortunes of the petrochemical industry will have withered. IPE

Refinery throughputs

	1977	1978	1979	1980	1981	1982	1983	1984	1985	1986	1987
						(Thousand barrels daily)					
U.S.	14,625	14,780	14,465	13,520	12,470	11,775	11,685	12,045	12,005	12,715	12,855
Canada	1,805	1,820	1,960	2,085	1,945	1,650	1,545	1,570	1,495	1,475	1,605
Latin America	5,715	5,975	6,220	6,460	6,575	6,080	5,825	5,715	5,450	5,480	5,530
Western Europe	12,775	12,855	13,665	12,330	10,970	10,075	9,710	9,790	9,490	10,060	9,795
Middle East	2,240	2,535	2,365	2,445	2,205	2,245	2,515	2,665	2,795	2,950	2,970
Africa	1,165	1,280	1,355	1,460	1,525	1,735	1,920	1,975	2,015	2,095	2,170
Japan	4,285	4,240	4,340	4,015	3,630	3,360	3,255	3,355	3,120	2,990	2,910
Southeast Asia	2,030	2,330	2,450	2,370	2,480	2,335	2,380	2,455	2,650	2,765	2,750
South Asia	605	645	675	705	795	840	880	885	1,020	1,120	1,145
Australasia	680	690	700	640	655	625	605	615	580	590	570
Total Non-Communist	**45,925**	**47,150**	**48,195**	**46,030**	**43,250**	**40,720**	**40,320**	**41,070**	**40,620**	**42,240**	**42,300**
Communist	11,930	12,875	13,230	13,320	13,420	13,575	13,675	13,885	14,165	14,500	14,700
Total World	**57,855**	**60,025**	**61,425**	**59,350**	**56,670**	**54,295**	**53,995**	**54,955**	**54,785**	**56,740**	**57,000**

Source: British Petroleum

Historical reserve data (billion bbl oil)

	1960	1965	1970	1975	1980	1985	1988	1989
Asia-Pacific								
Australia	—	—	2.0	1.7	2.4	1.4	1.7	1.7
Brunei	—	—	1.0	2.0	1.7	1.4	1.4	1.4
India	—	—	—	—	2.5	3.5	4.3	6.4
Indonesia	9.5	9.5	10.0	14.0	9.5	8.6	8.4	8.2
Malaysia	—	—	—	2.5	3.0	3.0	2.9	2.9
Europe								
Norway	—	—	1.0	7.0	5.5	8.3	14.8	10.4
United Kingdom	—	—	1.0	16.0	14.8	13.6	5.2	5.2
Middle East								
Abu Dhabi	—	10.0	11.8	29.5	29.0	30.5	92.2	92.2
Divided Zone	6.0	12.4	25.7	6.4	6.0	5.4	5.2	5.2
Dubai	—	—	—	1.0	1.4	1.4	4.0	4.0
Iran	35.0	40.0	70.0	64.5	57.5	48.5	92.8	92.8
Iraq	27.0	25.0	32.0	34.3	30.0	44.5	100.0	100.0
Kuwait	62.0	62.0	67.1	68.0	64.9	90.0	91.9	91.9
Oman	—	—	—	5.0	2.3	3.5	4.0	4.0
Qatar	2.5	3.0	3.5	5.8	3.6	3.4	3.2	3.1
Saudi Arabia	50.0	60.0	128.5	148.6	165.0	169.0	166.9	169.9
Syria	1.0	1.0	1.0	1.0	1.3	1.4	1.7	1.7
Africa								
Algeria	5.2	7.4	7.0	7.4	8.2	9.0	8.5	8.4
Angola-Cabinda	—	—	—	1.3	1.2	1.8	1.1	2.0
Egypt	—	2.0	4.5	3.9	2.9	3.2	4.3	4.3
Gabon	—	—	—	—	—	—	0.6	0.7
Libya	2.0	10.0	29.2	26.0	23.0	21.1	21.0	22.0
Nigeria	—	3.0	9.3	20.2	16.7	16.7	16.0	16.0
Tunisia	—	—	—	1.0	1.7	1.5	1.8	1.8
Western Hemisphere								
Argentina	2.2	1.9	4.5	2.5	2.5	2.3	2.3	2.3
Brazil	—	—	—	—	1.3	1.9	2.3	2.5
Canada	—	7.5	10.7	7.1	6.4	7.1	6.8	6.8
Colombia	—	1.5	1.7	1.7	1.0	0.624	1.6	2.0
Ecuador	—	—	—	2.5	1.1	1.4	1.6	1.4
Mexico	2.3	2.5	3.2	9.5	44.0	48.6	48.6	54.1
Venezuela	18.5	17.3	14.0	17.7	17.9	25.8	56.3	58.0
U.S.A.	33.5	35.4	37.0	33.0	26.4	27.3	25.3	26.5
Total Non-Communist	**268.5**	**319.6**	**511.4**	**555.7**	**562.2**	**614.6**	**808.1**	**823.6**
U.S.S.R.	33.5	33.5	77.0	80.4	63.0	63.0	59.0	58.5
China	—	—	—	—	—	19.1	18.4	23.6
Other	—	—	—	—	—	—	1.8	1.7
Total World	**302.0**	**353.1**	**611.4**	**658.7**	**648.5**	**698.7**	**887.3**	**907.4**

354

PHILLIPS PETROLEUM CO.'s ethylene complex at Sweeny, Tex.

World's refiners enjoy good demand, profits

THE WORLDWIDE REFINING INDUSTRY SAW A RE-
turn to healthy margins and profitability during 1988. Crude
oil prices remained below $18/bbl, and product demand
firmed.

Most world regions have continued to restructure their
refining industries, keeping supply and demand in better
balance than during previous years.

Improved economic conditions worldwide, coupled with
relatively low product prices, have maintained or increased
demand for refined products. Demand for higher-quality
products is also up.

In the U.S., the successful accomplishment of lead phase-
down provided refiners with improved octane producing
capability. This capability, combined with lower than world-
average gasoline prices, has triggered another push toward
increased gasoline octane.

Environmental concerns, however, remain the industry's
toughest challenge for the near future. Air-quality concerns
in both Europe and the U.S. will likely result in heavier
regulation of processing operations and the products they
produce.

Much of the air-quality improvement efforts in the U.S. will
be directed at gasoline. Already, two regions of the country
have passed legislation requiring oxygenates to be part of
the gasoline blend to reduce carbon monoxide (CO) emissions from motor vehicles.

Several states have introduced legislation requiring a
substantial reduction in gasoline volatility during summer
months. Some of these are more stringent than those
proposed at the federal level.

Efforts to legislate the use of alternatives to gasoline have
also received renewed attention by both the U.S. and state
governments.

In Europe, efforts continue to lower the lead content of
gasoline. But because rules requiring catalytic converters on
auto exhaust systems have not yet been strictly imposed,

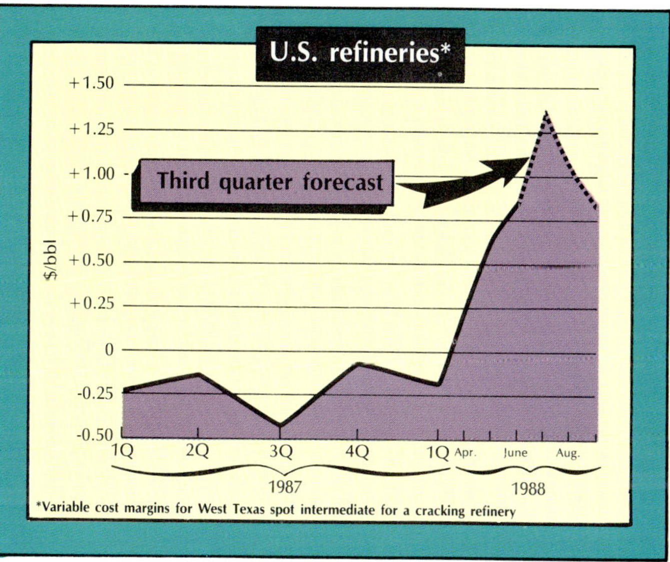

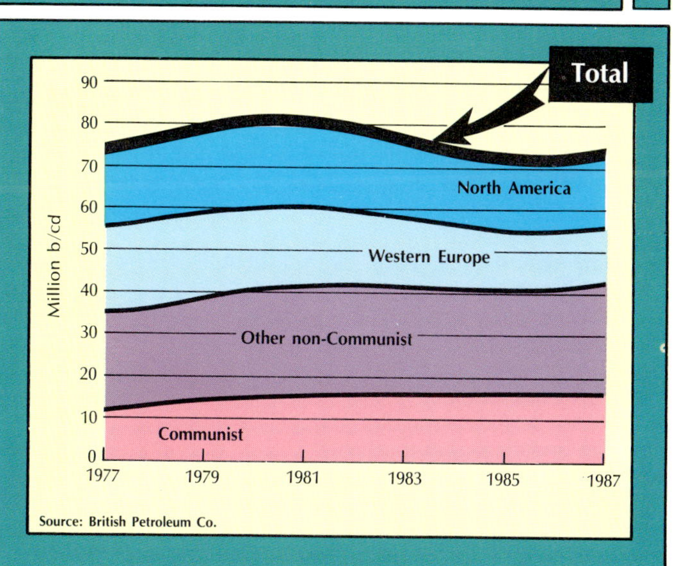

Demand and stocks for distillate and resid in the U.S. 

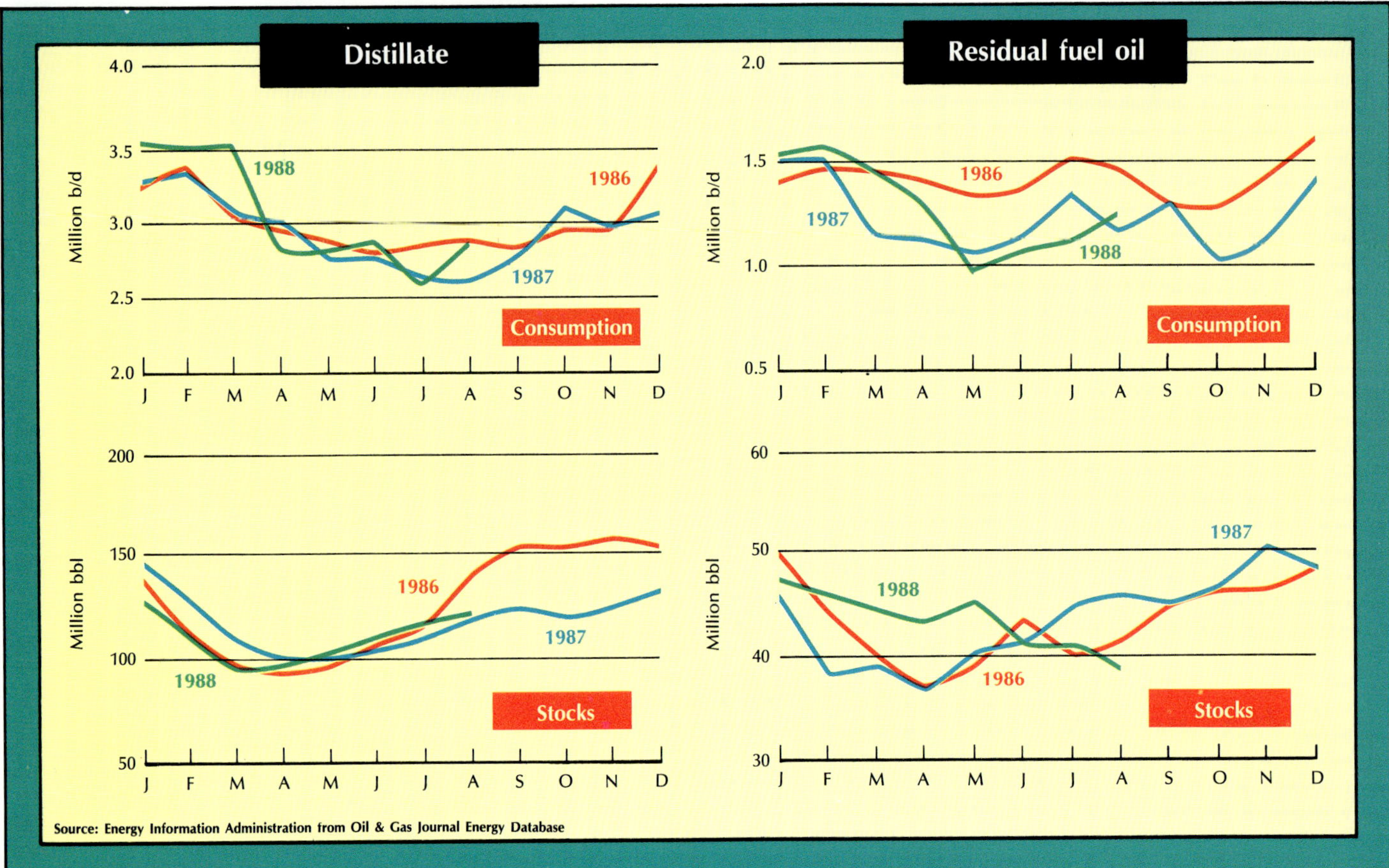

the introduction of unleaded gasoline is occurring at an uneven rate throughout the region.

The requirement for reduced sulfur content refined products could tighten the supply of light, sweet crude oils. The result will be growing utilization of distillation and conversion capacity and healthy economics for complex refiners later in this century.

During 1988, OPEC countries expanded further into the downstream refining operations of countries with good product markets. By gaining partial ownership of refining operations, these oil producers can partially stabilize the market for their crude supplies.

This expansion, meanwhile, has brought more competitive pressure on U.S. refiners. In the context of a global economy, U.S. refiners can no longer base investment decisions solely on the domestic market.

Although refined product supply and demand is in the best balance worldwide than it has been in recent years, the balance is tenuous at best. Construction projects, therefore, have been limited, for the most part, to revamps and improvements of existing facilities, not construction of new refineries.

Improved health of the industry

This improved health of the worldwide refining industry is expected to continue for the next few years, barring any sharp increases in crude oil prices and downturn in world economic conditions.

Strong demand and firm prices for petroleum products helped boost refining margins in the U.S. during 1988 (see Fig. 1).

In the first half of 1988, U.S. products demand rose 2.8% from first half 1987, while the average crude oil price in the first half was down about $2.30/bbl from the same period a

Europe's capacity, oil demand* 

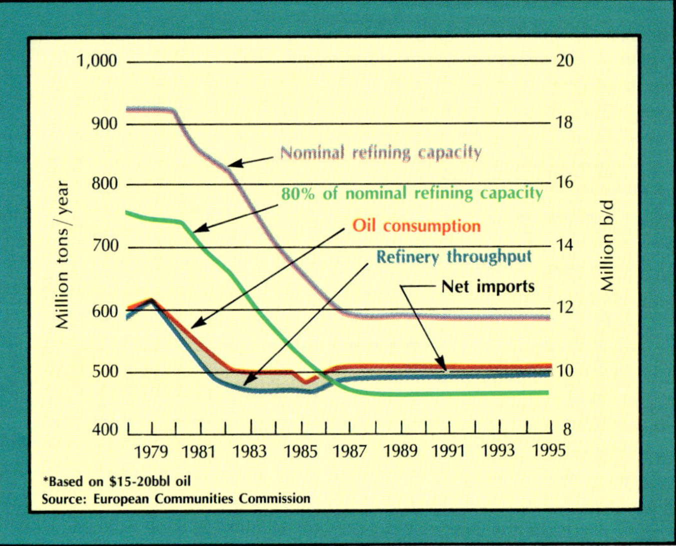

year ago. Sales of higher margin, high-octane gasoline also increased, helping buoy margins.

Refiners' cost-cutting efforts that meant survival during recent years, were a big key to improved profits last year.

Marketing is growing in importance as a downstream profit center, especially as companies diversify deeper into retail operations.

The outlook is for downstream profits to continue strong into 1989, with some leveling as companies can't expand further on several quarters of impressive, sometimes record

A sample of U.S. companies' downstream profits — Table 1

	Refining/marketing				Petrochemicals			
	2nd quarter		1st half		2nd quarter		1st half	
	1988	1987	1988	1987	1988	1987	1988	1987
	Million $							
Amoco	206	55	312	50	189	104	358	192
ARCO*	72	45	159	70	226	101	411	186
Chevron	123	24	322	8	76	67	185	119
Citgo	34	17	76	71	—	—	—	—
Coastal	21.3	26.7	29.9	39.8	—	—	—	—
Crown Central	7.2	(3)	7.6	(18.3)	—	—	—	—
Diamond Shamrock R&M	1.9	2.2	7.1	4.9	—	—	—	—
Exxon	281	19	712	55	314	225	622	411
Kerr-McGee	1.8	4.6	16	10	18.4	7.7	27.7	16.3
Mobil	260	84	477	162	160	75	294	118
Occidental†	—	—	—	—	221	51	339	212
Pacific Resources	8	2	24.2	4.7	—	—	—	—
Phillips	98	48	172	78	237	110	430	205
Shell	103	56	233	64	133	101	275	141
Sun	79	21	121	16	—	—	—	—
Texaco	215	18	368	24	44	19	91	27
Tosco	36.8	10.6	50.2	21.9	—	—	—	—
Unocal	28	19	63	43	17	19	27	29
Valero	5.8	(7.3)	10.5	(13.3)	—	—	—	—
Vista§	—	—	—	—	27.8	15.7	87.7	35.5
Total	**1,581.8**	**441.8**	**3,163.2**	**691**	**1,663.2**	**895.4**	**3,147.4**	**1,691.8**

* Petrochemical earnings represent ARCO's 80.4% interest in ARCO Chemical Co. and Lyondell Petrochemical Co., which entails a large refinery and petrochemical complex in Houston. †Includes earnings from Cain Chemical Co. as of May 2. §Fiscal 1988 third quarter and 9 months.

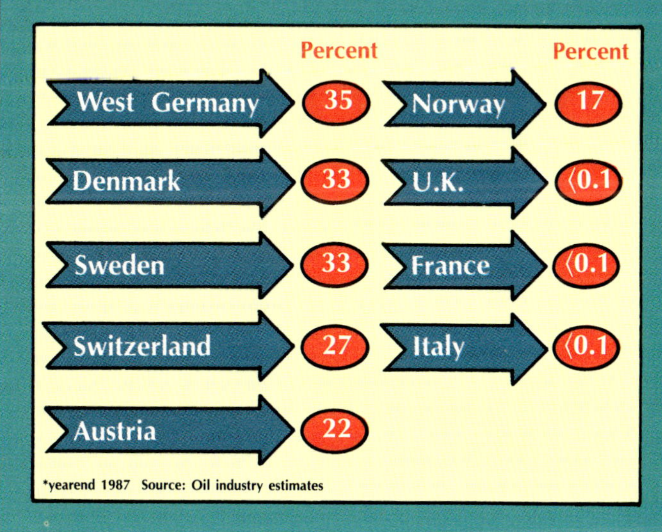

earnings.

Refining/marketing profits were up for almost all U.S. companies during 1988 (see table 1). Chevron Corp., for instance, recorded the highest margins in 2 years because of lower crude costs and reductions in processing and marketing costs. Chevron's results were limited, however, by $48 million of additions to reserves for environmental programs.

Chevron implemented a wide range of measures that involve recycling waste heat to process units, such as a boiler feedwater circulation system that collects heat formerly rejected by cooling towers. Plant improvements, such as one new refinery cogeneration unit that saves about $30 million/year, or about $0.35/bbl at the refinery, have contrib-uted to Chevron's improved profitability.

Improved product demand, especially for high-octane gasoline, contributed to improved profits for Amoco Oil Co. Amoco's first half product sales volume rose to 1.04 million b/d from 978,000 b/d during 1987. Its first half product sales was up 3% from 1987 figures.

Mobil Oil Corp. saw a 2% increase in motor gasoline sales, and a 17% rise in lubricant sales.

U.S. refiners have also increased emphasis on lubricant markets. Exxon Corp., for instance, introduced a new marine lubricant to replace older marine lube types. It also developed new engine oil additives intended to meet new industry-wide lube oil quality specifications introduced in the U.S. and Europe.

Retail lube sales was also an increasing market in 1988. Pennzoil Products Co. and Jiffy Lube International signed a long-term contract in 1988 to make Pennzoil the oil of choice at Jiffy Lube's outlets.

Refiners in the European Economic Community saw a substantial recovery in margins during 1988. Meager first-quarter margins rose considerably during the second and third quarters partially because of strong gasoline demand in the U.S.

Margins were helped greatly by the return to market related pricing by producer/exporters in their efforts to expand crude oil sales. Most Middle Eastern crudes now trade on a realization basis with North Sea Brent crude prices setting the pace.

European margins increased from about $1/bbl during the first quarter to about $2.15/bbl in the second quarter, and $2.25/bbl during the third quarter.

North American refining capacity grew by about 1.2% to about 17.7 million b/d. Throughput for the region rose 1.1% to about 12.9 million b/d in the U.S. and rose 8.8% to 1.6 million b.d in Canada (see table 2 and Fig. 2).

Western Europe still leads surplus

Western Europe still leads the world in surplus refining capacity, despite a capacity decline in 1987. Capacity in the region stood at 14.01 million b/d in 1987, down 2.2% from 1986.

Refinery throughput dropped to about 9.8 million b/d, down 2.6%. Capacity as of Jan. 1, 1988 stood at 11.84 million b/d for the 12 nations making up the European Economic Community.

Japan's throughput fell 2.7% to 2.91 million b/d, and capacity fell 4.3% to about 4.5 million b/d. Australian throughput fell 3.4% to 570,000 b/d, and capacity there dropped 1.9% to 730,000 b/d.

Middle East capacity rose 6.9% to 4.12 million b/d, while throughput was up about 0.7% to 2.97 million b/d. Capacity increases in Saudi Arabia were 22% to 1.375 million b/d.

U.S. demand for distillate, residual fuel oil, and gas is expected to be up for the 1988-1989 heating season because of continued economic growth and low product prices (see Fig. 3). The supply of liquid fuels to meet the winter surge in demand now comes from increased refining output, imports, and stocks. In the past, refiners met demand mainly from inventories.

Demand for distillates is expected to be up about 3.3%

Regional breakout of world refinery throughput — Table 2

	1977	1978	1979	1980	1981	1982	1983	1984	1985	1986	1987	% change 1986–87	1987 % share of total
						1,000 b/cd							
U.S.	14,625	14,780	14,465	13,520	12,470	11,775	11,685	12,045	12,005	12,715	12,885	−1.1	22.6
Canada	1,805	1,820	1,960	2,085	1,945	1,650	1,545	1,570	1,495	1,475	1,605	+8.8	2.8
Latin America	5,715	5,975	6,220	6,460	6,575	6,080	5,825	5,715	5,450	5,480	5,530	+0.9	9.7
Western Europe	12,775	12,855	13,665	12,330	10,970	10,075	9,710	9,790	9,490	10,060	9,795	−2.6	17.2
Middle East	2,240	2,535	2,365	2,445	2,205	2,245	2,515	2,665	2,795	2,950	2,970	+0.7	5.2
Africa	1,165	1,280	1,355	1,460	1,525	1,735	1,920	1,975	2,015	2,095	2,170	+3.6	3.8
Japan	4,285	4,240	4,340	4,015	3,630	3,360	3,255	3,355	3,120	2,990	2,910	−2.7	5.1
Southeast Asia	2,030	2,330	2,450	2,370	2,480	2,335	2,380	2,455	2,650	2,765	2,750	−0.5	4.8
South Asia	605	645	675	705	795	840	880	885	1,020	1,120	1,145	+2.2	2.0
Australasia	680	690	700	640	655	625	605	615	580	590	570	−3.4	1.0
Total non-Communist	45,925	47,150	48,195	46,030	43,250	40,720	40,320	41,070	40,620	42,240	42,300	+0.1	74.2
Communist	11,930	12,875	13,230	13,230	13,420	13,575	13,675	13,885	14,165	14,500	14,700	+1.4	25.8
Total world	57,855	60,025	61,425	59,350	56,670	54,295	53,995	54,955	54,785	57,740	57,700	+0.5	100.00

Source: British Petroleum Co. plc

during the 1988-89 winter at 3.44 million b/d. Low prices and good economic growth will strengthen demand and discourage fuel conservation.

Demand for residual fuel oil will be above the 1987-88 winter by about 2% to 111.415 million b/d. Table 3 shows the near-term forecast for winter fuels in the U.S.

Despite improved conditions in Europe, some refineries are expected to close during the next 6 years, but the closure rate will be much slower than during the period 1980-1987.

European refiners plan to shut down an additional 280,000 b/d of surplus primary distillation capacity by 1995. The plant closure rate slowed since 1986. About 480,000 b/d has been shut down since then. Only about 60,000 b/d was shut down during 1987.

Demand on refineries in the region is expected to be at about 10.4 million b/d for $15-20/bbl crude oil or 9.7 million b/d at $25-30/bbl crude. Closure plans to cut capacity in the region to 11.5 million b/d would still provide ample capacity for these demand scenarios (see Fig. 4).

Air-quality issues pose perhaps the biggest concern for U.S. refiners ability to compete in a global economy. U.S. refiners could bear a significant share of a burden estimated at $35-50 billion/year incurred by a broad expansion of the U.S. Clean Air Act.

Included in the expansion are stiffer rules on the volatility of gasoline, on the sulfur aromatics content of diesel fuel, and on the requirement to add oxygenates to the gasoline blend to reduce exhaust emission of carbon monoxide.

Gasoline will be the prime target for U.S. regulatory efforts to cut air pollution in the 1990s. The first main push will be to reduce gasoline volatility.

U.S. Environmental Protection Agency (EPA) rules on gasoline volatility have been expected for 2 years, and it is expected that the rules will be issued in early 1989. The rules will limit the Reid vapor pressure (Rvp) of gasoline blends during the summer months to 9 psi. The proposed EPA timetable is for refiners to be able to meet the lower Rvp by 1992.

Part of the delay in issuing volatility limits is that EPA wants to issue rules on vehicle onboard refueling emission control systems simultaneously with gasoline Rvp limits. That is complicated further because EPA must prepare a notice of proposed rulemaking to allow further comment on onboard control safety issues.

Additionally, nine northeastern states will limit gasoline to 9 psi Rvp during the summer of 1989. New rules that require the addition of oxygenates to reduce CO emissions were put into effect in the Denver, Phoenix, Tucson, and Albuquerque areas.

GM's variable fuel vehicle

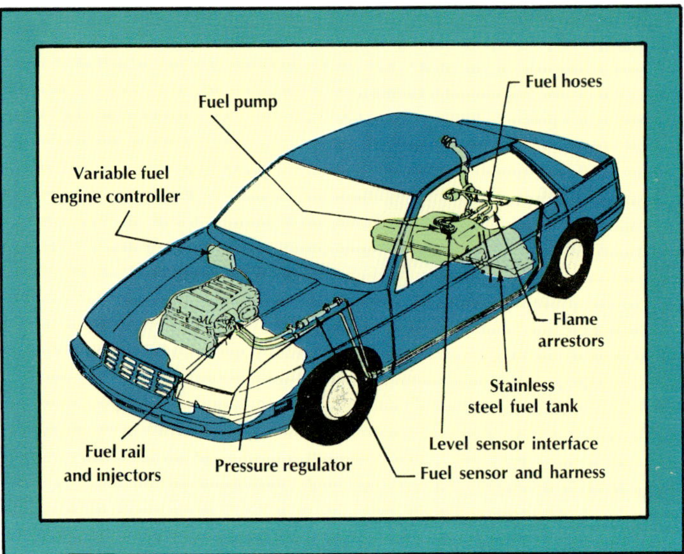

Refiners claim the industry couldn't immediately meet the demand for a lower volatility, high-octane gasoline that would come with Rvp limits.

Gasoline imports figure heavily in meeting current U.S. demand, but European refiners and traders contend they don't have the capability to produce a 9 psi, 92 octane gasoline for the U.S. market.

Alternate fuels

Tougher rules on U.S. fuels have spurred additional activity in alternate fuels. Ethanol interests are pushing the use of both ethanol and its ether, ethyl-tertiary-butyl-ether (ETBE) as CO reducers and octane increasers.

ETBE would be a substitute for MTBE, already in strong demand as both an octane booster and CO reducer. Of course, its success depends heavily on tax exemptions for ethanol.

Increased use of oxygenates for CO emission reduction may be short lived, however, with the introduction of new, adaptive catalytic converters. The new design sharply reduces CO emissions, independent of the type of gasoline burned.

In the Northeastern U.S., state rules to limit gasoline volatility came under fire from refining companies in the

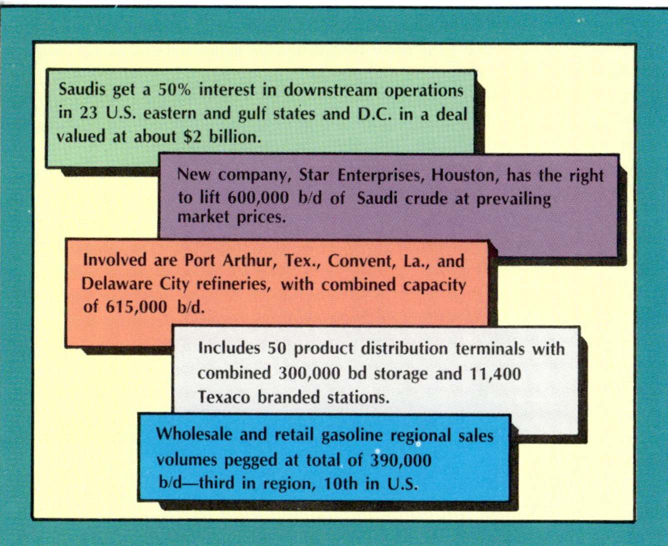

region. Of the nine states to issue proposed rules, New Jersey's rules were contested by Texaco Inc. as being too costly to consumers.

Texaco estimated that consumers would pay an additional $55 million/year without measurable benefits. Texaco made similar claims about a proposed rule in the state of New York.

New Jersey's rule would cut the Rvp of gasoline sold in the state to 9 psi from 11.5 psi during May 1-Sept. 15 each year, beginning in 1989. The proposal would reduce gasoline volume by about 5%, according to Texaco, resulting in the need to replace about 90 million gal/year.

That would increase the cost by as much as $0.03/gal because of lower gasoline yields and higher refinery costs.

Proposed rules in Pennsylvania were contested by Sun Refining & Marketing Co.

Pennsylvania's rules would limit Rvp to 9 psi for summer gasoline.

Sun contends that the rule would cut regional gasoline supply by about 5% and hike prices by $0.06-$0.09/gal.

Environmental changes in Europe are being made over a wide front. After Oct. 1, 1989, all European Community states must have unleaded gasoline available

Unleaded has made a major penetration in many Northwest European markets, but still has not taken off in Britain, France, or Italy (see Fig. 5). In addition, France and Italy have not yet reduced lead content of leaded gasoline from 0.4 g/l to the 0.15g/l required level.

Unleaded gasoline sold at the end of 1987 amounted to about 35% of all sold in West Germany, the leader in cleaning up air pollution in Europe. Most service stations in West Germany sell two grades of unleaded gasoline.

Denmark's volume of unleaded gasoline has jumped to 33% of sales in 1987, compared to 10% the preceding year.

In contrast, sales in Britain, France, and Italy were less than 0.1% of gasoline deliveries. Stations are sparse in these countries, and usually only on the most heavily traveled tourist routes.

Sulfur content limits of automotive and heating gas oil will be limited to 0.3 wt% after Jan.1, 1989, and individual EC members are empowered to impose a lower level of 0.2 wt%.

The heavy fuel oil market is divided into high sulfur product (3.5-4 wt% sulfur) and low sulfur (1 wt% sulfur). Individual countries have the right to impose a requirement for low sulfur fuel oil.

Companies have complained that meeting the low sulfur requirements of parts of West Germany, Switzerland, and Sweden places considerable strain on refiners. For instance, Rafinerie de Cressier SA in Switzerland has to run its 60,000 b/d crude unit on a single grade of low sulfur crude to meet local regulations, removing its ability to seek more competitively priced crudes.

Another octane race is possible. Process installations to meet the octane requirements of U.S. gasoline in

Outlook for U.S. winter fuels*   Table 3

	1985–86	1986–87	1987–88	Percent change 1988–87	Forecast 1988–89	Percent change 1989–99
		1,000 b/d			1,000 b/d	
MIDDLE DISTILLATE:						
Demand						
Domestic	3,136	3,148	3,329	+5.7	3,440	+3.3
Exports	112	79	78	−1.3	75	−3.8
Total	**3,248**	**3,227**	**3,407**	**+5.6**	**3,515**	**+3.2**
Supply						
Refined	2,879	2,718	2,910	+7.1	3,010	+3.4
Imported	257	268	288	+7.5	305	+5.9
From stocks	112	241	209	−13.3	200	−4.3
Total	**3,248**	**3,227**	**3,407**	**+5.6**	**3,515**	**+3.2**
Stocks (million bbl)	99	109	89	−18.3	98	+10.1
RESIDUAL FUEL OIL						
Demand						
Domestic	1,356	1,424	1,387	−2.6	1,415	+2.0
Exports	203	180	204	+13.3	190	−6.9
Total	**1,559**	**1,604**	**1,591**	**−0.8**	**1,605**	**+0.9**
Supply						
Refined	918	900	962	+6.9	985	+2.4
Imported	618	675	629	−6.8	600	−4.6
From stocks	23	29	0	—	20	—
Total	**1,559**	**1,604**	**1,591**	**−0.8**	**1,605**	**+0.9**
Stocks (million bbl)†	40	39	44	+12.8	40	−9.1
		Bcf		% change	Bcf	% change
Residential. 3,199	3,158	3,393	+7.4	3,450	+1.7	
Commercial	1,666	1,595	1,749	+9.7	1,800	+2.9
Industrial	3,262	2,907	3,488	+20.0	3,625	+3.9
Electric utilities	1,201	1,112	1,192	+7.2	1,195	+0.3
Pipeline and plant fuel	760	745	774	+3.9	780	+0.8
Total	**10,088**	**9,517**	**10,596**	**+11.3**	**10,850**	**+2.4**

* October-March heating season. †At end of March.

an unleaded gasoline environment provided U.S. refiners with an extra measure of octane producing capability. That capability, coupled with low product prices, has U.S. refiners looking at the possibility of another octane race.

Consumer demand for high octane, premium grade gasoline is growing at a greater rate overall gasoline demand. Estimated total demand in 1988 was about 3% higher than at the same time in 1987, while premium demand ran about 12% higher than a year earlier.

Because profits have been higher on premium unleaded gasoline, several refiners began to market a 93 or 94 (R + M)/2 octane premium grade.

There is some question whether there is enough octane capacity to meet the increasing demand.

Some industry officials contend the inherent constraints of the industry's octane capability eventually will drive the cost of high octane premium to a retail level than will crimp consumer demand. And higher octane will reduce overall liquid yields, causing an increase in gasoline imports.

EPA's rules on volatility could also restrict the ability of refiners to meet octane demand.

Auto makers consider redesign

The apparent availability of higher octane gasoline has caused the auto industry in the U.S. to eye additional octane as one answer to increased engine output while meeting present fuel economy standards.

The supply of higher octane grades has caused U.S. auto makers to consider redesigning engines to obtain higher compression ratios. That would allow the autos equipped with the new engines to more easily meet the corporate average fuel economy (CAFE) standards.

General Motors Corp. says that higher compression ratio engines would also allow larger cars that consumers seem to demand at present. But before an auto company can implement an engine design program, it would have to be assured that an adequate supply of high-octane gasoline would be available for an adequate period into the future.

Refiners can provide the higher octane by utilizing new octane catalysts in fluid catalytic crackers, running reforming units at higher severity, increasing throughput through alkylation units, and by blending ethers such as MTBE and ETBE as octane enhancers.

Although there is an appearance of unlimited octane for premium grades, it is estimated by some industry experts that the limit will be about two octane numbers above the present 92 (R + M)/2 premium level.

The heavy push to improve air quality in the U.S. continues to spur efforts, both technically and politically, to design vehicles that can use alternate fuels.

During 1988, the U.S. government and state governments joined automakers and the American Gas Association in promoting alternate fuels vehicles as solutions to energy and pollution problems. Bills have been approved in the U.S. congress to promote methanol, ethanol, and compressed natural gas as alternatives to gasoline.

The legislation will give an automaker credits against its average fleet fuel economy for producing alternate fuel vehicles designed to run on alternative fuels.

As a result, General Motors Corp. introduced its variable fuel vehicle (VFV) that has modified engine parts and fuel storage and pumping components (see diagram 6). Ford Motor Co. offers a 1987 Crow Victoria model than runs on methanol. The American Gas Association has a Ford Ranger modified to run on compressed natural gas.

GM's VFV is designed to run on any combination of alcohol fuel and gasoline. Gm also has other development programs underway, such as a bus operated by the Golden Gate Bridge and Transit Authority in San Francisco.

Efforts by OPEC nations to guarantee placement of their crude oil in equity or other ventures with foreign refiners continued to expand during 1988. These moves are intended primarily to assure steady revenues in a volatile market.

Venezuela and Kuwait have led OPEC in the push downstream into foreign refineries (see table 4). Venezuela owns substantial interests in Citgo Petroleum Co. and Champlin Petroleum Co. in the U.S. Kuwait has made large investments into operations in Europe.

Other OPEC countries have also begun to increase their efforts downstream. Nigeria, for instance, is pursuing an interest in Sun Co.'s U.S. refining and marketing operations as well as a refinery in Ireland.

Saudi Arabia formalized a downstream venture with Texaco Inc. in November 1988. In the agreement, Saudi Arabian Oil Co. acquired a 50% interest in Texaco's refining and marketing business in 23 U.S. East and Gulf coast states and the District of Columbia.

The nearly $2 billion joint venture agreement includes shared inventory, crude oil purchase rights, product distribution terminals, and wholesale and retail outlets (see Fig. 7). Purchase price of the Saudi Arabia's interest was $812 million.

A shortfall of light, sweet crude supplies could be on the horizon during the latter part of this century because of

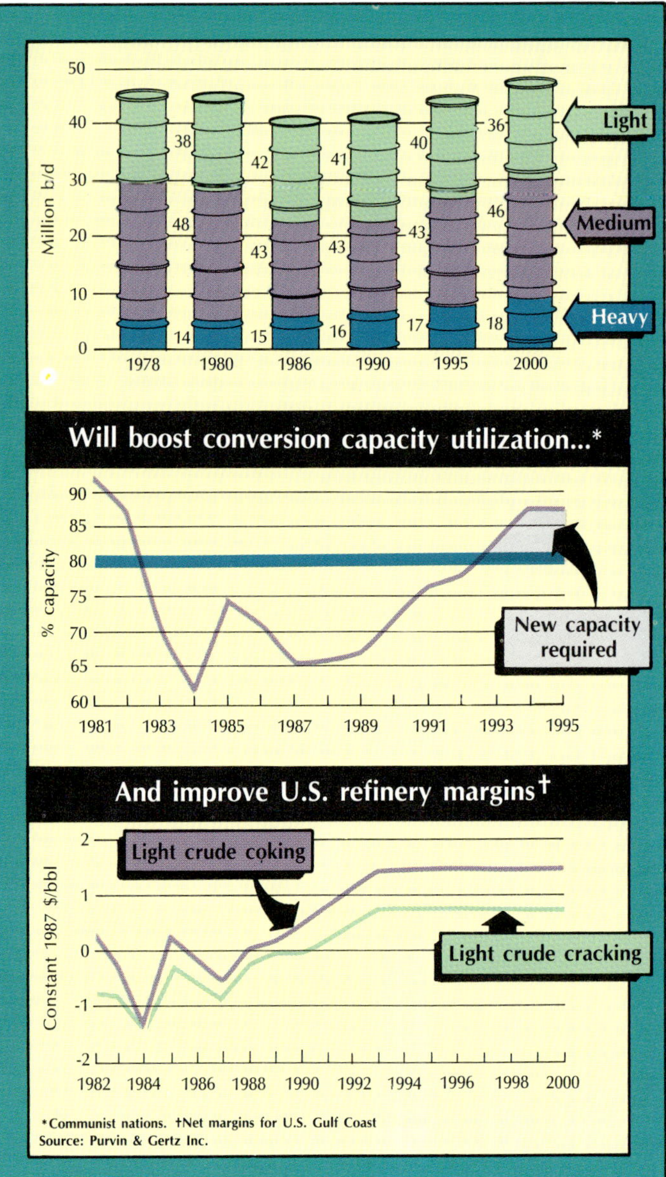

How OPEC's downstream capacity has grown

Table 4

	Refining capacity/(access)* 1,000 b/d			Percent change 1983-87	Oil production 1,000 b/d			Percent change 1983-87
	1983	1985	1987		1983	1985	1987	
Algeria	471.2	471.2	471.2	—	685.0	647.1	652.4	−4.8
Ecuador	94.6	94.6	94.6	—	233.0	275.7	173.9	−25.4
Gabon	44.0	44.0	21.0	−52.3	148.3	152.5	157.3	+6.1
Indonesia	473.4	837.0	837.0	+76.8	1,301.1	1,242.3	1,185.8	−8.9
Iran	670.0	615.0	615.0	−8.2	2,466.7	2,241.7	2,309.2	−6.4
Iraq	215.5	365.5	365.5	+69.6	883.3	1,447.5	2,135.0	+141.7
Kuwait	594.0	614.0	720.0	+21.2	887.5	862.5	1,120.8	+26.3
Access	—	132.5	132.5		—	—	—	—
Libya	117.0	315.0	342.0	+192.3	1,018.3	1,047.5	1,093.8	+7.4
Access	105.0	105.0	105.0	—	—	—	—	—
Neutral Zone	—	—	—	—	391.2	352.5	387.1	−1.0
Nigeria	234.0	234.0	243.0	+3.8	1,233.8	1,481.7	1,254.5	+1.7
Access†	—	—	(165)		—	—	—	—
Qatar	13.0	63.0	63.0	+384.6	294.4	298.8	280.3	−4.8
Saudi Arabia	765.0	1,190.0	1,490.0	+94.8	4,850.0	3,283.3	4,025.0	−17.0
Access	—	—	(886)		—	—	—	—
United Arab Emirates	121.5	162.0	162.0	+33.3	1,109.2	1,141.7	1,399.6	+26.2
Access	—	—	346.0		—	—	—	—
Venezuela	1,323.1	1,224.2	1,224.2	−7.5	1,775.8	1,645.8	1,580.8	−11.0
Access	294.0	614.0	1,354.9	+360.8	—	—	—	—
Subtotal	**5,136.3**	**6,229.5**	**6,648.5**	**+29.4**	—	—	—	—
Access subtotal	**399.0**	**851.5**	**1,938.4**	**+385.8**	—	—	—	—
Total	**5,535.3**	**7,081.0**	**8,586.9**	**+55.1**	**17,277.6**	**16,120.6**	**17,755.5**	**+2.8**

* Generally, access to refinery capacity outside country; refers to total capacity of outside refineries, even if equity interest or crude supply deal represents only a portion of total capacity. †Hypothetical, not included in total. Represents Nigeria National Petroleum Co. overtures for Sun Oil Co. refining assets in U.S. and Bantry Bay refinery in Ireland. §Hypothetical, not included in total. Represents current Saudi negotiations for possible 50% interest in Texaco's U.S. refining/marketing assets.

increased demand to meet environmental regulations, particularly in Europe, according to Purven & Gertz Inc., Houston in a report early in 1988. Although supplies are adequate to meet demand through 1995, declines in light, sweet crude's share of production in the U.S. and Canada, and a reduction in production in the North Sea, will parallel the growth in demand for light, sweet crude after 1995.

To meet higher crude runs and a heavier crude slate, refinery conversion units will be operating at capacity by the early 1990s, according to Purvin & Gertz. Because conversion capacity is operating at a higher rate than distillation capacity, conversion additions will be needed before 1995 to accommodate the reduction in light, sweet crude production.

New conversion capacity could be required by late 1992, and with those additions, U.S. margins would improve (see Fig. 8).

Construction projects

Refinery projects continue to be revamps and modernizations to improve efficiency and lower operating costs. Although worldwide demand for refined products during 1988 was up, and forecasts for the rest of the 20th century show increased demand, the levels won't support much grassroots refinery construction.

Some of the notable construction projects during 1988 include the announced expansion of two refineries in Thailand. The National Energy Policy Committee of Thailand has approved a $150 million project by Thai Oil Co. that would boost capacity of its 65,000 b/d refinery at Si Racha, by another 100,000 b/d. This project, and an ongoing project will result in the refinery having a total capacity of 183,500 b/d. The committee also approved a debottlenecking plan by Bangchak Petroleum Co. that will boost capacity of 65,000 b/d Bangkok refinery by 20,000 b/d.

Also under construction in Thailand were the expansion of Esso Standard Thailand Ltd.'s 63,000 b/d refinery at Si Racha to 110,000 b/d, and construction by Shell Co. of Thailand Ltd. of an 85,000 b/d refinery on the eastern seaboard.

Construction projects in Thailand are spurred by the country's substantial growth in consumption. Estimated at about 283,000 b/d, annual growth is estimated at 10% annually. In France, a $27 million revamp was completed at Shell Francaise's Etang de Berre refinery. Shell ordered the revamp to accommodate a fluid catalytic cracking unit it moved to Berre from Pauillac, France.

The existing distillation unit needed upgrading to bring it up to the operational level of the relocated FCCU. The revamp increased the vacuum unit capacity, including installation of a new fractionation tower, transfer line, and vacuum pumps. Total capacity of the refinery is 130,000 b/d

To meet the increased demand for high-octane gasoline, several U.S. refiners are spending millions of dollars to revamp refineries to provide higher octane gasoline.

Ashland Oil Co. will spend $250 million during the early 1990s to upgrade its three refineries to meet higher octane demand and stricter environmental rules. About $50 million will be spent during 1990-1992 to add MTBE capacity and boost recycle in the isomerization units at its refineries in Catlettsburg, Ky, St. Paul Park, Minn., and Canton, Ohio. It also expects to spend about $200 million at the refineries to meet new EPA rules calling for 0.05 wt% sulfur in diesel fuel.

Sohio Oil Co. has contracted for one of the world's largest continuous catalyst regeneration reforming units at it 130,000 b/d Toledo, Ohio refinery. Estimated cost is $60,000. The new unit, expected to start in 1991, will replace two older, less efficient cyclic reformers.

Marathon Petroleum Co. plans to build an MTBE unit and modernize controls at its 68,500 b/d Detroit refinery. The MTBE unit will have a capacity of 1,140 b/d. Control improvements will include converting conventional electronic controls to a computerized distributed control system. Completion of both projects is scheduled for mid 1989. IPE

Worldwide production statistics

NORTH AMERICA

Upper figures = 1,000 b/d
Lower figures = million t/y

	1940	1950	1960	1965	1970	1980	1981	1982	1983	1984	1985	1986	1987	1988
Canada	23.5 / 1.2	79.6 / 4.0	525.6 / 26.2	801.4 / 39.9	1,263.6 / 62.9	1,412.0 / 70.3	1,287.0 / 64.1	1,233.0 / 61.4	1,396.0 / 69.5	1,430.0 / 71.2	1,453.0 / 72.4	1,475.3 / 73.5	1,508.3 / 75.1	1,604.6 / 79.9
United States	3,707.4 / 184.6	5,407.1 / 269.3	7,054.6 / 351.3	7,804.1 / 388.6	9,630.0 / 479.6	8,569.0 / 426.7	8,588.0 / 427.7	8,655.0 / 431.0	8,669.0 / 431.7	8,750.0 / 435.8	8,919.0 / 444.2	8,790.0 / 437.7	8,276.7 / 412.2	8,165.9 / 406.7
Total	3,730.9 / 185.8	5,486.7 / 273.3	7,580.2 / 377.5	8,605.5 / 428.5	10,893.6 / 542.5	9,981.0 / 497.0	9,875.0 / 491.8	9,888.0 / 492.4	10,065.0 / 501.2	10,180.0 / 507.0	10,372.0 / 516.6	10,265.3 / 511.2	9,785.0 / 487.3	9,770.5 / 486.6

MIDDLE EAST

Upper figures = 1,000 b/d
Lower figures = million t/y

	1940	1950	1960	1965	1970	1980	1981	1982	1983	1984	1985	1986	1987	1988
Abu Dhabi	… / …	… / …	… / …	281.9 / 14.0	693.8 / 34.6	1,350.0 / 67.2	1,145.0 / 57.0	883.0 / 44.0	757.0 / 37.7	750.0 / 37.4	705.1 / 35.1	949.0 / 47.3	969.4 / 48.3	1,012.6 / 50.4
Bahrain	19.3 / 1.0	30.2 / 1.5	45.1 / 2.2	57.0 / 2.8	76.6 / 3.8	48.0 / 2.4	44.0 / 2.2	45.0 / 2.2	41.0 / 2.0	41.0 / 2.0	41.9 / 2.1	44.0 / 2.2	43.2 / 2.2	42.3 / 2.1
Dubai	… / …	… / …	… / …	… / …	… / …	349.0 / 17.4	358.0 / 17.8	357.0 / 17.8	327.0 / 16.3	324.0 / 16.1	348.0 / 17.3	350.0 / 17.4	381.2 / 19.0	355.3 / 17.7
Iran	181.2 / 9.0	664.3 / 33.1	1,067.6 / 53.2	1,885.5 / 93.9	3,328.8 / 165.8	1,467.0 / 73.1	1,375.0 / 68.5	1,896.0 / 94.4	2,606.0 / 129.8	2,166.0 / 107.9	2,279.0 / 113.5	1,806.3 / 90.0	2,341.7 / 116.6	2,207.5 / 109.9
Iraq	60.7 / 3.0	136.2 / 6.8	1,004.2 / 50.0	1,315.2 / 65.5	1,548.6 / 77.1	2,638.0 / 131.4	892.0 / 44.4	914.0 / 45.5	905.0 / 45.1	1,218.0 / 60.7	1,396.8 / 69.6	1,787.7 / 89.0	2,095.8 / 104.4	2,679.2 / 133.4
Israel	… / …	… / …	2.5 / 0.1	4.0 / 0.2	70.2 / 3.5	1.0 / …	0.3 / …	0.3 / …	0.2 / …	0.1 / …	0.1 / …	0.1 / …	1.0 / …	0.3 / …
Jordan	… / …	… / …	… / …	… / …	… / …	… / …	… / …	… / …	… / …	… / …	… / …	… / …	4.0 / 0.2	0.5 / …
Kuwait	… / …	344.4 / 17.2	1,628.2 / 81.1	2,169.5 / 108.0	2,734.5 / 136.2	1,382.0 / 68.8	916.0 / 45.6	675.0 / 33.6	912.0 / 45.4	925.0 / 46.1	823.3 / 41.0	1,202.0 / 59.9	1,095.8 / 54.6	1,254.2 / 62.5
Neutral Zone	… / …	… / …	136.2 / 6.8	360.9 / 18.0	500.6 / 24.9	540.0 / 26.9	370.0 / 18.4	315.0 / 15.7	398.0 / 19.8	420.0 / 20.9	357.8 / 17.8	325.7 / 16.2	399.2 / 19.9	316.3 / 15.8
Oman	… / …	… / …	… / …	… / …	332.4 / 16.6	283.0 / 14.1	317.0 / 15.8	328.0 / 16.3	378.0 / 18.8	404.0 / 20.1	484.3 / 24.1	540.8 / 26.9	565.0 / 28.1	596.7 / 29.7
Qatar	… / …	33.6 / 1.7	175.1 / 8.7	231.1 / 11.5	362.4 / 18.0	472.0 / 23.5	414.0 / 20.6	340.0 / 16.9	270.0 / 13.4	395.0 / 19.7	297.0 / 14.8	332.3 / 16.5	284.4 / 14.2	349.2 / 17.4
Ras Al Khaimah	… / …	… / …	… / …	… / …	… / …	… / …	… / …	… / …	… / …	5.8 / 0.3	9.1 / 0.5	11.0 / 0.5	10.0 / 0.5	10.0 / 0.5
Saudi Arabia	13.9 / 0.7	546.7 / 27.2	1,247.1 / 62.1	2,024.9 / 100.8	3,548.9 / 176.7	9,630.0 / 479.6	9,642.0 / 480.2	6,484.0 / 322.9	4,872.0 / 242.6	4,545.0 / 226.3	3,295.3 / 164.1	4,719.7 / 235.0	4,054.2 / 201.9	4,708.3 / 234.5
Sharjah	… / …	… / …	… / …	… / …	… / …	10.0 / 0.5	9.0 / 0.4	7.0 / 0.3	35.0 / 1.7	62.0 / 3.1	66.1 / 3.3	65.0 / 3.2	66.9 / 3.3	65.0 / 3.2
Syria	… / …	… / …	… / …	… / …	83.1 / 4.1	165.0 / 8.2	166.0 / 8.3	175.0 / 8.7	165.0 / 8.2	161.0 / 8.0	161.7 / 8.1	185.0 / 9.2	231.7 / 11.5	273.3 / 13.6
Turkey	… / …	0.3 / …	7.0 / 0.3	29.7 / 1.5	67.9 / 3.4	44.0 / 2.2	46.0 / 2.3	45.0 / 2.2	45.0 / 2.2	41.0 / 2.0	41.5 / 2.1	46.8 / 2.3	52.0 / 2.6	50.0 / 2.5
Yemen, North	… / …	… / …	… / …	… / …	… / …	… / …	… / …	… / …	… / …	… / …	… / …	10.0 / 0.5	10.0 / 0.5	159.9 / 8.0
Yemen, South	… / …	… / …	… / …	… / …	… / …	… / …	… / …	… / …	… / …	… / …	… / …	… / …	… / …	12.5 / 0.6
Total	275.1 / 13.7	1,755.7 / 87.5	5,313.0 / 264.5	8,359.7 / 416.2	13,347.8 / 664.7	18,379.0 / 915.3	15,694.3 / 781.5	12,464.3 / 620.5	11,711.2 / 583.0	11,457.9 / 570.6	10,307.0 / 513.4	12,375.4 / 616.1	12,605.5 / 627.8	14,093.1 / 701.8

EUROPE

Upper figures = 1,000 b/d
Lower figures = million t/y

	1940	1950	1960	1965	1970	1980	1981	1982	1983	1984	1985	1986	1987	1988
Austria	7.7	27.9	45.4	53.1	53.3	29.0	25.0	25.0	24.0	23.0	21.7	22.2	21.6	23.2
	0.4	1.4	2.3	2.6	2.7	1.4	1.2	1.2	1.2	1.1	1.1	1.1	1.1	1.2
Denmark	...	...	...	...	...	6.0	16.0	38.0	46.0	45.0	58.0	74.7	94.0	95.5
	...	...	...	...	...	0.3	0.8	1.9	2.3	2.2	2.9	3.7	4.7	4.8
France	1.4	2.5	38.6	59.5	45.8	26.0	34.0	33.0	32.0	35.0	49.0	59.0	63.5	68.7
	0.1	0.1	1.9	3.0	2.3	1.3	1.7	1.6	1.6	1.7	2.4	2.9	3.2	3.4
Germany, Fed. Rep.	20.1	22.2	107.3	155.0	147.2	90.0	86.0	84.0	81.0	79.0	80.5	80.0	73.8	77.9
	1.0	1.1	5.3	7.7	7.3	4.5	4.3	4.2	4.0	3.9	4.0	4.0	3.7	3.9
Greece	...	...	...	...	...	...	8.0	21.0	25.0	27.0	26.4	26.4	24.4	21.8
	...	...	...	...	...	...	0.4	1.0	1.2	1.3	1.3	1.3	1.2	1.1
Italy	0.2	0.2	37.5	47.0	26.3	39.0	37.0	28.0	37.0	45.0	41.7	49.0	50.0	92.9
	...	...	1.9	2.3	1.3	1.9	1.8	1.4	1.8	2.2	2.1	2.4	2.5	4.6
Netherlands	...	13.4	36.0	37.0	35.8	25.0	27.0	28.0	42.0	61.0	69.0	79.0	91.3	83.7
	...	0.7	1.8	1.8	1.8	1.2	1.3	1.4	2.1	3.0	3.4	3.9	4.5	4.2
Norway	...	...	...	...	...	528.0	508.0	488.0	600.0	688.0	771.0	823.3	973.3	1,069.1
	...	...	...	...	...	26.3	25.3	24.3	29.9	34.3	38.4	41.0	48.5	53.2
Spain	...	...	...	...	3.1	32.0	24.0	30.0	59.0	45.0	45.3	37.2	32.8	30.4
	...	...	...	...	0.2	1.6	1.2	1.5	2.9	2.2	2.3	1.9	1.6	1.5
United Kingdom	0.3	0.9	1.7	1.7	1.7	1,619.0	1,790.0	2,050.0	2,260.0	2,452.0	2,519.8	2,602.0	2,445.9	2,376.3
	...	...	0.1	0.1	0.1	80.6	89.1	102.1	112.5	122.1	125.5	129.6	121.8	118.3
Total	29.7	67.1	266.5	353.3	313.2	2,394.0	2,555.0	2,825.0	3,206.0	3,500.0	3,682.4	3,852.8	3,870.6	3,939.5
	1.5	3.3	13.3	17.5	15.7	119.1	127.1	140.6	159.5	174.0	183.4	191.8	192.8	196.2

ASIA-PACIFIC

Upper figures = 1,000 b/d
Lower figures = million t/y

	1940	1950	1960	1965	1970	1980	1981	1982	1983	1984	1985	1986	1987	1988
Australia	...	...	...	7.2	175.6	379.0	382.0	353.0	405.0	481.0	556.0	472.0	566.1	552.3
	...	...	...	0.4	8.7	18.9	19.0	17.6	20.2	24.0	27.7	23.5	28.2	27.5
Bangladesh	...	...	...	...	...	...	...	...	...	...	...	...	0.5	0.9
	...	...	...	...	...	...	...	...	...	...	...	...	...	...
Brunei	19.3	84.8	92.9	80.4	148.0	230.0	166.0	155.0	155.0	160.0	149.5	170.0	133.7	138.3
	1.0	4.2	4.6	4.0	7.4	11.5	8.3	7.7	7.7	8.0	7.4	8.5	6.7	6.9
Burma	21.1	1.5	11.1	12.1	16.4	30.0	30.0	30.0	30.0	30.0	30.0	30.0	24.0	15.0
	1.1	0.1	0.6	0.6	0.8	1.5	1.5	1.5	1.5	1.5	1.5	1.5	1.2	0.7
China, Taiwan	...	...	0.4	0.4	1.8	5.0	3.0	3.0	2.3	2.6	2.7	1.9	2.6	2.6
	...	...	...	...	0.1	0.2	0.1	0.1	0.1	0.1	0.1	0.1	0.1	0.1
India	3.9	5.1	9.0	61.8	138.6	182.0	298.0	384.0	390.0	543.0	614.0	622.7	608.7	631.8
	0.2	0.3	0.4	3.1	6.9	9.1	14.8	19.1	19.4	27.0	30.6	31.0	30.3	31.5
Indonesia	169.4	132.6	411.2	485.0	853.6	1,576.0	1,607.0	1,341.0	1,292.0	1,332.0	1,219.5	1,243.8	1,185.8	1,137.5
	8.4	6.6	2.05	24.2	42.5	78.5	80.0	66.8	64.3	66.3	60.7	61.9	59.1	56.6
Japan	7.2	5.6	10.2	11.4	16.3	10.0	7.0	6.0	6.9	6.4	10.0	12.8	12.1	12.3
	0.4	0.3	0.5	0.6	0.8	0.5	0.3	0.3	0.3	0.3	0.5	0.6	0.6	0.6
Malaysia	...	...	...	...	...	288.0	259.0	306.0	370.0	462.0	432.8	503.3	467.4	540.0
	...	...	...	...	...	14.3	12.9	15.2	18.4	23.0	21.6	25.1	23.3	26.9
New Guinea	...	4.8	4.2	...	...	...	...	...	...	...	...	...	...	...
	...	0.2	0.2	...	...	...	...	...	...	...	...	...	...	...
New Zealand	...	...	...	...	...	7.0	10.0	15.0	15.0	18.0	18.0	29.3	28.3	28.0
	...	...	...	...	...	0.3	0.5	0.7	0.7	0.9	0.9	1.5	1.4	1.4
Pakistan	2.4	3.5	7.2	11.2	9.9	10.0	10.0	12.0	13.0	18.0	34.3	41.0	41.6	47.0
	0.1	0.2	0.4	0.6	0.5	0.5	0.5	0.6	0.6	0.9	1.7	2.0	2.1	2.3
Philippines	...	...	...	...	...	15.0	2.0	7.0	15.0	12.0	9.3	6.3	5.1	9.2
	...	...	...	...	...	0.7	0.1	0.3	0.7	0.6	0.5	0.3	0.3	0.5
Thailand	...	...	...	...	0.4	...	...	6.0	11.0	19.0	34.3	35.6	31.3	38.4
	...	...	...	...	...	...	...	0.3	0.5	0.9	1.7	1.8	1.6	1.9
Total	223.3	237.9	546.2	669.5	1,360.6	2,732.0	2,774.0	2,618.0	2,705.2	3,084.0	3,110.4	3,168.7	3,107.2	3,153.3
	11.2	11.9	27.2	33.5	67.7	136.0	138.0	130.2	134.4	153.5	154.9	157.8	154.9	156.9

OFF AFRICA, 1 West Manambolo is drilled near the coast of Madagascar's Morondava basin. Photo by Petro-Canada Inter. Assistance Corp.

AFRICA

Upper figures = 1,000 b/d
Lower figures = million t/y

	1940	1950	1960	1965	1970	1980	1981	1982	1983	1984	1985	1986	1987	1988
Algeria	...	...	182.7	551.6	1,029.1	1,016.0	750.0	750.0	686.6	608.0	643.3	600.7	647.9	666.8
	...	...	9.1	27.5	51.2	50.6	37.4	37.4	34.2	30.3	32.0	29.9	32.3	33.2
Angola (inc. Cabinda)	...	...	1.3	13.0	13.7	150.0	140.0	122.0	174.0	207.0	228.5	280.7	341.3	449.3
	...	...	0.1	0.6	0.7	7.5	7.0	6.1	8.7	10.3	11.4	14.0	17.0	22.4
Cameroon	...	...	...	...	...	58.0	87.0	109.0	114.0	125.0	133.5	180.0	170.0	170.0
	...	...	...	...	...	2.9	4.3	5.4	5.7	6.2	6.6	9.0	8.5	8.5
Congo	...	...	...	...	...	56.0	79.0	87.0	95.0	118.0	115.1	115.0	118.0	134.8
	...	...	...	...	...	2.8	3.9	4.3	4.7	5.9	5.7	5.7	5.9	6.7
Egypt	18.1	44.9	61.6	122.7	327.3	596.0	578.0	667.0	690.0	790.0	885.8	773.7	899.3	851.3
	0.9	2.2	3.1	6.1	16.3	29.7	28.8	33.2	34.4	39.3	44.1	38.5	44.8	42.4
Gabon	...	...	16.1	25.2	108.8	180.0	147.0	130.0	150.0	150.0	150.0	146.2	155.8	175.0
	...	...	0.8	1.3	5.4	9.0	7.3	6.5	7.5	7.5	7.5	7.3	7.8	8.7
Ivory Coast	...	...	...	...	...	...	7.0	9.0	24.0	22.0	28.0	20.0	16.0	12.9
	...	...	...	...	...	...	0.3	0.4	1.2	1.1	1.4	1.0	0.8	0.6
Libya	...	...	...	1,220.0	3,318.0	1,785.0	1,063.0	1,127.0	1,020.0	1,090.0	1,045.3	1,030.7	1,019.7	1,012.5
	...	...	...	60.8	165.2	88.9	52.9	56.1	50.8	54.3	52.1	51.3	50.8	50.4
Nigeria	...	...	17.3	272.2	1,083.3	2,057.0	1,369.0	1,324.0	1,232.0	1,414.0	1,445.5	1,464.0	1,238.6	1,358.3
	...	...	0.9	13.6	53.9	102.4	68.2	65.9	61.4	70.4	72.0	72.9	61.7	67.6
Tunisia	...	...	3.0	3.6	87.7	110.0	118.0	106.0	115.0	114.0	125.0	106.0	104.3	102.8
	...	...	0.1	0.2	4.4	5.5	5.9	5.3	5.7	5.7	6.2	5.3	5.2	5.1
Zaire	...	...	...	...	...	20.0	20.0	22.0	26.0	27.0	30.5	32.1	31.8	29.5
	...	...	...	...	...	1.0	1.0	1.1	1.3	1.3	1.5	1.6	1.6	1.5
Others	...	0.8	3.0	3.6	1.5	4.0	2.2	1.7	6.6	7.7	8.9	7.6	12.3	5.0
	...	...	0.1	0.2	0.1	0.2	0.1	0.1	0.3	0.4	0.4	0.4	0.6	0.2
Total	18.1	45.7	285.0	2,211.9	5,969.4	6,032.0	4,360.2	4,454.7	4,333.2	4,672.7	4,839.4	4,756.7	4,755.0	4,968.2
	0.9	2.2	14.2	110.3	297.2	300.5	217.1	221.8	215.9	232.7	240.9	236.9	237.0	247.3

LATIN AMERICA

Upper figures = 1,000 b/d
Lower figures = million t/y

	1940	1950	1960	1965	1970	1980	1981	1982	1983	1984	1985	1986	1987	1988
Argentina	56.3	64.0	171.9	269.3	382.9	487.0	497.0	483.0	481.0	467.0	447.5	430.2	419.1	449.6
	2.8	3.2	8.6	13.4	19.1	24.3	24.8	24.1	24.0	23.3	22.3	21.4	20.9	22.4
Bolivia	0.8	1.7	9.8	9.2	16.3	30.0	24.0	24.0	22.0	20.0	20.0	18.7	18.5	18.8
	...	0.1	0.5	0.5	0.8	1.5	1.2	1.2	1.1	1.0	1.0	0.9	0.9	0.9
Brazil	...	0.9	80.9	94.0	160.5	182.0	215.0	252.0	315.0	437.0	540.5	578.0	562.8	555.6
	...	...	4.0	4.7	8.0	9.1	10.7	12.5	15.7	21.8	26.9	28.8	28.0	27.7
Chile	...	1.7	19.8	34.9	34.6	29.0	40.0	41.0	39.0	38.0	32.5	34.4	28.3	24.8
	...	0.1	1.0	1.7	1.7	1.4	2.0	2.0	1.9	1.9	1.6	1.7	1.4	1.2
Colombia	69.9	93.3	152.7	210.0	214.0	125.0	125.0	140.0	155.0	165.0	178.8	324.2	387.8	346.7
	3.5	4.6	7.6	10.5	10.7	6.2	6.2	7.0	7.7	8.2	8.9	16.1	19.3	17.3
Ecuador	6.4	7.2	7.7	8.0	4.1	222.0	204.0	215.0	236.0	254.0	273.3	270.1	157.2	310.1
	0.3	0.4	0.4	0.4	0.2	11.1	10.2	10.7	11.8	12.6	13.6	13.5	7.8	15.4
Guatemala	...	...	...	...	...	5.0	5.0	6.4	7.5	5.2	3.1	4.9	3.9	3.7
	...	...	...	...	...	0.2	0.2	0.3	0.4	0.3	0.2	0.2	0.2	0.2
Mexico	120.3	198.5	270.6	323.1	430.2	1,936.0	2,390.0	2,734.0	2,702.0	2,743.0	2,797.0	2,468.0	2,537.8	2,527.3
	6.0	9.9	13.5	16.1	21.4	96.4	119.0	136.2	134.6	136.6	139.3	122.9	126.4	125.9
Peru	33.2	41.1	52.6	63.2	72.2	191.0	184.0	198.0	171.0	201.0	189.0	179.0	166.7	141.7
	1.7	2.0	2.6	3.1	3.6	9.5	9.2	9.9	8.5	10.0	9.4	8.9	8.3	7.1
Trinidad & Tobogo	60.9	56.5	115.7	133.9	139.8	211.0	240.0	182.0	158.0	169.0	179.0	167.9	163.1	149.3
	3.0	2.8	5.8	6.7	7.0	10.5	12.0	9.1	7.9	8.4	8.9	8.4	8.1	7.4
Venezuela	507.0	1,498.0	2,846.1	3,472.9	3,708.0	2,167.0	2,093.0	1,826.0	1,791.0	1,724.0	1,669.0	1,664.9	1,591.9	1,658.0
	25.2	74.6	141.7	173.0	184.7	107.9	104.2	90.9	89.2	85.9	83.1	82.9	79.3	82.6
Others	...	...	...	...	...	...	0.7	0.6	0.5	2.2	2.9	3.6	3.6	3.7
	...	...	...	...	...	...	...	...	...	0.1	0.1	0.2	0.2	0.2
Total	854.8	1,962.9	3,727.8	4,618.5	5,162.6	5,585.0	6,017.7	6,102.0	6,078.0	6,225.4	6,332.6	6,143.9	6,040.7	6,189.3
	42.5	97.7	185.7	230.1	257.2	278.1	299.7	303.9	302.8	310.1	315.3	305.9	300.8	308.3
Total non-Communist	5,131.9	9,556.0	17,718.7	24,818.4	37,047.7	45,103.0	41,276.2	38,352.0	38,098.6	39,120.0	38,643.8	40,562.8	40,164.0	42,113.9
	255.6	475.9	882.4	1,236.1	1,845.0	2,246.0	2,055.2	1,909.4	1,896.8	1,947.9	1,924.5	2,019.7	2,000.6	2,097.1

COMMUNIST AREAS

Upper figures = 1,000 b/d
Lower figures = million t/y

	1940	1950	1960	1965	1970	1980	1981	1982	1983	1984	1985	1986	1987	1988
Bulgaria	...	...	4.0	4.6	6.7	6.0	6.0	3.0	3.0	3.0	3.0	3.0	6.0	6.0
	...	...	0.2	0.2	0.3	0.3	0.3	0.1	0.1	0.1	0.1	0.1	0.3	0.3
China	...	4.0	100.0	220.0	500.0	2,119.0	2,024.4	2,042.0	2,120.0	2,250.0	2,496.0	2,620.0	2,700.0	2,733.0
	...	0.2	5.0	11.0	24.9	105.5	100.8	101.7	105.6	112.1	124.3	130.5	134.5	136.1
Czechoslovakia	...	0.8	2.7	3.9	4.1	1.9	1.7	2.0	2.0	2.0	2.0	2.0	2.0	3.0
	...	...	0.1	0.2	0.2	0.1	0.1	0.1	0.1	0.1	0.1	0.1	0.1	0.1
German Dem. Rep.*	...	1.0	1.0	1.0	1.2	10.0	10.0	10.0	10.0	10.0	10.0	10.0	10.0	8.0
	...	...	...	...	0.1	0.5	0.5	0.5	0.5	0.5	0.5	0.5	0.5	0.4
Hungary	...	10.1	24.3	36.1	38.7	40.6	40.5	41.0	40.0	40.0	40.0	40.0	40.0	40.0
	...	0.5	1.2	1.8	1.9	2.0	2.0	2.0	2.0	2.0	2.0	2.0	2.0	2.0
Poland	...	3.3	3.9	6.8	8.5	6.6	6.0	4.0	5.0	5.0	5.0	4.0	4.0	4.0
	...	0.2	0.2	0.3	0.4	0.3	0.3	0.2	0.2	0.2	0.2	0.2	0.2	0.2
Romania	...	87.7	231.0	252.4	268.6	230.9	232.0	234.0	240.0	229.0	226.0	220.0	215.0	188.0
	...	4.4	11.5	12.6	13.4	11.5	11.6	11.7	12.0	11.4	11.3	11.0	10.7	9.4
U.S.S.R.*	624.9	760.6	2,969.1	4,877.3	7,089.2	12,031.0	12,180.0	12,260.0	12,320.0	12,260.0	11,900.0	12,300.0	12,480.0	12,446.0
	31.1	37.9	147.9	242.9	353.0	599.1	606.6	610.5	613.5	610.5	592.6	612.5	621.5	619.8
Yugoslavia	0.3	2.1	18.8	40.7	57.8	84.5	87.6	87.0	82.0	80.0	81.0	83.0	84.0	76.0
	...	0.1	0.9	2.0	2.9	4.2	4.4	4.3	4.1	4.0	4.0	4.1	4.2	3.8
Others	...	2.7	14.9	19.0	37.1	50.0	57.2	81.0	86.0	96.0	86.0	77.0	80.0	85.0
	...	0.1	0.7	0.9	1.8	2.5	2.8	4.0	4.3	4.8	4.3	3.8	4.0	4.2
Total	625.2	872.3	3,369.7	5,461.8	8,011.9	14,580.5	14,645.4	14,764.0	14,908.0	14,975.0	14,849.0	15,359.0	15,621.0	15,589.0
	31.1	43.4	167.7	271.9	398.9	726.0	729.4	735.1	742.4	745.7	739.4	764.8	778.0	776.3

*Includes gas liquids

	1940	1950	1960	1965	1970	1980	1981	1982	1983	1984	1985	1986	1987	1988
World Total	5,757.1	10,428.3	21,088.4	30,280.2	45,059.1	59,683.5	55,921.6	53,116.0	53,006.6	54,095.0	53,492.8	55,921.8	55,785.0	57,702.9
	286.7	519.3	1,050.1	1,508.0	2,243.9	2,972.0	2,784.6	2,644.5	2,873.4	2,693.6	2,663.9	2,784.5	2,778.6	2,873.4

World oil consumption

Upper figures = 1,000 b/d Lower figures = million t/y	1985	1986	1987		1985	1986	1987
U.S.A.	15,170 721.7	15,670 751.3	15,955 763.4	Switzerland	255 12.0	280 13.2	265 12.4
Canada	1,490 68.5	1,475 68.1	1,500 69.4	Turkey	345 16.8	375 18.3	445 21.4
Total North America	16,660 790.2	17,145 819.4	17,455 832.8	United Kingdom	1,630 77.4	1,645 77.4	1,610 75.2
Latin America	4,425 209.7	4,530 215.3	4,645 220.6	Other	45 2.3	50 2.5	55 2.8
Total Western Hemisphere	**21,085** **999.9**	**21,675** **1,034.7**	**22,100** **1,053.4**	Total Western Europe	11,900 566.4	12,380 589.1	12,390 587.4
Austria	200 9.8	215 10.4	220 10.7	Middle East	2,150 108.3	2,150 108.3	2,175 109.6
Belgium & Luxembourg	430 20.8	485 23.5	485 23.5	Africa	1,700 82.1	1,695 81.8	1,750 84.4
Denmark	220 10.7	215 10.5	205 10.0	South Asia	1,160 56.4	1,210 58.5	1,260 61.0
Finland	220 10.8	230 11.3	230 11.2	Southeast Asia	2,275 110.7	2,450 119.1	2,585 125.0
France	1,790 84.3	1,830 86.0	1,835 86.1	Japan	4,380 203.4	4,435 205.3	4,510 208.1
Germany, Fed. Rep.	2,390 112.9	2,530 119.9	2,430 114.6	Australasia	670 30.8	695 32.2	720 32.8
Greece	245 12.0	250 12.2	260 12.6	U.S.S.R.	9,020 445.8	9,045 447.2	9,090 449.2
Ireland	85 4.0	100 4.8	85 4.2	Eastern Europe	2,525 123.4	2,535 123.9	2,575 125.9
Italy	1,730 84.4	1,770 86.5	1,845 89.8	China	1,785 88.9	1,990 99.2	2,085 103.9
Netherlands	635 29.2	700 32.4	695 32.0	**Total Eastern Hemisphere**	**37,565** **1,816.2**	**38,585** **1,864.6**	**39,140** **1,887.3**
Norway	200 9.0	200 9.3	205 9.5	World excl. U.S.S.R. China & E. Europe	45,320 2,158.0	46,690 2,229.0	47,490 2,261.7
Portugal	180 8.8	195 9.5	195 9.4	**World Total**	**58,650** **2,816.1**	**60,260** **2,899.3**	**61,240** **2,940.7**
Spain	925 42.9	925 42.7	935 43.2				
Sweden	375 18.3	385 18.7	390 18.8				

Source: British Petroleum
South Asia—Afghanistan, Bangladesh, Burma, India, Nepal, Pakistan, Sri Lanka Southeast Asia—Brunei, Malaysia, Indonesia, Hong Kong, South Korea, Philippines, Singapore, Thailand, Taiwan, Papua New Guinea, SW Pacific Islands, Australasia—Australia, New Zealand

Worldwide refining capacity

EUROPE

Upper figures = 1,000 b/d
Lower figures = million t/y

	1940	1950	1960	1970	1980	1981	1982	1983	1984	1985	1986	1987	1988	1989
Austria	...	26	47	98	280	280	244	269	268	273	204	204	204	204
	...	1.3	2.3	4.9	13.9	13.9	12.2	13.4	13.3	13.6	10.2	10.2	10.2	10.2
Belgium	12	16	171	699	1,064	1,056	1,035	693	694	693	652	648	631	631
	0.6	0.8	8.5	34.8	53.0	52.6	51.5	34.5	34.6	34.5	32.5	32.3	31.4	31.4
Cyprus	...	...	...	...	16	16	16	16	15	16	16	16	16	17
	...	...	...	...	0.8	0.8	0.8	0.8	0.7	0.8	0.8	0.8	0.8	0.8
Denmark	1	1	...	173	214	215	215	215	174	176	166	166	177	177
	...	...	...	8.6	10.7	10.7	10.7	10.7	8.7	8.8	8.3	8.3	8.8	8.8
Finland	...	...	25	176	336	336	299	299	299	299	241	241	241	241
	...	...	1.2	8.8	16.7	16.7	14.9	14.9	14.9	14.9	12.0	12.0	12.0	12.0
France	152	306	769	2,405	3,385	3,342	3,291	2,871	2,670	2,386	1,947	1,834	1,941	1,876
	7.6	15.2	38.3	119.8	168.6	166.4	163.9	143.0	133.0	118.8	97.0	91.3	96.7	93.4
Germany, Fed. Rep.	48	81	573	2,351	2,986	3,021	2,937	2,471	2,386	2,172	1,933	1,720	1,648	1,518
	2.4	4.0	28.5	117.1	148.7	150.4	146.3	123.1	118.8	108.2	96.3	85.7	82.1	75.6
Greece	...	...	30	95	431	428	427	422	369	390	390	385	385	385
	...	...	1.5	4.7	21.5	21.3	21.3	21.0	18.4	19.4	19.4	19.2	19.2	19.2
Ireland	11	...	40	54	56	56	56	56	56	56	56	56	56	56
	0.5	...	2.0	2.7	2.8	2.8	2.8	2.8	2.8	2.8	2.8	2.8	2.8	2.8
Italy	41	107	773	3,214	4,131	4,092	4,003	3,283	3,050	3,095	2,738	2,679	2,563	2,450
	2.0	5.3	38.5	160.1	205.7	203.8	199.3	163.5	151.9	154.1	136.4	133.4	127.6	122.0
Netherlands	15	77	359	1,456	1,828	1,827	1,708	1,552	1,552	1,499	1,468	1,401	1,381	1,381
	0.7	3.8	17.9	72.5	91.0	91.0	85.1	77.3	77.3	74.7	73.1	69.8	68.8	68.8
Norway	1	1	2	113	264	253	243	243	243	244	240	240	240	239
	...	...	0.1	5.6	13.1	12.6	12.1	12.1	12.1	12.2	12.0	12.0	12.0	11.9
Portugal	4	8	23	82	378	341	365	365	282	290	296	294	294	313
	0.2	0.4	1.1	4.1	18.8	17.0	18.2	18.2	14.0	14.4	14.7	14.6	14.6	15.6
Spain	11	17	144	734	1,455	1,464	1,517	1,522	1,493	1,493	1,367	1,305	1,305	1,285
	0.5	0.8	7.2	36.6	72.5	72.9	75.5	75.8	74.4	74.4	68.1	65.0	65.0	64.0
Sweden	4	26	48	253	458	451	471	453	453	439	429	437	437	427
	0.2	1.3	2.4	12.6	22.8	22.5	23.5	22.6	22.6	21.9	21.4	21.8	21.8	21.3
Switzerland	...	...	...	106	137	102	137	137	137	127	137	65	132	132
	...	...	...	5.3	6.8	5.1	6.8	6.8	6.8	6.3	6.8	3.2	6.6	6.6
United Kingdom	133	197	962	2,420	2,527	2,629	2,482	2,260	2,092	2,007	1,792	1,780	1,803	1,803
	6.6	9.8	47.9	120.5	125.8	130.9	123.6	112.5	104.2	99.9	89.2	88.6	89.8	89.8
Yugoslavia	5	8	10	228	296	297	297	297	297	297	302	483	608	609
	0.2	0.4	0.5	11.4	14.7	14.8	14.8	14.8	14.8	14.8	15.0	24.1	30.3	30.3
Total Europe	**438**	**871**	**3,976**	**14,657**	**20,242**	**20,206**	**19,743**	**17,424**	**16,530**	**15,952**	**14,374**	**13,954**	**14,062**	**13,744**
	21.5	**43.1**	**197.9**	**730.1**	**1,007.9**	**1,006.2**	**983.3**	**867.8**	**823.3**	**794.5**	**716.0**	**695.1**	**700.5**	**684.5**

NORTH AMERICA

Upper figures = 1,000 b/d
Lower figures = million t/y

	1940	1950	1960	1970	1980	1981	1982	1983	1984	1985	1986	1987	1988	1989
Canada	203	329	1,002	1,355	2,222	2,168	2,200	2,020	1,807	1,869	1,856	1,760	1,869	1,856
	10.1	16.4	49.9	67.5	110.7	107.8	109.6	100.6	90.0	93.1	92.4	87.6	93.1	92.4
United States	4,460	6,696	10,400	12,079	17,720	18,470	18,700	16,800	15,930	15,400	15,182	15,258	15,288	15,557
	222.1	333.5	517.9	601.5	882.5	919.8	931.3	836.6	793.3	766.9	756.1	759.8	761.3	774.7
Total	**4,663**	**7,025**	**11,402**	**13,434**	**19,942**	**20,638**	**20,900**	**18,820**	**17,737**	**17,269**	**17,038**	**17,018**	**17,157**	**17,413**
	232.2	**349.9**	**567.8**	**669.0**	**993.2**	**1,027.8**	**1,040.9**	**937.2**	**883.3**	**860.0**	**848.5**	**847.4**	**854.4**	**867.1**

LATIN AMERICA

Upper figures = 1,000 b/d
Lower figures = million t/y

	1940	1950	1960	1970	1980	1981	1982	1983	1984	1985	1986	1987	1988	1989
Argentina	103	152	238	502	676	679	677	676	678	678	667	670	690	690
	5.1	7.6	11.9	25.0	33.7	33.8	33.7	33.7	33.8	33.8	33.2	33.4	34.4	34.4
Bahamas	...	...	...	...	500	500	500	500	500	350	...	...	...	...
	...	...	...	...	24.9	24.9	24.9	24.9	24.9	17.4	...	...	...	...
Barbados	...	...	...	3	3	3	3	3	3	3	3	3	3	3
	...	...	...	0.1	0.1	0.1	0.1	0.1	0.1	0.1	0.1	0.1	0.1	...
Bolivia	1	7	12	21	74	74	61	61	47	47	47	47	47	58
	...	0.3	0.6	1.0	3.7	3.7	3.0	3.0	2.3	2.3	2.3	2.3	2.3	2.9
Brazil	4	12	156	575	1,205	1,402	1,407	1,219	1,301	1,305	1,305	1,321	1,407	1,407
	0.2	0.6	7.8	28.6	60.0	69.8	70.1	60.7	64.8	65.0	65.0	65.8	70.1	70.1
Chile	1	4	24	115	139	139	141	141	141	141	141	150	141	147
	...	0.2	1.2	5.7	6.9	6.9	7.0	7.0	7.0	7.0	7.0	7.5	7.0	7.3
Colombia	15	24	51	140	193	199	199	214	211	211	211	226	226	227
	0.7	1.2	2.5	7.0	9.6	9.9	9.9	10.7	10.5	10.5	10.5	11.3	11.3	11.3
Costa Rica	...	...	...	8	11	15	15	16	17	16	15	16	16	16
	...	...	...	0.4	0.5	0.7	0.7	0.8	0.8	0.8	0.7	0.8	0.8	0.8
Cuba	4	7	87	93	160	160	160	160	160	160	160	160	160	160
	0.2	0.3	4.3	4.6	8.0	8.0	8.0	8.0	8.0	8.0	8.0	8.0	8.0	8.0
Ecuador	2	5	7	37	86	87	90	79	84	82	88	88	122	123
	0.1	0.2	0.3	1.8	4.3	4.3	4.5	3.9	4.2	4.1	4.4	4.4	6.1	6.1
El Salvador	...	...	...	14	16	16	16	16	16	16	16	17	17	17
	...	...	...	0.7	0.8	0.8	0.8	0.8	0.8	0.8	0.8	0.8	0.8	0.8
Guatemala	...	...	...	21	16	16	16	16	16	16	16	16	16	16
	...	...	...	1.0	0.8	0.8	0.8	0.8	0.8	0.8	0.8	0.8	0.8	0.8
Honduras	...	...	...	10	14	14	14	14	14	14	14	14	14	14
	...	...	...	0.5	0.7	0.7	0.7	0.7	0.7	0.7	0.7	0.7	0.7	0.7
Jamaica	...	...	...	28	34	22	22	36	36	36	36	36	36	34
	...	...	...	1.4	1.7	1.1	1.1	1.8	1.8	1.8	1.8	1.8	1.8	1.7
Mexico	98	160	357	515	1,394	1,394	1,470	1,289	1,269	1,269	1,269	1,349	1,354	1,354
	4.9	8.0	17.8	25.6	69.4	69.4	73.2	64.2	63.2	63.2	63.2	67.2	67.4	67.4
Netherlands Antilles	495	617	680	795	792	782	782	782	740	740	320	320	320	320
	24.7	30.7	33.9	39.6	39.4	38.9	38.9	38.9	36.9	36.9	15.9	15.9	15.9	15.9
Nicaragua	...	...	...	21	14	14	14	14	15	15	15	15	15	15
	...	...	...	1.0	0.7	0.7	0.7	0.7	0.7	0.7	0.7	0.7	0.7	0.7
Panama	...	...	...	140	100	100	100	100	100	100	100	100	100	100
	...	...	...	7.0	5.0	5.0	5.0	5.0	5.0	5.0	5.0	5.0	5.0	5.0
Paraguay	...	...	...	5	7	8	8	8	8	8	8	8	8	8
	...	...	...	0.2	0.3	0.4	0.4	0.4	0.4	0.4	0.4	0.4	0.4	0.4
Peru	22	35	49	104	170	169	162	168	169	176	176	176	182	172
	1.1	1.7	2.4	5.2	8.5	8.4	8.1	8.4	8.4	8.8	8.8	8.8	9.1	8.6
Puerto Rico	...	...	84	155	284	284	284	123	121	121	121	121	123	123
	...	...	4.2	7.7	14.1	14.1	14.1	6.1	6.0	6.0	6.0	6.0	6.1	6.1
Trinidad & Tobago	49	104	182	417	456	456	456	375	375	320	260	300	300	300
	2.4	5.2	9.1	20.8	22.7	22.7	22.7	18.7	18.7	15.9	12.9	14.9	14.9	14.9
Uruguay	5	16	28	40	40	40	45	45	45	45	45	45	45	33
	0.2	0.8	1.4	2.0	2.0	2.0	2.2	2.2	2.2	2.2	2.2	2.2	2.2	1.6
Venezuela	55	254	886	1,324	1,446	1,349	1,323	1,284	1,224	1,224	1,230	1,225	1,201	1,201
	2.7	12.6	44.1	65.9	72.0	67.2	65.9	63.9	61.0	61.0	61.3	61.0	59.8	59.8
Virgin Islands	...	...	...	220	728	728	640	600	600	545	600	600	545	545
	...	...	...	11.0	36.3	36.3	31.9	29.9	29.9	27.1	29.9	29.9	27.1	27.1
Others	...	...	...	31	62	62	59	77	55	57	57	57	61	60
	...	...	...	1.5	3.1	3.1	2.9	3.8	2.7	2.8	2.8	2.8	3.0	3.0
Total	854	1,397	2,841	5,334	8,620	8,712	8,664	8,016	7,945	7,695	6,920	7,080	7,149	7,143
	42.3	69.4	141.5	265.3	429.2	433.7	431.3	399.1	395.6	383.1	344.4	352.5	355.8	355.5

FEED CONTROL VALVES on ethylene furnaces in Japanese facility (photo by Larry Hackemesser, courtesy of M.S. Kellogg Co.).

ASIA-PACIFIC

Upper figures = 1,000 b/d
Lower figures = million t/y

	1940	1950	1960	1970	1980	1981	1982	1983	1984	1985	1986	1987	1988	1989
Australia	4	14	237	581	725	743	742	716	722	697	623	626	637	644
	0.2	0.7	11.8	28.9	36.1	37.0	37.0	35.7	36.0	34.7	31.0	31.2	31.7	32.1
Bangladesh	...	...	...	...	31	31	31	31	22	31	31	31	31	31
	...	...	...	...	1.5	1.5	1.5	1.5	1.1	1.5	1.5	1.5	1.5	1.5
Brunei-Malaysia-Singapore	10	10	46	302	1,093	1,246	1,271	1,271	1,313	1,286	1,240	1,183	1,080	1,071
	0.5	0.5	2.3	15.0	54.4	62.1	63.3	63.3	65.4	64.0	61.8	58.9	53.8	53.3
Burma*	...	...	...	...	26	26	26	26	26	26	26	26	26	26
	...	...	...	...	1.3	1.3	1.3	1.3	1.3	1.3	1.3	1.3	1.3	1.3
China, Taiwan	...	...	28	129	425	515	515	515	515	543	543	543	600	570
	...	...	1.4	6.4	21.2	25.6	25.6	25.6	25.6	27.0	27.0	27.0	29.9	28.4
Guam*	...	...	...	...	44	44	44	44	44	44	...	...	...	...
	...	...	...	...	2.2	2.2	2.2	2.2	2.2	2.2	...	...	...	...
India	10	6	112	436	557	557	557	753	779	705	867	991	1,059	1,051
	0.5	0.3	5.6	21.7	27.7	27.7	27.7	37.5	38.8	35.1	43.2	49.4	52.7	52.3
Indonesia	170	203	274	268	528	445	498	341	387	631	636	636	714	714
	8.5	10.1	13.6	13.3	26.3	22.2	24.8	17.0	19.3	31.4	31.7	31.7	35.6	35.6
Japan	53	34	640	2,796	5,509	5,454	5,601	5,548	5,020	4,813	4,613	4,790	4,567	4,363
	2.6	1.7	31.9	139.2	274.3	271.6	278.9	276.3	250.0	239.7	229.7	238.5	227.4	217.3
Korea, South	...	...	...	175	601	607	755	755	776	776	782	862	820	880
	...	...	...	8.7	29.9	30.2	37.6	37.6	38.6	38.6	38.9	42.9	40.8	43.8
New Zealand*	...	...	...	...	74	74	74	74	74	53	53	54	82	88
	...	...	...	...	3.7	3.7	3.7	3.7	3.7	2.6	2.6	2.7	4.1	4.4
Okinawa*	...	...	...	...	198	208	208	183	153	153	110	...	...	...
	...	...	...	...	9.9	10.4	10.4	9.1	7.6	7.6	5.5	...	...	...
Pakistan	...	5	6	115	98	122	133	133	126	129	130	130	130	130
	...	0.2	0.3	5.7	4.9	6.1	6.6	6.6	6.3	6.4	6.5	6.5	6.5	6.5
Philippines	...	...	22	180	253	246	286	286	286	286	216	286	286	254
	...	...	1.1	9.0	12.6	12.3	14.2	14.2	14.2	14.2	10.8	14.2	14.2	12.6
Sri Lanka*	...	...	...	...	38	50	50	50	50	50	50	50	50	50
	...	...	...	...	1.9	2.5	2.5	2.5	2.5	2.5	2.5	2.5	2.5	2.5
Thailand	...	...	1	62	186	186	176	176	176	172	192	192	192	191
	...	...	...	3.1	9.3	9.3	8.8	8.8	8.8	8.6	9.6	9.6	9.6	9.5
Others	27	4	9	137	...	...	...	...	...	...	...	...	...	...
	1.3	0.2	0.4	6.8	...	...	...	...	...	...	...	...	...	...
Total	274	276	1,375	5,181	10,386	10,554	10,967	10,902	10,469	10,395	10,112	10,400	10,274	10,063
	13.6	13.7	68.4	257.8	517.2	525.7	546.1	542.9	521.4	517.4	503.6	517.9	511.6	501.1

*Prior to 1980, included in others

AFRICA

Upper figures = 1,000 b/d
Lower figures = million t/y

	1940	1950	1960	1970	1980	1981	1982	1983	1984	1985	1986	1987	1988	1989
Algeria	...	...	...	47	122	122	122	137	137	464	465	465	465	465
	...	...	...	2.3	6.1	6.1	6.1	6.8	6.8	23.1	23.2	23.2	23.2	23.2
Angola-Cabinda	...	...	2	20	36	31	32	32	32	32	32	32	32	32
	...	...	0.1	1.0	1.8	1.5	1.6	1.6	1.6	1.6	1.6	1.6	1.6	1.6
Cameroon*	...	...	...	...	...	...	43	43	41	43	43	43	43	42
	...	...	...	...	...	...	2.1	2.1	2.0	2.1	2.1	2.1	2.1	2.1
Congo*	...	...	...	...	...	...	...	...	21	21	21	21	21	21
	...	...	...	...	...	...	...	...	1.0	1.0	1.0	1.0	1.0	1.0
Egypt	17	39	87	193	234	292	292	341	369	369	434	452	452	489
	0.8	1.9	4.3	9.6	11.7	14.5	14.5	17.0	18.4	18.4	21.6	22.5	22.5	24.4
Ethiopia*	...	...	...	...	14	14	14	14	15	14	18	18	18	18
	...	...	...	...	0.7	0.7	0.7	0.7	0.7	0.7	0.9	0.9	0.9	0.9
Gabon*	...	...	...	...	20	20	20	20	20	20	23	16	23	24
	...	...	...	...	1.0	1.0	1.0	1.0	1.0	1.0	1.1	0.8	1.1	1.2
Ghana	...	...	...	29	26	26	27	27	28	28	28	28	28	27
	...	...	...	1.4	1.3	1.3	1.3	1.3	1.4	1.4	1.4	1.4	1.4	1.3
Ivory Coast	...	...	...	19	50	50	50	90	90	90	90	73	60	60
	...	...	...	0.9	2.5	2.5	2.5	4.5	4.5	4.5	4.5	3.6	3.0	3.0
Kenya	...	...	...	44	95	95	79	79	79	95	95	95	90	90
	...	...	...	2.2	4.7	4.7	3.9	3.9	3.9	4.7	4.7	4.7	4.5	4.5
Liberia*	...	...	...	...	15	15	15	15	15	15	15	15	15	15
	...	...	...	...	0.7	0.7	0.7	0.7	0.7	0.7	0.7	0.7	0.7	0.7
Libya	...	...	...	10	138	142	130	130	125	330	330	329	329	329
	...	...	...	0.5	6.9	7.1	6.5	6.5	6.2	16.4	16.4	16.4	16.4	16.4
Madagascar*	...	...	...	...	11	11	11	16	16	16	16	16	16	16
	...	...	...	...	0.5	0.5	0.5	0.8	0.8	0.8	0.8	0.8	0.8	0.8
Morocco	...	...	2	35	72	72	72	74	74	80	81	81	155	155
	...	...	0.1	1.7	3.6	3.6	3.6	3.7	3.7	4.0	4.0	4.0	7.7	7.7
Mozambique	...	...	...	20	16	16	16	16	17	17	17	17	...	...
	...	...	...	1.0	0.8	0.8	0.8	0.8	0.8	0.8	0.8	0.8	...	...
Nigeria	...	...	...	40	160	160	254	260	247	250	250	250	270	415
	...	...	...	2.0	8.0	8.0	12.6	12.9	12.3	12.5	12.5	12.5	13.4	20.7
Senegal	...	...	...	13	20	20	18	18	18	30	30	30	30	30
	...	...	...	0.6	1.0	1.0	0.9	0.9	0.9	1.5	1.5	1.5	1.5	1.5
Sierra Leone*	...	...	...	...	10	10	10	10	10	10	10	10	10	10
	...	...	...	...	0.5	0.5	0.5	0.5	0.5	0.5	0.5	0.5	0.5	0.5
South Africa	1	...	25	178	478	469	424	424	389	389	389	389	434	434
	...	...	1.2	8.9	23.8	23.4	21.1	21.1	19.4	19.4	19.4	19.4	21.6	21.6
Sudan	...	...	...	20	26	24	24	24	24	24	24	24	21	21
	...	...	...	1.0	1.3	1.2	1.2	1.2	1.2	1.2	1.2	1.2	1.0	1.0
Tanzania	...	...	...	14	17	12	14	17	17	14	14	14	14	14
	...	...	...	0.7	0.8	0.6	0.7	0.8	0.8	0.7	0.7	0.7	0.7	0.7
Togo*	...	...	...	...	20	20	20	20	20	20	20	20	20	...
	...	...	...	...	1.0	1.0	1.0	1.0	1.0	1.0	1.0	1.0	1.0	...
Tunisia	...	...	...	23	34	34	34	34	34	34	34	34	34	34
	...	...	...	1.1	1.7	1.7	1.7	1.7	1.7	1.7	1.7	1.7	1.7	1.7
Zaire*	...	...	...	...	16	17	17	17	17	17	17	17	17	17
	...	...	...	...	0.8	0.8	0.8	0.8	0.8	0.8	0.8	0.8	0.8	0.8
Others	...	...	...	87	35	35	35	35	35	35	35	35	34	31
	...	...	...	4.3	1.7	1.7	1.7	1.7	1.7	1.7	1.7	1.7	1.7	1.5
Total	**18**	**39**	**116**	**792**	**1,665**	**1,707**	**1,773**	**1,893**	**1,890**	**2,457**	**2,531**	**2,524**	**2,631**	**2,789**
	0.8	**1.9**	**5.7**	**39.2**	**82.9**	**84.9**	**88.0**	**94.0**	**93.8**	**122.2**	**125.8**	**125.5**	**130.8**	**138.8**

*Prior to 1980, included in others

MIDDLE EAST

Upper figures = 1,000 b/d
Lower figures = million t/y

	1940	1950	1960	1970	1980	1981	1982	1983	1984	1985	1986	1987	1988	1989
Abu Dhabi*	...	...	...	...	14	14	128	135	185	185	185	185	180	180
	...	...	...	...	0.7	0.7	6.4	6.7	9.2	9.2	9.2	9.2	9.0	9.0
Bahrain	33	150	187	265	250	250	250	250	250	250	250	250	250	243
	1.6	7.5	9.3	13.2	12.5	12.5	12.5	12.5	12.5	12.5	12.5	12.5	12.5	12.1
Iran	333	502	495	702	921	1,121	530	530	530	530	530	530	530	530
	16.6	25.0	24.7	35.0	45.9	55.8	26.4	26.4	26.4	26.4	26.4	26.4	26.4	26.4
Iraq	4	9	56	83	168	169	169	169	319	319	319	319	319	319
	0.2	0.4	2.8	4.1	8.4	8.4	8.4	8.4	8.4	15.9	15.9	15.9	15.9	15.9
Israel	...	83	87	115	195	190	190	190	190	170	170	180	180	180
	...	4.1	4.3	5.7	9.7	9.5	9.5	9.5	9.5	8.5	8.5	9.0	9.0	9.0
Jordan*	...	...	...	...	21	36	91	100	100	100	100	100	100	100
	...	...	...	...	1.0	1.8	4.5	5.0	5.0	5.0	5.0	5.0	5.0	5.0
Kuwait†	...	25	270	569	645	645	623	623	623	669	634	618	628	817
	...	1.2	13.4	28.3	32.1	32.1	31.0	31.0	31.0	33.3	31.6	30.8	31.3	40.7
Lebanon	...	6	24	37	53	52	52	52	52	17	17	37	37	37
	...	0.3	1.2	1.8	2.6	2.6	2.6	2.6	2.6	0.8	0.8	1.8	1.8	1.8
Qatar*	...	...	...	...	11	13	13	12	63	56	56	62	62	62
	...	...	...	...	0.5	0.6	0.6	0.6	3.1	2.8	2.8	3.1	3.1	3.1
Saudi Arabia†	...	140	189	465	487	487	487	705	860	840	1,115	1,125	1,375	1,375
	...	7.0	9.4	23.2	24.3	24.3	24.3	35.1	42.8	41.8	55.5	56.0	68.5	68.5
Syria	...	...	...	54	223	223	223	229	229	228	229	229	229	244
	...	...	...	2.7	11.1	11.1	11.1	11.4	11.4	11.4	11.4	11.4	11.4	12.2
Turkey	...	1	7	209	356	356	367	467	472	460	460	460	676	725
	...	...	0.3	10.4	17.7	17.7	18.3	23.3	23.5	22.9	22.9	22.9	33.7	36.1
Yemen, South	...	...	120	178	175	175	178	178	130	178	162	170	170	162
	...	...	6.0	8.9	8.7	8.7	8.9	8.9	6.5	8.9	8.1	8.5	8.5	8.1
Others	...	...	1	9	...	...	...	46	48	50	48	58	87	87
	...	...	...	0.4	...	...	...	2.3	2.4	2.5	2.4	2.9	4.3	4.3
Total	370	916	1,436	2,686	3,519	3,730	3,301	3,686	3,901	4,052	4,275	4,323	4,823	5,061
	18.4	45.5	71.4	133.7	175.2	185.8	164.5	183.7	194.3	201.9	213.0	215.4	240.4	252.2
Total, exclu. U.S.S.R., China, E. Europe	6,617	10,524	21,146	42,084	64,374	65,547	65,348	60,741	58,472	57,820	55,250	55,299	56,096	56,213
	328.8	523.5	1,052.7	2,095.1	3,205.6	3,264.1	3,254.1	3,024.7	2,911.7	2,879.1	2,751.3	2,753.8	2,793.5	2,799.2

*Prior to 1980, included in others. †Includes Neutral Zone.

U.S.S.R.-EASTERN EUROPE-CHINA

Upper figures = 1,000 b/d
Lower figures = million t/y

	1940	1950	1960	1970	1980	1981	1982	1983	1984	1985	1986	1987	1988	1989
China	...	8	98	300	1,600	1,810	1,810	2,000	2,050	2,150	2,150	2,200	2,200	2,200
	...	0.4	4.9	14.9	79.7	90.1	90.1	99.6	102.1	107.1	107.1	109.6	109.6	109.6
Czechoslovakia	...	16	46	220	455	455	455	455	455	455	455	455	455	455
	...	0.8	2.3	11.0	22.7	22.7	22.7	22.7	22.7	22.7	22.7	22.7	22.7	22.7
Hungary	...	20	58	130	290	312	312	311	312	242	242	242	220	220
	...	1.0	2.9	6.5	14.4	15.5	15.5	15.5	15.5	12.1	12.1	12.1	11.0	11.0
Poland	...	4	20	156	385	385	385	385	385	385	385	385	385	385
	...	0.2	1.0	7.8	19.2	19.2	19.2	19.2	19.2	19.2	19.2	19.2	19.2	19.2
Romania	...	182	260	356	608	617	617	617	617	617	617	617	617	617
	...	9.1	12.9	17.7	30.3	30.7	30.7	30.7	30.7	30.7	30.7	30.7	30.7	30.7
U.S.S.R.	...	900	2,800	5,640	10,950	11,400	11,600	11,750	12,000	12,200	12,260	12,260	12,260	12,300
	...	44.8	139.4	280.9	545.3	567.7	577.7	585.2	597.6	607.6	610.5	610.5	610.5	612.5
Others	...	4	40	332	852	852	852	898	894	894	894	894	852	910
	...	0.2	2.0	16.5	42.4	42.4	42.4	44.7	44.5	44.5	44.5	44.5	42.4	45.3
Total	...	1,134	3,322	7,134	15,140	15,831	16,031	16,416	16,713	16,943	17,003	17,053	16,989	17,087
	...	56.5	165.4	355.3	754.0	788.3	798.3	817.6	832.3	843.9	846.8	849.3	846.1	851.0
World Total	6,617	11,658	24,468	49,218	79,514	81,378	81,379	77,157	75,185	74,763	72,253	72,352	73,085	73,300
	328.8	580.0	1,218.1	2,450.4	3,959.6	4,052.4	4,052.4	3,842.3	3,744.0	3,723.0	3,598.1	3,603.1	3,639.6	3,650.2

LARGE DIVERTER GATE VALVE for FCC unit at Statoil's Rafinor A.S. refinery in Mongstad, Norway (Photo courtesy Tapco Inc., Houston).

Refining survey

YOUR USE OF THE INTERNATIONAL REFINing survey may be aided by the following information:

All country-by-country volumes are presented in calendar-day figures except those for the U.S. In that tabulation, only crude is reported in calendar-day figures. Cat cracking and cat reforming are given in stream-day figures.

In the U.S. tabulation, when an asterisk (*) appears beside a number in the crude column, this indicates that the figure was reported as a stream-day figure and has been converted to a calendar-day figure using the crude conversion factor of 0.95.

Calendar-day figures are the average volume a refinery unit processes each day including downtime used for turnarounds. This is actual total volume for the year divided by 365.

Steam-day figures represent the amount a unit can process when running full capacity for short periods. [IPE]

LEGEND

CAT CRACKING
1. Fluid
2. Other

CAT REFORMING
Semiregenerative:
1. Conventional catalyst
2. Bimetallic catalyst
Cyclic:
3. Conventional catalyst
4. Bimetallic catalyst
Other:
5. Conventional catalyst
6. Bimetallic catalyst

Cat reforming definitions are:

Semiregenerative. Charactierized by shutdown to the reforming unit at specified intervals or at the operator's convenience for regeneration of catalyst in situ.

Cyclic. Charactierized by continuous or continual regeneration of catalyst in situ in any one of several reactors that can be isolated from and returned to the reforming operation. This is accomplished without changing feed rate or octane.

Other includes:

Nonregenerative—catalyst replaced by fresh catalyst.

Continuous—continuous regeneration of a part of the catalyst in a special regenerator followed by continuous addition to the reactor.

Moving-bed catalyst systems.

Survey of operating refineries worldwide

(capacities of January 1, 1989)

Country	No. plants	Crude	Catalytic cracking	Catalytic reforming
Abu Dhabi	2	180,000	.....	30,055
Algeria	4	464,700	.....	55,600
Angola	1	32,100	.....	1,900
Argentina	11	690,400	124,800	41,000
Australia	10	644,100	190,300	159,600
Austria	1	204,000	24,000	32,000
Bahrain	1	243,000	39,000	18,000
Bangladesh	1	31,200	.....	1,650
Barbados	1	3,000	.....	.....
Belgium	4	630,500	102,000	80,300
Bolivia	3	57,500	.....	14,900
Brazil	13	1,407,300	350,500	24,800
Brunei	1	10,000	.....	.....
Burma	2	26,300	.....	.....
Cameroon	1	42,000	.....	7,000
Canada	27	1,855,800	387,600	369,600
Chile	3	146,800	36,670	10,000
China, Taiwan	2	570,000	22,500	56,250
Colombia	4	227,400	91,000	6,000
Congo	1	21,000	.....	2,000
Costa Rica	1	16,200	.....	1,200
Cote d'Ivoire	2	60,000	.....	16,600
Cyprus	1	17,100	.....	4,250
Denmark	3	176,500	.....	31,700
Dominican Republic	2	47,000	.....	7,500
Ecuador	3	123,300	16,000	2,780
Egypt	8	489,203	.....	30,940
El Salvador	1	17,000	.....	3,800
Ethiopia	1	18,000	.....	2,270
Finland	2	241,000	41,800	42,900
France	14	1,875,970	319,200	247,000
Gabon	1	24,000	.....	1,400
Germany	15	1,518,200	179,400	275,100
Ghana	1	26,600	.....	5,850
Greece	4	384,500	50,500	51,300
Guatemala	1	16,000	.....	3,000
Honduras	1	14,000	.....	1,800
Hungary	3	220,000	20,000	23,000
India	12	1,051,441	133,953	26,030
Indonesia	6	714,200	12,600	61,500
Iran	4	530,000	.....	63,845
Iraq	8	318,500	.....	43,500
Ireland	1	56,000	.....	11,000
Israel	2	180,000	20,000	26,000
Italy	19	2,450,200	278,300	284,300
Jamaica	1	34,200	.....	3,240
Japan	41	4,362,750	596,950	551,650
Jordan	1	100,000	4,410	8,640
Kenya	1	90,000	.....	9,000
Korea	6	880,000	.....	62,450
Kuwait	4	817,000	42,000	33,000

Country	No. plants	Crude	Catalytic cracking	Catalytic reforming
Lebanon	2	37,000	7,250	7,492
Liberia	1	15,000	.....	2,000
Libya	3	329,400	.....	13,982
Madagascar	1	16,350	.....	2,600
Malaysia	4	209,300	.....	21,980
Martinique	1	12,800	.....	7,410
Mexico	9	1,354,000	267,000	157,800
Morocco	2	154,600	5,600	26,800
Netherlands	7	1,380,700	125,500	166,400
Netherlands Antilles	1	320,000	42,000	15,000
New Zealand	1	88,400	.....	21,900
Nicaragua	1	14,500	.....	2,800
Nigeria	4	414,500	84,000	71,000
Norway	3	239,400	.....	30,100
Oman	1	76,932	.....	15,386
Pakistan	3	130,050	.....	5,450
Panama	1	100,000	.....	7,500
Paraguay	1	7,500	.....	.....
Peru	6	172,438	22,540	1,680
Philippines	4	254,000	23,100	39,400
Portugal	3	313,300	10,100	52,200
Puerto Rico	2	123,000	12,000	5,800
Qatar	1	62,000	.....	12,030
Saudi Arabia	7	1,375,000	76,000	180,900
Senegal	1	29,800	.....	2,870
Sierra Leone	1	10,000	.....	.....
Singapore	5	852,000	.....	58,300
Somalia	1	10,000	.....	.....
South Africa	4	433,500	84,400	62,800
Spain	10	1,285,000	164,000	171,600
Sri Lanka	1	50,000	.....	3,750
Sudan	1	21,440	.....	1,700
Sweden	5	426,500	25,000	71,000
Switzerland	2	132,000	.....	26,000
Syria	2	243,744	.....	18,994
Tanzania	1	13,500	.....	3,000
Thailand	3	191,045	8,100	26,280
Trinidad	2	300,000	28,000	31,000
Tunisia	1	34,000	.....	3,300
Turkey	5	724,654	34,128	63,127
United Kingdom	15	1,803,100	442,000	307,500
Uruguay	1	33,000	4,100	2,500
Venezuela	6	1,201,100	193,100	6,000
Virgin Islands	1	545,000	.....	125,000
Yemen, N.	1	10,000	.....	2,500
Yemen, S.	1	161,500	.....	10,800
Yugoslavia	7	609,135	52,050	74,612
Zaire	1	16,700	.....	3,500
Zambia	1	20,594	.....	5,070
Total	**411**	**40,713,446**	**4,793,451**	**4,796,013**

Country-by-country refining survey

WORLDWIDE REFINING	Crude capacity	b/cd Catalytic cracking	Cat reforming
ABU DHABI			
Abu Dhabi National Oil Co.–Ruwais	120,000	....	[2]19,150
Umm Al-Nar 2	60,000	....	[4]10,905
Total	**180,000**	**....**	**30,055**
ALBANIA			
Government-owned refineries at Ballsh, Cerrik, Stalin.			
Total	**40,000**		
ALGERIA			
Sonatrach: Arzew	60,000	....	8,600
Hassi Messaoud	23,700	....	2,000
Maison Carree	58,000	....	15,000
Skikda	323,000	....	30,000
Total	**464,700**	**....**	**55,600**
ANGOLA			
Petrangol SARL–Luanda	**32,100**	**....**	**[1]1,900**
ARGENTINA			
Destileria Argentina de Petróleo SA–			
Lomas de Zamora	2,000	....	....
Esso SAPA–Campana	94,500	[1]29,700	[2]8,500
Galvan	18,600	....	....
Isaura SA–Bahia Blanca	12,000	....	....
Shell Cia. Argentina de Petróleo SA–			
Buenos Aires	121,700	[1]29,200	[1]12,000
Yacimientos Petroliferos Fiscales–			
Campo Duran	32,000	....	....
Dock Sud	4,000	....	....
La Plata	216,000	[1]45,000	[1]9,000
Lujan de Cuyo	129,000	[1]20,900	[1]9,000
Plaza Huincul	23,000	....	[2]2,500
San Lorenzo	37,600	....	....
Total	**690,400**	**124,800**	**41,000**
AUSTRALIA			
Ampol Refineries Ltd.–Lytton	72,000	[1]27,500	[1]13,000
Australian Lubricating Oil Refinery Ltd.–Kurnell	....	....	....
BP Australia–Brisbane	42,000	[1]15,800	[2]8,500
Kwinana	100,000	[1]24,000	[1]6,400 [2]16,400
Caltex Refining CPL–Kurnell[1]	100,000	[1]39,000	[1]25,600
Mobil Oil Australia Ltd.–Adelaide	....	....	....
Petroleum Refineries Australia PL–			
Adelaide	46,500	....	[2]20,000
Altona	98,600	[2]24,000	[2]28,000
Shell Refining (Australia) PL–Clyde	75,000	[1]35,000	[2]13,700
Geelong	110,000	[1]25,000	[2]8,000 [6]20,000
Total	**644,100**	**190,300**	**159,600**

Solvent extraction: [1]4,800 b/cd.

WORLDWIDE REFINING	Crude capacity	b/cd Catalytic cracking	Cat reforming
AUSTRIA			
OeMV–Schwechat	204,000	[1]24,000	[2]14,000 [4]18,000
Total	**204,000**	**24,000**	**32,000**
BAHRAIN			
Bahrain Petroleum Co. BSC (Closed) –Sitra	**243,000**	**[1]39,000**	**[2]18,000**
BANGLADESH			
Eastern Refinery Ltd.–Chittagong	**31,200**	**....**	**[1]1,650**
BARBADOS			
Mobil Oil Barbados Ltd.–Bridgetown	**3,000**	**....**	**....**
BELGIUM			
Belgian Refining Corp. NV–Antwerp	91,500	....	[2]12,000
Esso Inc.–Antwerp	218,000	[1]25,500	[4]38,100
Nynas Petroleum NV—Antwerp	15,000	....	....
Fina Raffinadeij Antwerpen–Antwerp	306,000	[1]76,500	[2]30,200
Total	**630,500**	**102,000**	**80,300**
BOLIVIA			
Yacimientos Petroliferos			
Fiscales Bolivianos–Cochabamba[1]	39,500	....	[1]8,500
Santa Cruz	15,000	....	[2]6,400
Sucre	3,000	....	....
Total	**57,500**	**....**	**14,900**

[1]Solvent extraction: 166 b/cd.

WORLDWIDE REFINING	Crude capacity	b/cd Catalytic cracking	Cat reforming
BRAZIL			
Petróleo Brasileiro SA: Araucária,			
Paraná	151,000	[1]47,200	....
Betim, Minas Gerais	144,600	[1]34,100	....
Canoas, Rio Grande do Sul	72,300	[1]15,100	....
Cubatão, São Paulo	163,600	[1]53,400	[1]11,000
Duque de Caxias, Rio de Janeiro	226,600	[1]47,200	[1]13,800
Fortaleza, Ceará	....	....	....
Manaus, Amazonas	10,000	[1]2,000	....
Mataripe, Bahia	121,400	[1]27,700	....
Mauá, Santo André, São Paulo	32,500	[2]18,200	....
Paulínia, São Paulo	302,000	[1]44,000	....
São José dos Campos, São Paulo	164,000	[1]59,100	....
Refinaria de Petróleo Ipiranga SA–Rio Grande do Sul	9,300	[1]2,500	....
Refinaría de Petroleos de Manguinhos SA–Rio de Janeiro	10,000	....	....
Total	**1,407,300**	**350,500**	**24,800**
BRUNEI			
Brunei Shell Petroleum CL–Seria	**10,000**	**....**	**....**
BULGARIA			
Government-owned refineries at Burgas, Pleven, Ruse.			
Total	**300,000**		
BURMA			
Petrochemical Industries Corp.–Chauk	6,300	....	....
Syriam	20,000	....	....
Total	**26,300**	**....**	**....**
CAMEROON			
Sonara–National Refining CL– Cape Limboh, Limbe	**42,000**	**....**	**[2]7,000**
CANADA			
Alberta			
Husky Oil Operations Ltd.–Lloydminster	23,500	....	....
Imperial Oil Ltd.–Edmonton	164,900	[1]47,000	[4]20,900
Petro-Canada Products Inc.– Edmonton	115,500	[1]36,000	[2]12,600

WORLDWIDE REFINING	Crude capacity	b/cd Catalytic cracking	Cat reforming
Shell Canada Ltd.–Scotford	59,700	....	[6]21,500
Turbo Resources Ltd.–Balzac	27,500	[1]8,800	[2]7,550
Total	**391,100**	**91,800**	**62,550**
British Columbia			
Chevron Canada Ltd.–North Burnaby	36,000	[1]10,500	[1]10,000
Husky Oil Operations Ltd.–			
Prince George	9,500	[1]3,300	[2]1,400
Imperial Oil Ltd.–Ioco	42,800	[1]12,300	[3]6,700
Petro-Canada Products Inc.–			
Port Moody	37,200	....	[2]8,600
Taylor	17,400	[1]6,300	[2]3,000
Shell Canada Ltd.–Shellburn, Burnaby	24,000	[1]5,400	[2]3,600
Total	**166,900**	**37,800**	**33,300**
New Brunswick			
Irving Oil Ltd.–St. John	237,500	[1]17,100	[2]34,650
Total	**237,500**	**17,100**	**34,650**
Newfoundland			
Newfoundland Processing Ltd.–			
Come By Chance	105,000	....	[2]26,000
Total	**105,000**	**....**	**26,000**
Northwest Territories			
Imperial Oil Ltd.–Norman Wells	3,500	....	....
Nova Scotia			
Imperial Oil Ltd.–Dartmouth	82,300	[1]23,000	[4]8,600
Texaco Canada Ltd.–Halifax	20,000	[1]7,200	[2]3,600
Total	**102,300**	**30,200**	**12,200**
Ontario			
Imperial Oil Ltd.–Sarnia	122,600	[1]23,500	[2]13,300
			[4]14,300
Petro-Canada Products Inc.–Clarkson	41,500	....	[6]9,800
Oakville	80,500	[1]25,400	[1]5,000
			[2]11,500
Shell Canada Ltd.–Sarnia	71,000	[1]14,400	[2]21,000
Suncor Inc.–Sarnia	70,000	[2]16,000	[2]26,000
Texaco Canada Ltd.–Nanticoke	95,000	[1]40,000	[6]24,000
Total	**480,000**	**119,300**	**124,900**
Quebec			
Petro-Canada Products Inc.–Montreal	87,400	[1]17,200	[2]31,300
Shell Canada Ltd.–Montreal	120,000	[1]22,000	[2]20,700
Ultramar Canada Inc.–St. Romuald———	103,000	[1]34,600	[2]15,000
Total	**310,400**	**73,800**	**67,000**
Saskatchewan			
Consumers' Cooperative			
Refineries Ltd.–Regina	45,200	[1]17,600	[2]9,000
Petro-Canada Products Inc.–			
Moose Jaw	13,300	....	....
Total	**58,500**	**17,600**	**9,000**
Total Canada	**1,855,800**	**387,600**	**369,600**

CHILE

	Crude capacity	Catalytic cracking	Cat reforming
ENAP–Gregorio-Magallanes	8,800	....	....
Petrox SA–Talcahuano	72,000	[1]16,670	[2]4,000
Refinería de Concón–Concón	66,000	[1]20,000	[1]6,000
Total	**146,800**	**36,670**	**10,000**

CHINA

Government-owned refineries at Anshan, Beijing, Dalian, Daqing, Fushun, Hangzhou, Jinxi, Karamai-Dushanzi, Lanchow, Lenghu, Maoming, Nanchong, Nanjing, Shanghai, Shengli, Tianjin, Yumen.

Total	**2,200,000**		

CHINA, TAIWAN

	Crude capacity	Catalytic cracking	Cat reforming
Chinese Petroleum Corp.–Kaohsiung	446,500	[1]22,500	[2]36,000
Tao-Yuan	123,500	....	[2]20,250
Total	**570,000**	**22,500**	**56,250**

WORLDWIDE REFINING	Crude capacity	b/cd Catalytic cracking	Cat reforming
COLOMBIA			
Empresa Colombiana de Petróleos			
–Barrancabermeja-Santander[1]	150,000	[1]62,000	[2]6,000
Cartagena, Bolivar	70,000	[1]29,000	....
Orito, Putumayo	2,400	....	....
Tibu, N. de Santander	5,000	....	....
Total	**227,400**	**91,000**	**6,000**

Solvent extraction: [1]35,000 b/cd.

CONGO

	Crude capacity	Catalytic cracking	Cat reforming
Coraf–Pointe-Noire	21,000	....	[2]2,000

COSTA RICA

	Crude capacity	Catalytic cracking	Cat reforming
Refinadora Costarricense de Petroleo			
SA–Limón	16,200	....	[2]1,200

COTE d'IVOIRE

	Crude capacity	Catalytic cracking	Cat reforming
Ste. Ivoirienne de Raffinage–Abidjan	50,000	....	[2]16,600
Ste. Multinationale de Bitumes–			
Abidjan	10,000	....	....
Total	**60,000**	**....**	**16,600**

CUBA

Government-owned refineries at Cabaiguan, Havana, Santiago de Cuba.

Total	**160,000**		

CYPRUS

	Crude capacity	Catalytic cracking	Cat reforming
Cyprus Petroleum Refinery Ltd.–			
Larnaca	17,100	....	[2]4,250

CZECHOSLOVAKIA

Government owned refineries at Bratislava, Kolin, Kralupi, Pardubice, Strazke, Zaluzi, Zyolen.

Total	**455,000**		

DENMARK

	Crude capacity	Catalytic cracking	Cat reforming
AS Dansk Shell–Fredericia	55,000	....	[2]12,000
Statoil AS–Kalundborg	65,000	....	[2]9,200
Kuwait Petroleum Refining			
(Danmark) A/S–Gulfhavn	56,500	....	[2]10,500
Total	**176,500**	**....**	**31,700**

DOMINICAN REPUBLIC

	Crude capacity	Catalytic cracking	Cat reforming
Falconbridge Dominicana C por À–La			
Peguera	16,000	....	....
Refineria Dominicana de Petróleo			
SA–Haina	31,000	....	[1]7,500
Total	**47,000**	**....**	**7,500**

ECUADOR

	Crude capacity	Catalytic cracking	Cat reforming
CEPE–Esmeraldas	90,000	[1]16,000	[2]2,780
Anglo Ecuadorian Oilfields Ltd.–			
Sta. Elena Peninsula	32,300	....	....
Texaco–Lago-Agrio	1,000	....	....
Total	**123,300**	**16,000**	**2,780**

EGYPT

	Crude capacity	Catalytic cracking	Cat reforming
Alexandria Petroleum Co.–Alexandria			
(El-Mex)	99,250	....	....
Cairo Oil Refining Co.–Mostorod	125,060	....	[2]9,000
Tanta	21,800	....	....
El Ameria Oil Refining Co.–			
Alexandria	60,340	....	[2]9,400
El-Nasr Petroleum Co.–El-Suez	58,400	....	....
Wadi-Feran	9,680	....	....
Suez Oil Processing Co.–El-Suez	62,530	....	[2]12,540
Assiout	52,143	....	....
Total	**489,203**	**....**	**30,940**

WORLDWIDE REFINING

	Crude capacity	b/cd Catalytic cracking	Cat reforming
EL SALVADOR			
Refinería Petrólera Acajutla SA–Acajutla	17,000	[1]3,200	....
Total	**17,000**	**3,200**	**....**
ETHIOPIA			
Ethiopian Petroleum Corp.–Assab	**18,000**	....	[1]**2,270**
FINLAND			
Neste Oy–Naantali	44,000	[2]13,300	[5]6,800
Porvoo	197,000	[1]28,500	[5]36,100
Total	**241,000**	**41,800**	**42,900**
FRANCE			
CRD—Total France—			
Gonfreville L'Orcher[1]	309,000	[1]31,500	[2]45,400
La Mède	136,000	[1]29,700	[4]23,400
Mardyck	122,000	[1]29,800	[4]20,500
Cie. Rhénane de Raffinage—			
Reichstett-Vendenheim	85,000	[1]14,000	[2]13,000
Elf France–Donges	200,000	[1]44,000	[2]25,400
Feyzin	175,070	[1]21,400	[2]8,400
Grandpuits	92,000	[1]28,800	[2]12,900
Esso SAF–Fos sur Mer	100,000	[1]17,000	[1]16,000
Port Jerome	136,000	[1]20,000	[1]16,000
Mobil Oil Francaise—			
Notre Dame de Gravenchon	57,000	....	[2]13,300
Shell Francaise–Berre l'Etang	128,000	[1]38,000	[2]19,000
Petit Couronne	154,000	[1]19,000	[2]28,000
Ste. Francaise des Petroles BP—			
Dunkirk	....	....	....
Lavera	181,900	[1]26,000	[1]5,700
Total	**1,875,970**	**319,200**	**247,000**

[1]Solvent extraction: 9,800 b/cd.

	Crude capacity	b/cd Catalytic cracking	Cat reforming
GABON			
Ste. Gabonaise de Raffinage–Port Gentil	**24,000**	....	[3]**1,400**
GERMANY, E.			
Government-owned	**470,000**		
GERMANY, WEST			
Deutsche BP-Erdol Ingolstadt–Vohburg†	102,000	[1]17,200	[1]17,200
Deutsche Marathon Petroleum GmbH–Burghausen	70,000	....	....
Deutsche Shell AG–Godorf	170,000	....	[2]17,000 [6]21,000
Harburg-Grasbrook	86,000	[1]15,000	[6]16,000
Deutsche Texaco AG–Heide	80,000	[2]9,000	[6]18,000
Erdoel Raffinerie Neustadt GmbH–Neustadt-Donau	144,000	[1]22,700	[1]17,700
Esso AG–Ingolstadt	95,000	[1]24,500	[2]12,800
Karlsruhe	150,000	....	[2]27,900
Mobil Oil AG–Worth	93,000	[1]19,000	[2]15,900
Oberrheinische Mineralolwerke GmbH–Karlsruhe	142,000	[1]66,000	[4]11,400 [6]20,100
Ruhr Oel GmbH–Gelsenkirchen	215,400	[1]6,000	[1]11,900 [5]27,000
Union Rheinische Braunkohlen Kraftstoff AG–Wesseling	103,000	....	[4]14,000
Wintershall AG–Lingen	65,000	....	[2]12,200 [4]15,000
Mannheim	....	....	....
Salzbergen	2,800	....	....
Total	**1,518,200**	**179,400**	**275,100**

†Operating on a semi-integrated basis.

	Crude capacity	b/cd Catalytic cracking	Cat reforming
GHANA			
Ghanaian Italian Petroleum CL–Tema	**26,600**	....	[1]**5,850**
GREECE			
Hellenic Aspropyrgos Refinery SA–Aspropyrgos	110,000	[1]23,300	[2]14,700 [6]20,000

	Crude capacity	b/cd Catalytic cracking	Cat reforming
WORLDWIDE REFINING			
Motor Oil (Hellas) Corinth Refineries SA–Aghii Theodori	100,000	[1]27,200	[1]8,500
Petrola Hellas SA–Elefsis	108,000	....	....
Thessaloniki Refining Co. AE–Thessaloniki	66,500	....	[2]8,100
Total	**384,500**	**50,500**	**51,300**
GUATEMALA			
Texas Petroleum Co.–Escuintla	**16,000**	....	[2]**3,000**
HONDURAS			
Refinería Texas de Honduras SA–Puerto Cortes	**14,000**	....	[1]**1,800**
HUNGARY			
Dunai KV–Százhalombatta	150,000	[1]20,000	[2]23,000
Tiszai KV–Leninváros	60,000	....	....
Zalai KV–Zalaegerszeg	10,000	....	....
Total	**220,000**	**20,000**	**23,000**
INDIA			
Bharat Petroleum CL–Mahul, Bombay	140,000	[1]31,000	[1]5,200
Bongaigaon Refinery & Petrochemicals Ltd.–Bongaigaon, Assam	27,392	....	[2]1,930
Cochin Refineries Ltd.–Kerala	90,000	[1]20,000	[1]5,000
Hindustan Petroleum CL—Mahul, Bombay[1]	110,000	[1]12,000	....
Visakhapatnam	92,699	[1]19,583	....
Indian Oil CL–Barauni	66,000	....	....
Digboi	10,050	....	....
Gauhati	60,400	....	....
Gujarat	165,200	[1]19,800	[3]7,000
Haldia	55,000	....	[3]4,500
Mathura	120,000	[1]19,800	....
Madras Refineries Ltd.–Madras	114,700	[1]11,770	[1]2,400
Total	**1,051,441**	**133,953**	**26,030**

[1]Solvent extraction: 10,000 b/cd.

	Crude capacity	b/cd Catalytic cracking	Cat reforming
INDONESIA			
Pertamina–Balikpapan, Kalimantan	221,500	....	[6]18,000
Cilacap, Central Java	271,200	....	[1]12,000 [6]18,000
Dumai, Central Sumatra	99,400	....	[1]5,600 [6]7,900
Musi, South Sumatra	81,400	[1]12,600	....
Pangakalan Brandan, North Sumatra	4,500	....	....
Sungai Pakning, Central Sumatra	36,200	....	....
Total	**714,200**	**12,600**	**61,500**
IRAN			
National Iranian Oil Co.–Esfahan	200,000	....	29,600
Shiraz	40,000	....	6,245
Tabriz	80,000	....	....
Tehran	210,000	....	28,000
Total	**530,000**	**....**	**63,845**

Note: Present status uncertain due to Iran-Iraq war.

	Crude capacity	b/cd Catalytic cracking	Cat reforming
IRAQ			
Oil Refineries Administration–Baiji	150,000	....	22,000 16,500
Basra	70,000	....	....
Daura	71,000	....	5,000
K3-Haditha	7,000	....	....
Khanaqin	12,000	....	....
Mufthia	4,500	....	....
Qaiyarah, Mosul	2,000	....	....
Iraqi Company for Oil Operations–Kirkuk	2,000	....	....
Total	**318,500**	**....**	**43,500**

Note: Present status uncertain due to Iran-Iraq war.

IRELAND

WORLDWIDE REFINING	Crude capacity	Catalytic cracking	Cat reforming
Irish Refining plc–Whitegate	56,000	....	[4]11,000

ISRAEL

WORLDWIDE REFINING	Crude capacity	Catalytic cracking	Cat reforming
Oil Refineries Ltd.–Ashdod	70,000	....	[2]10,000
Haifa	110,000	[1]22,000	[2]16,000
Total	**180,000**	**22,000**	**26,000**

ITALY

WORLDWIDE REFINING	Crude capacity	Catalytic cracking	Cat reforming
Agip Plus SpA–Livorno	100,000	....	[2]9,500
Agip Raffinazione–Rho, Milan	79,800	[2]14,300	[2]17,000
Sannazzaro, Pavia	210,000	[1]27,000	[2]10,000 [5]16,700
Taranto	83,800	....	[2]17,000
Venezia	90,000	....	[2]16,000
Anonima Petroli Italiana–Falconara, Marittima	78,900	....	[2]10,500
Arcola Petrolifera SpA–La Spezia	18,000	....	....
Esso Italiana SpA–Augusta, Siracusa	180,000	[1]45,300	[2]9,300 [4]7,300
Industrie Chimiche Italiane de Petrolio SpA–Frassino, Mantova	55,000	....	[2]7,000
Iplom SpA–Busalla	46,500	....	....
Isab–Melilli	220,000	....	[3]34,500
Mobil Oil Italiana SpA–Naples	100,000	[2]18,400	[2]12,500
Raffineria di Roma SpA–Rome	80,500	....	[2]14,200
Raffineria Mediterranea SrL–Milazzo	160,000	[1]28,000	[2]11,000
Raffineria Siciliana SrL–Gela	80,000	[1]34,000	[2]10,000
Saras SpA–Sarroch	285,000	[1]63,000	[5]27,000
Sarpom–Trecate, Novara	248,200	[1]16,300	[2]18,500 [4]8,100
Selm-Soc. Energia Montedison–Priolo	240,000	[1]32,000	[2]9,000
Tamoil Italia SpA–Cremona	95,000	....	[1]3,800 [2]6,400
Total	**2,450,200**	**278,300**	**284,300**

JAMAICA

WORLDWIDE REFINING	Crude capacity	Catalytic cracking	Cat reforming
Petrojam Ltd.–Kingston	34,200	....	[1]3,240

JAPAN

WORLDWIDE REFINING	Crude capacity	Catalytic cracking	Cat reforming
Asia Oil CL–Sakaide, Kagawa	13,500	[1]13,500	[2]9,000
Cosmo Oil CL–Chiba	209,000	[1]26,600	[2]31,800
Sakai	104,500	[1]17,550	[2]6,650
Yokkaichi City	166,300	[1]23,700	[2]5,400 [4]9,900
Fuji Oil CL–Sodegaura	126,600	[1]14,000	[2]13,600
General Sekiyu Seisei KK–Sakai	140,400	[1]32,500	[2]26,100
Idemitsu Kosan CL–Chita, Aichi	123,500	[1]27,000	[4]16,200
Himeji, Hyogo	104,500	....	[2]12,600
Ichihara, Chiba	199,500	[1]37,800	[2]15,300
Tokuyama, Yamaguchi	95,000	[1]20,700	....
Tomakomai, Hokkaido	85,500	[2]16,200	....
Kainan Petroleum Refining CL–Kainan	64,000	....	....
Kashima Oil CL–Kashima, Ibaragi	150,000	[1]17,500	[4]14,000
Koa Oil CL–Marifu	110,000	[1]20,000	[2]9,700
Osaka	80,000	[1]18,000	[2]9,000
Kyokuto Petroleum Ltd.–Chiba	120,000	....	[2]19,000
Kyushu Oil CL–Oita	130,000	[1]16,500	[2]8,000
Mitsubishi Oil CL–Kawasaki	55,000	....	[2]8,500
Mizushima	220,000	[1]34,000	[2]25,000 [6]18,500
Nansei Sekiyu KK–Nishihara, Okinawa	45,000	....	[1]8,600
Nichimo Sekiyu Seisei KK–Kawasaki	76,000	....	[1]7,600
Nihonkai Oil CL–Toyama	50,000	....	4,000
Nippon Mining CL–Chita	85,000	[1]13,000	[2]22,000
Funakawa	6,000	....	....
Mizushima	190,200	[1]38,000	[2]20,000
Nippon Oil CL–Niigata	26,000	[2]4,000	....
Nippon Petroleum Refining CL–			
Muroran	150,000	[1]21,000	[2]21,000
Nakagusuku, Okinawa	....	....	[2]2,500
Negishi	305,000	[1]39,000	[2]20,000 [6]30,000
Yokohama	....	....	[2]1,300
Okinawa Sekiyu Seisei–Yonashiro, Okinawa	57,000	....	....
Seibu Oil CL–Yamaguchi	110,100	[1]18,100	[2]17,800
Showa Shell Sekiyu KK–Kawasaki	136,800	....	[2]19,100
Niigata	27,500	....	[2]4,300
Showa Yokkaichi Sekiyu CL–Yokkaichi	215,700	[1]22,600	[2]38,100
Taiyo Oil CL–Ehime	61,750	....	[2]5,400
Toa Nenryo Kogyo–Kawasaki	190,000	[1]74,100	[2]15,200
Wakayama	157,700	[1]31,600	[2]34,200
Toa Oil CL–Kawasaki	65,000	....	[2]11,000
Toho Oil CL–Owase	35,000	....	....
Tohoku Oil CL–Sendai	75,700	....	[2]11,300
Total	**4,362,750**	**596,950**	**551,650**

JORDAN

WORLDWIDE REFINING	Crude capacity	Catalytic cracking	Cat reforming
Jordan Petroleum Refinery–Zerka	100,000	[1]4,410	[2]8,640

KENYA

WORLDWIDE REFINING	Crude capacity	Catalytic cracking	Cat reforming
Kenya Petroleum Refineries Ltd.–Mombasa	90,000	....	[1]3,800 [2]5,200
Total	**90,000**	**....**	**9,000**

KOREA, N.

WORLDWIDE REFINING	Crude capacity	Catalytic cracking	Cat reforming
Government-owned, Unggi	42,000		

KOREA, S.

WORLDWIDE REFINING	Crude capacity	Catalytic cracking	Cat reforming
Honam Oil Refinery CL–Yosu	355,000	....	[2]15,300
Kukdong Oil CL–Busan[1]	10,000	....	....
–Daesan	60,000	....	[2]3,000
Kyung In Energy CL–Inchon	60,000	....	[2]3,600
Ssangyong Oil Refining CL–Onsan[2]	90,000	....	[1]4,000
Yukong Ltd.–Ulsan	305,000	....	[2]17,650 [6]18,900
Total	**880,000**	**....**	**62,450**

Solvent extraction: [1]2,100 b/d. [2]8,000 b/d.

KUWAIT

WORLDWIDE REFINING	Crude capacity	Catalytic cracking	Cat reforming
Getty Oil Co. (subsidiary of Texaco Inc.)–Mina Al-Zour	70,000	....	....
Kuwait National Petroleum Co.–Mina Abdulla	190,000	....	....
Mina Al-Ahmadi	370,000	[1]28,000	[2]33,000
Shuaiba	187,000	[2]14,000	....
Total	**817,000**	**42,000**	**33,000**

LEBANON

WORLDWIDE REFINING	Crude capacity	Catalytic cracking	Cat reforming
Tripoli Oil Installations–Tripoli	20,000	[1]7,250	[2]4,392
Mediterranean Refining Co.–Sidon	17,000	....	[1]3,100
Total	**37,000**	**7,250**	**7,492**

LIBERIA

WORLDWIDE REFINING	Crude capacity	Catalytic cracking	Cat reforming
Liberia Petroleum Refining–Monrovia	15,000	....	[1]2,000

LIBYA

WORLDWIDE REFINING	Crude capacity	Catalytic cracking	Cat reforming
Azzawiya Oil Refining Co.–Azzawiya	120,000	....	[2]12,982
Ras Lanuf Oil & Gas Processing Co.–Ras Lanuf	201,000	....	....
Sirte Oil Co.–Brega	8,400	....	[1]1,000
Total	**329,400**	**....**	**13,982**

MADAGASCAR

WORLDWIDE REFINING	Crude capacity	Catalytic cracking	Cat reforming
Solima–Managareza, Tamatave	16,350	....	[1]2,600

MALAYSIA

WORLDWIDE REFINING	Crude capacity	Catalytic cracking	Cat reforming
Esso Malaysia Berhad–Port Dickson	47,300	....	[2]7,980
Petronas Penapisan Sdn. Berhad–Kerteh, Kemaman, Terengganu	27,000	....	....
Sarawak Shell Berhad–Luton	45,000	....	....
Shell Refining Co. Berhad–Port Dickson	90,000	....	[1]4,000 [2]10,000
Total	**209,300**	**....**	**21,980**

MARTINIQUE

WORLDWIDE REFINING	Crude capacity	Catalytic cracking	Cat reforming
Ste. Anonyme de la Raffinerie des Antilles–Fort-de-France	12,800	....	[2]7,410

WORLDWIDE REFINING	Crude capacity	b/cd Catalytic cracking	Cat reforming
MEXICO			
Petroleos Mexicanos–Azcapotzalco	110,000	[1]24,000	....
Cadereyta	235,000	[1]40,000	[2]20,000
Ciudad Madero[1]	195,000	[1]43,000	[2]15,000
		[4]9,000	
Minatitlan	200,000	[1]24,000	[2]48,000
		[2]16,000	
Poza Rica	50,000	....	....
Reynosa	9,000	....	....
Salamanca	235,000	[1]40,000	[2]24,800
Salina Cruz	165,000	[1]40,000	[6]20,000
Tula, Hidalgo	155,000	[2]40,000	[2]30,000
Total	**1,354,000**	**267,000**	**157,800**

Solvent extraction: [1]38,000 b/cd.

WORLDWIDE REFINING	Crude capacity	b/cd Catalytic cracking	Cat reforming
MOROCCO			
Samir–Mohammedia	129,000	....	[2]24,000
Ste. Cherifienne des Petroles–Sidi Kacem	25,600	[2]5,600	[1]2,800
Total	**154,600**	**5,600**	**26,800**
NETHERLANDS			
BP Raffinaderij Nederland NV–Rotterdam	437,000	[1]47,500	[2]27,000
Esso Nederland BV–Rotterdam	155,000	....	[4]25,200
Kuwait Petroleum Europoort BV–Rotterdam	75,500	....	[2]8,200
Shell Nederland Raffinaderij BV–Pernis	348,000	[1]78,000	[2]17,000
			[6]42,000
Smid & Hollander Raffinaderij BV–Amsterdam	8,200	....	....
Texaco Petroleum Mij. (Nederland) BV–Pernis	207,000	....	[2]29,000
Total Raffinaderij Nederland NV–Vlissingen	150,000	....	[6]18,000
Total	**1,380,700**	**125,500**	**166,400**
NETHERLANDS ANTILLES			
Refineria Isla Curazao SA–Emmastad	320,000	[1]42,000	[2]15,000
NEW ZEALAND			
New Zealand Refining–Whangarei	88,400	....	[2]21,900
NICARAGUA			
Esso Standard Oil SA Ltd.–Managua	14,500	....	[2]2,800
NIGERIA			
Nigerian Petroleum Refining CL–Kaduna[1]	104,500	[1]18,000	[2]15,300
Port Harcourt, Alesa Eleme	60,000	....	[1]6,000
Port Harcourt, Rivers State	150,000	[1]40,000	[6]33,000
Warri	100,000	[1]26,000	[2]16,700
Total	**414,500**	**84,000**	**71,000**

Solvent extraction: [1]9,492 b/cd.

WORLDWIDE REFINING	Crude capacity	b/cd Catalytic cracking	Cat reforming
NORWAY			
Esso Norge AS–Slagen-Valloy	90,000	....	[2]10,000
Norske Shell AS–Sola	60,000	....	[2]10,000
Rafinor AS-Mongstad	89,400	....	[2]10,100
Total	**239,400**	**....**	**30,100**
OMAN			
Oman Refinery Co.–Mina Al Fahal	76,932	....	[2]15,386
PAKISTAN			
Attock Refinery Ltd.–Rawalpindi	36,000	....	....
National Refinery Ltd.–Korangi, Karachi	47,750	....	[2]2,750
Pakistan Refinery Ltd.–Karachi	46,300	....	[1]2,700
Total	**130,050**	**....**	**5,450**

WORLDWIDE REFINING	Crude capacity	b/cd Catalytic cracking	Cat reforming
PANAMA			
Refinería Panama SA–Las Minas	100,000	....	[2]7,500
PARAGUAY			
Petroleos Paraguayos–Villa Elisa	7,500	....	....
PERU			
Petróleos del Peru:			
Conchan	5,670	....	....
Iquitos, Loreto	9,450	....	....
La Pampilla, Lima	93,900	[1]7,600	[3]1,680
Marsella, Loreto	1,953	....	....
Pucallpa	2,565	....	....
Talara	58,900	[1]14,940	....
Total	**172,438**	**22,540**	**1,680**
PHILIPPINES			
Caltex (Philippines) Inc.–Batangas	63,000	[1]8,100	[2]8,100
Petron Corp.–Limay, Bataan	130,000	[2]15,000	[1]23,800
Philippine Petroleum Corp.–Pililla	....	....	....
Pilipinas Shell Petroleum–Tabangao	61,000	....	[1]7,500
Total	**254,000**	**23,100**	**39,400**

POLAND

Government-owned refineries at Czechowice, Gdansk, Glinik Mariampolski, Jasto, Jealicze, Kralaty, L. Warynski, Plock, Trzebinia. Of these, Plock is by far the largest, accounting for roughly 80% of Polish refinery production. Its crude capacity is nearly 300,000 b/d.

	Crude capacity		
Total	**385,000**		

WORLDWIDE REFINING	Crude capacity	b/cd Catalytic cracking	Cat reforming
PORTUGAL			
Petrogal–Lisbon	....	[2]10,100	....
Leça da Palmeira, Porto	88,300	....	[2]14,200
			[4]10,500
Sines	225,000	....	[5]27,500
Total	**313,300**	**10,100**	**52,200**
PUERTO RICO			
Bayamón Refinery–Bayamón	38,000	[1]12,000	[2]5,800
Yabucoa Sun Oil Co.–Yabucoa	85,000	....	....
Total	**123,000**	**12,000**	**5,800**
QATAR			
National Oil Distribution Co.–Umm Said	62,000	....	[2]2,200
			[6]9,830
Total	**62,000**	**....**	**12,030**

ROMANIA

Government-owned refineries at Bacau, Borzesti, Brazi, Brazov, Cimpina, Darmanesti, G. Gheorghiu Dej, Navodari, Onesti, Pitesti, Ploesti, Sulpacu, and Telaejen.

	Crude capacity		
Total	**617,000**		

WORLDWIDE REFINING	Crude capacity	b/cd Catalytic cracking	Cat reforming
SAUDI ARABIA			
Arabian American Oil Co.–Ras Tanura	450,000	....	[4]60,000
Arabian Oil CL–Ras Al Khafji	30,000	....	....
Jeddah Oil Refinery–Jeddah	91,000	[1]10,000	[2]2,900
Petromin-Mobil–Yanbu	250,000	[1]66,000	[6]32,000
Petromin-Shell–Al-Jubail[1]	250,000	....	[6]17,000
Riyadh Oil Refinery–Riyadh[2]	134,000	....	[2]28,600
			[5]5,400
Yanbu Domestic Refinery–Yanbu	170,000	....	[2]35,000
Total	**1,375,000**	**76,000**	**180,900**

Solvent extraction: [1]12,500 b/cd. [2]8,800 b/cd.

WORLDWIDE REFINING	Crude capacity	b/cd Catalytic cracking	Cat reforming
SENEGAL			
Ste. Africaine de Raffinage–M'Bao (Dakar)	29,800	....	[4]2,870

WORLDWIDE REFINING	Crude capacity	b/cd Catalytic cracking	Cat reforming
SIERRA LEONE			
Sierra Leone Petroleum Refining Co.–Freetown	10,000	....	....
SINGAPORE			
BP Refinery Singapore PL Pasir Panjang	28,000	....	....
Esso Singapore PL–Pulau Ayer Chawan	170,000	....	[2]8,600
Mobil Oil Singapore PL–Jurong	193,000	....	[2]18,000
Shell Eastern Petroleum Ltd.– Pulau Bukom	300,000	....	[2]6,800 [6]13,900
Singapore Refinery CPL†– Pulau Merlimau	161,000	....	[4]11,000
Total	**852,000**	....	**58,300**

†Singapore Petroleum CPL shares percentages of refinery capacities.

WORLDWIDE REFINING	Crude capacity	b/cd Catalytic cracking	Cat reforming
SOMALIA			
Iraqsoma Ref. Co.–Mogadishu	10,000	....	....
SOUTH AFRICA			
Caltex Oil SA PL–Cape Town	90,000	[1]25,000	[2]10,500
Mobil Oil Refining Co.– Southern Africa PL–Durban	65,000	[1]15,000	[2]15,000
†Durban	....	....	....
National Petroleum Refiners of South Africa PL–Sasolburg OFS	78,500	[1]17,400	[2]11,300
Shell and BP South Africa Petroleum Refineries PL–Durban	200,000	[1]27,000	[2]26,000
Total	**433,500**	**84,400**	**62,800**

†South African Oil Refinery PL is owner; Mobil is operator.

WORLDWIDE REFINING	Crude capacity	b/cd Catalytic cracking	Cat reforming
SPAIN			
Asfaltos Españoles SA–Tarragona	21,000	....	....
Cia. Española de Petróles–San Roque (Cádiz)	160,000	[1]36,000	[2]35,000
Tenerife	130,000	....	[2]13,000
Petróleos del Mediterraneo– Castellón de la Plana	112,000	[1]14,000	[2]14,000
Petronor SA–Somorrostro, Vizcaya	220,000	[1]40,000	[2]30,500
Repsol Petroleo SA–Cartagena, Murcia	112,000	....	[2]25,000
La Coruña	135,000	[1]28,000	[2]22,000
Puertollano, Ciudad Real	135,000	[1]30,000	[2]18,000
Tarragona	180,000	[2]16,000	....
Unión Explosivos Rio Tinto SA– La Rábida, Huelva	80,000	....	[2]14,100
Total	**1,285,000**	**164,000**	**171,600**
SRI LANKA			
Ceylon Petroleum Corp.– Sapugaskanda	50,000	....	[2]3,750
SUDAN			
Port Sudan Refinery Ltd.–Port Sudan	21,440	....	[1]1,700
SWEDEN			
AB Nynas Petroleum–Gothenburg	12,500	....	....
Nynäshamn	28,000	....	....
BP Raffinaderi–Gothenburg	106,000	....	[2]20,500
Shell Raffinaderi BV–Gothenburg	80,000	....	[2]16,000
Skandinaviska Raffinaderi AB– Brofjorden-Lysekil	200,000	[1]25,000	[2]34,500
Total	**426,500**	**25,000**	**71,000**
SWITZERLAND			
Raffinerie de Cressier SA–Cressier	60,000	....	[2]16,000
Raffinerie du Sud-Ouest SA– Collombey	72,000	....	[2]10,000
Total	**132,000**	....	**26,000**

WORLDWIDE REFINING	Crude capacity	b/cd Catalytic cracking	Cat reforming
SYRIA			
Banias Refining Co.–Banias	126,350	....	[4]16,728
Homs Refinery Co.–Homs	117,394	....	[2]2,266
Total	**243,744**	....	**18,994**
TANZANIA			
Tanzanian & Italian Petroleum Refining CL–Kigamboni, Dar es Salaam	13,500	....	[2]3,000
THAILAND			
Esso Standard Thailand Ltd.–Sriracha	64,220	....	[5]7,200
Petroleum Authority of Thailand– Bangchak, Bangkok	61,750	....	[2]9,450
Thai Oil CL–Sriracha	65,075	[1]8,100	[1]9,630
Total	**191,045**	**8,100**	**26,280**
TRINIDAD			
Trinidad and Tobago Oil CL– Pointe-a-Pierre	220,000	[1]28,000	[4]25,000
Point Fortin	80,000	....	[3]6,000
Total	**300,000**	**28,000**	**31,000**
TUNISIA			
Ste. Tunisienne Industries des Raffinage–Bizerte	34,000	....	3,300
TURKEY			
Anadolu Tasfiyehanesi AS–Mersin	90,000	....	[2]10,500
Turkish Petroleum Refineries Corp.: Aliaga-Izmir[1]	249,000	[1]14,000	[2]10,700
Batman, Siirt	24,350	....	[2]1,300
Izmit	248,455	[1]20,128	[2]20,285
Kirikkale	112,849	....	[2]20,342
Total	**724,654**	**34,128**	**63,127**

Solvent extraction: [1]10,273 b/cd (furfural), 6,038 b/cd (propane deasphalting).

WORLDWIDE REFINING	Crude capacity	b/cd Catalytic cracking	Cat reforming
UNITED KINGDOM			
England			
Conoco Ltd.–South Killingholme	130,000	[1]38,000 [2]50,000	....
Eastham Refinery Ltd.–Eastham, Cheshire	12,000	....	....
Esso Petroleum CL–Fawley	300,200	[1]76,500	[2]27,900 [4]22,500
Lindsey Oil Refinery Ltd.– Killingholm, South Humberside	190,000	[1]40,000	[2]33,600
Mobil Oil CL–Coryton, Essex	145,000	[1]47,000	[2]27,000
Phillips Imperial Petroleum Ltd.– Port Clarence	100,000	....	....
Shell U.K. Ltd.–Shell Haven	92,000	....	[6]34,000
Stanlow	262,000	[1]54,000	[2]29,000 [6]27,000
Total	**1,231,200**	**305,500**	**201,000**
Scotland			
BP Refinery Grangemouth Ltd.– Grangemouth	178,500	[1]19,000	[2]32,000
William Briggs & Sons Ltd.—Dundee	6,400	....	....
Total	**184,900**	**19,000**	**32,000**
Wales			
Amoco UK Ltd.–Milford Haven	102,000	[1]31,500	[2]17,500
BP Refinery Llandarcy Ltd.– Llandarcy, Neath	....	....	....
Gulf Oil-GB–Milford Haven	105,000	....	[2]18,000
Pembroke Cracking Co. (65% Texaco, 35% Gulf Oil-GB)	....	[1]86,000	....
Texaco Ltd.–Pembroke, Dyfed	180,000	....	[6]39,000
Total	**387,000**	**117,500**	**74,500**
Total United Kingdom	**1,803,100**	**442,000**	**307,500**

REFORMER UNIT is part of Citgo Peteroleum Corp.'s refinery in Lake Charles, La. (photo courtesy Citgo Petroleum Corp.).

WORLDWIDE REFINING	Crude capacity	b/cd Catalytic cracking	Cat reforming
URUGUAY			
Ancap–La Teja, Montevideo	33,000	[1]4,100	[1]2,500

U.S.S.R.

Government-owned refineries at Achinsk, Angarsk, Baku, Batumi, Chimkent, Drogobych, Fergana, Gorki, Grozny, Guryev, Ishimbai, Khabarovsk, Kherson, Kirishi, Komsomolsk, Krasnovodsk, Kremenchug, Kuibyshev, Lisichansk, Mazheikiai, Moscow, Mozyr, Nadvornaya, Nizhnekamsk, Odessa, Omsk, Orsk, Pavlodar, Perm, Polotsk, Ryazan, Saratov, Syzran, Tuapse, Ufa, Ukhta, Vannovskiy, Volgograd, and Yaroslavl.

	Crude capacity		
Total	12,300,000		

VENEZUELA

	Crude capacity	Catalytic cracking	Cat reforming
Corpoven–El Palito, Carabobo	105,000	[1]45,000	[1]6,000
El Toreño, Barinas	4,500	....	....
Puerto La Cruz, Anzoategui	195,100	[1]11,400	....
San Roque, Anzoategui	5,200	....	....
Lagoven–Judibana, Falcon	600,000	[1]76,100	....
Maraven–Punta Cardon, Falcon	291,300	[1]60,600	....
Total	1,201,100	193,100	6,000

VIRGIN ISLANDS

	Crude capacity	Catalytic cracking	Cat reforming
Hess Oil Virgin Islands Corp.–St. Croix	545,000	....	[2]125,000

WORLDWIDE REFINING	Crude capacity	b/cd Catalytic cracking	Cat reforming
YEMEN, N.			
Yemen Hunt Oil Co.–Marib	10,000	....	[1]2,500
YEMEN, S.			
Aden Refinery Co.–Little Aden	161,500	....	[1]10,800

YUGOSLAVIA

	Crude capacity	Catalytic cracking	Cat reforming
Bosanski Brod[1]	112,835	....	[2]7,612 [6]13,500
Lendava	14,600	....	....
Novi Sad	63,000	....	[2]8,200
Pančevo	108,000	[1]20,750	[2]8,800
Rijeka	160,000	[1]21,300	[2]17,500
Sisak	150,000	[1]10,000	[2]19,000
Zagreb	700	....	....
Total	609,135	52,050	74,612

Solvent extraction: [1]3,300 b/cd (deasphalting).

ZAIRE

	Crude capacity	Catalytic cracking	Cat reforming
Sozir–Muanda	16,700	....	[2]3,500

ZAMBIA

	Crude capacity	Catalytic cracking	Cat reforming
Indeni Petroleum Refinery CL–Bwana Nkubwa Area, Ndola	20,594	....	[2]5,070

UNITED STATES

Survey of operating refineries in the U.S. (state capacities as of January 1, 1989)

State	No. plants	Crude capacity b/cd	Cat cracking Fresh feed	Cat reforming
Alabama	3	138,750	......	27,000
Alaska	6	217,000	......	12,000
Arizona	1	5,710	......	......
Arkansas	3	58,570	18,500	9,000
California	31	2,278,593	645,500	561,500
Colorado	2	72,000	23,000	16,500
Delaware	1	140,000	65,000	56,000
Georgia	2	36,800	......	......
Hawaii	2	129,500	22,000	12,000
Illinois	6	920,600	342,000	294,300
Indiana	4	422,900	159,000	99,000
Kansas	8	344,025	127,000	98,600
Kentucky	2	218,900	100,000	53,000
Louisiana	17	2,274,515	736,500	494,300
Michigan	4	120,100	45,500	33,500
Minnesota	2	285,600	78,000	55,500
Mississippi	5	358,600	72,000	95,800
Montana	4	137,200	50,600	35,500
Nevada	1	4,500	......	......
New Jersey	6	464,250	256,000	89,500
New Mexico	3	72,800	27,100	17,400
New York	1	42,750	......	......
North Dakota	1	58,000	26,000	12,100
Ohio	4	482,650	176,000	160,600
Oklahoma	6	387,500	136,500	91,000
Oregon	1	15,000	......	......
Pennsylvania	8	730,400	239,300	208,820
Tennessee	1	57,000	30,000	10,000
Texas	31	4,058,350	1,616,500	1,150,900
Utah	6	154,500	55,400	29,100
Virginia	1	51,000	27,500	9,750
Washington	7	469,675	102,500	119,800
West Virginia	2	15,500	......	4,900
Wisconsin	1	32,000	11,000	8,000
Wyoming	5	163,500	64,500	36,850
Total	**188**	**15,418,738**	**5,252,900**	**3,901,920**

U.S. refineries

Company and refinery location	Crude	Catalytic cracking	Catalytic reforming
ALABAMA			
Belcher Refining Co.—Mobile Bay	*14,250	......	[2]6,000
Hunt Refining Co.—Tuscaloosa	44,500	......	[2]6,000
Louisiana Land & Exploration Co.—Saraland	80,000	......	[2]21,000
Total	**138,750**	**......**	**27,000**
ALASKA			
Arco Alaska Inc.—Kuparuk	12,000	......	......
Prudhoe Bay	14,000	......	......
Chevron U.S.A. Inc.—Kenai	22,000	......	......
Mapco Petroleum Inc.—North Pole	90,000	......	......
Petro Star Inc.—North Pole	7,000	......	......
Tesoro Petroleum Corp.—Kenai	72,000	......	[4]12,000
Total	**217,000**	**......**	**12,000**
ARIZONA			
Intermountain Refining Cl—Fredonia	5,710	......	......
ARKANSAS			
Berry Petroleum Co.—Stevens	3,800	......	......
Cross Oil & Refining Co. of Arkansas—Smackover	6,770	......	......
Lion Oil Co.—El Dorado[1]	48,000	[1]18,500	[2]9,000
Total	**58,570**	**18,500**	**9,000**

Solvent extraction: [1]5,500 b/d.

Company and refinery location	Crude	Catalytic cracking	Catalytic reforming
CALIFORNIA			
Anchor Refining Cl—McKittrick	10,000	......	......
Atlantic Richfield Co.—Carson	214,000	[1]82,000	[2]48,000
Chemoil Refining Corp.—Signal Hill	14,400	......	......
Chevron U.S.A. Inc.—El Segundo	330,000	[1]60,000	[2]60,000
Richmond[1]	270,000	[1]63,000	[2]68,000
Conoco Inc.—Santa Maria	9,500	......	......
Edgington Oil Cl—Long Beach	41,600	......	......
Exxon Co.—Benicia	128,000	[1]64,000	[3]28,000
Fletcher Oil & Refining Co.—Carson	26,500	[1]10,500	[4]5,000
Golden Bear Division, Witco Chemical Corp.—Oildale	10,818	......	......
Golden West Refining Co.—Santa Fe Springs	40,600	[1]13,500	[2]19,000
Huntway Refining Co.—Benicia	8,400	......	......
Wilmington	5,500	......	......
Kern Oil & Refining Co.—Bakersfield	20,000	......	[2]3,000
Lunday-Thagard Co.—South Gate	7,700	......	......
Mobil Oil Corp.—Torrance	123,000	[1]63,000	[2]36,000
Newhall Refining Co. Inc.—Newhall	18,800	......	[3]2,500
Oxnard Refinery—Oxnard	*4,275	......	......
Pacific Refining Co.—Hercules	*52,250	......	[2]15,000
Paramount Petroleum Corp.—Paramount	42,700	......	[2]10,500
Powerine Oil Co.—Santa Fe Springs	*46,550	[1]12,500	[2]8,500
San Joaquin Refining Cl—Bakersfield	18,000	......	......
Shell Oil Co.—Martinez	140,100	[1]67,000	[3]28,000
Wilmington	144,000	[1]42,000	[2]24,000
Sunland Refining Corp.—South Gate	15,000	......	[2]1,500
Texaco Refining & Marketing Inc.—			
Bakersfield	47,000	......	[2]23,000
Wilmington	75,000	[1]28,000	[2]38,000
Tosco Corp.—Martinez	126,000	[1]55,000	[2]20,000 [6]23,000
Union Pacific Resources—Wilmington	68,000	[1]38,000	[6]14,500
Unocal Corp.—Los Angeles	108,000	[2]47,000	[2]52,000
Rodeo	112,900	......	[2]34,000
Total	**2,278,593**	**645,500**	**561,500**

Solvent extraction: [1]50,000 b/sd.

Company and refinery location	Crude	Catalytic cracking	Catalytic reforming
COLORADO			
Colorado Refining Co.—Commerce City	28,000	[1]8,000	[2]9,000
Conoco Inc.—Commerce City	44,000	[1]15,000	[2]7,500
Total	**72,000**	**23,000**	**16,500**
DELAWARE			
Star Enterprise—Delaware City	140,000	[1]65,000	[2]18,000 [5]38,000
Total	**140,000**	**65,000**	**56,000**

GEORGIA

Company and refinery location	Crude	Catalytic cracking	Catalytic reforming
Amoco Oil Co.—Savannah	28,000	......	......
Young Refining Corp.—Douglasville	8,800	......	......
Total	**36,800**	**......**	**......**

HAWAII

Company and refinery location	Crude	Catalytic cracking	Catalytic reforming
Chevron U.S.A. Inc.—Barber's Point	48,000	[1]22,000	......
Hawaiian Independent Refinery Inc.—Ewa Beach	81,500	......	[2]12,000
Total	**129,500**	**22,000**	**12,000**

ILLINOIS

Company and refinery location	Crude	Catalytic cracking	Catalytic reforming
Clark Oil & Refining Corp.—Blue Island	64,600	[1]25,000	[2]30,500
Hartford	60,000	[1]26,000	[2]12,000
Marathon Petroleum Co.—Robinson	195,000	[1]41,000	[4]38,000 [5]41,000
Mobil Oil Corp.—Joliet	180,000	[1]98,000	[2]46,000
Shell Oil Co.—Wood River	274,000	[1]94,000	[2]22,000 [3]75,000
Unocal Corp.—Lemont	147,000	[1]58,000	[2]29,800
Total	**920,600**	**342,000**	**294,300**

INDIANA

Company and refinery location	Crude	Catalytic cracking	Catalytic reforming
Amoco Oil Co.—Whiting	350,000	[1]140,000	[3]85,000
Indiana Farm Bureau Cooperative Association Inc.—Mt. Vernon	20,600	......	[2]4,000
Laketon Refining Corp.—Laketon	8,300	......	......
Rock Island Refining Corp.—Indianapolis[1]	44,000	[1]19,000	[2]10,000
Total	**422,900**	**159,000**	**99,000**

Solvent extraction: [1]4,500.

KANSAS

Company and refinery location	Crude	Catalytic cracking	Catalytic reforming
Coastal Refining and Marketing Inc.—Augusta	......	......	[2]10,000
Derby Refining Co.—El Dorado[1]	*28,500	[1]14,500	[2]4,500
Wichita	*29,925	[1]19,000	[2]6,500
Farmland Industries Inc.—Coffeyville	56,500	[1]23,000	[2]16,000
Phillipsburg	26,400	......	[2]5,300
National Cooperative Refinery Association—McPherson	70,900	[1]20,000	[2]15,000
Texaco Refining & Marketing Inc.—El Dorado	78,200	[1]31,500	[2]25,000
Total Petroleum Inc.—Arkansas City	53,600	[1]19,000	[2]16,000
Total	**344,025**	**127,000**	**98,300**

Solvent extraction: [1]5,600 b/sd.

KENTUCKY

Company and refinery location	Crude	Catalytic cracking	Catalytic reforming
Ashland Petroleum Co.—Catlettsburg	213,400	[1]60,000 §[2]40,000	[2]25,000 [5]27,000
Somerset Refinery Inc.—Somerset	5,500	......	[2]1,000
Total	**218,900**	**100,000**	**53,000**

§Reduced crude converter.

LOUISIANA

Company and refinery location	Crude	Catalytic cracking	Catalytic reforming
Atlas Processing Co., Division of Pennzoil—Shreveport	46,200	......	[2]10,000
Calcasieu Refining Co.—Lake Charles	12,000	......	......
Calumet Refining Co.—Princeton	4,100	......	......
Canal Refining Co.—Church Point	9,865	......	[1]1,900
Citgo Petroleum Corp.—Lake Charles	320,000	[1]150,000	[2]46,000 [5]45,000
Conoco Inc.—Westlake	156,500	[1]40,000	[2]15,900 [5]11,500
Exxon Co.—Baton Rouge[1]	455,000	[1]155,000	[4]90,000
Hill Petroleum Co.—Krotz Springs	55,300	[1]28,000	[2]12,500
St. Rose	33,000	......	......
Kerr-McGee Refining Corp.—Cotton Valley	7,800	......	......
Marathon Petroleum Co.—Garyville	255,000	[1]85,000	[5]48,000
Mobil Oil Corp.—Chalmette	145,000	[1]51,000	[2]28,000
Murphy Oil USA Inc.—Meraux[2]	92,500	[1]35,000	[6]23,000
Placid Refining Co.—Port Allen[3]	*47,500	[1]18,500	[2]10,000
Shell Oil Co.—Norco	215,000	[1]90,000	[2]18,000 [3]38,000
Sohio Oil Co.(BP America)—Belle Chasse	*194,750	[1]89,000	[2]37,500
Star Enterprise—Convent	225,000	[1]85,000	[2]40,000
Total	**2,274,515**	**736,500**	**494,300**

Solvent extraction: [1]28,500 b/sd. [2]12,000 b/sd. [3]6,000 b/sd.

MICHIGAN

Company and refinery location	Crude	Catalytic cracking	Catalytic reforming
Crystal Refining Co.—Carson City	4,000	......	......
Lakeside Refining Co.—Kalamazoo	5,600	......	[1]1,000
Marathon Petroleum Co.—Detroit	68,500	[1]27,000	[2]18,500
Total Petroleum Inc.—Alma	42,000	[1]18,500	[2]14,000
Total	**120,100**	**45,500**	**33,500**

MINNESOTA

Company and refinery location	Crude	Catalytic cracking	Catalytic reforming
Ashland Petroleum Co.—St. Paul Park	67,100	[1]23,000	[2]23,500
Koch Refining Co.—Rosemount	*218,500	[1]55,000	[6]26,000 [2]6,000
Total	**285,600**	**78,000**	**55,500**

MISSISSIPPI

Company and refinery location	Crude	Catalytic cracking	Catalytic reforming
Amerada-Hess Corp.—Purvis	30,000	[2]16,000	[2]5,800
Chevron U.S.A. Inc.—Pascagoula	295,000	[1]56,000	[2]90,000
Ergon Refining Inc.—Vicksburg	16,800	......	......
Southland Oil Co.—Lumberton	5,800	......	......
Sandersville	11,000	......	......
Total	**358,600**	**72,000**	**95,800**

MONTANA

Company and refinery location	Crude	Catalytic cracking	Catalytic reforming
Cenex—Laurel[1]	40,400	[1]12,000	[2]12,000
Conoco Inc.—Billings[2]	48,500	[1]15,500	[2]12,500
Exxon Co.—Billings	42,000	[1]21,000	[1]10,000
Montana Refining Co.—Great Falls	6,300	[1]2,100	[2]1,000
Total	**137,200**	**50,600**	**35,500**

Solvent extraction: [1]4,000 b/sd. [2]8,500 b/sd (PDA).

NEVADA

Company and refinery location	Crude	Catalytic cracking	Catalytic reforming
Nevada Refining Co.—Tonopah	4,500	......	......

NEW JERSEY

Company and refinery location	Crude	Catalytic cracking	Catalytic reforming
Amerada-Hess Corp.—Port Reading	......	[1]50,000	......
Chevron U.S.A. Inc.—Perth Amboy	80,000	......	......
Coastal Eagle Point Oil Co.—Westville	*109,250	[1]50,000	[2]38,000
Exxon Co.—Linden	130,000	[1]120,000	[4]28,000
Mantua Oil Co. LP—Thorofare	45,000	......	......
Mobil Oil Corp.—Paulsboro[1]	100,000	[1]36,000	[2]23,500
Total	**464,250**	**256,000**	**89,500**

Solvent extraction: [1]9,500 b/sd.

NEW MEXICO

Company and refinery location	Crude	Catalytic cracking	Catalytic reforming
Bloomfield Refining Co.—Bloomfield	16,800	[1]5,400	[2]2,800
Giant Industries Inc.—Gallup	18,000	[1]7,200	[2]6,800
Navajo Refining Co.—Artesia	*38,000	[1]14,500	[1]7,800
Total	**72,800**	**27,100**	**17,400**

NEW YORK

Company and refinery location	Crude	Catalytic cracking	Catalytic reforming
Cibro Petroleum Products Co.—Albany	42,750	......	......

NORTH DAKOTA

Company and refinery location	Crude	Catalytic cracking	Catalytic reforming
Amoco Oil Co.—Mandan	58,000	[1]26,000	[4]12,100
Total	**58,000**	**26,000**	**12,100**

OHIO

Company and refinery location	Crude	Catalytic cracking	Catalytic reforming
Ashland Petroleum Co.—Canton	66,000	[1]25,000	[5]20,000

Company and refinery location	Crude	Catalytic cracking	Catalytic reforming
Sohio Oil Co. (BP America)—Lima	*171,000	[1]36,000	[4]53,000
Toledo	*120,650	[1]55,000	[3]23,000
			[4]19,000
Sun Co. Inc.-Toledo[1]	125,000	[1]60,000	[2]45,600
Total	**482,650**	**176,000**	**160,600**

Solvent extraction: [1]9,000 b/sd.

OKLAHOMA

Company and refinery location	Crude	Catalytic cracking	Catalytic reforming
Barrett Refining Corp.—Thomas	13,000	......	......
Conoco Inc.—Ponca City	136,000	[1]46,000	[2]34,500
Kerr-McGee Refining Corp.—Wynnewood[1]	43,000	[1]20,000	[2]8,500
Sinclair Oil Corp.—Tulsa	50,000	[1]18,000	[2]12,000
Sun Co. Inc.—Tulsa[2]	85,000	[1]30,000	[2]24,000
Total Petroleum Inc.—Ardmore	60,500	[1]22,500	[6]12,000
Total	**387,000**	**136,500**	**91,000**

Solvent extraction: [1]5,000 b/sd. [2]5,900 b/sd.

OREGON

Company and refinery location	Crude	Catalytic cracking	Catalytic reforming
Chevron U.S.A. Inc.—Portland	**15,000**	......	......

PENNSYLVANIA

Company and refinery location	Crude	Catalytic cracking	Catalytic reforming
Atlantic Refining & Marketing Corp.—			
Philadelphia	125,000	[1]29,000	[2]60,000
Chevron U.S.A. Inc.—Philadelphia	174,100	[1]53,300	[2]34,000
Kendall-Amalie Division,			
Witco Chemical Co.—Bradford	8,500	......	[2]3,300
Pennzoil Products Co.—Rouseville	15,700	......	[2]5,820
Quaker State Oil Refining Corp.—			
Smethport[1]	6,500	......	[2]2,100
Sohio Oil Co. (BP America)—Marcus Hook	*171,000	[1]50,000	[6]48,000
Sun Co. Inc.—Marcus Hook	165,000	[1]87,000	[2]39,600
United Refining Co.—Warren	*64,600	[1]20,000	[2]16,000
Total	**730,400**	**239,300**	**208,820**

Solvent extraction: [1]730 b/sd

TENNESSEE

Company and refinery location	Crude	Catalytic cracking	Catalytic reforming
Mapco Petroleum Inc.—Memphis	*57,000	[1]30,000	[2]10,000
Total	**57,000**	**30,000**	**10,000**

TEXAS

Company and refinery location	Crude	Catalytic cracking	Catalytic reforming
Amoco Oil Co.—Texas City	420,000	[1]195,000	[4]160,000
Champlin Refining Co.—Corpus Christi	150,000	[1]70,000	[1]52,000
Chevron U.S.A. Inc.—El Paso	66,000	[1]22,000	[2]25,000
Port Arthur	329,000	[1]110,000	[2]23,000
			[4]44,100
Coastal Refining & Marketing Inc.—			
Corpus Christi	*90,250	[1]18,500	[2]11,000
			[6]17,500
Crown Central Petroleum Corp.—Houston	100,000	[1]56,000	[2]14,000
			[5]22,000
Diamond Shamrock Corp.—Sunray[1]	105,000	[1]45,000	[2]29,000
Three Rivers[2]	50,000	[1]19,000	[2]11,000
El Paso Refining Co. Ltd.—El Paso	21,000	[1]7,000	[2]5,600
Exxon Co. U.S.A.—Baytown[3]	493,000	[1]155,000	[3]60,000
			[4]63,000
Fina Oil & Chemical Co.—Big Spring[4]	55,000	[1]22,000	[2]20,000
Port Arthur[5]	100,000	[1]36,000	[5]34,000
Hill Petroleum Co.—Houston[6]	67,500	[1]55,000	[2]11,500
Texas City	119,600	[1]43,500	[2]11,000
			[5]12,000
Howell Hydrocarbons Inc.—San Antonio	2,750	......	[1]1,000
Koch Refining Co.—Corpus Christi	125,000	[1]40,000	[3]15,000
			[6]33,500
LaGloria Oil & Gas Co.—Tyler	55,000	[1]17,000	[2]4,000
			[5]11,700
Liquid Energy Corp.—Bridgeport	10,000	......	......
Lyondell Petrochemical Co.—Houston[7]	265,000	[1]90,000	[2]110,000
Marathon Petroleum Co.—Texas City	69,500	[1]38,000	[1]9,000
Mobil Oil Corp.—Beaumont	275,000	[1]102,000	[2]57,000
			[5]46,000
Phillips 66 Co.—Borger	105,000	[1]60,000	[2]26,000
Sweeny	175,000	[1]87,000	[2]36,000
Pride Refining Inc.—Abilene	42,750	......	......
Shell Oil Co.—Deer Park	215,900	[1]65,000	[2]20,000
			[3]43,000
Odessa	28,600	[1]10,500	[1]11,000
Southwestern Refining CI—Corpus Christi	104,000	[1]50,000	[2]30,000
Star Enterprise—			
Port Arthur and Port Neches[8]	250,000	[1]110,000	[5]40,000
Trifinery—Corpus Christi	28,500	......	......
Unocal Corp.—(Beaumnont), Nederland	120,000	[1]39,000	[1]12,000
			[2]20,000
Valero Refining Co.—Corpus Christi	20,000	†[2]54,000	......
Total	**4,058,350**	**1,616,500**	**1,150,900**

Solvent extraction: [1]15,000 b/sd. [2]7,000 b/sd. [3]53,000 b/sd. [4]10,000 b/sd. [5]18,000 b/sd. [6]11,000 b/sd. [7]10,200 b/sd. [8]4,000 b/sd.
†Heavy oil cracker.

UTAH

Company and refinery location	Crude	Catalytic cracking	Catalytic reforming
Amoco Oil Co.—Salt Lake City	40,000	[1]18,000	[4]7,600
Big West Oil Co.—Salt Lake City	24,000	[2]5,000	[2]5,000
Chevron U.S.A.—Salt Lake City	45,000	[1]11,000	[2]5,500
		[2]7,000	
Crysen Refining Inc.—Woods Cross	12,500	......	[2]3,000
Pennzoil Products Co.—Roosevelt	8,000	[1]6,000	[2]2,000
Phillips 66 Co.—Woods Cross[1]	25,000	[2]8,400	[4]6,000
Total	**154,500**	**55,400**	**29,100**

Solvent extraction: [1]5,000 b/sd.

VIRGINIA

Company and refinery location	Crude	Catalytic cracking	Catalytic reforming
Amoco Oil Co.—Yorktown	51,000	[1]27,500	[2]9,750
Total	**51,000**	**27,500**	**9,750**

WASHINGTON

Company and refinery location	Crude	Catalytic cracking	Catalytic reforming
Atlantic Richfield Co.—Ferndale	162,000	......	[2]56,000
Chevron U.S.A. Inc.—Seattle	5,000	......	......
Shell Oil Co.—Anacortes	84,000	[1]41,000	[3]25,000
Sohio Oil Co. (BP America)—Ferndale	77,000	[2]25,500	[4]11,800
Sound Refining Inc.—Tacoma	11,900	......	......
Texaco Refining & Marketing Inc.—			
Anacortes[1]	97,000	[1]36,000	[1]7,000
			[2]14,000
U.S. Oil & Refining Co.—Tacoma	*32,775	......	[2]6,000
Total	**469,675**	**102,500**	**119,800**

Solvent extraction: [1]4,500 b/sd.

WEST VIRGINIA

Company and refinery location	Crude	Catalytic cracking	Catalytic reforming
Mid Atlantic Fuels Inc.			
—St. Mary's—	5,000		[1]1,500
Quaker State Oil Refining Corp.—			
Newell	10,500	......	[2]3,400
Total	**15,500**	**......**	**4,900**

WISCONSIN

Company and refinery location	Crude	Catalytic cracking	Catalytic reforming
Murphy Oil USA Inc.—Superior	32,000	[1]11,000	[2]8,000
Total	**32,000**	**11,000**	**8,000**

WYOMING

Company and refinery location	Crude	Catalytic cracking	Catalytic reforming
Amoco Oil Co.—Casper	40,000	[1]13,500	[4]7,000
Frontier Oil & Refining Co.—Cheyenne	35,000	[1]12,000	[2]6,600
Little America Refining Co.—Casper	22,000	[1]14,000	[4]6,000
Sinclair Oil Corp.—Sinclair	54,000	[1]21,000	[2]14,500
Wyoming Refining Co.—Newcastle	12,500	[2]4,000	[1]2,750
Total	**163,500**	**64,500**	**36,850**

WORLDWIDE GAS PROCESSING

Capacities as of Jan. 1, 1988, and average production

Country	No. plants	Gas capacity	Gas through-put	Ethane	Prop.	Isobut.	Process or unsplit butane	LP-gas mix	Raw NGL mix	Debut. nat. gaso.	Other	Total products
		MMcfd			Production—1,000 gpd (Average based on the past 12 months)							
United States												
Alabama	6	177.0	161.5	16.0	178.5	44.8	100.4	....	1.2	71.1	....	412.0
Alaska	4	3,385.5	3,340.3	....	3.5	....	....	....	2,168.3	....	16.6	2,188.4
Arkansas	5	943.0	534.1	....	4.0	....	4.0	....	57.4	8.0	3.0	76.4
California	36	1,028.9	753.1	....	459.9	40.9	255.2	71.5	188.4	282.6	0.6	1,299.1
Colorado	37	925.2	498.4	....	191.5	....	42.0	17.7	452.1	41.3	25.5	770.1
Florida	2	790.0	647.0	105.4	138.0	14.0	77.0	....	....	75.0	....	409.4
Kansas	25	5,592.5	3,291.0	468.1	764.2	91.6	266.0	0.9	821.1	202.8	1.1	2,615.8
Kentucky	2	71.0	66.9	....	....	....	....	....	....	....	162.6	162.6
Louisiana	79	19,492.5	10,766.6	1,590.0	1,135.4	348.9	370.1	5.4	5,658.8	668.8	420.5	10,197.9
Michigan	28	4,744.1	1,439.5	....	71.5	....	....	15.7	730.5	103.8	17.6	939.1
Mississippi	6	901.0	396.7	....	9.2	....	8.0	....	26.1	11.5	....	54.8
Montana	6	22.0	8.2	....	23.8	2.2	....	22.1	25.0	2.1	0.8	76.0
Nebraska	2	3.5	2.2	....	....	....	....	....	7.2	....	....	7.2
New Mexico	27	3,310.8	2,107.1	213.6	200.7	16.9	65.2	....	3,687.8	76.0	21.3	4,281.5
North Dakota	9	308.0	213.8	....	196.0	....	88.1	....	83.6	102.6	13.1	483.4
Oklahoma	106	4,571.4	2,785.0	748.1	732.9	108.0	276.0	291.1	3,040.1	367.7	223.1	5,787.0
Pennsylvania	2	14.0	7.9	....	4.1	....	3.3	....	....	3.9	....	11.3
Texas	325	17,802.2	11,422.7	3,474.4	3,595.0	531.9	1,524.0	556.5	11,904.0	1,459.1	1,936.8	24,981.7
Utah	11	485.5	246.1	....	86.1	....	26.6	....	146.0	37.2	103.9	399.8
West Virginia	7	399.0	326.0	223.9	140.5	24.0	45.4	80.6	214.8	1.7	....	730.9
Wyoming	39	2,864.0	2,282.7	....	279.7	10.7	195.0	77.9	1,701.9	1,020.8	1,762.0	5,048.0
Total U.S.	**764**	**67,831.1**	**41,296.8**	**6,839.5**	**8,214.5**	**1,233.9**	**3,346.3**	**1,139.4**	**30,914.3**	**4,536.0**	**4,708.5**	***68,240.4**
Abu Dhabi	5	1,602.0	1,269.0	....	....	....	....	....	....	546.0	....	546.0
Algeria	4	5,335.2	4,268.2	....	....	....	....	632.8	....	....	5,449.2	6,082.0
Argentina	13	1,344.8	1,322.1	694.9	406.8	....	267.0	....	....	425.3	....	1,794.0
Australia	5	2,573.0	1,342.0	279.0	441.0	....	249.0	738.0	....	1,062.0	850.0	3,619.0
Austria	2	19.7	14.4	....	....	....	....	....	....	....	10.7	10.7
Bahrain	1	168.0	155.0	....	115.0	....	100.0	....	....	152.0	....	367.0
Bangladesh	2	240.0	178.0	....	....	....	....	....	....	....	8.2	8.2
Bolivia	2	474.0	177.3	....	1.4	2.5	....	161.7	....	61.5	136.9	364.0
Brazil	10	437.8	414.7	6.1	....	....	....	618.9	239.4	142.5	49.4	1,056.3
Brunei	2	1,032.0	845.0	....	7.7	....	....	....	54.4	242.0	....	304.1
Canada												
Alberta	415	21,905.8	12,328.2	4,894.3	1,762.0	115.4	928.4	905.1	4,877.1	2,504.7	196.3	16,183.3
British Columbia	13	2,294.4	1,250.3	....	147.3	....	92.0	....	....	75.3	52.5	367.1
Northwest Territories	2	196.0	22.1	....	....	....	....	....	....	....	....	....
Saskatchewan	19	260.4	166.9	39.2	....	....	....	....	81.4	0.3	12.4	133.3
Chile	2	470.0	400.0	....	194.0	....	124.0	....	....	102.8	....	420.8
China, Taiwan	4	271.0	105.7	41.2	29.8	1.9	16.9	....	1.6	14.9	....	106.3
Colombia	5	323.1	201.0	102.5	49.0	....	57.0	48.3	....	64.5	....	321.3
Denmark	1	450.0	240.0	....	....	....	....	....	....	....	100.0	100.0
Dominican Republic	1	....	....	....	....	....	....	....	....	....	277.2	277.2
Dubai	1	330.0	330.0	....	....	....	....	....	....	....	772.0	772.0
Ecuador	2	41.0	20.5	....	....	....	....	97.1	....	37.3	....	134.4
Egypt	9	1,449.0	1,338.7	....	....	....	....	796.5	....	....	1,176.8	1,973.3
France	1	570.0	563.0	....	139.0	....	132.0	....	....	270.0	....	541.0
Germany	5	399.0	341.0	....	....	....	....	....	....	....	....	....
Greece	1	10.5	9.4	....	....	....	....	....	86.6	....	....	86.6
Hungary	12	1,078.1	683.6	....	50.7	5.8	51.8	124.8	332.6	172.8	415.3	1,153.8
Indonesia	8	630.5	454.5	....	129.0	....	....	51.6	285.2	209.1	305.7	980.6
Italy	17	7,166.0	1,497.0	....	....	....	....	2.7	53.4	7.6	....	63.7
Kuwait	1	560.0	448.0	....	1,722.0	....	980.0	....	....	811.8	....	3,513.8
Libya	7	1,851.0	1,227.0	....	....	....	....	37.8	5,577.0	....	551.0	6,165.8
Malaysia	2	282.0	120.0	....	140.0	120.0	....	55.0	....	150.0	....	465.0
Mexico	10	4,334.0	3,249.0	3,487.7	5,657.2	3.5	153.0	385.4	....	316.5	14.0	10,017.3
Netherlands	1	440.0	96.0	....	....	....	....	....	....	....	4.9	4.9
New Zealand	3	852.0	94.0	....	39.0	....	....	....	....	9.0	....	48.0
Norway	2	950.0	455.0	....	785.0	102.0	262.0	....	251.0	197.0	....	1,597.0
Oman	8	626.0	309.0	....	....	....	....	....	202.0	43.0	....	245.0
Pakistan	6	1,095.0	923.0	....	....	....	....	37.0	....	....	12.1	49.1
Peru	2	70.0	37.0	....	3.4	....	1.0	5.6	28.9	....	4.0	42.9
Qatar	2	579.0	229.0	....	452.0	....	278.2	....	....	202.0	....	932.2
Saudi Arabia	5	4,300.0	....	....	....	....	....	....	1,176.0	....	....	1,176.0
Sharjah	2	990.0	960.4	....	220.5	....	251.2	....	....	....	1,341.4	1,813.1
Spain	3	428.0	93.0	....	....	....	....	....	....	....	94.0	94.0
United Kingdom	10	4,070.2	1,535.5	....	332.0	....	291.0	....	36.0	1,748.5	82.3	2,489.8
Venezuela	14	2,913.0	2,557.0	213.3	197.4	....	113.4	231.3	3,274.6	171.9	....	4,201.9
Total, outside U.S.	**643**	**75,411.5**	**42,270.5**	**9,758.2**	**13,021.2**	**351.1**	**4,347.9**	**4,929.6**	**16,557.2**	**9,740.3**	**11,916.3**	**70,621.8**

*Includes data reported only in aggregate form and, therefore, excluded from state-by-state figures above.

INDEX

ADVERTISERS' INDEX
INTERNATIONAL PETROLEUM ENCYCLOPEDIA
1989

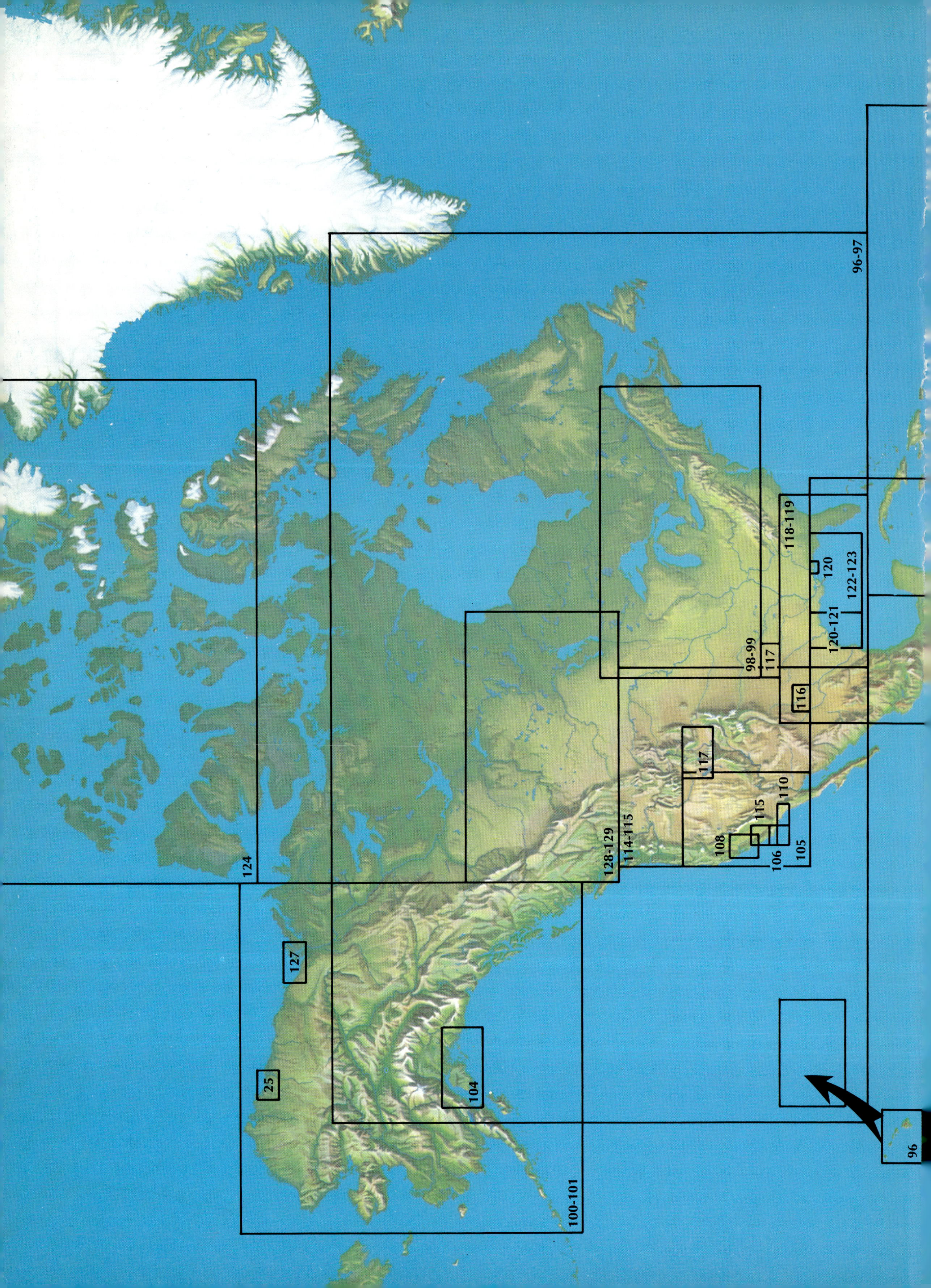

96-97
118-119
120
122-123
120-121
98-99
117
116
117
110
115
128-129
114-115
108
106
105
124
127
25
104
100-101
96